T0073958

Protocells

Protocells

Bridging Nonliving and Living Matter

edited by Steen Rasmussen, Mark A. Bedau, Liaohai Chen, David Deamer, David C. Krakauer, Norman H. Packard, and Peter F. Stadler

The MIT Press
Cambridge, Massachusetts
London, England

This book was set in Times New Roman and Syntax on 3B2 by Asco Typesetters, Hong Kong.

Library of Congress Cataloging-in-Publication Data

Protocells : bridging nonliving and living matter / edited by Steen Rasmussen ... [et al.].
p. ; cm.
Includes bibliographical references and index.
ISBN 978-0-262-18268-3 (hardcover : alk. paper) 1. Artificial cells. 2. Life (Biology) I. Rasmussen, Steen.

ISBN 978-0-262-54588-4 (paperback)

[DNLM: 1. Cells. 2. Biogenesis. 3. Cell Physiology. 4. Models, Biological. QU 300 P967 2008]
QH501.P76 2008
576.8′3—dc22 2007049243

Contents

Preface

The idea for this book grew out of two international protocell workshops in September 2003. One meeting, at Los Alamos National Laboratory and the Santa Fe Institute, was organized by Steen Rasmussen, Liaohai Chen, David Deamer, David Krakauer, Norman Packard, and Peter Stadler. The other meeting, at the European Conference on Artificial Life (ECAL) in Dortmund, Germany, was organized by Steen Rasmussen and Mark Bedau. We published a short summary of the state of the art of protocell research as reflected in those workshops in early 2004 (Rasmussen et al., 2004), and we planned to collect more details about this research in a longer volume. That plan was the seed for this book. But a series of events intervened, changing and delaying the book.

Those events grew out of the Seventh Artificial Life Conference in Portland, Oregon, organized by Mark Bedau, John McCaskill, Norman Packard, and Steen Rasmussen in August 2000. Coinciding with the millennium, the conference aimed to take stock of the young field of artificial life. Out of the Oregon meeting came a community consensus of specific grand challenges in artificial life. One of these challenges is to create wet artificial life from scratch.

Over the next three years, our activities were a portfolio of projects, most involving, in one way or another, the creation of life from scratch. In 2001 we coined the term *living technology* as an umbrella for our activities. The next year we realized how computer-controlled microfluidics could act as life support for the evolution of minimal chemical systems, and two months later we started creating a new roadmap to protocells. Our meetings led to a proposal for a new Center for Living Technology at which scientific developments in this area could be nurtured and developed along the way to producing practical applications.

The European Commission's program on complex systems funded the first phase of these plans. Just before the first protocell workshops in 2003, we learned that our EC proposal on Programmable Artificial Cell Evolution (PACE) was funded. John McCaskill led the PACE project, which consisted of fourteen European and U.S. partners and included plans for a European Center for Living Technology in Venice.

One of the members of the PACE consortium was the new startup company, ProtoLife Srl., of which Norman Packard and Mark Bedau became CEO and COO. Soon afterwards Los Alamos National Laboratory awarded a complementary grant for a project on Protocell Assembly, led by Steen Rasmussen. While these new activities absorbed our time for a couple of years, this book was on the back burner.

In 2005 Emily Parke agreed to manage the editorial process of producing the book. Our vision of the book had grown in the intervening years, so we solicited chapters from many who had missed the original workshops.

Though it had a convoluted gestation, we hope this book will be both a resource and an inspiration for the exciting and important quest to create life from scratch.

Mark Bedau
Norman Packard
Steen Rasmussen

Venice, Italy, June 2007

Reference

Rasmussen, S., Chen, L., Deamer, D., Krakauer, D., Packard, N., Stadler, P., & Bedau, M. (2004). Transitions between nonliving and living matter. *Science*, *303*, 963.

Acknowledgments

This book has been produced only because of the time and energy of a large group of people. The editors are extremely grateful for the various kinds of help provided, and we are pleased to have the opportunity here to acknowledge them.

Editorial Manager: Emily C. Parke

Editorial Board

James Bailey
James Boncella
James Cleaves
Stirling Colgate
Gavin Collis
Michael DeClue
Andrew Ellington
Goran Goranović
Martin M. Hanczyc
Takeshi Ikegami
Martin Nilsson Jacobi
Yi Jiang
Günter von Kiedrowski
Chad Knutson
Natalio Krasnegor
James La Clair
Doron Lancet
Pierre-Alain Monnard
Harold Morowitz
Ole Mouritsen
Andreea Munteanu
Peter Nielsen
Andrew Shreve
Eric Smith
Ricard Solé
Pasquale Stano
Eörs Szathmáry
Bryan Travis
Woody Woodruff
Hanz Ziock
Jinsuo Zhang

First, we want to convey special thanks to Emily Parke. As editorial manager, Emily coordinated and oversaw the complex process of soliciting chapters, coordinating reviews of the chapters, checking revisions, copy editing the final manuscripts, and generally keeping the project on track. None of the editors wants even to imagine what producing this book would have been like without Emily's unflagging expert assistance.

We are also especially grateful to the thoughtful critical attention devoted to early drafts of the chapters by our Editorial Board.

We want to single out John McCaskill for thanks. John has been a key intellectual partner and friend in the creation of the diverse portfolio of protocell-related activities, especially the European Commission's Programmable Artificial Cell Evolution (PACE) project, the Los Alamos Protocell Assembly project, the European Center for Living Technology, and ProtoLife Srl. All of those projects, and so indirectly this book, show the benefit of John's wit, intelligence, and vision.

We would also like to thank the MIT Press for critical early support that helped launch this project and critical later patience when the timeline slipped. We thank the Santa Fe Institute and the European Center for Living Technology for hospitality while some of the work on this volume was completed. For financial support for some of this work, we thank EU-supported PACE integrated project, EC-FP6-IST-FET-IP-002035, and the Los Alamos National Laboratory LDRD-DR Protocell Assembly project.

Introduction

I.1 What Are Protocells?

All life forms are composed of molecules that are not themselves alive. But how exactly do living and nonliving matter differ? Is there a fundamental difference at all? How could we possibly turn a collection of nonliving materials into something that is, at least operationally speaking, alive? This book is a compendium of the state of the art on attempts to answer these questions by building bridges between nonliving chemistry and emergent living states of matter. It focuses on attempts to create very simple life forms in the laboratory. These new forms of life might be quite unfamiliar.

In this book a living system is operationally defined as a system that integrates three critical functionalities (figure I.1, outer ring): First, it maintains an identity over time by localizing all its components. Second, it uses free energy from its environment to digest environmental resources in order to maintain itself, grow, and ultimately reproduce. Third, these processes are under the control of inheritable information that can be modified during reproduction. These properties enable selection and thus evolution as part of the reproduction process. Living systems are sometimes said to include various further essential properties, such as autonomous information processing, sensitivity to the environment, self-organization, and purposeful behavior. We agree that these are central properties of living systems, but we hold that they are derivative in the sense that they are effectively implied by the functionally integrated triad described previously.

This book aims to provide a very general understanding of how one could obtain entities with these lifelike properties from experiments that begin with nonliving materials. The book generally reflects the perspective that chemical instances of such forms of life must embody the three operational functionalities in three integrated chemical systems: a *metabolism* that extracts usable energy and resources from the environment, *genes* that chemically realize informational control of living functionalities, and a *container* that keeps them all together (see figure I.1, inner

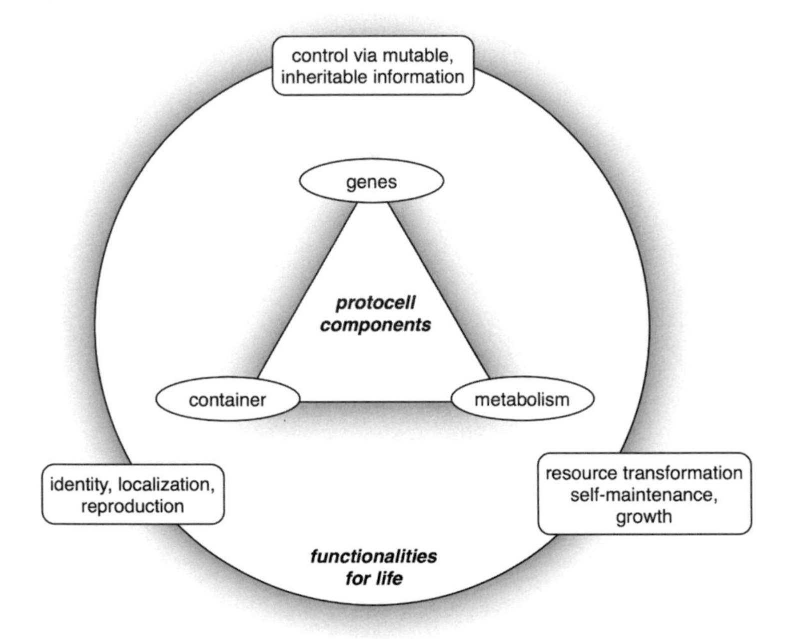

Figure I.1
The three essential operational functionalities of living systems (outer circle), together with the corresponding realizations of these functionalities in protocell architectures (inner triangle).

triangle). We will use the term *protocell* to refer to any realization of these three functional components. Our usage is close to the engineering word *prototype*, that is, an artificial structure that represents the first simple working model of a designed system. The importance of cooperative structures for minimal life was initially stressed in Eigen's hypercycle concept (Eigen, 1971), which focused on the cooperation between informational structures (genes). Gánti (1975) first identified the minimal cooperative structure capable of forming a cell as the triad of genes, metabolism, and container.

We use the terms *container*, *metabolism*, and *gene* quite generally, with minimal presupposition concerning their chemical details. In most contexts, the container will be an amphiphilic structure such as a vesicle or micelle, but immobilizing chemicals on a surface can also achieve the required spatial localization. Similarly, metabolism could harvest redox energy or light, it could work with more or less complex material precursors, and it might or might not use adenosine triphosphate (ATP) and complex enzymes. Furthermore, genes might achieve informational control and inheritance with some nucleotide other than DNA, or even without using any biopolymers.

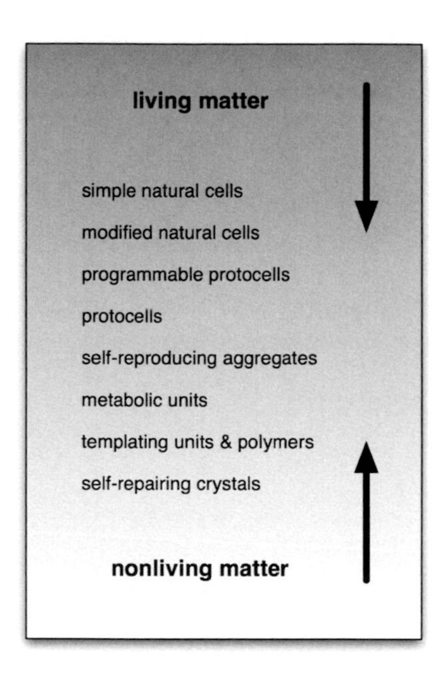

Figure I.2
The bottom-up approach focuses on assembling a minimal protocell from simple inorganic and organic components. The top-down approach focuses on the simplification of modern cells. Eventually these two approaches will meet in the middle.

One way to produce a protocell is to combine an appropriate mix of nonliving molecules and then let them react and self-assemble into a living protocell. This process begins with molecular constituents and so may be considered a *bottom-up* approach to protocells. Most of the work in this book falls within this bottom-up approach. Complementing this is a *top-down* approach based on the recent success of genomic research. The top-down approach begins with an existing contemporary living cell, typically a very simple one, and reduces its genome by successive removal of genes, to arrive at a minimal cell with just enough genes to maintain itself and reproduce (Hutchison et al., 1999). One long-range aim of this top-down research is to make an artificial cell by destroying a cell's original genome and inserting a new minimal genome that is synthesized externally from nucleotides using genomic technology (Glass et al., 2006). A spectrum of intermediate levels of functional organization is spanned by top-down approaches beginning with living matter (contemporary cells) and bottom-up approaches beginning with nonliving matter, as illustrated in figure I.2.

Once a population of protocells exists, their functional effectiveness might cause some to be selected over others. If there is a combinatorially large family of possible forms of informational control, and if information inheritance is neither too perfect nor too imperfect, then the process of evolution might be able to improve those functionalities over time. Evolvability is often thought to be an essential property of life (e.g., Maynard Smith, 1975), but it remains elusive in existing protocell realizations. In fact, it is an open question under which conditions such a system will be able to exhibit open-ended evolution, by which we mean a system's ability to create new properties in an open-ended manner over time as a result of selection and environmental pressures (Bedau et al., 2000). Statistical analysis of evolutionary systems, including artificial life models, indicates that the only systems known to exhibit open-ended evolution are natural life (the biosphere) and human technology, which is itself generated by living entities (Bedau et al., 1997; Skusa and Bedau, 2002). One key to achieving open-ended evolvability is the ability of evolving systems to generate novel properties. Novel properties in molecular systems can be obtained not only by genetic rearrangement but also by aggregation of existing systems. For example, combining a lipid membrane system with a photosynthesizer might produce a simple light-driven metabolic system. Aggregation-generated novelty could be as important as genetic variation in the evolution of protocells.

Most of the protocell architectures presented in this book rely crucially on the process of self-assembly. This means that the resulting protocell objects are not designed or constructed piece by piece, as happens with the products of traditional engineering; rather, the requisite materials are brought together under the appropriate laboratory conditions and the protocells spontaneously form. Protocells are themselves emergent structures, so designing them and controlling their functionality will require the development of new techniques of emergent engineering.

I.2 Scientific Roots of Protocell Research

Contemporary protocell research has a number of scientific roots. One is the long tradition of work on the origin of life. The question of how life might be obtained from nonliving materials is similar to that of how contemporary life originated, since contemporary life presumably arose from nonliving materials, but these two questions lead us in very different directions. The methods and conditions used to make protocells might be radically unlike those actually involved in life's origins, and it is an entirely open question whether such new forms of life would or could ever evolve into anything like naturally existing life-forms. The understanding of life obtained from making protocells, however, should contribute to research on the origin of life as well as benefit from it.

Research on the origin of life naturally spawned several of the research threads that comprise current protocell efforts. A particularly important thread is the concept of the RNA world, which concentrates on RNA as the primary element in origin of life scenarios (see, e.g., Gilbert, 1986; Orgel, 1994). The connection with protocell research comes when RNA chemistry is integrated with container structures (cf. Szostak, Bartel, and Luisi, 2001; see also chapters 2 and 5).

Protocell research may also be seen as an endeavor within the field of artificial life. The phrase "artificial life" is much broader and refers to any attempt to synthesize the essential features of living systems. Artificial life traditionally falls into three branches, corresponding to three synthesis methods. "Soft" artificial life creates computer simulations or other purely digital constructions that exhibit lifelike behavior. "Hard" artificial life produces hardware implementations of lifelike systems, usually in robotics. "Wet" artificial life involves the creation of lifelike systems in a wet lab, in most cases based on carbon chemistry in water. The holy grail of wet artificial life is the construction of protocells, which is the focus of this volume.

Another strand of human activity that produces living cells that are not found in nature is represented by the recent success of synthetic biology efforts to alter metabolic pathways of existing contemporary cells by genetic manipulation (for an overview see Baker et al., 2006). These techniques may produce a range of more or less artificial cells, depending on the extent of genetic modification and the resulting distance from naturally existing life. If synthetic biology is characterized as the attempt to engineer new biological systems, then protocell research can be viewed as a branch of synthetic biology (see the introduction to part IV).

Astrobiology is concerned with the search for life elsewhere in the universe, so it and protocell research share an interest in the fundamental properties of living systems. While the artificial life community asks what is minimal life and how can it be useful, the astrobiology community asks where life comes from and whether it exists only on the Earth. In contrast with astrobiology, protocell research does not need to justify the cosmological or geochemical origins of its starting materials. However, the possibility of detecting alien life on other planets or moons adds many intriguing questions, some of which we will touch on in the last section of this book.

I.3 Purposes and Organization of This Book

This volume is a comprehensive general resource on protocell research intended to serve a number of functions. The first, of course, is simply to convey the state of the art in contemporary protocell research. Doing this comprehensively involves a rich multiplicity of perspectives: different disciplines (e.g., physics, chemistry, biology, biochemistry, geochemistry, materials science, computer science, biophysics, evolutionary theory, engineering, philosophy), different approaches to creating protocells, the employment of different scientific methods (observation, experiment, simulation, theory), and the understanding of processes happening on different spatial and temporal scales. A resource that collects key protocell research in one place can promote mutually beneficial crosstalk with related and overlapping scientific and engineering projects, such as studies of the origin of life and extraterrestrial life, efforts in synthetic and systems biology, investigations in material science and nanotechnology, as well as the study of artificial life and artificial chemistry in general.

This snapshot of protocell research today can also promote, inform, and constructively focus future work in this area in at least two ways: by enhancing synergies among the various perspectives and approaches currently being pursued within protocell research, and by identifying and calling attention to milestones that represent the key tractable open questions that will most likely propel scientific progress.

Finally, being able to bridge the gap between the nonliving and the living will raise a number of new practical issues for society to face. The capacity to engineer truly living technology will galvanize a wealth of practical projects and surely spawn a diverse ecology of revolutionary applications. But in the wrong hands, or with the wrong designs, protocell technology could generate new kinds of risks to human health and the environment. In addition, the practice of creating novel forms of life from scratch could shock existing cultural norms, with the potential to fundamentally reshape our sense of who we are and where we come from. A comprehensive compendium on protocells could help those—whether scientists, journalists, public officials, or the general public—interested in gaining a scientifically informed perspective on this rapidly expanding new field.

These purposes for this volume have informed its overall organization. Part I contains four overviews of protocell science. These range from histories of experimental efforts to a conceptual framework for comparing experimental and theoretical protocell achievements and proposals. Together the chapters provide complementary views on how to evaluate and categorize the field. The eight chapters in part II present some of the leading approaches to integrating the core complementary functionalities of containment, metabolism, and informational control and transfer (genetics) into a fully functioning protocell. The work presented here includes experimental efforts, simulation of theoretical schemes, attempts to combine both approaches, and a method of achieving integration with the help of computer-controlled microfluidic and microelectrical devices.

Part III focuses on the foundational functional components of protocells provided by containment, metabolism, and genetics. These nine chapters dive into experimental and simulation details of replicating informational molecules, lipid membranes and compartments, and metabolic processes as well as energetics. This section also includes a chapter on protocell simulation methods.

Part IV situates protocell science in the broader contexts of theories of life, minimal forms of life, the related scientific investigation into the origin of life, and investigations of some practical applications and social and ethical implications. The book closes with a glossary of technical terms found in the chapters.

I.4 Why Now?

Modern studies of simple protocells began to appear in the 1980s, typically in the context of origin of life research. (For a historical account of earlier work, see chapter 1.) This modern research was inspired by the discovery of Bangham and coworkers in (1965) that phospholipids could not only assemble into the vesicular structures now called liposomes, but also could trap concentration gradients of ions and other solutes. From this work it immediately became apparent that the lipid bilayer was the primary permeability barrier to the free diffusion of ions and metabolites, and that specialized proteins must be present in the bilayer in the form of channels and pumps to allow communication between the intracellular volume and the external environment.

In the 1990s, other laboratories began to investigate encapsulated systems of macromolecules. The early goals were to develop efficient encapsulation methods, design lipid compositions and techniques that would permit external substrates to supply encapsulated enzymes, establish conditions in which lipid vesicles could undergo growth and division, and demonstrate catalyzed synthesis of polymers such as RNA, DNA, and peptides within the liposome volume. As these goals were achieved and reported in the literature, other laboratories began to think seriously about developing a variety of further protocell systems.

Much of the theoretical research on minimal life has its intellectual origin from the ideas presented by the experimentalist Manfred Eigen back in 1971, when the notions of quasispecies and hypercycle were created (Eigen, 1971). Cooperative structures that provide mutual support of genes, metabolism, and containers can be seen as generalizations of Eigen's ideas. Also, Ilya Prigogine's creation of the concept of dissipative structures in the late 1970s gripped the imagination of many theorists trying to understand the origins of life. The notion from the late 1980s of the RNA world inspired both theoretical and experimental investigations of minimal life, and even today this concept is probably still the most prevalent hypothesis about the bridge between early minimal life and contemporary life.

With the turn of the millennium, ever-increasing numbers of publications explicitly related to functional protocells began to appear. An Internet search in 2006 for the terms *protocell* and *artificial cell* produced approximately 50,000 references to each term. A similar search in the scientific literature revealed 435 citations for *protocell* and 1,310 citations for *artificial cell*. These numbers reveal exploding interest in these topics since the first modern publications about 25 years ago.

What might have given rise to this remarkable increase? First, of course, is the intrinsic interest in bridging the gap between nonliving and living matter. To truly understand what it would take to create a minimal living cell, the obvious challenge is to verify this understanding by assembling a minimal cell from its parts list. The first fabrication of a functioning protocell will be a major scientific breakthrough. It is also clear that protocell technology will have significant applications, particularly in the biotechnology, material science, computer and information technology, and environmental science industries.

All of these considerations make this book timely and of real use to the increasing number of investigators who see research on protocells as their primary scientific quest. We expect that this book will serve as a focus for such activity, calling attention to the progress that has been made so far, and identifying milestones to guide future research and development.

In an environment in which basic scientific research suffers from decreasing resources, society owes great thanks for the foresight and vision of national funding sources, such as the Department of Energy's support for Los Alamos National Laboratory and NASA programs in Astrobiology and Exobiology, as well as for the European Commission's program on Future and Emerging Technologies, the United Kingdom's Engineering and Physical Sciences Research Council, the Japan Society for the Promotion of Science, and the Japanese Minister of Education, Culture, Sports, Science, and Technology, all of which have provided significant financial support for protocell research well in advance of any proven applications. Without these kinds of resources, the research reported in this book would have been impossible.

The eventual applications of protocell research will come through the flowering of living technology. Living technology is one of the first concrete realizations of what the U.S. National Science Foundation (NSF) and the European Commission (EC) have termed *convergent technologies*. Convergent technologies are the emerging syntheses of nano-bio-info-cognitive (NBIC) knowledge production, and the NSF and EC both believe that convergent technologies will have a very large socioeconomic impact in the next 25 years. We hope that this volume will help forge an international community of interested stakeholders working with protocell scientists to explore the broader global opportunities and challenges of protocell research and the emergence of living technology.

References

Baker, D., Church, G., Collins, J., Endy, D., Jacobson, J., Keasling, J., & Modrich, P. (2006). Engineering life: Building a fab for biology. *Scientific American, 294*, 44–51.

Bangham, A. D., Standish, M. M., & Watkins, J. C. (1965). Diffusion of univalent ions across the lamellae of swollen phospholipids. *Journal of Molecular Biology, 13*, 238–252.

Bedau, M. A., McCaskill, J. S., Packard, N. H., Rasmussen, S., Adami, C., Green, D. G., et al. (2000). Open problems in artificial life. *Artificial Life, 6*, 363–376.

Bedau, M. A., Snyder, E., Brown, C. T., & Packard, N. H. (1997). A comparison of evolutionary activity in artificial evolving systems and the biosphere. In P. Husbands & I. Harvey (Eds.), *Proceedings of the fourth European conference on artificial life, ECAL97* (pp. 125–134). Cambridge, MA: MIT Press.

Eigen, M. (1971). Self-organization of matter and the evolution of biological macromolecules. *Naturwissenschaften*, *58*, 465–523.

Gánti, T. (1975). Organization of chemical reactions into dividing and metabolizing units: The chemotons. *Biosystems*, *7*, 15–21.

Gilbert, W. (1986). The RNA world. *Nature*, *319*, 618.

Glass, J. I., Smith, H. O., Hutchinson III, C. A., Alperovich, N. Y., & Assad-Garcia, N. (2007). Minimal bacterial genome. U.S. Patent Application Publication 2007/0122826.

Hutchison, C. A. III, Peterson, S. N., Gill, S. R., Cline, R. T., White, O., Fraser, C. M., et al. (1999). Global transposon mutagenesis and a minimal mycoplasma genome. *Nature*, *286*, 2165–2169.

Maynard Smith, J. (1975). *The theory of evolution*, 3rd ed. New York: Penguin.

Orgel, L. E. (1994). The origin of life on the Earth. *Scientific American*, *271*, 76–83.

Skusa, A., & Bedau, M. A. (2002). Towards a comparison of evolutionary creativity in biological and cultural evolution. In R. Standish, M. A. Bedau, & H. A. Abbass (Eds.), *Artificial life VIII* (pp. 233–242). Cambridge, MA: MIT Press.

Szostak, J. W., Bartel, D. P., & Luisi, P. L. (2001). Synthesizing life. *Nature*, *409*, 387–390.

I OVERVIEW: BRIDGING NONLIVING AND LIVING MATTER

1 The Early History of Protocells: The Search for the Recipe of Life

Martin M. Hanczyc

1.1 Introduction

Vitalism is an old ideology that makes a clear distinction between chemicals comprising a living organism and chemicals from inanimate matter. The synthesis of urea by Friedrich Wöhler (1828) challenged the philosophical distinction between nonliving and living matter by demonstrating that a biological organic molecule can be synthesized in a laboratory from nonliving chemical precursors. Further development of organic chemistry demonstrated how most of the organic molecules that comprise a cell could be synthesized in a laboratory under detailed reaction conditions. Subsequent experiments such as the Buchner brothers' fermentation of sugar into ethanol (a biological process) by the nonliving yeast extracts in 1897 opened the door further to exploration of the conceptual gray space between chemistry and biology (Friedmann, 1997). The following brief historical introduction recounts the principal experimental endeavors to create a protocell that began more than 100 years ago.

1.2 The First "Cell Models"

The first studies from the latter half of the nineteenth century that reported lifelike behaviors from nonliving systems were not based on cells or even cell extracts. Beginning in 1867, Moritz Traube described a system that demonstrated the formation and growth of semipermeable membranes composed of copper ferrocyanide surrounding a seed crystal of copper sulfate. Such an artificial membrane superficially resembled a cell membrane in that it exhibited selective permeability to different solutes and responded to osmotic pressure (Traube, 1867). In 1892, Bütschli described structures with an amoebalike movement formed by combining olive oil and potash. The dynamic structures appeared to form pseudopodia and engulf other particles, behaviors seen with living amoebae (Bütschli, 1892). Leduc (1907) also produced "osmotic cells" similar to Traube's by placing a seed crystal of calcium chloride in a saturated potassium salt solution. As the crystal dissolved, a membrane of calcium phosphate appeared and formed a structure similar in morphological appearance to a cell. He reported the shape change of the cell-like structures in response to environmental changes in osmotic pressure that presumably modulated the surface of the structures. These early cell models showed that simple, nonliving materials such as salt solutions could organize into structures at least superficially resembling living cells (Oparin, 1965a). Even though the relevance of such models to actual living cells was not clearly established at that time, the tendency of nonliving matter to adopt some characteristics of life fascinated researchers interested in not only the origin of but also the synthesis of life.

1.3 Herrera's Sulphobes

In 1897 Alfonzo L. Herrera published a book, *Recueil des lois de ta Biologie Generale* (*Collection of Laws of General Biology*), in which he introduced the idea that the functions and structures of living cells can be attributed to physicochemical laws (Negrón-Mendoza, 1994). In 1904, Herrera founded a new branch of science, plasmogeny, which he defined as the study of the origin of protoplasm as an experimental science. Protoplasm was a term used to describe the living material inside a cell. It had the observable properties of flow and movement, was the site of metabolism, and was able to self-replicate. Herrera began to explore how such properties could be the result of physicochemical forces acting and interacting within living cells. His approach was to reconstitute such properties using nonbiological chemicals. By mixing substances with different chemical potentials and different phases, he observed lifelike patterns emerging. He is most famous for producing "sulphobes" by exposing thin films of formaldehyde in water to fumes of ammonium sulfide (Herrera, 1942). Through microscopy (figure 1.1), Herrera observed populations of many diverse structures reminiscent of protoplasm, including structures that appeared to undergo mitotic division (Herrera, 1912). Apparently Herrera's sulphobes were so lifelike that when presented to an "eminent microscopist," they were determined to be living creatures and were then classified as such (Young 1965, p. 357).

Plasmogeny offered a view of biology seen through the eyes of a synthetic chemist. With his sulphobes, Herrera demonstrated that the superficial structure and organization of the protoplasm could be recapitulated through purely chemical means. Furthermore, such structures produced chemistry that resulted in the fortuitous synthesis of two amino acids and dyes. Therefore the sulphobes not only resembled lifelike structures but also were capable of limited biologically relevant synthesis. Most notably Herrera's synthesis of protoplasm-like properties in the laboratory was demonstrated starting from simple, nonbiological constituents.

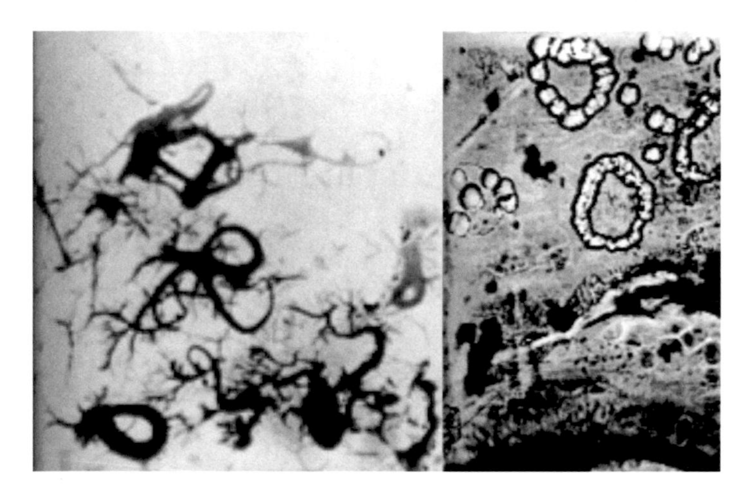

Figure 1.1
Sample micrographs from the work of Herrera showing various cell-like morphologies (×800 magnification) from Herrera (1912).

1.4 Bungenberg de Jong's Coacervates

Although Herrera began an extensive experimental research program to recapitulate and understand protoplasm through chemistry, the origin of the idea that the protoplasm is made up of chemicals subject to physicochemical laws can be found in early research on colloidal systems. Thomas Graham (1805–1869), an English chemist, studied the diffusion of nonbiological materials and biological extracts such as starch, gum, and gelatin. He termed the class of substances with very low diffusion rates *colloids*. In 1861, he wrote that the plastic parts of the animal body also behave like colloids (Graham, 1861).

The Dutch biochemist H. G. Bungenberg de Jong (1893–1977) coined the term *coacervate* as a special from of colloid. When certain organic substances such as gelatin and gum arabic are mixed with an aqueous solution, the fluids separate into two distinct phases. Upon agitation, the bottom, organic-rich layer breaks up into a population of small microscopic spherical structures or coacervates (see figure 1.2). These structures of a few microns in diameter do not readily dissolve because of a tight layer of water molecules that interacts with the charged organics at the surface surrounding the coacervate. This in effect delays the diffusion of concentrated material in the coacervate sphere into the surrounding media. Bungenberg de Jong (1932) noted the similarities between coacervates and living protoplasm, and like Graham before, held the idea that the physical conditions that produced coacervates may also explain something about the organization of the protoplasm.

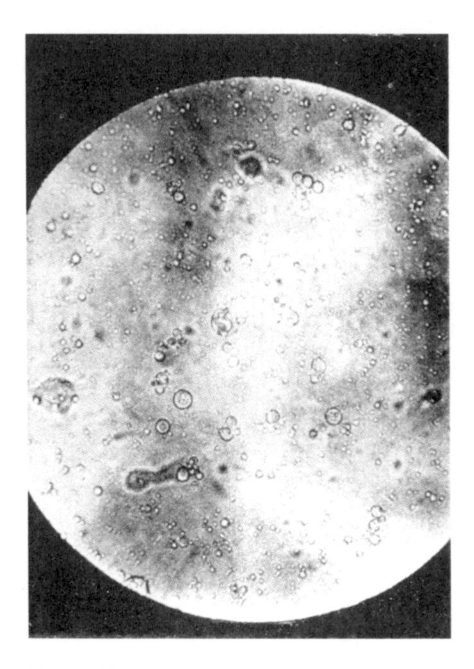

Figure 1.2
Coacervates containing bacterial polynucleotide phosphorylase enzyme able to synthesize RNA polymers, from Oparin, 1965b (magnification not specified).

Coacervates are interesting as model systems of simple cells in that they form spontaneously, are rich in organic material (as opposed to the aqueous environment around the coacervate), and provide a locally segregated environment with boundaries that allow selective absorption of exogenous organic molecules. Coacervates are most stable when formed from a mixture of positively and negatively charged substances resulting from an interplay of hydration and electrostatic forces. As a result, coacervates are dynamic, can respond to external manipulation, and at the same time are quite unstable, especially when the conditions under which they form are changed. As a result of such investigations into colloidal and coacervate systems, it was thought that living protoplasm consisted of different coacervate systems acting and interacting together to produce the various dynamical properties observed in living cells.

1.5 Crile's Autosynthetic Cells

At the Cleveland Clinic Foundation in the 1930s, Doctor George Crile and colleagues experimented with the reconstitution of artificial cells from living material. Two fractions were prepared from animal brain: one containing the lipid (lipoid) ma-

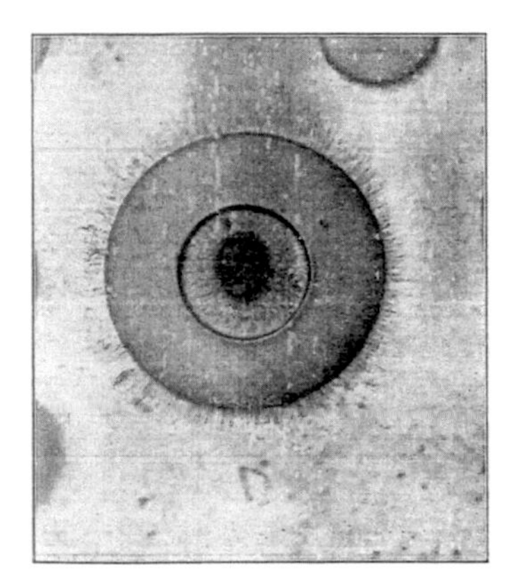
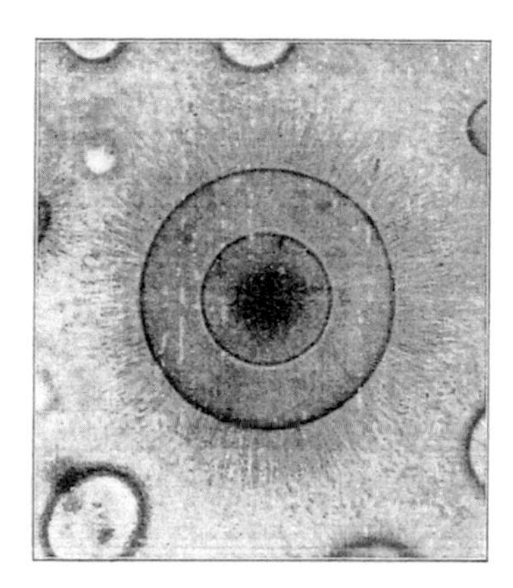

Figure 1.3
Crile's autosynthetic cells (×400 magnification; reproduced from Crile, Telkes, and Rowland, 1932).

terial and another containing sterilized protein material. When these fractions were mixed together with a salt solution representing the salinity found in the animal brain, structures resembling cells self-assembled. These structures, typically 50 to 150 microns in diameter, were termed *autosynthetic cells* (Crile, Telkes, and Rowland, 1932) (figure 1.3) and were scrutinized for lifelike behavior. Beyond their morphological similarities to living protozoa, Crile determined that the autocatalytic cell preparations consumed a certain amount of oxygen and produced carbon dioxide (in comparison with controls), and concluded that a certain amount of metabolism was taking place. However, it was never shown what this metabolism-like activity was, or if it was localized to the cell-like structures. In addition, they observed budding and division of the structures by microscopy. The autosynthetic cells could be maintained for months by feeding the structures with the sterile protein fraction, or destroyed by heat, poisons, radiation, lack of oxygen, or lack of food.

Crile and colleagues relied heavily on self-organized processes in living matter to reconstitute some of the properties of life from biological extracts. They then used currently available techniques to test for signs of life. The work of Noireaux and Libchaber (2004) recreated a similar approach of artificially reconstituting cells from extracts under more controlled conditions and using modern techniques.

1.6 Oparin and Coacervates

Although A. I. Oparin's interests were focused on understanding the origin of life, he made many contributions to the field of protocell research, primarily by introducing metabolism to coacervates. Oparin and colleagues continued the pioneering work of Bungenberg de Jong and made coacervates from gum arabic (an extract from the

Acacia tree, a complex and variable mixture of arabinogalactan oligosaccharides, polysaccharides, and glycoproteins) and gelatin (a protein product formed by partial hydrolysis of collagen extracted from animal skin, bones, cartilage, and ligaments). Such coacervates were prepared containing phosphorylase enzymes capable of polymerizing the substrate glucose-1-phosphate into starch. Once the coacervates containing the enzyme were prepared, the substrate was added to the external medium and adsorbed by the coacervates. It was found under such conditions that starch production proceeded within the coacervate. Taking this experimental design a step further, Oparin prepared coacervates containing two different enzymes, phosphorylase and β-amylase, which together could form a simple two-step enzymatic pathway. When the substrate glucose-1-phosphate was added exogenously, it was found that not only did starch form in the coacervate droplet but also maltose, the product of β-amylase acting on the starch, was found in the external media (figure 1.4). Reportedly Oparin and colleagues succeeded in encapsulating polyadenine synthesis (see figure 1.2), NAD oxidative-reductive reactions, and ascorbic acid oxidation within coacervates (Oparin, 1966). Oparin also understood that the rates of reaction for each step determine whether the coacervate assembly would grow by incorporating and converting the incoming substrates into retainable products or diminish by converting all available molecules into highly diffusible product (Oparin, 1965b).

This work with coacervates demonstrated a few essential points. First, structures with a superficial semblance to living protoplasm can be formed in the laboratory from organic material. Such structures can trap functional molecules such as enzymes. Molecules added to the external media can be absorbed by the coacervate structure. The absorption is selective and depends on the composition of the coacervate. Once absorbed, the molecules can be acted on by the trapped enzymes and the product accrues within the structure. Finally, the products of the reaction can also diffuse out of the structure into the environment. The essential role of selective permeability is clearly demonstrated. As Young stated in 1965, "It [a selectively permeable barrier] permits the cell to create its own internal environment necessary for its metabolic and reproductive activities" (Young 1965, p. 348). On the importance of metabolism, Oparin stated that "individual chemical reactions in living beings are strictly coordinated and proceed in a certain sequence, which as a whole forms a network of biological metabolism directed toward the perpetual self-preservation, growth, and self-reproduction of the entire system under the given environmental conditions" (Oparin 1965b, p. 331).

By engineering supramolecular structures capable of housing a simple metabolic pathway, Oparin demonstrated that the reconstitution of a population of structures with simple lifelike metabolism was possible. Although Oparin's main objective for this work was to understand fundamental processes related to chemical evolution and the origin of cellular life, his work with coacervates followed an approach more

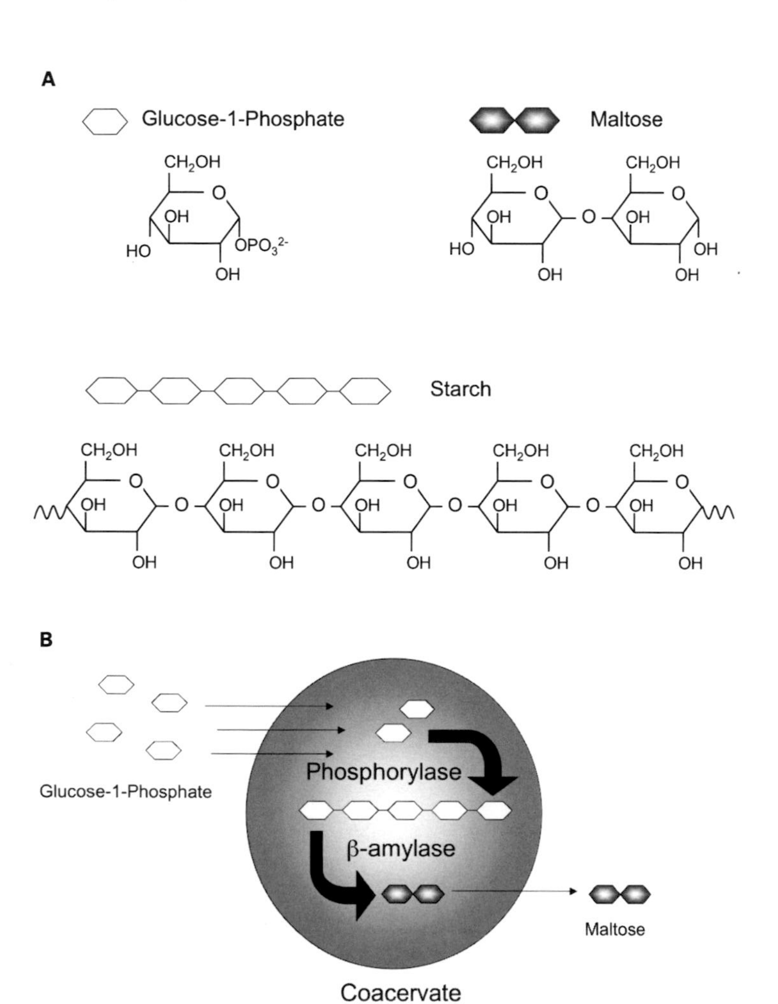

Figure 1.4
(A) Representation of molecules involved in the coacervate enzymatic pathway. (B) Schematic of the encapsulated enzymatic pathway. The enzymes, phosphorylase and β-amylase, are entrapped within the coacervate structure. The substrate, glucose-1-phosphate, is supplied exogenously and absorbed by the coacervate. The substrate is converted to starch within the coacervate droplet by phosphorylase. Then the β-amylase converts the starch to maltose, which diffuses out of the coacervate structure (Crile, Telkes, and Rowland, 1932).

akin to synthetic biology. This is because his work was based on organic molecules and enzymes extracted and purified from already living matter. (In contrast, Herrera's work involved the generation of structures directly from nonbiological chemicals.) Because of his success in reconstituting lifelike microscale structures in the 1950s and 1960s, Oparin inspired many to pursue the fundamental questions of the origin of life and to understand the basic physicochemical forces that underlie cellular life.

1.7 Jeewanu

In the 1960s Krishna Bahadur from the Department of Chemistry at the University of Allahabad reported on the synthesis of protocells called *Jeewanu* (Bahadur, 1966). The name given to these structures, Jeewanu, comes from a Sanskrit work for particles of life. Jeewanu were made using various protocols and components. Some were made by combining peptides, ascorbic acid, ammonium molybdate, and inorganic minerals. Others were made by mixing paraformaldehyde, molybdic acid, and ferric chloride in water and then placing this mixture on a nutrient-rich agar. After some time (hours to weeks) under exposure to a source of light (sunlight or artificial ultraviolet light), globular structures of about 0.5 to 15 microns in diameter appeared. Bahadur claimed that the Jeewanu had a similar morphology to living cells, grew over time, possessed weak metabolic activity, and could reproduce by budding. He also made the claim that Jeewanu were living cells. The creation of Jeewanu was an attempt to synthesize actual living cells using simple precursors with the goal of understanding the origin of life. The inadequate detail provided in his publications led to criticisms (see Caren and Ponnamperuma, 1967) and doubt was cast on his claims.

1.8 Fox's Microspheres

With the development of molecular biology in the 1950s through the 1970s, the fundamental role of self-organization or self-assembly in biological systems became increasingly apparent (Fox, 1968; Wald, 1954). It is this type of interaction among molecules that gives rise to higher-order structures of complexity in modern-day cells and is a main factor in the formation of sulphobes, autosynthetic cells, Jeewanu, and coacervates as described earlier. In addition, self-organization provides a tool that can be used to construct protocells without relying excessively on precise and often untenable manipulation on the molecular level.

In the 1960s and 1970s, Sidney Fox (1912–1998) and colleagues contributed to the pursuit of the synthesis of life by creating populations of peptide-based spherical structures in the laboratory. Specifically, by heating amino acids, mixing in water or

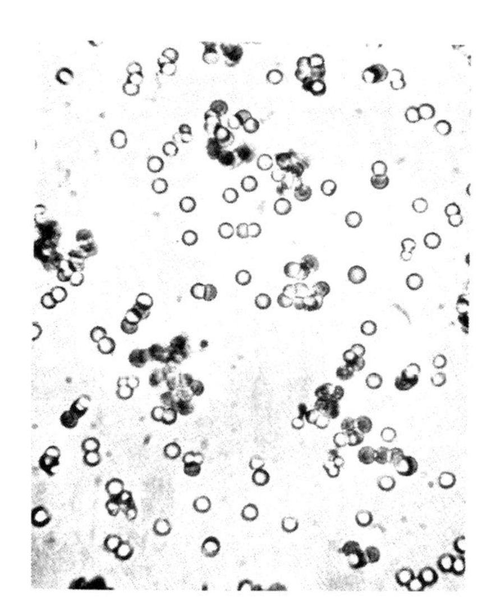

Figure 1.5
Microspheres of proteinoid in water from Young, 1965 (×900 magnification).

salt solution, and then allowing the solution to cool slowly, Fox found that spherical structures formed spontaneously (figure 1.5). These structures were similar in diameter (typically a few microns), with each microsphere containing an estimated 10^{10} proteinoid molecules. The proteinoid material formed during the heating step by linking and crosslinking of the amino acid monomers to form more elaborate and complex peptides. These peptides would then interact and self-assemble within minutes into larger macromolecular structures called microspheres (Fox and Dose, 1972). Fox's microspheres were found to be generally more stable than coacervates. Fox extensively characterized the dynamics of his microspheres, reporting that he observed many interesting phenomena including diffusion of material from the inside of the sphere to the outside, partial fission (due to dissolution of proteinoid and change in surface tension), generation of double-layer membrane, shrinkage or swelling resulting from osmotic changes, sensitivity to pH (depending on amino acid composition), selective permeation of polysaccharides and monosaccharides, motility (with zinc and ATP), budding, growth by accretion, and formation of junctions between microspheres. In addition, Fox and his colleagues reported some weak enzymelike activity in the proteinoid products (Fox, 1965, 1968). It is not clear whether microspheres exhibiting such behaviors were common or the conclusions were based on a few rare events in the population.

Like Oparin, Fox was searching for abiotic processes that could have led to the formation of cellular life. In much the same way, microspheres showed encapsulated metabolism and supramolecular dynamics. One fundamental difference between coacervates and microspheres is that Fox demonstrated the self-assembly of a population of complex microstructures from simple precursors.

1.9 Toward the Modern Concept of the Protoplasm

The cell model systems presented here were simplistic conceptualizations of biological cells and may not have been accurate representations of the important forces that define a living cell. Concurrent research into properties of living cells painted a different picture of the vital physicochemical properties of life. Although coacervates and microspheres were shown to have some type of selectively permeable barriers, other research has shown that all natural living cells use a particular kind of barrier composed of a lipid membrane. In 1900, Charles E. Overton showed that nonpolar, oily chemical substances were selectively absorbed by plant cells. He hypothesized that the cell membrane barrier is similar in some way to olive oil and is lipoid in composition (Overton, 1900). Further investigation into the physicochemical nature of lipids by Irving Langmuir (1917) led to the idea that fatty acid molecules form a monolayer at the air-water interface with the fatty hydrocarbon chains oriented away from the water. Gorter and Grendel (1925) then performed experiments similar to Langmuir's, but with lipid extracts from erythrocytes. They were able to calculate that enough lipid was present to cover approximately twice the surface area of the cells, concluding that the cells were covered in a layer of lipids "two molecules thick." This research may have inspired Crile and colleagues to use two fractions to construct their autosynthetic cells: the protein fraction to reconstitute the cytosol and the lipoid fraction to reconstitute the membrane.

Danielli and Davson (1935) proposed a bilayer model of lipid cell membrane in which lipids were arranged such that the hydrocarbon chains formed the interior of the membrane sandwich, and proteins covered the lipid headgroups on both sides of the membrane. This hypothetical structure was further supported by electron microscopy studies in the 1950s, and renamed the *unit membrane model* (Robertson, 1957). Further evolution in the conceptualization of the cell membrane led to the current fluid mosaic model of the cell membrane (Singer and Nicholson, 1972). In this model the cell is enclosed in a bilayer of lipids as before, but now the lipids as well as integral proteins move within the membrane. The fluid mosaic model allows for a detailed understanding of processes in real cells such as osmotic response, selective permeability, and specific cell-cell and cell-environment communication. Bangham demonstrated in 1965 that dispersions of phospholipid extracts self-assemble into spherical structures reminiscent of cell membranes, and these "liposomes"—like cell membranes—are osmotically active (Bangham, Standish, and Watkins, 1965). This provided the first in vitro model of a real cell membrane. Since then, liposomes have been developed extensively as models of natural cell membranes and used in constructing protocells.

In sulphobes, autosynthetic cells, coacervates, Jeewanu, and microspheres, the morphology of the structures bears a superficial resemblance to living cells. The electrostatic forces of interacting polymers and the crosslinking of peptides may explain their shape and consistency. However, the real structure of the cytoplasm is of a much different complexity. While Graham, Bungenberg de Jong, and others were investigating the properties of coacervates, other contemporaneous models of the protoplasm were being developed. One included a view in which the internal contents of the cell are physically organized in three dimensions by a protein network of fibers. In the eighteenth and nineteenth centuries, an intracellular fiber network was proposed to explain contraction of animal muscles (see Frixione, 2000). Sigmund Freud, the father of psychoanalysis, was one of the original proponents of the fibrillary theory of protoplasm structure as it applied to his research into nerve cells (Freud, 1882; Triarhou and Del Cerro, 1987). During the early part of the twentieth century, similar conceptualizations were applied to explain the intracellular organization and movement in many types of animal cells. Such hypotheses were supported in the 1950s through the visualization of protein fiber networks by electron microscopy. Today, we understand that the consistency of the eukaryotic cytoplasm is not due to colloidal properties of the constituents after all, but rather to an elaborate protein cytoskeleton that is responsible for the cell shape as well as internal organization and cell movement. Evidence also exists for similar protein networks in bacteria (see Carballido-Lopez, 2006).

Returning now to the attempts to synthesize protocells, those cell models did not, after all, represent the true complexity of living protoplasm beyond superficial morphological similarity. Nevertheless, the early research into the recipe of life has shown that much simpler aggregate constructions can exhibit some lifelike behaviors. Furthermore, some structures have been engineered to contain primitive metabolisms. So although such works may not have successfully explained the physicochemical forces of cellular life, they can be seen as useful attempts to synthesize a simplistic form of artificial life.

1.10 Summary

The ideology of vitalism has been challenged, and its once-defined lines blurred by the research programs that ambitiously tried to synthesize life and lifelike behavior from nonliving components. These early attempts at synthesizing life from nonlife fueled the debate about what life essentially is. Such research projects attracted the interest of many researchers and thinkers from a wide range of disciplines, and also of religious leaders. Indeed, Sidney Fox was invited on multiple occasions to the Vatican for discussions about the origin and synthesis of life with the Pope and his scientists (Fox, 1997). Self-organization of structure, response to environmental changes, uptake of nutrients, internal metabolism, expulsion of waste, movement, and self-replication: These concepts persist in modern-day attempts to synthesize protocells, as described in the following chapters of this book. Although these early attempts did not satisfactorily represent the real physicochemical forces that shape a modern living cell, they did show the tendency of matter, under the proper conditions, to organize itself into structures possessing similar qualities to living cells. And in that way, they imply that there may be many paths toward synthesizing primitive life.

In 1966, at the national meeting of the American Chemical Society (ACS), a symposium on the synthesis of living systems was held to present the major steps in the developing field of molecular biology. The topics ranged from the laboratory synthesis of bovine insulin and other proteins to the synthesis and replication of nucleic acids. The symposium prompted the publication of a paper in 1967 suggesting that "[i]t is the accomplishments of the past decade and the confidence in the creative powers of scientists which have led to speculation that these accomplishments may lead, before the end of the century to the synthesis of new types of rudimentary living systems" (Price, 1967, p. 144).

In an earlier optimistic view published in *Science* in 1960, Simpson noted that "[a]t a recent meeting in Chicago, a highly distinguished international panel of experts was polled. All considered the experimental production of life in the laboratory imminent, and one maintained that this has already been done ..." (Simpson 1960, p. 969).

Despite predictions, most researchers would agree that the goal of a synthetic cell has not yet been achieved. The question remains: When will science create synthetic life? This is difficult to answer. Recently, Jack Szostak remarked that $20 million dollars and three more years of research would produce a living and evolving protocell (Zimmer, 2004). Indeed, a well-funded, consistent research program may reach the desired goal in a relatively short time. Protocell history captured in manuscripts and texts over the past hundred years shows us that progress in chemistry, physics, biology, biochemistry, and other areas inspires an enthusiasm that makes the goal of a synthetic cell seem closer. Descriptive and exploratory research in many fields, along with technological advancements, is intimately linked with the vision of turning our *knowledge of life* toward the *synthesis of life*.

Looking back over the history of protocell development, we can see that there are many often related avenues that one can take to create a living cell through chemistry. We need not only a fundamental knowledge of a variety of subjects, but also our creativity. A few scientists spent a career exploring the benefits and limitations of the chosen path described here. In the end, none of these efforts has produced a living cell. But they all asked the important questions, highlighted the challenges, and shaped our thinking. These attempts to synthesize living cells from chemistry show that a viable experimental model of a living cell will depend on more than the co-localization of various essential characters such as the type of molecule, source of information, and functional metabolism. Higher-order interactive networks must also be in place to coordinate the form and function of all components in a living cell with high reliance on the self-organizing properties of the system.

The creation of a self-replicating protocell is only one step toward the synthesis of life. Our predecessors were interested in not only creation of a self-replicating protocell through chemistry and physics but also the synthesis of a "protoplasm" capable of evolving into a diverse array of cell types and a communal network of living entities. Although current efforts often focus on the creation of a single synthetic cell, our history envisions an even larger goal of an interwoven community of synthetic structures acting and interacting to produce a rich environment suitable not only for the creation but for the evolution of synthetic life.

Acknowledgments

My sincere thanks and gratitude to Christopher Barbour, Coordinator of Special Collections, at the Tisch Library Special Collections of Tufts University, and Patricia Killiard, Head of Electronic Services and Systems, of the Cambridge University Library, for their assistance in acquiring many of the older references cited in this chapter.

References

Bahadur, K. (1966). *Synthesis of Jeewanu: The protocell.* Allahabad-2, India: Ram Narain Lal Beni Prasad, Gyanodaya Press.

Bangham, A. D., Standish, M. M., & Watkins, J. C. (1965). Diffusion of univalent ions across the lamellae of swollen phospholipids. *Journal of Molecular Biology, 13*, 238–252.

Bungenberg de Jong, von H. G. (1932). Die koazervation und ihre bedeutung für die biologie. *Protoplasma, 15*, 110–176.

Bütschli, O. (1892). *Untersuchungen über microscopische Schäume und das Protoplasma.* Leipzig: Engelmann.

Carballido-Lopez, R. (2006). The bacterial actin-like cytoskeleton. *Microbiology and Molecular Biology Reviews, 70* (4), 888–909.

Caren, L. D., & Ponnamperuma, C. (1967). *A review of some experiments on the synthesis of "Jeewanu"* (NASA Document TM X-1439). Springfield, VA: Clearinghouse for Federal Scientific and Technical Information.

Crile, G., Telkes, M., & Rowland, A. F. (1932). Autosynthetic cells. *Protoplasma, 15*, 337–360.

Danielli, J. F., & Davson, H. (1935). A contribution to the theory of permeability of thin films. *Journal of Cellular and Comparative Physiology, 5*, 495–508.

Fox, S. W. (1965). Simulated natural experiments in spontaneous organization of morphological units from proteinoid. In S. W. Fox (Ed.), *The origins of prebiological systems and of their molecular matrices* (pp. 361–382). New York: Academic Press.

Fox, S. W. (1968). Self-assembly of the protocell from a self-ordered polymer. *Journal of Scientific and Industrial Research, 27*, 267–274.

Fox, S. W. (1997). My scientific discussions of evolution for the Pope and his scientists. *The Harbinger*, May 27. Mobile.

Fox, S. W., & Dose, K. (1972). *Molecular evolution and the origin of life*. San Francisco: W. H. Freeman and Company.

Freud, S. (1882). Über den bau der nervenfasern und nervenzellen beim flusskrebs. *Sitzungberichte der Kaiserliche Akademie der Wien, mathematische-naturwissenschaftliche*, Classe 85 (Abth) 3, 9–46.

Friedmann, H. C. (1997). From Friedrich Wöhler's urine to Eduard Buchner's alcohol. In A. Cornish-Bowden (Ed.), *New beer in an old bottle: Eduard Buchner and the growth of biochemical knowledge* (pp. 67–122). Valencia: Universitat de València.

Frixione, E. (2000). Recurring views on the structure and function of the cytoskeleton: A 300-year epic. *Cell Motility and the Cytoskeleton, 46*, 73–94.

Gorter, E., & Grendel, F. (1925). On bimolecular layers of lipoids on the chromocytes of the blood. *Journal of Experimental Medicine, 41*, 439–443.

Graham, T. (1861). Liquid diffusion applied to analysis. *Philosophical Transactions of the Royal Society of London, 151*, 183–224.

Herrera, A. L. (1912). Resume des recherches de plasmogenie 1898–1912. *Archives de Plasmologie. Générale, 1* (1), 55–110.

Herrera, A. L. (1942). A new theory of the origin and nature of life. *Science, 96* (2479), 14.

Langmuir, I. (1917). The constitution and fundamental properties of solids and liquids. II. Liquids. *Journal of the American Chemical Society, 39*, 1848–1906.

Leduc, S. (1907). *Les bases physiques de la vie*. Paris: Masson.

Negrón-Mendoza, A. (1994). Alfonzo L. Herrera: A Mexican pioneer in the study of chemical evolution. *Journal of Biological Physics, 20*, 11–15.

Noireaux, V., & Libchaber, A. (2004). A vesicle bioreactor as a step toward an artificial cell assembly. *Proceedings of the National Academy of Sciences of the United States of America, 101* (51), 17669–17674.

Oparin, A. I. (1965a). *Origin of life*. New York: Dover Publications, Inc.

Oparin, A. I. (1965b). The pathways of the primary development of metabolism and artificial modeling of this development in coacervate drops. In S. W. Fox (Ed.), *The origins of prebiological systems and of their molecular matrices* (pp. 331–345). New York: Academic Press.

Oparin, A. I. (1966). *The origin and initial development of life*. Moscow: Meditsina Publishing House. English translation: NASA TT F-488 (1968). Springfield, VA: Clearinghouse for Federal Scientific and Technical Information.

Overton, E. (1900). Studien über die Aufnahme der Anilinfarben durch die lebenden Zelle. *Jahrbucher fur wiss Botanik, 34*, 669–701.

Price, C. C. (Ed.) (1967). The synthesis of living systems. *Chemical & Engineering News*, August 7, 1967, 144–156.

Robertson, J. D. (1957). New observations on the ultrastructure of the membranes of frog peripheral nerve fibers. *Journal of Biophysical and Biochemical Cytology, 3*, 1043–1047.

Simpson, G. G. (1960). The world into which Darwin led us. *Science, 131* (3405), 966–974.

Singer, S. J., & Nicolson, G. L. (1972). The fluid mosaic model of the structure of cell membranes. *Science, 175* (4023), 720–731.

Triarhou, L. C., & Del Cerro, M. M. (1987). The histologist Sigmund Freud and the biology of intracellular motility. *Biology of the Cell, 61*, 111–114.

Traube, M. (1867). *Archives D Anatomie Physiologie U. Wiss. Med.*, 87–128; 129–165.

Wald, G. (1954). The origin of life. *Scientific American*, August 1954, 45–53.

Wöhler, F. (1828). On the artificial production of urea. *Annalen der Physik und Chemie*, *88*, Leipzig. (online English translation at: http://dbhs.wvusd.k12.ca.us/webdocs/Chem-History/Wohler-article.html, last accessed March 13, 2007)

Young, R. S. (1965). Morphology and chemistry of microspheres from proteinoid. In S. W. Fox (Ed.), *The origins of prebiological systems and of their molecular matrices* (pp. 347–357). New York: Academic Press.

Zimmer, C. (2004). What came before DNA? *Discover*, *25* (6), 1–5.

2 Experimental Approaches to Fabricating Artificial Cellular Life

David Deamer

2.1 Introduction

Recent scientific advances suggest that it may be possible to fabricate an artificial cell (i.e., a protocell) with most, if not all, of the properties associated with the living state. At first glance, this might seem to be an impossible task, but life appears to have arisen spontaneously on the early Earth, so perhaps we can be optimistic. Yet life did not spring into existence with a full complement of genes, ribosomes, membrane transport systems, metabolism and the DNA → RNA → protein information transfer that dominates all life today. There must have been something simpler, a kind of scaffold-life that was left behind in the evolutionary rubble. Can we reproduce that scaffold? One possible approach is to incorporate one or a few genes into artificial vesicles to produce molecular systems that display the properties of life. The properties of the system may then provide clues to the process by which life began in a natural setting of the early Earth, and perhaps lead to a second origin of life, but this time in a laboratory setting.

What would such a system do? We can answer this question by listing the basic functions that would be required of artificial cellular life:

- Self-assembled lipidlike molecules form membrane-bounded compartments that encapsulate internal molecular machinery.
- Macromolecules are encapsulated, but smaller nutrient molecules cross the membrane barrier by diffusion or through polymer pores.
- Energy is captured by a pigment system (light energy), or from chemical energy or oxidation-reduction reactions.
- A primitive version of metabolism is initiated within the boundary membrane.
- The energy is coupled to the synthesis of activated monomers, and macromolecules grow by polymerization of the monomers.

• The membrane-bounded compartment grows by addition of amphiphilic molecules.

• Macromolecular catalysts speed the growth process, and primitive regulatory mechanisms evolve to control metabolism and polymerization processes.

• Information is captured in the sequence of monomers in one set of polymers, and used to direct the synthesis of a second set of catalytic polymers, thereby reproducing the catalysts during growth.

• The membrane-bounded system of macromolecules can divide into smaller structures.

• Genetic information is passed between generations by duplicating the informational polymers and sharing them between daughter cells.

• Occasional mistakes (mutations) are made during replication or transmission of information so that the system can evolve through natural selection.

Looking at this list, one is struck by the complexity of even the simplest form of cellular life. This is why it has been so difficult to "define" life in the usual sense of a definition, that is, boiled down to a few sentences in a dictionary. Cellular life is a complex system that cannot be captured in a few sentences, so perhaps listing its observed properties is the best we can ever hope to do. Despite the apparent complexity, it is also significant that most of the functions have been reproduced in the laboratory. We can now describe how this set of functions might be integrated into artificial cellular life.

2.2 Self-Assembly Processes in Cellular Life

All modern cellular life incorporates two processes, which we will refer to as self-assembly and directed assembly. Spontaneous self-assembly occurs when certain compounds associate through noncovalent hydrogen bonds, electrostatic forces, and nonpolar interactions that stabilize orderly arrangements of small and large molecules. A classic example is the manner by which amphiphilic molecules in aqueous phases form micelles and bimolecular structures (figure 2.1, top panel). Another version of self-assembly requires the formation of covalent bonds between similar molecular species, such as the random polymers of amino acids that can be produced by energy-dependent condensation reactions (figure 2.1, center). In contrast, the directed assembly processes characteristic of today involve the formation of covalent bonds by energy-dependent synthetic reactions, but also require that a coded sequence in one type of polymer in some way directs the sequence of monomer addition in a second polymeric species (figure 2.1, lower panel).

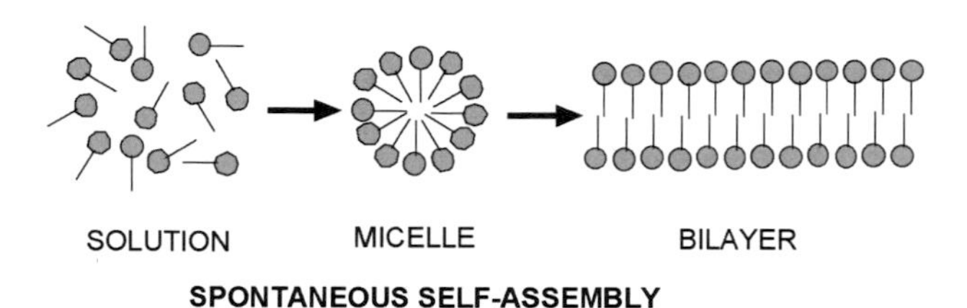

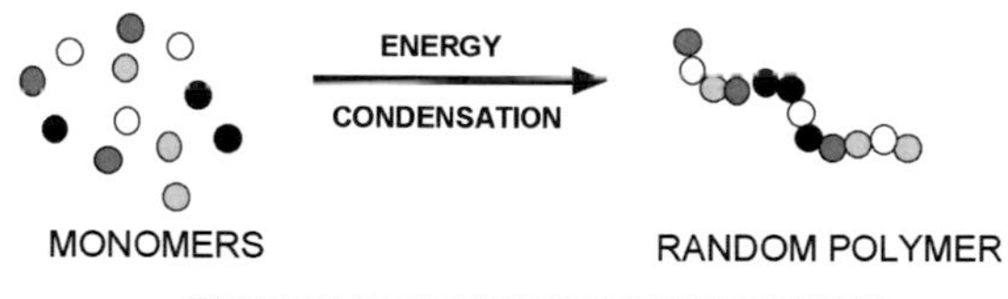

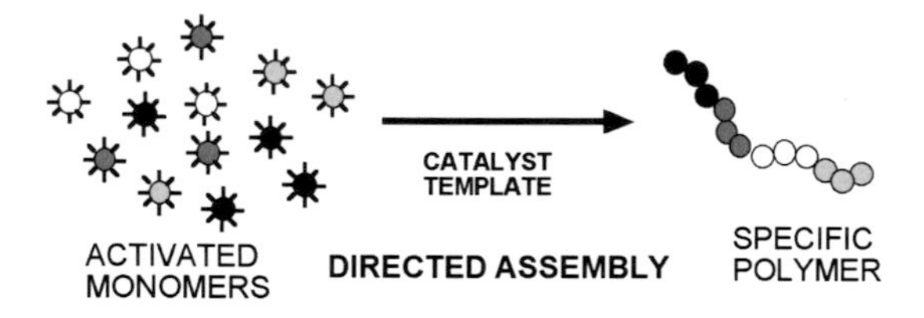

Figure 2.1
Cellular life today uses both self-assembly and directed assembly processes to grow. Self-assembly (upper panel) is essential for synthesis and stability of membrane structures and protein folding, whereas directed assembly (lower panel) underlies the synthesis of proteins according to the base sequences in DNA and mRNA. We assume that on the early Earth, random polymers similar to peptides and nucleic acids were produced by a yet unknown synthetic pathway (center). The random polymers, if capable of growth in a membrane-bounded microenvironment, would be subjected to selection and thereby begin biological evolution.

Spontaneous self-assembly of organic compounds in aqueous phases was presumably common on the prebiotic Earth, and likely involved certain compounds that can form closed membrane-bound microenvironments. Such boundary structures, and the compartments they produce, have the potential to make energy available in the form of ion gradients, and can provide a selective inward transport of nutrients. Furthermore, membranous compartments, in principle, are capable of containing unique systems of macromolecules. If a yet unknown macromolecular replicating system of polymers could be encapsulated within a membrane-bounded compartment, the components of the system would share the same microenvironment (figure 2.2) and the result would be a major step toward cellularity, speciation, and true cellular function (Cavalier-Smith, 1987; Conde-Frieboes and Blochliger, 2001; Deamer et al., 2002; Dyson, 1999; Hutchison et al., 1999; Kock and Schmidt, 1991; Morowitz,

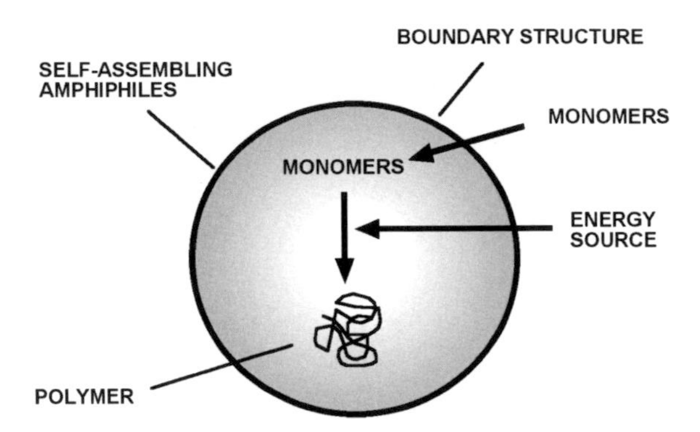

Figure 2.2
A protocell would have a minimal set of functional properties, including self-assembly of boundary membranes, transport of monomers, capture of energy to drive polymerization reactions, and encapsulation of polymer systems capable of growth.

1992; Ourisson and Nakatani, 1994; Segré, Deamer, and Lancet, 2001; Szostak, Bartel, and Luisi, 2001).

2.2.1 Self-Assembly Processes in Organic Mixtures

What physical properties are required for a molecule to become incorporated into a stable bilayer? All bilayer-forming molecules are amphiphiles, with a hydrophilic "head" and a hydrophobic "tail" on the same molecule. If amphiphilic molecules were present in the mixture of organic compounds available on the early Earth, it is not difficult to imagine that their self-assembly into molecular aggregates was a common process.

It is reasonable to conclude that a variety of simpler amphiphilic molecules can participate in the formation of membrane structures. The long-chain fatty acids and alcohols that contribute the amphiphilic property of contemporary membrane lipids are possible components. Significantly, such simple amphiphiles readily form vesicles (Apel, Deamer, and Mautner, 2002; Monnard et al., 2002). Stability of the vesicles is strongly dependent on chain length, pH, ionic strength, amphiphile composition and concentration, temperature, and head group characteristics. For example, even a 9-carbon monocarboxylic acid—nonanoic acid—can form vesicles at concentrations of 85 mM and pH 7.0, which is the pK of the acid in bilayers. At this pH, half of the carboxylic acid groups are protonated, half are anions, and hydrogen bonding between the protonated and anionic head groups stabilizes the bilayer configuration. Addition of small amounts of a nonanol can also stabilize the bilayers as a result of hydrogen bonding between the alcohol and acid head groups, so that vesicles form at lower concentrations ($\sim$20 mM) over a pH range from 7 to

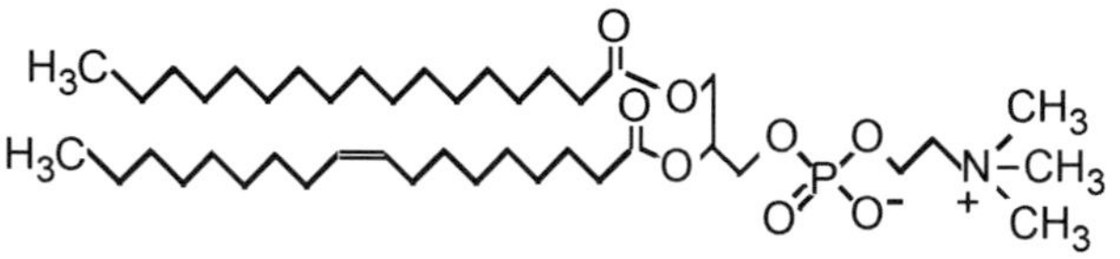

Figure 2.3
A phospholipid (1-palmitoyl, 2-oleoyl phosphatidylcholine). This molecule is amphiphilic as it has a "water-hating" (hydrophobic) tail and a "water-loving" (hydrophilic) head.

11. The vesicles provide a selective permeability barrier, as indicated by osmotic activity and entrapment of polar dyes. As chain length increases, stability also increases, and vesicles form at lower concentrations (Apel et al., 2002; Monnard et al., 2002).

2.2.2 What Amphiphiles Self-Assemble into Membranes?

Amphiphilic molecules are among the simplest of life's molecular components, and are readily synthesized by nonbiological processes. Virtually any normal alkane having 10 or more carbons in its chain takes on amphiphilic properties if one end of the molecule incorporates a polar or ionic group (see below). The simplest common amphiphiles are therefore molecules such as monocarboxylic acids (anions), monoamines (cations), and alcohols (neutral polar groups).

$CH_3\text{-}(CH_2)_n\text{-}COOH \rightarrow H^+ + CH_3\text{-}(CH_2)_n\text{-}COO^-$ (anion)

$CH_3\text{-}(CH_2)_n\text{-}NH_2 + H^+ \rightarrow CH_3\text{-}(CH_2)_n\text{-}NH_{3+}$ (cation)

$CH_3\text{-}(CH_2)_n\text{-}OH$ (neutral amphiphile)

Lipids are far more diverse chemically than other typical biomolecules such as amino acids, carbohydrates, and nucleotides. The definition of lipids includes simple fatty acids and their glycerol esters, sterols such as cholesterol, and phospholipids, sphingolipids, and cerebrosides. Lipids are generally defined by their common hydrophobic character, which makes them soluble in organic solvents such as chloroform. Virtually all lipids also have a hydrophilic group, which makes them surface-active.

Eukaryotic phospholipids typically have two fatty acid chains esterified to a glycerol, with the third position of the glycerol esterified to a phosphate group. Most phospholipids also have a head group such as choline, ethanolamine, or serine attached to the phosphate. One such lipid is shown in figure 2.3. The precise function of the variable head groups has not yet been established.

The other lipid commonly present in eukaryotic membranes is cholesterol, a polycyclic structure produced from isoprene by a complex biosynthetic pathway.

2.3 The Fluid Mosaic Model of Membrane Structure

In the 1970s, the fluid mosaic concept emerged as the most plausible model to account for the known structure and properties of biological membranes (Singer and Nicolson, 1972). The fact that membranes exist as two-dimensional fluids (liquid-disordered), rather than in a gel state (solid-ordered), was clearly demonstrated by Frye and Edidin (1970), who showed that the lipid and protein components of two separate membranes diffused into each other when two different cells were fused. Since that time, numerous studies have measured the diffusion coefficient of lipids and proteins in membranes, and the diffusion rates were found to correspond to those expected of a fluid with the viscosity of olive oil, rather than a gel phase resembling wax.

The lipid components of membranes must be in a fluid state to function as membranes in living cells. Straight-chain fatty acids have relatively high melting points because of the ease with which van der Waals interactions can occur along the hydrocarbon chains. Any discontinuity in the chains interrupts these interactions and markedly decreases the melting point. As an example, stearic acid contains 18 carbons in its alkane chain and melts at 68°C, while oleic acid, with a *cis*-double bond between carbons 9 and 10, has a melting point of 16°C. If cellular life requires fluid membranes, it is reasonable to assume that the membranes of artificial cells could also be composed of amphiphilic molecules in a fluid state. However, alternative boundary structures could also be incorporated, such as a self-assembling protein coat resembling that of viruses.

The idea that the proteins of biological membranes are embedded in a fluid sea of lipids arose from our increasing understanding of membrane structure. It has been demonstrated in numerous ways that most of the proteins associated with membranes are embedded in the lipid bilayer phase, rather than simply adhering to the surface. As a general rule, membrane proteins have stretches of hydrophobic amino acids in their sequences, and these are threaded back and forth through the bilayer multiple times, thereby anchoring the protein to the membrane. The hydrophobic proteins often are involved in production of pores, or transmembrane channels, that are essential for ion and nutrient transport processes.

Could channels capable of nutrient transport be produced in the bilayer membranes of artificial cells? In fact, channel-like defects do appear when nonpolar peptides interact with a lipid bilayer. For instance, polyleucine or polyalanine have been induced to fuse with planar lipid membranes, and the bilayers exhibited transient bursts of ionic conductance (Oliver and Deamer, 1994). More complex synthetic peptides have also been demonstrated to produce ion-conducting channels in lipid bilayers (Lear, Wasserman, and DeGrado, 1988). It is fair to expect that a variety of polymers are likely to be able to penetrate bilayer membranes and produce channels that bypass the permeability barrier. This is an area that is ripe for further investigations, as described in a recent review by Pohorille, Schweighofer, and Wilson (2005).

2.4 Function of Membranes in Artificial Cells

Membranes have many functions in addition to acting as containers for the macromolecular polymers of life. Three primary membrane functions associated with an artificial cell might include selective inward transport of nutrients from the environment, capture of the energy available in light or oxidation-reduction potentials, and coupling of that energy to some form of energy currency such as ATP to drive polymer synthesis.

The simplest of these functions is that of a permeability barrier, which limits free diffusion of solutes between the cytoplasm and the external environment. Although such barriers are essential for cellular life to exist, a mechanism by which selective permeation allows specific solutes to cross the membrane must also exist. In contemporary cells, such processes are carried out by transmembrane proteins, which act as channels and transporters. Examples include the proteins that facilitate the transport of glucose and amino acids into the cell, channels that allow potassium and sodium ions to permeate the membrane, and active transport of ions by enzymes that use ATP as an energy source.

If we are to assemble an artificial cell, it will be necessary to overcome the membrane permeability barrier. One possible way to accomplish this could be simple diffusion across the bilayer. To give a perspective on permeability and transport rates by diffusion, we can compare the fluxes of relatively permeable and relatively impermeable solutes across contemporary lipid bilayers. The measured permeability of lipid bilayers to small, uncharged molecules such as water, oxygen, and carbon dioxide is greater than the permeability to ions by a factor of $\sim 10^9$. For instance, the permeability coefficient of water is approximately 10^{-3} cm s^{-1}, and the permeability coefficient of potassium ions is 10^{-11} cm s^{-1}. By themselves, these values mean little, but they make more sense in the context of time required for exchange across a bilayer. Measurements show that half the water in a liposome exchanges in milliseconds, whereas potassium ion half-times of exchange are measured in days.

We can now consider some typical nutrient solutes like amino acids and phosphates. Such molecules are ionized, which means that they would not readily cross the permeability barrier of a lipid bilayer. Permeability coefficients of liposome membranes to phosphate and amino acids have been determined (Chakrabarti and Deamer, 1994) and were found to be in the range of 10^{-11} to 10^{-12} cm s^{-1}, similar to ionic solutes such as sodium and chloride ions. From these figures one can estimate that if an artificial cell depended on passive transport of phosphate across a lipid bilayer composed of a typical phospholipid, it would require several years to accumulate phosphate sufficient to double its DNA content, or pass through one cell cycle. In contrast, a modern bacterial cell can reproduce in as short a time as 20 minutes.

Given that lipid bilayers are so impermeable to typical polar and ionic solutes, how can we design artificial cells so that they will have access to essential nutrients? One clue may be that highly evolved modern lipids are products of several billion years of evolution, and typically contain hydrocarbon chains 16 to 18 carbons in length. These chains provide an interior "oily" portion of the lipid bilayer that represents a nearly impermeable barrier to the free diffusion of ions such as sodium and potassium. The reason is related to the common observation that "oil and water don't mix." That is, ion permeation of the hydrophobic portion of a lipid bilayer faces a very high energy barrier called Born energy, which is associated with the difference in energy for an ion in a high dielectric medium (water with a dielectric constant of 80) compared to the same ion in a low dielectric medium (hydrocarbon with a dielectric constant of 2). This energy barrier is immense, up to 40 kcal $mole^{-1}$ (Parsegian, 1969).

However, recent studies have shown that permeability is strongly dependent on chain length (Paula et al., 1996). For instance, shortening phospholipid chains from 18 to 14 carbons increases permeability to ions by a thousandfold (figure 2.4). The reason is that thinner membranes have increasing numbers of transient defects that open and close on nanosecond time scales, so that ionic solutes can get from one side of the membrane to the other without dissolving in the oily interior phase of the bilayer. Ionic solutes even as large as ATP can diffuse cross a bilayer composed of dimyristoylphosphatidylcholine, a 14-carbon phospholipid (Monnard and Deamer, 2001). We conclude that initial approaches to fabricating an artificial cell do not necessarily depend on peptide channels to provide nutrient transport. It may be sufficient simply to prepare the membranes with a lipid composition that permits relatively fast diffusion of small substrate molecules, yet can maintain macromolecular components in the internal volume.

2.5 Growth Processes in Artificial Cells

Earlier reports (Walde, Goto, et al., 1994; Walde, Wick, et al., 1994) showed that vesicles composed of oleic acid can grow and "reproduce" as oleoyl anhydride spontaneously hydrolyzed in the reaction mixture, thereby adding amphiphilic components (oleic acid) to the vesicle membranes. This approach has recently been extended by Hanczyc and coworkers (Hanczyc, Fujikawa, and Szostak, 2003; Hanczyc and Szostak, 2004; see also chapter 5), who prepared myristoleic acid membranes under defined conditions of pH, temperature, and ionic strength. The process by

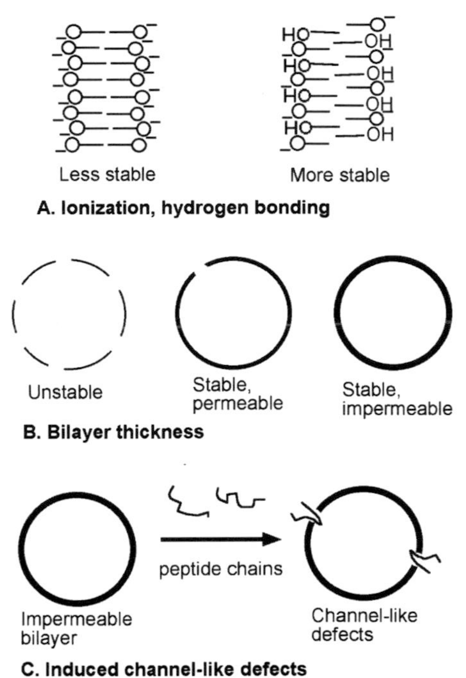

Figure 2.4
Stability and permeability of self-assembled amphiphilic structures. Amphiphilic molecules such as fatty acids having carbon chain lengths of 9 or more carbons form bilayer membranes when sufficiently concentrated. (A) Pure bilayers of ionized fatty acid are relatively unstable, but become markedly more stable in long-chain alcohols are added. (B) Dimensions of the amphiphile also play a role. Shorter-chain amphiphiles (9–10 carbons) are less able to form bilayers, while those of intermediate chain length (12–14 carbons) produce stable bilayers that also are permeable to ionic and polar solutes. Longer chain lengths (16–18 carbons) produce bilayers that are increasingly less permeable to solutes (Paula et al., 1996).

which the vesicles formed from micellar solutions required several hours, apparently with a rate-limiting step related to the assembly of "nuclei" of bilayer structures. However, if a mineral surface in the form of clay particles was present, the surface in some way catalyzed vesicle formation, reducing the time required from hours to a few minutes. The clay particles were spontaneously encapsulated in the vesicles. The authors further found that RNA bound to the clay was encapsulated as well, and remained within the vesicles for extended periods of time.

In a second series of experiments, Hanczyc, Fujikawa, and Szostak (2003; see also chapter 5) found that the myristoleic acid vesicles could be induced to grow by adding fatty acid to the medium, presumably by incorporating fatty acid molecules into the membrane rather than by fusion of vesicles. If the resulting suspension of large vesicles was then filtered through a polycarbonate filter with pores 0.2 μm in diameter, the larger vesicles underwent a kind of shear-induced division to produce smaller

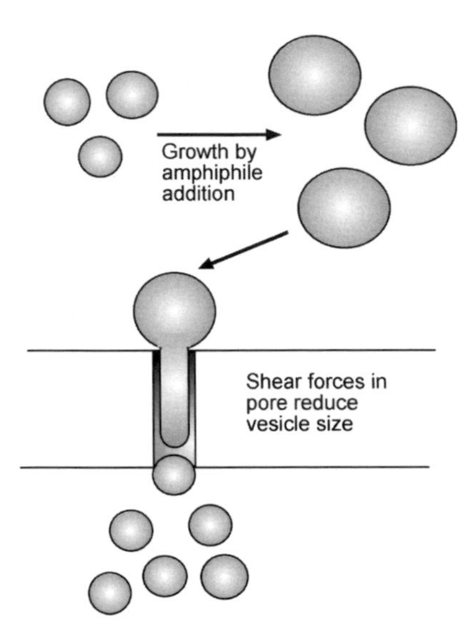

Figure 2.5
Fatty acid vesicles can grow by addition of fatty acid molecules to the membrane, and are then dispersed into smaller vesicles by passage through a porous filter. The cycle of growth and dispersion can be repeated several times (Hanczyc, Fujikawa, and Szostak, 2003) and presumably could go on indefinitely as long as a source of amphiphiles was available for vesicle growth. However, any encapsulated solutes would inevitably be diluted or lost in the process unless they were undergoing some form of growth and replication.

vesicles. This process could be repeated several times (figure 2.5). This remarkable series of experiments clearly demonstrated the relative simplicity of producing a complex system of lipid, genetic material, and mineral catalyst in a model protocellular structure that can undergo a form of growth and division.

2.6 Encapsulation Mechanisms

Even if membranous vesicles can be prepared with sufficient permeability to permit nutrient transport to occur, these structures would be virtually impermeable to larger polymeric molecules that were necessarily incorporated into molecular systems on the pathway to cellular life. The encapsulation of macromolecules in lipid vesicles has been demonstrated by hydration-dehydration cycles that simulate an evaporating lagoon (Shew and Deamer, 1983) or by freeze-thaw cycles (Pick, 1981). Molecules as large as DNA can be captured by such processes. For instance, when a dispersion of DNA and fatty acid vesicles is dried, the vesicles fuse to form a multilamellar sandwich structure with DNA trapped between the layers (figure 2.6). Upon rehydration, vesicles reform that contain highly concentrated DNA, a process that can be visualized by staining with a fluorescent dye.

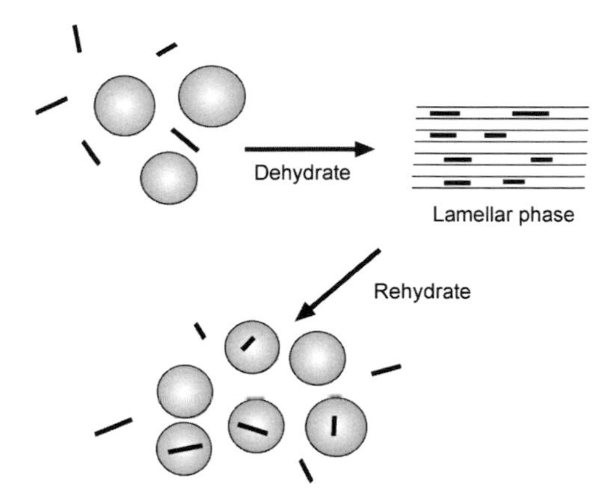

Figure 2.6
Macromolecules are readily encapsulated in lipid vesicles in a single cycle of dehydration-hydration (Shew and Deamer, 1983). Such wetting-drying cycles would commonly occur in the prebiotic environment at intertidal zones.

2.7 Environmental Constraints on Artificial Cell Membranes

Although self-assembly of amphiphilic molecules promotes the formation of complex molecular systems, the physical and chemical properties of an aqueous phase can significantly inhibit such processes, possibly constraining the environments in which cellular life first appeared. One such constraint is that temperature strongly influences the stability of vesicle membranes. A hot environment has the advantage that it provides activation energy to drive desired reactions. However, because the intermolecular forces that stabilize self-assembled molecular systems are relatively weak, it is difficult to imagine how lipid bilayer membrane constituents would be stable under these conditions.

A second concern is related to the ionic composition of the medium. High salt concentrations potentially exert significant osmotic pressure on any closed membrane system (figure 2.7). All marine organisms today have evolved highly developed membrane transport systems that allow them to maintain osmotic equilibrium against substantial salt gradients across their membranes. Furthermore, divalent cations bind to the anionic head groups of amphiphilic molecules, strongly inhibiting their ability to form stable membranes (Monnard et al., 2002).

2.8 Encapsulated Transcription and Translation Systems

In contemporary cells, there is a cycle represented by protein catalysts (enzymes) and nucleic acids that store genetic information, which in turn can transmit that

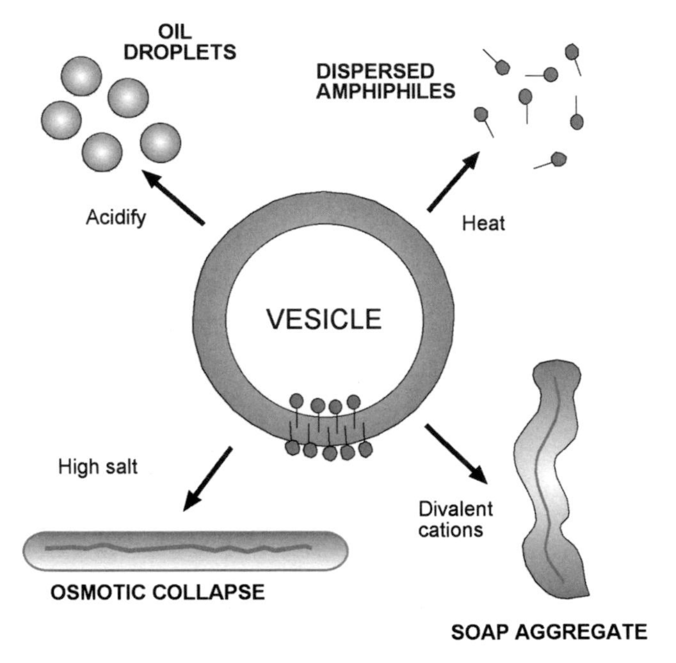

Figure 2.7
Vesicles produced by single-chain amphiphiles such as fatty acids tend to be destabilized by certain environmental factors. If the fatty acid is protonated at low pH ranges, the membranes collapse into droplets, while at high pH ranges only micelles can form. The vesicles also become increasingly unstable as temperature increases. In the presence of high salt concentrations, the vesicles undergo osmotic collapse, and may also form nonmembranous aggregates if divalent cations react with the carboxylate head groups.

information to a second molecule by replication or transcription. However, in an artificial cell, the same molecule could play both catalytic and information-containing roles, as suggested by recent studies of ribozymes (Johnston et al., 2001). Several approaches to artificial cells have been proposed to test various scenarios for the origin of cellular life (Luisi 1996, 2002; Luisi, Ras, and Mavelli, 2004; Pohorille and Deamer, 2002; Rasmussen et al., 2004; Szostak, Bartel, and Luisi, 2001). An ideal model cell would incorporate an encapsulated polymerase activity together with a template of some sort, so that sequence information in the template can be transcribed to a second molecule. The membrane must be sufficiently permeable to allow the polymerase to have access to externally added substrates. Furthermore, the membrane itself should be able to grow in order to accommodate the growth of the encapsulated polymers. Finally, in an ideal cell model, the polymerase itself would be reproduced from information in the template, so that the entire system is able to grow and evolve.

To demonstrate polymerase activity in a model cell, Chakrabarti and colleagues (1994) encapsulated polynucleotide phosphorylase in vesicles composed of dimyris-

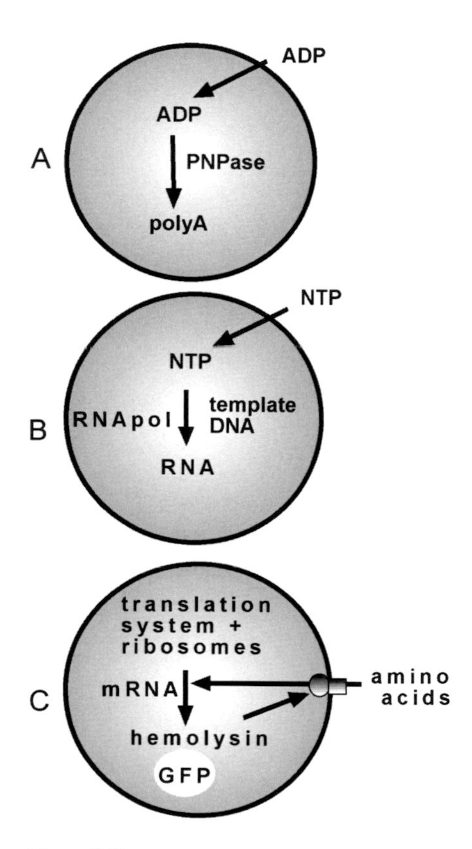

Figure 2.8
Model protocell systems. (A) An encapsulated polymerase (polynucleotide phosphorylase) can synthesize RNA from nucleoside diphosphates such as ADP (Chakrabarti et al., 1994; Walde, Goto, et al., 1994; Walde, Wick, et al., 1994). (B) RNA can be synthesized by a template-dependent T7 RNA polymerase (Monnard, Luptak, and Deamer, 2007), by the addition of nucleoside triphosphates (NTP), as they are able to diffuse into the vesicle. (C) Proteins such as green fluorescent protein (GFP) can be synthesized by an encapsulated translation system (Yu et al., 2001). If mRNA coding for hemolysin is also present, the hemolysin forms a pore in the lipid bilayer. Amino acids then permeate the bilayer and protein synthesis can continue for several days (Nomura et al., 2003).

toylphosphatidylcholine (DMPC). This enzyme can produce RNA from nucleoside diphosphates such as adenosine diphosphate (ADP) and does not require a template, so it has proven useful for initial studies of encapsulated polymerase activity (figure 2.8a). Furthermore, DMPC liposomes are sufficiently permeable to allow 5 to 10 ADP molecules per second to enter each vesicle. Under these conditions, measurable amounts of RNA in the form of polyadenylic acid were synthesized and accumulated in the vesicles after several days' incubation. The enzyme-catalyzed reaction could be carried out in the presence of a protease external to the membrane, demonstrating that the vesicle membrane protected the encapsulated enzyme from hydrolytic degradation. Similar behavior has been observed with monocarboxylic acid vesicles (Walde, Goto, et al., 1994; Walde, Wick, et al., 1994), and it follows that complex phospholipids are not required for an encapsulated polymerase system to function.

The Q-beta replicase (Oberholzer, Albrizio, and Luisi, 1995a) and the components of the polymerase chain reaction (Oberholzer et al., 1995b) have also been encapsulated in liposomes, together with templates and substrates in the form of nucleoside triphosphates. Both of these enzyme systems use templates, so it is clear that template-dependent polymer synthesis can occur in an encapsulated environment. The phospholipids used in these studies were relatively impermeable, so that substrates were necessarily encapsulated along with enzyme and template. This limited the amount of nucleic acid replication that could occur to a few molecules per vesicle. However, the permeability barrier can in principle be overcome by introducing transient defects in the membranes of lipid vesicles. For instance, a template-directed reaction can be encapsulated in DMPC liposomes in which externally added substrates were used to supply the enzyme (Monnard, Luptak, and Deamer, 2007). In this study, T-7 RNA polymerase and a circular 4,000-bp plasmid template were encapsulated, and substrates were provided by addition of ribonucleotide triphosphates (figure 2.8b). The system was subjected to temperature cycles of 23°C and 37°C in a polymerase chain reaction (PCR) apparatus. DMPC membranes are relatively permeable at the phase transition temperature of 23°C, permitting substrate ribonucleotides to enter the vesicles, while at 37°C the membranes become much less permeable but the polymerase is activated. RNA synthesis was monitored by incorporation of radiolabeled uridine triphosphate (UTP), and transcription was confirmed by reverse PCR. Figure 2.9 shows a micrograph of the resulting structures containing RNA synthesized within the vesicle volume.

Most recently, functioning translation systems that included ribosomes have been encapsulated in lipid vesicles (figures 2.8c and 2.10). Oberholzer, Nierhaus, and Luisi (1999) made the first attempt to assemble a translation system in a lipid vesicle system. However, only very small amounts of peptides were synthesized, largely because the lipid bilayer was impermeable to amino acids, so that ribosomal translation was limited to the small number of amino acids that were encapsulated within the vesicles. Yu and coworkers (2001) improved the yield substantially by using larger vesicles, demonstrating that green fluorescent protein (GFP) can be synthesized by an encapsulated translation system. Noireaux and Libchaber (2004) took this approach one step further by incorporating two genes into a similar encapsulated translation system, one for alpha hemolysin, a pore-forming protein, and a second for GFP. The investigators reasoned that if the system was, in fact, capable of synthesizing proteins, the newly translated hemolysin would insert into the membrane and produce a pore. This would allow externally added solutes (i.e., amino acids) to permeate the lipid bilayer barrier and supply the substrates required for protein synthe-

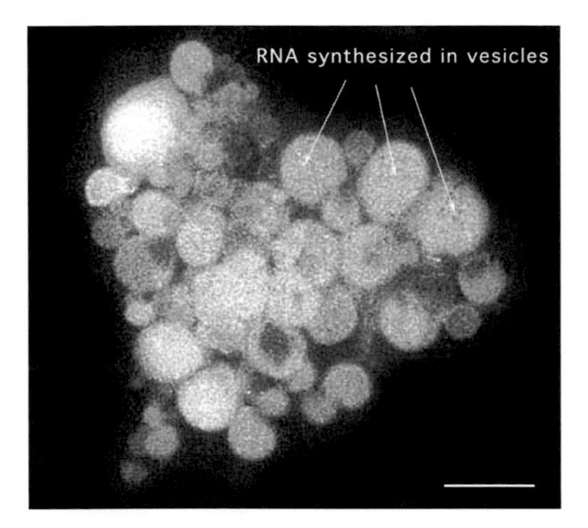

Figure 2.9
Lipid vesicles with encapsulated T7 RNA polymerase and DNA template. A mixture of four NTPs was added, and these diffused into the vesicles and were used by the polymerase to synthesize RNA with DNA as a template. The RNA was stained with ethidium bromide and appears as fluorescent material within the vesicles. Note that some of the vesicles do not contain fluorescent RNA, presumably because they lacked sufficient enzyme or template. Scale bar shows 20 μm.

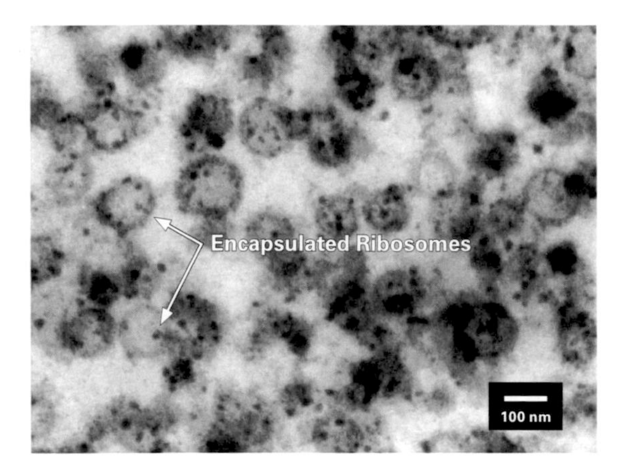

Figure 2.10
Ribosomes from *Escherichia coli* encapsulated in phospholipid vesicles. The vesicles were reconstituted from a detergent-lipid solution in the presence of cytoplasmic extracts from *E. coli* bacteria, and 10 or so ribosomes were present in each vesicle. Micrograph courtesy of Z. Martinez.

sis. The production of GFP would then indicate that synthesis was proceeding at appreciable rates. This worked very well, and GFP could be visualized accumulating in the vesicular volume for up to four days (figure 2.8c). Further similar systems were reported by Nomura and coworkers (2003) and Ishikawa and coworkers (2004), who managed to assemble a two-stage genetic network in liposomes, in which the gene for an RNA polymerase was expressed first, and the polymerase was then used to produce mRNA required for GFP synthesis.

2.9 An Approach to Synthetic Cellular Life

All that remains to be done to recreate a minimal cell based on the familiar chemistry of existing life, it seems, is to add up the individual processes described here and integrate them into a complete system, and we might be able to achieve a second origin of life, 3.8 billion years after the first origin, but this time in a laboratory setting. But would the system really be alive? No, because all of the systems we have listed depend on polymerase enzymes or ribosomes. Even though every other part of the system can grow and reproduce, these get left behind.

This, then, is the final challenge: to encapsulate a system of macromolecules that can make more of themselves. There is now good evidence that life passed through a stage in which RNA served both as catalyst and carrier of genetic information, so several laboratories are now attempting to produce a ribozyme that has polymerase activity. If this can be found, an obvious approach to fabricating an artificial cell will be to encapsulate an evolving ribozyme system (Beaudry and Joyce, 1992; Wilson and Szostak, 1994) within vesicles that are optimized for stability and permeability. Replication and molecular evolution could then potentially occur in immensely large numbers of microscopic volumes represented by the vesicle interiors, rather than in the macroscopic volume of a test tube. Under these conditions, relatively rare systems of replicating catalytic polymers would be selected by their ability to grow and reproduce. In contrast, such systems could not occur in a test tube volume where the individual components are dispersed among trillions of other molecules.

David Bartel and coworkers, using a technique developed for selection and molecular evolution called SELEX (Systematic Evolution of Ligands by Exponential Enrichment), have produced a ribozyme that can grow by polymerization, in which the ribozyme copies a sequence of bases (Johnston et al., 2001). So far, the polymerization has copied only a string of 14 nucleotides, but this is a good start. If a ribozyme system can be found that catalyzes its own complete synthesis using genetic information encoded in its structure, it could rightly be claimed to have the essential properties that are lacking so far in artificial cell models: reproduction of the catalyst itself. Given such a ribozyme, it is not difficult to imagine its incorporation into a system of lipid vesicles that would have the basic properties of the living state.

2.10 Conclusion

Any attempt to fabricate an artificial cell (i.e., a protocell) in the laboratory must incorporate a variety of amphiphilic hydrocarbon derivatives that can self-assemble into bilayer boundary structures and encapsulate polymers that are being synthesized by a separate process. The vesicle membranes can be made sufficiently permeable to allow passage of smaller ionic substrates required for metabolism and biosynthesis, yet maintain larger molecules within. Encapsulated catalysts and information-bearing molecules will therefore have access to nutrients required for growth. Furthermore, specific groupings of macromolecules will be maintained in a given vesicle, rather than drifting apart. This would allow true Darwinian-type selection of such groupings to occur, a process that could not take place in mixtures of molecules free in solution. A small number of encapsulated molecular systems are likely to have the specific set of properties that will allow them to capture free energy and nutrients from their environment and undergo growth by polymerization. At some point, the growth may be catalyzed by the encapsulated polymers, so that a primitive genetic process will be in place. Such structures would be on the evolutionary path to the first forms of artificial cellular life.

Note

Portions of this chapter were adapted from a previous review by Deamer et al. (2002).

References

Apel, C. L., Deamer, D. W., & Mautner, M. (2002). Self-assembled vesicles of monocarboxylic acids and alcohols: Conditions for stability and for the encapsulation of biopolymers. *Biochimica et Biophysica Acta*, *1559*, 1–10.

Beaudry, A. A., & Joyce, G. F. (1992). Directed evolution of an RNA enzyme. *Science*, *342*, 255.

Cavalier-Smith, T. (1987). The origin of cells: A symbiosis between genes, catalysts and membranes. *Cold Spring Harbor Symposia on Quantitative Biology*, LII, 805–824.

Chakrabarti, A., Breaker, R. R., Joyce, G. F., & Deamer, D. W. (1994). Production of RNA by a polymerase protein encapsulated within phospholipid vesicles. *Journal of Molecular Evolution*, *39*, 555–559.

Chakrabarti, A., & Deamer, D. W. (1994). Permeation of membranes by the neutral form of amino acids and peptides: Relevance to the origin of peptide translocation. *Journal of Molecular Evolution*, *39*, 1–5.

Conde-Frieboes, K., & Blochliger, E. (2001). Synthesis of lipids on the micelle/water interface using inorganic phosphate and an alkene oxide. *Biosystems*, *1*, 109–114.

Deamer, D. W., Dworkin, J. P., Sandford, S. A., Bernstein, M. P., & Allamandola, L. J. (2002). The first cell membranes. *Astrobiology*, *2*, 371–382.

Dyson, F. (1999). *The origins of life*. Princeton, NJ: Princeton University Press.

Frye, L. D., & Edidin, M. (1970). The rapid intermixing of cell surface antigens after formation of mouse-human heterokaryons. *Journal of Cell Science*, *7*, 319–335.

Hanczyc, M. M., Fujikawa, S. M., & Szostak, J. W. (2003). Experimental models of primitive cellular compartments: Encapsulation, growth and division. *Science*, *302*, 618–622.

Hanczyc, M. M., & Szostak, J. W. (2004). Replicating vesicles and models of primitive cell growth and division. *Current Opinions in Chemical Biology*, *28*, 660–664.

Hutchison, C., Peterson, S., Gill, S., Cline, R., White, O., Fraser, C., et al. (1999). Global transposon mutagenesis and a minimal Mycoplasma genome. *Science*, *286*, 2165–2169.

Ishikawa, K., Sato, K., Shima, Y., Urabe, I., & Yomo, T. (2004). Expression of a cascading genetic network within liposomes. *FEBS Letters*, *576*, 387.

Johnston, W. K., Unrau, P. J., Lawrence, M. S., Glasner, M. E., & Bartel, D. L. (2001). RNA-catalyzed RNA polymerization: Accurate and general RNA-templated primer extension. *Science*, *292*, 1319–1325.

Koch, A. L., & Schmidt, T. M. (1991). The first cellular bioenergetic process: Primitive generation of a proton motive force. *Journal of Molecular Evolution*, *33*, 297–304.

Lear, J. D., Wasserman, Z. R., & DeGrado, W. F. (1988). Synthetic amphiphilic peptide models for protein ion channels. *Science*, *240*, 1177–1181.

Luisi, P. L. (1996). Self-reproduction of micelles and vesicles: Models for the mechanisms of life from the perspective of compartmented chemistry. *Advances in Chemistry and Physics*, *92*, 425–438.

Luisi, P. L. (2002). Toward the engineering of minimal living cells. *Anatomical Record*, *268*, 208–214.

Luisi, P. L., Ras, P. S., & Mavelli, F. (2004). A possible route to vesicle reproduction. *Artificial Life*, *10*, 297–308.

Monnard, P.-A., Apel, C. L., Kanavarioti, A., & Deamer, D. W. (2002). Influence of ionic solutes on self-assembly and polymerization processes related to early forms of life: Implications for a prebiotic aqueous medium. *Astrobiology*, *2*, 139–152.

Monnard, P.-A., Lupak, A., & Deamer, D. W. (2007). Models of primitive cellular life: Polymerases and templates in liposomes. *Philosophical Transactions of the Royal Society B*, *362*, 1741–1750.

Morowitz, H. J. (1992). *Beginnings of cellular life*. New Haven, CT: Yale University Press.

Noireaux, V., & Libchaber, A. (2004). A vesicle bioreactor as a step toward an artificial cell assembly. *Proceedings of the National Academy of Sciences United States of America*, *101*, 17669–17674.

Nomura, S., Tsumoto, K., Hamada, T., Akiyoshi, K., Nakatani, Y., & Yoshikawa, K. (2003). Gene expression within cell-sized lipid vesicles. *Chemistry and Biochemistry*, *4*, 1172–1175.

Oberholzer, T. M., Albrizio, M., & Luisi, P. L. (1995). Polymerase chain reaction in liposomes. *Current Biology*, *2*, 677–682.

Oberholzer, T. R., Wick, R., Luisi, P. L., & Biebricker, C. K. (1995). Protein expression in liposomes. *Biochemical and Biophysical Research Communications*, *207*, 250.

Oberholzer, T. K., Nierhaus, H., & Luisi, P. L. (1999). Protein expression in liposomes. *Biochemical and Biophysical Research Communications*, *261*, 238–241.

Oliver, A., & Deamer, D. W. (1994). Alpha helical hydrophobic polypeptides form proton-selective channels in lipid bilayers. *Biophysical Journal*, *66*, 1364–1379.

Ourisson, G., & Nakatani, T. (1994). The terpenoid theory of the origin of cellular life: The evolution of terpenoids to cholesterol. *Chemistry and Biology*, *1*, 11–23.

Parsegian, A. (1969). Energy of an ion crossing a low dielectric membrane: Solutions to four relevant electrostatic problems. *Nature*, *221*, 844–846.

Paula, S., Volkov, A. G., Van Hoek, A. N., Haines, T. H., & Deamer, D. W. (1996). Permeation of protons, potassium ions, and small polar molecules through phospholipid bilayers as a function of membrane thickness. *Biophysical Journal*, *70*, 339–348.

Pick, U. (1981). Liposomes with a large trapping capacity prepared by freezing and thawing of sonicated phospholipid mixtures. *Archives of Biochemistry and Biophysics*, *212*, 186–194.

Pohorille, A., & Deamer, D. W. (2002). Artificial cells: Prospects for biotechnology. *Trends in Biotechnology*, *20*, 123.

Pohorille, A., Schweighofer, K., & Wilson, M. A. (2005). The origin and early evolution of membrane channels. *Astrobiology*, *5*, 1–17.

Rasmussen, S., Chen, L., Deamer, D., Krakauer, D. C., Packard, N. H., Stadler, P. F., & Bedau, M. A. (2004). Transitions from nonliving to living matter. *Science*, *303*, 963–965.

Segré, S., Deamer, D. W., & Lancet, D. (2001). The Lipid World. *Origins of Life and Evolution of the Biosphere*, *31*, 119–145.

Shew, R., & Deamer, D. (1983). A novel method for encapsulating macromolecules in liposomes. *Biochimica et Biophysica Acta*, *816*, 1–8.

Singer, S. J., & Nicolson, G. L. (1972). The fluid mosaic model of the structure of cell membranes. *Science*, *175*, 720–731.

Szostak, J. W., Bartel, D. P., & Luisi, P. L. (2001). Synthesizing life. *Nature*, *409*, 387–390.

Walde, P., Goto, A., Monnard, P.-A., Wessicken, M., & Luisi, P. L. (1994). Oparin's reactions revisited: Enzymatic synthesis of poly(adenylic acid) in micelles and self-reproducing vesicles. *Journal of the American Chemical Society*, *116*, 7541–7547.

Walde, P., Wick, R., Fresta, M., Mangone, A., & Luisi, P. L. (1994). Autopoietic self-reproduction of fatty acid vesicles. *Journal of the American Chemical Society*, *116*, 11649–11654.

Wilson, C., & Szostak, J. W. (1994). In vitro evolution of a self-alkylating ribozyme. *Nature*, *374*, 777–782.

Yu, W., Sato, K., Wakabayashi, M., Nakaishi, T., K-Mitamura, E. P., Shima, Y., et al. (2001). *Journal of Bioscience and Bioengineering*, *92*, 590.

3 Semisynthetic Minimal Cells: New Advancements and Perspectives

Pasquale Stano, Giovanni Murtas, and Pier Luigi Luisi

3.1 The Minimal Cell: From the Origin of Life to Synthetic Biology

An enormous complexity characterizes even the simplest living cells, in which several hundred genes and their expressed proteins control and catalyze hundreds of reactions simultaneously within the same tiny compartment. Is such complexity really essential for life? Or is cellular life possible with a much smaller number of components? The present enormous complexity is the result of billions of years of evolution, during which time defense and self-repair mechanisms, redundancies, and metabolic loops developed. It follows that research on the simplest "minimal" cells is related to the origin of cellular life and early evolution when such cells could not have been so complex.

Under a certain set of environmental conditions, a minimal cell is broadly defined as the one with the fewest components needed to be called alive. How to define "alive" is a complex question open to scientific and philosophical debate. A general description of life at the cellular level must incorporate three basic properties: self-maintenance (metabolism), self-reproduction, and evolvability (considering the Darwinian aspects of evolvability—referring to populations rather than individual cells—a more accurate definition should take into account an entire family of minimal cells in the stream of environmental pressure).

Full-fledged cellular life arises when all three of these properties are present. Notice, however, that the trilogy may not be perfectly implemented, particularly in synthetic constructs, and one can envision several approximations to cellular life that have different degrees of function. For example, we can have protocells capable of self-maintenance but not self-reproduction, protocells in which self-reproduction is active for only a few generations, or systems that lack evolvability. In other words, the term *minimal cell* refers to a variety of possible constructs.

The issue of minimal cells has been considered for many years, and we should in particular cite the work of Morowitz (1967), who considered enzymatic components of primary metabolism and estimated that the size of a minimal cell should be perhaps a tenth the size of *Mycoplasma genitalium*. In the early 1970s, Tibor Gánti proposed the *chemoton* model, focusing on the functional organization of the chemical reactions involved in a self-reproducing cell (Gánti, 1975; see also chapter 22). Significant earlier insights into the field include those by Jay and Gilbert (1987), Woese (1983), and Dyson (1982).

In a joint paper entitled "Synthesizing Life," Szostak, Bartel, and Luisi (2001) proposed an elegant construct, the minimal RNA-based cell, which is discussed in more detail in section 3.1.1. More recently, the reviews by Pohorille and Deamer (2002), Deamer (2005), Noireaux and colleagues (2005), and Luisi's group (Luisi, 2002; Oberholzer and Luisi, 2002) have sharpened the question and brought it in the perspective of modern molecular tools, paving the way to a semisynthetic approach to minimal cells in which existing components such as modern genes, enzymes, and ribosomes are assembled in artificial compartments as lipid vesicles.

A different approach is currently in progress at Los Alamos by Steen Rasmussen and coworkers, who designed a micellelike construct that would be simpler than the biology-inspired minimal cell discussed in this review, but still containing a high degree of complexity (Rasmussen et al., 2004; see also chapter 6). The project aims to develop a fully synthetic living system on molecules based on self-assembled fatty acids and synthetic peptide-nucleic acids (PNAs) using a rudimentary metabolism to grow by polymerization.

The alternative approach of "synthetic" or "constructive" biology (Benner, 2003; Benner and Sismour, 2005; Hud and Lynn, 2004; Schwarz, 2001) involves theoreticians and experimentalists in biology, chemistry, and physics. (A recent International School on Complexity, in Erice, Italy, in 2004, focused on the topic of minimal cells; for a report, see Szathmáry, 2005.) Furthermore, synthetic and system biology appear strongly interconnected (Kaneko, 2004), when their goal is to elucidate the basic mechanisms of biological systems.

In this chapter, we review the theoretical and experimental approaches that have been recently developed. Beginning with a short discussion of the RNA minimal cell we examine, in particular, the more concrete aspects of the road map for the realization of protein/DNA minimal cells. In this survey, we outline conceptual aspects of the minimal genome issue, as well as practical aspects such as controlling the vesicle dynamics and performing biochemical reaction in compartments. A concluding section is devoted to possible future developments that are already in their initial stages. A more detailed discussion can be found elsewhere (Luisi, Ferri, and Stano, 2006), and Stano, Ferri, and Luisi (2006) have written a concise review that discusses autopoiesis in the context of artificial cells. Several theoretical and computational models of minimal life are omitted here; the interested reader should refer to the corresponding original works or the computational and theoretical reviews in this volume (chapters 6, 8–12, 14, 18, 19, 21–23, and 27).

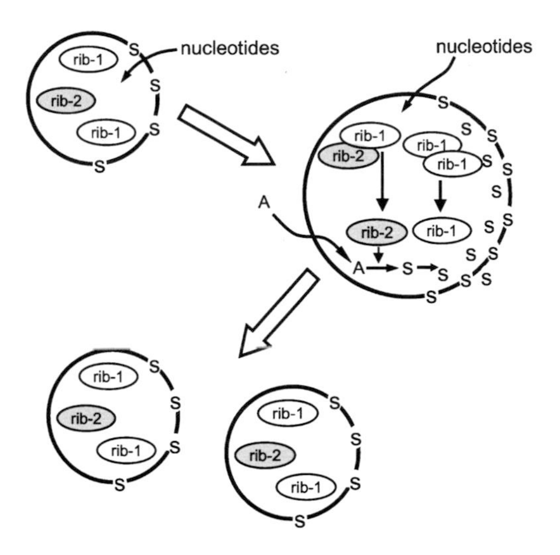

Figure 3.1
The RNA cell, containing two ribozymes. rib-2 is capable of synthesizing the cell membrane by converting precursor *A* to surfactant *S*; rib-1 is an RNA replicase capable of reproducing itself and making copies of rib-2. All necessary low-molecular components required for macromolecular synthesis are provided from the surrounding medium and are capable of permeating the membrane (adapted from Luisi, Oberholzer, and Lazcano, 2002). For the sake of simplicity, this figure represents an ideal cell division, in which all the core components are equally shared between the new vesicles.

3.1.1 The Minimal RNA Cell

One of the simplest constructs that incorporates criteria of evolvability, self-maintenance, and reproduction is the so-called *RNA cell* (figure 3.1). This purely theoretical object, presented some years ago by Szostak, Bartel, and Luisi (2001), extends the RNA World concept to an integrated system of replicating-catalytic RNA and cellular compartments. Here, the combined "genetic" and catalytic properties of ribozymes play a key role. The RNA cell consists of a vesicle containing two ribozymes, one with replicase activity, the other being a catalyst for the synthesis of the membrane component using an appropriate precursor. The first ribozyme (rib-1 in figure 3.1) replicates itself and the second ribozyme by a template-directed polymerization of nucleotides, which are supposed to be available. The second ribozyme (rib-2 in figure 3.1) transforms an available membranogenic precursor into a membrane-forming compound, allowing membrane growth and cell division. As a product of such processes new vesicles are formed, containing the same solutes (i.e., the ribozymes) that were originally in the parent vesicles. Consequently, a concerted core-and-shell reproduction of the entire construct (ribozymes *and* vesicle) can be obtained.

This hypothetical scheme is based on ribozymes that are not yet available and on a series of additional assumptions such as uptake of nutrients from the environment and cell division, and represents an extremely simple biochemical system that fulfills the three afore-mentioned criteria for life. Therefore, the RNA cell is a ribozyme-based minimal cell that is closely linked to the RNA World hypothesis and represents a primitive form of minimal life. On the other hand, if seen as a step in the origin of life, the RNA cell must eventually evolve into a protein/DNA cell, which also contains RNA as a functional and structural component, with genetic information stored in DNA and most of the processes catalyzed by protein enzymes.

For fabricating protein/DNA minimal cells in the laboratory, the molecular tools available to the experimentalist are existing and accessible genes, ribosomes, and enzymes. We can now turn to the question: What is the minimal number of genes needed to implement "minimal life"? And how can we proceed experimentally to the construction of minimal cells? The remaining part of this chapter deals with these questions, focusing on the experimental approaches.

3.2 Designing a Protein/DNA Minimal Cell

Here we discuss the theoretical and practical background for the realization of a minimal protein/DNA cell. We begin by characterizing the framework of possible strategic approaches for the experimental reconstruction of minimal cells. Traditionally, research approaches in the area of prebiotic chemistry use the so-called bottom-up approach, based on the notion that a continuous and spontaneous increase of molecular complexity has transformed inanimate matter into the first living cellular entities. For a number of reasons, this approach has not been particularly fruitful, and an alternative approach has been to encapsulate extant nucleic acids and enzymes in lipid vesicles to reconstitute a minimal cell (figure 3.2).

The terminology of this alternative route to the minimal cell can be confusing. The term *top-down* has been used to describe currently available molecular tools for assembling minimal cells (Luisi and Oberholzer, 2001). On the other hand, if we consider that simple molecular components are combined to form a cellular entity, the procedure also can be described as *bottom-up* because it produces an increase in com-

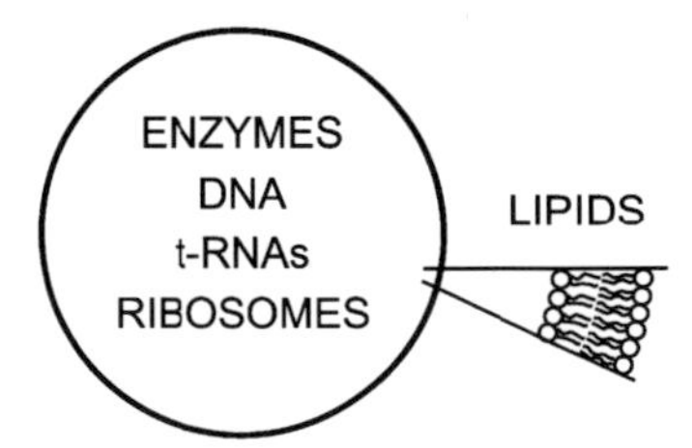

Figure 3.2
The semisynthetic approach to the construction of the minimal cell.

plexity. The term *reconstruction* is perhaps more appropriate, because it does not imply that the target is the construction of an extant cell. Since such cells do not necessarily match any known life form, the term artificial cells might also be used (Pohorille and Deamer, 2002) or better semiartificial/semisynthetic cells, because the molecular components are borrowed from extant cells.

We now present a selection of topics, all related to the minimal cell. We begin with studies on the minimal genome, as investigated recently by several authors. Experimental studies on self-reproduction of vesicles and other self-assembled structures are described, including general features of reactions occurring within compartments. Finally, we discuss recently published studies on the use of cellular extracts to incorporate one or more proteins into vesicles, followed by a summary discussion.

3.2.1 The Minimal Genome

Islas and coworkers (2004) computed the genome size distribution of free-living prokaryotes, obligate parasites, thermophiles, and endosymbionts. The resulting histogram (figure 3.3) shows that the DNA content of free-living prokaryotes can vary over a wide range, from the 1,450 kb of *Halomonas halmophila* to the 4,640 kb of *Escherichia coli* K-12, to the 9,700 kb genome of *Azospirillum lipoferum* Sp59b. Other species, such as thermophiles and obligate parasites, have a narrower genome size distribution, ranging from 1,000 to 5,000 kb. On the other hand, when endosymbionts are considered, it is evident that their DNA content is significantly smaller.

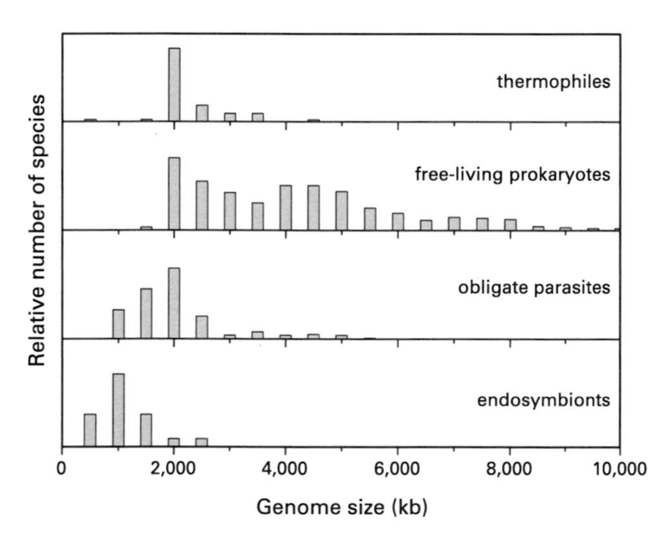

Figure 3.3
Prokaryotic genome size distribution ($N = 641$). Genome sizes, complete proteomes, and the number of ORFs (open reading frames) were all retrieved from NCBI (http://www.ncbi.nlm.nih.gov). (Adapted from Islas et al., 2004.)

Table 3.1
Core of a minimal bacterial gene set

DNA Metabolism		*16*
Basic replication machinery	13	
DNA repair, restriction and modification	3	
RNA Metabolism		*106*
Basic transcription machinery	8	
Translation: Aminoacyl-tRNA synthesis	21	
Translation: tRNA maturation and modification	6	
Translation: Ribosomal proteins	50	
Translation: Ribosome function, maturation and modification	7	
Translation factors	12	
RNA degradation	2	
Protein Processing, Folding, and Secretion		*15*
Protein post-translational modification	2	
Protein folding	5	
Protein translocation and secretion	5	
Protein turnover	3	
Cellular Processes		*5*
Energetic and Intermediary Metabolism		*56*
Poorly Characterized		*8*
TOTAL		*206*

Adapted from Gil et al., 2004.

Organisms such as *Mycoplasma genitalium* and *Buchnera* spp. reach the smallest sizes, with values that agree with the predictions of Shimkets (1998) for a minimal genome corresponding to about 600 kb. It is argued that these organisms have undergone massive gene losses and their limited encoding capacities are a result of their adaptation to the highly permissive (nutrient-rich) and stable intracellular environments provided by the hosts (Islas et al., 2004).

Other studies of minimal genomes have focused on the actual number of genes rather than the number of base pairs. For instance, Gil and colleagues (2004) used a functional approach to compare the genomes of five endosymbionts and other microorganisms. The results indicate 206 genes as the genomic core required for minimal living forms (table 3.1). This small number of genes code for the proteins that perform essential cell functions such as basic metabolism and self-reproduction.

These results generally agree with the values given by other authors (Kolisnychenko et al., 2002; Koonin, 2000; Mushegian, 1999; Mushegian and Koonin, 1996; Shimkets, 1998), as summarized in table 3.2. In particular, Mushegian and Koonin (1996) indicated an inventory of 256 genes, required to sustain a modern type of minimal cell under permissible conditions. Similar values, as indicated in a review by Koonin (2000), were inferred by Itaya (1995) and Hutchinson et al. (1999) on the

Table 3.2
Works on the minimal genome

Description of the System	Main Goal and Results	Reference
The complete nucleotide sequence (580,070 base pairs) of the *Mycoplasma genitalium* genome has been determined by whole-genome random sequencing and assembly.	Only 470 predicted coding regions were identified (genes required for DNA replication, transcription and translation, DNA repair, cellular transport, and energy metabolism).	Fraser et al., 1995
Site-directed gene disruption in *B. subtilis*.	The values of viable minimal genome size were inferred.	Itaya, 1995
The 468 predicted *M. genitalium* protein sequences were compared with the 1,703 protein sequences encoded by the other completely sequenced small bacterial genome, that of *Haemophilus influenzae*.	A minimal self-sufficient gene set: the 256 genes that are conserved in these two bacteria, Gram-positive and Gram-negative respectively, are almost certainly essential for cellular function.	Mushegian & Koonin, 1996
Computational analysis (quantification of gene content, of gene family expansion, of orthologous gene conservation, as well as their displacement).	A set of close to 300 genes was estimated as the minimal set sufficient for cellular life.	Mushegian, 1999
Global transposon mutagenesis was used to identify nonessential genes in *Mycoplasma* genome.	265 to 350 of the 480 protein-coding genes of *M. genitalium* are essential under laboratory growth conditions, including about 100 genes of unknown function.	Hutchison et al., 1999
Several theoretical and experimental studies are reviewed.	The minimal-gene-set concept	Koonin, 2000
The article focuses on the notion of a DNA minimal cell.	The conceptual background of the minimal genome is discussed.	Luisi et al., 2002
Full-length poliovirus complementary DNA (cDNA) was synthesized by assembling oligonucleotides of plus and minus strand polarity.	It is possible to create an infectious poliovirus, which is much simpler than a bacterium, by a synthetic approach.	Cello, Paul, and Wimmer, 2002
A technique for precise genomic surgery was developed and applied to deleting the largest K-islands of *E. coli*, identified by comparative genomics as recent horizontal acquisitions to the genome.	Twelve K-islands were successfully deleted, resulting in an 8.1% reduction in genome size, a 9.3% reduction of gene count, and elimination of 24 of the 44 transposable elements of *E. coli*. The goal was to construct a maximally reduced *E. coli* strain to serve as a better model organism.	Kolisnychenko et al., 2002
Buchnera genomes obtained from five aphid lineages are physically mapped.	They suggest that the *Buchnera* genome is still experiencing a reductive process toward a minimum set of genes necessary for its symbiotic lifestyle.	Gil et al., 2002
Buchnera and other organism genomes were compared.	206 genes were identified as the core of a minimal bacterial gene set.	Gil et al., 2004
Comparative genomics.	Estimates of the size of minimal gene complements were made to infer the primary biological functions required for a sustainable, reproducing cell nowadays and throughout evolutionary times.	Islas et al., 2004

basis of site-directed gene disruption and transposon-mediated mutagenesis knockout in *B. subtilis* and *M. genitalium*/*M. pneumonia*, respectively.

Having reached the small number of 200 to 300 genes as a minimal genome, we would like to go further, asking the question whether further reductions are possible. In particular, we consider the genomic core of very simple cells that may not be as efficient as modern cells but are still capable of carrying out basic life processes.

Obviously, only speculation can help at this point. One approach is to imagine a step-by-step knockout of the genome that reduces cellular complexity and nonessential functions (Luisi, Oberholzer, and Lazcano, 2002). The starting point of this analysis comes from the consideration that the "blueprint" of cellular life is primarily related to production of information-carrying molecules, of the cell boundary, and of the biochemical machinery for synthesis and replication (this is exactly represented by the ribozyme-based minimal cell of figure 3.1). According to this view, the large number of genes that are responsible for the endogenous synthesis of low-molecular-weight compounds can be eliminated, because—in principle—low-molecular-weight compounds could be present in the surrounding medium and be able to freely penetrate into the cell. This is a rather strong hypothesis, which assumes availability of a vast set of simple compounds (sometimes with complex chemical structures) and an almost fully permeable membrane. If such a model is to be the target of an experimental investigation, the first requirement must be taken as an ad hoc condition, and certain membrane properties will be adjusted to bypass low membrane permeability; as discussed later. Selective transport of compounds across a lipid bilayer membrane is also achievable if pore-forming agents are present.

In table 3.3 (first, second columns; from Luisi, Oberholzer, and Lazcano, 2002), we report a gene list of a minimal cell that is able to perform gene replication and protein and lipid biosynthesis. This cell would have about 25 genes for the entire DNA/RNA synthetic machinery, about 120 genes for the entire protein synthesis (including RNA synthesis and the 55 ribosomal proteins), and 4 genes for the synthesis of the membrane. This would come to a total of about 150 genes, somewhat less than the 206 genes proposed by Gil et al. (2004). Because an external supply of substrates is assumed to be available, such a minimal cell should be capable of self-maintenance (expression of enzymes and other proteins), self-reproduction (of all its components including DNA, t-RNA, and ribosomes), and synthesis of membrane components, which leads to growth and eventually division of the structure. However, it would not make low-molecular-weight compounds and would not have defensive, regulatory, and self-repair mechanisms. Also, cell division would occur by a statistically based physical process, rather than the controlled process used by cells today.

Proceeding with this thought experiment, the next reduction would involve ribosomal proteins (table 3.3, third column). There are some indications that ribosomal proteins may not be essential for protein synthesis (Zhang and Cech, 1998), and other

Table 3.3
A hypothetical list of gene products defining minimal cells (according to definitions in this chapter), sorted by functional category

Gene Product	Number of Genes		
	Minimal DNA Cell (*a*)	"Simple-Ribosome" Cell	Extremely Reduced Cell
DNA/RNA metabolism			
DNA polymerase III	4 (*b*)	4 (*b*)	1
DNA-dependent RNA polymerase	3 (*c*)	3 (*c*)	1
DNA primase	1	1	
DNA ligase	1	1	1
Helicases	2–3	2–3	1
DNA gyrase	2 (*d*)	2 (*d*)	1
ssDNA-binding proteins	1	1	1
Chromosomal replication initiator	1	1	
DNA topoisomerase I and IV	1 + 2 (*d*)	1 + 2 (*d*)	1
ATP-dependent RNA helicase	1	1	
Transcript. elong. factor	1	1	
RNases (III, P)	2	2	
DNases (endo/exo)	1	1	
Ribonucleotide reductase	1	1	1
Protein biosynthesis/translational apparatus			
Ribosomal proteins	51	0	0
Ribosomal RNAs	1 (*e*)	1 (*e*)	1 (*e*), self splicing
aa-tRNA synthetases	24	24	14 (*f*)
Protein factors required for biosynthesis and synthesis of membrane proteins	9–12 (*g*)	9–12 (*g*)	3
tRNAs	33	33	16 (*h*)
Lipid metabolism			
Acyltransferase 'plsX'	1	1	1
Acyltransferase 'plsC'	1	1	1
PG synthase	1	1	1
Acyl carrier protein	1	1	1
TOTAL	146–150	105–107	46

Notes: The hypothetical minimal genome, based on *Mycoplasma genitalium* (first and second columns) was first reduced by removing the ribosomal proteins (third column), and second, by reducing the number of different amino acids involved in protein formation, assuming the existence of less specific polymerases, and other assumptions (fourth column). (a) Based on *M. genitalium*; (b) subunits α, β, γ, and τ; (c) subunits a, b, b′; (d) subunits a, b; (e) one operon with three functions (rRNAs); (f) assuming a reduced code; (g) including the possible limited potential to synthesize membrane proteins; (h) assuming the third base to be irrelevant.
Adapted from Luisi, Oberholzer, and Lazcano, 2002.

suggestions that ancient translation systems may have been simpler (Nissen et al., 2000; for a review, see Calderone and Liu, 2004). In this context, the link with the scenario of early cells becomes stronger. There are, in fact, some claims that the first ribosomes consisted of rRNAs associated with basic peptides (Weiner and Maizels, 1987). If we accept this, and take out the 55 genes of the ribosomal proteins, the number of genes would be reduced to around 110. (A recent report of Mushegian, 2005, indicates 35–40 proteins as the components of minimal ribosomes; this number was calculated on the basis of a comparative genomic approach.)

Further reductions are possible by imagining a cell with a simplified replicating enzymatic suite (table 3.3, fourth column). The idea that a single polymerase could play multiple roles (for example, DNA *and* RNA-polymerase) has already been proposed as conceivable in early cells (Luisi, Oberholzer, and Lazcano, 2002). This concept was developed on the basis of previous theoretical accounts on the transition from RNA to DNA in early cells, on reverse-transcriptases and DNA primases (Frick and Richardson, 2001; Lazcano et al., 1988, 1992). Similarly, in order to decrease the number of t-RNA genes and the corresponding amino acyl-tRNA synthetase enzymes, one can claim that early cells did not use all 20 amino acids.

In this way it is possible to design—using 45 to 50 genes, as reported in table 3.3—a living, but inefficient minimal cell (Luisi, Oberholzer, and Lazcano, 2002). This number is significantly lower than that proposed by Moya in table 3.1 (see also chapter 16), and is, of course, highly speculative. Many authors would doubt that a cell with only 45 to 50 genes could be functional.

In conclusion, there is a general convergence of opinion about the minimal genome size being around 200 genes. This can be further reduced to about 50 genes if a biochemically rich environment, protein-poor ribosomes, and enzymes with a broader spectrum of activity are implemented.

3.3 The Use of Lipid Vesicles as Cell Models

Two important properties of lipid vesicles are closely related to the fabrication of semisynthetic minimal cells. Lipid vesicles can host reactions of biological significance and are capable of self-reproduction. These two points are discussed in the following sections, with particular emphasis on experimental results.

It is significant that vesicles can in principle simulate one of the most critical capacities of living cells, division, on the basis of physical and chemical properties alone, without sophisticated biochemical machineries. This capability is essential when we consider prebiotic scenarios involving protocells or laboratory versions of minimal cells. Investigations of reactions within lipid vesicles and the potential organizing effect of lipid matrices on chemical reactions are reviewed in the following sections, and the issue of membrane permeability is also briefly discussed.

3.3.1 Self-Reproduction of Self-Assembled Structures

In the context of this chapter, self-reproducing self-assembled structures are considered as models of protocell reproduction under the more general perspective of autopoietic systems (Luisi, 2003; Luisi and Varela, 1990; Varela, Maturana, and Uribe, 1974). Living cells generate structural components using their own macromolecular machinery, and these components in turn self-assemble into the system during growth and ultimately division into daughter cells. This consideration links vesicle self-reproduction and the chemical implementation of autopoiesis to studies of minimal cells.

Pioneering work on the subject began in the early 1990s by Luisi's group at ETH Zürich, and focused on the self-reproduction of reverse micelles (Bachmann et al., 1990, 1991; Bachmann, Luisi, and Lang, 1991). These results were subsequently extended to micelles (Bachmann et al., 1991; Bachmann, Luisi, and Lang, 1992), submicrometric vesicles (Berclaz, Mueller, et al., 2001; Berclaz, Bloechliger, et al., 2001; Morigaki et al., 1997; Rasi, Mavelli, and Luisi, 2003; Walde, Wick, et al., 1994), as well as giant vesicles (Wick, Walde, and Luisi, 1995). These reports, together with some recent developments (Takakura and Sugawara, 2004; Takakura, Toyota, and Suguwara, 2003), constitute our knowledge on this subject. Two cases discussed here in detail are the autocatalytic formation and self-reproduction of micelles driven by a chemical reaction, and the self-reproduction of fatty acid vesicles by increasing the concentration of the membranogenic precursor. In the latter case, we consider a new unexpected property of such systems—the matrix effect.

The chemical approach that Bachmann and coworkers (1992) used to form micelles from a precursor consists of the exploitation of ester hydrolysis to produce the corresponding acid and alcohol (figure 3.4). Consequently, the freshly formed long-chain fatty acids spontaneously self-assemble in alkaline water to form micelles (figure 3.4a). Figure 3.4b–c shows the experimental setup and the kinetics of fatty acid micelle formation, following the alkaline hydrolysis of the corresponding fatty acid ethyl ester. The initial state is a two-phase system, in which the water-insoluble ester is layered over an alkaline water solution. In this first stage, when the water-oil interface is small (macroscopic interface), the reaction is very sluggish. When enough ester has been converted to the corresponding fatty acid carboxylate, the first micelles form, and they act as a catalyst for the ester hydrolysis.

Notice that the formation of micelles corresponds to an increased area of the oil-water interface. The ester molecules can be incorporated into the micelles and are hydrolyzed at the boundary. In this way, increasing numbers of micelles are produced, which then catalyze their own formation. In other words, an autocatalytic process is initiated that is characterized by a pronounced sigmoidal yield-time profile. The process can be considered an autocatalytic self-reproduction of micelles, because the precursor is first incorporated in the "parent" micelle, then transformed in the

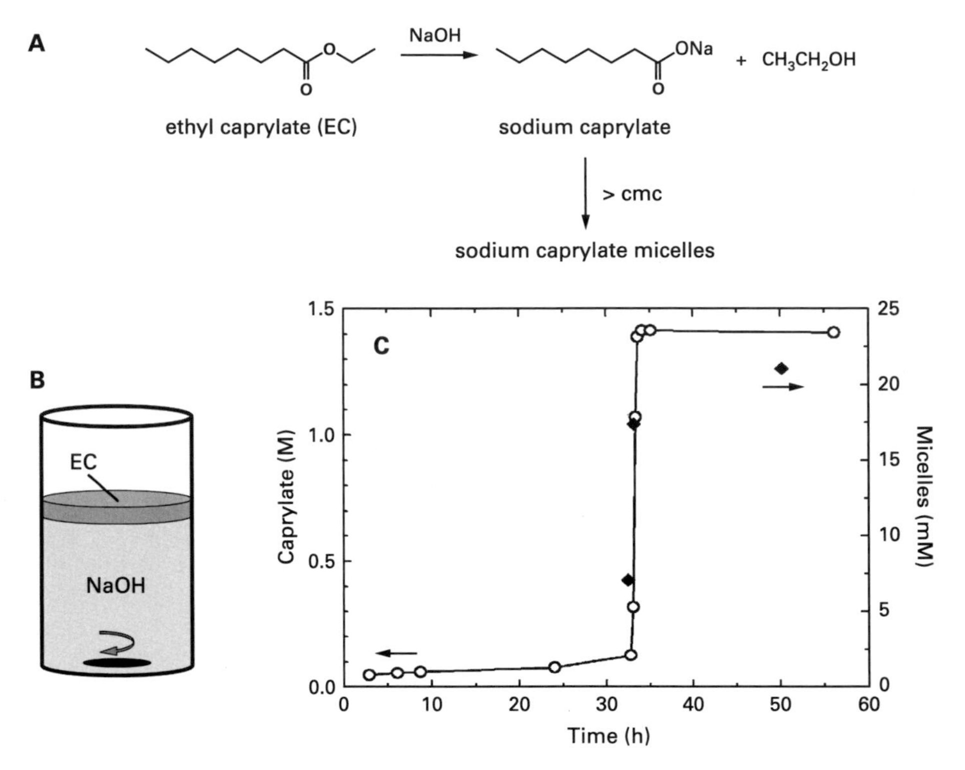

Figure 3.4
Autopoietic self-reproduction of caprylate micelles. (A) The alkaline hydrolysis of ethyl caprylate (EC) produces the surfactant sodium caprylate and ethanol. When the concentration of sodium caprylate exceeds a critical value (cmc = critical micelle concentration), caprylate micelles are formed. (B) Ethyl caprylate is layered on an alkaline aqueous solution; the reaction takes place at the interface between the two liquids and leads to the formation of sodium caprylate micelles in the aqueous phase, which solubilize new ethyl caprylate molecules, and these in turn are hydrolyzed at the micellar interface. (C) Time course of the reaction: The yield is very low in the first ~35 hours, but as soon as the cmc is reached (0.1 M) and micelles form, the remaining ethyl caprylate is quickly hydrolyzed. The quite pronounced S-shaped curve witnesses the autocatalytic process. (Adapted from Bachmann, Luisi, and Lang, 1992.)

micelle building block within the boundary of the micelle itself and used to form new micelles.

A structure that is self-bounded and able to self-generate due to reactions that take place within the boundary, indeed, meets the criteria of autopoiesis. The term, in fact, stems from the Greek *auto* (self) and *poiesis* (formation). In regard to the work with minimal living cells, we note that autopoiesis is the central feature of all cellular life, because it emphasizes the importance of self-generation "from within" (Luisi, 2003; Maturana and Varela, 1980, 1998; Varela, Maturana, and Uribe, 1974).

The second system discussed here is vesicle self-replication (figure 3.5). The reaction that is used to form vesicles is the so-called micelle-to-vesicle transformation,

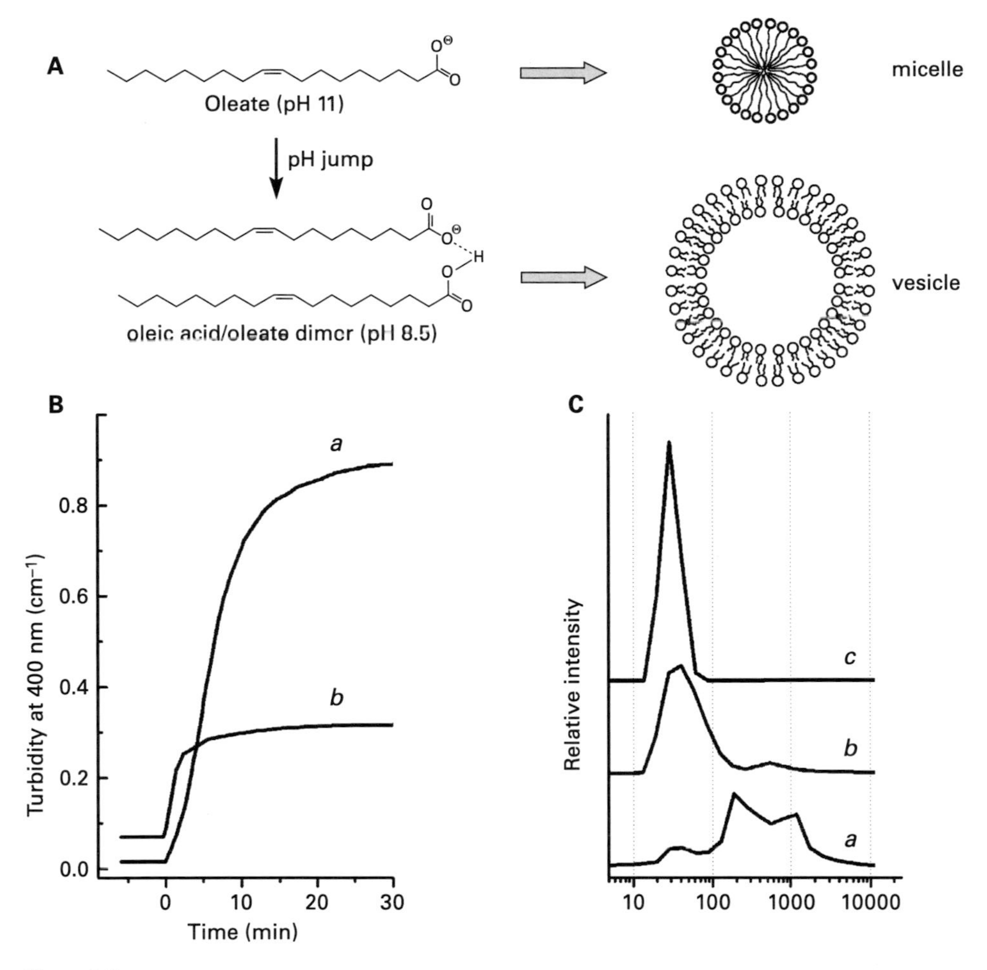

Figure 3.5
Oleate micelle-to-vesicle transformation and vesicle self-reproduction. (A) Injecting a concentrated solution of oleate micelles into a pH 8.5 buffer, the partial protonation of the carboxylate moiety allows the formation of a dimer, which permits the formation of oleate vesicles. (B) Time course of oleate vesiculation in the absence (a) and in the presence (b) of an equimolar amount of preformed extruded oleate vesicles. The process (b) is faster and brings to a lower turbidity value. (C) Dynamic light scattering analysis of vesicles obtained in the absence (a) and in the presence (b) of preformed extruded oleate vesicles. Curve (c) indicate the monomodal size distribution of the preformed vesicles, obtained by extrusion. Notice that curves (b) and (c) are almost coincident (the matrix effect). (Adapted from Rasi, Mavelli, and Luisi, 2003.)

which takes place when fatty acid micelles (stable at high pH, e.g., ~10 to 11) are diluted in a buffer with pH 8 to 9. Vesicles from oleic acid/oleate are widely studied, and we refer to them here. At high pH, oleate molecules are unprotonated anions, whereas at pH 8 to 9, the ionization degree of the oleic acid in the protonated (neutral) form is about 50%, which means that oleic acid and oleate molecules are present in approximately equimolar amounts. The different protonation state of oleate at different pH ranges, as shown in figure 3.5a, accounts for the micelle-to-vesicle transition, perhaps by varying the geometry from conical (high pH, which favors micelle formation) to approximately cylindrical (lower pH, which favors bilayer formation). This pH-jump experiment offers a method by which membranogenic material can be added to a lipid membrane so that vesicle growth occurs.

If the oleate vesicles form in the absence of any preformed vesicles, the vesiculation is relatively slow and leads to a heterogeneous vesicle sample with a broad size distribution, characterized by large vesicles (figure 3.5b–c, curve a). On the other hand, if the oleate vesicles form in the presence of preformed vesicles having a defined size, the process is faster and the size distribution of the final vesicles strongly shifts toward the size of the preformed vesicles (figure 3.5b–c, curve b). It is evident that the presence of preformed vesicles affects the kinetics and the mechanism of oleate vesiculation, and this has been named *the matrix effect* (Berclaz, Bloechliger, et al., 2001; Bloechliger et al., 1998; Lonchin et al., 1999; Rasi, Mavelli, and Luisi, 2003). Under these circumstances, the vesicle size distribution is maintained when the lipid concentration is increased, implying that an increment of the vesicle number occurs. Because the new vesicles resemble the old ones, vesicle self-reproduction takes place through a mechanism that could have been very relevant in prebiotic evolution of protocells (Luisi et al., 2004).

A possible mechanism, which derives from the analysis of kinetic, dynamic light scattering and electron microscopy data, invokes the initial interaction between freshly added oleate molecules and preformed vesicles that grow by incorporating membranogenic molecules, and finally divide to form new vesicles. In this progression, internal solutes are not lost (Berclaz, Bloechliger, et al., 2001). In other words, the process is a model for the self-reproduction of prebiotic vesicles, by means of a spontaneous and dynamic interaction between vesicles and membrane-forming molecules.

Although oleate vesiculation has been widely investigated (Chen and Szostak, 2004; Chungcharoenwattana and Ueno, 2004, 2005; Hanczyc, Fujikawa, and Szostak, 2003), a detailed description is still missing. Several theoretical approaches have been developed to explain vesicle self-reproduction (Bolton and Wattis, 2003a, 2003b; Bozic and Svetina, 2004; Chizmadzhew, Maestro, and Mavelli, 1994; Mavelli and Luisi, 1996).

In the case of vesicles, supramolecular assemblies are reproducing themselves and maintaining a constant size at the expense of the surfactant source. Furthermore, the process is localized within the boundary of the aggregate, making this a second example of an autopoietic system in addition to the micelles described earlier. Autopoietic self-reproduction of vesicles based on anhydride hydrolysis has also been described (Morigaki et al., 1997; Walde, Wick, et al., 1994), as well as a chemically regulated homeostatic vesicle system (Zepik, Bloechliger, and Luisi, 2001).

In conclusion, the self-reproduction of supramolecular structures is very relevant to studies of minimal cells. It will be apparent in the next section that the self-reproduction of a complete protocellular system could incorporate a similar mechanism for growth and division, and that the cellular boundary is not at all a passive container. In other words, in any attempt to fabricate semisynthetic cells, the membranous boundary structure and its reactivity must be taken into account.

3.3.2 Reactivity in Vesicles

In the previous section we discussed important preliminary aspects related to using liposomes as models for cellular compartments, and analogies between lipid vesicles and cellular membranes in terms of self-reproduction. The second area of inquiry concerns the use of vesicles as hosts for complex molecular biological reactions.

In attempting to understand the origin of prebiotic protocells, we must now consider another relevant aspect of liposome chemistry, which is the role of lipid membranes in the synthesis of biopolymers. In fact, a lipid membrane can favor the polymerization of amino acids or other compounds either by catalyzing the formation of covalent bonds (in this case we can have a specific effect of the membrane components that act as catalysts) or by favoring the polymerization of lipophilic compounds (in this case the membrane acts as a hydrophobic matrix). In principle, a combination of the two modes is also possible. We begin by illustrating relevant physicochemical properties of lipid membranes such as solute permeability and liposome fusion, as well as some experimental results concerning the development of liposomes as bioreactors. In the last part, we discuss the issue of membrane-assisted polymerization.

The first important physicochemical property, which concerns the use of vesicles as bioreactors, is membrane permeability to solutes. Basic membrane chemistry tells us that the lipid bilayers offer a permeability barrier to the free movement of ions and large water-soluble molecules. For instance, small molecules such as water and carbon dioxide cross liposome membranes with half-times measured in milliseconds, whereas a potassium gradient might take several days to approach equilibrium. Small nonpolar molecules such as general anesthetics (chloroform, halothane, and diethyl ether) are even more permeant than water, but certain hydrophobic molecules like peptides do not pass the barrier because they become embedded in the membrane. It follows that there is a continuum of membrane permeability, related to the different molecular structures of solutes, their chemical compositions, and the membrane itself, for example, in relation to the length of lipid hydrophobic chains. Thus, the first emergent property of the vesicle membrane is differential solute permeability.

This property can be used when it is desired to select particular compounds, either favoring or inhibiting their interaction with, or passage across, the membrane. For instance, some chemical reactions in compartments may respond to a vesicle-selection mechanism. Walde and Marzetta (1998) reported that α-chymotrypsin–containing vesicles show selective incorporation of externally added substrates.

From a biological point of view, it is clear that modern cells, with their sophisticated transport systems, take advantage of the very low natural membrane permeability. In the case of primitive cells and the onset of life, the main advantage of low permeability (especially for polymeric species such as proteins and nucleic acids) is the conservation of a specific cell identity that would arise from a particular composition of its internal and boundary composition. A vesicle protects its internal content from dilution in the surroundings and from potentially damaging environmental factors. In addition, considering that the aqueous content must be distributed between the daughter vesicles during vesicle self-reproduction without release of biochemicals into the surrounding medium, the low permeability of lipid membranes facilitates vesicle proliferation by preserving its chemical composition.

The main drawback of a low membrane permeability is that it inhibits uptake of nutrients from the environment. There are, however, some physical and chemical factors that increase the membrane permeability. It has been demonstrated that the dehydration-rehydration of initially empty vesicles in the presence of certain solutes leads to the encapsulation of the solute molecules in the vesicles (Deamer and Barchfeld, 1982). This pathway of vesicle loading is of great interest because it is considered a possible prebiotic route to the formation of functionalized vesicles (Deamer, 1998). However, because it involves the rupture and the resealing of the lipid bilayer, other mechanisms must be exploited to "feed" intact vesicles.

It is known, for instance, that permeability increases when membrane lipids are at the chain melting temperature, defined as the transition temperature between gel and fluid states of the bilayer. This effect is a result of the transient presence of membrane packing defects, and was experimentally exploited by Monnard and Deamer (2001) to demonstrate the passive diffusion of ATP into vesicles. In later work (Monnard, Luptak, and Deamer, 2007), all four nucleoside triphosphates (NTPs) were added externally to vesicles containing a template-dependent RNA-polymerase, and synthesis of RNA was observed as the rNTPs diffused through the defects into the vesicle interior volume and provided substrates for the polymerase enzyme. The RNA cannot escape from the vesicles because of the large dimension of the molecules, a useful example of how a primitive cell or laboratory protocell could grow by internalized polymerization.

Following the initial report on detergent-induced liposome loading (Schubert et al., 1986, 1991), Oberholzer and Walde repeatedly applied the detergent-based approach to load vesicles in order to perform enzymatic reactions within the liposome compartments. These experiments showed that small and large molecules can cross the lipid barrier in the presence of variable amounts of the detergent sodium cholate. In particular, glucose-1-phosphate (Oberholzer, Abrizio, and Luisi, 1999), mononucleotides (Treyer, Walde, and Oberholzer, 2002), glucose (Yoshimoto et al., 2003), and derivatized tripeptides (Yoshimoto et al., 2004) have been fed to, and reacted inside, enzyme-containing vesicles. Larger molecules, such as DNase I (29 kDa) and tRNA (25 kDa), are also able to permeate through the membrane (Oberholzer, Abrizio, and Luisi, 1999). However, it has been suggested that the maximum limit for membrane permeation in detergent-permeabilized systems is around 70 kDa (Schubert et al., 1986, 1991).

In addition to these cited studies, carried out on lecithin liposomes, Szostak and coworkers (Sacerdote and Szostak, 2005) recently carried out an interesting investigation on sugar permeability of fatty acid vesicles, suggesting that ribose, an essential building block of nucleic acids, can be selectively taken up by fatty acid vesicles.

To realize a real compartment system that can accumulate simple biochemicals in its aqueous core, we can also consider other possibilities. The use of membrane channels offers one possibility, but until now has met only modest success. An exception has been the use by Noireaux and Libchaber (2004) of α-hemolysin, a pore-forming protein that allowed a vesicle bioreactor to function continuously for up to four days.

The general problem is how to reach a high local concentration and a high compositional complexity of biochemicals in the aqueous core of liposomes. This difficulty might be partially circumvented if two or more liposomes, each containing a given substrate, are fused to produce liposomes containing all of the desired components. Vesicle fusion is becoming an active area of research, and interesting results have already been obtained (Arnold, 1995; Marchi-Artzner et al., 1996; Pantazatos and MacDonald, 1999; Stamatatos et al., 1988; Thomas and Luisi, 2004). The classic approaches to vesicle fusion involves: (a) charged vesicles and oppositely charged small ions or molecules, (b) oppositely charged vesicles, (c) dehydrating agents, and (d) fusogenic peptides. The interested reader should refer to the original contributions and to several reviews on this subject (Bailey and Cullis, 1997; Blumenthal et al., 2003; Nir et al., 1983).

If vesicles could be fused by means of a hypothetical prebiotic route, we would circumvent the problem of nutrient uptake and complexity achievement. We can imagine several vesicles fusing together in stepwise fashion to produce a new vesicle that contains all the solutes initially present in the original vesicles. In addition, this strategy would simplify the experimental approaches to minimal cell construction, the complexity of which can be achieved by fusing different preformed vesicle populations, each containing different compounds (for example, genes, ribosomes, enzymes, low-molecular-weight compounds).

The strategy of liposome fusion has not been pursued yet because of the complex balance of forces the fusion process involves. Parasite processes of aggregation or

Table 3.4
Molecular biology reactions into liposomes

Description of the System	Main Goal and Results	Reference
Lecithin biosynthesis	First attempt to synthesize lecithin within liposomes	Schmidli, Schurtenberger, and Luisi, 1991
Enzymatic poly(A) synthesis	Polynucleotides phosphorylase producing poly(A) from ADP	Chakrabarti et al., 1994
Enzymatic poly(A) synthesis	Poly(A) produced inside simultaneously with the (uncoupled) self-reproduction of vesicles	Walde, Goto, et al., 1994
Oleate vesicles containing the enzyme Qβ replicase, an RNA template and ribonucleotides. The water-insoluble oleic anhydride was added externally	A first approach to a synthetic minimal cell: The replication of an RNA template proceeded simultaneously with the self-replication of the vesicles	Oberholzer et al., 1995
POPC liposomes containing all different reagents necessary to carry out a PCR reaction	DNA amplification by PCR inside the liposomes; a significant amount of DNA was produced	Oberholzer, Nierhaus, and Luisi, 1995
POPC liposomes incorporating the ribosomal complex together with the other components necessary for protein expression	Ribosomal synthesis of polypeptides can be carried out in liposomes; synthesis of poly(Phe) was monitored by TCA of the ^{14}C-labeled products	Oberholzer, Nierhaus, and Luisi, 1999
T7 DNA within cell-sized giant vesicles formed by natural swelling of phospholipid films	Transcription of DNA and transportation by laser tweezers; vesicles behaved as a barrier preventing the attack of RNase	Tsumoto et al., 2001
DNA template and the enzyme T7 RNA polymerase microinjected into a selected giant vesicle; nucleotide triphosphates added from the external medium	The permeability of giant vesicles increased in an alternating electric field; mRNA synthesis occurred	Fischer, Franko, and Oberholzer, 2002

rupture can reduce the fusion yield. On the other hand, by using compartments such as water-in-oil emulsions, it is possible to express a functional protein by mixing different emulsion droplets, each containing different components for the reactions (see section 3.2.3 for a deeper discussion).

Regarding biochemical reactions in liposomes, considerable experimental work has paved the way to recent developments. These are studies in which liposomes have been used as host systems for molecular biology reactions (table 3.4). Two groups (Chakrabarti et al., 1994; Walde, Goto, et al., 1994) independently reported the biosynthesis of poly(A), a model for RNA. In both cases the authors entrapped polynucleotide phosphorylase (PNPase) in vesicles and observed the synthesis of poly(A), which remains in the aqueous core of such vesicles. In one case (Walde, Goto, et al., 1994) the internal poly(A) synthesis was coupled with the reproduction of the vesicle shell, obtained by external addition of oleic anhydride, a membranogenic precursor.

Oberholzer and coworkers (Oberholzer et al., 1995) reported RNA replication by $Q\beta$ replicase contained within vesicles. This enzyme replicates an RNA template, and using an excess of $Q\beta$ replicase and feeding the vesicles with nucleotides demonstrated that RNA replication could proceed for several generations. The replication of a core component could also be coupled to replication of the vesicle boundary. This system, as well as that of Walde, Goto, and coworkers (1994), is interesting because it represents a case of "core and boundary replication," in which both the content of the core and the membrane boundary itself undergo duplication. In addition, it should be clear that the mechanism of division is only the ideal case. In real cases, the core components of such vesicle reactors do not distribute evenly between the two or more "daughter" vesicles. Moreover, since all cellular components except $Q\beta$ replicase are produced, the system will eventually undergo a "death by dilution." After a while the new vesicles will contain neither enzyme nor template, and therefore the construct cannot reproduce itself indefinitely.

Another complex biochemical reaction implemented in liposomes is the polymerase chain reaction (PCR) (Oberholzer, Albrizio, and Luisi, 1995). Liposomes are able to endure PCR conditions, with several temperature cycles up to 90°C (liposomes are practically unchanged at the end of the reaction). One of the most interesting observations is that all the components needed for the reaction are encapsulated by liposomes in the correct relative amount. While it is not obvious that all chemicals are simultaneously trapped within individual liposomes, PCR products are nevertheless observed.

By using poly(U) as mRNA, Oberholzer and coworkers (Oberholzer, Nierhaus, and Luisi, 1999) show the production of poly(Phe), starting from phenylalanine, ribosomes, $tRNA^{Phe}$, and elongation factors entrapped in lecithin vesicles. The authors point out that the yield obtained (5% of the control reaction without liposomes) is surprisingly high, also considering the low probability that all the ingredients needed for the reaction are present in the same vesicle. In table 3.4 we also report the results of Fischer, Franco, and Oberholzer (2002) on mRNA synthesis utilizing a DNA template and T7 RNA polymerase, and the transcription of DNA by Tsumoto, Nomura, Nakatani, and Yoshikawa (2001); both studies refer to giant vesicles. Further considerations on polymerase reactions inside vesicles are provided by Monnard (2003).

The concluding part of this short review concerns membrane-assisted polymerization. Pioneering work initiated in the 1980s (Folda, Gros, and Ringsdorf, 1982; Fukada, Shibasaki, and Nakahara, 1981; Kunieda et al., 1981; Laschewsky et al., 1987; Neumann et al., 1987) focused on the use of covalently modified amino acids, generally with long alkyl chains, to anchor the amino acids to the liposome membrane. Because of the increased local concentration on the surface, oligomerization to peptides is possible. A few years ago, Luisi and coworkers investigated the liposome-assisted oligomerization of amino acids without hydrophobic derivatization (Blocher et al., 1999). It is shown that tryptophan oligomers (up to 29-mers) can be synthesized in the presence of lecithin liposomes, suggesting that the lipid membrane is somehow involved in the synthesis of long oligomers that are sparingly soluble in water depending on their length. Moreover, liposomes are able to selectively promote the condensation of tryptophan dimers (to give oligo-tryptophan) over other single-tryptophan containing dipeptides. These results demonstrate that a lipid membrane can act as a lipophilic matrix for the formation of long hydrophobic polypeptides, a reaction that cannot be obtained in aqueous solution, and that such membranes can display selectivity among different starting monomers.

As a further step, the same authors (Blocher, Liu, and Luisi, 2000) demonstrated that amino acids such as glutamic acid and tryptophan react in the presence of positively charged liposomes to give block oligomers. Furthermore, dipeptides of histidine or arginine react with tryptophan in the presence of negatively charged liposomes to give tryptophan-histidine or tryptophan-arginine oligomers. More recent work investigated the stereochemistry of racemic amino acid condensation in the presence of liposomes (Hitz et al., 2001; Hitz and Luisi, 2001). It was found that, starting from a racemic mixture of L- and D-tryptophan, the homochiral 10-mer is 40 times overrepresented with respect to the theoretically expected value. The preferential formation of homochiral chains is also observed in the absence of liposomes, but the yield of higher oligomers is reduced. Similar results on homochiral oligomers were obtained by Lahav and coworkers (Rubinstein et al., 2003) by using a monolayer of lipid as a matrix and other activated amphiphilic amino acids. The presence of homochiral products is ascribed to the formation of two-dimensional crystalline domains of monomers. These results suggest that similar phenomena can occur at the membrane interface.

Although the feature of membrane-assisted polymerization is more pertinent to prebiotic chemistry than to minimal cell fabrication, it is useful to note again that vesicles are not passive compartments, but can play an active role in several chemical processes. Our knowledge of such a potential role remains limited.

3.3.3 Protein Expression in Liposomes

In the previous section, we summarized some of the biochemical reactions that have been realized and optimized in liposomes. Following these initial reports, several investigators have achieved the goal of expressing one or more proteins in the aqueous core of liposomes, starting from the corresponding gene. For reasons of convenience, the easily detectable Green Fluorescent Protein (GFP) was chosen for all of these studies.

How many genes are involved in GFP expression? This question is not easily addressed from current data, because no calculation of the genes involved (i.e., the corresponding enzymes and RNAs) has been done. Commercial kits are often used for protein expression, and these kits are generally cell extracts (in particular, *E. coli* extracts) of unspecified composition. It is fair to say, however, that for the combined transcription/translation processes leading to the expression of one single protein, only a minimal part of the *E. coli* genome is required. (We refer here to the number of genes that code for the enzymes, RNAs, and ribosome components involved in the entire transcription/translation process, about 100 to 120 genes.) Recently, Ueda's team from the University of Tokyo developed a cell-free translation system with purified components (Shimizu et al., 2001) that can be used to have complete control over the factors involved in protein expression.

An overview of the work concerning protein expression in liposomes is presented in table 3.5, where the advances in this field are listed in chronological order. The common strategy is encapsulation in lipid vesicles of all the components required for in vitro protein expression; for example, the GFP gene, an RNA polymerase,

Table 3.5
Protein expression in compartments

Description of the System	Main Goal and Results	Reference
Liposomes from EggPC, cholesterol, DSPE-PEG5000 used to entrap cell-free protein synthesis	Expression of a mutant GFP, determined with flow cytometric analysis	Yu et al., 2001
Small liposomes prepared by the ethanol injection method	Expression of EGFP evidenced by spectrofluorometry	Oberholzer and Luisi, 2002
Gene-expression system within cell-sized lipid vesicles	Encapsulation of a gene-expression system; high expression yield of GFP inside giant vesicles	Nomura et al., 2003
A water-in-oil compartment system with water bubbles up to 50 mm	Expression of GFP by mixing different compartments able to fuse with each other	Pietrini and Luisi, 2004
A two-stage genetic network encapsulated in liposomes	A genetic network in which the protein product of the first stage (T7 RNA polymerase) is required to drive the protein synthesis of the second stage (GFP)	Ishikawa et al., 2004
E. coli cell-free expression system encapsulated in a phospholipid vesicle, which was transferred into a feeding solution containing ribonucleotides and amino acids	The expression of the α-hemolysin inside the vesicle solved the energy and material limitations; the reactor could sustain expression for up to 4 days	Noireaux and Libchaber, 2004

ribosomes, enzymes, and all the low-molecular-weight components such as amino acids, ATP, and other nucleotides needed for protein expression. This is generally accomplished by preparing the liposomes in the presence of all these compounds.

In their first report (Yu et al., 2001), Yomo and Urabe described the expression of a mutant GFP in lecithin liposomes. Large vesicles expressing GFP were prepared by the film hydration method in the presence of all the previously mentioned components. External GFP synthesis was inhibited by addition of RNase, and the production of functional mutant GFP protein inside vesicles was verified by flow cytometry and confirmed by confocal microscopy. Although the fraction of liposomes that were able to host the GFP expression was small, the procedure shows the feasibility of the method.

Oberholzer and Luisi (2002) employed a slightly different approach to prepare liposomes in the presence of cell-free expression components. Here the reaction was allowed to initiate for a few minutes in the water phase, and then unilamellar liposomes were formed by the ethanol injection method. The addition of ethylenediaminetetraacetic acid (EDTA) after liposome formation inhibited the external protein activity. The reporter protein, the so-called enhanced GFP (EGFP), was revealed by batch fluorescence spectroscopy after correcting for turbidity and normalization. The disadvantage of this procedure can be a low entrapping efficiency, but it has the great advantage of producing small uni- or oligolamellar liposomes. These relatively small structures generally have a homogeneous size distribution and similar physical properties. Even if they are typically much smaller (diameter around 100–200 nm) than micrometric-sized vesicles, and cannot be detected by optical microscopy, the use of small liposomes allows the researcher to simultaneously examine a large number of compartments by batch analytical techniques.

Direct observation of protein expression was accomplished by the procedure utilized by Nakatani and Yoshikawa's team (Nomura et al., 2003). These authors prepared giant vesicles (around 5 mm in size) by the osmotic swelling method, using a complete transcription/translation mixture as the aqueous phase during lipid hydration. The synthesized protein, the red-shifted GFP (rsGFP), was revealed by fluorescence and confocal microscopy. This is the first clear report on the synthesis of functional protein in giant vesicles. By studying the kinetics of the protein expression, the authors remark that—in the first three hours—the rsGFP yield inside the vesicles is surprisingly higher than that in the external solution.

More recently, Yomo's team accomplished the expression of a cascading genetic network within liposomes (Ishikawa et al., 2004; see figure 3.6a). The authors implemented a sequential synthesis of two proteins in the same compartment, made possible by using two different genes activated by two different promoters, and by employing different RNA polymerases. In particular, a plasmid was designed and constructed containing a T7 RNA polymerase gene under the control of SP6 pro-

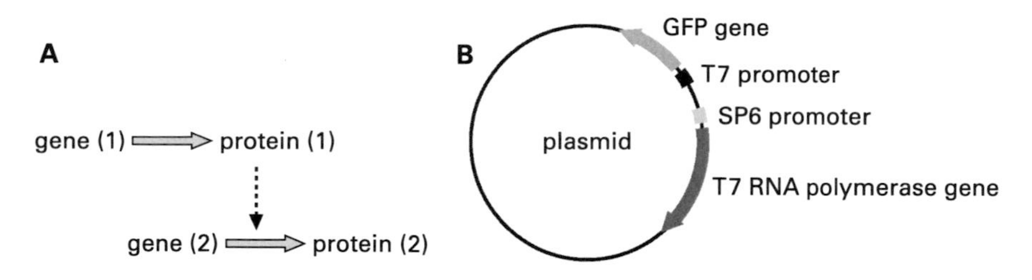

Figure 3.6
Expressing a two-level genetic network into vesicles. (A) General scheme. Gene (1) codifies for protein (1), which is actually an RNA polymerase. Protein (2), which is the GFP, can be expressed starting from gene (2), only if the RNA polymerase has been expressed in the first step. (B) Schematic drawing of the plasmid used by Ishikawa et al. (2004).

moter, and a mutant GFP gene under the control of T7 promoter (figure 3.6b). GFP is produced only when the first gene is activated, producing functional T7 RNA polymerase.

Entrapping the components for protein expression and SP6 RNA polymerase, Yomo and coworkers showed that the two-level cascade protein synthesis mentioned previously occurred within vesicles. External protein synthesis was inhibited by addition of RNase, and the GFP was detected by flow cytometry. Under these conditions, entrapped SP6 RNA polymerase triggers the expression of T7 RNA polymerase, which in turn induces the synthesis of GFP.

Of similar interest is the work by Noireaux and Libchaber (2004). In their approach, two proteins were expressed within vesicles. In addition to EGFP, the system was designed to produce α-hemolysin, a water-soluble protein that assembles as heptameric pores in the lipid bilayer. The pore is 1.4 nm wide (cutoff ~ 3 kDa), so that small solutes can freely enter and exit from the liposome internal volume according to concentration gradients, whereas large molecular components remain trapped inside. Giant vesicles were obtained by a recently described method, originally designed to produce asymmetric vesicles (Pautot, Frisken, and Weitz, 2003) that allowed essentially 100% entrapment of the expression kit, thus avoiding the use of an externally added inhibitor. Using this strategy, it was shown that the inner aqueous vesicle space can be supplied with monomers and energy-rich compounds such as amino acids and nucleotides needed for protein expression, which continues for up to four days. The α-hemolysin pore permits the uptake of small metabolites from the external medium and thus solves the energy and material limitations typical of the impermeable liposomes. Obviously, it also allows the release of internalized content, being an unselective and bidirectional pore. This offers an advantage in terms of unwanted byproduct accumulation and osmotic stress relief.

To conclude this review on protein expression in compartments, we describe water-in-oil (w/o) macroemulsions that were introduced and ingeniously used as selection and evolution tools by Tawfik and Griffiths (1998). Such cell-like compartments, formed by dispersion of a tiny aqueous volume (~0.5%) in a surfactant/hydrocarbon phase, have certain practical advantages with respect to liposomes. First, it is easy to reach 100% entrapment during the formation of compartments, because the water-soluble compounds are completely contained inside the w/o compartments. Second, it is possible to obtain solute exchange between different droplets, achieving a way to increase molecular complexity by mixing different compartments and feed the aqueous core of the emulsion droplets with fresh components. Pietrini and Luisi (2004) used a w/o emulsion to express EGFP by incorporating the translation/transcription kit in these microscopic water droplets. The work demonstrates that EGFP can be efficiently synthesized by mixing four different emulsions, each containing a subset of the molecular components (DNA, amino acids, RNA polymerase, cellular extracts). Following solute exchange, fusion, or both, some emulsion droplets were found to contain all the ingredients (in the correct relative amounts) needed to express a functional protein, which was detected by fluorescence microscopy and quantitative image analysis.

To summarize, a handful of pioneering studies have appeared in the last few years that demonstrated expression of water-soluble proteins within liposomes. With the possible exception of α-hemolysin, membrane-associated proteins have not yet been expressed, although Ueda's team succeeded in cell-free expression of membrane proteins in the presence of, but not inside, inverted membrane vesicles (Kuruma et al., 2005). This result is interesting because it shows that membrane proteins, which carry out fundamental processes of lipid biosynthesis and energy metabolism, can potentially be synthesized and inserted into vesicle membranes.

3.4 Outlining the Next Steps

Our account of current research, summarized in tables 3.4 and 3.5, makes clear that self-reproduction remains as an essential element in defining a living cell described in the introduction: None of the systems listed in table 3.5 reproduce themselves to give rise to a second generation of protein-expressing vesicles. To accomplish this result, researchers must focus on the self-reproduction of vesicle boundary structures. The synthesis of new lecithin molecules in lecithin liposomes would be a significant step forward in the roadmap to a minimal cell. Production of the cell boundary from within, as depicted in figure 3.7, corresponds to the notion of autopoiesis (Luisi, 2003; Luisi and Varela, 1990; Varela, Maturana, and Uribe, 1974).

The most obvious approach involves vesicle self-reproduction by the endogenous synthesis of lipids. Two strategies can be pursued: incorporation of the enzymes that synthesize the lipids, or incorporation of the corresponding genes that have those enzymes as gene products.

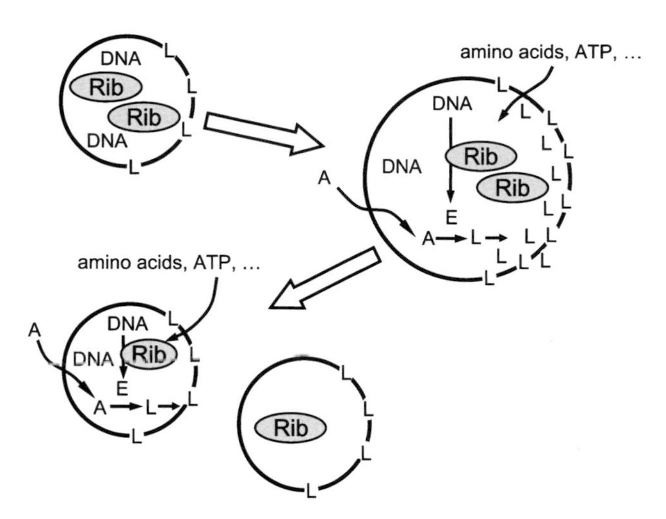

Figure 3.7
A cell that makes its own boundary. The complete set of biomacromolecules needed to perform protein synthesis (RNA polymerases, ribosomes, translation factors) is indicated as "Rib." The product of this synthesis (indicated as E) is the complete set of enzymes for the lipid (L) synthesis. After growth and division, some of the vesicles formed could undergo a "death-by-dilution."

One early attempt to produce lecithin in liposomes used the so-called salvage pathway, a metabolic pathway that converts glycerol-3-phosphate to phosphatidic acid, then diacylglycerol, and finally phosphatidylcholine (Schmidli, Schurtenberger, and Luisi, 1991). The four enzymes needed to accomplish these reactions—G3P-AT (*sn*-glycerol-3-phosphate acyltransferase), LPA-AT (1-acyl-*sn*-glycerol-3-phosphate acyltransferase), PA-P (phosphatidate phosphatase), and CDPC-PT (cytidinediphosphocholine phosphocholine transferase)—were simultaneously inserted in liposomes by the detergent depletion method, and the synthesis of new phosphatidylcholine (10% yield) was followed by radioactive labeling. Liposome transformation, followed by dynamic light scattering, showed that vesicles changed their size distribution during the process. This was a relatively complex system and it was realized later that the reaction of lipid biosynthesis could, in principle, be limited to the formation of phosphatidic acid, because this compound also forms stable liposomes. Further studies (Luci, 2003) involved overexpression in *E. coli* and reconstitution into liposomes of the first two enzymes of the phospholipid salvage pathway.

Returning now to minimal cells as models of prebiotic life, we note that all of the vesicular systems utilized phospholipids, which are rather complex (evolved) compounds. The reason is quite clear: phospholipids provide a stable and relatively inert membrane that is convenient to handle experimentally. From the point of view of potentially prebiotic compartments, a simpler surfactant must be involved in membrane formation. Most of the scientists involved in origin of life research agree that fatty acid vesicles are the best candidates, also including branched fatty acids such as oligoprenyl derivatives. The use of fatty acid vesicles, however, presents a challenge because of potential chemical incompatibilities, particularly the effects of magnesium ions, low pH, and high isoelectric point (pI) enzymes, all of which reduce the stability of fatty acid membranes. Therefore, the properties and self-reproduction of fatty acid–based vesicles as compartments for minimal cell studies is an objective that is worthy of continued research interest over the next few years.

In a more precise approach, protein expression within vesicles should be carried out utilizing, instead of commercial expression kits, the required number of enzymes involved in protein synthesis (*E. coli* extracts are in a sense "black boxes"). As introduced in section 3.2.3, a recently developed expression kit (Shimizu et al., 2001), composed of purified enzymes and ribosomes, is likely to fulfil these requirements. This system would correspond to the implementation of the minimal genome concept by means of specific incorporation of genes into liposomes, and can thereby enable the next steps toward a minimal cell.

Finally, looking at the systems of table 3.5, we are still dealing with ribosomal protein biosynthesis, and this implies 100 to 200 genes. Simplification of the ribosomal machinery, and of the enzymes devoted to RNA and DNA synthesis, can be one of the next targets. Is this feasible experimentally? For example: Can simple matrices be developed that are operative in vitro as ribosomes? Can a minimal cell work, at the expense of specificity, with only very few, less specific DNA/RNA polymerases? Also it might not be necessary at the beginning of life to have all possible specific tRNAs present, but instead work with a few less specific ones that use a limited number of amino acids.

3.5 Concluding Remarks

We have witnessed a significant increase in theoretical and experimental investigations focused on the minimal cell. A significantly greater control of molecular reactions within compartments has been achieved in the past few years, allowing a realistic approach to simple cell models. The notion of minimal cells, initially developed for studies on the origin of life, is now becoming an attractive target for constructive or synthetic biology. In fact, a deep understanding of basic cellular processes and their reconstruction in vitro will certainly increase our basic knowledge of life as an emergent property.

While the definition of a minimal cell appears simple, its experimental implementation has just begun. We have outlined the main difficulties involved in generating minimal cells; the clearest of which is efficient core-and-boundary self-reproduction, a goal that has not yet been accomplished. To self-reproduce all the components of the liposome core, one must consider implementing DNA duplication as well as in situ expression (starting from the corresponding genes) of all enzymes and RNAs composing the transcription/translation machinery. Likewise, the self-reproduction of the lipids constituting the membrane must proceed by the endogenous synthesis of membranogenic compounds, starting from an appropriate precursor.

All this may appear to be quite a formidable task, but a concerted effort seems likely to achieve minimal living cells. These cells represent, in our opinion, a very interesting goal for ongoing research on the origin of life, even though their current versions have obvious limitations. In the progression of studies on minimal cells we will encounter cells that produce proteins but do not reproduce, or cells that reproduce for a few generations and then die out by dilution of internal components, or cells characterized by very poor metabolic rates that cannot maintain cellular integrity. All of these constructs are significant because they are likely to be functionally similar to those intermediates that were involved in evolution before full-fledged biological cells appeared.

According to our definition of semisynthetic minimal cells, we are now making these constructs by using currently available macromolecular components, such as evolved nucleic acids and enzymes. As we noted earlier, this would correspond to a sort of top-down approach, in which complex systems are assembled starting from many simpler components. Though the use of modern genes and proteins cannot be considered prebiotic, their applications in constructing minimal living cells do demonstrate that life is indeed an emergent property. It is from this perspective that minimal cell studies reveal their major connection to research on the origin of life. "Synthesizing Life"—to cite the title of a programmatic review (Szostak, Bartel, and Luisi, 2001)—will be of fundamental importance in understanding the real essence of cellular life, and perhaps the evolutionary pathways that led to present-day life.

Acknowledgments

Research on vesicle properties and functionalization takes place within the European COST D27 Action "Prebiotic Chemistry and Early Evolution." We thank the Enrico Fermi Centre (Rome, IT) for financial support. The manuscript was greatly improved thanks to the suggestions of the reviewers, to whom we are grateful.

References

Arnold, K. (1995). Cation-induced vesicle fusion modulated by polymers and proteins. In R. Lipowsky & E. Sackmann (Eds.), *The structure and dynamics of membranes*, volume 1 (pp. 865–916). Amsterdam: Elsevier.

Bachmann, P. A., Luisi, P. L., & Lang, J. (1991). Self-replication of reverse micelles. *Chimia*, *45*, 266–268.

Bachmann, P. A., Luisi, P. L., & Lang, J. (1992). Autocatalytic self-replicating micelles as models for prebiotic structures. *Nature*, *357*, 57–59.

Bachmann, P. A., Walde, P., Luisi, P. L., & Lang, J. (1990). Self-replicating reverse micelles and chemical autopoiesis. *Journal of the American Chemical Society*, *112*, 8200–8201.

Bachmann, P. A., Walde, P., Luisi, P. L., & Lang, J. (1991). Self-replicating micelles: Aqueous micelles and enzymatically driven reactions in reverse micelles. *Journal of the American Chemical Society*, *113*, 8204–8209.

Bailey, A. L., & Cullis, P. R. (1997). Liposome fusion. In R. Epand (Ed.), *Lipid polymorphism and membrane properties* (pp. 359–373). San Diego: Academic Press.

Benner, S. A. (2003). Act natural. *Nature*, *421*, 118–118.

Benner, S. A., & Sismour, A. M. (2005). Synthetic biology. *Nature Reviews Genetics*, *6*, 533–543.

Berclaz, N., Bloechliger, E., Mueller, M., & Luisi, P. L. (2001). Matrix effect of vesicle formation as investigated by cryotransmission electron microscopy. *Journal of Physical Chemistry B*, *105*, 1065–1071.

Berclaz, N., Mueller, M., Walde, P., & Luisi, P. L. (2001). Growth and transformation of vesicles studied by ferritin labelling and cryotransmission electron microscopy. *Journal of Physical Chemistry B*, *105*, 1056–1064.

Blocher, M., Liu, D., & Luisi, P. L. (2000). Liposome-assisted selective polycondensation of alpha-amino acids and peptides: The case of charged liposomes. *Macromolecules*, *33*, 5787–5796.

Blocher, M., Liu, D., Walde, P., & Luisi, P. L. (1999). Liposome-assisted selective polycondensation of α amino acids and peptides. *Macromolecules*, *32*, 7332–7334.

Bloechliger, E., Blocher, M., Walde, P., & Luisi, P. L. (1998). Matrix effect in the size distribution of fatty acid vesicles. *Journal of Physical Chemistry*, *102*, 10383–10390.

Blumenthal, R., Clague, M. J., Durell, S. R., & Epand, R. M. (2003). Membrane Fusion. *Chemical Reviews*, *103*, 53–69.

Bolton, C. D., & Wattis, J. A. D. (2003a). The size-templating matrix effect in vesicle formation. 1. A microscopic model and analysis. *Journal of Physical Chemistry B*, *107*, 7126–7134.

Bolton, C. D., & Wattis, J. A. D. (2003b). Size-templating matrix effect in vesicle formation. 2. Analysis of a macroscopic model. *Journal of Physical Chemistry B*, *107*, 14306–14318.

Bozic, B., & Svetina, S. (2004). A relationship between membrane properties forms the basis of a selectivity mechanism for vesicle self-reproduction. *European Biophysical Journal*, *33*, 565–571.

Calderone, C. T., & Liu, D. R. (2004). Nucleic-acid-templated synthesis as a model system for ancient translation. *Current Opinions in Chemical Biology*, *8*, 645–653.

Cello, J., Paul, A. V., & Wimmer, E. (2002). Chemical synthesis of poliovirus cDNA: Generation of infectious virus in the absence of natural template. *Science*, *297*, 1016–1018.

Chakrabarti, A. C., Breaker, R. R., Joyce, G. F., & Deamer, D. W. (1994). Production of RNA by a polymerase protein encapsulated within phospholipid vesicles. *Journal of Molecular Evolution*, *39*, 555–559.

Chen, I. A., & Szostak, J. W. (2004). A kinetic study of the growth of fatty acid vesicles. *Biophysical Journal*, *87*, 988–998.

Chizmadzhew, Y., Maestro, M., & Mavelli, F. (1994). A simplified kinetic model for an autopoietic synthesis of micelles. *Chemical Physics Letters*, *22*, 656–662.

Chungcharoewattana, S., & Ueno, M. (2004). Size control of mixed egg yolk phosphatidylcholine (eggPC)/oleate vesicles. *Chemical and Pharmaceutical Bulletin*, *52*, 1058–1062.

Chungcharoewattana, S., & Ueno, M. (2005). New vesicle formation upon oleate addition to preformed vesicles. *Chemical and Pharmaceutical Bulletin*, *53*, 260–262.

Deamer, D. (1998). Membrane compartments in prebiotic evolution. In A. Brack (Ed.), *The molecular origin of life: Assembling the pieces of the puzzle* (pp. 189–205). Cambridge: Cambridge University Press.

Deamer, D. (2005). A giant step towards artificial life? *Trends in Biotechnology*, *23*, 336–338.

Deamer, D., & Barchfeld, G. L. (1982). Encapsulation of macromolecules by lipid vesicles under simulated prebiotic conditions. *Journal of Molecular Evolution*, *18*, 203–206.

Dyson, F. J. (1982). A model for the origin of life. *Journal of Molecular Evolution*, *18*, 344–350.

Fischer, A., Franco, A., & Oberholzer, T. (2002). Giant vesicles as microreactors for enzymatic mRNA synthesis. *ChemBioChem*, *3*, 409–417.

Folda, T., Gros, L., & Ringsdorf, H. (1982). Polyreactions in oriented systems 29: Formation of oriented polypeptides and polyamides in monolayers and liposomes. *Makromoleculare Chemie Rapid Communications, 3*, 167–174.

Fraser, C. M., Gocayne, J. D., White, O., Adams, M. D., Clayton, R. A., Fleischmann, R. D., et al. (1995). The minimal gene complement of Mycoplasma genitalium. *Science, 270*, 397–403.

Frick, D. N., & Richardson, C. C. (2001). DNA Primases. *Annual Reviews of Biochemistry, 70*, 39–80.

Fukada, K., Shibasaki, Y., & Nakahara, H. (1981). Polycondensation of long-chain esters of alpha-amino-acids in monolayers at air-water-interface and in multilayers on solid-surface. *Journal of Macromolecular Science, A15* (5), 999–1014.

Gánti, T. (1975). Organization of chemical reactions into dividing and metabolising units: The chemotons. *Biosystems, 7*, 189–195.

Gil, R., Sabater-Munoz, B., Latorre, A., Silva, F. J., & Moya, A. (2002). Extreme genome reduction in Buchnera spp: Toward the minimal genome needed for symbiotic life. *Proceedings of National Academy of Sciences of the United States of America, 99*, 4454–4458.

Gil, R., Silva, F. J., Peretó, J., & Moya, A. (2004). Determination of the core of a minimal bacteria gene set. *Microbiology and Molecular Biology Reviews, 68*, 518–537.

Hanczyc, M. M., Fujikawa, S. M., & Szostak, J. W. (2003). Experimental models of primitive cellular compartments: Encapsulation, growth, and division. *Science, 302*, 618–621.

Hitz, T., Blocher, M., Walde, P., & Luisi, P. L. (2001) Stereoselectivity aspects in the condensation of racemic NCA-amino acids in the presence and absence of liposomes. *Macromolecules, 34*, 2443–2449.

Hitz, T., & Luisi, P. L. (2001). Liposome-assisted selective polycondensation of alpha-amino acids and peptides. *Biopolymers* (including *Peptide Science*), *55*, 381–390.

Hud, N. V., & Lynn, D. G. (2004). From life's origins to a synthetic biology. *Current Opinions in Chemical Biology, 8*, 627–628.

Hutchinson, C. A., Peterson, S. N., Gill, S. R., Cline, R. T., White, O., Fraser, C. M., et al. (1999). Global transposon mutagenesis and a minimal Mycoplasma genome. *Science, 286*, 2165–2169.

Ishikawa, K., Sato, K., Shima, Y., Urabe, I., & Yomo, T. (2004). Expression of a cascading genetic network within liposomes. *FEBS Letters, 576*, 387–390.

Islas, S., Becerra, A., Luisi, P. L., & Lazcano, A. (2004). Comparative genomics and the gene complement of a minimal cell. *Origins of Life and Evolution of the Biosphere, 34*, 243–256.

Itaya, M. (1995). An estimation of the minimal genome size required for life. *FEBS Letters, 362*, 257–260.

Jay, D., & Gilbert, W. (1987). Basic protein enhances the encapsulation of DNA into lipid vesicles: Model for the formation of primordial cells. *Proceedings of National Academy of Sciences of the United States of America, 84*, 1978–1980.

Kaneko, K. (2004). Constructive and dynamic systems approach to life. In A. Deutsch, M. Falcke, J. Howard, & W. Zimmerman (Eds.), *Function and regulation of cellular systems: Experiments and models* (pp. 213–224). Basel: Birkhaeuser Verlag.

Kolisnychenko, V., Plunkett III, G., Herring, C. D., Fehér, T., Pósfai, J., Blattner, F. R., & Pósfai, G. (2002). Engineering a reduced Escherichia coli genome. *Genome Research, 12*, 640–647.

Koonin, E. V. (2000). How many genes can make a cell: The minimal-gene-set concept. *Annual Reviews of Genomics and Human Genetics, 1*, 99–116.

Kunieda, N., Watanabe, M., Okamoto, K., & Kinoshida, M. (1981). Polycondensation of thioglycine s-dodecyl ester hydrobromide in water. *Macromolecular Chemistry and Physics—Makromolekulare Chemie, 182*, 211–214.

Kuruma, Y., Nishiyama, K., Shimizu, Y., Muller, M., & Ueda, T. (2005). Development of a minimal cell-free translation system for the synthesis of presecretory and integral membrane proteins. *Biotechnology Progress, 21*, 1243–1251.

Laschewsky, A., Ringsdorf, H., Schmidt, G., & Schneider, J. (1987). Self-organization of polymeric lipids with hydrophilic spacers in side groups and main chain-investigation in monolayers and multilayers. *Journal of the American Chemical Society, 109*, 788–796.

Lazcano, A., Guerriero, R., Margulius, L., & Oró, J. (1988). The evolutionary transition from RNA to DNA in early cells. *Journal of Molecular Evolution*, *27*, 283–290.

Lazcano, A., Valverde, V., Hernandez, G., Gariglio, P., Fox, G. E., & Oró, J. (1992). On the early emergence of reverse transcription: Theoretical basis and experimental evidence. *Journal of Molecular Evolution*, *35*, 524–536.

Lonchin, S., Luisi, P. L., Walde, P., & Robinson, B. H. (1999). A matrix effect in mixed phospholipid/fatty acid vesicle formation. *Journal of Physical Chemistry B*, *103*, 10910–10916.

Luci, P. (2003). Gene cloning expression and purification of membrane proteins. Unpublished doctoral dissertation (Nr. 15108), ETH, Zurich.

Luisi, P. L. (2002). Toward the engineering of minimal living cells. *The Anatomical Record*, *268*, 208–214.

Luisi, P. L. (2003). Autopoiesis: A review and a reappraisal. *Naturwissenschaften*, *90*, 49–59.

Luisi, P. L., Ferri, F., & Stano, P. (2006). Approaches to a semi-synthetic minimal cell: A review. *Naturwissenschaften*, *93*, 1–13.

Luisi, P. L., & Oberholzer, T. (2001). Origin of life on earth: Molecular biology in liposomes as an approach to the minimal cell. In F. Giovanelli (Ed.), *The bridge between the big bang and biology: Stars, planetary systems, atmospheres, volcanoes and their link to life* (pp. 345–355). Rome: CNR Publications.

Luisi, P. L., Oberholzer, T., & Lazcano, A. (2002). The notion of a DNA minimal cell: A general discourse and some guidelines for an experimental approach. *Helvetica Chimica Acta*, *85*, 1759–1777.

Luisi, P. L., Stano, P., Rasi, S., & Mavelli, F. (2004). A possible route to prebiotic vesicle reproduction. *Artificial Life*, *10*, 297–308.

Luisi, P. L., & Varela, F. J. (1990). Self-replicating micelles—A chemical version of minimal autopoietic systems. *Origin of Life and Evolution of the Biosphere*, *19*, 633–643.

Marchi-Artzner, V., Jullien, L., Belloni, L., Raison, D., Lacombe, L., & Lehn, J. M. (1996). Interaction, lipid exchange, and effect of vesicle size in systems of oppositely charged vesicles. *Journal of Physical Chemistry*, *100*, 13844–13856.

Maturana, H., & Varela, F. (1980). *Autopoiesis and cognition: The realization of the living*. Dordrecht: Reidel.

Maturana, H., & Varela, F. (1998). *The tree of knowledge (Rev. Ed.)*. Boston: Shambala.

Mavelli, F., & Luisi, P. L. (1996). Autopoietic self-reproducing vesicles: A simplified kinetic model. *Journal of Physical Chemistry*, *100*, 16600–16607.

Monnard, P. A. (2003). Liposome-entrapped polymerases as models for microscale/nanoscale bioreactors. *Journal of Membrane Biology*, *191*, 87–97.

Monnard, P. A., & Deamer, D. W. (2001). Nutrient uptake by protocells: A liposome model system. *Origins of Life and Evolution of the Biosphere*, *31*, 147–155.

Monnard, P. A., Luptak, A., & Deamer, D. W. (2007). Models of primitive cellular life: Polymerases and templates in liposomes. *Philosophical Transactions of the Royal Society B*, *362*, 1741–1750.

Morigaki, K., Dallavalle, S., Walde, P., Colonna, S., & Luisi, P. L. (1997). Autopoietic self-reproduction of chiral fatty acid vesicles. *Journal of the American Chemical Society*, *119*, 292–301.

Morowitz, H. J. (1967). Biological self-replicating systems. *Progress in Theoretical Biology*, *1*, 35–58.

Mushegian, A. (1999). The minimal genome concept. *Current Opinions in Genetics and Development*, *9*, 709–714.

Mushegian, A. (2005). Protein content of minimal and ancestral ribosome. *RNA*, *11*, 1400–1406.

Mushegian, A., & Koonin, E. V. (1996). A minimal gene set for cellular life derived by comparison of complete bacterial genomes. *Proceedings of the National Academy of Sciences of the United States of America*, *93*, 10268–10273.

Neumann, R., Ringsdorf, H., Patton, E. V., & O'Brien, D. F. (1987). Preparation and characterization of long-chain amino-acid and peptide vesicle membranes. *Biochimica Biophysica Acta*, *898*, 338–348.

Nissen, P., Hansen, J., Ban, N., Moore, P. B., & Steitz, T. A. (2000). The structural basis of ribosome activity in peptide bond synthesis. *Science*, *289*, 920–930.

Nir, S., Bentz, J., Wilschut, J., & Duzgunes, N. (1983). Aggregation and fusion of phospholipid vesicles. *Progress in Surface Science*, *13*, 1–124.

Noireaux, V., Bar-Ziv, R., Godefroy, J., Salman, H., & Libchaber, A. (2005). Toward an artificial cell based on gene expression in vesicles. *Physical Biology*, *2*, 1–8.

Noireaux, V., & Libchaber, A. (2004). A vesicle bioreactor as a step toward an artificial cell assembly. *Proceedings of the National Academy of Sciences of the United States of America*, *101*, 17669–17674.

Nomura, S. M., Tsumoto, K., Hamada, T., Akiyoshi, K., Nakatani, Y., & Yoshikawa, K. (2003). Gene expression within cell-sized lipid vesicles. *ChemBioChem*, *4*, 1172–1175.

Oberholzer, T., Albrizio, M., & Luisi, P. L. (1995). Polymerase chain reaction in liposomes. *Chemistry and Biology*, *2*, 677–682.

Oberholzer, T., & Luisi, P. L. (2002). The use of liposomes for constructing cell models. *Journal of Biological Physics*, *28*, 733–744.

Oberholzer, T., Meyer, E., Amato, I., Lustig, A., & Monnard, P. A. (1999). Enzymatic reactions in liposomes using the detergent-induced liposome loading method. *Biochimica Biophysica Acta*, *1416*, 57–68.

Oberholzer, T., Nierhaus, K. H., & Luisi, P. L. (1999). Protein expression in liposomes. *Biochemical Biophysical Research Communications*, *261*, 238–241.

Oberholzer, T., Wick, R., Luisi, P. L., & Biebricher, C. K. (1995). Enzymatic RNA replication in self-reproducing vesicles: An approach to a minimal cell. *Biochemical Biophysical Research Communications*, *207*, 250–257.

Pantazatos, D. P., & MacDonald, R. C. (1999). Directly observed membrane fusion between oppositely charged phospholipid bilayers. *Journal of Membrane Biology*, *170*, 27–38.

Pautot, S., Frisken, B. J., & Weitz, D. A. (2003). Engineering asymmetric vesicles. *Proceedings of the National Academy of Sciences of the United States of America*, *100*, 10718–10721.

Pietrini, A. V., & Luisi, P. L. (2004). Cell-free protein synthesis through solubilisate exchange in water/oil emulsion compartments. *ChemBioChem*, *5*, 1055–1062.

Pohorille, A., & Deamer, D. (2002). Artificial cells: Prospects for biotechnology. *Trends in Biotechnology*, *20*, 123–128.

Rasi, S., Mavelli, F., & Luisi, P. L. (2003). Cooperative micelle binding and matrix effect in oleate vesicle formation. *Journal of Physical Chemistry B*, *107*, 14068–14076.

Rasmussen, S., Chen, L., Deamer, D., Krakauer, D. C., Packard, N. H., Stadler, P. F., & Bedau, M. A. (2004). Transitions from nonliving to living matter. *Science*, *303*, 963–965.

Rubinstein, I., Bolbach, G., Weygand, M. J., Kjaer, K., Weissbuch, I., & Lahav, M. (2003). Amphiphilic homochiral oligopeptides generated via phase separation of nonracemic alpha-amino acid derivatives and lattice-controlled polycondensation in a phospholipid environment. *Helvetica Chimica Acta*, *86*, 3851–3866.

Sacerdote, M. G., & Szostak, J. W. (2005). Semipermeable lipid bilayers exhibit diastereoselectivity favoring ribose. *Proceedings of the National Academy of Sciences of the United States of America*, *102*, 6004–6008.

Schmidli, P. K., Schurtenberger, P., & Luisi, P. L. (1991). Liposome-mediated enzymatic synthesis of phosphatidylcholine as an approach to self-replicating liposomes. *Journal of the American Chemical Society*, *113*, 8127–8130.

Schubert, R., Beyer, K., Wolburg, H., & Schmidt, K. H. (1986). Structural-changes in membranes of large unilamellar vesicles after binding of sodium cholate. *Biochemistry*, *25*, 5263–5269.

Schubert, R., Wolburg, H., Schmidt, K. H., & Roth, H. J. (1991). Loading of preformed liposomes with high trapping efficiency by detergent-induced formation of transient membrane holes. *Chemistry and Physics of Lipids*, *58*, 121–129.

Schwarz, J. (2001). A pure approach to constructive biology. *Nature Biotechnology*, *19*, 732–733.

Shimkets, L. J. (1998). Structure and sizes of genomes of the archaea and bacteria. In F. J. De Bruijn, J. R. Lupskin, & G. M. Weinstock (Eds.), *Bacterial genomes: Physical structure and analysis* (pp. 5–11). Boston: Kluwer Academic Publishers.

Shimizu, Y., Inoue, A., Tomari, Y., Suzuki, T., Yokogawa, T., Nishikawa, K., & Ueda, T. (2001). Cell-free translation reconstituted with purified components. *Nature Biotechnology*, *19*, 751–755.

Stamatatos, L., Leventis, R., Zuckermann, M. J., & Silvius, J. R. (1988). Interactions of cationic lipid vesicles with negatively charged phospholipid vesicles and biological membranes. *Biochemistry*, *27*, 3917–3925.

Stano, P., Ferri, F., & Luisi, P. L. (2006). From the minimal genome to the minimal cell: Theoretical and experimental investigation. In J. Seckbach (Ed.), *Life as we know it*. Berlin: Springer.

Szathmáry, E. (2005). Life: In search of the simplest cell. *Nature*, *433*, 469–470.

Szostak, J. W., Bartel, D. P., & Luisi, P. L. (2001). Synthesizing life. *Nature*, *409*, 387–390.

Takakura, K., & Sugawara, T. (2004). Membrane dynamics of a myelin-like giant multilamellar vesicle applicable to a self-reproducing system. *Langmuir*, *20*, 3832–3834.

Takakura, K., Toyota, T., & Suguwara, T. (2003). A novel system of self-reproducing giant vesicles. *Journal of the American Chemical Society*, *125*, 8134–8140.

Tawfik, D. S., & Griffiths, A. D. (1998). Man-made cell-like compartments for molecular evolution. *Nature Biotechnology*, *16*, 652–656.

Thomas, C. F., & Luisi, P. L. (2004). Novel properties of DDAB: Matrix effect and interaction with oleate. *Journal of Physical Chemistry B*, *108*, 11285–11290.

Treyer, M., Walde, P., & Oberholzer, T. (2002). Permeability enhancement of lipid vesicles to nucleotides by use of sodium cholate: Basic studies and application to an enzyme-catalyzed reaction occurring inside the vesicles. *Langmuir*, *18*, 1043–1050.

Tsumoto, K., Nomura, S. M., Nakatani, Y., & Yoshikawa, K. (2001). Giant liposome as a biochemical reactor: Transcription of DNA and transportation by laser tweezers. *Langmuir*, *17*, 7225–7228.

Varela, F., Maturana, H. R., & Uribe, R. B. (1974). Autopoiesis: The organization of living system, its characterization and a model. *Biosystems*, *5*, 187–196.

Walde, P., Goto, A., Monnard, P. A., Wessicken, M., & Luisi, P. L. (1994). Oparin's reactions revisited: Enzymatic synthesis of poly(adenylic acid) in micelles and self-reproducing vesicles. *Journal of the American Chemical Society*, *116*, 7541–7544.

Walde, P., & Marzetta, B. (1998). Bilayer permeability-based substrate selectivity of an enzyme in liposomes. *Biotechnology and Bioengineering*, *57*, 216–219.

Walde, P., Wick, R., Fresta, M., Mangone, A., & Luisi, P. L. (1994). Autopoietic self-reproduction of fatty acid vesicles. *Journal of the American Chemical Society*, *116*, 11649–11654.

Weiner, A. M., & Maizels, N. (1987). tRNA-Like structures tag the 3′ ends of genomic RNA molecules for replication: Implications for the origin of protein synthesis. *Proceedings of the National Academy of Sciences of the United States of America*, *84*, 7383–7387.

Wick, R., Walde, P., & Luisi, P. L. (1995). Light microscopic investigations of the autocatalytic self-reproduction of giant vesicles. *Journal of the American Chemical Society*, *117*, 1435–1436.

Woese, C. R. (1983). The primary lines of descent and the universal ancestor. In D. S. Bendall (Ed.), *Evolution from molecules to man* (pp. 209–233). Cambridge: Cambridge University Press.

Yoshimoto, M., Wang, S., Fukunaga, K., Walde, P., Kuboi, R., & Nakao, K. (2003). Preparation and characterization of reactive and stable glucose oxidase-containing liposomes modulated with detergent. *Biotechnology and Bioengineering*, *81*, 695–704.

Yoshimoto, M., Wang, S. Q., Fukunaga, K., Treyer, M., Walde, P., Kuboi, R., & Nakao, K. (2004). Enhancement of apparent substrate selectivity of proteinase K encapsulated in liposomes through a cholate-induced alteration of the bilayer permeability. *Biotechnology and Bioengineering*, *85*, 222–233.

Yu, W., Sato, K., Wakabayashi, M., Nakatshi, T., Ko-Mitamura, E. P., Shima, Y., et al. (2001). Synthesis of functional protein in liposome. *Journal of Bioscience and Bioengineering*, *92*, 590–593.

Zepik, H. H., Bloechliger, E., & Luisi, P. L. (2001). A chemical model of homeostasis. *Angewandte Chemie International Edition*, *40*, 199–202.

Zhang, B., & Cech, T. R. (1998). Peptidyl-transferase ribozymes: Trans reactions, structural characterization and ribosomal RNA-like features. *Chemistry and Biology*, *5*, 539–553.

4 A Roadmap to Protocells

Steen Rasmussen, Mark A. Bedau, John S. McCaskill, and Norman H. Packard

4.1 Introduction

The diversity of research presented in this volume illustrates many of the possible approaches to synthesizing protocells. In this chapter, we present a simple graphical language that enables one to characterize and classify the essential chemical features of many different protocell projects. The representation achieves several ends: It allows for comparison of different achievements, it enables articulation of how targets of ongoing research fit into a larger picture containing both those targets and the work of other researchers, and it also enables a comparison between experimental results, simulation results and hypothetical scenarios. After presenting the basic vocabulary, we show how our graphical representations may be used to construct a roadmap of different paths that lead to the creation of protocells. We then use these representations to depict the important functional features of several current protocell research efforts.

Protocell research consists of a mixture of experimental creations in the laboratory, computer simulations of chemical systems, and conceptual schemes that guide experimental and computational efforts. Our graphical language can represent each of these kinds of achievements and express them in a common language that enables direct comparisons among all these facets of protocell research.

Our representation scheme aims to focus on the essential features of protocell systems, and to abstract away most of their details. The general protocell vision that is expressed in the following diagrams is that life is the functional integration of metabolism, containment, and information processing, and the graphical language has primitives corresponding to these three functionalities. The diagrams provide a practical and easy method for generating natural milestones by which to judge progress in protocell research. They also enable us to side-step the seemingly interminable debates about the definition of life. However, it should be noted that the vision of life as the functional integration of metabolism, containment, and information processing is not a presumption of our graphical language. There are a range of visions of life that concentrate on particular subsets of these functionalities, many of which are presented here; all of these may be diagramed to facilitate precise comparisons between different protocell visions. Likewise, the language may grow to include other functionalities if and when they become important in protocell research.

Our graphical language represents the key causal and functional interactions chemically realized in protocell systems. Accordingly, more complicated chemical systems require more complicated diagrams. Thus, our method is cleanest when applied to the very simple protocells characteristic of pure bottom-up efforts (see, e.g., chapters 5 and 6), for they yield suitably simple diagrams. We have made some attempt to generalize our language to represent bottom-up methods that use material from existing living cells, such as cell-free extract containing a myriad of structures like ribosomes, enzymes, and the basic building blocks out of which nucleic acids and proteins are constructed (see chapter 7). But these attempts must sacrifice the fundamental clarity and precision of our diagrammatic language in order to represent much more complicated chemical systems.

Some top-down approaches to protocells begin with fully functioning natural living cells, such as bacteria, and modify them, creating even more complicated chemical systems (see chapter 16). We have not tried to represent these top-down schemes here; they are complicated enough that our diagrams might prove useless.

4.2 Functional Components and Functional Interactions

Protocells are chemical systems, but the chemical interactions (i.e., chemical reactions) depend on other important functional interactions stemming from non-chemical physical processes and effects. Several kinds of functional interactions are important both between the protocell subcomponents and between the protocell and its environment. The most important additional interactions affecting chemical reactions and material supply are catalysis, self-assembly (including template chemistry), energy transfer, and spatial concentration (e.g., induced by enclosure within a membrane).

While traditional graphical representation in chemistry deals with reversible or irreversible chemical reactions of the form $A_1 + A_2 + \cdots \rightleftharpoons B_1 + B_2 \ldots$, we use unidirectional arrows only to represent the dominant direction of a process, even if it is inherently reversible. Historically, some diagrammatic enhancements have been found useful to represent catalytic networks, for example, including arrows from a catalyst (chemical symbol) to a reaction (another arrow), which saves repeating a catalyst on both sides of the reaction. We now propose to expand the traditional chemical diagrammatic vocabulary by representing an even wider range of functional interactions.

A	aggregate
I	informational structure
E	energy harvester
F/hν	fuel, e.g., light or chemical
M, L,...	material components
pX	precursor of X
X*	energized form of X

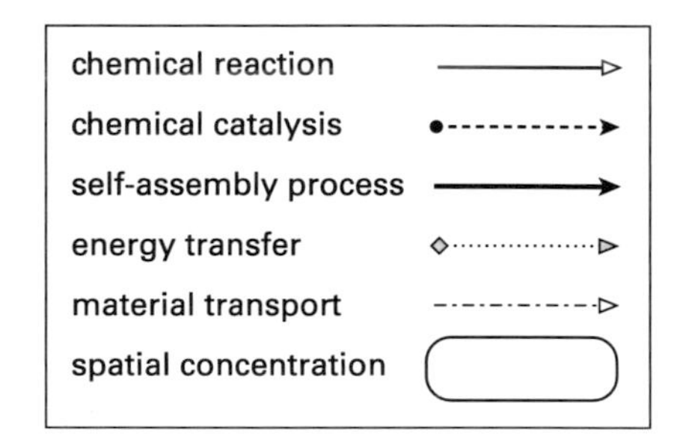

Figure 4.1
Above: key classes of functional elements involved in our graphical representations. Below: key causal links or interaction processes between protocellular subcomponents and between the protocell and its environment.

Our diagrams employ letters to represent different functional elements that may be found as components of protocells. The set of possible functional elements is quite general; some may be molecules, some may be molecular aggregates, and some are nonmaterial (e.g., energy). The functional elements used in our diagrammatic grammar are listed in the top part of figure 4.1, and the most important are the first three. A molecular aggregate or other structure functioning as a container is represented with the letter "A." We use the letter "I" to represent a chemical or other structure that contains combinatorial information that plays some causal role in the operation of the protocell. In all cases of interest, the information does not just exist by virtue of its combinatorial identity but also regulates the proper functioning of the system and provides the potential for evolvability. We use the letter "E" to represent a structure that makes energy from the environment available to power the chemical reactions in the protocell. Other letters represent fuels or material components of functional elements, and further symbolic conventions represent their precursors and energized forms.

In addition to material precursors, the following discussion also refers to building blocks of protocell structures. Building blocks are typically smaller components out of which something is constructed, while its precursors are molecules from which it is derived. Some molecules are small and simple, while others are large and complex. This difference is important to chemists, and our diagrams do not explicitly include this information (thought one can imagine easy ways to add it).

The physical entities represented by the letters A, I, or E might consist of one molecule, or they might consist of many different molecules. In fact, these three protocell functional components could be realized in quite unusual ways, as we will see later. The letters represent functional items, not necessarily physically separate entities that are performing those functions; the same physical structure could even play two different functional roles. What matters is that the critical functionality is achieved, not that the functions are realized in ways that are familiar from ordinary cellular life.

We sometimes use more specific symbols than A, I, or E to represent functional components. For example, when this helps to clarify a specific diagram, "M" might represent a monomer element that participates in polymerization, "L" might represent lipids that self-assemble into an aggregate, "DNA" or "RNA" might represent specific informational components, and "P" might represent a functional protein produced by other informational molecules. In addition, our diagrams can indicate more or less detail about the different functional components in protocells. For example, we sometimes explicitly represent multiple kinds of informational structures, and we sometimes explicitly represent transcription and translation. In general, we can represent more or fewer of the intermediate stages in a sequence of functional interactions, by expanding or collapsing the sequence of interactions. Furthermore, we sometimes choose to suppress a lot of detail with a labeled box. In these ways, the diagrams can represent more or less detail, as desired.

Our diagrams use arrows and rectangles to represent six types of interactions (figure 4.1, bottom). Chemical reactions that involve forming covalent bonds are represented with a black arrow with an open head. Instances in which one entity chemically catalyzes another are represented with a dashed arrow originating at a circle. We represent the self-assembly process by a bold arrow. The process of energy flow in a protocell is represented by a dotted arrow that originates with a diamond and ends with an open head. Protocells typically also involve a material flow, which involves transporting material both into and out of the protocell. When we want to represent the process of material transport, we use a dot–dashed arrow with an open head, although for simplicity we often omit this detail (especially material flow out of the protocell). Finally, to represent the process of spatial concentration of reagents, we use a rounded rectangular enclosure. Often (but not always) this spatial concentration is achieved by an aggregate that may constitute a container or an even more primitive structure that serves to localize other components; we indicate this with the letter "A," representing the aggregate in the middle of the line.

Of course, these six categories of interactions are partially overlapping. Energy transfer can be employed in catalysis. Self-assembly arises from reversible association interactions, and where the number of molecules participating in structures is small, this may also be represented by traditional binding equilibria $A_1 + A_2 + \cdots \rightleftharpoons B$. However, supramolecular and mesoscale self-assembly can involve extended aggregate structures, which may have a variable number of molecular components better described by physical phase properties. Membrane and micelle self-assembly, for example, entail complex dynamical, structural, and combinatorial relationships between the components that make a purely chemical notation cumbersome, so we prefer separate arrows to describe physical self-assembly.

The process of spatial concentration in our diagrams typically (though not always) occurs by encapsulation of reagents within a vesicle membrane or within the interior of a micelle. Because vesicles and membranes are self-assembled structures, self-assembly in these cases is implicated in spatial concentration. Furthermore, self-assembly typically involves a kind of catalysis, in which an existing assembled structure physically aids in the continued assembly of more of the structure. This form of catalysis is assumed to be represented by the self-assembly arrow without the need to add a separate catalysis arrow. Whereas the chemical reactions indicated in our diagrams involve the formation of covalent bonds, the self-assembly processes typically create aggregate structures through the operation of weak chemical associations such as the hydrophobic effect. We include under "catalysis" the process by which a nucleic acid template speeds up the formation of covalent bonds between complementary monomers organized appropriately along the template. The spatial organization of monomers by a template is a kind of self-assembly, so we represent the replication of informational polymers with a self-assembly arrow, although it also obviously involves covalent bond formation. There is an important difference between these two kinds of self-assembly. The formation (and breaking) of covalent bonds can exact a high energetic cost and require a substantial contribution from the metabolism. On the other hand, self-assembly processes that are free of covalent bonding are energetically downhill, and can proceed without the help of metabolism.

An important chemical difference exists between micelles, which are extremely small lipid structures with an oily interior, and vesicles, which are many orders of magnitude larger and consist of a lipid bilayer membrane surrounding an aqueous interior. Yet our diagrams might represent them in the same way, with the letter "A." Nevertheless, our diagrams are not hiding critical information about the functional difference between micelles and vesicles in protocells. Any important functional difference between a micelle in one protocell and a vesicle in another will be represented by different functional arrows. However, if the diagrams look essentially the same, then no important functional difference exists between micelles and vesicles in those protocells. This chapter's central thesis is that our causal diagrams capture the important functional components and interactions in protocells.

4.3 Routes to Protocells

Using this vocabulary of causal interactions, we can now represent the causal diagrams of most of the protocellular schemes currently being explored. Most of the bottom-up approaches to making protocells involve various ways of functionally coupling genes, metabolism, and a container. Figure 4.2 summarizes these qualitatively different kinds of interactions in a sequence starting with the individual elementary functional-diagrammatic modules in figure 4.2a and proceeding to possible couplings of the aggregate to other components (figures 4.2b and c), the coupling of the informational component I in figure 4.2d, and the coupling of the energetic component 2e. In these elementary diagrams, we can see some of the ambiguities that still exist in the diagrammatic grammar. We have chosen intuitive simplicity over detailed correctness at times.

In figure 4.2a, the three component functionalities—replicating genes, metabolism, and reproducing container—are represented as their separate diagrammatic modules. Note that in the middle of figure 4.2a for energy metabolism, energy is denoted as being provided either by a chemical reaction (F/F) or by a photon making a transition to another photon ($h\nu^*/h\nu$). Already in the diagram for informational polymer (I) synthesis from constituent monomers (M), in the top line of figure 4.2a, we see some ambiguity deriving from the fact (discussed earlier) that template-directed replication is actually a combination of chemical reaction (represented by solid black arrows) and self-assembly (represented by the red double arrows). We have chosen to represent the combination as illustrated; other more detailed choices might also be possible, depending on which aspects of the phenomenon need to be emphasized in an explanation or comparison, as illustrated in figure 4.5.

Figures 4.2b and c illustrate two ways of coupling aggregation to other processes; both by simple proximity enforced by either the container composed of the aggregate in 4.2b or the catalytic interaction in 4.2c. Parts d and e of figure 4.2 illustrate coupling of informational components, I, and energetic components, E, respectively.

By combining successively more of these functional interactions in various sequences, we can construct a map spanning the simplest to the most complex of these interaction networks (figure 4.3). This collection of diagrams constitutes a roadmap from the simplest protocell antecedents (top row) to a functionally fully integrated protocell (bottom row). We should note that the diagrams in figures 4.2 and 4.3 abstract away from all but the central functional processes, and not all material components are represented separately. Furthermore, unlike some of the diagrams to fol-

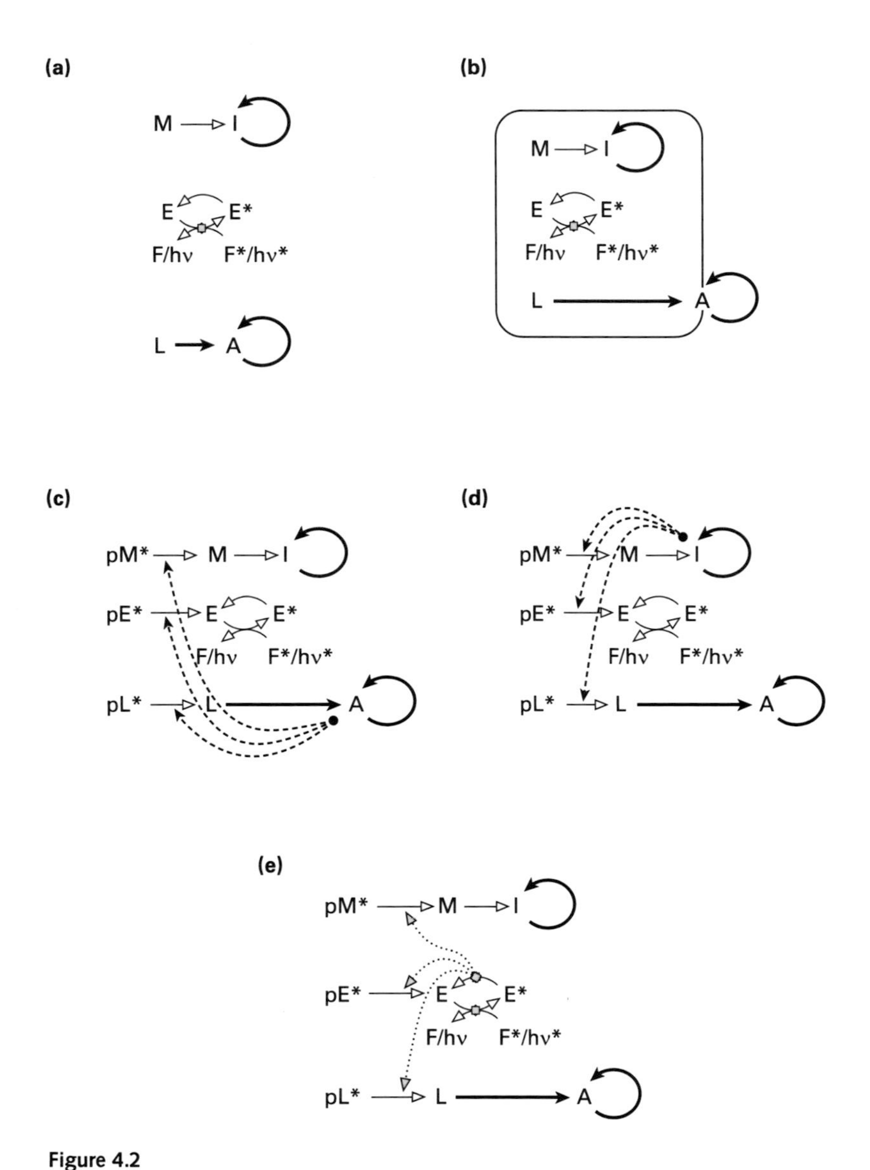

Figure 4.2
Each of the main qualitatively different causal interactions in a protocell (ignoring material transport). (a) The three components: replicating genes, metabolism, and reproducing container. (b) The container enables all components to be concentrated in close physical proximity and keeps environmental poisons and parasites away. (c) Aggregates can act as catalysts in different ways; for example, the thermodynamic conditions within a lipid aggregate, or at the water-aggregate interface, are different from the thermodynamics of bulk water properties. (d) The gene sequence or structure can also have catalytic properties. (e) Production of the three main building blocks can be coupled to an energy source through the metabolic process.

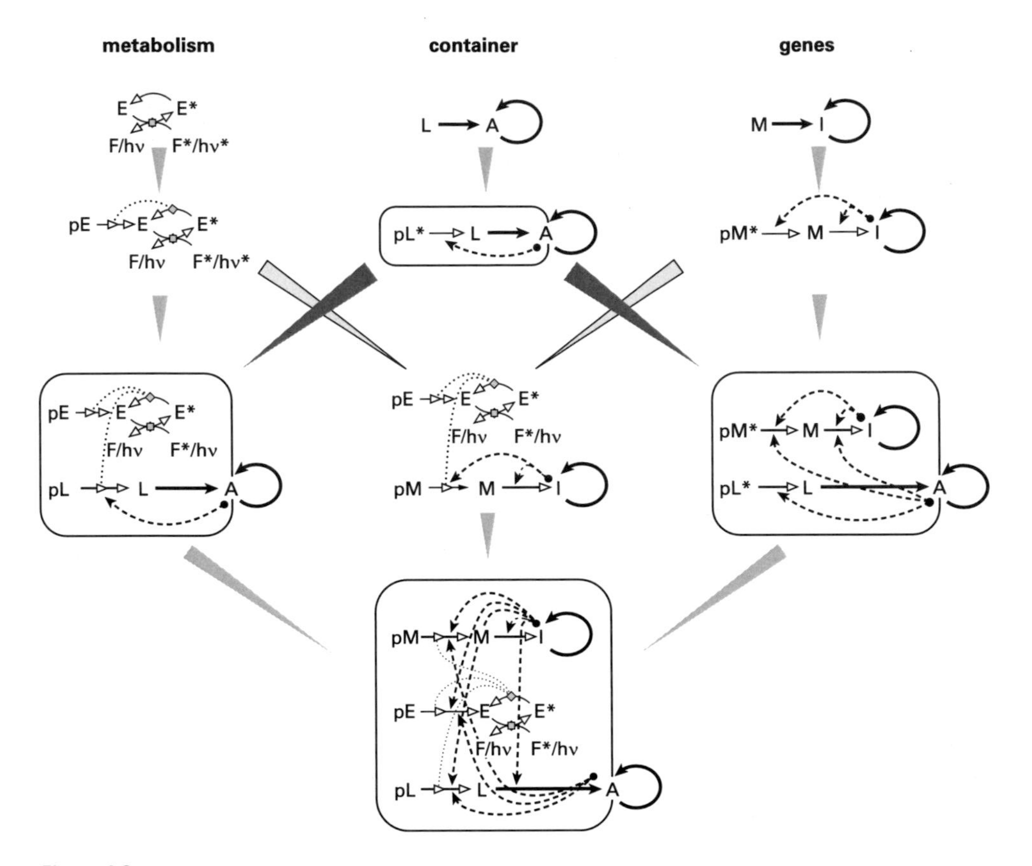

Figure 4.3
A bottom-up map for constructing protocells from nonliving materials, achieved by combining in various sequences the different key functional interactions present in protocells. (Top row) One can start by constructing a metabolic system, a reproducing container, or replicating genes. (Second row) Ideally each component in such a system would be produced from energized precursors, in a way facilitated by the metabolism, container, or genes. (Third row) A further step toward protocells is achieved by functionally coupling a metabolism and a container, a metabolism with genes, or a container with genes. (Bottom row) Functionally coupling the metabolism, container, and genes finally yields a complete protocell. Most bottom-up protocell models currently under construction are special cases or combinations of the idealized steppingstones in this map, and many also exploit the functionality of complex biological components.

Table 4.1
Possible system realizations in the pure bottom-up route map

Unary	A A* E E* I I*
Binary	AE AE* AI AI* A*E A*E* A*I A*I* EI EI* E*I E*I*
Ternary	AEI AEI* AE*I AE*I* A*EI A*EI* A*E*I A*E*I*

Note: Combining container (A), metabolic (E), and genetic (I) components, externally (X) or internally energized (X*), yields 26 possibilities, or $3^3 - 1$ (three possibilities {none, X, X*} for each of the three species types, omitting {none, none, none}). In section 4.3 we show that many current protocell models and implementations are derivatives of those depicted in this table. Other implementations include a variety of modern biological components, which of course make the set of possibilities much larger.

low, the diagrams in figures 4.2 and 4.3 ignore the issue of transport of material into (and out of) protocells for simplicity.

Figure 4.3 shows only a central set of possible routes to the protocell; the full spectrum of possibilities requires consideration of the role of energetics. With and without metabolic conversion of component precursors, we have one externally energized and one metabolically energized case for each of the three fundamental subcomponents (table 4.1). Self-assembly and metabolic processes obviously require available free energy. Ideally, the protocell metabolism should generate or transform every subcomponent building block employed in the protocell from environmental resources. The alternative is that energized building blocks are provided directly. The driving force behind the self-assembly of the lipid container is usually the hydrophobic effect, which lowers free energy by separating water from the other components. The free energies of the template-directed arrangement of the genetic building blocks, and possibly the metabolic and genetic component assembly (e.g., within the lipid phase), also run energetically downhill. The metabolic processes are nonequilibrium processes, since free energy is pumped into the system in the form of light or redox energy, which is converted into the chemical energy used to break or form covalent bonds.

4.4 Elementary Protocell Schemes

To understand how our causal diagrams can represent different kinds of protocell systems, it is useful to consider some of the simplest schemes first. Subsequent sections apply the causal diagrams to more complex achievements, including experimental results, simulation results, and merely hypothetical schemes that shape experimental and simulation efforts.

The seminal contributions of Morowitz, Deamer, and Luisi have increased awareness of the functional importance of containers, which provide a protected microenvironment, physical proximity, and a persisting individual identity. Of course, the importance of containers to the origin of cellular life has a long history (Fox, 1960;

Oparin, 1924). Some authors argue that Gánti (1975) produced the first complete model of a protocell, based on stoichiometric coupling of lipid synthesis with genetic replication, although the genes in his model seem more like an afterthought than an essential source of information for the protocell. Later work has argued that stoichiometric coupling of genes to membrane production is neither a feature of modern cells nor essential in simpler protocells (Rocheleau et al., 2007). Maturana and Varela's model (1980) of an autopoeitic protocell also captured the essence of a homeostatic protocell without attributing a central role to genetic information.

Figure 4.4 represents the most elementary diagrammatic container module, where we see the process of container (A) self-assembly from building blocks (L), and the subsequent production of further containers (e.g., from division). Note, however, that while the aggregate, A, makes more aggregates, it does not necessarily produce an actual lineage of aggregates, because A might simply catalyze the production of further aggregates rather than grow and divide itself. Thus, the circular arrow from A to A represents production of A from A, but not necessarily the growth and division of A. Rather than simply assuming a supply of building blocks, a more autonomous scheme would have the self-assembled container itself catalyze the production of building blocks from their precursors, as depicted in figure 4.4. This diagram also shows the aggregate providing the spatial contiguity needed to elevate reagent concentration and keep parasites and poisons at bay. Luisi and colleagues (Bachmann, Luisi, and Lang, 1992; Berclaz, Bloechliger, et al., 2001; Berclaz, Mueller, et al., 2001; Walde, Goto et al., 1994; Walde, Wick, et al., 1994; see also chapter 3) have demonstrated the autocatalytic nature of this scheme experimentally for micellar and vesicular aggregates, for example, using oleic acid chemistry.

For containers to be functional components of protocells, they must be coupled with the two other component functionalities, metabolism and information (replicator) chemistry. In fact, apart from the receptacle role containers play, spatial compartmentation also plays various key roles in stabilizing the chemical kinetics of replicator chemistry. Before discussing the diagrammatic representation of

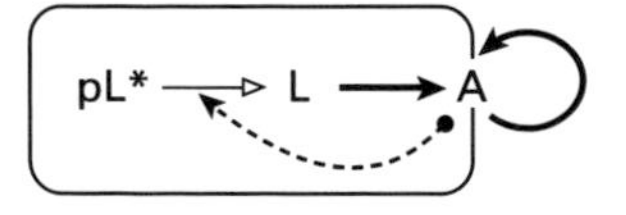

Figure 4.4
Container (A) self-assembly (straight bold arrow) and reproduction (circular bold arrow) driven by container-catalyzed transformation (dashed catalysis arrow) of activated precursor lipids (pL) to functional lipids (L). Note that the container can produce a catalytic concentration effect (rounded rectangle) if the precursors or building blocks are preferentially associated with the aggregate, for example, by being embedded in a micelle or encapsulated within the aqueous interior of a vesicle. Versions of this diagram have been experimentally realized by Luisi, Walde, and coworkers (see text for discussion; see also chapter 3).

container–replicator coupling, we set the stage with a brief historical overview. This container–replicator history illustrates how each of the binary functional couplings has its own detailed history.

Since the 1970s, many authors have asserted the importance of some kind of compartmentation for the evolution of stable protocells. While Eigen (1971) and Eigen and Schuster (1977, 1978a, 1978b) argued that hypercyclic organization is necessary for protocell evolution, and spatial compartmentation is necessary for their evolutionary stability, Harnasch and Bresch (Niesert, Harnasch, and Bresch, 1981) demonstrated the strong constraints on the path between genetic fluctuations and cooperative selection. Maynard Smith and Szathmáry (Maynard Smith and Szathmáry, 1995; Szathmáry and Demeter, 1987) have argued that stochastic correction of genetic composition in dividing protocells is sufficient for cooperative selection, subject to well-known small copy number constraints on group selection. Boerlijst and Hogeweg (1991) demonstrated that hypercyclic organization could be evolutionarily stabilized in spiral waves without other compartments. McCaskill (1994) and Cronhjort (Cronhjort and Blomberg, 1997; Cronhjort and Nyberg, 1996) showed that the simplest cooperative replication processes could be stabilized in self-replicating spot patterns in reaction-diffusion systems. Füchslin and McCaskill (2001) demonstrated that stochastic spatial systems could even stably evolve genetic coding of translation without compartmentation and, together with Altmeyer (McCaskill, Füchslin, and Altmeyer, 2001), they developed an analytic framework, PRESS, to describe the ability of spatial replication systems to evolve functionally differentiated genetic components. More formally, automata models (Fontana, 1991; Ikegami, 1994; McCaskill, 1992; Thürk, 1993; Tangen, 1994) have revealed the generic importance of spatial structure for cooperative evolution. Munteanu, Rasmussen, Ziock, and Solé (2007) demonstrated how the protocellular growth laws depend on the compartment and its kinetic interaction with the genes and metabolism. McCaskill and coworkers (McCaskill et al., 2007) have explored the explicit coupling of self-assembly and replicator kinetics in a lattice model, with coevolution of structure and replicator population.

Experimentally, Gebinoga and Eigen (Eigen et al., 1991) demonstrated viral-based hypercycles in vivo using "zombie" cells as bioactive compartments; Ehricht, Ellinger, and McCaskill (Ehricht, Ellinger, and McCaskill, 1997; Ellinger, Ehricht, and McCaskill, 1998) constructed the first cell-free evolvable experimental system for molecular cooperation; and Joyce (Paul and Joyce, 2002) constructed a ribozyme-based evolvable cooperation system. Evolving systems of proteins coupled to nucleic acids by in vitro protein translation in separate compartments have been investigated by Yomo (Matsuura et al., 2002). Some of the modern experimental model systems for protocells, particularly those not making use of protein enzymes extracted from natural cells, may also be criticized for the lack of significant evolvability

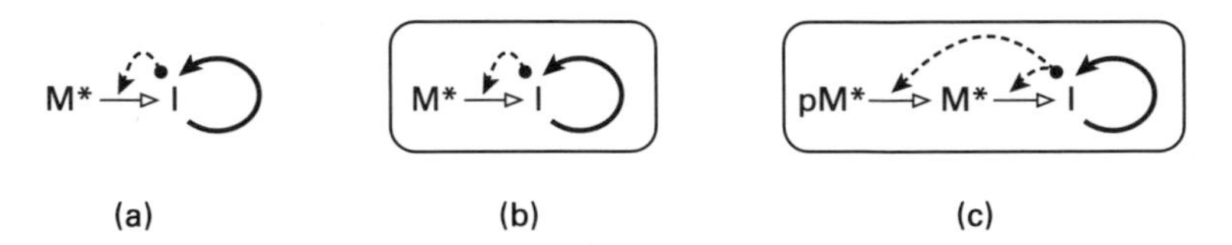

Figure 4.5
(a) Template-directed replication with activated monomers as investigated by many groups; see text for discussion and references. Note that the templating process is shown by the self-assembly arrow, while optional specific (higher order, e.g., 2X → 3X) catalysis by the biopolymers is shown by the dotted catalytic arrow. (b) With spatial control, this is the causal structure of the evolvable systems investigated by McCaskill and coworkers for spatial replicator models (see text for discussion and references). (c) Templating polymers capable of evolution of both building blocks (M), catalyzed from precursors (pM), and of the polymeric sequences (I). Because of exploitation by parasitic replicators (see chapter 17), this system is evolutionarily stable only with spatial colocalization, indicated by the rounded rectangles, here achieved by an unspecified mechanism.

stemming from minimal (sometimes single-bit) genetic information. This downgrading of the importance of sequence information is in strong contrast with the Eigen-Schuster school of physicochemical evolution of life and notable attempts by Bartel and colleagues (Johnston et al., 2001) to evolve an RNA polymerase for general copy functionality.

The causal diagram for template-directed self-replicating molecules is depicted in figure 4.5. Earlier experimental work used external enzymes as catalysts to achieve evolvable systems of this type, starting with Spiegelman's work with the Qß replicase enzyme (Mills, Peterson, and Spiegelman, 1967). Such systems have been implemented without enzymes using RNA oligomers (Acevedo and Orgel, 1987; Joyce, 1984), DNA oligomers (von Kiedrowski, 1986; see also chapter 13), artificial organic templates (Tijvikua, Ballester, and Rebek, 1990), and peptides (Lee et al., 1996); similar systems are currently being developed for PNA (see chapter 15). The formation of the monomers themselves is discussed in chapter 26. Spatial localization is illustrated in figure 4.5, parts b and c, and investigated by McCaskill and coworkers for spatial replicator models (Füchslin and McCaskill, 2001; McCaskill, 1994; McCaskill, Füchslin, and Altmeyer, 2001).

Metabolic coupling is diagrammed in figure 4.6. At a minimum, protocell metabolism consists of a molecule, E, that is able either to capture light or to react with another molecule to reach a higher energetic state, E^*, associated with a change in its redox potential. This energized metabolic molecule E^* can be utilized to drive chemical reactions, usually by donating or receiving an electron. In figure 4.6, the essence of this mechanism is shown (a) together with an example (b) in which the metabolic reaction produces more metabolic molecules from appropriate precursors.

Deamer (1992) developed a functional system with pigments inside a vesicle membrane that can generate a gradient of protons (pH) across the membrane, which are utilized for the production of lipids from precursors that self-assemble to form more

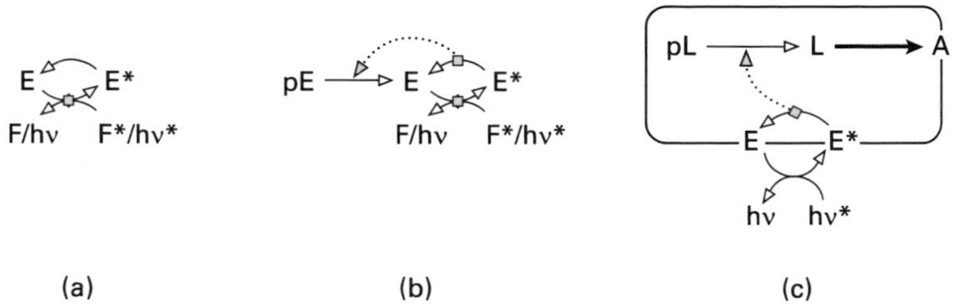

Figure 4.6
(a) A metabolic molecule, E, is energized by either light or an energy-rich compound, F*. (b) E* can release its free energy by driving another chemical reaction, for example, producing more metabolic compounds from a precursor pE. (c) In the membrane of a vesicle (A) Deamer (1992) located a metabolic molecule (E) that utilizes light ($h\nu^*$) as an energy source (therefore called pigment or sensitizer) and produces more membrane molecules (L) from precursors (pL) via the buildup of a proton or pH gradient across the membrane.

membrane (see figure 4.6c). Note how this simple system makes the metabolism and the container functionally dependent on each other. The container locates the sensitizers (E) in spatial proximity to the container chemistry and enables the metabolism to function, and the metabolism provides the energy to form the aggregate building blocks (lipids, L) from their precursors (pL). This is a simple example of a cross-coupled container and metabolism system.

The following sections show how our causal diagrams can be applied to a diverse range of protocell research, including experimental achievements, simulation results, and hypothetical schemes. We hasten to point out that our discussion is not a complete atlas of existing protocell accomplishments. Rather, it illustrates how our causal diagrams apply to many of the accomplishments discussed in this volume, as well as some other work from the literature.

4.5 Experimental Achievements

This section diagrams experimental realizations of some important systems on the route to bottom-up protocells, as well as some experimental realizations that are intermediate between bottom-up and top-down, in that they use higher-order components either synthesized or obtained from natural cells (e.g., cell extract). These systems realize various parts of a protocell's functional requirements for containment, metabolism, and genetics, using a variety of components and biochemical aids. They also achieve various kinds and degrees of functional coupling of containment, metabolism, and genetics. Further details about these and other important experimental achievements can be found in chapters 2 and 3, as well as the experimental chapters in part II of this volume.

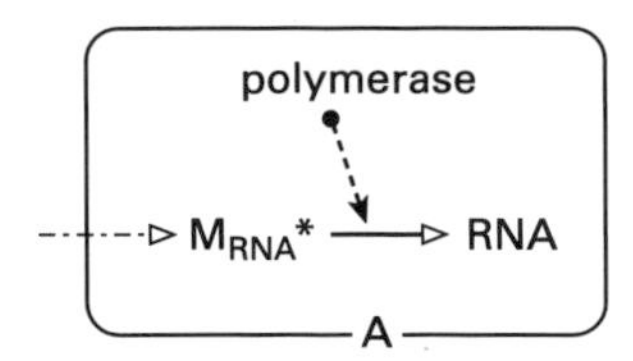

Figure 4.7
Deamer and coworkers (Chakrabarti et al., 1994; see also chapters 2 and 3) demonstrated that a polymerase (DNAse) trapped inside vesicles (A) composed of the amphiphile DMPC could produce RNA without a template (template independent or "random") from nucleoside diphosphates that slowly diffuse through the vesicle membrane.

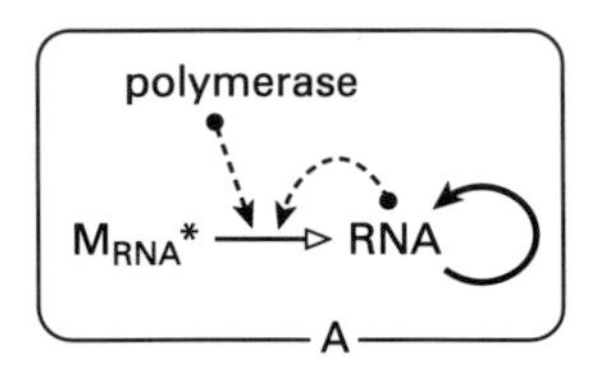

Figure 4.8
Oberholzer, Albrizio, and Luisi (1995), Oberholzer et al. (1995; see also chapter 3) demonstrated that it is possible to replicate RNA trapped inside vesicles (A) using activated nucleotides (M^*_{DNA}) and a polymerase, Qβ replicase.

Four different protocell realizations based on RNA chemistry contained in vesicles are depicted in figures 4.7 through 4.14, with the last including DNA and enzymes as well. Figure 4.7 illustrates containment of chemistry, and uses a polymerase to produce RNA without a template from nucleoside diphosphates that slowly diffuse through the vesicle membrane. Figure 4.8 illustrates the efforts of Luisi and coworkers to replicate RNA trapped inside vesicles using activated nucleotides (M^*_{DNA}) and a polymerase, Qβ replicase. The vesicle membranes were relatively impermeable to the nucleotides, so this reaction can run only until the entrapped reagents are consumed. In similar work, Oberholzer, Albrizio, and Luisi (1995; see also chapter 3) were able to run the polymerase chain reaction (PCR) inside liposomes and observe the production of PCR products. Figure 4.9 illustrates a similar experiment that uses ADP as an activated RNA monomer (designated M^*_{RNA}) that is first entrapped, and then added externally to the vesicles, while oleic anhydride is simultaneously hydrolyzed to form oleic acid, which feeds aggregate growth.

The first spatially resolved in vitro amplification and evolutionary selection experiment based on CATCH (cooperative amplification by cross hybridization; Ehricht, Ellinger, and McCaskill, 1997; Ellinger, Ehricht, and McCaskill, 1998) is shown in figure 4.10. In the experiments, spatial localization was assisted by a low-dimensional open microfluidic environment. Ellington (Kim and Joyce, 2004; Levy and Ellington, 2003) has proposed a more recent cooperative evolution system involving ribozymes.

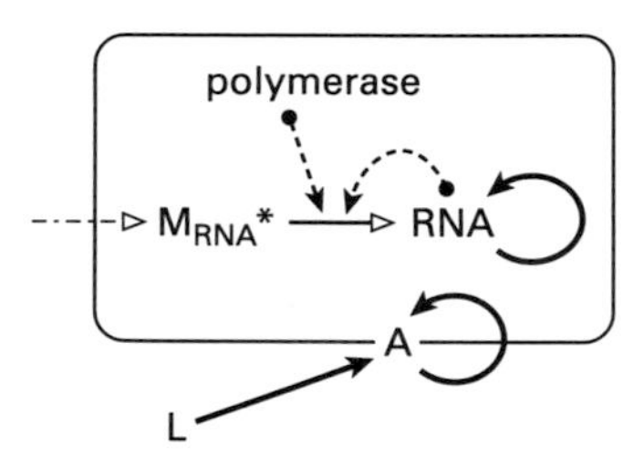

Figure 4.9
Luisi and coworkers (Walde, Goto, et al., 1994; Walde, Wick, et al., 1994; see also chapter 3) demonstrated the copying of an RNA template inside an oleic acid vesicle (A) with the aid of a polymerase, and the simultaneous growth of the population of vesicles. No functional coupling was demonstrated between those two processes in this system, nor was it established whether the new vesicles contained the replicating RNA system.

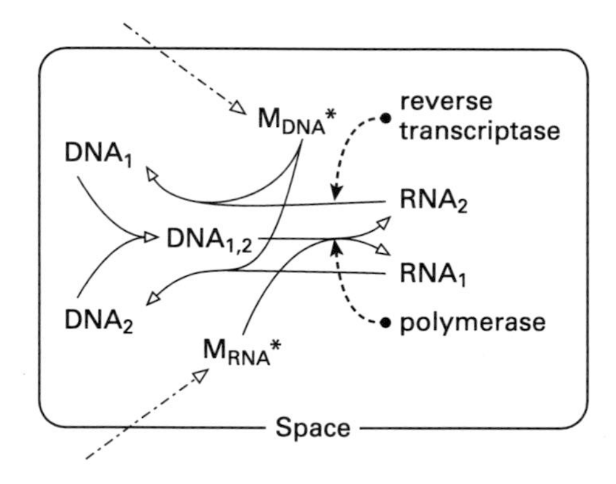

Figure 4.10
The first facultative cooperation system CATCH (Ehricht, Ellinger, and McCaskill, 1997; Ellinger, Ehricht, and McCaskill, 1998) that could combine two different informational molecules (and functions) in an evolutionary stable replication system, which profits from spatial colocalization.

Another example of vesicle-contained RNA synthesis is shown in figure 4.11, where we have added an ad hoc diagrammatic notation to indicate temperature cycling needed to enable the transportation of activated ribonucleotides across the membrane through defects created at high temperature.

Sugawara and colleagues (Takakura, Toyota, and Sugawara, 2003) have produced a giant mulitlamellar vesicle system composed of synthetic amphiphiles (bolamphiphiles) in a system illustrated in figure 4.12. The vesicles are loaded with a catalyst that converts a lipid precursor to lipids at their initial self-assembly, and the lipids from this reaction then feed the vesicle growth and division process. The lipids self-assemble into new vesicles, which grow and are shed as long as entrapped catalysts are present. Since the giant vesicles are large enough to be seen with a light microscope, one can verify that daughter vesicles are being shed by a single parent, and

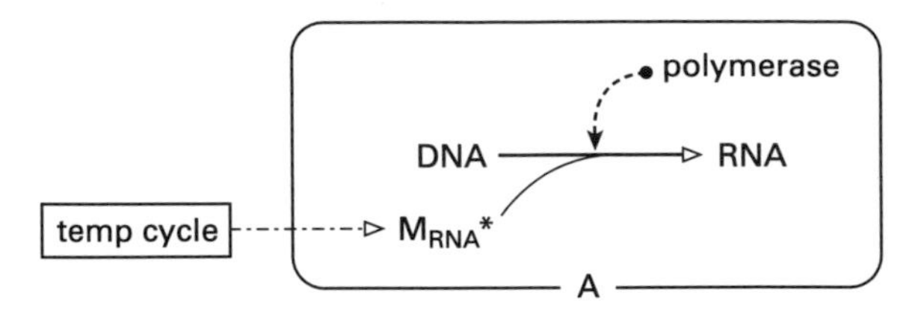

Figure 4.11
Monnard and Deamer (2002; see also chapters 2 and 3) created a (DMPC) vesicle (A) system that sustains transcription of RNA from DNA, with the help of T7 RNA polymerase. An externally generated temperature cycle creates transitory membrane defects that allow activated ribonucleotides (M^*_{RNA}) used in transcription to enter the system continually.

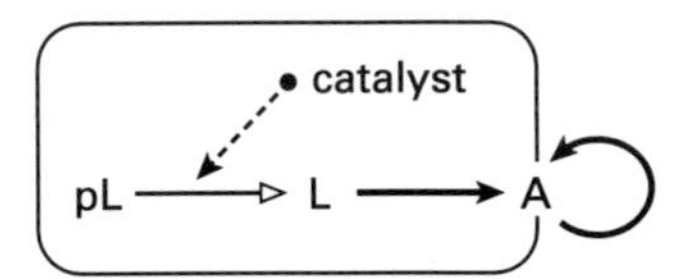

Figure 4.12
Sugawara and colleagues (Takakura, Toyota, and Sugawara, 2003) have produced a giant mulitlamellar vesicle system (A) that produces new vesicles with a catalyst that converts a lipid precursor (pL) to a lipid (L).

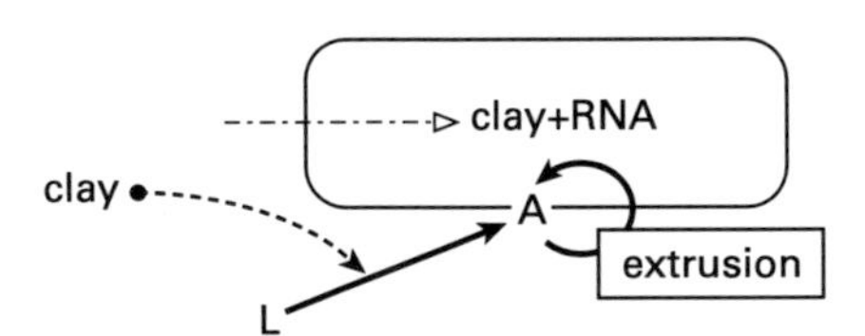

Figure 4.13
Hanczyc and coworkers (Hanczyc, Fujikawa, and Szostak, 2003; Hanczyc, Mansy, and Szostak, 2007; see also chapter 5) demonstrated that clay both catalyzes the formation and growth of vesicles and facilitates the incorporation of RNA into vesicles.

so a true lineage of vesicles is being produced. The division process eventually halts because of catalyst dilution.

Figure 4.13 illustrates experiments by Hanczyc and coworkers (Hanczyc, Fujikawa, and Szostak, 2003; Hanczyc, Mansy, and Szostak, 2007; see also chapter 5) that incorporate two novel elements requiring ad hoc diagrammatic extension: (1) the use of clay to catalyze the formation and growth of vesicles and act as a substrate for RNA inside vesicles, and (2) the use of extrusion for vesicle division. The same clay is also known to catalyze the polymerization of RNA from activated ribonucleotides (Erten et al., 2000; Ferris, 2006; Monnard, 2005). Hanczyc and coworkers also demonstrated that manual extrusion caused vesicle division with preservation of

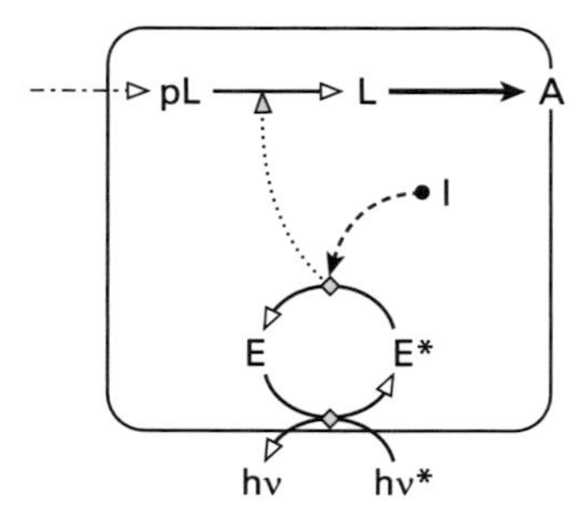

Figure 4.14
DeClue et al. (2007) (see also chapter 6) experimentally obtained the informational molecule catalyzed metabolic production of self-assembling containers (A).

internal contents, thus enabling vesicles containing informational polymers, which are perhaps replicating, to grow and divide indefinitely.

Figure 4.14 diagrams an experiment that uses an informational molecule to catalyze metabolic production of a lipid, which then assembles into an aggregate. The redox properties of the informational molecules (I) determine whether the sensitizer (E) is able to catalyze the metabolic production of fatty acids (L) from their precursors (pL). In this system, it is debatable whether the so-called informational molecules really have any functionally relevant information content. For example, their modulation of the metabolism has not been shown to be sequence dependent. So, one could argue that it is more appropriate to consider them simply part of the metabolic system.

The next four figures represent experimental realizations of protocell systems that are not, strictly speaking, bottom-up protocell architectures, because they utilize cell-free extract obtained from natural cells. The cell extract provides a chemically rich and reactive substrate because it contains ribosomes, many enzymes, and other complex biochemicals. A fully living version of this class of protocell architecture would require a replication process that could continuously produce the cell extract, a functionality that is currently attained only by existing living cells. These experiments are of interest for their encapsulation of significant metabolic functionality.

Figure 4.15 illustrates encapsulation of green fluorescent protein synthesis (Oberholzer, Nierhaus, and Luisi, 1999; Yu et al., 2001; see also chapter 7). Figure 4.16 illustrates the experiment of Noireaux and Libchaber (2004), which encapsulates the green fluorescent protein synthesis but also the synthesis of a pore-forming protein. The pore-forming protein enabled the cell to be continually resupplied with RNA and amino acid building blocks, which extended the time of the system's functioning for an order of magnitude. Since the membrane pore enables the container to resupply building blocks, and thus to function better as a container, this diagram

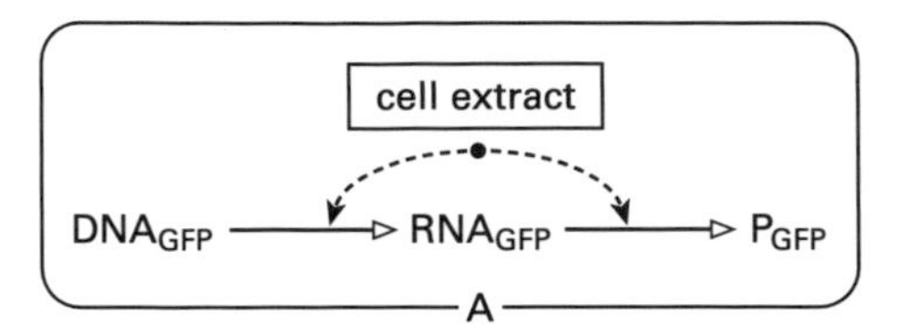

Figure 4.15
Building on the work of Oberholzer, Nierhaus, and Luisi (1999), Yu et al. (2001; see also chapter 7) demonstrated the synthesis of green fluorescent protein (P_{GFP}) inside a vesicle that contained the gene for green fluorescent protein (DNA_{GFP}) along with a cell extract containing all of the components for in vitro protein expression, including ribosomes, polymerase, and miscellaneous other enzymes and building blocks of RNA and proteins.

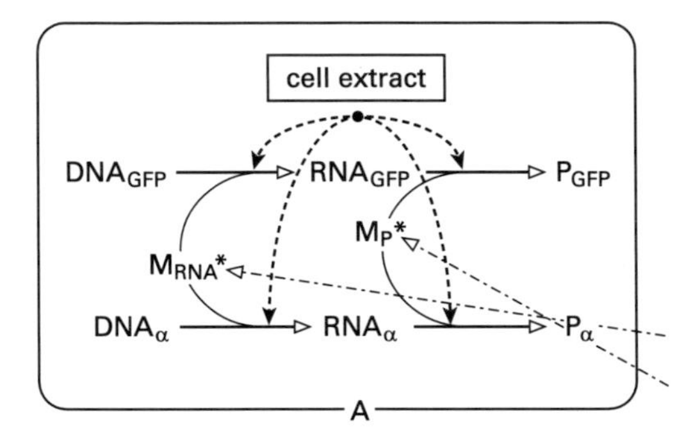

Figure 4.16
Noireaux and Libchaber (2004) obtained a functional vesicle-encapsulated system of two transcribed and translated genes, one coding for the easily detectable green fluorescent protein (P_{GFP}) and one coding for a membrane pore-forming protein, alpha hemolysin (P_α). The pores enabled the ongoing supply of RNA and amino acid building blocks.

represents one particular kind of functional support of the container that genes can provide.

Figure 4.17 represents the encapsulation of a two-layer cascading genetic network (Ishikawa et al., 2004). One gene codes for green fluorescent protein (P_{GFP}), and the other for T7 RNA polymerase (P_{T7}) to drive the synthesis of the former. The transcription and translation products are constructed from activated monomers (M^*_{RNA} and M^*_P) using the complex mixture of ingredients available in the cell extract. Note that here the operation of one part of the information system (the expression of P_{T7}) functionally supports the operation of another part of the information system (the expression of P_{GFP}), thus enhancing the production of P_{GFP} over that seen in the previous two experiments described.

To show how these causal diagrams could apply to a very simple biological system, figure 4.18 illustrates a virus's life cycle within our diagrammatic language. As

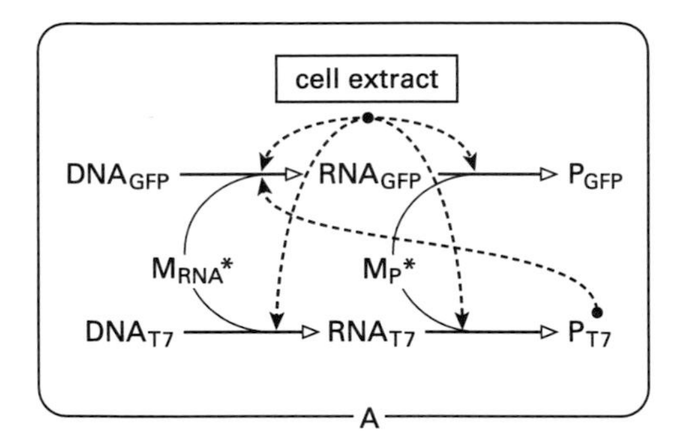

Figure 4.17
Yomo and colleagues (Ishikawa et al., 2004; see also chapters 3 and 7) realized a vesicle-encapsulated system cascading genetic network of two transcribed and translated genes. One gene coded for green fluorescent protein (P_{GFP}) and the other coding for T7 RNA polymerase (P_{T7}), which is required to drive the synthesis of the fluorescent protein.

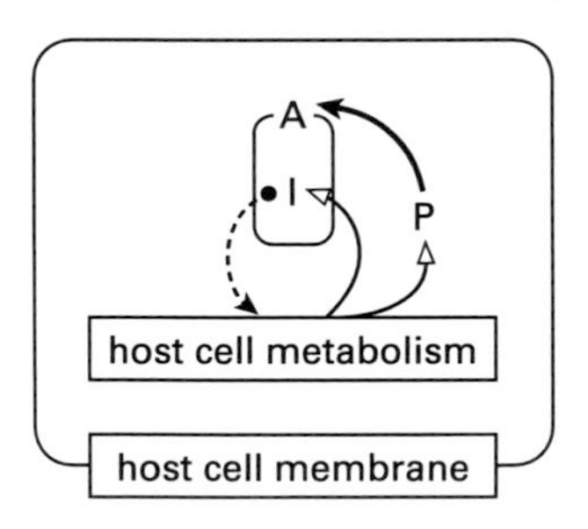

Figure 4.18
A representation of a virus. For a comparison with protocells, we represent a virus (A) as an encapsulated genome (I) that can utilize the complex metabolic machinery of its host cell to make more copies both of its genome and of its self-assembling capsule proteins (P).

with the representation of cell extract in figures 4.15 through 4.17, clean representation of the dependence of the virus on the host cell metabolism involves the introduction of another ad hoc "black box" that encompasses the myriad biochemical details of that metabolism. Some viruses replicate their genome only after using the host cell metabolism to produce a special polymerase for this purpose. The diagram shows how the virus is similar in some ways to artificial systems that use highly evolved polymerases or cell extracts containing complex structures such as ribosomes.

4.6 Simulation Results

Exactly the same kind of functional diagrams that we have used to represent protocell achievements in the laboratory can also be used to represent simulations of

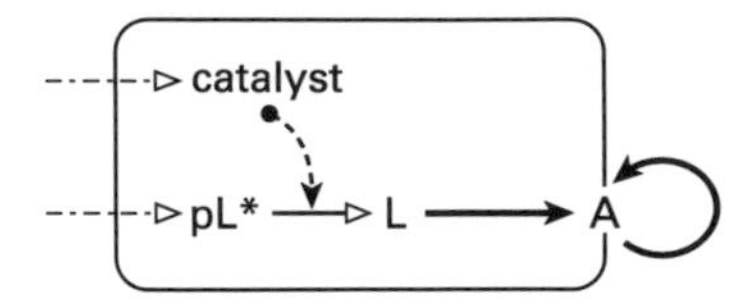

Figure 4.19
Ikegami and coworkers (see chapter 9) develop an abstract computational system (lattice gas simulation) capable of spontaneously generating a semipermeable container (A) that encapsulates a catalyst capable of converting activated precursor lipids (pL*) into the lipids (L) that self-assemble into the container.

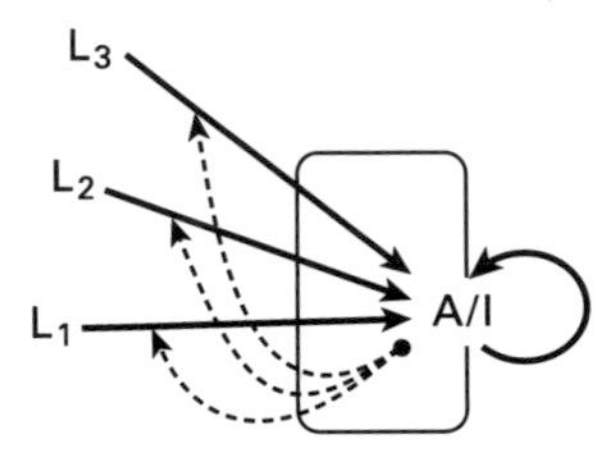

Figure 4.20
Segré and coworkers (2000; see also chapter 11) developed a model system in which aggregates (A) self-assemble from many types of lipids. In addition, the composition of the aggregate (a mixture of different types of lipids) is itself compositional information (I) that catalytically influences the selective incorporation of additional lipids (L_1, L_2, L_3).

protocell architectures. Different simulations capture different aspects of a protocell's functional requirements and achieve functional coupling of containment, metabolism, and genetics to different degrees. The chapters in part II describe in detail some of these models. We do not attempt here to cover protocell modeling efforts comprehensively; rather we pick just a few examples to illustrate how simulations may be diagrammed, thus enabling a structural comparison between them and, perhaps more important, between them and experimental efforts. All but one of the simulation results we diagram here are further detailed in other chapters.

Figure 4.19 describes a minimalist protocell model with nontrivial container self-assembly that may be driven to make the container structures divide (see chapter 9). Figure 4.20 represents a "lipid world" model consisting of lipids that are assumed to be in a self-assembled state such that they cross-catalyze each other's production (Segré, Ben-Eli, and Lancet, 2000; see also chapter 11). In this model, the functional roles of aggregate (A) and information (I) are realized by exactly the same material substrate (which we represent by A/I), and the information contained in the aggregate is termed a *composome*. At a predetermined point in the aggregate's growth, it reproduces by dividing. Since the composome achieves the functions of container and informational chemistry simply with the energetically downhill self-assembly of lipid

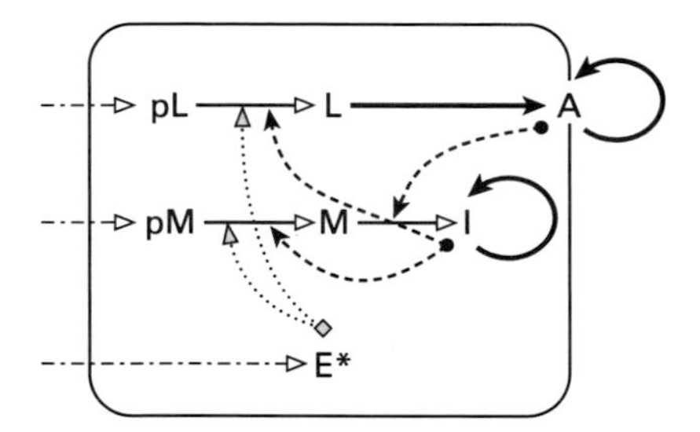

Figure 4.21
Fellermann and coworkers (2007; see also chapter 6) developed a computational system with a replicating gene (I) that catalyzes both the metabolic production of lipids (L) and the gene components (M). The precursors (pL, pM) of these reactions and light sensitizer molecules are continuously provided from the environment. The lipids self-assemble into an aggregate (A) that grows and divides.

aggregates, it has no need for a significant metabolism that harvests energy from the environment. Thus, the composome model shows how it might be possible for a protocell to avoid the need for a metabolism altogether, as long as the system is provided with the energy-rich supply of amphiphiles. With lipids continuously provided from the environment, simulations (Segré, Ben-Eli, and Lancet, 2000) have shown that the compositional information in a population of composomes evolves without any informational polymers functioning as genes.

Fellermann, Rasmussen, Ziock, and Solé (2007; see also chapter 6) have produced a dissipative particle dynamics (DPD) simulation of each of the steps in the life cycle of a simple protocell (figure 4.21). The model represents various amphiphilic and hydrophobic elements in a water environment. The precursors (pL, pM) of lipids (L) and dimers (M) are hydrophobic, as is the metabolic molecule (E^*). The precursors are converted into amphiphilic building blocks (lipids and monomers) when they are in close proximity to the metabolic molecule, E^*, which is interpreted as a sensitizer that converts light energy into chemical energy. The creation of building blocks also depends on the close proximity of amphiphilic genes (I); this is interpreted as the genes catalyzing those reactions. Since the genes and their dimer components are amphiphilic, they can align properly to achieve ligation only if they are at the interface of the micelle (they tend to form micelles in bulk water). In this way, the micelle aggregate, A, catalyzes the production genes from the dimers. The production of new lipids causes the aggregate to grow and divide.

Figure 4.22 illustrates a model that couples self-assembly with replication kinetics (McCaskill et al., 2007). Self-assembly takes place on a lattice whose site values are binary (or three-valued), governed by an appropriate Hamiltonian, and replicators are represented by a reaction-diffusion system on a parallel lattice, with bidirectional coupling between the two lattices. The presence of the replicating molecules locally

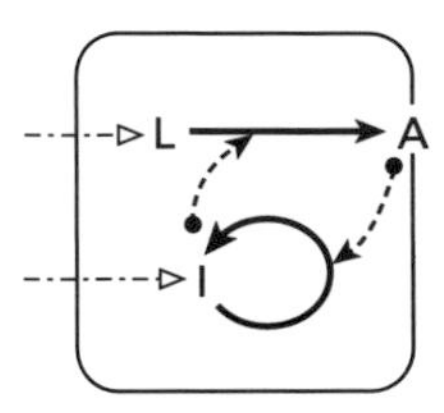

Figure 4.22
McCaskill and coworkers (2007) developed a model of the interaction between the self-assembly process and the process of evolution. The model consists of a two- or three-dimensional system of replicating combinatorial templates in a ternary fluid of hydrocarbons, amphiphiles, and water.

modulates the phases of the complex fluid, and the replication rate and mobility of the templates is influenced by the local amphiphilic configuration through an energetic coupling. Consequently, the replicators can potentially modify their environment to enhance their own replication. Through this coupling, the system can associate hereditary properties with self-assembling mesoscale structures in the complex fluid. The template replication process drives the thermodynamics of the coupled template/ternary fluid system out of equilibrium, as each replication step is associated with an energy cost. In the diagram shown, the presence of the templates (I) changes the local phases of the ternary fluid to produce micelles (A), a functional interaction that is represented by a catalysis arrow from I to the self-assembly process that produces A. In turn, the blue catalytic arrow from A to the replication of I indicates that the presence of micelles creates an abundance of the oil-water interface that the templates need to replicate. Energetics are determined primarily by the grand canonical ensemble of the amphiphilic spins. Because the statistical mechanics is taking place in the grand canonical ensemble, material and energy are exchanged freely in order to maintain a free energy equilibrium with an implicit "external reservoir," which comprises the environment. The local amphiphilic spin energetics have a direct catalytic influence on the ability of the replicators to replicate. New replicators are supplied and removed as determined by the replicator dynamics.

4.7 Hypothetical Schemes

In this section we move from the realm of actual achievements in experiments and simulations to the realm of the merely hypothetical. We represent a handful of examples with the goal of illustrating how the diagrammatic language can be used to abstract essential elements of hypothetical protocell architectures for comparison with each other and with experimental implementations and simulation. Each of the architectures include hypothetical couplings among all three protocell functional components—containment, metabolism, and information—with the aim of sustaining the whole as an entire living system.

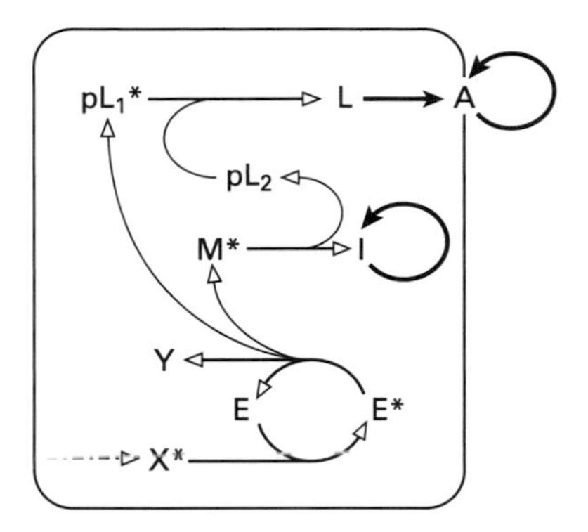

Figure 4.23
The chemoton (Gánti, 1975; see also chapter 22) couples a metabolism with a genetics and containment.

Figure 4.23 illustrates the chemoton model, a term coined by its creator, Tibor Gánti (1975; see also chapter 22). In this model, metabolism, containment, and genetic information functionalities are coupled as shown. The metabolism converts energized food (X^*) into energized monomers (M^*), energized precursors for lipids (pL_1^*), and waste (Y). Copies of genetic polymers (I) and a second kind of lipid precursor (pL2) are made from the energized monomers. The two lipid precursors combine to form lipids, which self-assemble into a self-reproducing container. When a gene is copied, the aggregate reproduces by an extrinsic process.

This representation of the chemoton is schematic, in that it abstracts away various details about the component substances and the stoichiometric coupling of the chemical reactions. An important hallmark of the chemoton is the fact that it stoichiometrically couples the functional elements. This is shown in figure 4.23 by the prevalence of reaction arrows (black) in the diagram. Note that the genes do not modulate the metabolism or the container. Another notable aspect of the chemoton is that the externally supplied "food," X^*, must be a large, energetically activated molecule. The size and complexity of X^* can be inferred from the fact that it is the source of all of the precursors for all other components. These last aspects indicate that the chemoton, as originally conceived, might be practically unfeasible. However, it occupies an important historical role as the first hypothetical architecture to explicitly integrate the three functional elements of containment, metabolism, and information.

Figure 4.24 illustrates a protocell based primarily on RNA chemistry (Szostak, Bartel, and Luisi, 2001; see also chapters 3 and 5). This protocell architecture is a natural follow-on to the first four experiments described in section 4.6, all of which aim to encapsulate RNA chemistry in vesicles, and it illustrates one way in which the "RNA world" approach to protocells could be protected from molecular parasites by containers (see chapter 17). The design has two RNA elements, one that is a

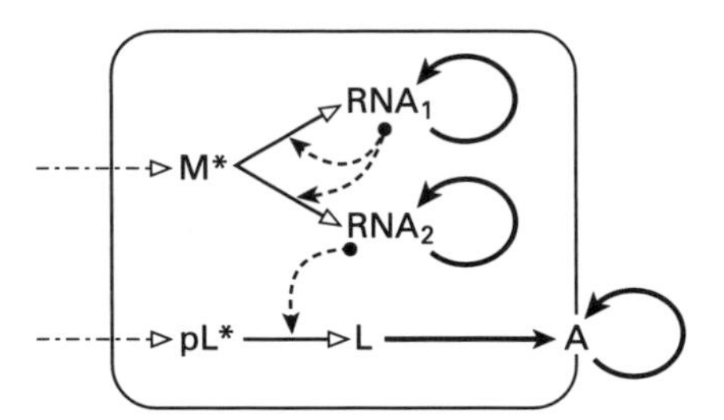

Figure 4.24
The RNA cell (Szostak, Bartel, and Luisi, 2001; see also chapter 3). A vesicle encapsulates two RNA ribozymes; one (RNA_1) acts as a polymerase (RNA replicase) and facilitates the replication of both RNA strings, and the other (RNA_2) catalyzes the production of lipids (L) from appropriate activated precursors (pL*).

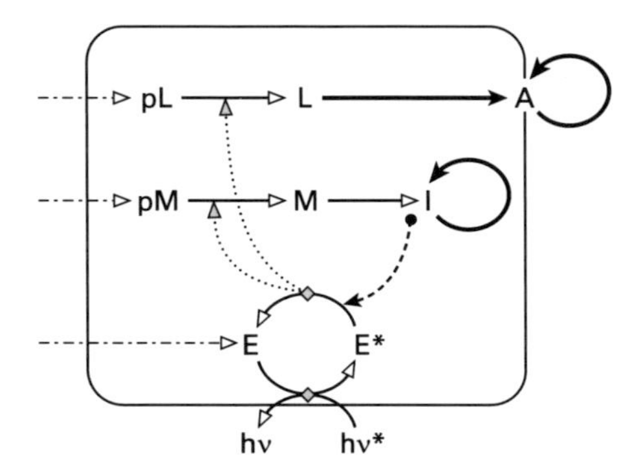

Figure 4.25
Los Alamos bug (Rasmussen et al., 2003; see also chapter 6). Light energy is harvested by a sensitizer (E), and this energy is used in the production of monomer components (M) of genetic molecules (I) from precursors (pM). The energy is also used to convert precursor lipids (pL) into lipids (L). These lipids self-assemble into a micelle or vesicle aggregate (A). Genetic molecules catalyze both the metabolic cycle and the production of further genetic molecules. The continual production of lipids causes the aggregate to grow and ultimately divide.

polymerase and one that catalyzes the production of lipids to feed container growth. Those lipids self-assemble into a vesicle, which then grows and eventually divides. Transporting energized precursors into the cell is experimentally problematic, but a temperature cycle could possibly generate enough membrane defects to do so. This scheme contains no explicit metabolism that creates the building blocks for cellular processes; instead, they are assumed to enter the cell from the environment (by an unspecified process) in an energized form.

Figure 4.25 illustrates the hypothetical "Los Alamos bug" (Rasmussen et al., 2003). This protocell architecture strives to be realistic, in the sense that it does not involve complex molecules such as polymerase, and the precursors needed to feed the

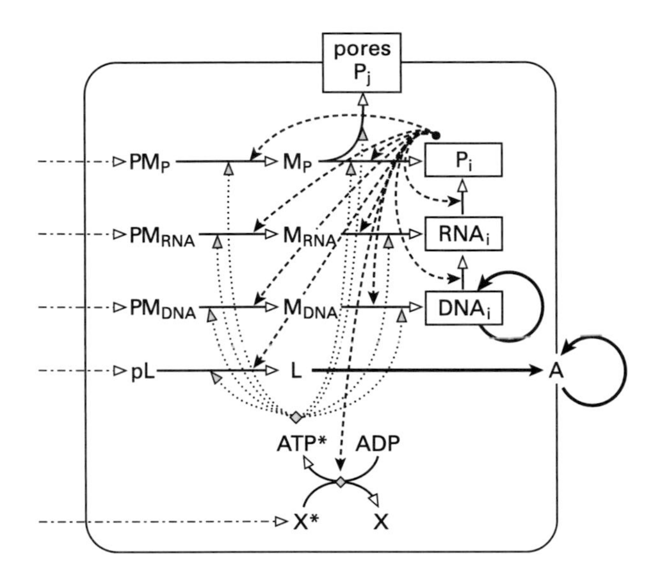

Figure 4.26
The protein/DNA cell (see chapter 3). The metabolism extracts energy from an energized food source (X^*) to produce ATP^*, which drives all the chemical reactions in the cell. Those reactions ultimately produce various kinds of DNA, RNA, and proteins (DNA_i, RNA_i, and P_i) from their components (M_{DNA}, M_{RNA}, M_P), which in turn are produced by their precursors (pM_{DNA}, pM_{RNA}, pM_P) lodged in vesicle membranes.

protocell are relatively simple. The lipid aggregate may not necessarily be a vesicle, but could be something much smaller, such as a micelle or small lipid aggregate. The functionality of the informational component (I) is believed to be the catalysis of the energy metabolism through a sequence-dependent charge transfer.

A rather complex proposed protocell architecture based on DNA, RNA, and proteins is illustrated in figure 4.26 (see chapter 3). This semisynthetic cell incorporates much of the complexity of actual living cells. The architecture uses 30 or more genes to code for a minimal set of proteins, including some with enzymatic or membrane pore functionality, and also some that assemble into a primitive ribosome. In some versions, the DNA, RNA, and protein components might enter the cell directly from the environment through membrane pores, rather than being synthesized in the cell from precursors.

Figure 4.27 illustrates a protocell architecture based on technological complementation, that is, the replacement of certain key functionalities with technological support, in this case microfluidic and microelectronic support (see chapter 12). This fully synthetic but hybrid electronic-chemical cell can survive only in an artificial microfluidic environment created by the experimenter. Different versions have varying degrees of cell functionality and regulation assumed by the microfluidic system,

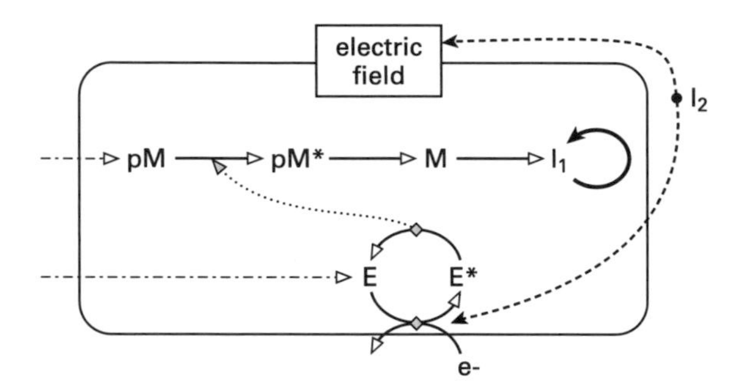

Figure 4.27
In the electronic cell (see chapter 12) containment no longer needs lipid aggregation; instead, spatial localization can be achieved through programmed electrical fields, as in this diagram. Likewise, the informational functionality need not be carried by an informational polymer (like I_1 in this diagram), but can be achieved by the external program that is part of the experimental environment (as shown here by I_2, which is outside the containment provided by the electric field) controlling both the electrical fields for localization and the injection of energy. See text for further discussion.

resulting in different levels of technological complementation. This is not a simulated cell, but a real hybrid chemical-electronic cell. In a simple version of this scheme, the container is replaced by an electronically regulated field barrier (artificial membrane) that can be reproduced by the computer at different locations. In a more radical version, the genes are replaced by electrode control sequences that direct the local chemical processes, and can be copied by the computer to operate locally at different sites. Finally, photometabolism can be replaced with electrochemical activation by individual cell microelectrodes, interacting either directly with redox chemistry or indirectly, for example, by induced pH gradients. The version shown employs both chemical and electronic genes. Only one example of technologically complemented protocell architecture is shown here, but there are numerous variations on this theme, altering the components and functionalities that are complemented.

4.8 Conclusion

The central thesis of this chapter is that the causal diagrams introduced herein capture the important functional components and functional interactions in protocells and their precursors. This makes comparison of different approaches to making protocells straightforward. In addition, the diagrams can play a constructive role in identifying key milestones, highlighting key scientific advancements, and calling attention to key gaps in present accomplishments.

Although certainly not ideal, the causal diagramming method has some notable advantages. The figures in sections 4.5, 4.6, and 4.7 show that the method applies uniformly to experiments, simulations, and hypothetical schemes. Figures 4.15

through 4.17 show that it also applies, with minor modification, to protocell constructions that use complex cell-free extract. This illustrates how the method can represent more or less chemical detail, according to the purposes at hand. The method should also apply to top-down protocell constructions, provided that sufficient chemical detail is suppressed, as it was in diagrams of protocells using cell extracts and the virus causal diagram (see figure 4.18).

The causal diagramming method makes explicit the essential functionally distinct interactions while abstracting away from many extraneous details, and it highlights which high-level properties are present and which are missing. The central challenge for this kind of representation scheme is to strike a balance between functional abstractness and chemical concreteness. The appropriate balance depends on the purposes at hand, and different purposes can be achieved by different kinds of diagrams. The diagramming method presented here is a work in progress; we expect it to be augmented in the future and to evolve through further use.

Acknowledgments

The bulk of this work was supported by the EU in the PACE integrated project, EC-FP6-IST-FET-IP-002035. Additional support for S.R. came from the LDRD-DR Protocell Assembly (PAs) project. For hospitality while working on this project, we thank the European Center for Living Technology (ECLT) and the Santa Fe Institute. We also thank the audiences at presentations of this material at the ECLT for helpful comments.

References

Acevedo, O. L., & Orgel, L. E. (1987). Non-enzymatic transcription of an oligodeoxynucleotide 14 residues long. *Journal of Molecular Biology*, *197*, 187–193.

Bachmann, P. A., Luisi, P. L., & Lang, J. (1992). Autocatalytic self-replicating micelles as models for prebiotic structures. *Nature*, *357*, 57–59.

Berclaz, N., Bloechliger, E., Mueller, M., & Luisi, P. L. (2001). Matrix effect of vesicle formation as investigated by cryotransmission electron microscopy. *Journal of Physical Chemistry B*, *105*, 1065–1071.

Berclaz, N., Mueller, M., Walde, P., & Luisi, P. L. (2001). Growth and transformation of vesicles studied by ferritin labeling and cryotransmission electron microscopy. *Journal of Physical Chemistry B*, *105*, 1056–1064.

Boerlijst, M., & Hogeweg, P. (1991). Self-structuring and selection: Spiral waves as a substrate for prebiotic evolution. In C. Langton, C. Taylor, J. D. Farmer, & S. Rasmussen (Eds.), *Artificial life II* (pp. 255–276). Redwood, CA: Addison-Wesley.

Chakrabarti, A., Breaker, R. R., Joyce, G. F., & Deamer, D. W. (1994). Production of RNA by a polymerase protein encapsulated within phospholipid vesicles. *Journal of Molecular Evolution*, *39*, 555–559.

Cronhjort, M. B., & Blomberg, C. (1997). Cluster compartmentalization may provide resistance to parasites for catalytic networks. *Physica D*, *101*, 289–298.

Cronhjort, M. B., & Nyberg, A. M. (1996). 3D hypercycles have no stable spatial structure. *Physica D*, *90*, 79–83.

Deamer, D. W. (1992). Polycyclic aromatic hydrocarbons: Primitive pigment systems in the prebiotic environment. *Advances in Space Research, 12*, 183–189.

DeClue, M. S., Monnard, P.-A., Bailey, J. A., Boncella, J. M., Rasmussen, S., & Ziock, H. (2007). Towards artificial life: Catalytic coupling of information, metabolism, and container in a minimal protocell. Submitted.

Eigen, M. (1971). The molecular quasispecies. *Naturwissenschaften, 58*, 465–523.

Eigen, M., Biebricher, C. K., Gebinoga, M., & Gardiner, W. C. (1991). The hypercycle: Coupling of RNA and protein biosynthesis in the infection cycle of an RNA bacteriophage. *Biochemistry, 30*, 11005–11018.

Eigen, M., & Schuster, P. (1977). The hypercycle: A principle of natural self-organization. Part A: Emergence of the hypercycle. *Naturwissenschaften, 64* (11), 541–565.

Eigen, M., & Schuster, P. (1978a). The hypercycle: A principle of natural self-organization. Part B: The abstract hypercycle. *Naturwissenschaften, 65*, 7–41.

Eigen, M., & Schuster, P. (1978b). The hypercycle: A principle of natural self-organization. Part C: The realistic hypercycle. *Naturwissenschaften, 65*, 341–369.

Ehricht, R., Ellinger, T., & McCaskill, J. S. (1997). Cooperative amplification of templates by cross-hybridization (CATCH). *European Journal of Biochemistry, 243*, 358–364.

Ellinger, T., Ehricht, R., & McCaskill, J. S. (1998). In vitro evolution of molecular cooperation in CATCH, a cooperatively coupled amplification system. *Chemistry and Biology, 5*, 729–741.

Fellermann, H., Rasmussen, S., Ziock, H., & Solé, R. (2007). Life cycle of a minimal protocell: A dissipative particle (DPD) study. *Artificial Life, 13*, 319–345.

Ferris, J. P. (2006). Montmorillonite-catalyzed formation of RNA oligomers: The possible role of catalysis in the origins of life. *Philosophical Transactions of the Royal Society, 361*, 1777–1786.

Fontana, W. (1991). Algorithmic chemistry. In C. Langton, C. Taylor, J. D. Farmer, & S. Rasmussen (Eds.), *Artificial life II* (pp. 159–209). Redwood, CA: Addison-Wesley.

Fox, S. (1960). How did life begin? *Science, 132*, 200–208.

Füchslin, R. M., & McCaskill, J. S. (2001). Evolutionary self-organization of cell-free genetic coding. *Proceedings of the National Academy of Science of the United States of America, 98*, 9185–9190.

Gánti, T. (1975). Organization of chemical reactions into dividing and metabolizing units: The chemotons. *Biosystems, 7*, 15–21.

Hanczyc, M. M., Fujikawa, S. M., & Szostak, J. W. (2003). Experimental models of primitive cellular compartments: Encapsulation, growth and division. *Science, 302*, 618–622.

Hanczyc, M. M., Mansy, S. S., & Szostak, J. W. (2007). Mineral surface directed membrane assembly. *Origin of Life and Evolution of the Biosphere, 37* (1), 67–82.

Ikegami, T. (1994). From genetic evolution to emergence of game strategies. *Physica D, 75*, 310–327.

Ishikawa, K., Sato, K., Shima, Y., Urabe, I., & Yomo, T. (2004). Expression of a cascading genetic network within liposomes. *FEBS Letters, 576*, 387–390.

Johnston, W. K., Unrau, P. J., Lawrence, M. S., Glasner, M. E., & Bartel, D. L. (2001). RNA-catalyzed RNA polymerization: Accurate and general RNA-templated primer extension. *Science, 292*, 1319–1325.

Joyce, G. F. (1984). Non-enzymatic template-directed synthesis of RNA copolymers. *Origins of Life, 14*, 613–620.

Kim, D.-E., & Joyce, G. F. (2004). Cross-catalytic replication of an RNA ligase ribozyme. *Chemistry and Biology, 11*, 1505–1512.

Lee, D. H., Granja, J. R., Severin, K., & Ghadiri, M. R. (1996). A self-replicating peptide. *Nature, 382*, 525–526.

Levy, M., & Ellington, A. D. (2003). Exponential growth by cross-catalytic cleavage of deoxyribozymogens. *Proceedings of the National Academy of Science of the United States of America, 100*, 6416–6421.

Matsuura, T., Yamaguchi, M., Ko-Mitamura, E. P., Shima, Y., Urabe, I., & Yomo, T. (2002). Importance of compartment formation for a self-encoding system. *Proceedings of the National Academy of Science of the United States of America, 99*, 7514–7517.

Maturana, H., & Varela, F. (1980). *Autopoiesis and cognition: The realization of the living*. Boston: D. Reidel.

Maynard Smith, J., & Szathmáry, E. (1995). *The major transitions in evolution*. Oxford: Oxford University Press.

McCaskill, J. S. (1992). How is genetic information generated? In S. Andersson, Å. E. Andersson, & U. Ottoson (Eds.), *Theory and control of dynamical systems: Applications to systems in biology (Huddinge, Stockholm, Sweden)* (pp. 137–164). Singapore: World Scientific Press.

McCaskill, J. S. (1994). The origin of molecular cooperation. Inaugural professorial lecture, Institut für Molekulare Biotechnologie IMB, Friedrich-Schiller-Universtät, Jena, Germany.

McCaskill, J. S., Füchslin, R. M., & Altmeyer, S. (2001). The stochastic evolution of catalysts in spatially resolved molecular systems. *Biology and Chemistry*, *382*, 1343–1363.

McCaskill, J. S., Packard, N. H., Rasmussen, S., & Bedau, M. A. (2007). Evolutionary self-organization in complex fluids. *Philosophical Transactions of the Royal Society B*, *362*, 1763–1779.

Mills, D. R., Peterson, R. L., & Spiegelman, S. (1967). An extracellular Darwinian experiment with a self-duplicating nucleic acid molecule. *Proceedings of the National Academy of Science of the United States of America*, *58*, 217–224.

Monnard, P.-A. (2005). Catalysis in abiotic structured media: An approach to selective synthesis of biopolymers. *Cellular and Molecular Life Sciences*, *62*, 520–534.

Monnard, P.-A., & Deamer, D. W. (2002). Membrane self-assembly processes: Steps toward the first cellular life. *Anatomical Record*, *268*, 196–207.

Munteanu, A., Attolini, C. S., Rasmussen, S., Ziock, H., & Solé, R. V. (2007). Generic Darwinian selection in catalytic protocell assemblies. *Philosophical Transactions of the Royal Society B*, *362*, 1847–1855.

Niesert, U., Harnasch, D., & Bresch, C. (1981). Origin of life between Scylla and Charybdis. *Journal of Molecular Evolution*, *17*, 348–353.

Noireaux, V., & Libchaber, A. (2004). A vesicle bioreactor as a step toward an artificial cell assembly. *Proceedings of the National Academy of Science of the United States of America*, *101*, 17669–17674.

Nomura, S., Tsumoto, K., Hamada, T., Akiyoshi, K., Nakatani, Y., & Yoshikawa, K. (2003). Gene expression within cell-sized lipid vesicles. *Chemistry and Biochemistry*, *4*, 1172–1175.

Oberholzer, T. K., Albrizio, M., & Luisi, P. L. (1995). Polymerase chain reaction in liposomes. *Current Biology*, *2*, 677–682.

Oberholzer, T. K., Nierhaus, H., & Luisi, P. L. (1999). Protein expression in liposomes. *Biochemical and Biophysical Research Communications*, *261*, 238–241.

Oberholzer, T. K., Wick, R., Luisi, P. L., & Biebricher, C. K. (1995). Enzymatic RNA replication in self-reproducing vesicles: An approach to a minimal cell. *Biochemical Biophysical Research Communications*, *207*, 250–257.

Oparin, A. I. (1924; 1957). *The origin of life on the Earth* (3rd Ed.). Edinburgh: Oliver and Boyd.

Paul, N., & Joyce, G. F. (2002). A self-replicating ligase ribozyme. *Proceedings of the National Academy of Science of the United States of America*, *99*, 12733–12740.

Rasmussen, S., Chen, L., Nilsson, M., & Shigeaki, A. (2003). Bridging nonliving and living matter. *Artificial Life*, *9*, 269–316.

Rocheleau, T., Rasmussen, S., Nielson, P. E., Jacobi, M. N., & Ziock, H. (2007). Emergence of protocellular growth laws. *Philosophical Transactions of the Royal Society B*, *362*, 1841–1845.

Segré, D., Ben-Eli, D., & Lancet, D. (2000). Compositional genomes: Prebiotic information transfer in mutually catalytic noncovalent assemblies. *Proceedings of the National Academy of Sciences of the United States of America*, *97*, 4112–4117.

Szathmáry, E., & Demeter, L. (1987). Group selection of early replicators and the origin of life. *Journal of Theoretical Biology*, *128*, 463–486.

Szostak, J. W., Bartel, D. P., & Luisi, P. L. (2001). Synthesizing life. *Nature*, *409*, 387–390.

Takakura, K., Toyota, T., & Sugawara, T. (2003). A novel system of self-reproducing giant vesicles. *Journal of the American Chemical Society*, *125*, 8134–8140.

Tangen, U. (1994). A computational model for functional evolution. In H. Ritter, H. Cruse, & J. Dean (Eds.), *Prerational intelligence: Adaptive behavior and intelligent systems without symbols and logic, Volume 2*, (pp. 299–320). Dordrecht: Kluwer.

Thürk, M. (1993). A model of self-organizing automata algorithms describing molecular evolution. PhD thesis (J. S. McCaskill & G. Wechsung, advisors). Jena: Friedrich-Schiller-Universität.

Tijvikua, T., Ballester, P., & Rebek, J. Jr. (1990). A self-replicating system. *Journal of the American Chemical Society, 112*, 1249–1250.

von Kiedrowski, G. (1986). A self-replicating hexadeoxynucleotide. *Angewandte Chemie, 25*, 932–935.

Walde, P., Goto, A., Monnard, P.-A., Wessicken, M., & Luisi, P. L. (1994). Oparin's reactions revisited: Enzymatic synthesis of poly(adenylic acid) in micelles and self-reproducing vesicles. *Journal of the American Chemical Society, 116*, 7541–7547.

Walde, P., Wick, R., Fresta, M., Mangone, A., & Luisi, P. L. (1994). Autopoietic self-reproduction of fatty acid vesicles. *Journal of the American Chemical Society, 116*, 11649–116454.

Yu, W., Sato, K., Wakabayashi, M., Nakaishi, T., Ko-Mitamura, E. P., Shima, Y., et al. (2001). Synthesis of functional protein in liposome. *Journal of Bioscience and Bioengineering, 92*, 590–593.

II INTEGRATION

Existing living cells are understood to be composed of four primary chemical components: proteins, nucleic acids, lipids, and carbohydrates. These are integrated into a system that achieves the functions of containment and catalyzed metabolism and that has the capacity to grow, reproduce, and evolve. The physical structure of many of the molecular components has been described in great detail, primarily through x-ray crystallography and nuclear magnetic resonance studies. The functions of many of those structures, such as the mechanism by which a particular enzyme catalyzes a particular reaction, have also been discerned. However, for the most part these structure-function relationships are explained for each individual component, rather than for the integrated system of components.

Recently computationally based systems biology has been integrating individual structure-function relationships into working subsystems that can be investigated in the laboratory (Kitano, 2001; *Science*, 2002). The first such systems to be analyzed (although not yet reproduced experimentally) were the metabolic pathways established by biochemists. Next the relationship between nucleic acids and proteins was elucidated, which was characterized as a system involving multiple molecular species ranging from the DNA of chromosomes to ribosomal translation of genetic information into amino acid sequences of proteins. More recently, we have begun to understand the functional relationships between thousands of proteins in a cell, which form patterns called *interactomes* (Bader et al., 2003; Park, Lappe, and Teichmann, 2001).

These biochemical and cell biological systems are far more complex than the systems physicists traditionally tried to model and understand. The theoretical perspective on such studies of complex systems changed dramatically in the 1970s and 1980s, with the creation of the personal computer and the explosion of work in nonlinear dynamics. Now that fractals, chaotic dynamics, bifurcations, and self-organizing phenomena could be simulated on an inexpensive computer in real time, new computational methods were increasingly used to model complex systems involving life. This new perspective led to the quest to model the essential minimal condition for life, which became a holy grail for the emerging artificial life community (Farmer et al., 1986; Langton, 1989).

Artificial life is concerned with more than computer models; it focuses equally on attempts to synthesize the essential features of living systems in hardware and in the laboratory (Bedau, 2003). In fact, one of artificial life's grand challenges is to generate a molecular protoorganism in vitro (Bedau et al., 2000)—the subject of this volume. Protocell research has succeeded in creating self-replicating (templating) molecules and self-reproducing compartments (micelles, lipid aggregates), as well as simple metabolic systems. However, combining these capabilities in an autonomous, evolvable, self-replicating autopoietic system remains a grand challenge.

The foundation for current attempts to produce protocells in the laboratory is 30 years of past experimental experience in which a variety of partial systems has been reconstituted from disassembled parts (see chapters 2 and 3). In parallel, theoretical and computational studies of simple lifelike systems have shed light on the intricate interactions among their many functional components. This foundation knowledge has led to increasing confidence that we might now achieve the ultimate goal of assembling a system of molecules that displays the basic properties of the living state.

The first targets in this quest should be the simplest possible forms of life. These simple molecular life forms would be self-reproducing molecular systems that continually reconstitute themselves and can evolve. Their environment, whether natural or artificial, should involve only simple materials and forms of energy. These localized biochemical systems, which might be held together by lipid aggregates, would derive free energy from simple chemicals or light available in the environment. They would use information carried in primitive genes to transform material resources and free energy from their environment into building blocks, and thereby to repair themselves, grow, and divide. Since the growth and division process would be controlled by inheritable information that controls protocell functioning, and since this information transmission is imperfect, protocells could be subject to natural selection and thus could evolve.

Such an ambitious goal requires application of powerful tools. A primary aim of this volume is to assemble a set of tools and show how their integrated application advances the goal of fabricating wet carbon-based artificial life. The choice of authors and topics in this part reflects this view that protocell research will be most productive if experimental work, theory, and computational simulation all guide and inform each other.

In general, research strategies for making protocells in the laboratory can be roughly categorized into two main approaches (Rasmussen et al., 2004). The top-down approach starts with the simplest existing living cells and attempts to simplify them further by a variety of methods, with the goal of determining the minimal set of necessary components for life as it exists today. The bottom-up approach starts with nonliving components and puts these into a context in which they assemble themselves into systems that display the emergent properties associated with the living state. The simplest version of the bottom-up approach uses components such as ribosomes and genes derived from living cells. A chemically more challenging alternative is to use molecules that are not derived from living cells, though they are inspired by what we know of the living state.

Both the top-down and bottom-up approaches are being pursued by various combinations of three independent methodologies. One approach is the traditional rational design method that uses chemical first principles and chemical intuition honed by experience to work out the details of a predefined design of the protocell components and how they are integrated into a single coherent system. Rational design is the most common method used in protocell research.

A second evolutionary design method is inspired by the way in which the complex functional designs in natural living systems have been spontaneously created by the process of natural selection. If protocell precursors could be intentionally fabricated with the capacity to evolve, for example, then desired protocell qualities could perhaps be coaxed to emerge through a process of artificial selection. Perhaps the clearest example of this is the artificial selection of RNA or proteins by means of in vitro or directed evolution (for reviews, see Breaker and Joyce, 1994; Lorisch and Szostak, 1996). Iterations of rounds of selection, amplification, and mutagenesis have time and again produced dramatic improvements of target functionality, well beyond what one could have hoped to achieve by rational design. Alternatively, even with protocell precursors that cannot themselves evolve, one could create a population of such precursors in the laboratory and grade them according to some artificial selection criterion, and then make a new population of experiments inspired by and generalizing from the earlier partial successes. Iterating this artificial selection process many times could produce more and more functional protocell designs, even if the chemical mechanisms employed in such designs were discovered by chemical analysis only after the fact (Theis et al., 2008).

With modern techniques of computer controlled microfluidics and micro-electrico-mechanical devices (MEMS), it is possible to conceive of a living system that is seamlessly integrated with an entirely synthetic environment. This opens the door to a new machine-complementation methodology for creating protocells (see chapter 12). The functionalities of spatial localization and isolation (containment), of energy and raw material harvesting (metabolism), or of heritable informational control of metabolism and containment could be partly realized in human engineered technology. This could then enable cycles of catalyzed metabolism, growth, and division to be controlled by a computer instead of by internal regulatory chemistry.

The chapters in this part describe a variety of combinations of these various approaches to making protocells. Chapter 5 introduces some of the methods of the bottom-up approach, in which RNA is encapsulated in lipid membrane vesicles that are themselves capable of growth and even division. The authors describe recent experimental results showing how simple fatty acid membranes have desirable properties in this regard. They also discuss possible roles for evolution within simple vesicular systems, and the use of combinatorial chemistry to explore optimal mixtures of components.

Chapter 6 describes a novel experimental approach that is based on chemical principles and does not include biological components. It is a significant first step toward capturing light energy in order to drive synthetic reactions, one producing a PNA polymer and the second an amphiphile that permits the boundary membrane to grow in concert with the polymer.

A working translational system encapsulated within vesicles that can produce substantial numbers of a desired protein molecule is described in some detail in chapter 7. The translational apparatus is obtained from cell-free extracts of bacteria, and contains all the components required for protein synthesis. This system can incorporate what the authors refer to as a cascading genetic network, since the gene controlling protein synthesis is actually part of a DNA plasmid and must be transcribed to mRNA before being expressed.

Chapter 8 discusses both theory and experimental design of an evolving protocell. The theory portion introduces new concepts such as minority-controlled state, recursive production, and its relation to evolvability. These concepts are then related to experimental results from Yomo's group, the authors of the previous chapter.

Chapter 9 discusses a computational model called a "lattice artificial chemistry." In this, a set of simple rules are incorporated into particles designed to simulate certain properties of amphiphilic compounds, catalysts, and resource molecules, which are then allowed to interact *in silico*. Remarkably, the computer model reproduces the aggregation of such molecules into cell-like vesicular structures, which encapsulate catalytic components that fuel the production of boundary building blocks.

Chapter 10 extends computational modeling to modes of replication and fissioning. They show that a proper choice of permeability, concentration gradients, and resulting osmotic energy can induce division of a protocell into two daughter cells. How to implement this in chemical protocells is an important open issue.

What is perhaps the simplest protocell model yet conceived is described in chapter 11. This system is entirely free of polymers such as nucleic acids; instead, the composition of molecular components provides a combinatorially rich source of functional and heritable information. Specifically, mixtures of phase-separated amphiphilic components in a bulk medium undergo alterations of composition over time that depend on energy input, local interactions (free energy), and nutrient resources. Aggregates with different compositions compete with one another for resources to support their growth and division. Thus, the aggregates undergo selection, and this can be interpreted as evolution of their compositional information.

Chapter 12 describes a system that departs entirely from the requirement for a membranous container. It discusses an "electronic protocell" that achieves the basic reactions involved in protocell growth through combining computer control of microfluidic flow and electronic field control of reactant diffusion processes.

References

Bader, G. D., Donaldson, I., Wolting, C., Ouellette, B. F. F., Pawson, T., & Hogue, C. W. V. (2003). BIND: The biomolecular interaction network database. *Nucleic Acids Research, 31*, 248–250.

Bedau, M. A. (2003). Artificial life: Organization, adaptation, and complexity from the bottom up. *Trends in Cognitive Science, 7*, 505–512.

Bedau, M. A., Catskill, J. S., Packard, N. H., Rasmussen, S., Adam, C., Green, D. C., et al. (2000). Open problems in artificial life. *Artificial Life, 6*, 363–376.

Breaker, R. R., & Joyce, G. F. (1994). Inventing and improving ribosome function: Rational design versus iterative selection methods. *Trends in Biotechnology, 12*, 268–275.

Farmer, J. D., Lipids, A., Packard, N., & Wendoff, B. (Eds.) (1986). *Evolution, games, and learning: Models for adaptation for machines and nature*. Amsterdam: North Holland.

Kitano, H. (Ed.) (2001). *Foundations of systems biology*. Boston: MIT Press.

Langton, C. (Ed.) (1989). *Artificial life*. Redwood City, CA: Addison-Wesley.

Lorisch, J. R., & Szostak, J. W. (1996). Chance and necessity in the selection of nucleic acid catalysts. *Accounts of Chemical Research, 29*, 103–110.

Park, J., Lappe, M., & Teichmann, S. A. (2001). Mapping protein family interactions: Intramolecular and intermolecular protein family interaction repertoires in the PDB and yeast. *Journal of Molecular Biology, 307* (3), 929–938.

Rasmussen, S., Chen, L., Deamer, D., Krakauer, D. C., Packard, N., Stadler, P. F., & Bedau, M. A. (2004). Transition from nonliving to living matter. *Science, 303*, 963–965.

Science. (2002). Special issue on systems biology. *295* (5560), 1589–1780.

Theis, M., Gazzola, G., Forlin, M., Poli, I., Hanczyc, M. M., & Bedau, M. A. (2008). Optimal formulation of complex chemical systems with a genetic algorithm. *ComPlexUs*.

5 Steps Toward a Synthetic Protocell

Martin M. Hanczyc, Irene A. Chen, Peter Sazani, and Jack W. Szostak

5.1 Introduction

The synthesis of artificial living systems is one of the grand challenges of modern chemistry. This goal goes beyond simple autocatalysis or self-replication to include the essential biological dimension of Darwinian evolution, defined here as natural selection acting on populations that have heritable variations, so that the genetic composition of the population changes over time. For a supramolecular, replicating chemical system to exhibit Darwinian evolution, the system must embody heritable transmission of variation from generation to generation. Moreover, this inherited variation must lead to variations in phenotype that affect the reproduction or survival of the system itself. Many possible phenotypes may be selectively advantageous, from improved rates or accuracy of replication, to greater stability of the system, to the ability to adapt to changing environments, to interactions among members of the population.

In our view, the complexities of Darwinian cellular evolution can be manifested in relatively simple replicating systems consisting of two major components: a genetic polymer capable of encoding heritable information in its sequence, and a means of compartmentalization such as a membrane boundary. To be consistent with terminology used throughout this book, we use the term *protocell precursor* to describe a self-replicating genome inside a self-replicating compartment and *protocell* to denote a fully integrated evolving genome-compartment system. In this chapter, we focus on work from our laboratory that is grounded in this conception of minimal life; experimental and theoretical work from other laboratories is presented in other chapters of this volume.

Both nucleic acids and lipids exhibit remarkable qualities of self-assembly and self-organization (e.g., the spontaneous folding of nucleic acids into complex secondary and tertiary structures, and the spontaneous assembly of lipids into bilayer membranes), which form the basis for their critical roles in biology. However, the complexity of modern biological machinery obscures the underlying simplicity of the chemical and physical phenomena that gave rise to the first living systems. In attempting to design simple living systems, we have directed our attention to spontaneous processes such as membrane self-assembly and templated nucleic acid synthesis, and searched for ways to integrate these phenomena into a complete cycle of reproduction. Membranes composed of single-chain amphiphiles such as fatty acids exhibit many properties not seen in the more familiar phospholipid membranes, and there is much to be learned about such membranes and their dynamics. Similarly, while we know most about the replication of RNA and DNA, many other related and perhaps simpler nucleic acids such as TNA (threose nucleic acid) and GNA (glycerol nucleic acid) can be considered as possible genetic polymers for a protocell.

We must also consider plausible sources of energy. Life is an intrinsically nonequilibrium process that requires fluxes of matter and energy, and in fact planetary environments are far from equilibrium and provide numerous possibilities for inputs of energy required by primitive protocells. Thus, our generic protocell model (figure 5.1) includes potential inputs of energy from activated nucleotides for genomic synthesis, from phase transitions as amphiphiles are incorporated into bilayer membranes, from mechanical energy for cell division, and from the generation of energy-rich gradients as the system responds to nonequilibrium environments. A simplified metabolic pathway is also necessary to process energy inputs to the system. Strong selective pressures would favor the evolution of internal generation and storage of energy and the synthesis of useful compounds, making the cell progressively less vul-

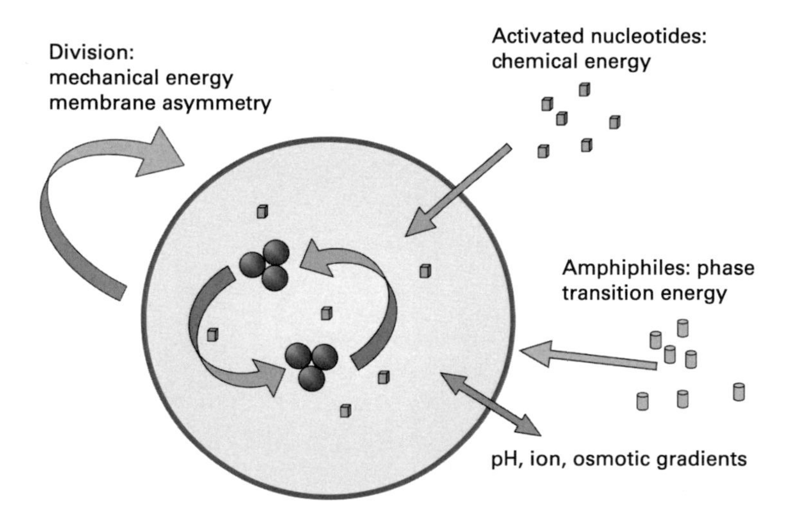

Figure 5.1
Cross-sectional schematic of our protocell model. The spheres represent self-replicating polymers encapsulated within the membrane.

nerable to fluctuations in the environment as the internal metabolism becomes more sophisticated.

Our efforts to design artificial living systems have been driven by two distinct motivations: the challenge of devising a self-reproducing, evolving chemical system, and the desire to learn something about the origin of life. The first motivation drives experimental work unconstrained by consistency with the prebiotic environment or chemistry, while the goal of the second is to find a plausible pathway from prebiotically feasible chemistry to the first cells. In practice, these approaches are distinct yet highly synergistic. We find that it is often beneficial to step back from the constraints of prebiotic conditions, which in any case are highly uncertain, and simply focus on finding any solution to a particular aspect of the protocell life cycle. Finding one solution—for example, the use of extrusion to mediate vesicle division—often leads to ideas for other possible solutions, such as division driven by membrane asymmetry, or propagation through internal vesicle synthesis. The experimental examples discussed in this chapter support our contention that efforts to synthesize protocell precursors will shed light on the origin of life on earth. It is our hope that, through modern chemistry and physics, we can devise self-reproducing and evolving systems that embody the simplicity of the first living cells.

5.2 Nucleic Acid Replication

In principle, there are two possibilities for replicating the nucleic acid genome of a synthetic protocell. The process could be spontaneous, that is, independent of macromolecular catalysis. The more commonly discussed possibility is that a folded nucleic acid, such as a ribozyme, might catalyze its own replication. We are pursuing both possibilities.

5.2.1 Spontaneous Replication

Orgel and his colleagues have explored spontaneous template-directed synthesis of RNA for many years. Perhaps the best studied system involves the use of 2-methylimidazole–activated nucleotide monomers, which extend a primer with good regiospecificity (i.e., almost exclusive synthesis of 3′-5′ phosphodiester bonds) on certain templates (Ferris et al., 1996). Unfortunately, despite considerable effort, the efficient copying of mixed-sequence templates longer than a few nucleotides has remained out of reach, and the idea that such chemistry could lead to complete cycles of replication has fallen from favor. However, the situation may be more promising if the constraints of prebiotic chemistry are ignored. For example, Orgel examined 3′-amino nucleotides as possible monomers, and observed significantly faster reactivity due to the greater nucleophilicity of the 3′-amino group compared to the standard 3′-hydroxyl group.

Other work has shown that modifications to the nucleobases can increase the stability of the A:U pair. We are exploring templated primer-extension reactions using such modified nucleotides, in the hope of finding a practical system that will allow for the replication of sequences in the range of 30 to 50 nucleotides.

Of course, complete cycles of replication require strand separation (presumably thermal) or strand displacement synthesis, both of which may require less stable duplexes. We are therefore exploring modified backbones such as TNA (threose nucleic acid) and GNA (glycerol nucleic acid), in the hope of finding a backbone chemistry that will be compatible with complete cycles of replication (see figure 5.2). Success in this endeavor would have one particularly interesting implication in that the first genome-encoded function to emerge in the context of cellular replication would not necessarily be related to nucleic acid replication. For example, the genome

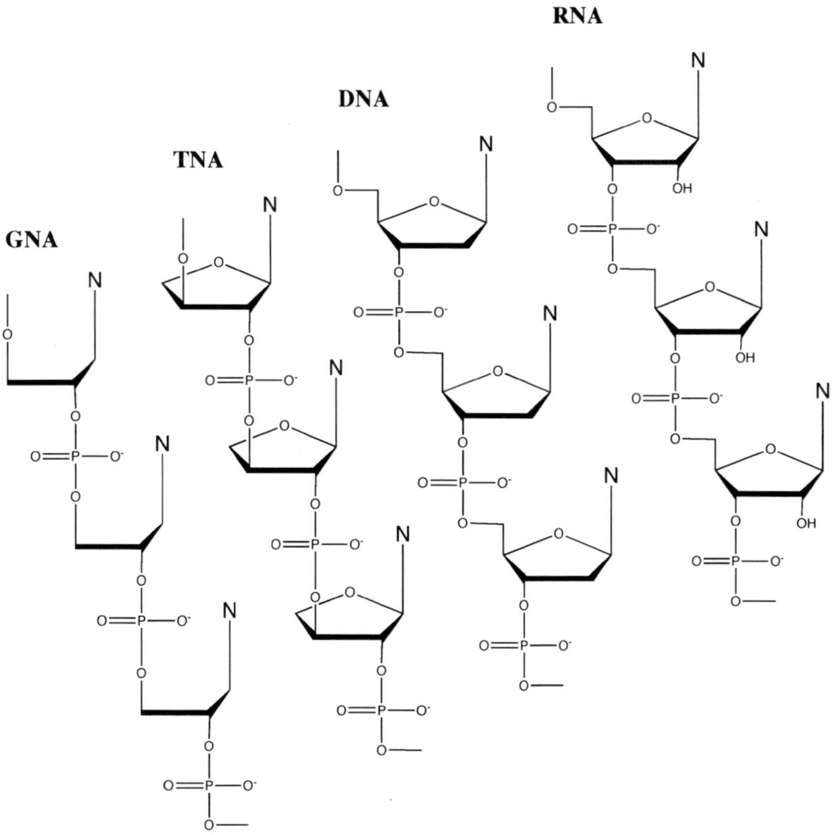

Figure 5.2
Two-dimensional representation of RNA, DNA, TNA, and GNA showing the similarity in the position of the phosphate backbones and the nucleotide bases (N).

could code for sequences with a structural role, a catalytic role in metabolism, or some unanticipated function, as long as it improved overall reproductive success.

5.2.2 The RNA Replicase

Given the potential difficulty in discovering a spontaneous nucleic acid replication, we have devoted more effort to the alternative "replicase" model. Our work in this area began almost twenty years ago with an exploration of the capacity of the natural group I self-splicing introns to act as ribozymes that catalyze replication-related reactions. This early work (Bartel et al., 1991; Doudna, Couture, and Szostak, 1991; Doudna, Usman, and Szostak, 1993; Green and Szostak, 1992) produced two promising leads. First, we showed that the template-directed ligation of short oligonucleotides was a viable strategy. Second, we showed that the active ribozyme could be assembled from short oligonucleotides 30 to 50 nucleotides (nts) in length. Ultimately, however, the lack of a strong energetic driving force for the transesterification chemistry catalyzed by the group I introns limited the yield of full-length product strands, and convinced us that further progress would require the use of chemically activated substrates.

This consideration led us to develop the evolutionary methods based on combinatorial chemistry needed to explore sequence space in search of novel RNA catalysts that could polymerize activated substrates. Because the selection of ligase ribozymes is extremely straightforward compared to the direct selection of polymerases, we began by selecting for a ribozyme ligase that could catalyze the same chemistry modern protein polymerases use, for example, to attach the 3′-hydroxyl of the growing chain on the α-phosphate of the triphosphate of the incoming nucleotide (which in this case was the first nucleotide of an oligonucleotide). The combinatorial search began with a large RNA pool of 10^{15} variants with a randomized region of 220 nts. Through subsequent iterations of selection for RNA variants best able to perform the ligation reaction, we isolated a ribozyme now known as the class I ligase (Bartel and Szostak, 1993). The compact, highly structured ribozyme was unique among other ligases isolated in that selection in that it showed superb 5′-3′ (as opposed to 5′-2′) regiospecificity. The class I ligase was subsequently evolved and engineered to exhibit true multiple-turnover reactions, with a catalytic rate of $\sim$100 min^{-1} (Ekland, Szostak, and Bartel, 1995). Later work carried out in the Bartel lab at the Whitehead Institute showed that the class I ligase itself could catalyze limited sequence-specific nucleotide additions to a primer (Ekland and Bartel, 1996). The subsequent evolution of a 90-nucleotide RNA motif enabled the class I ligase to extend a primer in a sequence-specific fashion for up to 14 nucleotides, using an external primer-template complex (Johnston et al., 2001). The reaction is quasi-processive, with several nucleotide additions occurring during one primer-template/ribozyme binding event, depending on the sequence being synthesized (Lawrence and Bartel, 2003). This ribozyme represents a major advance, showing definitively that RNA can act as an RNA polymerase on supplied substrates. However, significant improvements in rate, extent of synthesis, and fidelity are required to reach the goal of generating a polymerase capable of copying its own sequence.

The class I polymerase ribozyme, and other ribozyme polymerases that have been evolved from ligases, all use nucleoside triphosphates (NTPs) as substrates. This is a significant difficulty for any protocell design that relies on the spontaneous diffusion of substrates across the cell membrane. Substrates with less polar leaving groups (e.g., 2-methyl-imidazole [Ferris, Hill, and Liu, 1996; Weimann et al., 1968] dimethylaminopyridine [DMAP; Prabahar, Cole, and Ferris, 1994] or 1-methyladenine [Prabahar and Ferris, 1997]) have all been shown to polymerize nonenzymatically on a template with good 5′-3′ regiospecificity, and should be much more membrane permeable. A polymerase ribozyme that used such substrates would not need great catalytic rate enhancement, and might therefore occur more frequently in sequence space than a polymerase that uses NTPs; furthermore, such a ribozyme polymerase could potentially operate inside membrane vesicles, with substrates supplied from the outside. Our current efforts in the directed evolution of a replicase are largely focused on this strategy. We have recently (Sazani and Szostak, unpublished) evolved a new ribozyme that extends the 3′ end of an RNA primer with pA, using 2-methylimidazole-activated AMP (adenosine monophosphate) as a substrate. Thus, it is indeed possible to evolve new ribozymes that catalyze the desired chemistry. However, the development of more selective enrichment schemes will be required to allow for the evolution of efficient template-directed polymerases that use such substrates. We are currently exploring a number of approaches to replicase evolution, including the use of different activating chemistries and di- or trinucleotides as substrates in place of mononucleotides. We are also considering the use of small molecule cofactors that might facilitate either catalysis or binding to the primer-template complex. Finally, we expect that new selection schemes involving compartmentalization will allow for the direct selection of multicomponent ribozymes that self-assemble from oligonucleotides.

5.3 The Dynamics of Simple Membranes

The other primary direction of our protocell precursor research program involves the engineering of a suitable dynamic membrane compartment. Compartments constructed from simple, single-chain fatty acids form spontaneously under appropriate aqueous conditions (i.e., low salt and slightly basic pH; see figure 5.3). Such simple membranes are themselves dynamic, as individual membrane molecules are in constant flux between the lamellar bilayer and the aqueous environment. Perturbations of the environment can result not only in the formation or dissolution of vesicles

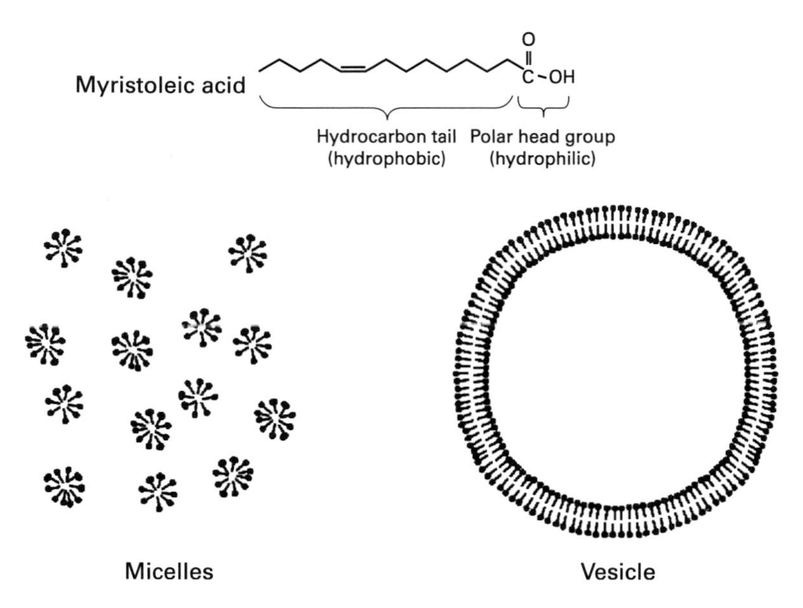

Figure 5.3
Representations of myristoleic acid showing the hydrophobic tail and the hydrophilic head group. Fatty acids can assemble into different supramolecular structures such as spherical micelles and vesicles (represented here in cross section). In the lower panels, the hydrophobic tails are represented as sticks and the head groups as balls. The head groups are facing toward the water molecules (not shown).

but also in a wide range of other dynamic behaviors including growth, shape changes, and division. Based on these principles and on previous work with both fatty acid and phospholipid vesicles, we engineered a system of vesicle growth and division. By adding myristoleate micelles to a population of preformed myristoleate/myristoleic acid vesicles (figure 5.3), we were able to observe vesicle growth. Growth occurs through the uptake of the additional micellar material into the preexisting membranes. We determined that this uptake process is efficient (up to 90% of incoming material is incorporated into the preexisting membranes) by dynamic light scattering (DLS), flow field–flow fractionation (FFF) coupled with multiangle laser light scattering (MALLS) and fluorescence resonance energy transfer (FRET). In our system, the division step was executed by extruding the grown vesicles through small 100-nm pores in polycarbonate membranes. This system of vesicle growth and division can operate for multiple cycles (Hanczyc, Fujikawa, and Szostak, 2003), and there is no obvious reason why it might not run indefinitely.

Our system of vesicle growth and division has certain properties that are suitable for incorporation into the design of a protocell. The compartments are permeable and allow the passive diffusion of water and dissolved salts and buffer. This allows the internal volume to increase as the surface area grows. However, it is not desirable for all molecules to be able to pass through the membrane. For example, RNA catalysts need to be retained within the compartment. We have shown that water and bicine buffer quickly equilibrate through the membrane, whereas larger molecules such as RNA oligomers are retained during vesicle growth and division (Hanczyc and Szostak, unpublished results). For future applications in which a rudimentary metabolism is encapsulated, the diffusion rates of metabolites become critical. For example, when energy and precursors in the form of small molecules are supplied externally to the system, such molecules must be able to pass through the compartment membrane at rates compatible with the encapsulated metabolism (Deamer et al., 2002).

Our method of division is executed through the input of mechanical energy, using an extrusion process that is efficient but artificial. No living cells today use such a process for cell division. We would therefore like to develop alternative methods for compartment fission or vesicle propagation that are more spontaneous. Such mechanisms may be based on a change in membrane curvature that can lead to fission. Membrane curvature can be modulated by the addition of external factors to the system such as electrolytes, enzymes, or molecules that interact with only one leaflet of the bilayer, thereby producing an asymmetric force across the bilayer that is relieved by fission into two or more smaller vesicles. Alternatively, spontaneous curvature leading to vesicle fission may be intrinsic to the membrane composition and effected through phase separation and lipid domain formation. All these mechanisms have elements of biological realism. For further discussion of such alternative processes of vesicle fission, see our recent review (Hanczyc and Szostak, 2004).

Finally, our system was designed to be executed at the population level. Large numbers of vesicles are grown and divided in parallel by our methodology. In a typical experiment, 10^{14} individual compartments are manipulated through multiple growth and division cycles (see figure 5.4). This approach is not limited to the manipulation and analysis of small numbers of individual vesicles, and therefore can be used to evolve large populations of replicating protocell precursors and even full protocells in which strong selective pressures can be imposed.

5.3.1 Mechanistic Considerations

Mechanistic studies of vesicle growth have revealed that fatty acid membranes exhibit surprisingly complex dynamics. When micelles are added to buffered solution (pH 8.5), they are immediately destabilized and tend to aggregate. If preformed vesicles are present, some micelles quickly associate with the vesicles, and the remaining micelles form metastable aggregates. The growth of oleate vesicles after rapid micelle addition therefore proceeds through two pathways with distinct timescales (Chen and Szostak, 2004a). The amount of material incorporated into preformed vesicles through the fast ($k \sim 3\ s^{-1}$) pathway is stoichiometrically limited by the sur-

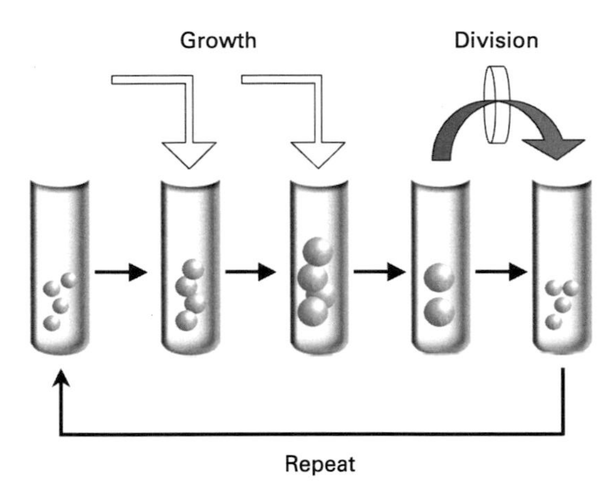

Figure 5.4
A population of vesicles manipulated through sequential growth and division cycles.

face area of the vesicles, suggesting the formation of a "shell" of micelles around each vesicle, with a micelle fatty acid to vesicle fatty acid molar ratio of ~0.4. Although both micelles and vesicles are negatively charged, correlations among counter ions can decrease the electrostatic repulsion between similarly charged surfaces (Allahyarov, D'Amico, and Lowen, 1998; Angelini et al., 2003; Butler et al., 2003; Ha, 2001; Levin, 1999; Linse and Lobaskin, 1999). Micelle-vesicle interactions might be further stabilized by the increase in entropy caused by the displacement and release of bicine anions from the outer Helmholtz plane of the membrane (Dinsmore et al., 1998; Kaplan et al., 1994). The rate-limiting step for incorporation by this pathway appears to be sodium ion (Na^+)-oleate flip-flop across the membrane. The remaining micelles form metastable aggregates that either contribute to a slower pathway of vesicle growth ($k \sim 0.1\ s^{-1}$) or assemble into new vesicles. The slow pathway for growth is probably rate-limited by the dissociation of oleate monomers from the aggregates (Chen and Szostak, 2004a; Thomas et al., 2002; Zhang, Kamp, and Hamilton, 1996).

The improved understanding of the mechanism of vesicle growth attained through these detailed kinetic experiments is important for optimizing the growth procedure for protocells in the laboratory. Because de novo vesicle formation competes for micelles in the slow pathway but not the fast pathway, the yield of growth is essentially stoichiometric for the fast addition of ~0.4 equivalents of micelles, but drops thereafter and appears to plateau beyond 1.5 equivalents. The slow, continuous addition of micelles avoids micelle-to-vesicle ratios greater than 0.4, resulting in a high yield of vesicle growth (Hanczyc, Fujikawa, and Szostak, 2003). Since electrostatic interactions are important for micelle-vesicle interactions, the buffer and ionic conditions are likely to influence growth, possibly accounting for the lack of growth observed when a borate buffer is used instead of bicine (Fujikawa and Szostak, unpublished data).

Vesicle growth is driven by the energy difference between the micelle and bilayer phases when the solution pH is close to the pK_a of the fatty acid. Can a protocell capture and store the energy released during this phase transition, making it available for cellular processes? At a molecular level, growth consists of the incorporation of fatty acid molecules into the outer leaflet of the vesicle membrane, followed by flipping of half of the molecules into the inner leaflet to maintain mass balance between the leaflets. Since the protonated form of the fatty acid is electrically neutral and can therefore cross the hydrophobic membrane more easily, most fatty acid molecules incorporated into the inner leaflet are protonated. When these reach the interior of the vesicle membrane, roughly half of these molecules will dissociate to release protons, because the solution is buffered near the pK_a of the fatty acid. The end result is the formation of a transmembrane pH gradient (figure 5.5), comparable to the chemical gradients used to store energy in modern cells (Chen and Szostak, 2004b).

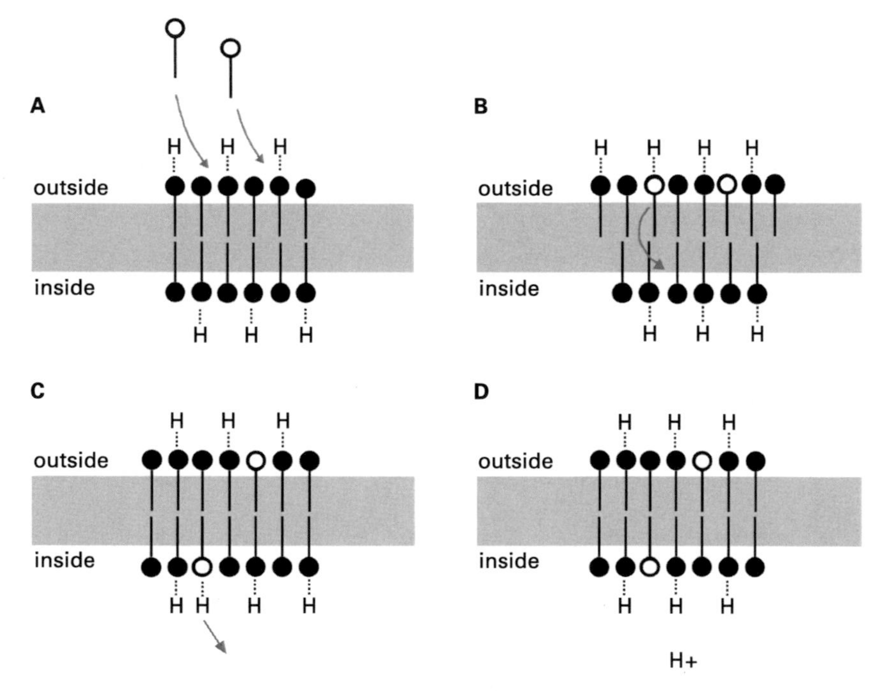

Figure 5.5
Membrane growth by micelle addition creates a pH gradient. The ball and stick represent the head and tail of a fatty acid, respectively. Newly added amphiphiles are depicted as having white head groups. The inside and outside of the vesicles are indicated. (Adapted from figure 2, Chen and Szostak, 2004b.)

This mechanism of energy transduction begins with a system that is out of thermodynamic equilibrium as soon as micelles have been added to a buffered solution, at which point they are a thermodynamically unstable phase. This energy is translated into a pH gradient by the differential permeability of the protonated and deprotonated forms of the fatty acid, capturing approximately 12% of the released free energy under conditions in which the pH gradient is maintained. Because the counterpart of cations can dissipate the pH gradient, its lifetime is determined by the permeability of the cations in solution. For fatty acid membranes, alkali metal cations such as Na^+ cross the membrane in a few seconds, most likely through the formation and translocation of an ion pair (e.g., Na^+-oleate) (Zeng et al., 1998). If other cations, such as arginine, are used instead, the pH gradient can be trapped for several hours (Chen and Szostak, 2004b). From the perspective of the cellular system, the spontaneous growth of one component, the membrane, could in principle be coupled to energetically uphill activities of other components, such as replication.

5.3.2 Integrating the Nucleic Acid and Membrane Components

Once the replication of both the membrane compartment and the nucleic acid component has been engineered independently, the two systems will be integrated to produce a population of protocell precursors that consists of a self-replicating nucleic acid polymer encapsulated within a self-replicating membrane. The superposition of these two self-replicating components (membrane and genome) does not constitute a true "cell" unless the components are coupled to form a unified system out of two independently replicating systems (Szostak, Bartel, and Luisi, 2001). The fundamental question is how to formulate a complete protocell as a unit of selection, that is, how to couple the activities of the genome and membrane such that Darwinian competition acts at the level of the cell. One possibility is to increase the complexity of the system by invoking additional genome-encoded enzymes (ribozymes), for example, catalyzing the synthesis of membrane constituents (Bartel and Unrau, 1999). We have recently found a simpler solution that relies only on the physical properties of an ionic polymer and a semipermeable membrane (Chen, Roberts, and Szostak, 2004).

In our original blueprint for the design of an artificial cell, we distinguished between two levels of organization: a protocell precursor level in which the nucleic acid genome evolved within the membrane compartment without affecting that compartment in any phenotypically relevant way, and a true protocell level in which the genomic and compartmental functions were constructively intertwined. (We used somewhat different terminology in the original blueprint.) We imagined that the evolution of distinct genome-encoded functions would be required to mark the transition from the protocell precursor level to the protocellular level. Examples of linking functions that would integrate the compartment with the genome include ribozymes that help synthesize the molecular components of the membrane from some precursor. For example, a ribozyme that synthesized the phosphate ester of a fatty alcohol, or the glycerol ester of a fatty acid, could confer membrane stability over a greater range of pH or salt conditions. Alternatively, structural RNAs might stabilize the compartment boundary by providing an internal framework (cytoskeleton), or help regulate cell division by controlling the shape of the cell.

The distinction between protocell precursor and protocell has been blurred, however, by our recent demonstration of the potential for a direct physical link between genome replication and membrane growth. While nucleotide monomers can pass through fatty acid membranes with a timescale of hours, RNA polymers as small as a decamer are essentially impermeable (Chen, Salehi-Ashtiani, and Szostak, 2005). The polymerization of anionic RNA monomers inside a vesicle increases the concentration of counter-ions associated with the backbone, causing the internal osmotic pressure and membrane tension to rise. If unstressed (e.g., empty) vesicles are also present, the membrane tension of the swollen vesicle drives transfer of membrane material from the unstressed vesicles to the swollen vesicles. Relief of this tension makes the incorporation of additional amphiphiles more thermodynamically favorable, allowing membrane growth at the expense of other nearby membranes that are under less tension. Using a FRET assay to monitor membrane surface area, we verified that membrane components were transferred from empty vesicles to vesicles encapsulating high concentrations of RNA (Chen, Roberts, and Szostak, 2004). The mechanism of transfer probably involves the dissociation and uptake of monomers by the vesicles. Thus in this scenario, genomic mutations that result in faster replication, and thus a faster increase in internal nucleic acid concentration, will lead to faster membrane growth and faster growth of the vesicle as a whole (figure 5.6). If vesicle division is either stochastic or occurs at a critical size, the average cell cycle

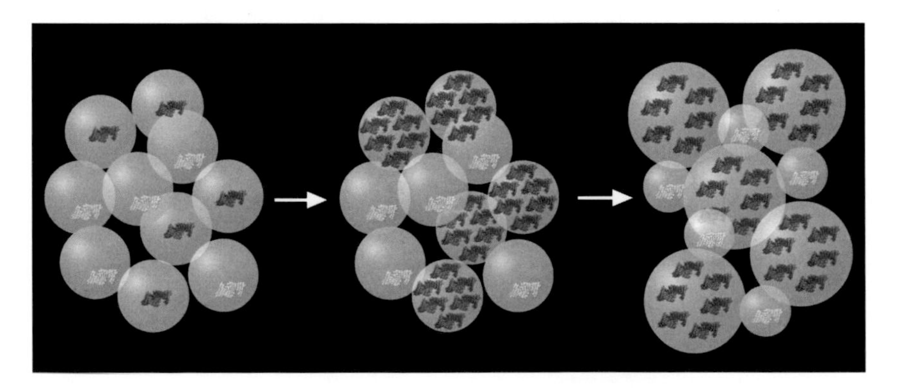

Figure 5.6
An illustration of vesicle competition. RNA replicases are shown encapsulated within spherical compartments. The dark gray molecules represent variants that are more efficient replicators.

will be shorter. This unexpected effect of mutations that primarily influence nucleic acid replication efficiency on overall cellular replication highlights the potential for an intrinsic linkage between genomic and cellular replication. Although distinct protocellular levels of organization remain possible, it is no longer necessary to postulate these as distinct, sequential steps toward the origin of life.

5.3.3 Are Fatty Acids the "Right" Membrane Material for Protocells?

Fatty acids do form surprisingly strong membranes, as measured by the maximum sustainable membrane tension (t^*). The t^* for oleic acid vesicles (10 mN/m) falls well within the range of values measured for phospholipid membranes (3–40 mN/m), indicating that they have comparable tolerance of osmotic fluctuations (Chen, Roberts, and Szostak, 2004). The dynamic behaviors of fatty acid vesicles, including growth, membrane transfer, and the buildup of pH gradients, are observable in fatty acid vesicles because of the relatively small energetic differences and fast timescales of transfer between different phases and structures (seconds to minutes). In comparison, typical phospholipids have very low critical aggregate concentrations and long dissociation and flip-flop timescales (hours to days) (Jones and Thompson, 1989; McLean and Phillips, 1981; van der Meer, 1993) due to the presence of two acyl chains and a large head group. Phospholipid membranes are also relatively impermeable to solutes that are important for enzyme activity, such as Mg^{2+} (permeability coefficient, $P_{POPC} \ll 10^{-11}$ cm/s), compared to vesicles made from fatty acid membranes ($P = 2 \times 10^{-7}$ cm/s) (Chen, Salehi-Ashtiani, and Szostak, 2005). In the absence of transporters or special machinery, high permeability to solutes like Mg^{2+} is necessary to maintain internal concentrations after multiple cycles of growth and division. However, the advantageous dynamics of fatty acid vesicles come at the expense of stability. Fatty acid vesicles are relatively sensitive to disruption by dilution and are also disrupted by low concentrations of cations, including Mg^{2+} (Monnard et al., 2002). The addition of the cognate acyl glycerol ester or acyl alcohol to fatty acid vesicles can increase membrane stability, presumably by enhancing the hydrogen bonding network (Apel, Deamer, and Mautner, 2001; Monnard and Deamer, 2003), although this addition also dampens the dynamics of such vesicles (Chen et al., submitted). Alternative amphiphiles and other combinations of components may yield even better candidates for protocell membranes, but these observations indicate that the membrane composition of a protocell will represent a compromise between stability and dynamic behavior.

5.4 The Role of Clays

Investigations into the origin of life have elucidated a role for clay minerals in the prebiotic synthesis of organic molecules and polymers (Ferris et al., 1996; Rode,

1999; Sowerby et al., 2001; Sowerby, Petersen, and Holm, 2002; Wächtershäuser, 1988). We discovered that clays also participate in the assembly of vesicles. When the acidity of a fatty acid micellar solution is dropped from pH greater than 10 to pH 8.5 near the pK_a of the carboxylate head group in the membrane, vesicles form spontaneously. This phase transition can be monitored simply as the change in turbidity of the solution over time. With dilute solutions of micelles (but still above the critical aggregate concentration), this transition proceeds slowly, on the order of minutes to hours. By adding a dispersion of mineral clay particles to the system, the vesicle formation process is accelerated as much as 100-fold (Hanczyc, Fujikawa, and Szostak, 2003). In addition, clay particles often become encapsulated within the resulting vesicles.

We have shown that clay particles with RNA adsorbed to the surface become encapsulated within the newly formed vesicles. An intriguing speculation resulting from this investigation is that mineral clays may have been involved in the formation of the first cells through both the synthesis of organic polymers and the concomitant encapsulation of these polymers in a primitive membrane. Although such a construct is not a functional cell, the product of these experimental investigations contains three essential components of life: a compartment boundary, a potential information-containing polymer, and a site for catalysis.

In constructing a protocell, mineral clays can play a role in bringing together and concentrating organic components on the surface of the particle and perhaps also provide a catalytic surface for the production of larger molecules and polymers (Ferris et al., 1996; Rode, 1999; Sowerby, Petersen, and Holm, 2001, 2002; Wächtershäuser, 1988). A diverse pool of polymers can be produced though clay-mediated synthesis and may serve as the basis of selection experiments. In addition, clays can be used for compartmentalizing adsorbed molecules. These compartments are composed of the very same fatty acids used in our growth and division protocol. It may therefore be feasible to use clay minerals to organize the encapsulation of biological molecules at the start of a protocell replication cycle. Through random segregation of internal components during vesicle fission, the clay particles are lost during successive cycles of replication, unless they can be reintroduced through vesicle fusion or perhaps through repeated internal synthesis of mineral surfaces (Mao et al., 2003, 2004; Sweeney et al., 2004; Whaley et al., 2000). This search for clues to the origin of life has led to the elucidation of the role for a nonliving mineral surface in the organization and synthesis of biological materials. Such materials and processes can be used in the present day as tools to synthesize life.

5.5 Conclusions

We have already demonstrated a functioning system of replicating vesicles, in which growth occurs by the slow addition of micelles to preformed vesicles, and division is mediated by extrusion through small pores. This is the first system in which the continuity of vesicle contents and membrane components, from generation to generation, has been experimentally demonstrated. It is highly likely that continued research will lead to the identification of other means of vesicle propagation, some of which may be more feasible from a prebiotic perspective. For example, we have recently demonstrated that, in principle, nucleic acid replication could drive growth of the membrane compartment. On the other hand, at this point we lack a functioning system of nucleic acid replication that is independent of additional complex machinery of biological origin. We are pursuing three approaches to solving this problem: identification of a nucleic acid capable of spontaneous templated replication, of small molecule catalysts that enhance spontaneous templated replication, and of a nucleic acid sequence that encodes an effective polymerase. Furthermore, we are pursuing strategies to produce spontaneous vesicle division in order to produce a dynamic container capable of spontaneous replication. If spontaneous nucleic acid replication can be achieved, this system will be incorporated into the replicating vesicle system. To construct the system as a whole, both the nucleic acid system and the vesicle will need further engineering to provide mutual compatibility. For example, the membrane may need to be modified not only to allow the diffusion of externally added substrates, but also to retain its integrity under the conditions that promote nucleic acid replication. The resulting self-reproducing system should begin to evolve spontaneously as a result of Darwinian selection for nucleic acid sequences that replicate faster and more accurately, and that interact more effectively or in novel ways with the membrane compartment. Once such integrated evolving systems have been constructed, it will of course be possible to experimentally manipulate the selective pressures experienced by the replicating protocell system. It should be possible to use the protocell as a means for evolving novel catalysts (e.g., of metabolic transformations), or of evolving novel supramolecular structures that confer enhanced stability or controllable division. The evolving populations of protocells will be amenable to analysis at many levels of description, from complete nucleic acid sequences to supramolecular compositions, bringing applied molecular evolution to molecular cellular systems.

References

Allahyarov, E., D'Amico, I., & Lowen, H. (1998). Attraction between like-charged macroions by coulomb depletion. *Physical Review Letters, 81*, 1334–1337.

Angelini, T. E., Liang, H., Wriggers, W., & Wong, G. C. L. (2003). Like-charge attraction between polyelectrolytes induced by counterion charge density waves. *Proceedings of the National Academy of Sciences of the United States of America, 100*, 8634–8637.

Apel, C. L., Deamer, D. W., & Mautner, M. N. (2001). Self-assembled vesicles of monocarboxylic acids and alcohols: Conditions for stability and for the encapsulation of biopolymers. *Biochimica et Biophysica Acta, 1559*, 1–9.

Bartel, D. P., Doudna, J. A., Usman, N., & Szostak, J. W. (1991). Template-directed primer extension catalyzed by the Tetrahymena ribozyme. *Molecular and Cellular Biology*, *11*, 3390–3394.

Bartel, D. P., & Szostak, J. W. (1993). Isolation of new ribozymes from a large pool of random sequences. *Science*, *261*, 1411–1418.

Bartel, D. P., & Unrau, P. J. (1999). Constructing an RNA world. *Trends in Cell Biology*, *9*, M9–M13.

Butler, J. C., Angelini, T. E., Tang, J. X., & Wong, G. C. L. (2003). Ion multivalence and like-charge polyelectrolyte attraction. *Physical Review Letters*, *91*, 028301.

Chen, I. A., Roberts, R. W., & Szostak, J. W. (2004). The emergence of competition between model protocells. *Science*, *305*, 1474–1476.

Chen, I. A., Salehi-Ashtiani, K., & Szostak, J. W. (2005). RNA catalysis in model protocell vesicles. *Journal of the American Chemical Society*, *127*, 13213–13219.

Chen, I. A., & Szostak, J. W. (2004a). A kinetic study of the growth of fatty acid vesicles. *Biophysical Journal*, *87*, 988–998.

Chen, I. A., & Szostak, J. W. (2004b). Membrane growth can generate a transmembrane pH gradient in fatty acid vesicles. *Proceedings of the National Academy of Sciences of the United States of America*, *101*, 7965–7970.

Deamer, D., Dworkin, J. P., Sandford, S. A., Bernstein, M. P., & Allamandola, L. J. (2002). The first cell membranes. *Astrobiology*, *2*, 371–381.

Dinsmore, A. D., Wong, D. T., Nelson, P., & Yodh, A. G. (1998). Hard spheres in vesicles: Curvature-induced forces and particle-induced curvature. *Physical Review Letters*, *80*, 409–412.

Doudna, J. A., Couture, S., & Szostak, J. W. (1991). A multisubunit ribozyme that is a catalyst of and template for complementary strand RNA synthesis. *Science*, *251*, 1605–1608.

Doudna, J. A., Usman, N., & Szostak, J. W. (1993). Ribozyme-catalyzed primer extension by trinucleotides: A model for the RNA-catalyzed replication of RNA. *Biochemistry*, *32*, 2111–2115.

Ekland, E. H., & Bartel, D. P. (1996). RNA-catalysed RNA polymerization Using nucleoside triphosphates. *Nature*, *383*, 192.

Ekland, E. H., Szostak, J. W., & Bartel, D. P. (1995). Structurally complex and highly active RNA ligases derived from random RNA sequences. *Science*, *269*, 364–370.

Ferris, J. P., Hill Jr., A. R., Liu, R., & Orgel, L. E. (1996). Synthesis of long prebiotic oligomers on mineral surfaces. *Nature*, *381*, 59–61.

Green, R., & Szostak, J. W. (1992). Selection of a ribozyme that functions as a superior template in a self-copying reaction. *Science*, *258*, 1910–1915.

Ha, B.-Y. (2001). Modes of counterion density fluctuations and counterion-mediated attractions between like-charged fluid membranes. *Physical Review E*, *64*, 031507.

Hanczyc, M. M., Fujikawa, S. M., & Szostak, J. W. (2003). Experimental models of primitive cellular compartments: Encapsulation, growth, and division. *Science*, *302*, 618–622.

Hanczyc, M. M., & Szostak, J. W. (2004). Replicating vesicles as models of primitive cell growth and division. *Current Opinion in Chemical Biology*, *8*, 660–664.

Johnston, W. K., Unrau, P. J., Lawrence, M. S., Glasner, M. E., & Bartel, D. P. (2001). RNA-catalyzed RNA polymerization: Accurate and general RNA-templated primer extension. *Science*, *292*, 1319–1325.

Jones, J. D., & Thompson, T. E. (1989). Spontaneous phosphatidylcholine transfer by collision between vesicles at high lipid concentration. *Biochemistry*, *28*, 129–134.

Kaplan, P. D., Rouke, J. L., Yodh, A. G., & Pine, D. J. (1994). Entropically driven surface phase separation in binary colloidal mixtures. *Physical Review Letters*, *72*, 582–585.

Lawrence, M. S., & Bartel, D. P. (2003). Processivity of ribozyme-catalyzed RNA polymerization. *Biochemistry*, *42*, 8748–8755.

Levin, Y. (1999). When do like charges attract? *Physica A*, *265*, 432–439.

Linse, P., & Lobaskin, V. (1999). Electrostatic attraction and phase separation in solutions of like-charged colloidal particles. *Physical Review Letters*, *83*, 4208–4211.

Mao, C., Flynn, C. E., Hayhurst, A., Sweeney, R., Qi, J., Georgiou, G., Iverson, B., & Belcher, A. M. (2003). Viral assembly of oriented quantum dot nanowires. *Proceedings of the National Academy of Sciences of the United States of America, 100*, 6946–6951.

Mao, C., Solis, D. J., Reiss, B. D., Kottmann, S. T., Sweeney, R. Y., Hayhurst, A., et al. (2004). Virus-based toolkit for the directed synthesis of magnetic and semiconducting nanowires. *Science, 303*, 213–217.

McLean, L. R., & Phillips, M. C. (1981). Mechanism of cholesterol and phosphatidylcholine exchange or transfer between unilamellar vesicles. *Biochemistry, 20*, 2893–2900.

Monnard, P. A., Apel, C. L., Kanavarioti, A., & Deamer, D. W. (2002). Influence of ionic inorganic solutes on self-assembly and polymerization processes related to early forms of life: Implications for a prebiotic aqueous medium. *Astrobiology, 2*, 139–152.

Monnard, P. A., & Deamer, D. W. (2003). Preparation of vesicles from nonphospholipid amphiphiles. *Methods in Enzymology, 372*, 133–151.

Prabahar, K. J., Cole, T. D., & Ferris, J. P. (1994). Effect of phosphate activating group on oligonucleotide formation on montmorillonite: The regioselective formation of 3′,5′-linked oligoadenylates. *Journal of the American Chemical Society, 116*, 10914–10920.

Prabahar, K. J., & Ferris, J. P. (1997). Adenine derivatives as phosphate-activating groups for the regioselective formation of 3′,5′-linked oligoadenylates on montmorillonite: Possible phosphate-activating groups for the prebiotic synthesis of RNA. *Journal of the American Chemical Society, 119*, 4330–4337.

Rode, B. M. (1999). Peptides and the origin of life. *Peptides, 20*, 773–786.

Sowerby, S. J., Cohn, C. A., Heckl, W. M., & Holm, N. G. (2001). Differential adsorption of nucleic acid bases: Relevance to the origin of life. *Proceedings of the National Academy of Sciences of the United States of America, 98*, 820–822.

Sowerby, S. J., Petersen, G. B., & Holm, N. G. (2002). Primordial coding of amino acids by adsorbed purine bases. *Origin of Life and Evolution of the Biosphere, 32*, 35–46.

Sweeney, R. Y., Mao, C., Gao, X., Burt, J. L., Belcher, A. M., Georgiou, G., & Iverson, B. L. (2004). Bacterial biosynthesis of cadmium sulfide nanocrystals. *Chemistry and Biology, 11*, 1553–1559.

Szostak, J. W., Bartel, D. P., & Luisi, P. L. (2001). Synthesizing life. *Nature, 409*, 387–390.

Thomas, R. M., Baici, A., Werder, M., Schulthess, G., & Hauser, H. (2002). Kinetics and mechanism of long-chain fatty acid transport into phosphatidylcholine vesicles from various donor systems. *Biochemistry, 41*, 1591–1601.

van der Meer, B. W. (1993). Fluidity, dynamics and order. In: Shinitzky M. (Ed.), *Biomembranes: Physical aspects* (pp. 97–158). Weinheim: VCH.

Wächtershäuser, G. (1988). Before enzymes and templates: Theory of surface metabolism. *Microbiology Review, 52*, 452–484.

Weimann, B. J., Lohrmann, R., Orgel, L. E., Schneider-Bernloehr, H., & Sulston, J. E. (1968). Template-directed synthesis with adenosine-5′-phosphorimidazolide. *Science, 161*, 387.

Whaley, S. R., English, D. S., Hu, E. L., Barbara, P. F., & Belcher, A. M. (2000). Selection of peptides with semiconductor binding specificity for directed nanocrystal assembly. *Nature, 405*, 665–668.

Zeng, Y., Han, X., Schlesinger, P., & Gross, R. W. (1998). Nonesterified fatty acids induce transmembrane monovalent cation flux: Host-guest interactions as determinants of fatty acid-induced ion transport. *Biochemistry, 37*, 9497–9508.

Zhang, F., Kamp, F., & Hamilton, J. A. (1996). Dissociation of long and very long chain fatty acids from phospholipid bilayers. *Biochemistry, 35*, 16055–16060.

6 Assembly of a Minimal Protocell

Steen Rasmussen, James Bailey, James Boncella, Liaohai Chen, Gavin Collis, Stirling Colgate, Michael DeClue, Harold Fellermann, Goran Goranović, Yi Jiang, Chad Knutson, Pierre-Alain Monnard, Fouzi Mouffouk, Peter E. Nielsen, Anjana Sen, Andy Shreve, Arvydas Tamulis, Bryan Travis, Pawel Weronski, William H. Woodruff, Jinsuo Zhang, Xin Zhou, and Hans Ziock

6.1 Introduction

The bottom-up approach we have adopted to assemble a minimal self-replicating protocell borrows ideas from a diverse set of theoretical and experimental traditions. The approach described in *Artificial Life* (Bedau et al., 2000) focuses on the *systemic and functional* properties of living systems, and this idea has from the outset been the primary driver in our protocellular design. We did not begin with a set of biochemical reactions historically associated with life and then attempt to integrate them into an artificial form of life. Instead, we first defined the functional requirements of a living system and then searched for their simplest possible physicochemical implementation (Rasmussen et al., 2003, 2004). This approach involved an iterative design process dictated by the physicochemical constraints that became increasingly apparent in our experimental studies.

Historically, the notion of a hypercycle has been a central element for theory-driven approaches to the origin of life (Eigen, 1971). Such approaches include plausible chemical model systems with autocatalytic sets of polymers as developed by, for example, Farmer, Kauffman, and Packard (1986), Kauffman (1986), Bagley and Farmer (1991); Segré, Ben-Eli, and Lancet (2000); see also chapter 11. To this mix of ideas, Wächterhäuser (1997) added mineral surfaces as catalysts for a primitive metabolism of the first self-reproducing systems. More abstract random graph generalizations of the cooperative hypercycle feedback concept for protogenetic systems were developed by Rasmussen (1985, 1988), and spatial hypercycle generalizations were developed by Hogeweg and coworkers (Boerlijst and Hogeweg, 1991). McCaskill (1993, 1997) showed that hypercycles, or other reaction-based cyclic cooperative feedback structures, are not necessary for stabilizing distributed catalysis when proximity in space is used. Thus, proximity in space itself might act as a cooperative agent. Gánti (1997) early on articulated further integration of genes, metabolism, and container as a model for minimal protocells. Ikegami and coworkers (Ono and Ikegami, 1999; see also chapter 9) developed abstract self-reproducing, computational protocells, which also include certain realistic physical properties. Inspired by von Neumann's original conjectures, Langton (1984) and Sayama (1998) proposed even more abstract self-reproducing structures.

Our protocell design also benefits from the experimental concepts for a protocell developed by Walde and coworkers (Walde, 1994; Walde, Goto et al., 1994; Walde, Wick, et al., 1994), Morowitz, Heinz, and Deamer (1988), Deamer (1997), and Monnard and Deamer (2002). But at the same time, it differs from these and other proposals in several aspects. First, our focal point is a minimal thermodynamic coupling between the three functional structures: container, metabolism, and genes. We do not start with a self-replicating container and a self-replicating gene, which are then combined. Rather, in our case, the function and reproduction of each individual component is directly dependent on the functions of the other components. Second, instead of RNA, we envision molecules that are simpler to synthesize prebiotically, such as peptide nucleic acids (PNA) (Nielsen, 1993; Nelson, Levy, and Miller, 2000; see also chapter 15) because PNA's nonionic backbone might more easily couple with the lipid layer than RNA. Third, instead of sophisticated ribozymes, we consider very simple short polymers that are capable of self-replication by means of a ligation mechanism (von Kiedrowski, 1986, 1993; see also chapter 13). Fourth, as in the other protocell proposals, we utilize lipid containment to maintain the cooperative system, but on or within the surface layer of a lipid aggregate rather than inside a vesicle. We can therefore work with smaller and more dynamic lipid structures such as micelles. Fifth, we make extensive use of the differences in the thermodynamic properties within the lipid phase compared to the water phase, as well as the lipid/water interface, and as a result we obtain quite different chemistries in these three environments. Sixth, a key component in our protocellular design is the choice of metabolic processes, which are selected from a variety of possible redox and photochemical reaction schemes and are directly coupled to the genetic and container systems.

Minimal protocells have intrinsic scientific interest, but we can also ask whether such systems have potential practical applications. One such application discussed by McCaskill (chapter 12) and by Goranović, Rasmussen, and Nielsen (2006), is that it might be possible to "program" protocells by computer-controlled microfluidics technology. We also note that the assembly of novel self-replicating systems raises a number of ethical and perhaps even religious issues, as well as safety concerns. A discussion of these issues can be found in Bedau and Parke (chapter 28).

6.2 Systemic Design Principle

6.2.1 Practical (Functional) Definition of Minimal Life

It is notoriously difficult to define minimal life in a strict sense. However, an operational definition of minimal life in this context is the integration of certain key functions localized in a container, which can metabolically transform nutrients into building blocks to grow and divide, and finally to have the capacity to undergo evolution. Thus, from a practical point of view we need to connect a metabolic system with a genetic system in a container (see figure 6.1b).

6.2.2 Systems Approach

The design principle behind our protocell is simple: *Minimize the number and complexity of the physicochemical structures for the required cooperative functionalities.* This means that our approach to a physicochemical protocell implementation from the outset has been guided by the end goal, which has guided our choice of functional and structural organization, as well as the component molecules. Because simplicity is our main goal, we are not restricted to contemporary cellular structures or building blocks. Two ideas have been important to designing a protocell:

1. We have simplified the notion of a container by allowing the metabolic and genetic complexes to operate at the external interface of a lipid (fatty acid) aggregate, as well as within the lipid itself. The genes incorporate amphiphilic peptide nucleic acid (PNA) and the metabolic molecules are hydrophobic photosensitizers, either organic dyes or organometallic structures. All of the protocell components are designed to self-assemble in water.

2. We have simplified the cooperative genetics, metabolism, and container by integrating genes as a component of a charge transfer process, which means that the genes directly catalyze the metabolic processes. The metabolism produces the container components (as well as the other building blocks), which in turn self-assemble into an aggregate, which in turn contains the metabolism and the genes as well as catalyzing the gene self-replication and the metabolic processes.

Accepting these two ideas allows us to explore protocell containers as small as a few nanometers in diameter, which is orders of magnitude smaller than most contemporary cells (see figure 6.1a).

Experimental investigations and theoretical exploration are mutually supportive in fabricating the component subsystems, as we shall see in the following sections. However, it should be emphasized that we seek to utilize computational explorations in two quite different ways. Traditionally, theoretical models and computational simulations are designed to predict the corresponding experimental systems. However, the complexity of the molecular systems we are operating with and the questions we need to address about these systems are, in general, beyond strict model predictability. Certain aspects of these systems can be modeled rather precisely, and when possible we seek to do just that. In most cases, however, we are left with another benefit of modeling. Our models clearly expose our ignorance as they force us to bring

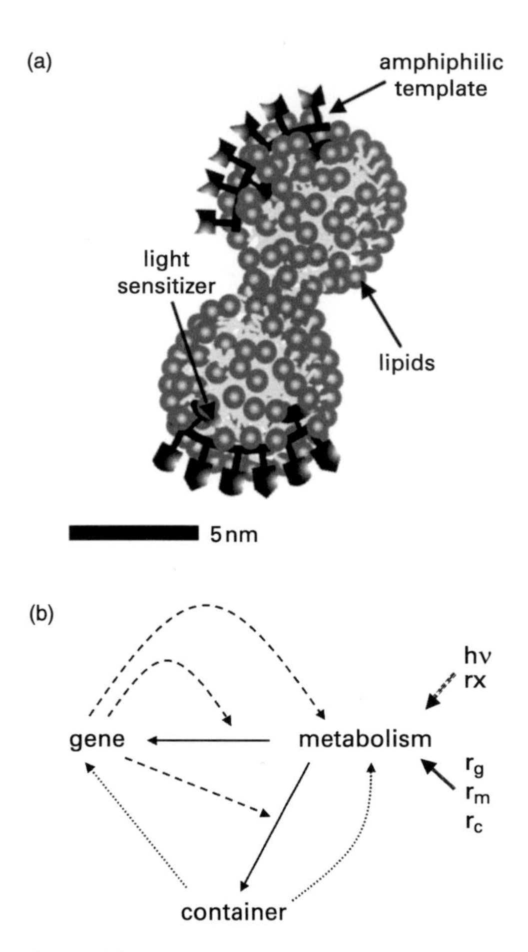

Figure 6.1
(a) The Los Alamos protocell consists of fatty acid container molecules and a peptide nucleic acid (PNA) gene with photosensitizer complexes attached to the backbone (organic dye or organometal). (b) The lipophilic inheritable replicator (PNA) catalyzes (dashed arrows) the metabolic formation (solid arrows) of both gene and container building blocks. The amphiphilic container molecules self-assemble and increase the container size. The container ensures a high local concentration as well as better thermodynamic reaction conditions (dotted arrows) of both the hydrophobic metabolic molecules and the amphiphilic replicator polymers. The necessary free energy is supplied by light in our system, but could just as well be based on redox reactions. All three protocellular components are internally synthesized or directly available from the environment: the resources for the gene monomers r_g, the metabolic complexes r_m, (r_g and r_m covalently coupled), and the container monomers r_c.

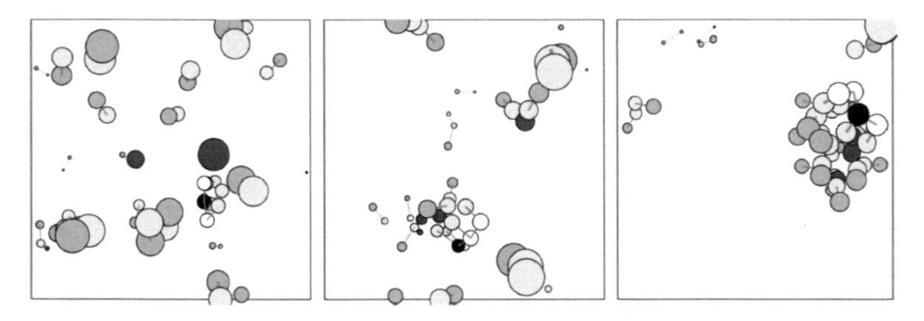

Figure 6.2 (color plate 1)
Simplified dissipative particle dynamics (DPD) simulation (Fellermann et al. 2007) of protocellular component self-assembly in water (water not shown). Lipid molecules (L) are represented as amphiphilic dimers with a hydrophilic head (green) and a hydrophobic tail (yellow), the (hydrophobic) sensitizer molecules (Z) are represented as red particles, and the gene (a 4-mer template, T) is represented by a polymer with black and white monomers, each representing different bases, and yellow hydrophobic anchors on each monomer. The self-assembly dynamics is shown starting from random initial conditions (left), initial assembly (middle), and the fully assembled protocell (right). Note how the yellow lipid hydrocarbon tails define the micellar interior while the green lipid head groups orient themselves toward the water. The hydrophobic sensitizer molecules are located in the nonpolar micellar interior, while the amphiphilic gene is located at the water lipid interface. A hydrophobic (or amphiphilic) sensitizer could also be covalently linked to the backbone of the gene (not shown in this simulation).

together what we know about the system in a coherent manner. In certain cases, modeling can also be used to qualitatively predict systemic issues and emergent properties when, for instance, subsystems are merged.

6.3 Metabolism-Genetics-Container

6.3.1 Photofragmentation: Basic Concept

For the protometabolism to operate, the photosensitizer and the catalytic gene must be located within the same container. By using a hydrophobic or amphiphilic sensitizer and amphiphilic genes, thermodynamic principles will ensure that collocation occurs. Figure 6.2 (color plate 1) shows a dissipative particle dynamics (DPD) simulation of the self-assembly of the lipid container, the sensitizer, and a gene (Fellermann et al., 2007). It should be noted that the DPD simulation is a mesoscale simulation technique that, for systems of this level of complexity, has very little predictive capability. We are utilizing the DPD simulations to uncover possible consequences of subsystem couplings as well as a test-bed. For a broad range of realistic parameters, self-assembly of the container molecules, the hydrophobic or amphiphilic metabolic center, and the amphiphilic genes can occur.

The basic metabolic processes are discussed in figure 6.3. To optimize the reaction conditions and identify the best lipid precursor and sensitizer structures suitable for the protocell, we investigated two types of photofragmentation reactions to generate

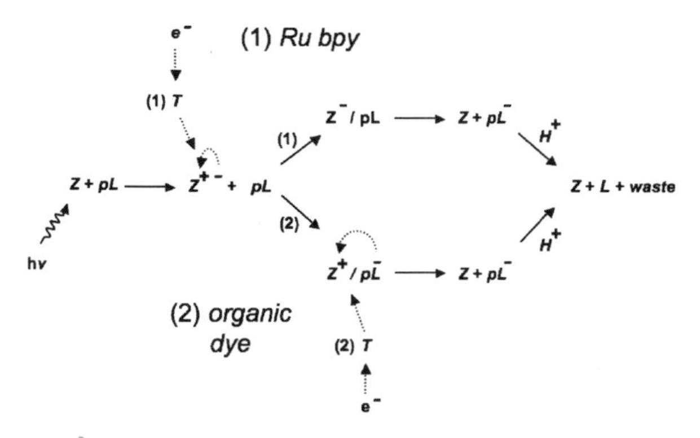

Figure 6.3
Summary of the two metabolic schemes for fragmentation of a precursor lipid (pL) into a functional lipid (L) in the presence of a sensitizer (Z) and a gene (T). After the photon (light energy) has excited Z, a charge separation, that is, an electron-hole/pair is formed on the sensitizer. In the organometal-based scheme (1), a gene (T) is introduced at this stage to donate an electron that neutralizes the positive separated charge on Z and thus prevents the back electron transfer (curved dotted arrow); if successful, the process enables the transfer of the remaining electron from Z to pL, causing the irreversible production of L. Note that the electron donation by T comes directly from T, thereby leaving a T^+, which is later neutralized by an electron from a reducing agent in the solution. In the organic dye–based scheme (2), the energy of the excited electron is sufficient to cause it to jump onto a precursor lipid, pL, at which point a gene (T) donates an electron that neutralizes the charged sensitizer and prevents the back transfer from pL^- to Z^+. It is critical that without the presence of the gene, the back electron transfer would be fairly robust since the functional lipids are otherwise formed (spontaneously) without any practical influence of the gene, T. The "/" in Z^-/pL and Z^+/pL^- indicates the close proximity of the contact ion species. To enhance the charge transfer efficiency between T and Z, Z can be covalently bonded to the backbone of T. Electrons, e^-, and protons, H^+, are supplied by the environment to ensure neutral end-products.

surfactant molecules in the organic phase, one based on organic dyes and the other on a transition metal compound. These same metabolic schemes can also generate functional gene components from precursor gene molecules because they are based on identical photofragmentation reactions.

The first reaction uses tris(2,2′-bipyridine) ruthenium ($Ru(bpy)_3$) as a sensitizer and pyridinium ester as a lipid precursor. A general discussion of the possible origin of a biological metabolism can be found in chapter 20. See chapter 21 for a general discussion of biological and protocellular energetics.

6.3.2 Ru-bpy Sensitizer Scheme

As described previously, the metabolic process that our protocell will use is composed of a reductive cleavage of a fatty acid precursor, *pL*. This process is energetically unfavorable and it requires nearly 1.0 V of reductive potential to affect release of the carboxylic acid *L*. In our scheme, the source of this energy is a suitably high-energy photon (~540 nm or shorter wavelength visible light) by a photosensitizer/catalyst, *Z*. This photosensitizer (*Z*) is a ruthenium tris(bipyridine) dication or one of its derivatives and pyridinium ester as lipid precursor (pL). The chemical and physical properties of ruthenium di-imine excited state have been the focus of extensive experimental and theoretical studies that explore, for example, the localization of the excited electron and the resulting redox properties of this relatively long-lived state (Liu et al., 2005; Wallin et al., 2005).

Figure 6.4b presents a simplified diagram of the energetics of the ruthenium tris(bipyridine) dication used as the sensitizer; the idea can be generally applied to a range of ruthenium derivatives possessing differing ligands or even differing metal complexes (Hoffman et al., 1989; Kalyanasundaram, 1982; Vlcek et al., 1995). One important derivative is that which directly links the ruthenium center to the information system so that the transfer of the electron from Z to pL can be modulated by the information system (see following). To accomplish this linkage, one of the bipyridine ligands is substituted with a link to a PNA chain (see figure 6.4a). This substitution will perturb the energetics from that depicted, but this need not be detrimental if the method of substitution is chosen carefully. The transfer of the electron from Z to pL must be modulated by the information system (T) such that the gene base sequence directly influences the rate of production of L.

The complex depicted at the bottom of figure 6.4b shows the ground-state sensitizer Ru^{II} with three neutral bpy ligands. As discussed earlier regarding the theoretical calculations, absorption of a photon results in a relatively long-lived triplet metal-to-ligand charge-transfer state. The abbreviation $Ru^{III}(bpy^0)_2(bpy^{-I})^{2+}$ designates the formally oxidized metal and reduced ligand in the excited state. This excited state will decay via several pathways including luminescence and various nonradiative pathways with a lifetime that is characteristic of the species and influenced to some extent by the solvent. For $Ru^{III}(bpy^0)_2(bpy^{-I})^{2+}$, the lifetime ranges from several hundred nanoseconds to near 1 μs (Kalyanasundaram, 1982). This duration provides a long time interval and an opportunity for transition via another channel if one is provided. In our case, the intention is to have an electron transfer from the PNA act as the other channel. Because this transfer is a relatively slow process, having a long lifetime for the $Ru^{III}(bpy^0)_2(bpy^{-I})^{2+}$ state increases the probability for the desired electron transfer to take place. The long lifetime also provides an opportunity for detailed spectrographic analysis of the processes discussed in more detail later.

The excited state has stored some fraction of the energy of the absorbed photon. For $Ru^{III}(bpy^0)_2(bpy^{-I})^{2+}$, the excited state is approximately 2.1 eV higher in energy than the ground state (Vlcek et al., 1995). It is this energy that we need to harness to drive the reductive cleavage of the pL. To make use of the stored energy, we can intercept the excited state and react via two possible pathways, following either an oxidative- or a reductive-quenching route.

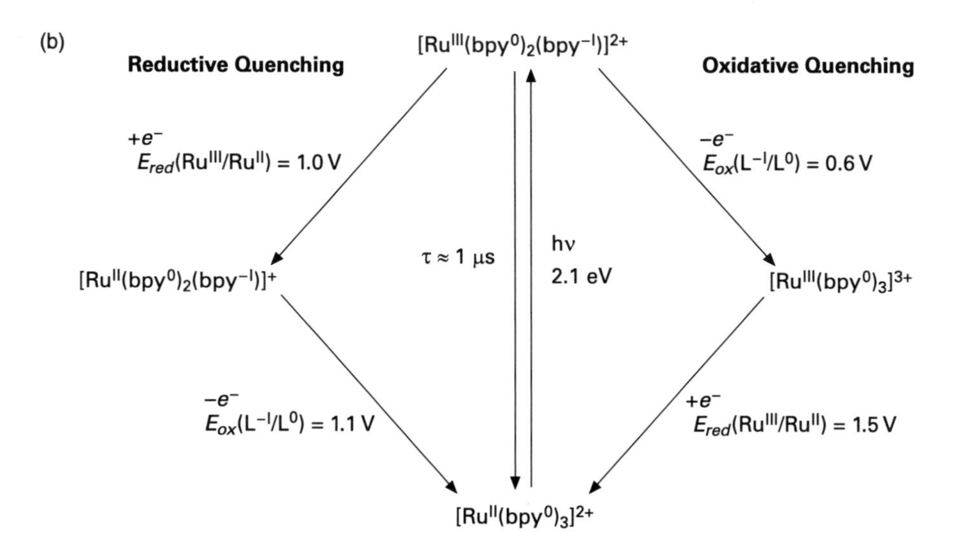

Figure 6.4
Ru bpy–based sensitizer scheme. (a) Structure of the photometabolic center. A PNA (see chapter 15) (a gene, T) is covalently linked to the Ru-bpy complex (Z). The precursor lipid (pL) is reduced through a bimolecular reaction. (b) We utilize the left-hand side of the overall reaction. The upper left part is the up-stream electron donation (from the gene), and the lower downstream part is where the photoexcited electron is donated to the precursor that fragments into a functional building block.

The oxidative pathway, shown on the right side of figure 6.4b, transfers an electron from the reduced ligand (bpy^{-I}) to the acceptor (pL) to produce the intermediate $Ru^{III}(bpy^0)_3{}^{3+}$ and the reduced pL^-, which in turn will irreversibly cleave to form the desired product L, provided no back electron transfer occurs. While this might appear to be a desirable pathway, there are a few drawbacks. First, the reductive potential of 0.6 V is insufficient to reduce pL and initiate the fragmentation, although this might be adjusted by engineering the enzyme system through choice of the ligand or even the metal complex (e.g., $Re(bpy)(CO)_2Cl$). Second, the rate of this electron transfer must compete with the rate of excited state decay to result in any pL fragmentation, which requires a high concentration (mM) of pL. Most important, the production of L proceeds before the contribution of the information system (the gene), which in essence decouples the metabolic process from the gene. For our purposes, the mechanism must proceed down the left-hand side. The low potential of 0.6 V for oxidation of the (bpy^{-I}) ligand on the oxidative quenching (right-hand) side is therefore advantageous because it precludes this route.

On the reductive quenching (left) side of figure 6.4b, the process proceeds by reduction of the Ru^{III} center giving the intermediate $Ru^{II}(bpy^0)_2(bpy^{-I})^+$. Therefore, the remaining key issues to complete our metabolism are (1) to provide an electron to accomplish the reductive quenching step and (2) to couple the process to the protogene. We call this part of the overall metabolic process the *upstream* process. Our favored solution to address these points is to design the information system with the appropriate properties to provide the electron itself; that is, in addition to the requirements of a nonperfect genome replication, the gene needs to have a reversible redox couple of the appropriate energy to quench the Ru excited state through electron donation. For this electron transfer to compete effectively with the excited state decay, we plan to covalently attach the gene to this metabolic protoenzyme, thereby removing the requirement of high concentrations.

The molecular system for the protogene construction that we have focused on is based on a PNA backbone structure (see chapter 15) in which the specifics of the nucleobases attached to the backbone (and their sequence) will ultimately determine the efficiency of the metabolic process. The identity of the bases is preferably similar to those employed in DNA for the obvious reasons of replication. However, the oxidizing potential of the Ru excited state ($\sim$1.0 V) is insufficient to oxidize any of the nucleosides (ribose + nucleobase) (Bi et al., 2005; Langmaicr ct al., 2004). The most easily oxidized standard nucleoside is guanosine at about -1.16 V (slightly less favorable than the nucleobase guanine, -1.03 V; see table 6.1). The unfavorable oxidation is reflected in the excited state quenching dynamics. The lifetime of the Ru excited state can be measured easily by monitoring the phosphorescence decay at 610 nm. Quenching is detected by a decrease in the luminescent lifetime. Ascorbate and ferrocene have both been successfully employed as reductive quenchers with oxidation

Table 6.1
Observed redox potential compared to normal hydrogen electrode (NHE) of electron donor for the Ru tris bpy complex

Electron Donor (Gene Substitute)	Redox States Involved	Reduction Potential (vs. NHE)	Reductively Quench Ru Complex?
$Ru(bpy)_3{}^{2+}$	$[Ru^{2+}]^* + e^- \rightarrow Ru^+$	+1.0	—
Ferrocene	$Fe^{3+} + e^- \rightarrow Fe^{2+}$	+0.59	Yes
Ascorbate	$Asc^{\cdot -} + H^+ + e^- \rightarrow AscH^-$	+0.282	Yes
Guanine	$G^{\cdot +} + e^- \rightarrow G$	+1.03	No
Guanosine	$Gs^{\cdot +} + e^- \rightarrow Gs$	+1.16	No
Guanosine-monophosphate	$GMP^{\cdot +} + e^- \rightarrow GMP$	+1.17	No
GT-PNA-oligomer	$G^{\cdot +} + e^- \rightarrow G$	+1.05	No
GG-PNA-oligomer	$G^{\cdot +} + e^- \rightarrow G$	+1.0	No
8-Oxoguanine	$oG^{\cdot +} + e^- \rightarrow oG$	+0.6	Yes

potentials of −0.28 V and −0.59 V, respectively. However, none of the guanine derivatives studied demonstrated any detectable luminescence quenching, which is not surprising given that all of these electron transfer reactions are endergonic. Our challenge then is to devise a strategy to generate a redox center with a low enough reversible oxidation that will undergo facile electron transfer to ruthenium. A number of potential solutions provide this lowered potential. Here, we discuss just two of these: base stacking and incorporation of a "synthetic" base.

In DNA it has been shown that short sequences of the repeated base guanine (G) result in localized oxidative damage at that sequence. It has been proposed that stacking of the bases lowers the oxidation potential through interaction of the neighboring ring systems (Prat, Houk, and Foote, 1998; Sistare et al., 2000), but evidence for this lowered oxidation potential is lacking. Our recent attempts to measure the short (two-base) stack have revealed a slight lowering (~50 mV) of the oxidation potential. It is possible that this interaction is enhanced in a duplex structure by enforcing the stacking arrangement. This has yet to be demonstrated.

The second method involves the incorporation of "synthetic" bases in the genome. For our purposes, a synthetic nucleobase is one that does not appear in naturally occurring DNA. These kinds of molecules are otherwise plentiful in biology and often have exactly the redox properties that we are interested in exploiting. For example, nicotinamide (from nicotinamide adenosine dinucleotide, NAD) and riboflavin have readily accessible reversible redox couples; however, they do not have an obvious complement for base pairing purposes to allow a replication mechanism. Fortunately, there is a derivative of guanine that might satisfy all of these requirements; 8-oxoguanine is a naturally occurring oxidation product of guanine. (It is

Table 6.2
Downstream Ru bpy sensitizer reduction of precursor lipid

Entry	Lipid Precursor (Anion)	Irradiation Time	Photosensitizer	Precursor Present after Photolysis	Products present after Photolysis
1	I^- (Iodide)	24 hours	None used	Yes	Yes—benzene, picolinium byproduct and acetic ion
2	$CF_3SO_3^-$ (Triflate)	12 hours	Comarin 6 [stoichoimetric]	Yes	No products observed (unreacted)
3	$CF_3SO_3^-$ (Triflate)	2 hours	$Ru(bpy)_2Cl_2.2H_2O$ [stoichiometric]	No	Yes—benzene, picolinium byproduct and acetic ion
4	$CF_3SO_3^-$ (Triflate) and ascorbate [stoichiometric]	12 hours	$Ru(bpy)_3Cl_2.6H_2O$ [catalytic]	No	Yes—benzene, picolinium byproduct and acetic ion
5	$CF_3SO_3^-$ (Triflate) and ferrocene [stoichiometric]	2 hours	$Ru(bpy)_3Cl_2.6H_2O$ [catalytic]	No	Yes—benzene, picolinium byproduct and acetic ion

Note: All reactions are performed in depurated dichloromethane as solvent and are degassed with nitrogen prior to irradiation.

worth noting that biology has devised a process to enzymatically remove this damaged base from DNA strands.) It is known to base pair with cytosine in a similar fashion to guanine (but with a higher propensity to mispair) and has an oxidation potential of only −0.6 V. These properties might lend themselves nicely to our construction of the protogene in which the electron transfer energetics are viable (see table 6.1) and rate modulation might be influenced by the base (genetic) sequence that links the electron source (8-oxoG) to the ruthenium center, hence providing a mechanism for evolutionary control.

In the final part of the overall metabolic process, which we might call the *downstream* portion, the product $Ru^{II}(bpy^0)_2(bpy^{-I})^+$ has a reductive potential of 1.1 V. This potential can reduce pL at a leisurely rate based on diffusion without requiring large concentrations and giving our desired product, L (see table 6.2). The efficiency with which the reductive quenching step competes with all other (nonproductive) excited state deactivation processes will ultimately determine the yield of L, and therefore offers the potential for evolutionary control of the metabolic process.

We can investigate separately the upstream gene-controlled electron donation process as well as the downstream irreversible precursor fragmentation. In figure 6.5, we

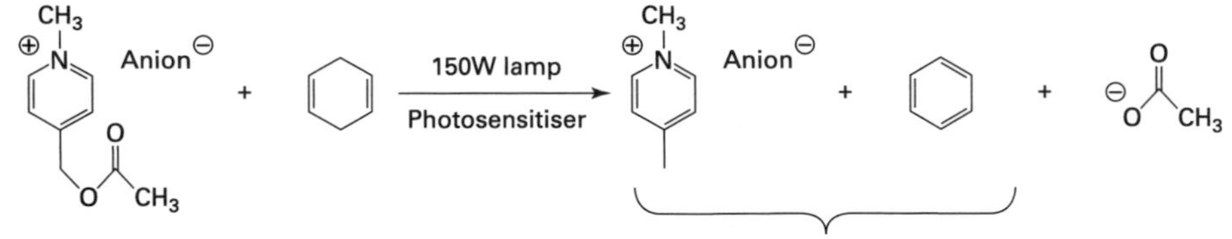

CH_3 Anion + 150W lamp Photosensitiser CH_3 Anion + + O CH_3

N-Methyl picolinium ester salt (Lipid precursor)

Hydrogen radical donor (1,4-cyclohexadiene)

Photochemical by-products (*N*-Methyl picolinium salt and benzene)

Acetate ion ("fatty acid" component)

Figure 6.5
Schematic of a systematic investigation of the downstream reductive production of fatty acid from precursor pyridine complex using ferrocene as the upstream electron donor and as a surrogate for the gene.

document a systematic investigation of the downstream reductive production of fatty acid from precursor pyridine using ferrocene as the upstream electron donor and as a surrogate for the gene. Sundararajan and Falvey (2004) have shown that *N*-methyl picolinium methyl ester iodide directly undergoes photorelease in the absence of a sensitizer or radical donor to afford high yields of the acetate ion when irradiated with a high-energy mercury or xenon lamp source. This reaction occurs as the picolinium/iodide ion pair exhibits a charge-transfer band, which, on irradiation, results in the iodide ion serving as the electron donor source. With the aim of developing milder conditions to produce the same transformation, we investigated whether a low-energy light source would be sufficient.

The direct reaction of the precursor lipid *pL* is shown in entry 1 of table 6.2, using a commercial 150 W/120 V visible light source. Analysis of the reaction mixture after a 24-hour period by proton NMR spectroscopy revealed only trace amounts of acetate ion and photochemical byproducts.

A second possible reaction was tested with the same light source using the *N*-methyl picolinium methyl ester triflate salt and a separate photosensitizer, Coumarin 6 (entry 2), in stoichiometric amounts, but also proved unfruitful, giving essentially unreacted methyl ester. As described previously, tris(bipridine)ruthenium complexes are known to absorb in the visible range, yet more important, they have the necessary reductive strength to release the acetate ion.

Including a stoichiometric amount of *cis*-dichlorobis(2,2′-bipyridyl)ruthenium (II) dihydrate (entry 3) under the conditions described for the previous entries, provides the first photorelease reaction with complete conversion of the *N*-methyl picolinium methyl ester triflate salt. In this case, the oxidative pathway (right-hand side of figure 6.4b is followed as the $Ru(bpy)_2Cl_2$ excited state has sufficient reducing power (1.1 V) to reduce the picolinium ester (Vlcek et al., 1995). Close examination of the proton NMR spectrum shows signals that can be attributed to the formation of photochemical byproducts and the desired acetate ion. Entry 3 gives the results when an

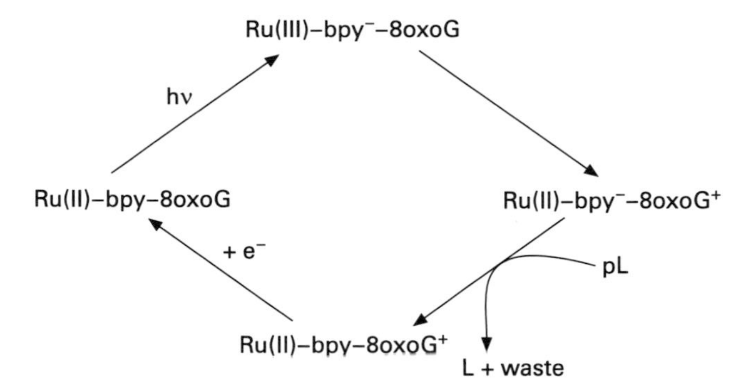

Figure 6.6
Summary of a protocellular metabolic cycle. This system demonstrates how an informational molecule (8-oxoguanine) can regulate the metabolic production of container components and establish the desired catalytic connection between genes, metabolism, and a container. The production of fatty acids from N-methyl picolinium precursors is observed both for covalently linked "informational molecule" and in solution. The process starts with the absorption of a photon that causes a charge separation within the Ru-bpy sensitizer molecule. Because the gene molecule (in this initial case, the 8-oxoguanine base) donates an electron, the sensitizer center is neutralized and the energy-rich electron can cleave the picolinium ester (resource molecule) to form fatty acid and waste. Finally, the "ground state" of the gene has to be regenerated. For details, see DeClue et al. (2007).

equivalent of sensitizer is employed per molecule of *N*-methyl picolinium methyl ester triflate salt; this result could essentially be explained as a lower-energy light version of the direct mechanism of Falvey and coworkers (Sundararajan and Falvey, 2004) for the reductive photocleavage reaction. The entry 3 results suggest that the ruthenium complex is acting as both sensitizer and electron donor.

To further expand this photochemistry, we seek to determine whether the reaction could be driven by an external electron donor (i.e., genetic material from an upstream process) and a catalytic photosensitizer, which are desirable for the protocellular system. Entry 4 indicates that ferrocene is sufficient as an external electron donor, initially providing the electron to follow the left-hand side of figure 6.4b, and catalytic tris(2,2′-bipyridine)ruthenium (II) chloride hexahydrate is employed as the photosensitizer to reductively photocleave *N*-methyl picolinium methyl ester triflate salt, as shown for entry 3.

The experimental results from entries 3 and 4 are encouraging. As we exchange the upstream electron donation from the gene surrogate, ferrocene, to 8-oxoguanine, the downstream photofragmentation process still works under the prevailing conditions. 8-Oxoguanine Ru-bpy provides a catalytic production of fatty acids, both as the 8-oxoguanine is covalently linked to the Ru-bpy complex, and as a bimolecular reaction in solution (DeClue et al., 2007) (see figure 6.6). These results open new avenues where we can study a multitude of parameters, such as different solvents, pH, different lipid precursors, alternative hydrogen radical donors, vesicle production, and so on, and thereby improve and optimize this process that is essential for the protocellular system. The next major challenge is to incorporate the 8-oxoguanine as a PNA monomer into a different PNA strands and investigate the sequence-dependent catalytic properties, as the possible basis for a later protocellular selection mechanism.

6.3.3 Organic Sensitizer Scheme

Our second metabolic reaction scheme based on organic dyes utilizes stilbene and aromatic amine as the sensitizers (Z) and phenacyl esters as the lipid precursor (pL). As shown in figure 6.7, we have established that phenacyl esters (precursor lipids, pL) can undergo photo-induced C-O bond scission to form acetophenone and the corresponding carboxylic acid (R-COOH, surfactant L) in the presence of stilbene

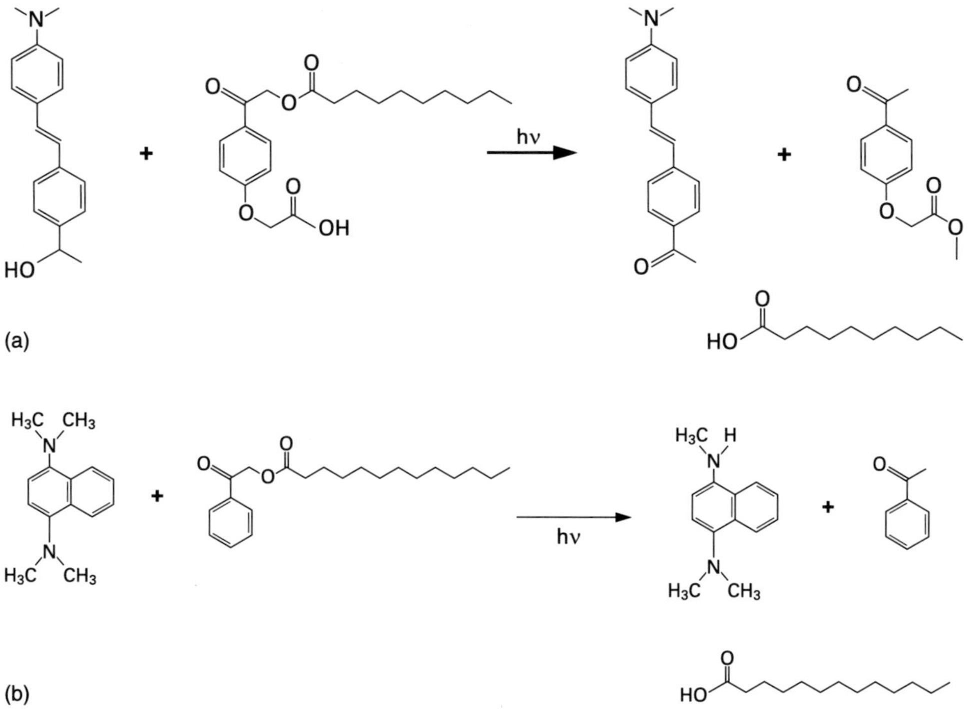

Figure 6.7
Fragmentation of a phenacyl ester or acid (pL) produces a fatty acid and "waste" (acetophenone) when (a) stilbene or (b) an aromatic amine (commercially available) is used as the photosensitizer (Z) (in later experiments the acid group at the phenacyl precursor can be used to further functionalize the precursor as necessary). The stilbene is sufficiently quenched at ~450 nm (visible light) whereas the aromatic amine is quenched at ~380 nm (slightly ultraviolet). Recall figure 6.2 and see the charge transfer calculations in figure 6.8.

or various aromatic amine sensitizer molecules (Z), similar to the results reported by Falvey's group (Banerjee and Falvey, 1997; Lee and Falvey, 2000; Sundararajan and Falvey, 2005).

In brief, a stilbene sensitizer is made by a samarium-mediated reductive coupling of the corresponding stilbene ketone, which is prepared from a Wittig reaction. Stilbene sensitizer has an absorption spectrum peak at 360 nm. The fluorescence of the stilbene sensitizer, at approximately 450 nm, is found to be efficiently quenched by the precursor molecules with a Stern-Volmer quenching constant (Mouffouk and Chen 2006, in preparation) of $1.5 \times 10^4\ M^{-1}$. A quenching rate of $5.7 \times 10^{10}\ M^{-1}\ s^{-1}$ is calculated based on the fluorescence lifetime of 0.2 ns, measured via a single-photon counting experiment, indicating the efficient electron transfer reaction between the sensitizer and precursor molecules. Consequently, we can follow the reaction by both ^{1}H-NMR, UV-Vis and fluorescence spectroscopy.

The charge separation between the sensitizer and the precursor lipid is believed to occur according to the quantum mechanical simulation in figure 6.8 (Tamulis et al., 2008), in this case with another organic sensitizer, an aromatic amine sensitizer. Once the electron jumps to (and stays at) the phenacyl ester it will cause the breakage of the C-O bond. However, a back electron transfer reaction can negate the charge separation initially by neutralizing the positive charge still on the sensitizer,

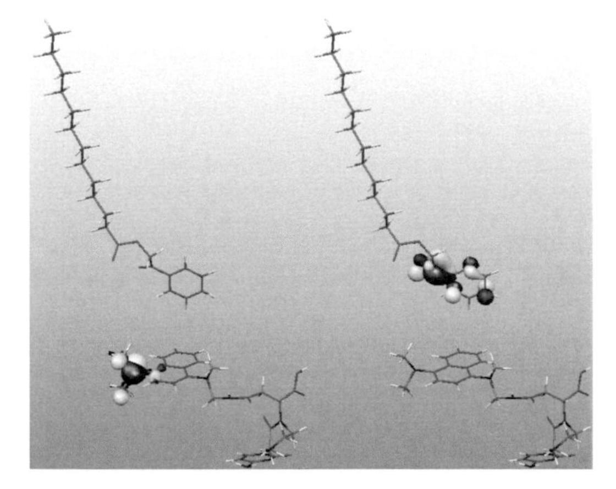

Figure 6.8
Simulation results for the electron charge transfer, π-π^*, transition associated with the first excited state, which occurs from the aromatic amine sensitizer 1,4-bis(N,N-dimethylamino) naphthalene molecule (recall figure 6.7), highest occupied orbital (HOMO) (lower, left) to the lipid precursor molecule's lowest unoccupied molecular orbital (LUMO) (upper, right) in a water solvent (not shown) (Tamulis et al., 2008). Note how the energy-rich electron initially localized at the sensitizer jumps to the precursor lipid and resides at the ester bond, which eventually breaks (not shown). The sensitizer is covalently bounded to a PNA backbone (only one PNA monomer shown). The charge transfer is calculated by the quantum mechanical TD-DFT PBEPBE/6-31+G* method.

as discussed in figure 6.3. To prevent this back reaction, an information molecule (a gene) could donate an electron to neutralize the sensitizer, leaving the excess electron on the lipid precursor and thereby allowing the fission of the phenacyl ester bond to proceed.

As indicated earlier, a direct autocatalytic feedback between the PNA protogenes and the production of lipid molecules could be implemented by using a modified PNA as a photocatalyst. Since the quantum yield of photolysis of phenacyl esters normally is very low (~0.01–0.05) because of the fast back electron transfers, a sensitizer coupled with an electron relay system could block the back electron transfer process, and thus increase the quantum yield of the surfactant production, just as discussed in the previous section, where we utilize the Ru-bpy sensitizer complex. Theoretical investigations based on a reaction kinetic analysis of this metabolic scheme (Knutson et al., 2008) indicate that the rate-limiting reaction is the photoexcitation process, which is of the order of one excitation per sensitizer molecule per second. To enhance the overall metabolic efficiency, the gene electron donation rate has to be of the same order of magnitude as the back reaction rate. The back reaction is estimated to be of the order of 10^4 electron transfers per sensitizer molecule per second.

6.3.4 Metabolic Flexibility

We plan to incorporate the same sensitizers (both $Ru(bpy)_3$ and stilbene molecules) to the PNA backbone and use an appropriate sequence of nucleobases as the electron relay system. In the presence of PNA with an appropriate base sequence and sensitizer attached, photo-induced electron transfer reaction occurs between the sensitizer and the phenacyl ester; consequently, the sensitizer cation radical can receive an electron from guanine (8-oxoguanine) forming a guanine cation radical. Note that the surfactant molecules can be synthesized only when the particular base sequence is present. As a result of the increased production of surfactant molecules, the associated lipid aggregate (a micelle or a vesicle) should grow, becoming unstable, and eventually divide.

6.4 Genetics-Container-Metabolism

6.4.1 Basic Concept of Amphiphilic Gene Replication

Genes are essential structures in our protocell because they provide both metabolic catalysis and inheritance. An imperfect replication of the metabolic catalyst also provides variation essential for selection, thereby enabling Darwinian evolution.

A review of prebiotic gene synthesis, nonenzymatic gene replication, and theoretical studies on the origins, organization, and dynamics of primitive gene dynamics is beyond the scope of this chapter. For a discussion of the possible molecular synthesis of genes, see chapter 26; for a review of the RNA world perspective, see, for example, Joyce and Orgel (1999) and chapter 17; for a discussion of simple replicators see, for example, Paul and Joyce (2002) and chapters 13 and 15. Experimental container-gene integration is discussed in chapter 5, and a discussion about container-gene-metabolism integration can be found in chapters 2 and 3. Later in this section we summarize a theoretical discussion of gene-container replication kinetics, and further discussion can be found in Rocheleau and coworkers (2007; see also chapter 14), while the coupled gene-metabolism-container replication kinetics are covered in Munteanu and coworkers (2007). Figure 6.9 presents the key steps in the protocellular gene replication dynamics.

6.4.2 PNA-Partition into Nonpolar Solvents at Water-Lipid Interface

It is generally recognized that standard PNA is fairly soluble in water and insoluble in octanol or other nonpolar liquids. PNA does not penetrate phospholipid membranes, so it can be contained in liposomes for an extended time (Wittung et al.,

(a)

R = Phe Val Leu

Peptide nucleic acid (PNA) with lipophilic backbone

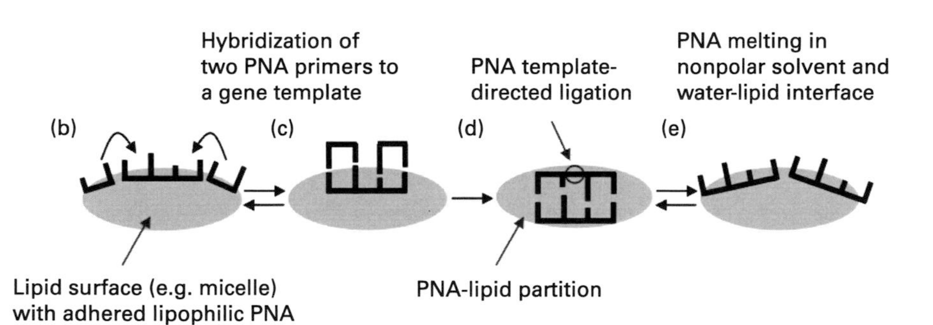

Figure 6.9
Picture of gene replication in a lipid aggregate surface. We use amphiphilic genes to ensure a close gene-container connection because amphiphilic genes will attach themselves to the surface of the lipid aggregates in water. The genes are backbone-modified PNA strings, (a), with attached hydrophobic peptide side groups as found in the analog to phenylalanine, isoleucine, or valine amino acid monomers. At the onset of the gene replication cycle (b), a template and, at a minimum, two complementary gene fragments that together complement the template strings are anchored at the water-lipid interface. Thermodynamics drive the template and the two smaller strings to hybridize (c) as a result of base recognition. The resulting three-component gene duplex is more hydrophobic than the individual gene pieces, because of its hydrophobic backbone and now-shielded polar sides, and the duplex will therefore tend to sink deeper into the nonpolar lipid environment. The nonpolar environment will enhance the ligation process (amide or peptide bond formation), possibly in the presence of 1-[3-(dimethylamino)propyl]-3-ethylcarbodiimide hydrochloride (EDC). The eventual dehybridization (e) of the two PNA strands could occur through a gentle temperature cycle, or, if the system is residing near the melting temperature, through a random thermodynamic fluctuation, although the latter also has implications for the melting of the unligated oligomers.

1995). However, molecular dynamics (MD) simulations indicate that even standard PNA at high concentrations and low ionic strength tends to accumulate at phospholipid interfaces (Weronski, Jiang, and Rasmussen, 2007) as can be seen in figure 6.10a. The driving force for this affinity is believed to be a slight free energy minimum at the interface as depicted in figure 6.10b.

Because the PNA backbone can be decorated with lipophilic amino acids, a PNA string can be made arbitrarily hydrophobic. However, it is not yet clear what appropriate decoration means in terms of creating an amphiphilic gene with the desired properties as discussed in section 4.1. Lipophilic PNAs, PNA amphiphiles (PNAA), are known to form micellar aggregates in water (Vernille and Schneider, 2004). The issues associated with PNA solvation, as well as PNA hybridization and melting in nonpolar environments, are further discussed in chapter 15.

6.4.3 PNA Template-Directed Ligation and Replication

Preliminary template-directed PNA ligation and replication experiments are demonstrated in water (Sen and Nielsen, 2006). The overall reaction consists of the steps presented in figure 6.9, but without the presence of lipid aggregates, and assisted by EDC-activated PNA oligomers. As predicted, we have issues with product inhibition in these experiments (von Kiedrowski, 1986, 1993; see also chapter 13).

DPD simulation studies of the lipophilic PNA template-directed ligation at lipid aggregate interfaces predict that the hybridization process is strongly enhanced by using micelles loaded with oil (lipid precursor), as a result of enhanced lipid aggregate stability (Fellermann et al., 2006). Furthermore, these DPD simulation studies indicate that as the hydrophobicity of the PNA backbone is strengthened, it increasingly hampers the standard PNA hybridization process because of competition between the free energies associated with the hybridization process and the hydrophobic anchoring. A DPD simulation–based discussion of the container-associated gene replication process is shown in figure 6.11 (color plate 2).

An interesting question is whether it is possible to eliminate the EDC for the amide bond preactivation. For that to be possible, the conditions for the amide bond formation reaction have to be enhanced as much as possible. Amide bond formation follows the reaction R-COOH + H_2N-R → Transition States → R-CO-NH-R + H_2O, where the equilibrium constant for this reaction is given by

$$K = \frac{[R - CO - NH - R][H_2O]}{[R - COOH][H_2N - R]}.$$

It is clear that an environment with a low water concentration and a very high concentration of the reactants will favor the forward reaction (i.e., bond formation). Locally, the R-COOH and H_2N-R concentration is very high as a result of the jux-

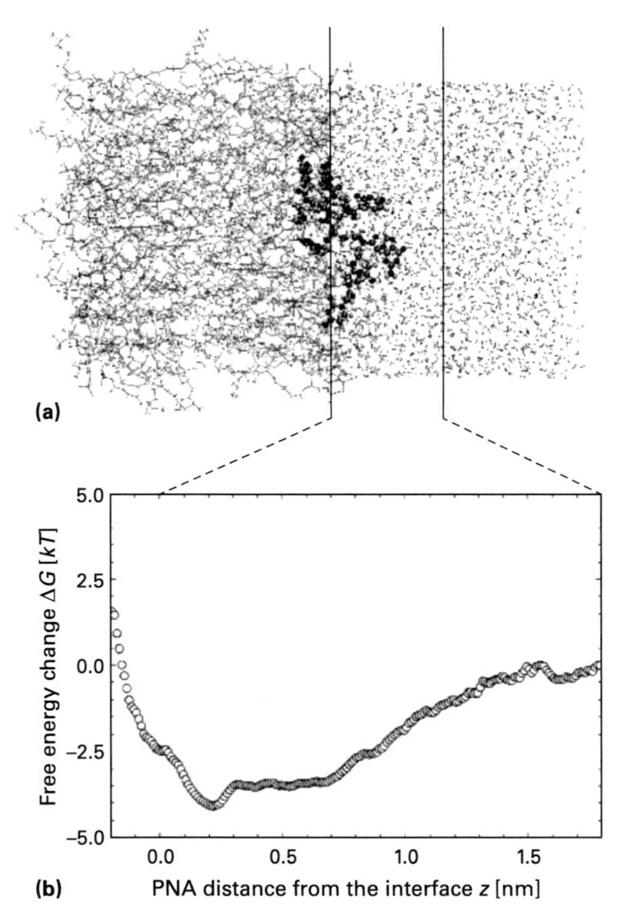

Figure 6.10
(a) Standard PNA molecule with glycine backbone (protein data bank identifier 1PUP) at the lipid-water interface. The lipid bilayer is composed of POPC. The picture presents the system at equilibrium (after about 350 ns). The left and right vertical lines denote the coordinates $z = 0$ and $z = 1.8$ nm, respectively. Simulations are conducted using NAMD (Kale et al., 1999) and VMD (Humphrey, Dalke, and Schulten, 1996) packages. (b) Profile of the free energy and average force acting on the PNA string near the membrane-water interface. A free energy minimum is seen at the membrane-water interface, which predicts a PNA-membrane attraction.

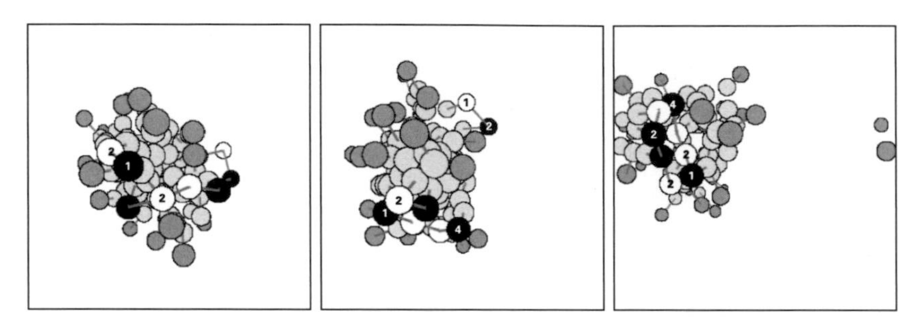

Figure 6.11 (color plate 2)
DPD simulation of the three steps of template-directed (gene) ligation at a micellar interface (water not shown), where the micelle is loaded with oil-like (lipid precursor) molecules. The amphiphilic fatty acids are represented as green and yellow dimers (green = hydrophilic, and yellow = hydrophobic), and the oil-like dimers consist of two yellow beads. The simplified lipophilic PNA gene template consists of four monomers, ABBA. The A base component monomer is white, whereas the B base component monomer is black. This template can hybridize with two small PNA dimers consisting of BA and AB sequences (monomer A base pair with monomer B). The hydrophobic component of the PNA backbone is modeled by a (yellow) hydrocarbon tail attached to each PNA base component monomer. For details, see Fellermann et al. (2007). In the left panel, the three pieces of PNA are associated to the lipid aggregate interface. In the middle panel, one of the dimers is hybridized, and in the right panel, both dimers are hybridized and the ligation process occurs.

taposition of these two groups because of the templating. If the ligation could occur in the lipid phase with little water, it should favor amide bond formation. However, preliminary *ab initio* calculations for the amide bond formation using the density function theory (DFT) method (Zhou, 2006) indicate that the activation barrier is higher in a nonpolar (oil) environment than in a polar (water-rich) environment. The quantum calculations also indicate that the thermodynamic properties are similar in the two media, and free energy differences between the bonded and nonbonded states are negligible.

6.5 Container-Metabolism-Genetics

6.5.1 Container Loading and Replication

One of the essential structural elements of a protocell is the container. Its key functions are to localize the genetic and metabolic elements within its boundaries, to permit an accumulation of reactive resource species and metabolic products, and to promote metabolic processes by offering favorable physicochemical environments.

Over the years, different types of containers have been envisioned: coacervates composed of a mixture of arabic gum (Oparin et al., 1976; see also chapter 1), histones or RNA fragments, proteinoid spheres composed of amino acid polymers (Fox and Dose, 1977; see also chapter 1), and amphiphilic self-assembled structures (Apel, Mautner, and Deamer, 2002; Hargreaves and Deamer, 1978; Monnard et al.,

2002; Walde, 1994; Walde, Wick, et al., 1994; see also chapters 2, 3, 5, and 18). Among all possible amphiphilic structures, liposomes and emulsions are the structures most frequently used to design chemical models of protocells because these containers are biomimetic (Karlsson et al., 2004; Monnard, 2003; Noireaux and Libchaber, 2004; Tawfik and Griffiths, 1998; Yu et al., 2001; see also chapters 2, 3 and 5). However, other amphiphilic structures such as micelles could fulfill the requirements of our system.

Our protocell model has the following requirements for a container: It must (1) self-assemble and (2) remain stable enough to contain the metabolic and genetic processes while (3) being dynamic enough for growth and division. Finally, it must (4) be composed of molecules that have precursors which are nonmiscible with water and do not build structures on their own. The choice of amphiphile is therefore crucial. Protocellular models based on liposomes have usually been composed of two types of amphiphiles: double-chained phospholipids or single-chained fatty acids. Phospholipids tend to form very stable structures (Oberholzer, Albrizio, and Luisi, 1995) but their chemical synthesis from simple precursor molecules is rather difficult (Oro et al., 1978). The stability of fatty acid structures is restricted to well-defined ranges of environmental conditions, such as pH (Apel et al., 2002) or ionic strength (Monnard et al., 2002), but simple precursor molecules such as anhydrides (Bachmann, Luisi, and Lang, 1992; Walde, 1994) and phenacyl esters are relatively hydrophobic. In addition, fatty acids also form micelles, and these might be good alternatives to liposomal containers.

For the container to grow and reproduce, fatty acid surfactants must be generated metabolically in situ (inside a micelle or container membrane), thus sensitizer molecules, precursor molecules, and genetic material must be incorporated within the lipid structures. This can be achieved by photosensitizers that absorb light energy to initiate the photofragmentation reactions of the precursor molecules, which in turn are catalyzed by the genes. A DPD simulation of these processes is illustrated in figure 6.12.

Micellar or micellelike systems might offer small, more dynamic containers that would facilitate a rapid growth and an autonomous division. However, these two processes can be observed only at the macroscopic (population) level. By contrast, it is possible with a microscope to observe the growth and division processes of individual vesicle containers, and eventually this would give more detailed insights.

6.5.2 Aggregate Association of Metabolic Complex

Not surprisingly, the positively charged Ru-pby complex associates nicely with negatively charged fatty acid vesicles as demonstrated in figure 6.13 (Monnard et al., 2007).

The interactions between the Ru-complexes and the fatty acid vesicles are probably mostly electrostatic, but as figure 6.13 shows, a significant amount of Ru-bpy is

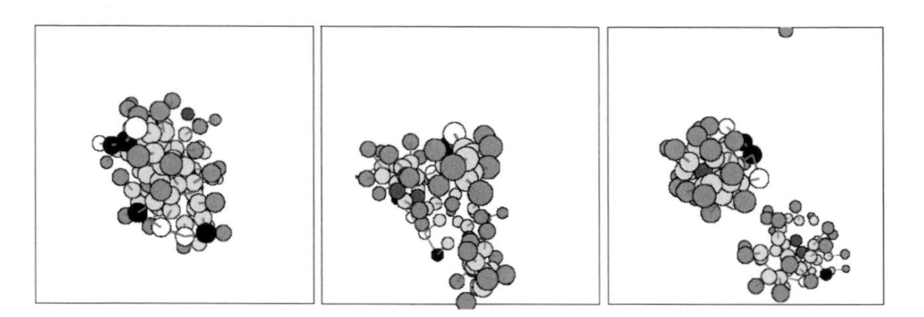

Figure 6.12 (color plate 3)
DPD simulation of the protocellular division starting from preloaded micelle, free photosensitizers, and two dehybridize, single-stranded genes (left panel). For details, see Fellermann et al. (2007). Water molecules are not shown, the lipids are represented as green-yellow (head-tail) dimers, the photosensitizers are red, and the oily precursor lipids are yellow-yellow dimers. As the oily precursors are "digested" and transformed into fatty acids driven by the sensitizers and catalyzed by the genes, the micellar size exceeds its equilibrium size and becomes unstable (middle), and it eventually splits into two micelles (right). Note the nice partition of sensitizers and gene molecules in this particular simulation, which does not always occur. Further it should be noted that reloading ("feeding") the micelles with the oil-like precursor lipids is nontrivial. Simulation studies show that a delicate balance between feeding and diffusion rates is necessary to prevent the formation of large oil droplets that can absorb multiple micelles and thus destroy their individuality.

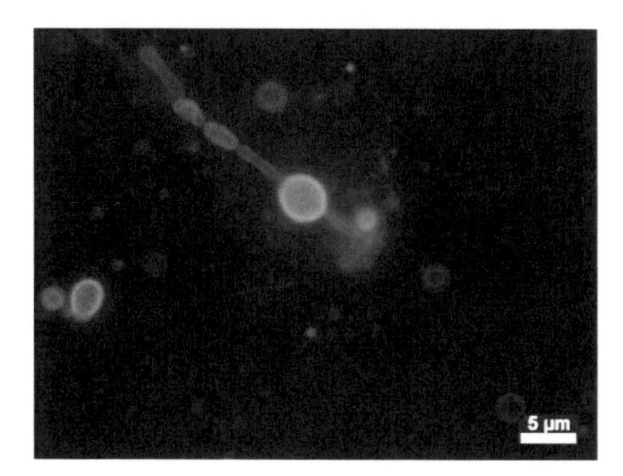

Figure 6.13
Metabolic Ru-bpy complex associated with fatty acid vesicles: 60 mM decanoic acid, 100 mM PO buffer, pH 7.0, where vesicles are made by pH vesiculation and 0.5 mM $Ru(bpy)_3(PF_6)_2$. Note the clear metabolic complex–fatty acid aggregate association. For details, see Monnard et al. (2007).

still solvated in the water phase. It should be noted that the Ru complex does not dissolve/partition in octanol or isooctane, but does in aqueous suspensions of fatty acid, as shown in figure 6.13. To enhance the association of the metabolic complex with the fatty acid container, hydrophobic ligands could be attached to the Ru-bpy structure.

6.5.3 Tracking Individual Fatty Acid Containers

A major challenge for protocell container research is to monitor the stability, dynamics, and replication processes of the containers (see chapter 7). In the past, such studies have been carried out with bulk suspensions of containers (e.g., light scattering and other diagnostic measurements of container systems). However, for research on protocell growth and division, bulk-phase samples have significant limitations, and a preferred approach would be to investigate the structure and dynamics of single containers. This approach would allow the growth and reproduction processes of protocells to be monitored on the single (proto)cell level. We have begun developing strategies for such studies, focusing, for example, on vesicle systems formed from amphiphiles.

The basic approach we use is to tether containers to surfaces using biomolecular recognition strategies. Briefly, we prepared oleic acid assemblies in buffered solution at pH ≈ 8.5. Under these conditions, oleic acid spontaneously forms large vesicles. We incorporated a small amount of fluorescent label, which allowed us to visualize them with a standard fluorescence microscope. Finally, to this mixture, we also added a small percentage of biotin-labeled phospholipid, which provides a means of attaching the vesicles to suitably derivatized surfaces. We have found it useful to attach vesicles to surfaces coated with polyethylene glycol, but with a small fraction of exposed biotin (figure 6.14; see also Dattelbaum and Anderson, unpublished work). Using such methods, we have been able to prepare surfaces on which isolated fatty acid assemblies are immobilized. We are now beginning exploration of the dynamics of such assemblies in order to examine such processes as growth, change of shape, incorporation of other components (e.g., metabolic or information containing molecules), and ultimately division. We believe that the opportunity for long-term visualization of single containers offered by these tethered vesicle approaches will be a powerful tool for the development of protocells.

6.6 Full Protocell Integration

6.6.1 Integrated Protocellular Reaction Kinetics

It is known that template-directed replication obeys a parabolic growth law because of product inhibition (Lee et al., 1996; Sievers and von Kiedrowski, 1994; Varga and Szathmáry, 1997). By coupling a template-directed replication to a

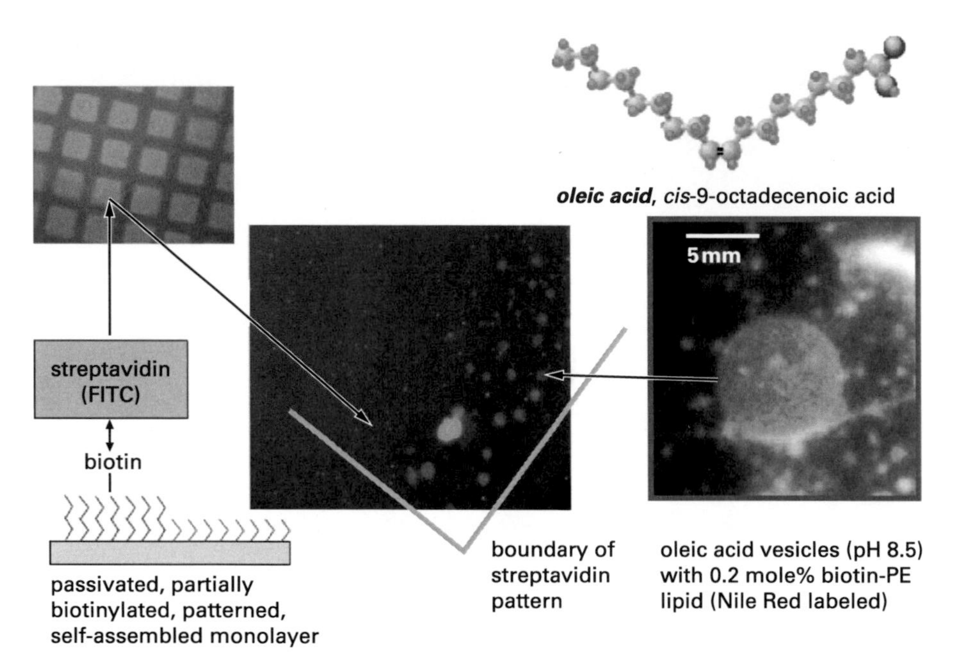

Figure 6.14
Example of attachment of biotin-containing fluorescently labeled oleic acid vesicles to patterned surfaces. The surfaces are patterned to expose a small amount of biotin in a matrix of polyethylene glycol. On addition of (fluorescently labeled) streptavidin, preferential binding of the streptavidin occurs in regions with exposed biotin (upper left corner). Then, subsequent addition of biotinylated oleic acid vesicles to this streptavidin derivatized surface leads to binding of the vesicles to the streptavidin-containing regions (center image). On the far right, a higher-resolution view of a tethered fatty acid vesicle is presented.

template-catalyzed lipid aggregate production, we can show analytically (Rocheleau et al., 2007) that the autocatalytic template-container feedback ensures balanced exponential replication kinetics after many generations, because the genes and the container grow exponentially with the same exponent. Our analysis also suggests that the exponential growth of most modern biological systems emerges from the inherent spatial localization of the container replication process. From this we can analytically show how the internal gene and metabolic kinetics determine the cell population's generation time (Burdett and Kirkwood, 1983; Novak et al., 1998; Tyson, Chen, and Novak, 2003). Actually, these results can be extended to include an explicit representation of both a metabolic production of gene components and new metabolic complex components (Munteanu et al., 2007). These theoretical results are encouraging for experimental groups currently engaged in assembling self-reproducing protocells. However, the theory does not guarantee a balance of the transient kinetics, which of course is the first issue confronting any experimental system. The results do prescribe ideal initial concentrations for system components in

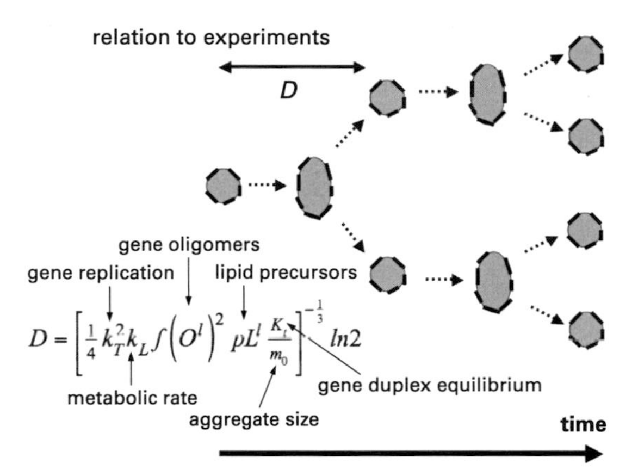

Figure 6.15
The predicted generation (doubling) time (D) for the integrated protocellular system can be expressed in terms of the constants that define the system. Note how the various constants determine the doubling time, but the overall population will always be exponential. The k_is denote the rate constants for gene replication k_T and metabolic production k_L, K_t is the equilibrium constant for gene hybridization-dehybridization, f is a function expressing the replication mechanism for gene replication typically proportional with the local oligomer concentration O^l, pL^l is the local concentration of the precursor lipid components, and m_0 is the number of lipids per aggregate. For details, see Rocheleau et al. (2007).

relation to gene replication and metabolic rates. Figures 6.15 and 6.16 illustrate these results.

Previous kinetic studies of replication reactions (Stadler and Stadler, 2003; Stadler, Stadler, and Schuster, 2000; see also chapter 14) have focused mainly on template replication and have not included coupling to metabolism and container production. Our results extend these investigations.

6.6.2 Open Experimental Integration Issues

We still have many open experimental issues to address before an experimental protocell can be assembled in the laboratory:

1. We need to demonstrate that the 8-oxoguanine–containing PNA sequence can catalyze metabolism. This task requires the synthesis of 8-oxoguanine PNA monomer, followed by the synthesis of a variety of PNA strings with the 8-oxoguanine located at different sites. The metabolic rate constant's dependency on the 8-oxoguanine's location on the PNA is then recorded.

2. We need to demonstrate that gene catalysis of the metabolism occurs primarily when the gene is a duplex structure. Otherwise the gene is not a true self-replicating catalyst, and it might be difficult to ensure a balance in the full protocellular replication process, as described in section 6.1. As with issue 1, the metabolic rate constant's dependency on duplex formation needs to be measured.

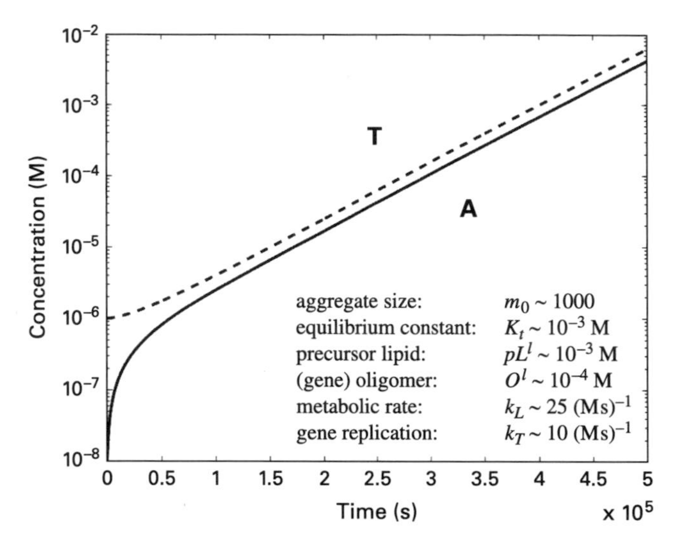

Figure 6.16
Predicted exponential growth for catalytically coupled components: replicating genes (T) and container aggregate (A). This coordinated component growth generalizes to include gene-catalyzed metabolic conversion of precursor genes, precursor metabolic complexes, and precursor lipids. For details, see Rocheleau et al. (2007) and Munteanu et al. (2007).

3. Another investigation must demonstrate the feasibility of PNA replication at lipid interfaces, which is not yet achieved. Ideally, the ligation (polymerization) process should occur without prior activation. This task is presumably the most complex of the remaining individual tasks. As a result of product inhibition, it is a priori difficult to obtain reasonable yields in this process (see, e.g., Rocheleau et al., 2007), and further, the process has to occur as the PNA is associated with the lipid aggregate. With the problems we have encountered so far, this task could be a temporary showstopper because of technical complications that could take time to resolve.

4. We need to demonstrate metabolic production of fatty acids with a gene-sensitizer complex attached to a fatty acid aggregate, resulting in aggregate growth. Although not reported in this chapter, it has already been established (see DeClue et al., 2007).

5. Division of micelles or vesicles as a result of the metabolic activity remains to be shown for our system. Vesicle extrusion is a relatively easy solution to this problem, as demonstrated by several groups (see, e.g., chapter 5). Alternatively, starting with a ternary solution should autonomously provide multiplication of containers as the oily lipid precursors are converted into lipids, turning the ternary solution into a micellar solution. Further, the utilization of two or more lipids (and corresponding precursors) could generate phase separation with budding and division as a result. Finally, differential flow stresses in a microfluidics system might also facilitate vesicle

division (see chapter 12). Many possible solutions to this problem exist, and some of them are currently being explored in simulation.

6. We need to combine milestones 3, 4, and 5 to demonstrate one full protocell life-cycle. Even after reaching these milestones, this final integration could pose unanticipated problems. This full dynamics integration has not even been achieved yet in simulation.

7. Finally, we must carry out step 6 for a number of generations to observe gene sequence selection in a protocell population. This would demonstrate a chemical implementation of a cooperative structure that is able to utilize external free energy to convert external resources into building blocks, thereby allowing the structure to grow, divide, and have inheritable components.

6.7 Conclusion

Many challenges confront any attempt to assemble a minimal life form from a set of simple components. From the very beginning, the approach described in this chapter has focused on component integration. Our main experimental achievement is to demonstrate a gene (8-oxoguanine)-controlled light-driven metabolic (Ru-bpy) production of container building blocks (fatty acids). This demonstrates the viability of establishing a direct catalytic connection from the genes to the container through the metabolism. Theoretical investigations of the integration kinetics are encouraging, in that they demonstrate coordinated component growth in a catalytically coupled protocell. Uncoordinated component growth is a valid concern, because the integrated protocell replication process is a highly complex experimental undertaking.

In this chapter we have summarized a number of partial experimental and computational results and problems related to achieving nonenzymatic template-directed PNA replication, Ru-bpy vesicle association, and single vesicle attachment to prepared surfaces. Our computational investigations have also directed our attention to a number of systemic issues, such as the supply of micellar or vesicle systems with precursor lipids, partition of metabolic complexes and genes during container replication, strong nonlinear influence of the interior oil phase on micellar stability, and the necessary balance between gene backbone hydrophobicity and hybridization energies. Finally, we have tallied a list of experimental milestones we have yet to meet. Judging by this list, it is clear that major technical challenges must still be overcome.

Acknowledgments

We are grateful for input and constructive criticism received from the external reviewers to the Protocell Assembly (PAs) and Programmable Artificial Cell

Evolution (PACE) projects, in particular David Deamer, David Whitten, Susan Atlas, Frank Alexander, and Peter Schuster. This work is supported in part by a Los Alamos National Laboratory LDRD-DR grant and a grant from the European Commission's 6th Framework on emerging information technologies.

References

Apel, C. L., Mautner, M. N., & Deamer, D. W. (2002). Self-assembled vesicles of monocarboxylic acids and alcohols: Conditions for stability and for encapsulation of biopolymers. *Biochimica et Biophysica Acta, 1559* (1), 1–9.

Bachmann, P. A., Luisi, P. L., & Lang, J. (1992). Autocatalytic self-replicating micelles as models for prebiotic structures. *Nature, 357*, 57–59.

Bagley, J. R., & Farmer, J. D. (1991). Spontaneous emergence of a metabolism. In C. Langton, C. Taylor, J. D. Farmer, & S. Rasmussen (Eds.), *Artificial life II* (pp. 93–140). Redwood City, CA: Addison-Wesley.

Banerjee, A., & Falvey, D. E. (1997). Protecting groups that can be removed through photochemical electron transfer: Mechanistic and product studies on photosensitized release of carboxylates from phenacyl esters. *Journal of Organic Chemistry, 62*, 6245–6251.

Bedau, M., McCaskill, S. J., Packard, N. H., Rasmussen, S., Adami, C., Green, D. G., Ikegami T., et al. (2000). Open problems in artificial life. *Artificial Life, 6*, 363–376.

Bi, S. P., Liu, B., Fan, F. R. F., & Bard, A. J. (2005). Electrochemical studies of guanosine in DMF and detection of its radical cation in a scanning electrochemical microscopy nanogap experiment. *Journal of the American Chemical Society, 127* (11), 3690–3691.

Boerlijst, M., & Hogeweg, P. (1991). Self-structuring and selection: Spiral waves as a substrate for prebiotic evolution. In C. Langton, C. Taylor, J. D. Farmer, & S. Rasmussen (Eds.), *Artificial life II*, SFI Volume X (pp. 255–276). Redwood, CA: Addison-Wesley.

Burdett, I. D. J., & Kirkwood, T. B. L. (1983). How does a bacterium grow during its cell cycle? *Journal of Theoretical Biology, 103*, 11–20.

Dattelbaum, A., & Anderson, A., unpublished work.

Deamer, D. (1997). The first living systems: A bioenergetic perspective. *Microbiology and Molecular Biology Reviews, 61*, 230–261.

DeClue, M., Monnard, P.-A., Boncella, J., Bailey, J., Collis, G., Shreve, A., et al. (2007). Towards artificial life—Nucleobase controlled light driven metabolic formation of amphiphilic containers for a minimal protocell. Los Alamos National Laboratories: Los Alamos Unclassified Report 07-2511.

Eigen, M. (1971). Self-organization of matter and the evolution of macromolecules. *Naturwissenshaften, 58*, 465–523.

Farmer, J. D., Kauffman, S., & Packard, N. H. (1986). Autocatalytic replication of polymers. *Physica D, 22*, 50–67.

Fellermann, H., Rasmussen, S., Ziock, H., & Solé, R. (2007). Life cycle of a minimal protocell—a dissipative particle dynamics (DPD) study. *Artificial Life, 13*, 319–345.

Fox, S. W., & Dose, K. (1977). *Molecular evolution and the origin of life*. New York: Marcel Dekker.

Gánti, T. (1997). Biogenesis itself. *Journal of Theoretical Biology, 187*, 583–593.

Goranović, G., Rasmussen, S., & Nielsen, P. E. (2006). Artificial life forms in microfluidic computers. *Proceedings of MicroTAS, 2*, 1408.

Hargreaves, W. R., & Deamer, D. W. (1978). Liposomes from ionic, single-chain amphiphiles. *Biochemistry, 17* (18), 3759–3768.

Hoffman, M. Z., Bolletta, F., Moggi, L., & Hug, G. L. (1989). Rate constants for the quenching of excited states of metal complexes in fluid solution. *Journal of Physical and Chemical Reference Data, 18* (1), 219–543.

Humphrey, W., Dalke, A., & Schulten, K. (1996). VMD: Visual molecular dynamics. *Journal of Molecular Graphics*, *14*, 33–38.

Joyce, G., & Orgel, L. (1999). Prospects for understanding the origins of the RNA world. *Cold Spring Harbor Monograph Series*, *37*, 49–78.

Kale, L., Skeel, R., Bhandarkar, M., Brunner, R., Gursoy, A., Krawetz, N., et al. (1999). NAMD2: Greater scalability for parallel molecular dynamics. *Journal of Computational Physics*, *151*, 283–312.

Kalyanasundaram, K. (1982). Photophysics, photochemistry and solar-energy conversion with tris(bipyridyl)ruthenium(II) and its analogs. *Coordination Chemistry Reviews*, *46* (Oct.), 159–244.

Karlsson, M., Davidson, M., Karlsson, R., Karlsson, A., Bergenholtz, J., Konkoli, Z., et al. (2004). Biomimetic nanoscale reactors and networks. *Annual Review of Physical Chemistry*, *55*, 613–649.

Kauffman, S. (1986). Autocatalytic sets of proteins. *Journal of Theoretical Biology*, *119*, 1–24.

Knutson, C., Benko, G., Rocheleau, T., Mouffouk, F., Maselko, J., Shreve, A. P., et al. (2008). Metabolic photo-fragmentation kinetics for a minimal protocell: Rate limiting factores, efficiency, and implications for evolution. *Artificial Life*, *14*, 1–14.

Langmaier, J., Samec, Z., Samcova, E., Hobza, P., & Reha, D. (2004). Origin of difference between one-electron redox potentials of guanosine and guanine: Electrochemical and quantum chemical study. *Journal of Physical Chemistry B*, *108* (40), 15896–15899.

Langton, C. G. (1984). Self-reproduction in cellular automata. *Physica D*, *10*, 135–144.

Lee, D. H., Granja, J. R., Martinez, J. A., Severin, K., & Ghadri, M. R. (1996). A self-replicating peptide. *Nature*, *382*, 525.

Lee, K., & Falvey, D. E. (2000). Photochemically removable protecting groups based on covalently linked electron donor-acceptor systems. *Journal of the American Chemical Society*, *122*, 9361–9366.

Liu, X. W., Li, J., Deng, H., Zheng, K. C., Mao, Z. W., & Ji, L. N. (2005). Experimental and DFT studies on the DNA-binding trend and spectral properties of complexes $[Ru(bpy)_2L]^{2+}$ (L = dmdpq, dpq, and dcdpq). *Inorganica Chimica Acta*, *358*, 3311–3319.

McCaskill, J. S. (1993). *Inhomogene molekulare evolution*. Jena, Germany: Jahresbericht 1992/1993, Institut fur Molekulare Biotechnologie.

McCaskill, J. S. (1997). Spatially resolved *in vitro* molecular ecology. *Biophysical Chemistry*, *66*, 145–158.

Monnard, P.-A. (2003). Liposome-entrapped polymerases, a model for microscale/nanoscale bioreactors. *Journal of Membrane Biology*, *191*, 87–97.

Monnard, P.-A., Apel, C. L., Kanavarioti, A., Deamer, D. W. (2002). Influence of ionic solutes on self-assembly and polymerization processes related to early forms of life: Implications for a prebiotic aqueous medium. *Astrobiology*, *2* (2), 139–152.

Monnard, P.-A., & Deamer, D. W. (2002). Membrane self-assembly processes: Steps towards the first cellular life. *Anatomical Record*, *268*, 196–207.

Monnard, P.-A., DeClue, M., Bailey, J., Boncella, J., & Shreve, A. (2007). In preparation.

Morowitz, H. J., Heinz, B., & Deamer, D. W. (1988). The chemical logic of a minimum protocell. *Origins of Life and Evolution of the Biosphere*, *18*, 281–287.

Mouffouk, F., & Chen, L. (2007). In preparation.

Munteanu, A., Attolini, C. S.-O., Rasmussen, S., Ziock, H., & Solé, R. (2007). Generic Darwinian selection in protocell assemblies. *Philosophical Transactions of the Royal Society of London B*, *362*, 1847–1855.

Nelson, K., Levy, M., & Miller, S. (2000). Peptide nucleic acid rather than RNA might have been the first genetic molecule. *Proceedings of the National Academy of Sciences of the United States of America*, *97*, 3868–3871.

Nielsen, P. E. (1993). Peptide nucleic acid (PNA): A model structure for the primordial genetic material? *Origins of Life and Evolution of the Biosphere*, *23*, 323–327.

Noireaux, V., & Libchaber, A. (2004). A vesicle bioreactor as a step toward an artificial cell assembly. *Proceedings of the National Academy of Sciences of the United States of America*, *101* (51), 17669–17674.

Novak, B. A., Csikasz-Nagy, B., Gyorffy, K., Chen, K., & Tyson, J. J. (1998). Mathematical model of the fission yeast cell cycle with checkpoint controls at the G1/S, G2/M and metaphase/anaphase transitions. *Biophysical Chemistry*, *72*, 185–200.

Oberholzer, T., Albrizio, M., & Luisi, P. L. (1995). Polymerase chain reaction in liposomes. *Chemistry and Biology*, *2* (Oct.), 677–682.

Ono, N., & Ikegami, T. (1999). Model of self-replicating cell capable of self-maintenance. In D. Floreano, J.-D. Nicoud, & F. Mondala (Eds.), *Advances in artificial life* (pp. 399–406). Berlin: Springer-Verlag.

Oparin, A. I., Orlovskii, A. F., Bukhlaeva, V. Y., & Gladilin, K. L. (1976). Influence of the enzymatic synthesis of polyadenylic acid on a coacervate system. *Doklady Akademii Nauk SSSR*, *226*, 972–794.

Oro, J., Sherwood, E., Eichberg, J., & Epps, D. (1978). Formation of phospholipids under primitive earth conditions and roles of membranes in prebiological evolution. In D. W. Deamer (Ed.), *Light transducing membranes* (pp. 1–22). London: Academic Press, Inc.

Paul, N., & Joyce, J. (2002). A self-replicating ligase ribozyme. *Proceedings of the National Academy of Sciences*, *99* (20), 12733–12740.

Prat, F., Houk, K. N., & Foote, C. S. (1998). Effect of guanine stacking on the oxidation of 8-oxoguanine in b-DNA. *Journal of the American Chemical Society*, *120* (4), 845–846.

Rasmussen, S. (1985). *Aspects of instabilities and self-organizing processes*. PhD thesis, Department of Physics, Technical University of Denmark, Lyngby, Denmark (in Danish).

Rasmussen, S. (1988). Toward a quantitative theory of the origin of life. In C. Langton (Ed.), *Artificial life* (pp. 79–104). Redwood City, CA: Addison-Wesley.

Rasmussen, S., Chen, L., Deamer, D., Krakauer, D., Packard, N., Stadler, P., & Bedau, M. (2004). Transitions from nonliving to living matter. *Science*, *303*, 963.

Rasmussen, S., Chen, L., Nilsson, M., & Abe, S. (2003). Bridging nonliving and living matter. *Artificial Life*, *9*, 269.

Rocheleau, T., Rasmussen, S., Nielsen, P., Jacobi, M., & Ziock, H. (2007). Emergence of protocellular growth laws. *Philosophical Transactions of the Royal Society of London B*, *362*, 1841–1845.

Sayama, H. (1998). Introduction of structural dissolution into Langton's self-replicating loops. In C. Adami, R. K. Belew, H. Kitano, & C. E. Taylor (Eds.), *Artificial life VI* (pp. 114–122). Cambridge, MA: MIT Press.

Segré, D., Ben-Eli, D., & Lancet, D. (2000). Compositional genomes: Prebiotic information transfer in mutually catalytic noncovalent assemblies. *Proceedings of the National Academy of Sciences of the United States of America*, *97*, 4112–4117.

Sen, A., & Nielsen, P. (2006). Personal communication.

Sistare, M. F., Codden, S. J., Heimlich, G., & Thorp, H. H. (2000). Effects of base stacking on guanine electron transfer: Rate constants for G and GG sequences of oligonucleotides from catalytic electrochemistry. *Journal of the American Chemical Society*, *122* (19), 4742–4749.

Sievers, D., & Von Kiedrowski, G. (1994). Self-replication of complementary nucleotide-based oligomers. *Nature*, *369*, 221.

Stadler, B., & Stadler, P. (2003). Molecular replicator dynamics. *Advances in Complex Systems*, *6*, 47.

Stadler, B., Stadler, P., & Schuster, P. (2000). Dynamics of autocatalytic replicator networks based on higher-order ligation reactions. *Bulletin of Mathematical Biology*, *62*, 1061–1086.

Sundararajan, C., & Falvey, D. E. (2004). C-O bond fragmentation of 4-picolyl- and N-methyl-4-picolinium esters triggered by photochemical electron transfer. *Journal of Organic Chemistry*, *69*, 5547–5554.

Sundararajan, C., & Falvey, D. E. (2005). Photorelease of carboxylic acids, amino acids, and phosphates from n-alkylpicolinium esters using photosensitization by high wavelength laser dyes. *Journal of the American Chemical Society*, *127*, 8000–8001.

Tamulis, A., Tamulis, V., Ziock, H., & Rasmussen, S. (2008). Influence of water and fatty acid molecules on quantum photoinduced electron tunneling in self-assembled photosynthetic centers of minimal protocells. In R. B. Ross & S. Mohanty (Eds.), *Multiscale simulation methods for nanomaterials* (pp. 9–28). Hoboken, NJ: Wiley-Interscience.

Tawfik, D. S., & Griffiths, A. D. (1998). Man-made cell-like compartments for molecular evolution. *Nature Biotechnology, 16*, 652–656.

Tyson, J. J., Chen, K. C., & Novak, B. (2003). Sniffers, buzzers, togglers, and blinkers: Dynamics of regulatory and signaling pathways in the cell. *Current Opinion in Cellular Biology, 15*, 221–231.

Varga, Z., & Szathmáry, E. (1997). An extremum principle for parabolic competition. *Bulletin of Mathematical Biology, 59*, 1145.

Vernille, J., & Schneider, J. (2004). Sequence-specific oligonucleotide purification using peptide nucleic acid amphiphiles in hydrophobic interaction chromatography. *Biotechnology Progress, 20* (6), 1776–1782.

Vlcek, A. A., Dodsworth, E. S., Pietro, W. J., & Lever, A. B. P. (1995). Excited-state redox potentials of ruthenium diimine complexes: Correlations with ground-state redox potentials and ligand parameters. *Inorganic Chemistry, 34* (7), 1906–1913.

von Kiedrowski, G. (1986). A self-replicating hexadeoxynucleotide. *Angewandte Chemie International Edition English, 25*, 932–935.

von Kiedrowski, G. (1993). Minimal replicator theory I: Parabolic versus exponential growth. In H. Dugas (Ed.), *Bioorganic chemistry frontiers, Volume 3* (pp. 115–146). Berlin: Springer-Verlag.

Wächterhäuser, G. (1997). The origin of life and its methodological challenges. *Journal of Theoretical Biology, 187*, 483–494.

Walde, P. (1994). Self-reproducing vesicles. In G. R. Fleischaker, S. Colonna, P. L. Luisi (Eds.), *Self-production of supramolecular structures: From synthetic structures to model of minimal living systems, Volume 1* (pp. 209–216). Netherlands: Kluwer Academic Publishers.

Walde, P., Goto, A., Monnard, P.-A., Wessicken, M., & Luisi, P. L. (1994). Oparin's reaction revisited: Enzymatic synthesis of poly(adenyl acid) in micelles and self-reproducing vesicles. *Journal of the American Chemical Society, 116*, 7541–7547.

Walde, P., Wick, R., Fresta, M., Mangone, A., & Luisi, P. L. (1994). Autopoietic self-reproduction of fatty acid vesicles. *Journal of the American Chemical Society, 116*, 11649–11654.

Wallin, S., Davidsson, J., Modin, J., & Hammarstrom, L. (2005). Femtosecond transient absorption anisotrophy study on $[Ru(bpy)_3]^{2+}$ and $[Ru(bpy)(py)_4]^{2+}$: Ultrafast interligand randomization of the MLCT state. *Journal of Physical Chemistry A, 109*, 4697–4704.

Weronski, P., Jiang, Y., & Rasmussen, S. (2007). Molecular dynamics study of small PNA molecules in lipid-water system. *Biophysical Journal, 92*, 3081–3091.

Wittung, P., Kajanus, J., Edwards, K., Nielsen, P., Nordén, B., & Malmström, B. G. (1995). Phospholipid membrane permeability of peptide nucleic acid. *Federation of the Societies of Biochemistry and Molecular Biology Letters, 365*, 27–29.

Yu, W., Sato, K., Wakabayashi, M., Nakaishi, T., Ko-Mitamura, E. P., Shima, Y., et al. (2001). Synthesis of functional protein in liposome. *Journal of Bioscience and Bioengineering, 92* (6), 590–593.

Zhou, X. (2006). Personal communication.

7 Population Analysis of Liposomes with Protein Synthesis and a Cascading Genetic Network

Takeshi Sunami, Kanetomo Sato, Keitaro Ishikawa, and Tetsuya Yomo

7.1 Introduction: The Necessity of Small Compartments

The simplest living creatures are small compartments called cells that contain one or several copies of a genome. These compartments prevent intracellular proteins and other essential components from diffusing into the bulk solution, thereby maintaining cellular integrity. Because typical cellular compartments are microscopic, it is interesting to consider whether the size of the compartment bears any relationship to the number of proteins that are encapsulated by the boundary membrane. A reasonable assumption is that a single copy of genetic information is used to synthesize 10^3 molecules of a given protein. This would produce a protein concentration of about 1 μM for a compartment on the order of 1 μm in diameter, and about 1 nM for a compartment on the order of 10 μm in diameter. Ordinary dissociation constants for protein-protein interactions and protein-DNA interactions range from 1 μM to 1 nM, but not to 1 pM, so a cell size on the order of 1 to 10 μm in diameter is reasonable for regulating metabolic and genetic networks through biopolymer interactions.

Primordial cells could have increased the genome copy number to produce larger numbers of biopolymers for larger compartments. However, experimental evolution with an in vitro self-encoding system using DNA polymerase and its gene showed that a replicating unit with a large copy number of genetic information was not able to sustain a population, whereas one with a single copy was able to survive even though mutations were occurring in the system (Matsuura et al., 2002). The failure to replicate units with a large number of copies of genetic information can be understood if less fit genes were passed on to the next generation by virtue of their coexistence with fitter genes in the same compartment. The fitness of each compartment was the overall average of the coexisting mutated genes. Based on the average fitness, the compartments propagated and passed all the mutated genes to the next generation regardless of the fitness of each mutated gene. Thus, through the accumulation of less fit genes, compartments with multiple copies of genetic information declined. On the other hand, compartments with single copies of genes showed fitness of the individual genes and were able to exclude less fit genes through selection at each generation, consequently their populations were maintained. One theoretical study has also confirmed that chemical networks can evolve a small number of copies of those genetic elements that are crucial for replication (Kaneko and Yomo, 2002).

There are two other possible ways to employ large compartments to maintain small numbers of copies of genetic information: (1) the efficiency of transcription and translation can be increased to maintain a suitably high protein concentration, or (2) dissociation constants of interactions among biopolymers can be reduced in controlled metabolic and genetic networks with low protein concentrations. Because these two processes require greater efficiency or molecular interactions than is achieved by modern cells over several billion years of evolution, it seems reasonable to assume that primordial cells began with small compartments on the order of not more than 10 μm in diameter.

Conducting experimental metabolic and genetic network reactions within small compartments would provide important insights on both the origin of life and the design of protocells. The compartments for achieving this aim should be simple yet still satisfy the essential functions of membranes of living cells. Liposomes, or lipid vesicles, have been used as compartments for protocells, and various types of biochemical reactions have been carried out in liposomes (Chakrabarti, Joyce, Breaker, and Deamer, 1994; Chen, Salehi-Ashtiani, and Szostak, 2005; Deamer, 2005; Luisi, 2002, 2003; Luisi, Ferri, and Stano, 2005; Monnard, 2003; Pohorille and Deamer, 2002; Walde and Ichikawa, 2001; see also chapters 2, 3, and 5). In particular, protein synthesis from DNA in liposomes has attracted a great deal of attention because it permits the investigation of the complex biochemical reactions involved in translation (Ishikawa et al., 2004; Noireaux and Libchaber, 2004; Nomura et al., 2003; Oberholzer, Nierhaus, and Luisi, 1999; Yu et al., 2001). In this chapter, we review the results of our efforts in this regard to address the following questions:

1. What is the extent of size diversity in liposome formation?
2. What are the average levels of protein synthesis per liposome?
3. How does selection work with a wide distribution in liposome size and protein synthesis?
4. Is it possible to run a cascading genetic network in liposomes?

7.2 Size Distribution of Liposomes

Liposomes ranging from 1 μm to 10 μm in diameter are heterogeneous populations, and even when prepared under controlled conditions they show variations in both shape and size (Svetina and Zeks, 2002). This variation can be an obstacle for reproducible experiments, but also indicates the potential for morphological changes such as division (Hanczyc, Fujikawa, and Szostak, 2003; Hanczyc and Szostak, 2004; Oberholzer et al., 1995) and fusion (Tanaka and Yamazaki, 2004). Thus, it would be worthwhile to examine the distribution of liposomes prepared without tight artificial controls such as extrusion, gel permeation chromatography, and centrifugation (Woodle and Papahadjopoulos, 1989). Flow cytometry (FACS) is a very useful method that applies the principle of light scattering and fluorescence to analyze particles in a fluid stream (Shapiro, 1995). Each individual particle can be measured in a continuous flow system at an analysis rate of up to 40,000 events per second; in a way, flow cytometry can be considered equivalent to a high-throughput fluorescence microscope that is able to detect and read forward-scattering light, side-scattering light, and the multiple colors of fluorescence simultaneously (Hai et al., 2004). We applied these powerful analytical functions of flow cytometry to the quantitative evaluation of liposome populations, and also to the selection of specific liposomes.

To investigate the structural properties of individual liposomes in regard to internal aqueous volume and lipid membrane volume, the freeze-dried empty liposome rehydration method was used to encapsulate a solution of green fluorescent protein (GFP) (Cormack, Valdivia, and Falkow, 1996; Kikuchi et al., 1999) into liposomes with a trace amount of red fluorescence–labeled fatty acid (BODIPY-RED-UA) (Yamada et al., 2001). The liposome prepared in our study consisted of a mixture of lipids, that is, POPC:PLPC:SOPC:SLPC:cholesterol:DSPE-PEG5000:BODIPY-RED-UA = 129:67:48:24:180:14:1 (molar ratio). The internal volume and lipid membrane volume of each liposome were estimated by flow cytometry based on the standard curves for the intensities of green and red fluorescence, respectively (figure 7.1a, Sato, 2005). The fluorescent standard calibration beads were introduced before the samples were measured for alignment of the 488-nm argon ion laser source, in order to set the half-peak coefficient of variation (CV) to less than 2% for better detection accuracy. The internal volume was shown to increase with increasing lipid membrane volume per liposome. Although prepared in a single test tube, the liposomes exhibited distributions of more than 1 order of magnitude in both internal volume and lipid membrane volume, indicating heterogeneous size and structure of individual liposomes ranging from 1 μm to 10 μm in diameter. Such a wide distribution occurs to some extent whenever liposomes are formed in ways that could occur naturally without artificial selection to control size and shape. The wide distribution in size implies that when a liposome changes its size, it might stay intact and persist with some level of stability. Such a process might have conferred the dynamic flexibility that primordial cells needed for cell division.

We studied a population consisting of two groups of liposomes. The first showed a linear increase in lipid membrane volume with increasing internal volume (*solid line*),

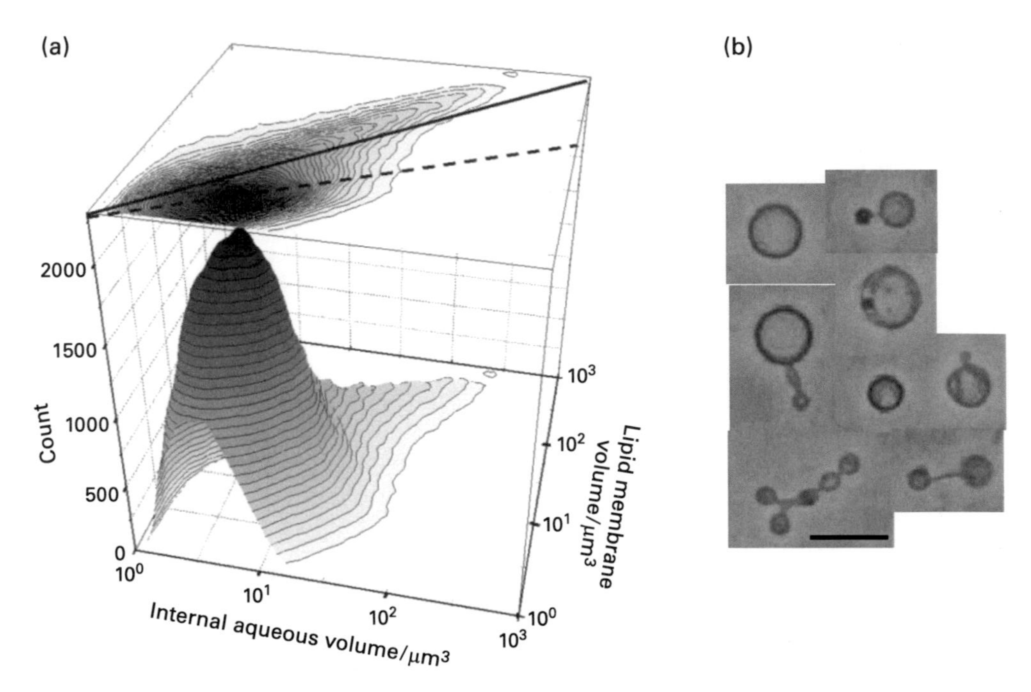

Figure 7.1
Liposome distribution in the internal aqueous volume and lipid membrane volume. (a) A three-dimensional histogram was drawn against the internal aqueous volume and lipid membrane volume plane. The contour lines were projected over the internal aqueous volume and lipid membrane volume plane. The dotted line with a slope of 2/3 lies in the liposome subpopulation with the lower lipid content, whereas the solid line with a slope of 1 lies in that with the higher lipid content. (b) Liposomes belonging to the subpopulation with the lower lipid content were collected by a cell sorter for microscopic observation. Scale bar is 10 μm.

and in the second group the lipid membrane volume increased as an exponential function of internal volume with a slope of 2/3 (*dotted line*). The former subpopulation seemed to have formed through aggregation of small liposomes maintaining a constant ratio between the lipid membrane volume and aqueous internal volume. The large lipid membrane content suggested that this subpopulation possessed multilamellar membrane structures. On the other hand, the latter subpopulation apparently grew by swelling after aggregation, maintaining a constant ratio between the surface or boundary volume and the internal volume. In contrast to the former subpopulation, the latter possessed a smaller number of lamellae, which allowed it to take up large amounts of water from the bulk solution during swelling.

The subpopulation with the slope of 2/3 possessed a higher degree of morphological diversity (see figure 7.1b). The first liposome population that differed in its internal aqueous volume and membrane volume was subjected to fractionation by FACS and to microscopic observation. We observed more than 200 liposomes, and concluded that the subpopulation with a slope of 1 possessed multilamellar membranes, whereas the second subpopulation with a slope of 2/3 had a small number of lamellae, which was consistent with the volume-surface ratio estimated from the internal aqueous volume and membrane volume. The subpopulation with a smaller number of lamellae showed a large diversity in shape compared with the subpopulation with multilamellar membranes. Furthermore, some liposomes in the subpopulation with a small number of lamellae exhibited a spherical structure, although other shapes were also seen (see figure 7.1b). The subpopulation with a smaller number of lamellae may be less stable than the multilamellar liposomes, but it seemed to have a greater potential for various types of morphology. In contrast, if liposomes are prepared by a physical process such as shearing forces (Hanczyc, Fujikawa, and Szostak, 2003; see also chapter 5), they tend to divide into small, relatively homogeneous vesicles.

7.3 Quantitative Analysis of Liposome Populations with Protein Synthesis

If proteins are transcribed and translated from DNA within liposomes in a population with a large diversity in size, as shown in figure 7.1, there will be a distribution in the efficiency of protein synthesis. Because liposomes encapsulate each of the components necessary for transcription and translation by chance, they possess different numbers of components and produce different amounts of protein molecules. The distribution of the liposomal protein synthesis efficiency must be taken into account to quantitatively understand protein synthesis in liposomes. Although microscopic measurements have imaged protein synthesis in liposomes (Noireaux and Libchaber, 2004; Nomura et al., 2003; Yu et al., 2001), they tend to provide higher estimates of efficiency by unintentionally focusing on liposomes with larger numbers of GFP molecules in the heterogeneous liposome population.

In our experiment, protein synthesis in liposomes was carried out under the following conditions. Plasmid DNA (0.15 copies/μm^3) encoding the GFP gene under the control of the T7 promoter, along with other components for protein synthesis (Shimizu et al., 2001), were encapsulated into liposomes to produce GFP. Internal aqueous volume was estimated based on the red fluorescent protein R-phycoerythrin, which was added to the solution before encapsulation as an internal aqueous marker. RNaseA was added to the solution outside the liposomes to ensure that GFP was transcribed only within the liposomes. We determined the distributions in both number of GFP molecules synthesized in liposomes and internal aqueous volume simultaneously by flow cytometric analysis based on the method described in section 1 (figure 7.2). Liposomes with a larger internal volume possessed larger numbers of GFP molecules. From the black dotted line with a slope of 1, we estimated that the averaged efficiency of GFP translation was 78 GFP molecules/μm^3. The internal volume and lipid membrane volume of each liposome were estimated by flow cytometry

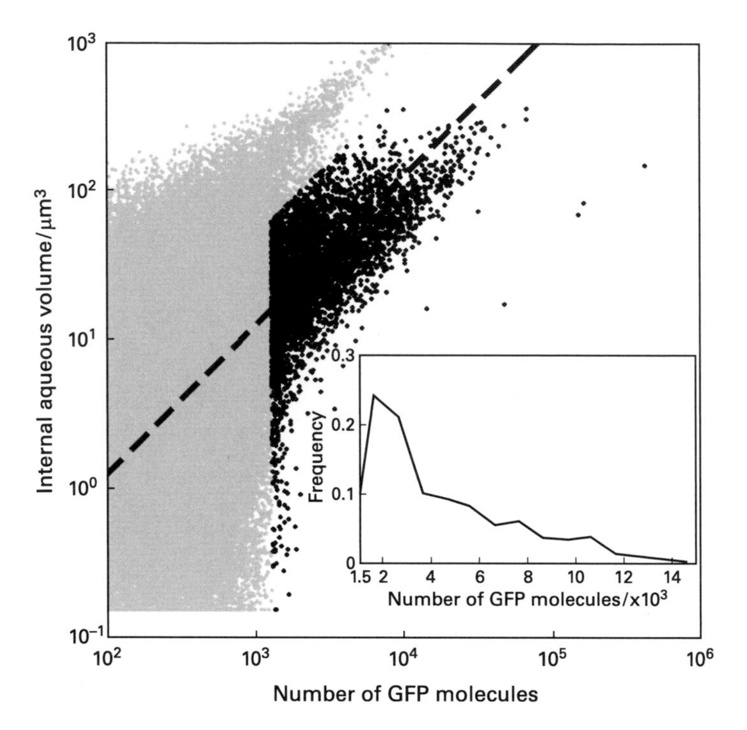

Figure 7.2
GFP synthesis from its gene in liposomes. The number of GFP molecules synthesized and internal aqueous volume for each liposome were estimated by flow cytometry (black dots). Internal aqueous volume was estimated based on the red fluorescent protein R-phycoerythrin, which was added to the solution before encapsulation as an internal aqueous marker. Gray dots represent background, which was determined by protein synthesis within liposomes without the GFP gene. The insert shows the distribution of GFP production among liposomes ranging in internal aqueous volume from 50 μm^3 to 100 μm^3.

based on the standard curves for the intensities of green and red fluorescence, respectively. Although liposomes encapsulate DNA at various efficiencies depending on the conditions used (Gregoriadis et al., 1999; Monnard, Oberholzer, and Luisi, 1997), provided that the GFP gene was encapsulated by chance without any repulsive effect due to its negative charge, the number of GFP molecules synthesized per gene was 520, which was one tenth of the efficiency for in vitro protein synthesis from the same plasmid. The low efficiency was mostly a result of inward leakage of RNaseA (Fischer, Oberholzer, and Luisi, 2000) added to the external bulk solution after encapsulation to prevent protein synthesis outside the liposomes.

Even among liposomes with the same internal volume, the efficiency of GFP synthesis varied to a large degree. For example, liposomes ranging in internal volume from 50 μm^3 to 100 μm^3 showed a wide distribution with a tail for higher GFP production per volume (see the insert histogram in figure 7.2). Liposomes with high efficiency were also observed for internal volumes smaller than 10 μm^3. When the

Table 7.1
Selectabilities of liposomes with different internal volumes

Fractionated Internal Volume (μm^3)	Number of Sorted Liposomes	Total Gene Copy Number per Liposome	Ratio of the Original Gene (%)
2.1–6.8	500	0.9	70
6.8–13	2,700	1.0	71
13–31	24,000	2.3	59
31–48	20,000	2.8	44
48–120	47,000	6.4	27

Note: Internal volume was estimated based on the marker for internal aqueous volume, a red fluorescent protein, as described in figure 7.2. The number of selected liposomes was determined from the number of counts sorted with a cell sorter. The total number of gene copies per liposome was estimated from the numbers of liposomes collected, the original GFP gene, and the mutant GFP gene.

components necessary for protein synthesis were encapsulated by chance, some liposomes excluded inhibitory factors (i.e., proteases or nucleases; Ueda et al., 1991), leading to high efficiency of protein synthesis. One report claims that the efficiency is higher than that of the bulk solution outside liposomes, though the mechanism responsible for this is not clear (Nomura et al., 2003). Their unexpected high efficiency probably was a result of selection of liposomes with high fluorescence by fluorescent microscopy. Quantitative studies on liposomes need to take into account their heterogeneity.

7.4 Selection of Liposomes Based on Protein Expression

Although a single genotype yielded a large degree of heterogeneity in GFP production among liposomes, as discussed in the previous section, fractionation based on internal aqueous volume allowed liposomes to be selected for genes of higher fitness. To observe this selectability, the gene for GFP and a mutant GFP gene with only one eighth of the original fluorescence were mixed at a molar ratio of 0.15 to 0.85. The mixed genes were encapsulated into liposomes, transcribed, translated, and subjected to flow cytometric analysis. Liposomes were separated into six groups according to internal aqueous volume. In each group, liposomes with a fluorescence intensity above the lower limit of detection were collected (table 7.1). The number of liposomes collected increased with internal volume because the larger liposomes contained more of the genes to produce more GFP, leading to a higher frequency of collected liposomes beyond the limit of detection. After collection, the average numbers of copies of the two genes were measured separately by quantitative polymerase chain reaction (PCR). The total gene copy number per liposome increased with internal volume. The ratio of the original *GFP* gene, which was 15% before selection, increased with reduction of the internal aqueous volume, indicating that the selection rates were higher for smaller than for larger liposomes. Although larger liposomes displayed higher levels of green fluorescence, as shown in figure 7.2, the ratio of the original GFP gene was close to the initial ratio before selection. Larger liposomes were rarely selected because mutant genes for weak green fluorescence coexisted with the original gene in the larger liposomes and selection was performed based on the average fluorescence over the two genes. On the other hand, smaller liposomes possessed nearly single copies of genes, so that selection based on sufficient green fluorescence per liposome excluded the mutant gene. Selection was therefore more efficient for smaller than for larger liposomes. In other words, liposomes or protocells should contain a small number of copies of genetic elements for retaining their sensitivity to mutation and ensuing selection. This effect is consistent with the experiment using the self-encoding system, confirming the importance of a small number of gene copies for the origin of life (Matsuura et al., 2002). To employ larger liposomes for artificial evolutionary engineering, one may dilute the concentration of DNA before encapsulation. However, this might not work for a genetic network that includes the binding of proteins synthesized in liposomes, because of the limited affinities, as discussed in the introduction of this chapter.

7.5 A Cascading Genetic Network

Since protein synthesis efficiency in liposomes has been optimized to 10^3 molecules per gene, which is comparable to that in a living cell, more complex genetic reactions can be investigated. For example, Noireaux and Libchaber (2003) constructed a functional cell-free genetic network in vitro. We recently conducted an experiment with a cascading genetic network within liposomes (Ishikawa et al., 2004). A plasmid encoding the T7 RNA polymerase gene under the control of the SP6 promoter and the *GFP* gene under the control of the T7 promoter shown in figure 7.3a was encapsulated into liposomes with other components for protein synthesis (Shimizu et al., 2001) before protein synthesis was initiated. In vitro kinetic experiments confirmed that T7 RNA polymerase was synthesized from its gene and bound to the T7 promoter, leading to GFP mRNA synthesis. In contrast to the single-stage GFP expression system, we observed a delay of 20 minutes for accumulation of T7 RNA polymerase before the production of GFP encoded after the T7 promoter. In liposomes, the cascading genetic network reaction produced GFP, whereas protein synthesis outside was inhibited by externally added RNaseA (Fischer, Oberholzer, and Luisi, 2000). A considerable number of liposomes exhibited green fluorescence above the limit of detection, while liposomes without the plasmid or with the same plasmid encoding the deletion mutant of T7 RNA polymerase did not (figure 7.3b). These observations indicate that as long as they are not too complex, genetic networks can proceed at rates comparable to protein synthesis by living cells.

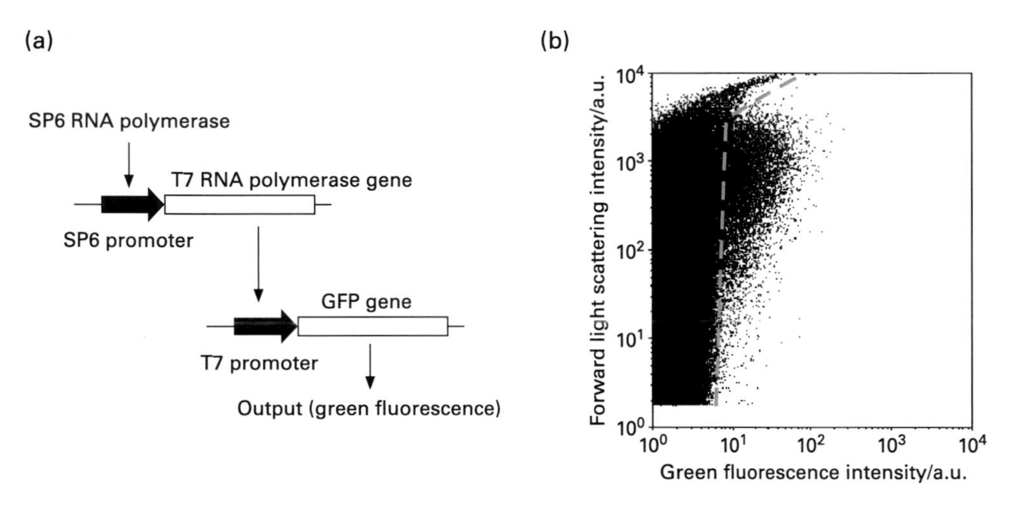

Figure 7.3
A cascading genetic network in liposomes. (a) Scheme for cascading genetic networks. In the cascading genetic reaction, the T7 RNA polymerase gene is first transcribed by SP6 RNA polymerase, and then the T7 RNA polymerase is translated. Second, the synthesized T7 RNA polymerase transcribes the GFP gene, then GFP is translated as an output signal. Both genes are encoded on the same plasmid DNA. (b) Liposomes encapsulating the plasmid DNA described in (a) were analyzed by flow cytometry. The gray dotted line represents the limit of detection. This line was based on the autofluorescence intensity in liposomes with the same plasmid in (a) except for encoding an inactivated T7 RNA polymerase deletion mutant gene. Dots beyond the limit of detection indicate the liposomes in which the cascading genetic network reaction occurred.

7.6 Conclusion

The results presented in this chapter can be summarized as follows. (1) Liposome populations show great diversity in internal volume. (2) Taking this diversity into account, the efficiency of protein synthesis in liposomes was estimated to be 520 GFP molecules per gene, on average, with a minor fraction showing very high efficiency. (3) When liposomes are fractionated according to their internal volume, smaller liposomes possessing virtually a single copy of the *GFP* gene were shown to be selectable. (4) A cascading genetic network was able to proceed in the liposomes. These results indicate that more complex reaction networks can evolve experimentally with mutation and selection as long as the diversity in both internal volume and efficiency of protein synthesis are taken into account. Otherwise, the diversity of a single genotype would overwhelm that among mutated genotypes, resulting in a loss of evolvability.

In the origin of life, primordial cells must have faced similar phenotypic diversity. Assuming that they employed compartments that were not too large to maintain evolvability, heterogeneity in phenotype and functions such as protein synthesis would be inevitable because of the small number of components necessary for reproduction. In the face of such diversity, only compartments that by chance possessed a balanced set of necessary components would have evolved into the forms of cellular life that exist today.

The next possible steps toward the synthesis of a protocell is the encapsulation of a self-encoding genetic network. An in vitro self-encoding system has been shown to possess sustainability with selectability (Matsuura et al., 2002). If a similar self-encoding genetic network is encapsulated, it might be possible to optimize the network through a cycle of mutation and selection, as discussed in section 3. Moreover, ribosomes and many other necessary cofactors, which in the work reported here were added manually, might be able to be encoded in the network through an evolutionary processes. Once a highly optimized self-encoding network is obtained, it seems plausible that it could be merged with replicating lipid compartments to fabricate a functional protocell (Hanczyc, Fujikawa, and Szostak, 2003; Oberholzer et al., 1995; Szostak, Bartel, and Luisi, 2001).

Acknowledgments

This study was supported by a grant-in-aid from the 21st Century COE program, "New Information Technologies for Building a Networked Symbiosis Environment," of the Japan Society for the Promotion of Science (JSPS).

Abbreviations

POPC	1-Palmitoyl-2-oleoyl-*sn*-phosphatidylcholine
PLPC	1-palmitoyl-2-linoleoyl-*sn*-phosphatidylcholine
SOPC	1-stearoyl-2-oleoyl-*sn*-phosphatidylcholine
SLPC	1-stearoyl-2-linoleoyl-*sn*-phosphatidylcholine
DSPE-PEG5000	distearoyl phosphatidyl ethanolamine-poly(ethylene glycol) 5000
BODIPI-RED-UA	11-{3″,5″-bis(4‴-methoxyphenyl)-4″,4″-difluoro-4″-bora-3a,4a-diaza-*s*-indacenyl}-3′,5′-dimethylphenoxyundecanoic acid

References

Chakrabarti, A., Joyce, G. F., Breaker, R. R., and Deamer, D. W. (1994). RNA synthesis by a liposome-encapsulated polymerase. *Journal of Molecular Evolution*, *39*, 555–559.

Chen, I. A., Salehi-Ashtiani, K., & Szostak, J. W. (2005). RNA catalysis in model protocell vesicles. *Journal of the American Chemical Society*, *127* (38), 13213–13219.

Cormack, B. P., Valdivia, R. H., & Falkow, S. (1996). FACS-optimized mutants of the green fluorescent protein (GFP). *Gene*, *173*, 33–38.

Deamer, D. (2005). A giant step towards artificial life? *Trends in Biotechnology*, *23*, 336–338.

Fischer, A., Oberholzer, T. & Luisi, P. L. (2000). Giant vesicles as models to study the interactions between membranes and proteins. *Biochimica et Biophysica Acta*, *1467* (1), 177–188.

Gregoriadis, G., McCormack, B., Obrenovic, M., Saffie, R., Zadi, B., & Perrie, Y. (1999). Vaccine entrapment in liposomes. *METHODS*, *19*, 156–162.

Hai, M., Bernath, K., Tawfik, D., & Magdassi, S. (2004). Flow cytometry: A new method to investigate the properties of water-in-oil-in-water emulsions. *Langmuir*, *20*, 2081–2085.

Hanczyc, M. M., Fujikawa, S. M., & Szostak, J. W. (2003). Experimental models of primitive cellular compartments: Encapsulation, growth, and division. *Science*, *302*, 618–622.

Hanczyc, M. M., & Szostak, J. W. (2004). Replicating vesicles as models of primitive cell growth and division. *Current Opinion in Chemical Biology*, *8* (6), 660–664

Ishikawa, K., Sato, K., Shima, Y., Urabe, I., & Yomo, T. (2004). Expression of a cascading genetic network within liposomes. *FEBS Letters*, *576* (3), 387–390.

Kaneko, K., & Yomo, T. (2002). On a kinetic origin of heredity: Minority control in a replicating system with mutually catalytic molecules. *Journal of Theoretical Biology*, *214*, 563–576.

Kikuchi, H., Suzuki, N., Ebihara, K., Morita, H., Ishii, Y., Kikuchi, A., et al. (1999). Gene delivery using liposome technology. *Journal of Controlled Release*, *62*, 269–277.

Luisi, P. L. (2002). Toward the engineering of minimal living cells. *The Anatomical Record*, *268*, 208–214.

Luisi, P. L. (2003). Autopoiesis: A review and a reappraisal. *Naturwissenschaften*, *90*, 49–59.

Luisi, P. L., Ferri, F., & Stano, P. (2005). Approaches to semi-synthetic minimal cells: A review. *Naturwissenschaften*, *15*, 1–13.

Matsuura, T., Yamaguchi, M., Ko-Mitamura, E. P., Shima, Y., Urabe, I., & Yomo, T. (2002). Importance of compartment formation for a self-encoding system. *Proceedings of the National Academy of Sciences*, *99* (11), 7514–7517.

Monnard, P.-A. (2003). Liposome-entrapped polymerases as models for microscale/nanoscale bioreactors. *Journal of Membrane Biology*, *191*, 87–97.

Monnard, P. A., Oberholzer, T., & Luisi, P. L. (1997). Entrapment of nucleic acids in liposomes. *Biochimica et Biophysica Acta*, *1329*, 39–50.

Noireaux, V., & Libchaber, A. (2003). Principles of cell-free genetic circuit assembly. *Proceedings of the National Academy of Sciences*, *100* (22), 12672–12677.

Noireaux, V., & Libchaber, A. (2004). A vesicle bioreactor as a step toward an artificial cell assembly. *Proceedings of the National Academy of Sciences*, *101* (51), 17669–17674.

Nomura, S. M., Tsumoto, K., Hamada, T., Akiyoshi, K., Nakatani, Y., & Yoshikawa, K. (2003). Gene expression within cell-sized lipid vesicles. *Chembiochem*, *4* (11), 1172–1175.

Oberholzer, T., Nierhaus, K. H., & Luisi, P. L. (1999). Protein expression in liposomes. *Biochemical and Biophysical Research Communications*, *261* (2), 238–241.

Oberholzer, T., Wick, R., Luisi, P. L., & Biebricher, C. K. (1995). Enzymatic RNA replication in self-reproducing vesicles: An approach to a minimal cell. *Biochemical and Biophysical Research Communications*, *207*, 250–257.

Pohorille, A., & Deamer, D. (2002). Artificial cells: Prospects for biotechnology. *Trends in Biotechnology*, *20*, 123–128.

Sato, K. (2005). *Fundamental research for constructing the artificial cell model with liposomes*. Unpublished doctoral dissertation, Osaka University, Japan.

Shapiro, H. M. (1995). *Practical Flow Cytometry*, 3rd ed. New York: John Wiley & Sons.

Shimizu, Y., Inoue, A., Tomari, Y., Suzuki, T., Yokogawa, T., Nishikawa, K., & Ueda, T. (2001). Cell-free translation reconstituted with purified components. *Nature Biotechnology*, *19*, 751–755.

Svetina, S., & Zeks, B. (2002). Shape behavior of lipid vesicles as the basis of some cellular processes. *The Anatomical Record*, *268*, 215–225.

Szostak, J. W., Bartel, D. P., & Luisi, P. L. (2001). Synthesizing life. *Nature*, *409*, 387–390.

Tanaka, T., & Yamazaki, M. (2004). Membrane fusion of giant unilamellar vesicles of neutral phospholipid membranes induced by La^{3+}. *Langmuir*, *20*, 5160–5164.

Ueda, T., Tohda, H., Chikazumi, N., Eckstein, F., & Watanabe, K. (1991). Phosphorothioate-containing RNAs show mRNA activity in the prokaryotic translation systems in vitro. *Nucleic Acids Research*, *19*, 547–552.

Walde, P., & Ichikawa, S. (2001). Enzymes inside lipid vesicles: Preparation, reactivity and applications. *Biomolecular Engineering*, *18*, 143–177.

Woodle, M. C., & Papahadjopoulos, D. (1989). Liposome preparation and characterization. *Methods in Enzymology*, *171*, 193–217.

Yamada, K., Toyota, T., Takakura, K., Ishimaru, M., & Sugawara, T. (2001). Preparation of BODIPY probes for multicolor fluorescence imaging studies of membrane dynamics. *New Journal of Chemistry*, *25*, 667–669.

Yu, W., Sato, K., Wakabayashi, M., Nakaishi, T., Ko-Mitamura, E. P., Shima, Y., et al. (2001). Synthesis of functional protein in liposome. *Journal of Bioscience and Bioengineering*, *92* (6), 590–593.

8 Constructive Approach to Protocells: Theory and Experiments

Kunihiko Kaneko

8.1 Introduction

To understand what life is, we need to reveal universal features that all life systems have to satisfy at minimum, irrespective of detailed biological processes. Present living organisms, however, include detailed and elaborate processes that have been captured through the history of evolution. For our purpose, then, it is desirable to set up a minimal biological system to understand the universal logic that organisms should necessarily obey. Hence, our approach will be constructive in nature. This is an approach that we call *constructive biology*, which has been carried out both experimentally and theoretically (Kaneko, 1998, 2003b, 2004; Kaneko and Tsuda, 2003).

The constructive approach to complex systems was proposed in the early 90s (Kaneko and Tsuda, 1994), whereas the term *constructive biology* was coined in the mid-90s (Kaneko, 1998). Nowadays the term *synthetic biology* prevails (Benner and Sismour, 2005; Sprinzak and Elowitz, 2005). Although there are some differences in emphases in the direction of research, the intention in both of these approaches is to artificially construct a system that does not exist in nature but that has some biological features. In synthetic biology, design- and engineering-oriented studies are stressed. In contrast, constructive biology does not aim at goal-oriented projects to design some function. Rather, by setting minimal conditions for basic properties of life, we try to unveil the universal logic that a biological system has to satisfy. In this sense, construction is for analyzing or understanding what life is (Kaneko, 2003b). Of course, some studies in synthetic biology also aim at unveiling the universal logic of life (Sprinzak and Elowitz, 2005); the difference between the two fields is, perhaps, the emphasis of their studies.

Constructive biology is not limited to the "wet" experiment. A synergetic experimental approach combines (1) gedanken experiments, (2) modeling, and (3) wet experiments. Gedanken experiments are theoretical study, to reveal a logic underlying universal features in life processes, which is essential to understanding the logic for "what life is." Still, we have not yet gained sufficient theoretical intuition to design a complex biological system in which the state of each part is constrained by the whole system. It is therefore relevant to make computer experiments to heuristically find such logic. This is the second approach mentioned—constructing an artificial world in a computer (Furusawa and Kaneko, 1998; Kaneko and Yomo, 1997, 1999).

Still, in a system with potentially huge degrees of freedom, like life, construction through computer modeling may miss some essential factors. Hence, we need the third experimental approach: construction in a laboratory. In this approach, one constructs a possible biological world in the laboratory by combining several procedures. For example, this experimental constructive biology has been pursued by Yomo and his collaborators (see, e.g., Kashiwagi et al., 2002; Ko, Yomo, and Urabe, 1994; Matsuura et al., 2002) at the levels of biochemical reactions, cells, and ensembles of cells.

Taking this three-part synergetic approach to constructive biology, we have been working on the construction of minimal replicating cells, multicellular organisms, adaptation, evolution, and symbiosis, both theoretically and experimentally (Kaneko 2003b, 2004). The construction of a replicating system with compartments, of course, is essential to considering the origin of a protocell, which is the theme of the present volume, and will be discussed here. Note, however, that we intend neither to reproduce what occurred in evolution on Earth, nor to guess the environmental condition of the past Earth. Rather, we try to construct such replicating systems from complex reaction networks under conditions preset by us.

8.2 Theoretical Issue 1: Origin of Bioinformation

8.2.1 Question: Origin of Heredity

In a cell, among many chemicals, only some (e.g., DNA molecules) are regarded as carriers of genetic information. Why do only some specific molecules play this role? How has distinction of roles in molecules, from genetic information to metabolism, progressed? Is it a necessary progression for a system with internal degrees and reproduction?

Eigen, following Spiegelman's experimental study on replication of RNA (Mills, Petersen, and Spiegelman, 1967), considered how reproduction of catalytic molecules is possible (Eigen and Schuster, 1979). For replication to progress, catalysts are necessary, and information must be preserved in a polymer to coordinate replication. However, error rates in replication must have been high at a primitive stage of life, and accordingly, it is recognized that the information to carry out catalytic activity would have been lost within a few generations.

To resolve this problem of inevitable loss of catalytic activities through replication errors, Eigen and Schuster (1979) proposed the hypercycle, whereby replicating chemicals catalyze each other to form a cycle: as A catalyzes B, B catalyzes C, C catalyzes D, and D catalyzes A. This hypercycle avoids the original problem of error accumulation. However, the hypercycle itself turned out to be weak against parasitic molecules—those that are catalyzed by a molecule in the cycle but do not in turn catalyze other molecules. One possibility for resolving the problem of parasitism is compartmentalization by a cell (Eigen 1992; Maynard-Smith, 1979; Szathmáry and Demeter, 1987; Szathmáry and Maynard-Smith, 1997). Consider a compartment ("cell") consisting of molecules, and assume that this compartment grows and divides as the number of molecules within increases. Such a compartment occupied by parasite molecules stops growing, while one not "infected" by parasites is able to continue. Similarly, spatially localized structures in reaction-diffusion systems may suppress the invasion of parasitic molecules (Altmeyer and McCaskill, 2001; Boerilst and Hogeweg, 1991).

Note, however, that compartmentalization requires some molecules (e.g., for the membrane) other than the replicating molecules (such as RNA). In this sense, some complexity in primitive cells is assumed. Dyson (1985) took this complexity for granted, and considered the possibility that a set of a large number of chemical species may continue reproduction, sustaining catalytic activity. In contrast to the picture of replicating molecules in a hypercycle, Dyson considered a set of chemicals that show inaccurate (loose) reproduction. Although accurate replication of such a variety of chemicals is not possible, chemicals, as a set, may continue reproducing themselves loosely, while maintaining catalytic activity.

It is important to consider whether such loose reproduction as a set is possible in a mutually catalytic reaction network. If it is, and if these chemicals also include molecules forming a membrane for compartmentalization, reproduction of a primitive cell would be possible. In fact, given the chemical nature of lipid molecules, it is not so surprising that a compartment structure is formed from lipids.

Still, in this reproducing system, particular molecules carrying information for reproduction do not exist, in contrast to the present-day cell, which has specific molecules (DNA) for this purpose. As for a transition from early, loose reproduction to later, accurate replication with genetic information, Dyson referred only to "genetic take-over" (Cairns-Smith, 1982), but does not discuss its mechanism. In other words, the question we have to address is how a specific molecular species within the mutually catalytic reaction network can take over the system to behave as a carrier of genetic information.

Now, it is important to study how recursive production of a cell is possible with the appearance of new molecules that may play a specific role for heredity. (By *recursive production* we mean that a state, that is, chemical composition, of the daughter cell repeatedly returns to the state of the mother cell. We are not referring to the mathematical term *recursive*.) Let us consider a simple protocell that consists of a mutually catalyzing molecular species whose increase in number leads to cell reproduction (Kaneko and Yomo, 2002). In this protocell, the molecules that carry the genetic information are not initially specified. The first question we discuss here is whether some specific molecules will start to carry information for heredity in order to realize the continual reproduction of such a protocell.

In present-day cells, it is generally believed that DNA is a carrier of heredity and controls the cells' behavior. Still, even in these cells, proteins and DNA both influence each other's replication processes. At this point, we need to first clarify what "heredity" really means. Here, one might point out that DNA molecules would be suited to encode many bits of information, and hence would be selected as an information carrier. Although this combinatorial capacity of a DNA molecule is important, what we are interested in here is a more basic property that must be satisfied prior to that—that of carrying heredity at the minimum, for which the following two features are necessary:

1. Such molecules are preserved well over generations. The number of such molecules exhibits smaller fluctuations than that of other molecules, and their chemical structure (such as polymer sequence) is preserved over a long time span. We refer to this as the *preservation property*.
2. If this molecule is replaced by some other type of molecule, there is a much stronger influence on the behavior of the cell than when other molecules are similarly replaced. We refer to this as the *control property*.

The question we address is as follows: Under what conditions does recursive production of a protocell continue maintaining catalytic activities? Does this require molecules carrying heredity? Under what conditions does one molecular species begin to satisfy conditions 1 and 2 so that the molecule carries heredity?

8.2.2 A Simple Model with Mutual Catalytic Process

A theoretical study to answer the question posed above was presented in Kaneko and Yomo (2002). By setting up a condition for a prototype of a cell consisting of mutually catalyzing molecules, they showed, under rather general conditions, that symmetry breaking between two kinds of molecules takes place. Through replication and selection, one kind of molecule comes to satisfy conditions 1 and 2 in section 8.2.1. Here, we outline the logic of the study.

First, assume a prototype of a cell consisting of molecules that catalyze each other (see figure 8.1). As the reaction progresses, the number of molecules will increase. Then, considering the physical nature of the membrane, this cell will divide when its volume (the total number of molecules) is beyond some threshold. Here we simply assume that if the total number of molecules in the cell is larger than some threshold, it divides into two, and the molecules are split between the two "daughter cells." We first consider the simplest case: Only two kinds of molecules, X and Y, exist in this

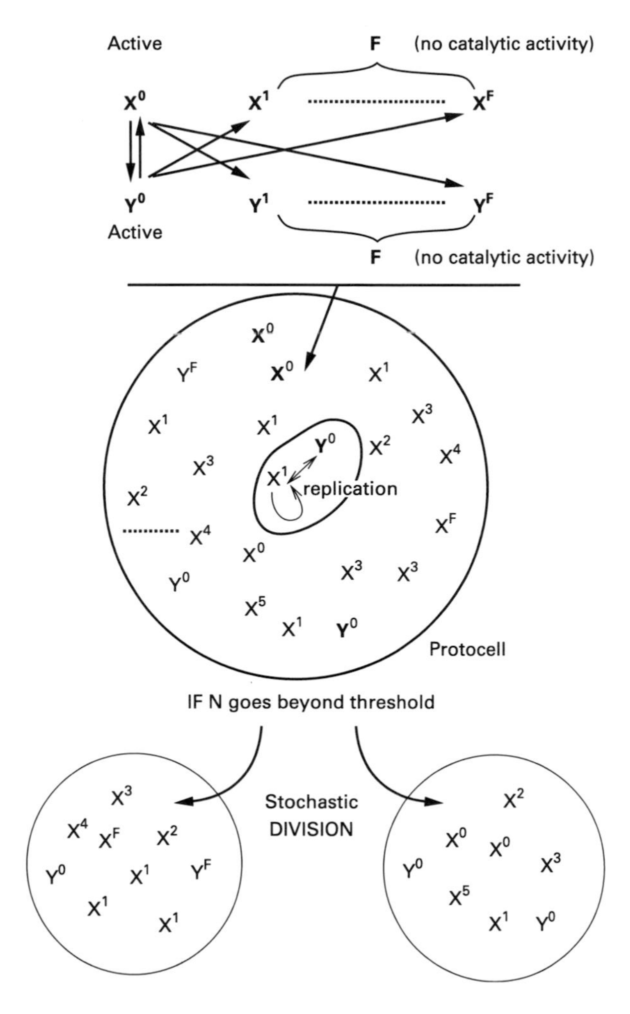

Figure 8.1
Schematic representation of our model.

protocell, and they catalyze each other for the synthesis of the molecules. Without losing generality, one can assume that the synthesis speed of X is faster than that of Y, or in other words, the catalytic activity of Y is much stronger than that of X.

In the replication process of complex polymers, some structural changes in molecules can occur, which may be termed *replication errors*. These structural changes in each kind of molecule often result in the loss of catalytic activity because molecules without catalytic activity are common. This protocell is then taken over by such noncatalytic molecules as are discussed in the parasite problem for the hypercycle. Hence, maintaining reproduction of this protocell is not easy.

As a simple "gendaken experiment" consider the following model:

1. There are two species of molecules, X and Y, which are mutually catalyzing.

2. For each species, there are active and inactive types. The active molecule type is rather rare. There are F types of inactive molecules per active type. There is only one type of active molecule for each species. Active types are denoted as X^0 and Y^0, and inactive types are denoted X^i and Y^i with $i = 1, 2, \dots, F$. In other words, X and Y are different kinds of molecules, while the index, X^j $(j = 0, 1, \dots, F)$, designates different polymers with different configurations of monomers.

The active type has the ability to catalyze the replication of both types (active and inactive) of the other species of molecules. Thus, the catalytic reactions for replication are assumed to take the form

$$X^j + Y^0 \to 2X^j + Y^0 \quad (\text{for } j = 1, 2, \dots, F)$$

and

$$Y^j + X^0 \to 2Y^j + X^0 \quad (\text{for } j = 1, 2, \dots, F).$$

3. The rates of synthesis (or catalytic activity) of the molecules X and Y differ. We stipulate that the rate of this replication process for Y, γ_Y, is much smaller than that for X, γ_X. This difference in the rates may also be caused by a difference in catalytic activity between the two molecular species.

4. In the replication process, structural changes may occur that alter the activity of the molecules. Therefore, the type (active or inactive) of a daughter molecule can differ from that of the mother. The rate of such structural change is given by μ, and is not necessarily small, due to thermodynamic fluctuations. This change can consist of the alternation of a sequence in a polymer or other conformational change, and may be regarded as "replication error." Thus, there are processes described by

$$X^i \to X^0; \quad \text{and} \quad Y^i \to Y^0 \quad (\text{with rate } \mu)$$

$$X^0 \to X^i; \quad \text{and} \quad Y^0 \to Y^i \quad (\text{with rate } \mu \text{ for each}),$$

for $i = 1, 2, \dots, F$. Hence, the probability for loss of activity is F times greater than for its gain, since there are F times more types of inactive molecules than active molecules.

5. When the total number of molecules in a protocell exceeds a given value $2N$, it divides into two, and the chemicals therein are distributed randomly into the two daughter cells, with N molecules going to each. Subsequently, the total number of molecules in each daughter cell increases from N to $2N$, at which point these divide.

6. To include competition, we assume that there is a constant total number M_{tot} of protocells, so that one protocell, randomly chosen, is removed whenever a (different) protocell divides into two.

8.2.3 Minority-Controlled State

The question we address here is how the active type 0 is maintained through reproduction, even if F is large.

Here, because of the difference in the speed of synthesis, the number of X molecules will get larger than that of Y. As long as the number of molecules in a cell is small, eventually, there comes a stage at which the catalytically active Y molecule (i.e., Y^0) is extinct. Then, X molecules are no longer synthesized. Inactive Y molecules (Y^i) may still be synthesized as long as X^0 molecules remain. However, after each division, the number of X^0 molecules becomes half, and sooner or later, the cell stops dividing. Hence, once the number of Y^0 becomes 0, the reproductivity of the cell will be lost.

Recall that the number of molecules in a (proto)cell is not huge. As a result of fluctuations in the number of molecules, some cells may keep active Y molecules. Since there is little room for Y molecules in protocells (because of their slow synthesis) the total number of Y^j molecules (for all $j = 0, 1, \ldots, F$) should be very small. Hence, typically, to keep the active Y molecules, the number of all other Y molecules should be suppressed to zero. Once the inactive Y molecules go extinct, they do not reappear as often, since both the number of Y molecules and the error rate are small. Hence, a cell state with few Y^0 molecules and almost zero Y^i molecules can keep reproducing.

Note that such a state does not exist if the total number of molecules is very large. In the continuum limit, the rate equation of the molecules is given by

$$dN_X^j/dt = \gamma_X N_X^j N_Y^0 / N^t; \quad dN_Y^j/dt = \gamma_Y N_X^0 N_Y^j / N^t. \tag{8.1}$$

From these equations, under mutational changes among types, it is expected that the relations $N_X^0/N_Y^0 = \gamma_X/\gamma_Y$, $N_X^0/N_X^i = 1/F$, and $N_Y^0/N_Y^i = 1/F$ are eventually satisfied.

We have carried out stochastic simulations by randomly colliding a pair of molecules in the preceding model. Indeed, even with our stochastic simulation, the above number distribution is approached as N is increased.

However, when N is small, under repeated division processes of cells with selection, a significant deviation from this distribution appears. When the total number of molecules N is small, a state occurs satisfying $\langle N_Y^0 \rangle \approx 2 - 10$, $\langle N_Y^i \rangle \approx 0$ ($i = 1, 2, \ldots, F$). Since $F \gg 1$, such a state with $\langle N_Y^0 \rangle / \langle N_Y^i \rangle > 1$ is not expected from the preceding estimate by the rate equation.

This state with few active Y molecules (Y^0) and almost zero inactive Y molecules (Y^i) is established under the following conditions: (1) the number of molecules in the cell is not too large, (2) the number of types of inactive molecules (F) is large, and (3) there is sufficient difference between the growth speeds of the two kinds of molecules

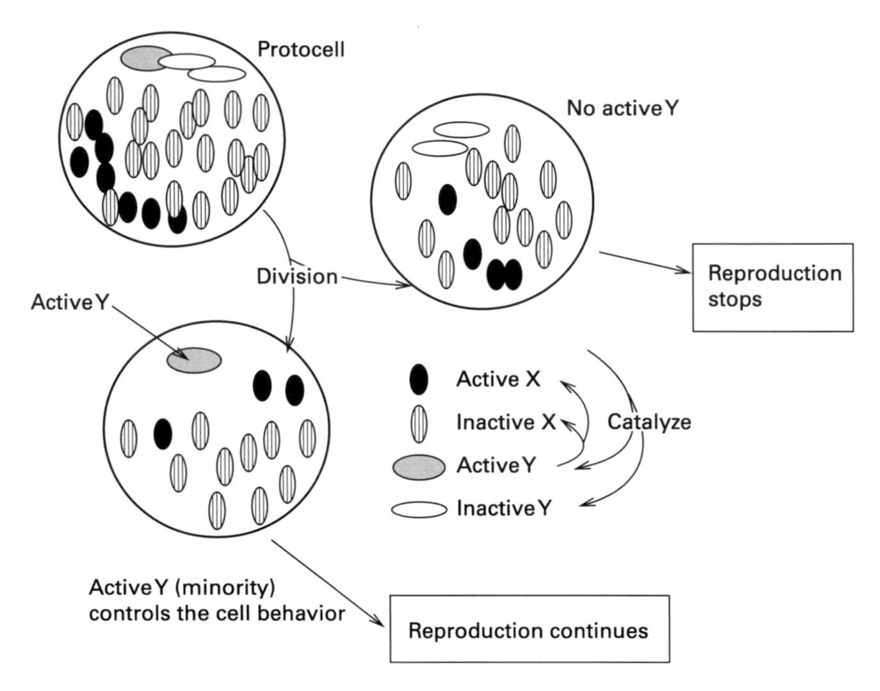

Figure 8.2
Schematic representation of the minority controlled state.

(X and Y). Because of the finiteness in molecule numbers, fluctuations arise to produce such initial conditions; such rare fluctuations, once they occur, are preserved, since a cell with such a composition can continue reproduction (see figure 8.2).

Here, the importance of the smallness in the number of Y molecules is twofold. First, large fluctuation in the number of Y molecules is essential to reach the above state. In spite of a large variety of inactive Y molecules, a rare state without such molecules is selected. This is because the smallness in the number of Y molecules allows for the possibility to reach such a state through fluctuation. On the other hand, once such a rare state is reached, it is preserved, because the probability of producing inactive Y molecules is rather low, since the number of active Y molecules is very small. Hence, a rare state that suppresses inactive Y molecules, once established, is preserved.

In this state with few Y^i molecules, the active Y molecule is a carrier for heredity, in the sense that the molecule has the following properties (Kaneko and Yomo, 2002):

Preservation Property The active Y molecules are preserved well over generations. Indeed, a state with a number of active Y molecules between ~2 and 6, with almost zero inactive Y molecules, is selected and preserved. The realization of such a state is very rare from the calculation of probability, but, once reached, it is preserved over generations. Also, the number fluctuation of Y molecules is much smaller than that of X, even after being normalized by the average number.

Control Property Consider a structural change in Y molecules, which may occur as a replication error and cause a change of catalytic activity. Since the number of active Y molecules is small, and all the X molecules are catalyzed by them, this influence is enormous. The synthesis speed of a protocell should change drastically. On the other hand, a change in X molecules has a weaker influence, since there are many active X molecules, and the influence of a change in each molecule is averaged out. Hence, a change in the Y molecules has a crucial influence on the cell behavior, in contrast to a change in the X molecules.

Summing up the argument in this section, the molecular species with slower replication speed and (accordingly) a minor population, comes to possess the properties for heredity. The state controlled by the minority molecular species is termed the minority-controlled state (MCS).

Note that the compartment is essential to the establishment of this MCS. Indeed, the rare cell state with no inactive Y molecules is selected, since a "cell" with inactive Y molecules cannot reproduce itself repeatedly. The necessity of compartmentalization for eliminating inactive states was already discussed by Szathmáry and Demeter (1987) in the *stochastic corrector model.* Koch (1984), on the other hand, has pointed out the importance of minority molecules with regard to evolvability which will be discussed later (but not with regard to controllability). Also, the kinetic suppression of the fluctuation of Y molecules is stronger than expected (Kaneko and Yomo, 2002) according to the standard statistical analysis adopted by Koch.

8.2.4 Significance of Minority-Controlled State

As the replication process is entirely facilitated by catalytic activity, the growth speed of the protocell depends on the catalytic activity of molecules inside it. With some change to the molecules, the protocells containing greater catalytic activity will be selected through evolution, leading to the selection of molecules with higher catalytic activity. Because only a few Y^0 molecules exist in the MCS, a change to one of them strongly influences the catalytic activity of the protocell, as supported by the control property. On the other hand, a change to X molecules has a weaker influence, on average: If N molecules change their activity randomly, the variance of the change in the average activity is reduced to $1/N$, as Koch (1984) also points out. Since the change of the minority molecules is not smeared out by this law of large numbers, an important characteristic of MCS is evolvability. (By evolvability we simply mean the ability for evolution. We are not discussing the phenotype-genotype relationship for evolution discussed in recent studies, e.g., de Visser et al., 2003; Kaneko and Furusawa, 2006; Wagner and Altenberg, 1996; West-Eberhard, 2003.)

This MCS has a positive feedback process to stabilize the state itself. Since the preservation of minority molecules is essential to a cell with MCS, a new selection pressure emerges to further ensure the preservation of the minority molecule in the offspring cells (otherwise the reproduction of the cell is highly damaged). A machinery to guarantee the faithful transmission of the minority molecule should evolve, further strengthening the preservation of the minority molecule. Hence, heredity evolves just as a result of kinetic phenomena in a reproducing protocell consisting of mutually catalytic molecules.

Once this faithful transmission of a minority molecule has evolved, then other chemicals that are synthesized in connection with it are also likely to be transmitted. Then, more molecules are transmitted in conjunction with the minority molecule. With more molecules catalyzed by the minority molecule in this evolution, the machinery to take better care of minority molecules will also evolve, since the latter are involved in reactions for the synthesis of many other molecules. Hence, the MCS allows for coevolution for better transmission of minority molecules and for coding of more information, leading to the separation of genetic information from other molecules carrying metabolism. This scenario provides one possible answer to the question of how genetic takeover progressed, which is unanswered in Dyson's study (1985).

Remark I In the model, we assumed that the replication rate of Y molecules was smaller. One might wonder whether, through the evolution of the catalytic activity and replication rates of the molecules, the difference in the catalytic activities between X and Y molecules may be lost. We have tested this problem by allowing for the evolution of catalytic activities of molecules, and found that the difference is preserved through evolution (Kaneko and Yomo, 2002). Hence, the MCS is stable, in contrast to the result from the stochastic corrector model (Grey, Hutson, and Szathmáry, 1995). This is because the evolution of the catalytic activity of Y molecules is accelerated because of its minority, so that the replication rate of X is increased and the difference between the replication rates of X and Y molecules is maintained.

Remark II Of course, even in present-day organisms, some genes exist in multiple copies—plasmids in bacteria, genes in the macronucleus in ciliates, and so forth. These genes are transferred to their descendants. In this sense, not all genes are in the minority. However, even in these cases, genes with multiple copies are often less important than the genes in the minority with regard to control. (In ciliates, micronucleus rather than macronucleus, is transferred through sexual recombination, and genes in macronucleus are "reset" with it.)

8.3 Theoretical Issue 2: Origin of Recursive Production and Evolvability

8.3.1 A Reaction Network Model with Compartment

A cell consists of several replicating molecules that mutually help in their synthesis and keep some synchronization for replication. How is such recursive production maintained while keeping a diversity of chemicals? This recursive production is not complete, and a slow "mutational" change occurs over generations, leading to evolution. How are evolvability and recursive production compatible?

In section 8.2, we considered a system consisting of two kinds of molecules for simplicity's sake. To study the general features of a system with mutually catalyzing molecules, however, it is important to consider a system with a variety of chemicals (k molecular species), forming a mutually catalyzing network (see figure 8.3). The molecules replicate through catalytic reactions, so that their numbers within a cell increase. Again, when the total number of molecules exceeds a given threshold (here we used $2N$), the cell divides into two, with each daughter cell inheriting half of the molecules of the mother cell, chosen randomly. Here we choose a random catalytic network, that is, a chemical species catalyzes the synthesis of some other randomly chosen chemical as

$$X^i + X^j \to 2X^i + X^j, \tag{8.2}$$

with $i, j = 1, 2, \ldots, k$. The connection rate of the catalytic paths is given by p for each chemical, while the rate of each reaction depends on the catalytic activity of the enzyme, denoted by $c(j)$. We assume that the catalytic activity $c(j)$ is a random number, chosen from the interval $[0, 1]$.

The replication is accompanied by some "error." Instead of the correct replication of the molecule X^i in the preceding equation, one of the other k molecular species is synthesized with an error rate μ.

Note that this type of reaction network has been studied following Eigen and Shuster's (1979) work on the hypercycle (see, e.g., Kauffman, 1993; Stadler, Fontana, and Miller, 1993; Stadler and Schuster, 1990). The most important difference from earlier studies is the use of cell division resulting from an increase in the number of molecules through catalytic reactions (see also Zintzaras, Mauro, and Szathmáry, 2002, for comparison with the stochastic corrector model and the hypercycle).

This model has four basic parameters: the total number of molecules N, the total number of molecular species k, the mutation rate μ, and the reaction path rate p. By carrying out simulations of this model, by choosing a variety of parameter values for N, k, μ, and p, and also by taking various random networks, we have found that the behaviors can be classified into the following three types (Kaneko, 2002, 2003a, 2005):

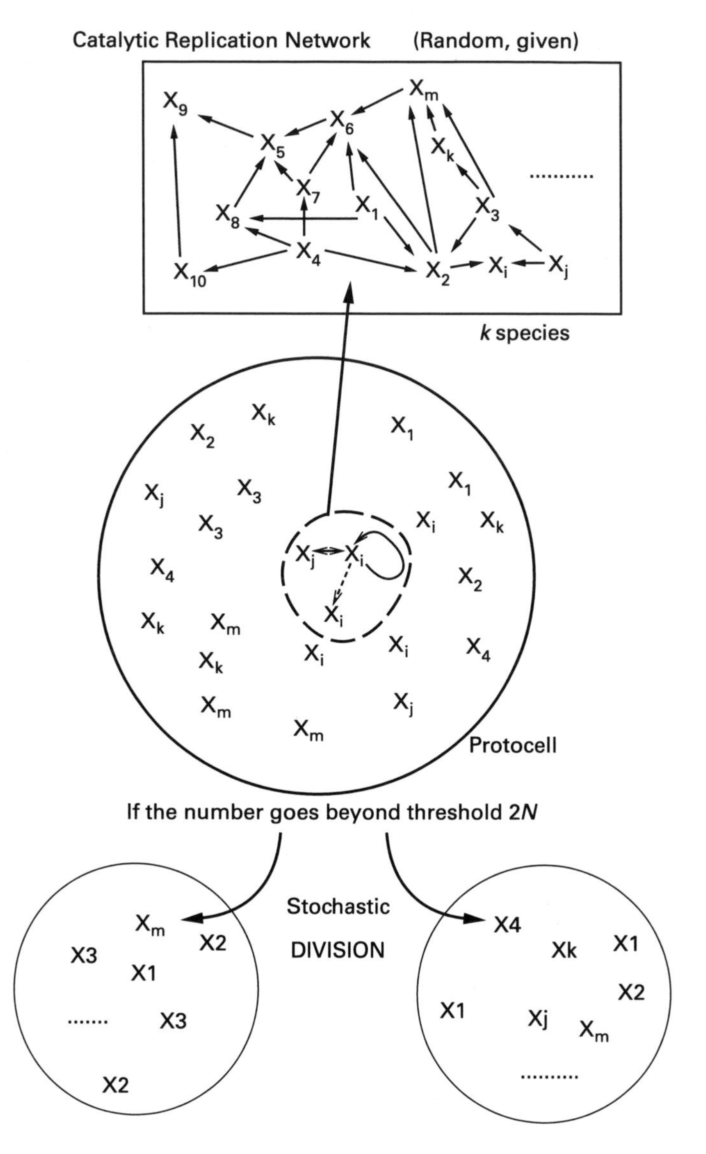

Figure 8.3
Schematic representation of our model with mutually catalyzing network.

A. fast switching states without recursiveness;

B. achievement of recursive production with similar chemical compositions; and

C. switching over several quasi-recursive states.

In phase A, there is no clear recursive production and the molecular species that is dominant in the population changes frequently. At one time step, some chemical species are dominant but only a few generations later, this information is lost, and the number of molecules in this species goes to zero (see figure 8.4a, color plate 4a). No stable mutual catalytic relationships are formed. Indeed, by autocatalytic reactions, the population of one dominant species can be amplified, but it is soon replaced by another chemical that is catalyzed by it.

Here, the time required for reproduction of a cell is huge compared with case B, and will go beyond the time scale accessible experimentally. Hence, this phase is not suitable for a protocell.

In phase B, on the other hand, a state with recursive production is established, and the chemical composition is stabilized so that it is not altered much by the division process. Once reached, this state lasts across all the generations ($O(10^4)$) in the simulation (see figure 8.4b, color plate 4b).

The recursive state ("attractor") here is not necessarily a fixed point, and the number of molecules may oscillate in time. Nevertheless, the overall chemical compositions remain within certain ranges: For example, the major species (i.e., those synthesized by themselves, not by an error) are not altered over the generations.

Generally, all the observed recursive states consist of 5 to 10 species, except for those species with low numbers of molecules, which exist only as a result of replication errors. These 5 to 10 chemicals form a mutually catalytic network, as will be discussed later (see figure 8.5). The members of these 5 to 10 species do not change over generations, and the chemical compositions are transferred to the offspring cells. Once reached, this state is preserved over the whole range of simulation steps.

In the phase C, after one recursive state lasts over many generations (typically a thousand), a fast switching state appears until a new (quasi-)recursive state appears. As shown in figure 8.4c (color plate 4b), for example, each (quasi-)recursive state is similar to that in the phase B, but in this case, after many generations, it is replaced by a fast switching state as in phase A. Then the same or a different (quasi-)recursive state is reached again, which lasts until the next switch occurs.

Although the behavior of the system depends on the choice of network, there is a general trend regarding phase changes from A, to C, and then to B with the increase of N, or with the decrease of k or p. We also note that the state with the (quasi-)recursive production corresponds to the "composome" in the model by Segré and Lancet (Segré, Ben-Eli, and Lancet 2000).

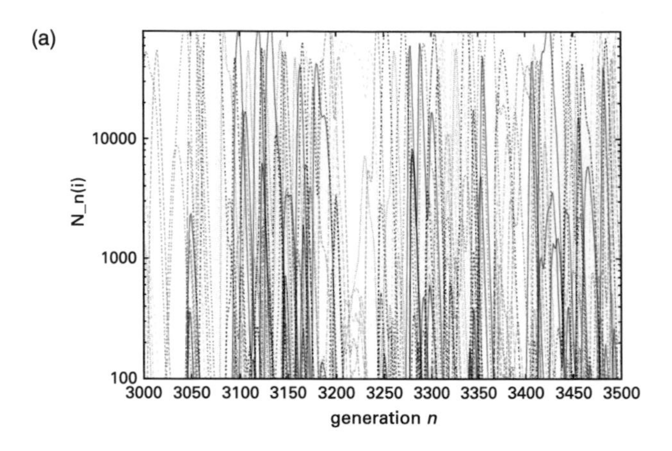

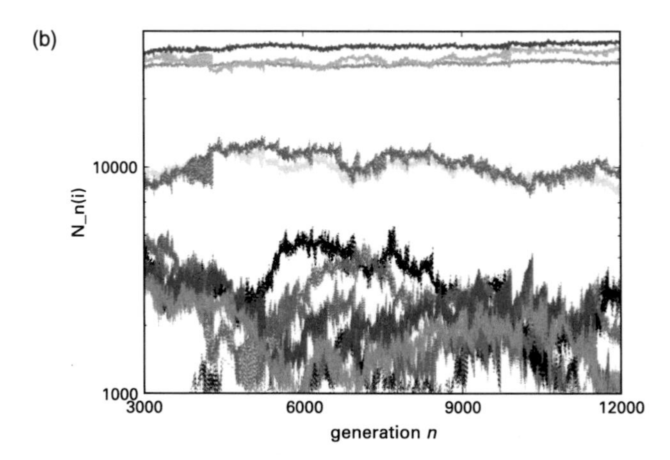

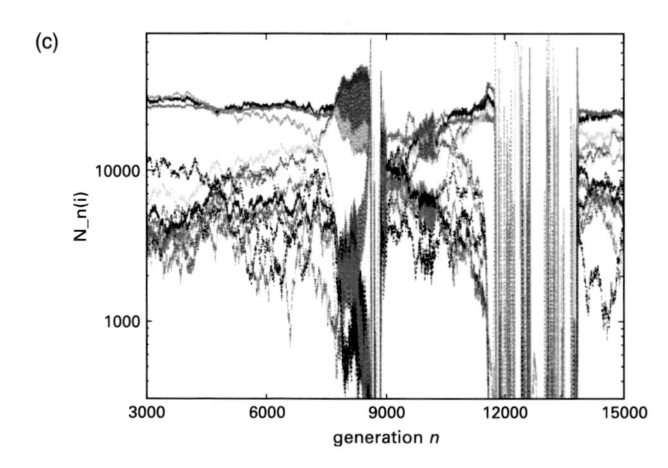

Figure 8.4 (color plate 4)
The number of molecules $N_n(i)$ for the species i is plotted as a function of the generation, n (after n division events). In (a), a random network with $k = 500$ and $p = 0.2$, and in (b) a random network with $k = 200$ and $p = 0.2$, was adopted, with $N = 64000$ and $\mu = 0.01$. Only some species (species whose population becomes large at some generation) are plotted. In (a), dominant species change successively by generation, whereas in (b) three quasi-recursive states are observed. Reproduced from Kaneko (2003a).

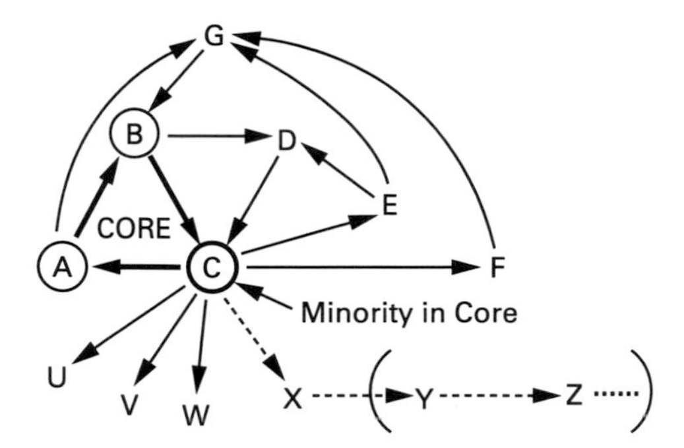

Figure 8.5
An example of mutually catalytic network in our model. The core network for the recursive state is shown by circles, whereas parasitic molecules ($X, Y, \ldots$), connected by broken arrows, are suppressed at the (quasi-)recursive state.

8.3.2 Maintenance of Recursive Production

How is recursive production sustained in the phase B? We have already discussed the danger of parasitic molecules that have lower catalytic activities and are catalyzed by molecules with higher catalytic activities. As discussed in section 2.1, such parasitic molecules can invade the hypercycle. Given the error in replication and the fluctuation of the species that make up the network, the recursive production state could be destabilized.

Of course, use of a compartment is essential to select a state maintaining recursive production. Still, it is not sufficient, as demonstrated by phase A, which exists commonly when the number of chemical species k and reaction path rate p are large. To answer the question of the stability of recursive states, we need to examine salient features for such recursive states by simulating several reaction networks. The following summarizes the unveiled logic for maintaining recursive states.

Stabilization by Intermingled Hypercycle Network The 5 to 12 species in the recursive state form a mutually catalytic network, such as that shown in figure 8.5. This network has a *core hypercycle network*, shown by thick arrows in the figure. This core hypercycle has a mutually catalytic relationship, as A catalyzes B, B catalyzes C, and C catalyzes A. However, these chemical species are also connected by other hypercycle networks such as $G \to D \to B \to G$, and $D \to C \to E \to D$, and so forth. The hypercylces are intermingled to form a network. Coexistence of core hypercycles and other attached hypercycles is common to the recursive states we have found in our model.

This intermingled hypercycle network (IHN) leads to stability against parasites and fluctuations. Assume that a molecule parasitic to one species in the members of IHN appears, say, X as a parasite to C in figure 8.5. The species X may reduce the number of species C. If only a single hypercycle ($A \rightarrow B \rightarrow C \rightarrow A$) existed, the population of all members—A, B, C—would be easily decreased by this invasion of parasitic molecules, resulting in the collapse of the hypercycle. In the present case, however, other parts of the network compensate for the decreased population of C by the parasite, so that the populations of A and B are not decreased as much. Hence, complexity in the hypercycle network leads to stability against the attack of parasite molecules. The analysis of dynamical systems also shows that IHN is relevant to stability against fluctuations (Kaneko, 2005).

Here, the minority molecule in the core network is most important for sustaining the recursive state. Consider a core catalytic cycle with $A \rightarrow B \rightarrow C \rightarrow A$. The number of molecules, N_j, of species j is in the inverse order of their catalytic activity, $c(j)$, that is, $N_A > N_B > N_C$ if $c(A) < c(B) < c(C)$. Because a molecule with higher catalytic activity helps the synthesis of others more, this inverse relationship is expected. Here, the C molecule is catalyzed by a molecular species with lower activity, but larger populations. Hence, the parasitic molecular species cannot easily invade to disrupt this mutually catalytic network. The stability in the minority molecule is also accelerated by the complexity in IHN. Within the IHN, this minority molecular species is involved in several hypercycles (see C in figure 8.5), because it has higher catalytic activity. This, on the one hand, demonstrates the prediction in section 8.2.3 about minority molecules, while on the other hand, it leads to the suppression of the fluctuation in the number of minority molecules.

8.3.3 Switching and Evolvability: Relevance of Minority Molecules in the Core Hypercycle

Next, we discuss the mechanism of switching. When the total population of molecules in a cell is not very large, the number of fluctuations is relatively large, especially with regard to the minority molecule. Although the amount of the minority molecule in the core may decrease as a result of fluctuations, the amount of some parasitic molecules will increase instead. Then, the minority molecule in the core network may be taken over by the parasitic molecular species. If this happens, all the other molecular species in the original network lose the main source of catalysis for their synthesis. Then successive switching of dominant parasitic molecular species occurs within a few generations, as in phase A. After the switching stage, another (or possibly the same) catalytic network is formed, where existence of a minority molecular species with a higher catalytic activity stabilizes this (quasi-)recursive state. This is the process that occurs in phase C.

Note that phase C gives a basis for evolvability, since a novel, (quasi-)recursive state with different chemical compositions is visited successively. To address this point, we have also studied the present model by including evolution (Kaneko, 2005). Starting from a network of molecules with low catalytic activity, the evolution to a network with a higher activity is found to occur, through successive switching of (quasi-)recursive states as realized in phase C. In this case, this evolution is triggered by the extinction of minority molecules in the core network. Once the amount of this chemical species goes to zero, the original IHN collapses, to be replaced by another IHN later. Evolution from a rather primitive cell consisting of low catalytic activities to that with higher activities is possible accordingly, with the aid of (relatively) minor molecules in the core network. This demonstrates the relevance of minority control to evolution.

Switching over several quasi-stationary states has been generally studied in dynamical systems as chaotic itinerancy (Kaneko, 1990; Kaneko and Tsuda, 2003; Tsuda, 1991), whereas Jain and Krishna (2002) analyzed transitions over autocatalytic sets in a linear reaction network by introducing the mutation of the network only after the temporal evolution of growth dynamics reaches a stationary state. (Such artificial separation between the growth stage and the mutation in the network paths is not introduced in our model.)

8.3.4 Statistical Law

Since the number of molecules is not huge, there are relatively large number fluctuations. Is there some statistical law for such fluctuations?

Since the total number of molecules N is not large, the number fluctuation of each molecule is inevitable even for a recursive production phase B. Although the fluctuations in the core network (A, B, C in figure 8.5) are typically small, for other species in the hypercycle network, the number fluctuation is much larger, and the distribution is close to log-normal distributions

$$P(N_i) \approx \frac{1}{N_i}\, exp\left(-\frac{(log\ N_i - \overline{log\ N_i})^2}{2\sigma}\right), \tag{8.3}$$

rather than the normal distribution. The origin of the log-normal distributions can be understood by the following rough argument: For a replicating system, the growth of the number (N_m) of molecules of the species m is given by

$$dN_m/dt = AN_m, \tag{8.4}$$

where A is the average effect of all the molecules that catalyze m. We can then obtain the estimate

$$d\log N_m/dt = \bar{a} + \eta(t) \tag{8.5}$$

by replacing A with its temporal average $\bar{a}$ plus fluctuations $\eta(t)$ around it. If $\eta(t)$ is approximated by a Gaussian noise, the log-normal distribution for $P(N_m)$ is suggested. Note that this argument is valid only if $\bar{a} > 0$ (since a random walk to the region with $N_m < 1$ is not allowed). Then, N_m in equation 5 diverges with time, but here, the cell divides in two before the divergence becomes significant. Hence, we need further elaboration of this rough argument; but the emergence of log-normal distribution itself is shown to be rather universal.

Indeed, Furusawa and the author (Furusawa and Kaneko, 2003; Furusawa et al., 2005) have studied several models of minimal cells consisting of catalytic reaction networks, without assuming the replication process itself. In this class of catalytic reaction networks, a huge number of chemical species coexists to form a recursive production. There again, the number of molecules of each chemical species over all cells generally obeys the log-normal distribution for a state with recursive production. The existence of such log-normal distributions is also experimentally confirmed in bacteria using fluorescent proteins (Furusawa et al., 2005). (Furthermore, there is a universal statistical law on the average number over all molecule species. The rank-abundance law obeys a universal power law, as is also confirmed in the data of gene expression of the present cells, see Furusawa and Kaneko, 2003.)

The fluctuations in the log-normal distribution, however, are generally very large, and range over orders of magnitude. This is in strong contrast to our naive impression that a process in a cell system must be well controlled. To have a more precise replication process, some molecules that may deviate from this log-normal distribution should be necessary.

Indeed, the minority control mechanism suggests the possibility of suppressing the fluctuation, as discussed in section 8.2.2. For a recursive production system, some mechanism to decrease the fluctuation in minority molecular species must be evolved. Indeed, the fluctuation in the chemicals among the core network in the IHN is highly suppressed, and deviates from the log-normal distribution. Within the core network, the number fluctuation of the minority species is lowest.

This suppression of the fluctuation in a reproducing system is an important issue, not only from a theoretical viewpoint, but also in constructing a stable protocell experimentally. In fact, if the number fluctuation of all the molecular species is large, stable reproduction of a protocell should be difficult. Suppression of fluctuation by the core network and minority species will be important in constructing a protocell.

8.4 Experiment

8.4.1 Steps to Replicating Protocells

Recently, some experiments have constructed minimal replicating systems in vitro. Let us consider synthesis of a protocell that reproduces itself. For this, we need the following steps:

1. A system consisting of chemicals (polymers) with some catalytic activity, which reproduces itself as a set, even though the reproduction may not be precise.

2. A compartment structure that uses a membrane to distinguish its inside from the outside environment. This membrane also grows through chemical reactions, so that it divides when its size is large.

3. Within the protocell, surrounded by the membrane, the reaction system (1) works, while the synthesis of membrane is coupled to this internal reaction system.

4. The internal reaction process and the synthesis of membrane work in some synchrony, so that a system with membrane and internal chemicals is reproduced recursively.

In our complex systems biology project, together with T. Yomo and T. Sugawara, we have achieved the following steps:

a. In vitro autonomous replication of DNA and protein for step 1;

b. Repetitive replication of liposomes for step 2;

c. Protein synthesis from RNA as well as the amplification of RNA within a liposome for step 3;

We discuss these three topics briefly.

8.4.2 In Vitro Autonomous Replication; Mutual Synthesis of DNA and Protein

As an experiment corresponding to this problem, we describe an in vitro replication system, constructed by Yomo's group (Matsuura et al., 2002).

In general, proteins are synthesized from the information in DNA through RNA, whereas DNA is synthesized through the action of proteins. As a set of chemicals, they autonomously replicate themselves. Now, simplifying this replication process, Matsuura and coworkers (2002) constructed a replication system including DNA and DNA polymerase. This DNA polymerase is synthesized by the corresponding genes in the DNA, while it works as the catalyst for the corresponding DNA. Through this mutual catalytic process, the chemicals replicate themselves. Roughly speaking, the polymerase in the experiment corresponds to X in our model in section 8.2, while the polymerase gene corresponds to Y.

As for the amplification of DNA, polymerase chain reaction (PCR) is a standard tool for molecular biology. In this case, however, enzymes that are necessary for the replication of DNA must be supplied externally, and it is not a self-contained autonomous replication system. In the experiment here, though the PCR is adopted as one step of the experimental procedure, the enzyme for DNA synthesis (DNA polymerase) also replicates in vitro within the system. Of course, some material, such as amino acids or adenosine triphosphate (ATP), have to be supplied, but otherwise, the chemicals are replicated by themselves (see figure 8.6 for the experimental procedure).

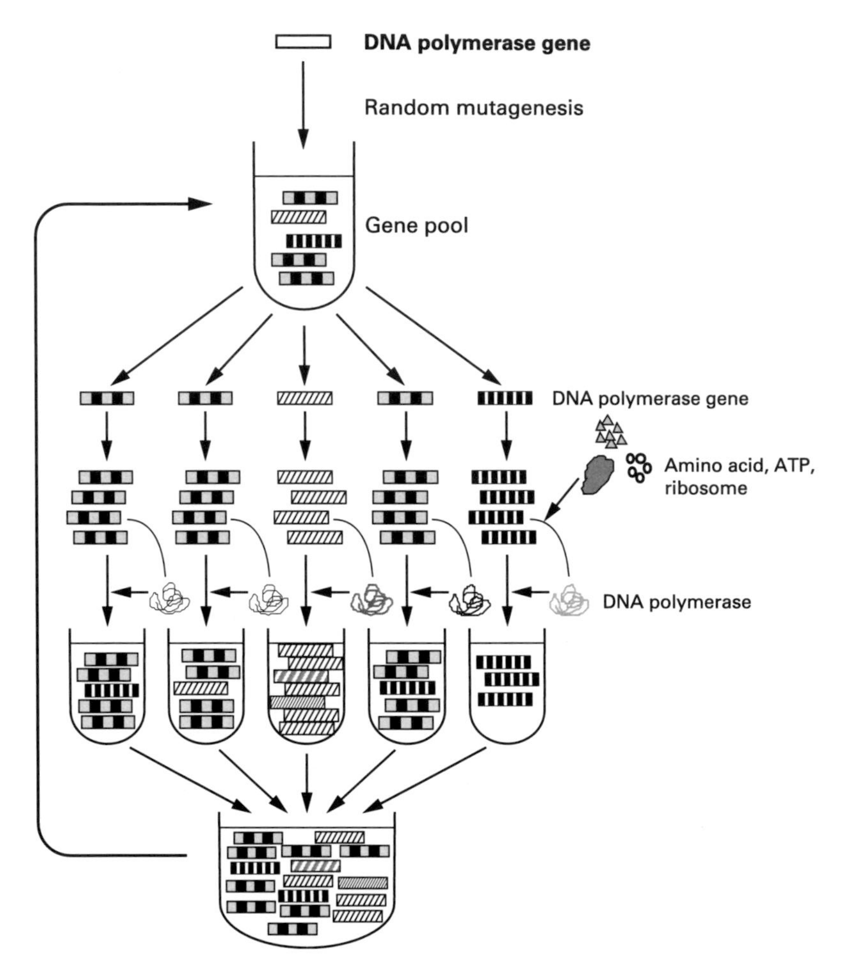

Figure 8.6
Schematic illustration of an in vitro autonomous replication system consisting of DNA and DNA polymerase. (See Matsuura et al., 2002, for details.) Supplied by courtesy of Matsuura et al. (2002).

Now, at each step of replication, about $2^{30} \sim 2^{40}$ DNA molecules are replicated. Here, of course, there are some errors. These errors can occur in the synthesis of enzymes, and also in the synthesis of DNA. With these errors, DNA molecules with different sequences appear. Now a pool of DNA molecules with a variety of sequences is obtained as a first generation.

From this pool, the DNA and enzymes are split into several tubes. Then, materials with ATP and amino acids are supplied, and this replication process is repeated. In other words, the "test tube" here plays the role of "cell compartmentalization." Instead of autonomous cell division, the split into several tubes is implemented externally, in the context of the experiment.

In this experiment, instead of changing the synthesis speed of molecules or N as in the theoretical model of section 8.2.1, one can control the number of DNA molecules by changing the condition of how the pool is split into several test tubes. Indeed, Yomo's group studied the two distinct cases—the pool split into tubes containing a single DNA molecule in each and the pool split into tubes containing 100 DNA molecules—and compared the results to test the minority control theory.

First, we describe the case with a single molecule of DNA in each tube. Here, the sequence of DNA molecules could be different in each of 10 separated tubes, since there are replication errors. If so, the activity of DNA polymerase, and accordingly, the copy number of DNA therein, would also be different. The contents of each tube are mixed, and this soup of chemicals is used for the next generation. In this soup, the DNA molecules (or their direct mutants) with higher replication rates occupy a larger fraction. Next, a (different) single DNA is selected from the soup for each of 10 tubes, and the same procedures are repeated. Accordingly, the probability that a DNA molecule with a higher reproduction activity will be selected for the next generation is higher. The self-replication activity from this soup is plotted successively over generations in figure 8.7. As shown, the self-replication activity is not lost (or in some cases can evolve), although it varies with each tube in each generation.

One might say that the maintenance of replication is not surprising at all, since a gene for the DNA polymerase is included in the beginning. However, an enzyme possessing such catalytic activity is rare. Indeed, with mutations, some proteins that lost such catalytic activity but are synthesized in the present system could appear, and might take over the system. Then the self-replication activity would be lost. In fact, this is nothing but Eigen's error catastrophe, discussed in section 8.2.1. So why is self-replication activity maintained in the present experiment?

The discussion in section 8.2 makes the answer clear. In the experiment, mutants that lose catalytic activity are much more common (i.e., F times larger according to section 8.2.1). Still, the number of such molecules is suppressed. This is possible, first, because the molecules are in a cell. In the experiment, also, they are in a test tube, that is, in a compartment. The selection works for the compartment, not for each molecule. So the tube (cell) with a lower activity produces a smaller number of offspring. Hence, compartmentalization is one essential factor in the maintenance of catalytic activity (Altmeyer and McCaskill, 2001; Boerilst and Hogeweg, 1991; Eigen, 1992; Maynard-Smith, 1979; Szathmary and Demeter, 1987; Szathmáry and Maynard-Smith, 1997) and another important factor is that each compartment (cell) contains a single or very few DNA molecules.

Recall that in the theory, if the number of Y molecules is larger, inactive Y molecules surpass the active ones in the population. To check this point, Matsuura and coworkers split the chemicals in the soup so that each tube had 100 DNA molecules instead of a single one. Otherwise, they adopted the same procedure. In other words, this corresponds to a cell with 100 copies of the genome. The change in self-replication

(a)

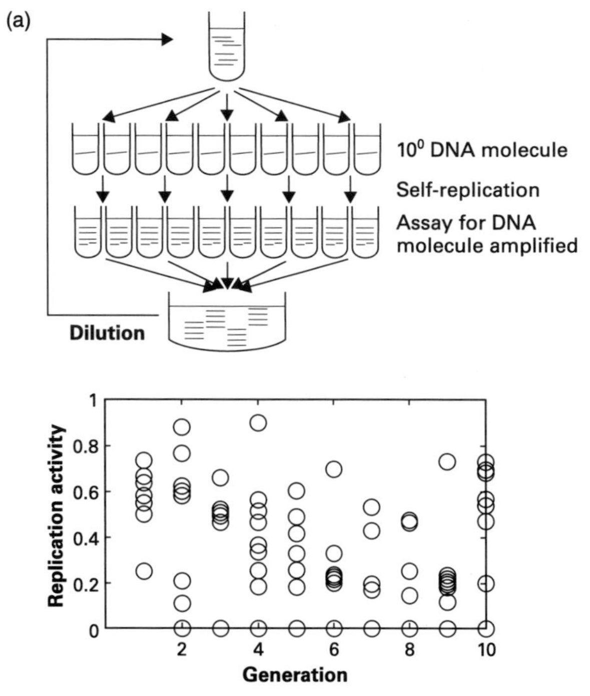
10^0 DNA molecule
Self-replication
Assay for DNA molecule amplified
Dilution
Replication activity
Generation
0
0.2
0.4
0.6
0.8
1
2
4
6
8
10

(b)

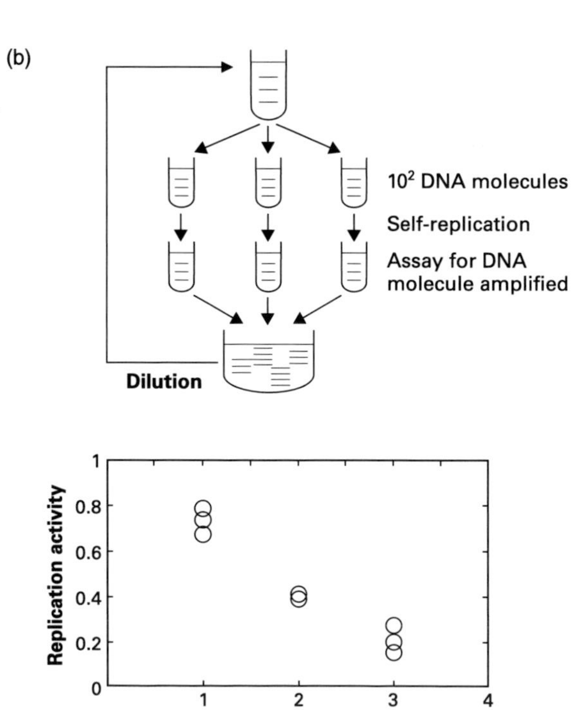
10^2 DNA molecules
Self-replication
Assay for DNA molecule amplified
Dilution
Replication activity
Generation
0
0.2
0.4
0.6
0.8
1
2
3
4

activity in the experiment is plotted in the lower panel of figure 8.7. As shown, each generation loses self-replication activity, and after the fourth generation, capacity for autonomous replication is totally lost.

Note that because these 100 DNA molecules can be different, a horizontal gene transfer occurs, as in a viral system with a high multiplicity of infection (Szathmáry, 1992). This point, however, is not essential to the following discussion on the relevance of minority DNA molecules to the maintenance and evolution of reproduction. Furthermore, as long as the error rate is high, 100 DNA molecules are statistically distributed even if there is no horizontal transfer.

When there are many DNA molecules, there should be variation in them. In each tube, the self-replication activity is given by the average of the enzyme activities from these 100 DNA molecules. Although the catalytic activity of each molecule is different, the variance of the average activity by tube is reduced, since the variance of the average of N variables decreases in proportion to $1/N$, according to the central limit theorem of probability theory. Hence, the average catalytic activity does not differ much by tube.

Furthermore, mutants with higher catalytic activity are rare, so most changes in the gene lead to reduced or null catalytic activity. On average, catalytic activity after mutations thus decreases, whereas the fluctuation around this average is very small, and selecting for a tube with a higher catalytic activity does not work. Deleterious mutations therefore remain in the soup, and the self-replication activity will be lost over generations. In other words, the selection works only when the number of information carriers in a replication unit is very small. This relevance of minority molecules was discussed by Koch (1984) and is consistent with our discussion in section 8.2.

In summary, the experiment found that replication is maintained even under deleterious mutations, but only when the number of DNA polymerase genes is small. The information contained in the DNA polymerase genes is preserved, and the system also has evolvability. This is made possible by the maintenance of rare fluctuations. These experimental results are consistent with the minority control theory described here.

8.4.3 Replicating Liposomes

All present-day cells are surrounded by membranes that consist of lipid bilayers. This membrane allows us to distinguish between the inside and outside of cells, while

Figure 8.7
Self-replication activities for each generation, measured as described in the text. The activities for 10 tubes are shown. (a) Result from a single DNA molecule where the next generation is produced mostly from the top DNA. Although activities vary by tube, those with higher activity are selected, so that the activities are maintained. (b) Result from 100 DNA molecules. Activity decays within three generations. Provided by courtesy of Matsuura et al. (2002).

catalytic reactions progress within the membrane. This unit, with lipid-layer structure, is called a liposome or vesicle (the terms are used interchangeably).

In general, oil molecules often form a bilayer membrane in an aqueous solution. This membrane often forms a closed spherical structure. If resource molecules are supplied, the membrane surface increases. Because of the balance with the surface tension, this growth cannot continue forever, and a large membrane will become destabilized. In some cases, this results in a division of this closed membrane liposome. Indeed, Luisi succeeded in synthesizing such division processes of liposomes (Bachmann, Luisi, and Lang, 1992), and the synthesis of replicating vesicles has been studied extensively since then (see Hanczyc, Fujikawa, and Szostak, 2003; Szostak, Bartel, and Luisi, 2004).

Quite recently, Sugawara's group succeeded in constructing a stable replication process that continues from daughter to granddaughter. With a suitable setup of chemical conditions, giant vesicles increase in size by absorbing nutrient chemicals, and then divide in two, and the daughters continue dividing (Takakura, Toyota, and Sugawara, 2003); further repetition of divisions is estimated from the analysis of flow cytometry (Toyota et al., 2005).

8.4.4 Toward the Synthesis of Artificial Replicating Cell

To achieve the next step, nucleic acids (such as RNA or DNA) as well as proteins have to be synthesized within the liposome. Since the environment within the liposome is very "oily," this step is not easy to achieve.

Recently, Yomo and his colleagues succeeded in amplifying RNA and synthesizing protein from mRNA (Yu et al., 2001; see also Noireaux and Libchaber, 2004, for discussion of a cell-free expression system encapsulated in a vesicle). Within the liposome of the diameter 1 μm, the transcription of RNA—that is, the synthesis of proteins from the mRNA—is shown to occur. By making the corresponding protein fluorescent (using Green Fluorescent Protein), Yu and coworkers (2001) have shown by the measurement of fluorescence that the synthesis indeed occurs within the liposome. This rate of protein synthesis is high enough to be measured by the cell-sorter (flow cytometry) using the fluorescence of protein (see chapter 7).

By selecting liposomes exhibiting higher fluorescence with the help of the cell sorter, one can now select a "better" RNA molecule with higher protein synthesis within the liposome. Thus, the evolution to a better RNA molecule is possible. On the other hand, this selection process can also work in choosing a better "oil" molecule for liposomes; therefore, evolution to liposomes that allow for robust protein synthesis is possible. Accordingly, by repeating this evolutionary process, one can achieve the third step (from section 8.4.1) for artificial cells.

As discussed in chapter 7, the distribution of protein abundances within the liposome is also measured. Since the fluorescent protein is synthesized with the aid of other molecules (such as RNA), the fluctuations should be large, and the distribution will be log-normal if the argument of section 8.3.4 is valid. Indeed, the distribution obtained from the experiment is close to being log-normal (or sometimes much broader).

This large fluctuation may introduce a barrier to achieving the last step for the synthesis of artificial replicating cells—unification of the replication of liposome and of protein. Indeed, to achieve this unification, it is important to balance the two replication processes: If the replication of membrane is faster, then the inside ingredients will be sparse, but if protein synthesis is faster, the density of molecules will be too high and destroy the liposome. The two processes must be balanced to achieve recursive production.

To accomplish this balanced replication, some link between liposome growth and the internal protein synthesis is required, but this has not been realized yet. Overly large fluctuations must be avoided. Regulation of fluctuations by minority control and intermingled network, as discussed in sections 8.2 and 8.3, are important.

Another difficulty in achieving this unification lies in the interference of several processes. Success of each of steps 1 through 3 in section 8.4.1 assumes that each process is separate. When we try to combine the processes, they interfere with each other, resulting in the collapse of each process. To achieve the separation of processes, minority control mechanisms and the switching process described in section 8.3.4 may be relevant.

One might think there is a long road to constructing an autonomous replicating artificial cell. However, once a system with loose reproduction in any form is realized, such a "cell" can be an object for Darwinian selection processes. By selecting a relatively "better" reproducing cell through a cell sorter, we obtain a more reliable replication system.

Acknowledgments

I would like to thank Tetsuya Yomo, Kanetomo Sato, and Chikara Furusawa for discussions and collaborations. I would also like to thank anonymous referees for pointing out some earlier studies important to the present issue.

References

Altmeyer, S., & McCaskill, J. S. (2001). Error threshold for spatially resolved evolution in the quasispecies model. *Physical Review Letters*, *86*, 5819–5822.

Cairns-Smith, A. G. (1982). *Clay minerals and the origin of life*. Cambridge University Press.

Bachmann, P. A., Luisi, P. L., & Lang, J. (1992). Autocatalytic self-replicating micelles as models for prebiotic structures. *Nature*, *357*, 57–59.

Benner, S. A., & Sismour, A. M. (2005). Synthetic biology. *Nature Reviews Genetics*, *6* (7), 533–543.

Boerlijst, M., & Hogeweg, P. (1991). Spiral wave structure in pre-biotic evolution: Hypercycles stable against parasites. *Physica D*, *48*, 17–28.

de Visser, J. A. G. M., Hermisson, J., Wagner, G. P., Meyers, L. A., Bagheri-Chaichian, H., Blanchard, J. L., et al. (2003). Evolution and detection of genetic robustness. *Evolution*, *57*, 1959–1972.

Dyson, F. (1985). *Origins of life*. Cambridge: Cambridge University Press.

Eigen, M. (1992). *Steps towards life*. Oxford: Oxford University Press.

Eigen, M., & Schuster, P. (1979). *The hypercycle*. Berlin: Springer.

Furusawa, C., & Kaneko, K. (1998). Emergence of rules in cell society: Differentiation, hierarchy, and stability. *Bulletin of Mathematical Biology*, *60*, 659–687.

Furusawa, C., & Kaneko, K. (2003). Zipf's law in gene expression. *Physical Review Letters*, *90* (8), 8102.

Furusawa, C., Suzuki, T., Kashiwagi, A., Yomo, T., & Kaneko, K. (2005). Ubiquity of log-normal distributions in intra-cellular reaction dynamics. *Biophysics*, *1*, 25.

Grey, D., Hutson, V., & Szathmary, E. (1995). A re-examination of the stochastic corrector model. *Proceedings of the Royal Society of London B*, *262*, 29–35.

Hanczyc, M., Fujikawa, S. M., & Szostak, J. W. (2003). Experimental models of primitive cellular compartments: Encapsulation, growth, and division. *Science*, *302*, 618–622.

Jain, S., & Krishna, S. (2002). Large extinctions in an evolutionary model: The role of innovation and keystone species. *Proceedings of the National Academy of Sciences of the United States of America*, *99*, 2055–2060.

Kaneko, K. (1990). Clustering, coding, switching, hierarchical ordering, and control in networks of chaotic elements. *Physica D*, *41*, 137–172.

Kaneko, K. (1998). Life as complex systems: Viewpoint from intra-inter dynamics. *Complexity*, *3*, 53–60.

Kaneko, K. (2002). Kinetic origin of heredity in a replicating system with a catalytic network. *Journal of Biological Physics*, *28*, 781–792.

Kaneko, K. (2003a). Recursiveness, switching, and fluctuations in a replicating catalytic network. *Physical Review E*, *68* (3), 1909.

Kaneko, K. (2003b). *What is life? A complex systems approach*. Tokyo: University of Tokyo Press (in Japanese; English version published by Springer, 2006).

Kaneko, K. (2004). Constructive and dynamical systems approach to life. In A. Deutsch, M. Falcke, J. Howard, & W. Zimmerman (Eds.), *Function and regulation of cellular systems: Experiments and models* (pp. 213–224). Basel: Birkhaeuser Verlag.

Kaneko, K. (2005). On recursive production and evolvability of cells: Catalytic reaction network approach. *Advances in Chemical Physics*, *130*, 543–598.

Kaneko, K. (2006). *Life: An introduction to complex systems biology*. New York: Springer.

Kaneko, K., & Furusawa, C. (2006). An evolutionary relationship between genetic variation and phenotypic fluctuation. *Journal of Theoretical Biology*, *240*, 78–86.

Kaneko, K., & Tsuda, I. (1994). Constructive complexity and artificial reality: An introduction. *Physica D*, *75*, 1–10.

Kaneko, K., & Tsuda, I. (2003). Chaotic itinerancy. *Chaos*, *13*, 926–936.

Kaneko, K., & Yomo, T. (1997). Isologous diversification: A theory of cell differentiation. *Bulletin of Mathematical Biology*, *59*, 139–196.

Kaneko, K., & Yomo, T. (1999). Isologous diversification for robust development of cell society. *Journal of Theoretical Biology*, *199*, 243–256.

Kaneko, K., & Yomo, T. (2002). On a kinetic origin of heredity: Minority control in replicating molecules. *Journal of Theoretic Biological*, *214*, 563–576.

Kashiwagi, A., Noumachi, W., Katsuno, M., Alam, M. T., Urabe, I., & Yomo, T. (2001). Plasticity of fitness and diversification process during an experimental molecular evolution. *Journal of Molecular Evolution*, *52*, 502–509.

Kauffman, S. A. (1993). *The origin of order*. Oxford: Oxford University Press.

Ko, E., Yomo, T., & Urabe, I. (1994). Dynamic clustering of bacterial population. *Physica D*, *75*, 81–88.

Koch, A. L. (1984). Evolution vs. the number of gene copies per primitive cell. *Journal of Molecular Evolution*, *20* (1), 71–76.

Matsuura, T., Yomo, T., Yamaguchi, M., Shibuya, N., Ko-Mitamura, E. P., Shima, Y., & Urabe, I. (2002). Importance of compartment formation for a self-encoding system. *Proceedings of the National Academy of Sciences of the United States of America*, *99*, 7514–7517.

Maynard-Smith, J. (1979). Hypercycles and the origin of life. *Nature*, *280*, 445–446.

Mills, D. R., Peterson, R. L., & Spiegelman, S. (1967). An extracellular Darwinian experiment with a self-duplicating nucleic acid molecule. *Proceedings of the National Academy of Sciences of the United States of America*, *58*, 217.

Noireaux, V., & Libchaber, A. (2004). A vesicle bioreactor as a step toward an artificial cell assembly. *Proceedings of the National Academy of Sciences of the United States of America*, *101*, 17669–17674.

Segré, D., Ben-Eli, D., & Lancet, D. (2000). Compositional genomes: Prebiotic information transfer in mutually catalytic noncovalent assemblies. *Proceedings of the National Academy of Sciences of the United States of America*, *97*, 4112.

Sprinzak, D., & Elowitz, M. B. (2005). Reconstruction of genetic circuits. *Nature*, *438*, 443–448.

Stadler, P. F., Fontana, W., & Miller, J. H. (1993). Random catalytic reaction networks. *Physica D*, *63*, 378.

Stadler, P. F., & Schuster, P. (1990). Dynamics of small autocatalytic reaction networks I: Bifurcations, permanence and exclusion. *Bulletin of Mathematical Biology*, *52*, 485–508.

Szathmáry, E. (1992). Natural selection and the dynamical coexistence of defective and complementing virus segments. *Journal of Theoretical Biology*, *157*, 383–406.

Szathmáry, E., & Demeter, L. (1987). Group selection of early replicators and the origin of life. *Journal of Theoretical Biology*, *128* (4), 463–486.

Szathmáry, E., & Maynard Smith, J. (1997). From replicators to reproducers: The first major transitions leading to life. *Journal of Theoretical Biology*, *187*, 555–571.

Szostak, J. W., Bartel, D. P., & Luigi Luisi, P. (2001). Synthesizing life. *Nature*, *409*, 387–390.

Takakura, K., Toyota, T., & Sugawara, T. (2003). A novel system of self-reproducing giant vesicles. *Journal of the American Chemical Society*, *125*, 8134–8140.

Toyota, T., Takakura, K., Kageyama, Y., Kurihara, K., Maru, N., Ohnuma, K., Kaneko, K., & Sugawara, T. (in press). Statistical study on sizes and components of self-reproducing giant multilamellar vesicles by flow cytometry. Langmuir.

Tsuda, I. (1991). Chaotic itinerancy as a dynamical basis of hermeneutics in brain and mind. *World Futures*, *32*, 167–185.

Wagner, G. P., & Altenberg, L. (1996). Complex adaptation and the evolution of evolvability. *Evolution*, *50*, 967–976.

West-Eberhard, M. J. (2003). *Developmental plasticity and evolution*. Oxford: Oxford University Press.

Yu, W., Sato, K., Wakabayashi, M., Nakaishi, T., Ko-Mitamura, E. P., Shima, Y., et al. (2001). Synthesis of functional protein in liposome. *Journal of Bioscience and Bioengineering*, *92* (6), 590–593.

Zintzaras, E., Mauro, S., & Szathmáry, E. (2002). Living under the challenge of information decay: The stochastic corrector model versus hypercycles. *Journal of Theoretical Biology*, *217*, 167–181.

9 Origin of Life and Lattice Artificial Chemistry

Naoaki Ono, Duraid Madina, and Takashi Ikegami

9.1 Introduction

A membrane is often more than merely a container of chemicals. Membrane surfaces are sometimes complex fabrics that perform a number of functions vital to maintaining a cell in the living state. From the very earliest stages of life, biological units require boundaries before they can perform complex functions such as recognizing and interacting with external units, in addition to maintaining self-coherency.

Once a self-replicating protocell is organized, it may be a unit of Darwinian selection and the functions of its membranes may evolve. However, how could this first protocell evolve from primitive, precellular metabolic systems? Recently, various attempts to artificially synthesize primitive cell-like systems have been reported. These typically incorporate various catalysts into artificially formed vesicles. For example, Nomura and others (Nomura et al., 2003) demonstrated protein synthesis inside lipid vesicles, and a pore protein that allows small molecules to permeate the membranes may be synthesized inside the vesicle so that it can sustain the reactions inside (Noireaux and Libchaber, 2004). Several authors (Takakura, Toyota, and Sugawara, 2003; Walde et al. 1994) have reported self-reproducing lipid vesicles. Membrane resources were supplied, producing vesicles that could themselves assimilate the resources, and as the vesicles grew, their structures became unstable, breaking into daughter vesicles.

A theoretical framework for the organization of protocell formation is provided by Maturana and Varela's systems theory *autopoiesis* (Maturana and Varela, 1980), which defines an autonomous system as a self-referential network of processes that produces itself, transforming itself, external components, or both as required. It follows that any physically realized autopoietic system must be a spatially extended unit with a boundary. Luisi and others implemented self-reproducing structures in a chemical system (Luisi, 2003). Computational models of autopoietic systems were first studied by Varela, and recently reexamined by McMullin and Varela (McMullin and Varela, 1997) and Breyer and coworkers (Breyer, Ackermann, and McCaskill,

1998). While these models demonstrate some aspects of autopoiesis, they are rather abstract two-dimensional cellular automata (CA), which lack physical or chemical plausibility. Further, we think it important that a membrane has some effect on the chemical network it encloses. Since this chemical network already plays a vital sustaining role for the membrane, there is a circular relationship between the microscale chemical activity and the macroscale membrane structure. Rasmussen and his colleagues have studied this circular relationship with a simplified molecular dynamics (MD) method (Rasmussen et al., 2004), their Lattice Molecular Automaton (Mayer, Köhler, and Rasmussen, 1997; Mayer and Rasmussen, 1998, 2000).

Here we present another approach: Lattice Artificial Chemistry (LAC), a system of abstract chemicals interacting on a given lattice. LAC simulates chemical reactions as interactions between particles distributed on a discrete lattice (Ono and Ikegami, 2001, 2002). In two dimensions, one may observe evolution of cellular structures from a random chemical soup, maintenance and self-reproduction of cell structures, and evolution of chemical networks through cellular selection. In three dimensions, new forms of membrane structure appear and the reproduction profile becomes more strongly dependent on the initial conditions (Madina, Ono, and Ikegami, 2003). In light of these various results, we believe that the emergence of cellular structure is a rather natural outcome of physicochemical dynamics. However, it is also true that such cell formation has evolutionary advantages. We will show that cells can survive in poor environments as a result of membrane formation. Also, it is important that cells maintain a nonequilibrium state, as this is a prerequisite for self-reproduction and evolution. Our model yields such dynamics, and indeed reproducing cell units will provide a global nonequilibrium environment for the cell assembly.

9.2 Modeling Protocells

Modeling an autopoietic protocell is challenging as one must take into account both chemical and physical aspects of self-maintenance, namely, metabolic cycles and self-organization of membranes—processes operating on rather different scales of time and space. Of course, these processes have been well studied independently. Aiming to understand the origin of metabolic systems, various theoretical models of primitive autocatalytic cycles have been proposed, such as analytic dynamical systems (Dyson, 1985) and artificial chemistries (Dittrich, Ziegler, and Banzhaf, 2001). These studies are motivated by the question of how a self-replicating chemical cycle can emerge from primitive chemical evolution.

Self-reproducing spatial patterns can be observed in a two-dimensional chemical system (Lee et al., 1994; Pearson, 1993). Consider a chemical system in which a set of catalysts reproduce themselves by consuming resources while continually decaying into waste. The spatial dynamics of such chemical systems strongly depends on the reaction and diffusion rate of the chemicals. When growth and decay rates are suitably balanced, a spot of catalysts grows in some regions while shrinking in others due to shortages of resources, and as a result breaks into smaller spots. These self-replicating spots may support the evolution of the autocatalysts (Cronhjort and Blomberg, 1997), however, the evolution of such a system would be very limited because the dynamics of the patterns is too sensitive to the parameters. Of more interest to us are autopoietic systems that are necessarily enclosed by membranes, and hence more robust targets of evolution.

Other interesting approaches to realize self-replicating spatial patterns have been constructed as CA (von Neumann, 1966). Along this line, various models of self-replicating patterns have been proposed (Chou and Reggia, 1997; Langton, 1984; Sayama, 1999). However, it is difficult to discuss these studies in the context of the evolution of protocells, because these models ignore physicochemical constraints such as mass conservation or free energy. As an abstract model built on a thermodynamically reasonable framework, our model aims to bridge the gap between studies of abstract self-replicating patterns and accurate biophysical phenomena.

In order to construct a plausible model, we need to know the membrane dynamics from a microscopic point of view. Recently, the dynamics of hydrophobic interaction between membrane molecules and water have been studied in detail using molecular dynamics (MD) methods (see chapter 19). However, such approaches demand large amounts of computation yet reveal little about what primitive membranes might have been, and how they might have evolved. Saving computation to study long-term and large-scale dynamics, coarse-grained models of membrane molecule interactions have been proposed (Edwards, Peng, and Reggia, 1998; Noguchi and Takasu, 2001a, 2001b). Discrete models, such as the Lattice Molecular Automaton (Mayer, Köhler, and Rasmussen, 1997; Mayer and Rasmussen, 2000) and Lattice Monte Carlo models (Drefahla et al., 1998) seem to be promising approaches because of their high scalability.

However, to investigate how, and under what conditions, an autopoietic protocell can emerge, it is important to realize both metabolic- and membrane-related processes within a single consistent framework. Our attempt to construct such a framework has led to a discrete, stochastic particle dynamics model. Chemical reactions are considered as probabilistic transitions between different particle states, whereas diffusion of molecules is simulated as a (biased) random walk over a lattice. As a result, transition probabilities for both the chemistry and the physics of the system are given by an integrated energy function. Since our aim is to observe generic principles underlying the emergence of autopoietic systems, we choose a simple, more abstract model without incorporating knowledge of, for example, existing biomolecules. This has the fortunate side effect of reducing the computational complexity of the model, which is, however, offset by the large scale of the simulations that must be performed, both in time (to observe multiple protocell generations) and in space (to contain reasonable cell populations).

9.2.1 Lattice Artificial Chemistry

Imagine a primordial soup in which primitive chemicals (e.g., segments of nucleotides, peptides, lipids and their complexes) react with each other.

Even though the variety of chemicals may be limited at first, the combination of such building blocks could create a diversity of chemicals, including various catalysts. Work along this line has been ongoing (Dyson, 1985; Segré, Ben-Eli, et al., 2001; Segré, Lancet, et al., 2001).

We assume that such basic chemical building blocks are provided by some inorganic sources, such as meteorites or hot springs, and that the earliest metabolisms used such building blocks rather than synthesizing all required substances from small molecules such as CO_2.

Figure 9.1 is a schematic drawing of the LAC protocell model (Ono and Ikegami, 2001, 2002). Chemicals are represented as coarse-grained particles that move across lattice sites. Unlike in many CA, any number of particles may exist on a single site. Therefore, the state of the system at time t is given by an array of vectors

$$n(x,t) = (n_1(x,t), \ldots, n_N(x,t)) \tag{9.1}$$

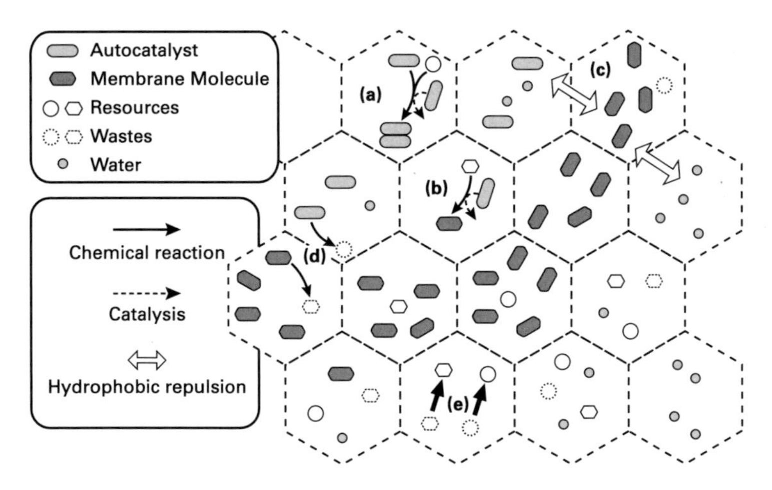

Figure 9.1
Illustration of the LAC model. (a) Autocatalysts replicate themselves. (b) Some autocatalysts additionally synthesize membrane particles. (c) Membrane particles form clusters as a result of the hydrophobic interaction. (d) Autocatalysts and membrane particles decay into wastes at a constant rate. (e) Resources are supplied by an external driving force that recycles waste into resources uniformly across the space.

where N is the number of chemical states; $n_\alpha(x, t)$ denotes the number of particles α on the site x at time t. On average, each site has 30 particles in the experiments reported here. The system's evolution is computed with a Monte-Carlo method, which models chemical reaction and transportation as stochastic transitions whose probabilities are biased by a function of potential difference,

$$p(\alpha \to \alpha') = k_{\alpha \to \alpha'} f(\Delta E) \tag{9.2}$$

$$f(\Delta E) = \frac{\Delta E}{e^{\beta \Delta E} - 1}, \tag{9.3}$$

where $k_{\alpha \to \alpha'}$ denotes the rate coefficient of the state change (e.g., a reaction rate or a diffusion rate), ΔE gives the potential difference of the particle, and β is the inverse of the product of the Boltzmann constant and the temperature.

Using this framework, we model the later stages of the emergence of primitive cells, that is, we assume that some autocatalytic molecules—be they self-replicating ribozymes, autocatalytic peptides, or something similar—have already begun to replicate (figure 9.1a). Let A_i represent the amount of a set of catalysts that catalyze each other's reproduction. We suppose that these autocatalysts (A_j) replicate using another autocatalyst (A_i) as a template, in which case their rate of reproduction, that is, the rate coefficient of the transition from resource particle X_A to A_j $(k_{X_A \leftrightarrow A_j})$, depends on the concentration of autocatalysts,

$$k_{X_A \leftrightarrow A_j}(x, t) = k_A + n_{A_j}(x, t) C_A \sum_i n_{A_i}(x, t), \tag{9.4}$$

where k_A denotes the base reaction coefficient and C_A denotes the catalytic reproduction activity. Catalysts catalyze each other's replication at this shared rate, whereas mutation occurs at a constant rate.

Further, we suppose that some autocatalysts have the ability to catalyze the production of primitive membrane molecules (figure 9.1b),

$$k_{X_M \leftrightarrow M}(x, t) = k_M + \sum_i C_{M_i} n_{A_i}(x, t), \tag{9.5}$$

where k_M denotes the base reaction rate and C_{M_i} denotes the catalysts' membrane production activity. In reality, it is difficult to synthesize the complex membrane molecules of present living organisms without using complex enzymes. However, it is well known that a wide range of surfactantlike substances form membranes and vesicles spontaneously (Lipowsky and Sackmann, 1995). For example, Vauthey and others demonstrated that short peptide chains form nanovesicles (Vauthey et al., 2002). Giant vesciles may also be artificially synthesized (Takakura, Toyota, Sugawara, 2003).

In our model, membrane molecules are represented by abstract particles, and membrane formation is driven by a very simple interaction. First, the diffusion of particles is biased according to the potential gradient sensed by the particle α,

$$\Psi_\alpha(x,t) = \sum_{|\Delta x| \leq 1} \sum_{\beta} \psi_{\beta\alpha}(\Delta x) n_\beta(x + \Delta x, t), \tag{9.6}$$

where α and β denote the particles involved and ψ denotes the coefficients of the repulsion between particle α and β that may also depend on their relative configuration, Δx. The repulsion coefficients represent the stability of the interface between particles. We assume that the repulsion between membrane particles and water particles is stronger than that between the same type of particles, so that they form separated clusters, like oil in water. Moreover, membrane particles have an axis in which direction the repulsion becomes particularly strong. This anisotropic repulsion of membrane particles represents the alignment of membrane molecules, and it leads to elongated clusters of membrane particles. We further assume that the autocatalytic particles are hydrophilic. The membranes create a potential barrier for the diffusion of autocatalytic particles (figure 9.1c). We assumed that the resource and waste particles are elemental components of autocatalysts or membranes, and therefore are smaller, thus diffusing more rapidly than these. Moreover, the repulsion between membrane and resource or waste particles is weaker than that between hydrophilic particles and membrane particles. Resource and waste particles thus diffuse through membranes more easily.

We assume that particles decay spontaneously (figure 9.1d). Autocatalysts and membrane particles change into waste particles at a constant rate. This implies that both catalysts and membranes must be continually reproduced in order to maintain cell structure. For simplicity, all particles share the same decay rate in the simulations hereafter. The synthesis of molecules is driven by the metabolic resources, that is, smaller compounds with high chemical potential. In this model, wastes are recycled into resources by an external energy source at a constant rate (figure 9.1e). This external source is represented by the difference in the rate coefficients of resource and waste particles, supplied as

$$k_{X_* \to Y_*}(x,t) \equiv k_Y \tag{9.7}$$

$$k_{Y_* \to X_*}(x,t) = k_Y + S_*(x,t), \quad (* \text{ is in } \{A, M\}) \tag{9.8}$$

where k_Y denotes the base rate, and S_A and S_M denote the rate of each resource X_A and X_M, respectively. Generally speaking, the supply of resources varies with various conditions, such as location in the environment. Imagine, for example, a case in which resources come from a hot spring on the sea floor. The resources will be plentiful near the spring but become scarcer with distance. Similarly, we suppose that the supply of whatever fueled the earliest metabolic systems would have varied with distance from the source (even if it is not clear exactly what these resources were).

Although the values of chemical and repulsion potential can be associated with thermodynamical parameters (see equation 9.3), some arbitrariness remains in the temporal and spatial scales, because the model is abstract and coarse-grained.

Although much of the model's parameter space yields trivial dynamics and may be easily ignored, many parameters remain. Choosing various sets from these parameters, we investigated the parameter space in detail and found regions where the spontaneous emergence of protocells from a random initial state can be observed. (For a detailed description of the algorithms and parameter values involved, see, e.g., Ono, 2005; Ono and Ikegami, 2002).

Figure 9.2 (color plate 5) shows snapshots of a typical run demonstrating the evolution of protocells from a primitive chemical system. The initial state contains autocatalysts that can reproduce themselves but that do not produce membrane particles (figure 9.2a, color plate 5), randomly distributed. In this experiment, we considered a gradient of resource supply. The supply rate of resources is highest at the upper boundary of the system and decreases linearly toward the lower boundary.

At first, the autocatalysts can survive only in the upper area where resources are plentiful (figure 9.2b, color plate 5). As they replicate, new species of catalysts are produced through random mutation. Some of the new catalysts may have the ability to produce membrane particles, though not efficiently. Membrane particles automatically cluster together as a result of their hydrophobic repulsion, and short fragments of membranes begin to appear (figure 9.2c, color plate 5).

Because of the gradient of the resource supply, there is a line where metabolism is no longer sustainable. When the membrane fragments appear, however, they prevent catalysts from diffusing away, enabling them to reproduce more easily. This leads to an increase in the number of autocatalysts whose ability to produce membranes is greater. If the average activity to produce membranes increases further, the autocatalysts produce enough membrane molecules to enclose themselves, and we observe the beginning of a cell-like structure, that is, a metabolic system enclosed by a membrane (figure 9.2d, color plate 5).

Such structures are maintained by their continuous production of membrane particles. However, if their enclosure is broken, the autocatalysts leak away and can no longer reproduce, since they are unlikely to find themselves in an area with a sufficient concentration of resource particles. Moreover, having lost its supply of membrane particles, the membrane ultimately decays. Therefore, these structures may be regarded as autopoietic systems, which we will refer to as "protocells" or simply "cells."

Protocells not only maintain themselves but grow in size by assimilating resources from their surroundings. Because resource particles can permeate membranes, they

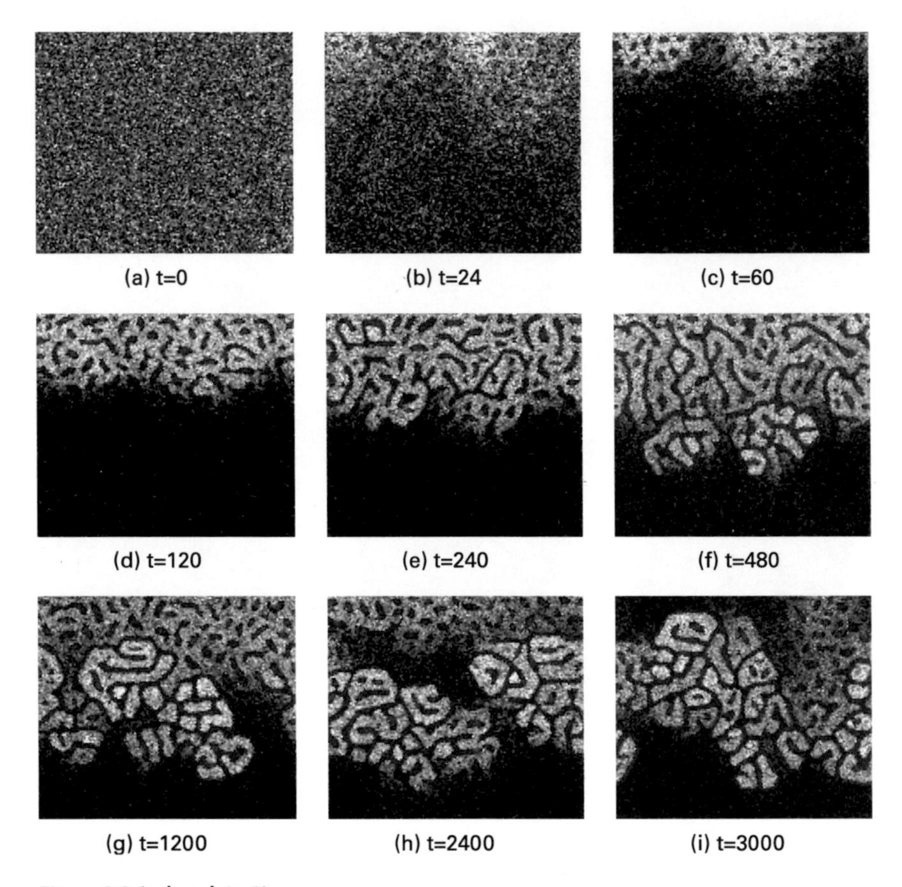

Figure 9.2 (color plate 5)
Evolution from precellular metabolism to protocells. Resource supply decreases gradually from top to bottom. In the beginning, autocatalysts produce few membrane particles and they survive only in the richest area (a, b, c). Due to the selection which occurs, more efficient catalysts multiply and the total amount of membrane grows. Once a protocell is organized, it begins to reproduce itself by growing and then dividing. Protocells become more stable through evolution until they finally replace the precellular autocatalysts.

diffuse into cells according to the density difference that arises from metabolism (since they are consumed inside cells, resources become scarce there). Membranes grow outward because of the pressure of the increasing number of particles inside each cell. As a protocell grows, it is able to produce membrane particles at an increasing rate, until the point when the supply of membrane particles (which grows according to the cell volume) more than offsets the loss of membrane particles due to decay (which grows according to the cell surface area). At this point, a new membrane is formed within the cell, and this formation lengthens until it completely divides the original cell into two daughter cells (figure 9.2e, color plate 5). In this sense, the protocell is not only self-maintaining but also self-reproducing.

When a cell divides, the daughter cells inherit the mother cell's distribution of autocatalysts. We may regard this identity as a primitive "unary genetics" (i.e., one in which genes are "accumulated" rather than coded) whereby the protocells realize simple evolution. Segré and Lancet proposed another form of protogenetic inheritance of membrane components (Segré, Ben-Eli, and Lancet, 2000).

Selection is based primarily on the stability of a cell's membrane, for if a cell failed to maintain its membrane, its catalysts, and thus its identity, would leak into the environment. Therefore, more stable cells, namely, cells that can produce enough membrane particles, are "selected" (figure 9.2f). As a result of this selection, the average activity of autocatalysts increases over time, making the protocells more stable as they evolve.

It is worth noting that, in contrast to typical evolutionary models, we do not explicitly impose a measure of fitness or system of reproduction; we merely define a physicochemical dynamics. Although there is no a priori distinction between the living and nonliving state, one may observe the birth, growth, and death of protocells. It seems that the evolutionary dynamics emerged from lower-level interactions through competition for resources.

As shown in figure 9.2e–g (color plate 5), the transition to protocells first takes place at the frontier of precellular metabolism. However, once protocells are established at the frontier, they can advance into the areas where the resources were otherwise too scarce, or too dense. This is due to the semipermeability of the membranes. As resources are consumed by metabolism inside a protocell, resource particles outside the cell diffuse into it according to the density difference, further promoting metabolism. On the other hand, the membrane keeps the autocatalysts suitably dense. As a result of this effect, protocells can sustain metabolism in poorer resource conditions (figure 9.2g, h, color plate 5), and as long as the concentration of autocatalysts within a protocell is sufficiently high and its membrane remains intact, it may survive indefinitely. The protocells also spread into the richer area. Moreover, because of the nonlinear nature of autocatalytic reactions, if the density of autocatalysts in some area is high, resource consumption is also increased. This results in a density difference that leads to the diffusion of further resources into this area, which in turn further increases the number of autocatalysts. On the other hand, in the area where resources are exploited, it becomes difficult to sustain metabolism. Thus, finally, protocells replace the precellular metabolic systems and come to dominate the space (figure 9.2i, color plate 5).

9.2.2 Conditions of Protocellular Emergence

To understand the process of protocellular takeover in more detail, we examined the evolution under a wide range of resource conditions. Suppose that autocatalysts and membrane particles are composed of different resource particles supplied at different

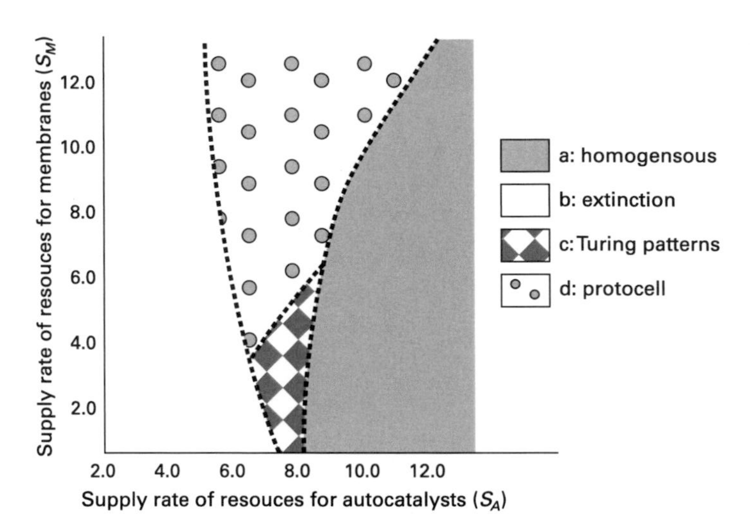

Figure 9.3
Phase diagram generated after varying the supply rates of the two resources. Data points are taken from three runs with different random seeds. There is some overlap between the regions, meaning that each of the three initial conditions evolve into different basins with different macrobehavior.

rates, S_A and S_M, respectively. Varying the two supply rates as the control parameters, we investigate the changes in evolutionary behavior. In the experiment presented in this section, the supply of resources is homogeneous and the initial condition consists of the ten species of autocatalysts randomly spread over the lattice.

We found four qualitatively different behaviors. Figure 9.3 shows the phase diagram plotted for the variables S_A and S_M. First, where S_A is very high, the autocatalysts can reproduce even outside membranes. Because there is no pressure to produce membranes, the activity to produce membranes remains very low. No protocells appear, but only small patches of membranes. On the other hand, in the second region where S_A is under a certain threshold, they can never maintain reproduction. In the third, region, when S_A is between these two regions and S_M is low, the autocatalysts may reproduce only in certain areas of the lattice because of the resource shortage. Indeed, they show patterns similar to those observed in models of self-replicating spots (e.g., Pearson, 1993). The shortage of resource particles makes it impossible for membrane structures to develop. On the other hand, in the fourth region where S_A is between the two regions and S_M is high, protocells evolve spontaneously because the formation of membranes is an advantage for the catalysts.

Note that if a cellular structure is given as an initial configuration, the parameter region where it can survive and reproduce is much wider than the fourth region of the phase diagram in figure 9.3 (where protocells emerge from random initial configurations), because the cell gathers resources from the surrounding environment, and the membrane keeps the density of the catalysts sufficiently high.

It should also be noted that a protocell is not merely a spatial structure, but also a unit of evolution. Because membranes obstruct diffusion—the spontaneous mixing of particles—the *effective* degrees of freedom of the chemical system increase when a membrane separates a protocell from its environment. Eventually, this separation of dynamics changes the evolutionary landscape. As shown in section 9.2.1, after protocells are formed, the selection pressure to maintain more stable membranes becomes dominant. And once cell structures are established, the focus of the evolution shifts to higher-order structures, that is, a transition from molecular to cellular evolution takes place even though the fundamental dynamics of the system remain unchanged.

9.2.3 Cell Structures

The sizes of cells depend on the ratio between the rates of production and diffusion of molecules. For example, under the conditions of the preceding experiments, an autocatalyst moves around 10 lattice units, on average, before it decays, whereas the diameter of protocells ranges from 7 to 20 lattice units. As the rate of membrane production increases relative to the rate of diffusion, the cells tend to become smaller because they can divide faster. On the other hand, if the diffusion rate is larger than the production rate, the evolution of protocells is more difficult as the catalysts become and remain strongly mixed. In such cases, protocells do not appear.

In a first attempt at studying the relationships between cell *shapes* and the chemical reactions occurring therein, we extend the LAC model to three dimensions, since this allows a vastly greater number of possible cellular morphologies. For example, we wonder if chemical reactions may proceed differently when constrained to occur in quasi–one-dimensional structures such as tubelike membranes. Such questions are of independent interest, regardless of the problem of the origin of cells.

9.2.4 Three-Dimensional LAC Model

In three dimensions, chemicals are represented by abstract particles moving on a cubic lattice. The model is essentially unchanged except for the forces involving membrane particles, which now have an anisotropic potential field in three dimensions such that sheetlike structures are able to form. Figure 9.4 illustrates an example of the repulsion around a membrane particle. As separation occurs due to the repulsion between hydrophilic and hydrophobic particles, thin, membranelike structures form as a result of this anisotropy.

9.2.5 Structures in Three Dimensions

When sufficient resource and catalytic particles are available to allow the formation of membrane particles, a number of different cellular structures may be formed. (We

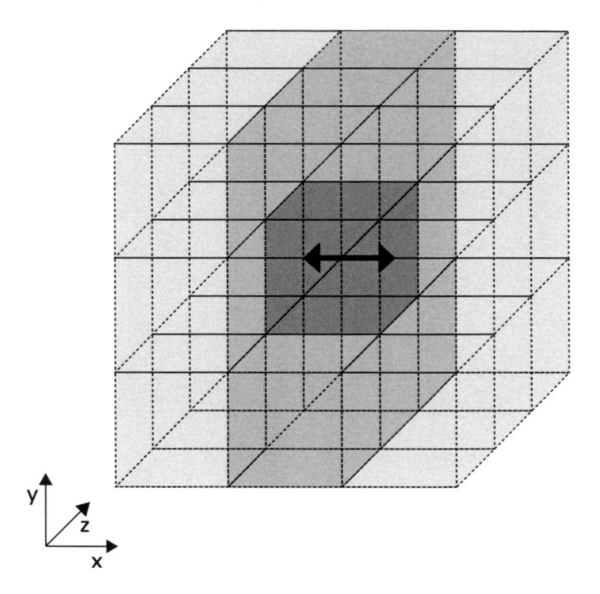

Figure 9.4
The anisotropic repulsion field around an membrane particle. Deeper gray indicates stronger repulsion.

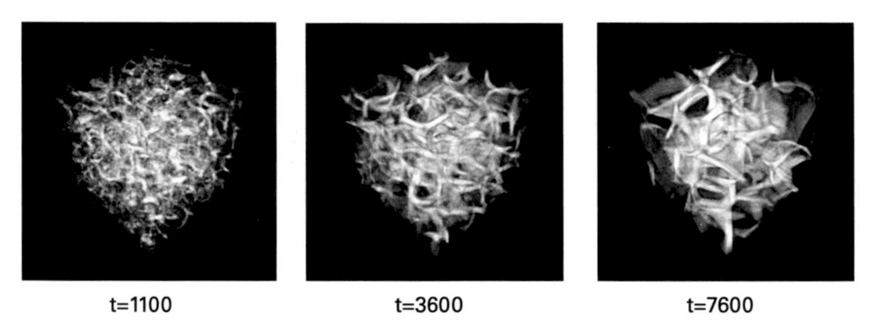

Figure 9.5
From a random homogeneous initial configuration, filaments form and begin to organize.

say that a cell has formed whenever a region of space is completely enclosed by a nonzero number of membrane particles.) From random initial conditions, the evolution typically consists of the spontaneous formation of thin filaments of membrane particles, which begin to grow in diameter. Once they grow sufficiently, thin membrane surfaces may begin to form between nearby and similarly aligned filaments (figure 9.5). Since the magnitude of the repulsion between two membrane particles is proportional to the angle between them, it is not energetically favorable for membranes to have sharp features, and eventually the original filaments are no longer apparent. Figure 9.6 is an example of tube formation given specific initial conditions.

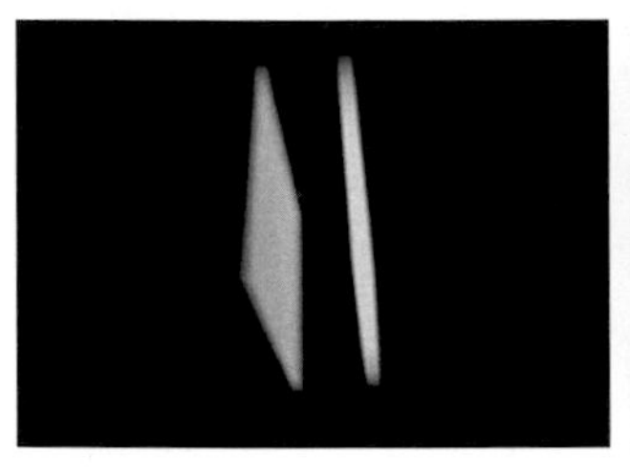
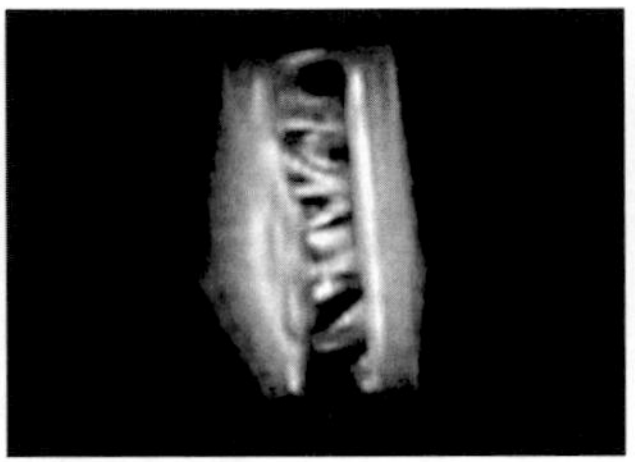
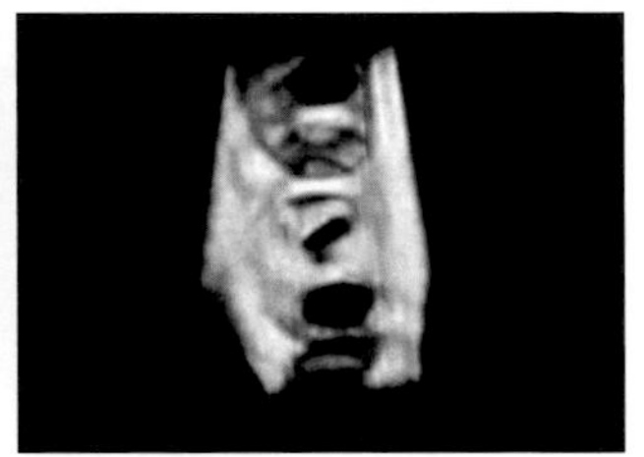

Figure 9.6
From an initial configuration of two parallel membranes, tubular structures form to connect them.

While initial structures are often preserved, a variety of new structures also form. Instead of spherical cell structures, tubes may grow to connect two parallel membranes. We must stress, however, that the process of cell division is not influenced by the details of cellular structure; there is no analog of complex division processes such as mitosis here. Indeed, even the dynamics of the two-dimensional (2D) replication process is not fully preserved in three dimensions (3D). Although it is possible to engineer initial conditions such that, for a fine parameter range, the model exhibits similar behavior in 2D and 3D, such similarities do not occur naturally. Because of the computational costs involved, we have investigated the 2D model more thoroughly; however, we expect that self-replicating 3D protocells may be more easily observed in the future because of the larger simulations that must be performed. (Informally, self-replication of protocells occurs more slowly in three dimensions because of the increased contribution of membrane particle repulsion, which tends to flatten membranes, making protocells larger. This larger size implies replication on longer timescales, since our model is not renormalizable with respect to the diffusion constant.)

9.2.6 Protocells in Three Dimensions

As filaments grow, they turn into sheets that ultimately join to form cells. These cells then continue to grow, compete for resources, and divide. A cell grows by absorbing resource particles from its neighborhood, which may or may not include other cells. As a cell grows, it may produce more membrane particles than it needs to maintain its structure. If this occurs, these excess particles can gather to seed the growth of a new membrane near the center of the cell. Eventually, this new membrane may grow to span the original cell completely, dividing it into two or more new, smaller cells. It is important for cells to maintain their structural integrity; catalysts may leave the cell by diffusing through any defect in the membrane. If too many catalysts escape, those that remain will be insufficient to maintain the cell membrane, and eventually the cell will decay completely. Thus, in this region, diffusion and reaction are balanced: The two interact as long as cellular structures exist. Figure 9.7 shows an

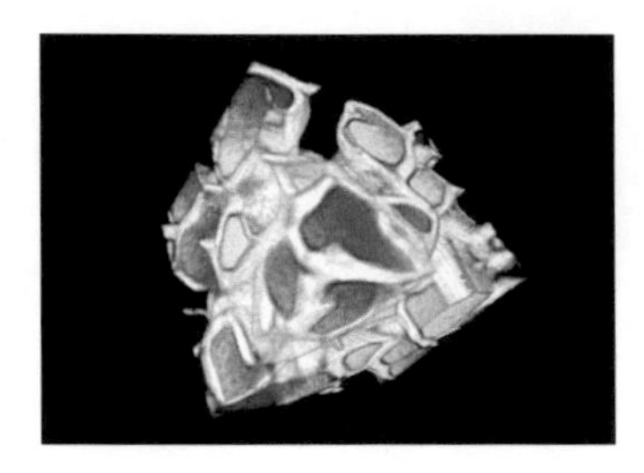
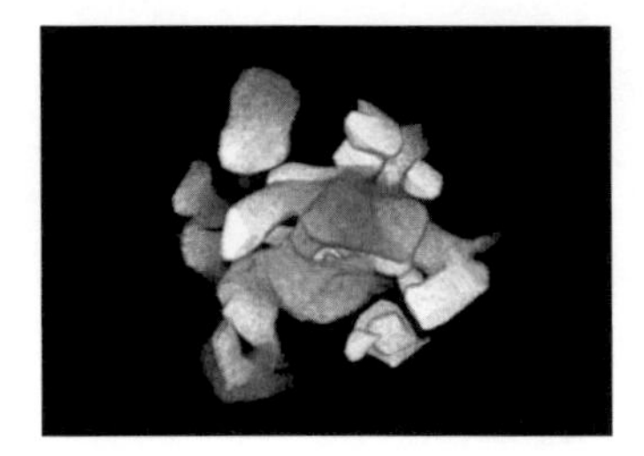

Figure 9.7
Fully formed cellular structures in a rich environment (left) and the distribution of catalytic particles in the same structure (right). The effect of competition for resources is evident: A previously homogeneous distribution of resources has given way to cellular "hoarding."

example of this behavior. One interesting aspect of cell division in this model is that cells do not have any mechanism with which to select an internal axis of division. As a result, cell division does not resemble mitosis: When cells divide, they do not displace each other, but instead share a newly formed membrane surface. Furthermore, unlike real cell division, in our model there is often a high degree of asymmetry between the "child" cells that are created. This leads to rapid disintegration of the smaller child cells, and the resources they would have otherwise consumed become available for another child cell to grow. This phenomenon leads to strongly fluctuating cell populations over time.

9.2.7 Further Work

We should point out that our consideration of metabolic processes is not necessarily restricted to lattice-based chemical simulation. Indeed, we have studied much the same metabolic system in a more conventional (and more physically accurate) MD framework (Madina and Ikegami, 2004). One can view this as a way of collapsing (we hesitate to use the word "approximating") relatively slow biochemical processes such as metabolism, bringing them within reach of the short timescales that MD simulation can handle (e.g., processes such as lipid aggregation). It is worth noting that despite a great increase in the accuracy with which molecules were modeled (and thus in spite of a decrease in the number of free simulation parameters), the model behavior was not markedly different. This raises the more general question of where and how to make abstractions when simulating protocells, which we hope to explore in future work.

9.3 Conclusions

In this chapter we have constructed autopoietic systems in two and three dimensions, with a minimal metabolic system comprising autocatalytic and membrane particles in a discrete, physically inspired system. We could observe in this system transitions from precellular chemistry to protocellular evolution. Importantly, we could observe that the formation of protocells allowed evolution to move into regions that would otherwise be too poor to sustain metabolism. In two and three dimensions, the model dynamics was broadly similar, though the latter case exhibited a variety of membrane forms and increased sensitivity to initial conditions.

Acknowledgments

This work has been supported by grants-in-aid from The 21st Century COE (Center of Excellence) program (Research Center for Integrated Science) of the Minister of Education, Culture, Sports, Science, and Technology, Japan, the "Academic Frontier, Intelligent Information Science" (AFIIS) project of Doshisha University, and the ECAgent project, part of the Future and Emerging Technologies program of the European Community (IST-1940).

References

Breyer, J., Ackermann, J., & McCaskill, J. (1998). Evolving reaction-diffusion ecosystems with self-assembling structures in thin films. *Artificial Life*, *4*, 25–40.

Chou, H.-H., & Reggia, J. A. (1997). Emergence of self-replicating structures in a cellular automata space. *Physica D*, *110*, 252–276.

Cronhjort, M. B., & Blomberg, C. (1997). Cluster compartmentalization may provide resistance to parasites for catalytic networks. *Physica D*, *101*, 289–298.

Dittrich, P., Ziegler, J., & Banzhaf, W. (2001). Artificial chemistries: A review. *Artificial Life*, *7*, 225–275.

Drefahla, A., Wahabb, M., Schillerb, P., & Mögel, H.-J. (1998). A Monte Carlo study of bilayer formation in a lattice model. *Thin Solid Films*, 327–329, 846–849.

Dyson, F. J. (1985). *Origins of life*. Cambridge: Cambridge University Press.

Edwards, L., Peng, Y., & Reggia, J. A. (1998). Computational models for the formation of protocell structures. *Artificial Life*, *4*, 61–77.

Langton, C. G. (1984). Self-reproduction in cellular automata. *Physica D*, *10*, 135–144.

Lee, K. J., McCormick, W. D., Pearson, J. E., & Swinney, H. L. (1994). Experimental observation of self-replication spots in a reaction diffusion system. *Nature*, *369*, 215–218.

Lipowsky, R., & Sackmann, E. (Eds.) (1995). *Structure and dynamics of membranes*, vol. 1A of *Handbook of biological physics*. Amsterdam: Elsevier.

Luisi, P. L. (2003). Autopoiesis: A review and a reappraisal. *Naturwissenschaften*, *90* (2), 49–59.

Madina, D., & Ikegami, T. (2004). Cellular dynamics in a 3D molecular dynamics system with chemistry. In J. Pollack et al. (Eds.), *Proceedings of the Ninth International Conference on Artificial Life (ALIFE9)* (pp. 461–465). Cambridge, MA: MIT Press.

Madina, D., Ono, N., & Ikegami, T. (2003). Cellular evolution in a 3D lattice artificial chemistry. In W. Banzhaf et al. (Eds.), *Proceedings of the Seventh European Conference on Artificial Life (ECAL2003)* (pp. 59–68). Berlin: Springer.

Maturana, H. R., & Varela, F. J. (1980). *Autopoiesis and cognition: The realization of the living*. Dordrecht, Holland: D. Reidel Publishing.

Mayer, B., Köhler, G., & Rasmussen, S. (1997). Simulation and dynamics of entropy-driven, molecular self-assembly processes. *Physical Review E*, *55* (4), 4489–4499.

Mayer, B., & Rasmussen, S. (1998). The lattice molecular automaton (LMA): A simulation system for constructive molecular dynamics. *International Journal of Modern Physics C*, *9*, 157–177.

Mayer, B., & Rasmussen, S. (2000). Dynamics and simulation of micellar self-reproduction. *International Journal of Modern Physics C*, *11* (4), 809–826.

McMullin, B., & Varela, F. J. (1997). Rediscovering computational autopoiesis. In P. Husbands & I. Harvey (Eds.), *Proceedings of the Fourth European Conference on Artificial Life (ECAL97)* (pp. 38–47). Cambridge, MA: MIT Press.

Noguchi, H., & Takasu, M. (2001a). Fusion pathways of vesicles: A Brownian dynamics simulation. *Journal of Chemical Physics* (DOI: 10.1063/1.1414314), *115*, 9547–9551.

Noguchi, H., & Takasu, M. (2001b). Self-assembly of amphiphiles into vesicles: A Brownian dynamics simulation. *Physiscal Review E*, *64*, 041913.

Noireaux, V., & Libchaber, A. (2004). A vesicle bioreactor as a step toward an artificial cell assembly. *Proceedings of the National Academy of Sciences of the United States of America*, *101* (51), 17669–17674.

Nomura, S. M., Tsumoto, K., Hamada, T., Akiyoshi, K., Nakatani, Y., & Yoshikawa, K. (2003). Gene expression within cell-sized lipid vesicles. *Chembiochem*, *4* (11), 1172–1175.

Ono, N. (2005). Computational studies on conditions of the emergence of autopoietic protocell. *Biosystems*, *81*, 223–233.

Ono, N., & Ikegami, T. (2001). Artificial chemistry: Computational studies on the emergence of self-reproducing units. In J. Kelemen & S. Sosik (Eds.), *Proceedings of the Sixth European Conference on Artificial Life (ECAL'01)* (pp. 186–195). Berlin: Springer.

Ono, N., & Ikegami, T. (2002). Selection of catalysts through cellular reproduction. In R. Standish et al. (Eds.), *Proceedings of the Eighth International Conference on Artificial Life (ALIFE8)* (pp. 57–64). Cambridge, MA: MIT Press.

Pearson, J. E. (1993). Complex patterns in a simple system. *Science*, *261*, 189–192.

Rasmussen, S., Chen, L., Deamer, D., Krakauer, D., Packard, N., Stadler, P., & Bedau, M. (2004). Transitions from nonliving to living matter. *Science*, *303*, 963–965.

Sayama, H. (1999). A new structurally dissolvable self-reproducing loop evolving in a simple cellular automata space. *Artificial Life*, *5*, 343–365.

Segré, D., Ben-Eli, D., Deamer, D., & Lancet, D. (2001). The lipid world. *Origin of Life and Evolution of the Biosphere*, *31*, 119–145.

Segré, D., Ben-Eli, D., & Lancet, D. (2000). Compositional genomes. *Proceedings of the National Academy of Sciences of the United States of America*, *97*, 4112–4117.

Segré, D., Lancet, D., Kedem, O., & Pilpel, Y. (2001). Graded autocatlysis replication domain (GARD). *Origin of Life and Evolution of the Biosphere*, *28*, 501–514.

Takakura, K., Toyota, T., & Sugawara, T. (2003). A novel system of self-reproducing giant vesicles. *Journal of the American Chemical Society*, *125* (27), 8134–8140.

Vauthey, S., Santosa, S., Gong, H., Watson, N., & Zhang, S. (2002). Molecular self-assembly of surfactantlike peptides to form nanotubes and nanovesicles. *Proceedings of the National Academy of Sciences of the United States of America*, *99*, 5355–5360.

von Neumann, J. (1966). *Theory of self-reproducing automata.* Champaign: University of Illinois Press.

Walde, P., Wich, R., Fresta, M., Mangone, A., & Luisi, P. L. (1994). Autopoietic self–reproduction of fatty acid vesicles. *Jounal of the American Chemical Society*, *116*, 11649–11654.

10 Models of Protocell Replication

Ricard V. Solé, Javier Macía, Harold Fellermann, Andreea Munteanu, Josep Sardanyés, and Sergi Valverde

10.1 Introduction

The transition to cellular life was marked by the emergence of a chemical coupling between simple autocatalytic processes (perhaps a primitive form of metabolism) and a container. The emergence of cells allowed the propagation of information with multiple selective advantages, from having the reactants closer in space to allowing division of labor. It also provided the conditions for escaping from molecular parasites, which are known to destroy cooperative dynamics in hyperbolic replicators (see chapters 13 and 14).

Generally cell biomass increases, usually doubling, before the cell cycle ends with cell division. The cell cycle can be defined as the orderly duplication of intracellular components, including the cell genome (DNA), followed by division of the cell into two cells (figure 10.1). Current cells have a number of sophisticated molecular mechanisms to control cell cycle dynamics, including checkpoints and regulatory functions. Unicellular life forms replicate by simple splitting of the cell into two cells, a process known as binary fission. The process takes place provided that external resources allow the template-based copy of genetic information and the building of all required molecular components necessary for the build-up of two daughter cells.

Although lipid aggregates, micelles, and vesicles are known to spontaneously form under a wide range of conditions, the problem of cell growth and replication is far from trivial. Several problems emerge when modeling an effective catalytic cooperation between components leading to a coherent replication cycle. In this chapter, we address some early and recent steps toward understanding possible paths toward simple replicating protocells. They all involve the problem of generating a spatially structured pattern generated through some dynamical process that leads to partial or complete reproduction of the whole structure.

The whole cycle of cell reproduction is strongly constrained by two main factors: the kinetics associated with reactions among different chemical components and the interactions between these components and the container. Early models (to be

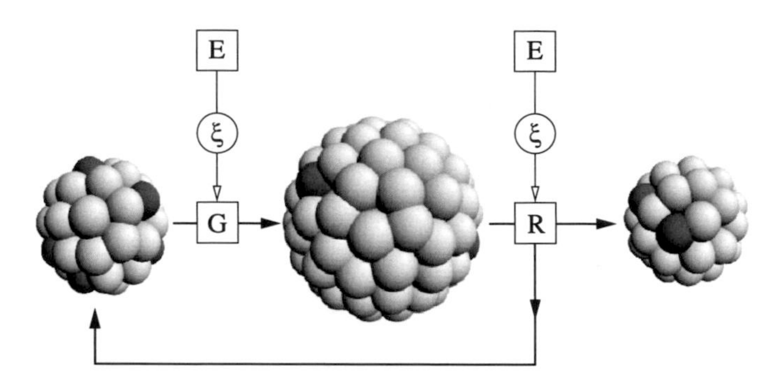

Figure 10.1
Two main events are required for a protocell to complete a life cycle. The first is somehow controlled by the interaction of chemical reactions and membrane growth (G) with an external environment, likely to be subject to noisy fluctuations (here indicated as x) given the small scale at which the cell lives. The second deals with replication (R). Splitting can be associated with external fluctuations (E) or somehow controlled by the cell dynamics, or both.

reviewed here) explored these two features separately. Only some attempts (such as Rashevsky, 1960) explored how toy models of metabolism could effectively couple them, and the conditions under which the compartment enclosing metabolites would experience growth and eventually splitting. From a different perspective, simple models of nanoscale organization of lipid aggregates can trigger fission events that are connected to the problem of how self-organized systems can spontaneously experience some type of fission.

The growth of protocell-like systems (or parts of them) is strongly tied to a compromise between forces driving the system toward growth and forces that trigger destabilization. Growth is easily achieved by accretion of material that forms aggregates or through membrane expansion under increasing osmotic pressures. These processes can end up in some stable structure (a large aggregate) or reach some rupture threshold leading to breaking the cell container. Under certain conditions (to be discussed here) growth is followed by some type of instability that triggers the formation of smaller subsystems. At both the mesoscopic and microscopic scales, a physical implementation of the rules dealing with compartment dynamics must be considered.

Here we explore some basic results dealing with the kinetic behavior of spatially extended, replicating systems. Our review includes (a) basic replicator dynamics (and how particular reaction processes lead to specific dynamical patterns of growth), (b) spatially extended, reaction-diffusion systems with no membrane leading to self-replicating spots, (c) physical models of amphiphile aggregates involving some type of spontaneous or induced fission mechanisms, and (d) models of self-replicating cells including both a closed membrane and a simple metabolic core.

10.2 Replicators and Producers

Primitive cellular life forms presumably involved the emergence of a catalytically coupled set of chemical reactions. In its simplest form, it might have included a vesicle or micelle coupled to a minimal metabolism. Such metabolism might have been favored by special, membrane-bound molecules acting as primitive enzymes. The so-called chemoton (Gánti, 1975; see also chapter 22) was suggested as a simple approach to this picture, where the three basic components of cellular organization—namely, metabolism, container, and information—would be tightly coupled. Metabolism would provide the building blocks for a population of replicating molecules, properly encapsulated by the membrane.

An important distinction exists between replicating and reproducing entities. Reproducers, in contrast to replicators, involve copy and development whereas replicators can be understood as informational molecules that can exist in several forms (A, B, C) and, when replicating, the new structures resemble the old ones. In the prebiotic scenario or chemical evolution context, natural selection is essentially the dynamics of replicators and thus dynamics and competition between replicators are key features to explore. In this sense, replicators can be units of selection (Szathmáry, 1997) since they have

1. multiplication—entities should give rise to more entities of the same kind;
2. heredity—A type entities produce A type entities, B type entities produce B type entities, and so on;
3. variability—heredity is not exact; occasionally A type objects give rise to A' type objects (it may be that $A' = B$).

10.3 Self-Replication Spots

This type of cell-like, spatially distributed system belongs to the large class of so-called reaction-diffusion models. The best studied examples of the two types of reaction-diffusion systems are the Meinhardt system (Gierer and Meinhardt, 1972) and the diffusive Gray-Scott system (Pearson, 1993), respectively. The complex interplay between activator and inhibitor or substrate chemical, aided by the reaction and diffusion components, creates most startling spatiotemporal (Turing) patterns, such as spots, stripes, traveling waves, or spot replication. The Turing patterns are characterized by the active role that diffusion plays in destabilizing the homogeneous steady state of the system. They emerge spontaneously as the system is driven into a state in which it is unstable toward the growth of finite-wavelength stationary perturbations. Interestingly enough, the replication characteristic is a particularity of the diffusive Gray-Scott model alone, which makes it the ideal model for developmental research.

In this case, cell-like localized structures grow, deform, and replicate themselves until they occupy the entire space.

The Turing patterns from the work of Pearson (1993) on the diffusive Gray-Scott model were confirmed experimentally by Lee and coworkers (Lee et al., 1993), including spot replication (Lee et al., 1994). Theoretically, extensive work exists in the literature on the dynamics of this model concerning the "spot replication" in one, two, and three dimensions (Muratov and Osipov, 2000). The model was originally introduced in Gray and Scott (1985) as an isothermal system with chemical feedback in a continuously fed, well-stirred tank reactor, where the last property implied the lack of diffusion. The analysis of the system revealed stationary states, sustained oscillations and even chaotic behavior. The model considers the chemical reactions

$$U + 2V \rightarrow U + 3V$$

$$V \rightarrow P,$$

which describe the autocatalytic growth of an activator, V, on the continuously fed substrate, U, and the decay of the former in the inert product, P, subsequently removed from the system. A major development was performed by Pearson (1993) who introduced the role of space by relaxing the constraint of a well-stirred tank and studied the system in two dimensions. In two dimensions, the concentrations of the two chemical components, $u(x, y; t)$ and $v(x, y; t)$ are given by

$$\wp_1(u, v) \equiv \frac{\partial u}{\partial t} = D_u \Delta u - uv^2 + F(1 - u)$$

$$\wp_2(u, v) \equiv \frac{\partial v}{\partial t} = D_v \Delta u + uv^2 - (F + k)v$$

where D_u and D_v are the diffusion coefficients, F is the dimensionless flow rate (the inverse of the residence time), and k is decay constant of the activator, V. The original study involved fixed diffusion coefficients, $D_u = 2 \times 10^{-5}$ and $D_v = 10^{-5}$, with F and k being the control parameters. As a typical Turing pattern, the system has a steady state stable with respect to homogeneous temporal oscillations, which becomes unstable toward standing, space-periodic perturbations when diffusion is taken into account (see Mazin et al., 1996, for a detailed linear analysis of the Gray-Scott model).

For the numerical study of the partial differential equations, we used the conditions initially employed in Pearson (1993), consisting in a system size of $R \times R$, with $R = 2.5$ discretized through $x \rightarrow (x_0, x_1, x_2, \ldots, x_N)$ and $y \rightarrow (y_0, y_1, y_2, \ldots, y_N)$ with $N = 256$. The numerical integration of the partial differential equations was

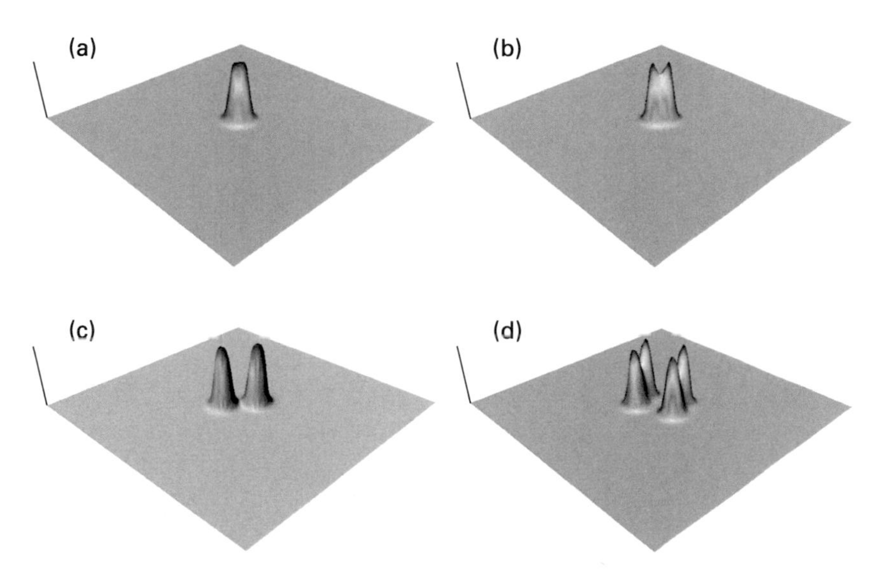

Figure 10.2
Replicating spots in a color map representation of the V concentration as a temporal process from (a) to (d). The values of the parameters are: $F = 0.04$, $k = 0.0655$, $N = 128$, $R = 1.25$.

performed by means of forward Euler integration, with a time step of $\tau \cong 0.9$ and spatial resolution $h = R/N$, and using the standard five-point approximation for the $2D$ Laplacian with periodic boundary conditions (figure 10.2). More precisely, the concentrations $(u_{i,j}^{n+1}, v_{i,j}^{n+1})$ at the moment $(n+1)\tau$ at the mesh position (i, j) are given by

$$u_{i,j}^{n+1} = u_{i,j}^{n} + \tau D_u \Delta_h u_{i,j}^{n} + \tau \wp_1(u_{i,j}^{n}, v_{i,j}^{n})$$

$$v_{i,j}^{n+1} = v_{i,j}^{n} + \tau D_u \Delta_h v_{i,j}^{n} + \tau \wp_2(u_{i,j}^{n}, v_{i,j}^{n})$$

with the Laplacian defined by

$$\Delta_h u_{i,j}^{n} = \frac{u_{i+1,j}^{n} + u_{i,j+1}^{n} + u_{i-1,j}^{n} + u_{i,j-1}^{n} - 4u_{i,j}^{n}}{h^2}.$$

The system was initialized with $(u_i; v_i) = (1; 0)$ with the exception of a small central square of initial conditions $(u_i; v_i) = (0.5; 0.25)$ perturbed with a 1% random noise. The works existent in the literature illustrate examples of patterns following a color map on the U-concentration, with the red color representing the $(u_R; v_R)$ steady state and the blue one a value in the vicinity of the $(u_B; v_B)$ state, for example, $(u; v) = (0.3; 0.25)$.

Spatial structures derived from spatially extended replicator dynamics are a first step toward localization of chemicals in a confined domain. Such a spatial localization is necessary to favor chemical reactions and selection processes. Although these kinds of spatial patterns have been found to have relevant, high-order selection properties, they are far from the real compartments defining cellular life. To gain insight into a more realistic scenario, we next consider two approaches largely based on considering the dynamics of membranes or aggregates.

10.4 Nanoscale Replicating Aggregates

If we consider aggregates of amphiphilic molecules with small numbers of components, we are actually looking at the nanoscale level. Here self-replication implies both growth and fission of amphiphile assemblies. Growth is understood as the outcome of an autocatalytic process: Amphiphiles are products of a chemical reaction that is driven by the presence of parent amphiphiles. As new monomers form, they are incorporated in the henceforth growing assembly. When the assembly reaches a critical size, it undergoes a fission process that results in two daughter assemblies. Here a microscopic approach to the underlying physics is needed.

Models that are used to study the dynamics of amphiphile aggregation are deduced from the technique of molecular dynamics (MD), in which forces derived from an assumed potential determine the motion of individual particles and hence the trajectory of the whole system in phase space.

Newton's Second Law is hereby applied to calculate the trajectory from the position r_i, mass m_i and the potential U_i of each individual particle:

$$\frac{d^2 r}{dt^2}(t) = -\frac{1}{m_i}\nabla U_i(t) \tag{10.1}$$

The potential is assumed to be the superposition of pair-wise potentials for all particle pairs within a certain interaction range:

$$U_i = \sum_{j \neq i} U(r_j - r_i)$$

Force field methods are well-established in the realms of molecular modeling, where the positions and forces of every single atom are calculated in each time step of a numerical integrator for equation 10.1. Unfortunately, the computational expense of fully atomistic molecular dynamics is considerably high when modeling the previously mentioned aggregation and fission processes.

Under the assumption that degrees of freedom of these lower scales do not affect amphiphile aggregation, coarse-graining techniques can de developed to reduce the unmanageable complexity of atomic interactions. In these coarse-grained approaches, amphiphile molecules are usually represented by two or more hydrophilic and hydrophobic particles or "beads" that represent sections of the molecule and are connected by elastic springs. Water particles are either modeled as structureless particles or completely excluded from the model, but are handled implicitly by adding their effects to the amphiphile interactions. Depending on the presence of water particles, models are called either explicit or implicit.

Usually, these particles are not meant to represent single molecules, but rather exemplary particles or "lumps" of fluid. Thus, the coarse-grained models assume an underlying medium with which the simulated particles can exchange energy due to friction and thermal noise. For example, Langevin equations can be used to achieve this energy flow:

$$\frac{d^2 r_i}{dt^2} = -\nabla U_i - \eta v_i + \xi,$$

where v_i is the particle velocity, η, the friction coefficient and ξ, a Gaussian-distributed random variable describing heat effects (Brownian motion). More sophisticated "thermostats" are used in the technique called dissipative particle dynamics (DPD) to ensure momentum conservation in the system.

In addition, particle interactions are usually modeled by a much simpler potential function than those used in fully atomistic MD. Most common is the so-called soft-core potential

$$U_{ij} = a_{ij}\left(1 - \frac{r_{ij}}{r_c}\right)^2 \quad \text{if} \quad r_{ij} < r_c, \quad \text{otherwise } U_{ij} = 0$$

where r_{ij} is the particle distance, r_c, the cutoff distance and a_{ij}, the repulsion strength, that depends on the hydrophilic/hydrophobic character of the interacting particles. This potential induces a solely repulsive force that decreases linearly with distance, until at $r_{ij} = r_c$ both potential and force are 0. In implicit models, the repulsion of water particles that drives hydrophobic interaction is mimicked by an additional adhesive potential between hydrophobic particles. These simple potential functions further reduce the computational complexity because they do not obey the singularities found in more realistic descriptions (like, e.g., the Lennard-Jones potential common in atomistic MD).

10.4.1 Simulations of Vesicle Fission

In early 2003, Yamamoto and Hyodo simulated the fission process of vesicles. They used DPD—described in chapters 18 and 19—to observe the deformation and eventual fission of a two-component vesicle. Their simulation consists of water and

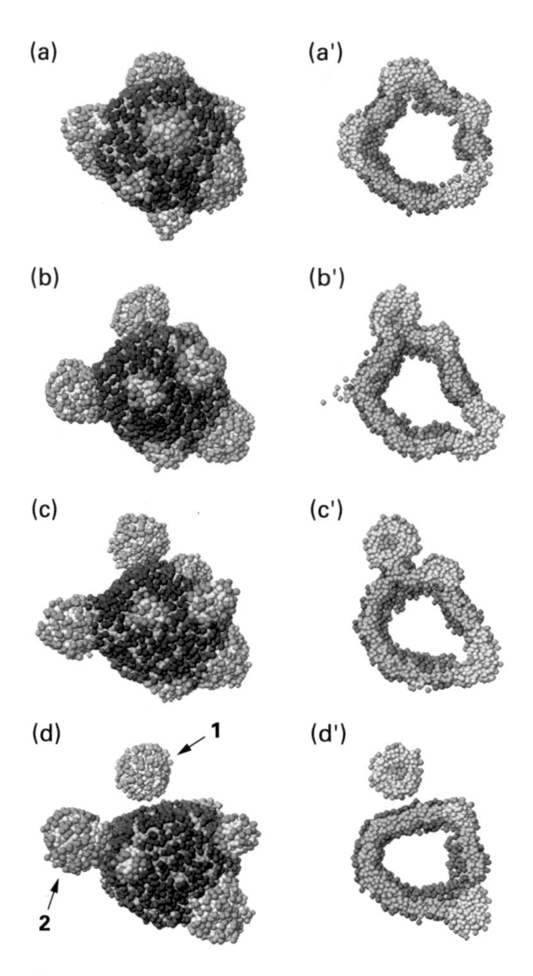

Figure 10.3
Fission of vesicles with diverse amphiphiles. Reprinted with permission from Yamamoto and Hyodo (2003).

amphiphile particles, whereby amphiphiles are composed of one hydrophilic head and three hydrophobic tail particles, which are covalently bonded by elastic springs. Interactions between these particles follow a softcore repulsion potential, a common approach in DPD simulations. Repulsion parameters are chosen so that the forces correspond to cylindrical amphiphiles that form bilayers.

The simulations are initialized with a bilayer structure that bends to form a vesicle consisting of 3,024 amphiphiles that encapsulate 5,389 water particles (figure 10.3). After the initial vesicle has formed, 35% of the amphiphiles are randomly exchanged for a second type of amphiphile. The second amphiphile possesses the same repulsion parameters as the former, but mutual repulsion of the monomers is supposed to be high (0.6 times stronger then the repulsion of identical monomers). Yamamoto and Hyodo (2003) observed that the amphiphiles drift within the bilayer and rearrange into separate phases to reduce the high mutual repulsion at domain edges. Depending on the ratio with which monomers are exchanged in the outer/inner layer of the membrane, the vesicle deforms. If the outer/inner ratio is greater than 3.0, a budding process is initiated, in which the exchanged amphiphiles form bulges on the membrane with the neck at the domain edge. Finally, these buds pinch off as a microvesicle from the parent vesicle.

Strictly speaking, their simulations do not model the spontaneous fission required for self-replication in the previously mentioned meaning: Their fission process is the result of an artificial exchange of amphiphiles after the initial vesicle is formed. However, their simulation sketches a possible way to an ongoing growth and fission process of vesicles. The system could be exposed to an ongoing influx of amphiphile monomers that get incorporated in the growing membrane. Successively added monomers rearrange within the membrane to form domains of identical amphiphiles. Once domains reach a critical size, the high potential at domain boundaries triggers the fission process.

Although this rough outline looks reasonable at first glance, some difficulties arise in the details of this scenario: Amphiphiles will enter only the outer membrane of the vesicle, hence increasing its surface tension. This might propagate the fission in the first place, but it is not clear how this will affect the fission process in the long term. The work of Yamamoto and Hyodo confirms that vesicle fission depends on both the number and rates of different amphiphiles in the outer and inner membrane. Thus, flipflop motion of amphiphiles from the outer to the inner membrane would be necessary to increase the surface tension needed for vesicle fission. It is not clear, however, which mechanism establishes flipflop motion against the energy gradient based on membrane curvature.

Noguchi and Takasu (2002) propose a fission process for vesicles that is mediated by a nanoparticle. They use an implicit model of the Langevin type to simulate budding, fission, and fusion of vesicles. Starting from a random initial condition with 1,000 amphiphiles, vesicles form spontaneously (figure 10.4). After vesicle formation, Noguchi and Takasu place what they call a nanoparticle at the center of mass of the vesicle: In their model, the nanoparticle represents a protein or colloid that acts as an initiator for morphological change, for example, vesicle fission. The nanoparticle is adhesive to the hydrophilic head groups of the amphiphiles, and therefore sticks to the inner membrane of the vesicle, where it introduces budding. The stronger the adhesion of the nanoparticle, the stronger will be the change of local curvature where the nanoparticle adheres to the membrane. For strong adhesion coefficients, the surface tension is induced but is so high that vesicle fission is energetically favored.

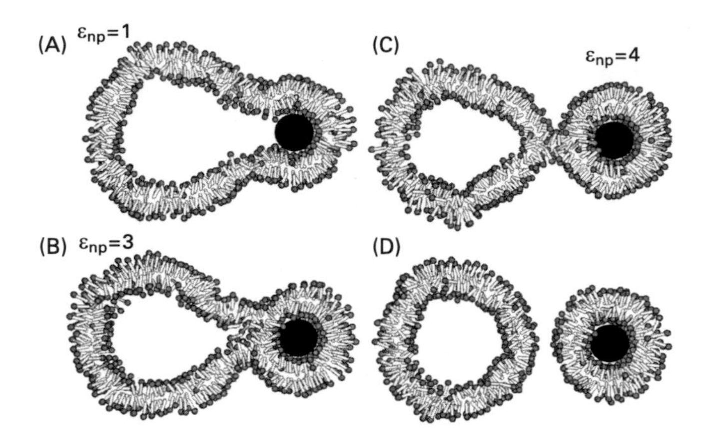

Figure 10.4
Vesicle fission in the presence of nanoparticle. Reprinted with permission from Noguchi and Takasu (2002).

While the nanoparticle might provide reliable vesicle fission, one now must find a mechanism for the nanoparticle to replicate. Otherwise, only the daughter vesicle that holds the nanoparticle is able to reproduce further.

In the previous examples we have not considered metabolism as an explicit part of the aggregation dynamics. Moreover, the importance of fluctuations (which might be huge) at this level has also been neglected. Both aspects are relevant in understanding the possible behavior exhibited by simple protocells. In the next section, we consider both a model membrane (at a large scale) and a toy metabolism together.

10.5 Mesoscopic, Mechanochemical Approximations

The objective of this section is to analyze the behavior of membranes from a mechanical point of view, with respect to the possible deformations that the membrane can experience and that can finally give rise to the division of the membrane into two structural daughters, which can grow and again return to divide themselves if the initial conditions can be regenerated in each daughter structure.

A first approximation (first applied by Rashevsky, 1960) to cell replication considers a minimal model of membrane physics, defined in terms of the average behavior of a continuous, closed membrane involving some sort of simple internal metabolism. Rashevsky presented the basic conditions required (under some constraints) to obtain replication that we summarize here. This is a mesoscopic approach that allows the incorporation of other components, such as chemical reactions able to create instabilities and symmetry breaking. The two membrane division mechanisms we consider are spontaneous and induced division.

10.5.1 Spontaneous Division

Different models of vesicle shape transformations have been widely studied. Essentially, the shape of the vesicles is analyzed in a continuum model under two variants (Seifert, Berndl, and Lipowsky, 1991). First, the spontaneous-curvature model (Helfrich, 1973) analyzes changes in vesicle shape minimizing the bending energy for a given area and volume. Second, the bilayer-coupling model (Svetina and Zeks, 1983) assumes that the two monolayers do not exchange lipids between them. This model imposes the minimization of the bending energy for a given area, volume, and ΔA, where ΔA is the difference between the areas of the external and internal layers. Both models lead to the same shape equations (Svetina and Zeks, 1989).

Here we summarize the theoretical calculation Rashevsky (1960) suggested to determine the critical radius beyond which the cellular membrane becomes unstable and increases the probability of a spontaneous division. This model focuses not explicitly on the spontaneous shape transformations, but on the energetic stability as a function of membrane size and, implicitly, on the conditions favorable to cellular division. We consider a simple structure formed by a spherical membrane in whose interior metabolic reactions take place. When a substance is produced or consumed by a metabolic system, the forces that act on each element of volume and of surface (membrane) are directed outwards or inwards. In this situation, it is possible to calculate the variation of energy of the system resulting only from the work of these forces. The spontaneous division takes place only when the work these forces make in the process of driving the system from the initial configuration to complete division into halves is positive, which means the change in energy is negative.

We assumed, as a first approximation, that the structure is formed of a spherical membrane and that the metabolic reactions can take place only within this membrane because of the presence of a number of Ω catalytic particles intervening in the consumption and production of metabolites.

In the simplest case, where the rate of reaction is constant and the particles distribution is uniform, we can calculate the energy balance. Basically, it is necessary to consider three aspects: the variation of energy associated to the increase of surface, the work due to the pressure that different substances exert on the membrane when crossing it, and finally the work the diffusion forces make on the particles present in the volume limited by the membrane. Rashevsky proposed a method of calculation based on assuming that the net variation of the system's energy is independent of the precise way the division occurs. The work made by the forces dividing the spherical membrane of radius r_o into two equal spheres of radius r_1 is equal to the work needed to expand the initial sphere to infinite radius and later to contract it into a sphere of radius r_1 (multiplying by 2 because there are two spheres).

Regarding energy associated to the increase of surface, if we suppose a spherical initial configuration of radius r_o, and assume that the volume remains constant throughout the entire division process (the sum of the volumes of two resulting halves is equal to the initial volume), we find the following relation between the final radius r_1 and r_o:

$$r_1 = \sqrt[3]{\frac{1}{2}} r_o,$$

with the total increase of surface being

$$\Delta S = 1.12\pi r_o^2.$$

Assuming that γ is the coefficient of superficial tension of the membrane, the division of the membrane implies an increase in the superficial energy by

$$\Delta W_S = 1.12\pi r_o^2 \gamma$$

independently of the mechanism used to obtain this division. From this point of view, the division of the membrane would not take place spontaneously because $\Delta W_S > 0$, unless other factors taking part in this process compensate this increase of energy.

The work on the membrane is a result of pressure of osmotic origin, produced by the different concentrations at either side of the membrane. Assuming a uniform distribution of particles inside and outside the membrane, the force that acts on each point of the spherical membrane is

$$F_m = 4\pi r_o^2 \frac{RT}{M}(c_i - c_e),$$

where c_i and c_e are the concentrations inside and outside the membrane, M is the molecular weight, T is the temperature, and R is the ideal gas constant (assuming that the substances are at low concentration in the dilution). This force can be expressed as

$$F_m = \frac{4\pi RTqr_o^3}{3Mh},$$

where q is the rate of metabolic reaction and h is the membrane permeability (see Rashevsky, 1960).

In the first theoretical approach, from a purely physical point of view and without consideration of real cellular membranes composition, Rashevsky (1960) assumed that permeability, h, is a function of the thickness of the membrane, d, and that increasing the radius of the membrane decreases its thickness in agreement with the following relation:

$$B \equiv 4\pi r_o^2 d \Rightarrow d = \frac{B}{4\pi r_o^2} \Rightarrow h = \frac{A}{d} = \frac{4\pi A r_o^2}{B}.$$

The work made by this force to expand the radius from r_o to infinity is

$$\int_{r_o}^{\infty} F_m \, dr_o = \frac{RTqB}{4\Pi AMr_o}.$$

In the calculus of the resulting spheres of radius r_1, it suffices to replace q by $0.5 \cdot q$ and r_o by r_1 ($r_1 = 0.8 \cdot r_o$) in the previous expression. The resulting work to contract the sphere from infinity to r_1 is

$$\frac{RTqB}{6.4\pi AMr_o},$$

and the net work of all the processes (multiplying by 2 the work of contraction because there are two resulting spheres) is

$$\Delta W_m = -\frac{\pi RTqr_o^4}{2Mh}.$$

Turning to the work done on catalytic particles present in the volume enclosed by the membrane, the existence of concentration gradients created by the metabolic reactions in the volume enclosed by the membrane implies the existence of diffusion forces acting on these particles. In particular, if we have n particles per volume unit (assuming a uniform distribution), and each one of these particles occupies a volume v, the force that acts per volume unit is

$$-\frac{3}{2}\alpha n \frac{RT}{M} v grad(c),$$

with

$$\alpha = 1 - \frac{3}{2}\frac{N\delta V_m}{M_o},$$

where δ is the density of the solvent, V_m is the volume of one molecule, and M_o is the molecular weight of the solvent (see Rashevsky, 1960, chapter VIII).

If we increase the volume by an amount dV, increasing the radius r_0 by an amount dr_0, it is possible to calculate the work made by the forces of diffusion:

$$\delta W_V = \frac{9}{40}\frac{RT\Omega^2 q_o v\alpha}{M\pi D_i r_o^2} dr_o,$$

where Ω is the number of catalytic particles, q_o is the reaction rate per catalytic particle, and D_i is the inner diffusion coefficient. If we expand the volume to infinity, we obtain

$$\int_{r_o}^{\infty} \delta W_V = \frac{9}{40} \frac{RT\Omega^2 q_o v\alpha}{MD_i \pi r_o}.$$

Similarly to the previous point, the work of expansion of two halves can be calculated with the same expression, replacing Ω with $0.5 \cdot \Omega$ and r_o with $r_1 = 0.8 \cdot r_o$ and multiplying by 2.

The final balance of work will consist of three terms, the first associated with modification of the total membrane surface, the second associated with the work done by the diffusion forces on the membrane, and finally the third associated with work done by the diffusion forces on the particles present in the inner volume.

$$\Delta W = 1.12\pi\gamma r_o^2 - \frac{\pi RTqr_o^4}{2Mh} - \frac{3}{20} \frac{\pi RT\alpha\mu qr_0^5}{D_i M},$$

where we have changed

$$\mu = \alpha \cdot n; \quad \Omega q_o = \frac{4}{3}\pi r_o^3 q; \quad \Omega v = \frac{4}{3}\pi r_o^3 \mu.$$

From this final expression we can obtain a criterion of spontaneity in the cellular division:

- If $\Delta W > 0$, spontaneous division is not possible.
- If $\Delta W < 0$, spontaneous division is possible.

The critical value of the radius r_o will be the one causing $\Delta W = 0$.

We can analyze the values of this expression based on some of its most significant parameters. It is important to emphasize that significant variations in the permeability and the diffusion coefficient do not imply significant variations in the value of the critical radius, having the same order of magnitude (see figure 10.5). This implies that the assumption of the variability of permeability with membrane thickness is not of fundamental importance. The values of r_o are in agreement with the experimental values of the radius of the actual cells.

10.5.2 Induced Division

It has been seen that the membrane can divide spontaneously when it reaches a critical radius value r_o, as it is favorable from an energetic point of view. But the fact that this can occur does not guarantee that this division is unavoidable. Rashevsky

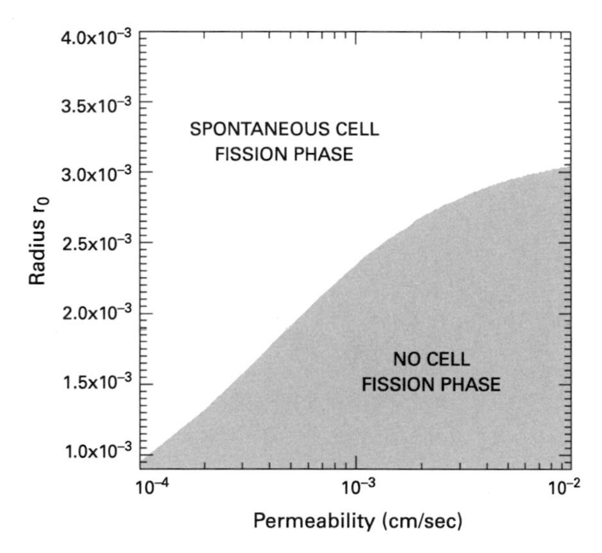

Figure 10.5
Range of values of the radius r_o versus the permeability h satisfying $\Delta W = 0$. For values of r_o leading to $\Delta W < 0$ (or $r > r_o$), spontaneous division can take place. For values of r_o leading to $\Delta W > 0$ ($r < r_o$), the configuration of the membrane remains stable.

(1960) and other authors have suggested that division of the membrane can be induced by different factors and more likely by a joint operation of several of them simultaneously. One of the factors that play an important role is the time-varying nature of the osmotic pressures. From a mechanical point of view, the action of time-varying osmotic pressures by itself can give rise to membrane division.

The origins of these variable osmotic pressures can be very diverse. For example, metabolic reactions can generate skew distributions of concentrations of metabolites. Thus, they generate an asymmetric distribution of osmotic pressures on the membrane as they diffuse outwards. Another example is given by metabolic reactions associated to localized metabolic centers (for example, molecules or clusters of molecules with catalytic properties) that can move in the membrane-enclosed volume.

We have simulated the situation generated by two localized metabolic centers (each metabolic center corresponds to a molecule acting as a catalyst) in the volume limited by the membrane (figure 10.6). We suppose that inward flux of substances exists. Additionally, in the presence of these metabolic centers, particular reactions take place in which the substances are partially consumed and new substances are generated on the surface of the metabolic centers. The latter ultimately spread outwards. The variation of the concentrations in time resulting from diffusion is given by the diffusion equations:

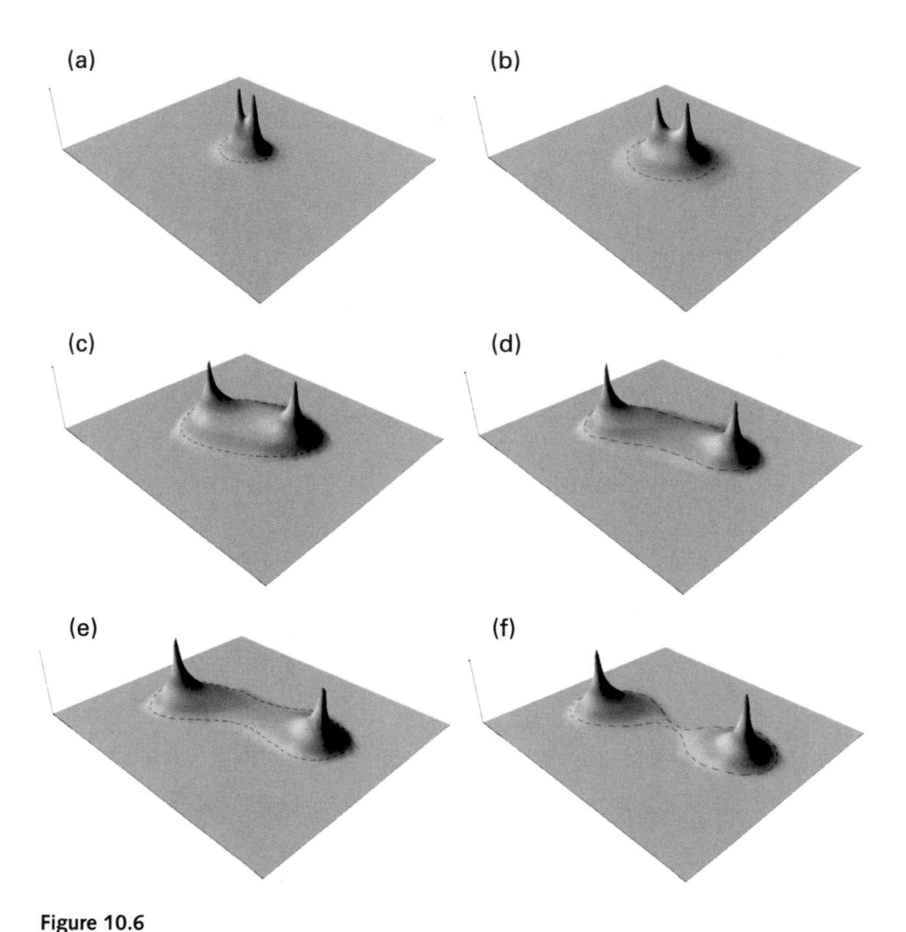

Figure 10.6
Different steps of membrane division. At each point of the membrane, the balance between pressures exerted by the incoming consumed substance, the outgoing produced substance and the surface tension are calculated. At the same time the metabolic centers are moving because of the electrostatic repulsion force. The model allows cell membranes to grow (with certain additional metabolic reactions that are able to produce membrane building blocks), and the volume of the cell can grow, too. The pictures show the concentration of the substance produced inside the volume limited by the membrane. Dashed line corresponds to the membrane boundary. For more details about the simulations, see Macia and Solé (2005).

$$\frac{\delta c_i}{\delta t} = D_i \Delta^2 c_i \quad \text{and} \quad \frac{\delta c_e}{\delta t} = D_e \Delta^2 c_e,$$

where c_i, and c_e are the concentrations inside and outside the membrane, and D_i and D_e are the inner and outer diffusion coefficients, respectively. In the membrane, the continuity of the boundary conditions must occur, supposing that the flow that arrives at the membrane leaves it (there is no accumulation of matter in the membrane):

$$D_i \frac{\delta c_i}{\delta \eta} = D_e \frac{\delta c_e}{\delta \eta} = h(c_i - c_e),$$

where h is the permeability of the membrane and η is the normal direction to the membrane.

The substances generate a certain pressure when they cross the membrane inwards and outwards. In a first approach, we can suppose that the pressure acting on each point of the membrane is the sum of the pressures generated by each substance independently. Also, we can suppose that the pressure generated by each substance is proportional to the difference in its concentrations on each side of the membrane. Thus, the pressure that acts on each point of the membrane can be calculated as

$$P_m = P_S + \sum_S k_s(c_i^s - c_e^s),$$

where s is the substance crossing the membrane, P_S is the pressure associated to the surface tension, and K_s is the proportionality coefficient that can be approximated by the following expression:

$$k_S = \frac{RT}{M_S},$$

with M_s being the molecular weight of the substance s.

If, under these conditions, the metabolic centers are displaced by any cause, variability in the osmotic pressures acting on each point of the membrane would be generated. For example, the molecules that form these metabolic centers might have an electrical charge and produce a repulsion phenomenon. We have simulated these conditions in order to understand one division cycle. The simulations have allowed us to verify that simply the variation in the osmotic pressures leads to division of the membrane. In this example, we have considered two metabolic centers that consume the input substance. The products resulting from the metabolic reactions are the same for both metabolic centers.

10.6 Conclusion

Modeling protocell replication dynamics requires a physics-inspired consideration of the membrane (container) together with an appropriate consideration of the kinetics of chemical reactions. So far, only a few models have been able to provide a full, reproducible set of steps leading to spontaneous replication. Most of these models incorporate the essential, bare bones of the underlying physics and consider some sort of active mechanism that can create instabilities in the container. Such deformations can result from energetic constraints, segregation of diverse amphiphiles, or active mechanisms. The latter can be associated with metabolic components that force the system to move out of equilibrium and split. These are, of course, the most relevant ones within the context of building artificial protocells, but they can be helped by considering other scenarios where splitting events are triggered by purely physical mechanisms.

In this chapter, we reviewed previous efforts in this direction. The limitations imposed by each approximation are obvious but also informative. The next steps toward the synthesis of artificial self-replicating cells require active processes that can regulate membrane dynamics in such a way that reaction kinetics dominates membrane growth. When thresholds of membrane stability are reached, splitting should be possible and the process can begin again. Steps taken in this direction reveal that such a scenario is feasible (Macia and Solé, 2005).

References

Gánti, T. (1975). Organization of chemical reactions into dividing and metabolizing units: The chemotons. *BioSystems, 7*, 189–195.

Gierer, A., & Meinhardt, H. (1972). Theory of biological pattern formation. *Kybernetik, 12* (1), 30–39.

Gray, P., & Scott, S. K. (1985). Sustained oscillations and other exotic patterns in isothermal reactions. *Journal of Physical Chemistry, 89* (1), 25–32.

Helfrich, W. (1973). Elastic properties of lipid bilayers: Theory and possible experiments. *Zeitschrift für Naturforschung, 28*, 693–703.

Lee, K. J., McCormick, W. D., Ouyang, Q., & Swinney, H. L. (1993). Pattern formation by interacting chemical fronts. *Science, 251*, 192–194.

Lee, K. J., McCormick, W. D., Swinney, H. L., & Pearson, J. E. (1994). Experimental observation of self-replicating spots in a reaction-diffusion system. *Nature, 369*, 215–218.

Macia, J., & Solé, R. V. (2005). Computational modeling of protocell division: A spatially extended metabolism-membrane system. Available at http://es.arxiv.org/abs/q-bio.CB/0511041 (accessed February, 26, 2007).

Mazin, W., Rasmussen, K. E., Mosekilde, E., Borckmans, P., & Dewel, G. (1996). Pattern formation in the bistable Gray-Scott model. *Mathematics and Computers in Simulation, 40*, 371–396.

Muratov, C. B., & Osipov, V. V. (2000). Static spike autosolitons in the Gray-Scott model. *Journal of Physics A: Mathematical and General, 33* (48), 8893–8916.

Noguchi, H., & Takasu, M. (2002). Adhesion of nanoparticles to vesicles: A Brownian dynamics simulation. *Biophysical Journal, 83* (1), 299–308.

Pearson, J. E. (1993). Complex patterns in a simple system. *Science*, *261*, 189.

Rashevsky, N. (1960). *Mathematical biophysics: Physico-mathematical foundations of biology*, *Vol. 1* (3rd ed.). Mineola, NY: Dover Publications.

Seifert, U., Berndl, K., & Lipowsky, R. (1991). Shape transformations of vesicles: Phase diagram for spontaneous-curvatures and bilayer-coupling models. *Physical Review*, *A44*, 1182–1202.

Svetina, S., & Zeks, B. (1983). Bilayer couple hypothesis of red cell shape transformations and osmotic hemolysis. *Biomedica Biochimica Acta*, *42* (86), 11–12.

Svetina, S., & Zeks, B. (1989). Membrane bending energy and shape determination of phospholipid vesicles and red blood cells. *European Biophysics Journal*, *17*, 101–111.

Szathmáry, E., & Maynard Smith, J. (1997). From replicators to reproducers: The first major transitions leading to life. *Journal of Theoretical Biology*, *187*, 555–571.

Yamamoto, S., & Hyodo, S. (2003). Budding and fission dynamics of two-component vesicles. *Journal of Chemical Physics*, *118*, 7937–7943.

11 Compositional Lipid Protocells: Reproduction without Polynucleotides

Doron Lancet and Barak Shenhav

11.1 Introduction

11.1.1 Life's Conundrums

A living cell is an utterly complex piece of molecular machinery. Contemplating prebiotic protocells or constructing artificial cells would require a thorough understanding of the intricacies of present-day living cell, and the methods of systems biology may become useful in this respect (Shenhav, Solomon et al., 2005; see also chapter 8). It is also necessary to employ divergent thinking, guided by an assessment of which of the cellular features are fundamental and which are consequential.

It is widely accepted that the most central property of living entities is a capacity to generate their own copies. This property underlies the ability of living organisms to undergo natural selection and evolution, and therefore must have been present very early in life's history. An often-stated conundrum is that life cannot reach even a most rudimentary level of complexity without self-replication, but only minimally complex chemical systems can multiply. The discovery of catalytic RNAs (Fedor and Williamson, 2005; Stark et al., 1978; Zaug and Cech, 1986), potentially capable of creating their own copies, has led to a wide belief that polynucleotides provide a solution for such incongruity.

An examination of the structure of RNA, however, reveals chemical intricacies that defy the notion that they might have been the first replicators. In the present-day setting, polynucleotides undergo replication only in the context of a complex cellular milieu. In a primordial highly heterogeneous chemical environment, the spontaneous emergence of sufficient quantities of the required carbohydrate moieties and nitrogen bases may be highly unlikely. Phosphodiester polymerization is thermodynamically unfavorable, and the repeated building of a long unbranched nucleic acid first and second strands would require intricate and highly specific catalysis (Bartel and Unrau, 1999; Ertem and Ferris, 1996). For these and other reasons it has been argued that an RNA-like polymer could not possibly serve as life's early seed (Schwartz, 1995; Shapiro, 1984). The main obstacle to offering an alternative scenario to the RNA world is conjuring a self-reproducing chemical system that is not based on polynucleotide information storage and templating.

It should be stressed that the origin of organic molecules and the origin of replication are two separate problems. In fact, the first may be viewed as a prerequisite for the second. The groundbreaking experiments of Miller and Urey (Miller, 1953; Miller and Urey, 1959) were specifically aimed at answering the first question. They asked how, under primitive earth conditions, a sufficient supply of organics could be generated, and found that surprisingly high yields of amino acids and other biologically relevant compounds are synthesized when electrical discharge is directed into a simulated primitive atmosphere. However, their choice of a global reducing atmosphere now seems implausible in light of more recent knowledge. The present chapter avoids this controversy, positing that organic compounds could have arisen by a wide variety of processes under a broad spectrum of planetary conditions (Basiuk and Navarro-Gonzalez, 1996; Chyba, Thomas, Brookshaw, and Sagan, 1990; Clark, 1988; Cody et al., 2000; Greenberg and Mendoza-Gomez, 1992; Keefe and Miller, 1995; Leif and Simoneit, 1995; Maurette, 1998; Miyakawa et al., 2002; Wächtershäuser, 1997). Rather, this discussion addresses only the ways by which more complex entities with lifelike properties could have emerged once such compounds were available.

11.1.2 Oparin Reconsidered

Oparin first published his pamphlet *The Origin of Life* in 1924 (Oparin, 1924, 1967), later expanding it into his well-known book with the same title (Oparin, 1936, 1938, 1953). Oparin dealt for the first time with the problem of life's origin based on a materialistic perspective. The standard Oparin-Haldane theory (Miller, Schopf, and Lazcano, 1997) invokes the generation of organic molecules on the early Earth followed by chemical reactions that produced increased organic complexity. This process is proposed to have led to organic life capable of reproduction, mutation, selection, and evolution. Oparin does not specify which molecules were first, or how self-replication came about. In fact, his portrayal predates by two decades the Watson-Crick DNA structure breakthrough, and Oparin is therefore unbiased by its implications.

According to Oparin's teachings the following steps occurred on the path to early life:

1. random synthesis of simple organic molecules from atmospheric gases;
2. formation of larger, more complex molecules from the simple organic molecules;
3. formation of coacervates—unique droplets containing the different organic molecules;

4. development of a capacity to take up molecules and discharge specific molecules, and maintain a characteristic chemical pattern or composition;

5. development of "organizers" that allowed controlled reproduction to ensure that daughter cells produced by division have the same chemical capabilities;

6. beginning of evolutionary development so that a group of cells could adapt to changes in the environment over time.

For decades, this clearcut hierarchical scenario was considered an acceptable view for how life began. In later years, Oparin guessed that the hypothesized "organizers" in step 5 might include nucleic acids (Oparin, 1976), but clearly he was not thinking in terms of a full-fledged transcription and translation apparatus, as these would have been too complex for the primitive stages he was considering. Oparin had the insight to propose simile generation based on a web of chemical interactions, which would be referred to today as a prebiotic metabolic network (Kauffman, 1993; Morowitz, 1999; Morowitz et al., 2000). Self-replication was thought to result from splitting of coacervate droplets, including their entire molecular repertoire, a process conceptually similar to modern cell division.

11.1.3 Polynucleotides Versus Assemblies

Following the DNA/RNA revolution in the 1950s and 1960s, origin of life research has undergone a radical change, no less significant than that which affected all of molecular biology. This led to the decline of the coacervates/molecular assemblies scenario ("garbage bag" in Dyson's terminology; see Dyson, 1999). This happened despite the fact that the older scenarios could potentially be more suitable for the harsh and unpredictable conditions that prevailed on early Earth.

The new scenario is simple and seemingly elegant. If only a single self-replicating polynucleotide could form in the prebiotic soup, life would readily emerge (Lifson, 1997). This is based on the generally applicable and indisputable notion that once self-replication is in place, selection and evolution can lead to increasing complexity, culminating in a protocell. This scenario envisions that the additional paraphernalia needed for functional protocells, such as metabolic pathways for the synthesis of many molecular components, membrane enclosures, passive and active transporters, energy-harvesting molecular complexes, ATP-like energy-storing compounds, primitive tRNAs and their associated sythetases, ribosomes, RNA polymerases, and much more, would all somehow appear and join the original nucleic acid. Typically, relatively little is said about how, if a single molecular replicator is in place, all the other components would appear and be recruited.

In this respect, the Oparin-style scenario has an advantage: It envisages early entities that include a diverse repertoire of components from the outset, in which full-fledged polynucleotide-based mechanisms would arise much later as a *consequence* of elaborate evolution of the early assemblies. Importantly, replication (or reproduction) is envisaged in these scenarios as an attribute of an entire molecular ensemble, so that evolution by gradual improvement would apply jointly to all components and not solely to a core replicating informational polymer.

11.2 Compositional Protocells

11.2.1 Templates and Catalysts

A common way to portray the chicken and egg nature of early evolution is to ask what came first: nucleic acid-based information-storage, needed to synthesize proteins, or macromolecular catalysts without which the DNA/RNA machinery would come to a grinding halt. One way to answer the question is to ask which of the two sets of chemical phenomena appears more basic. The RNA world protagonists would argue that, since ribonucleic acids are capable of both template-based replication *and* catalysis, it follows that polynucleotides must have been the primordial mover. However, an alternative view would probe the very basic definition of a templating reaction. As figure 11.1 suggests, a templating strand may be thought of merely as enhancing the incorporation of "correct" nucleotides into a growing strand, in comparison to the basal (slower) incorporation of "wrong" nucleotides.

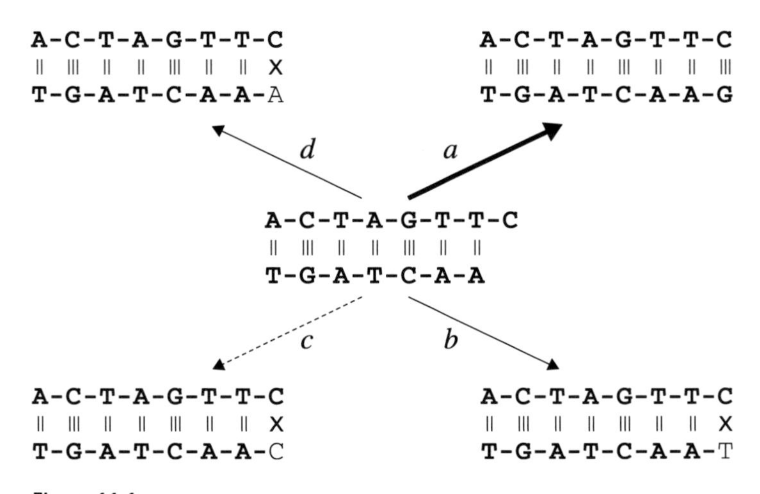

Figure 11.1
It is proposed that templating is a derived concept, and the more basic notion is catalysis. The elementary step of adding a base to an extending second strand in a double-stranded RNA/DNA-type biopolymer may be summarized in four competing reactions, one which involves legitimate Watson-Crick (WC) pairing (a), and the other three (b–d) forming illegitimate pairing with decreasing kinetic efficacy. The kinetic handicap of non-WC pairing is often a result of helix distortion leading to the termination of chain growth. This is despite the fact that non-WC base pairing in itself can be thermodynamically favorable. The top strand may thus be considered an "enzyme" that catalyzes chemical reaction (a) in favor of the side reactions (b), (c), and (d).

Thus, templating is a special case of catalysis, and DNA/RNA replication is, in some respects, part of a cell's (or protocell's) metabolic network. Consequently, one could envisage a primitive system in which Watson-Crick base-pairing has not yet evolved, and replication/reproduction-like processes are controlled by a more unwieldy set of organic catalysts. In other words, mutually catalytic networks (Bagley, Farmer, and Fontana, 1991; Dyson, 1999; Jain and Krishna, 2001; Kauffman, 1993; Morowitz, 2002) may, in principle, be sufficient for embodying a primitive form of self copying.

It is, of course, necessary to delineate how, through a gradual process of selection and evolution, templating polynucleotides could emerge from their "bag of catalysts" ancestors. Indeed, Oparin's original paper (Oparin, 1924, 1967) makes the argument that "slowly but surely, from generation to generation, over many thousands of years, there took place an improvement ... directed towards increasing the efficiency of the apparatus for absorption and assimilation of nutrient compounds ... (and) the ability to metabolize" (Oparin, 1924, p. 26). Part of that would be the appearance of RNA as a key molecule of life.

11.2.2 Compositional Information Storage

The coacervate theory describes qualitatively how structure and function might have been passed down through the generations, but a more accurate, rigorous, and quantitative description is needed to convince a modern molecular biologist that simple division of a molecular assembly is sufficient for information transfer. We have advanced the notion of compositional information, in an attempt to answer the question of how information could be propagated in an early molecular assembly devoid of informational biopolymers. The idea of sequence-based information is so widely accepted that it is hard for many to accept an alternative.

For compositional information, only the numbers of different molecule types are considered, while the spatio-organization is disregarded. To come to grips with how composition can carry a significant amount of information, consider the following example. A peptide containing ten amino acid monomers of 20 possible kinds can be constructed in 20^{10} different ways, hence has $\log_2(20^{10}) \approx 43$ bits of information. This can be compared to a small compositional assembly containing 10 amphiphilic amino acid derivatives selected from an "alphabet" of 20 kinds, which can be constructed in 2×10^7 different ways (figure 11.2; see Shenhav, Segré, and Lancet, 2003), and therefore has $\log_2(2 \times 10^7) \approx 24$ bits of information. One can thus see that, in this example, compositional information can amount to more than 50% of the equivalent sequential information! If the assembly/polymer size is considerably larger than 10, then for a small alphabet size (e.g., 20), the compositional information becomes much smaller than the sequential, but this can be compensated by increasing the alphabet (number of monomer types) (Shenhav et al., 2003). In a

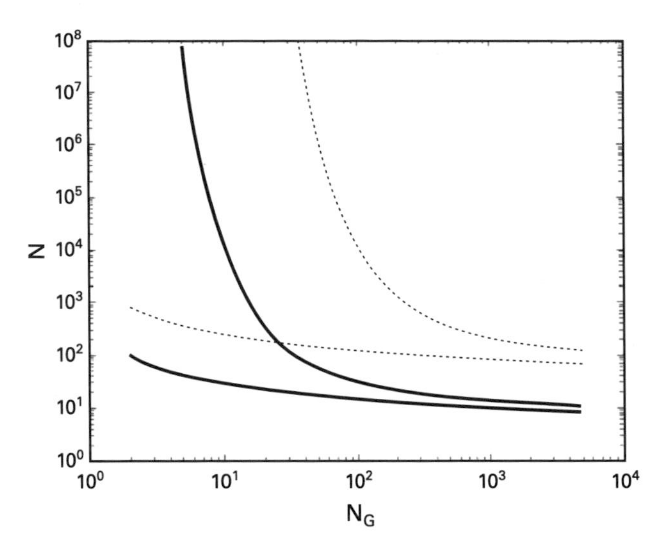

Figure 11.2
Required N (sequence length or assembly size) for encoding 100 (dark) and 800 (light) binary bits by a sequence (full line) or assembly (dashed line) as a function of the size of the molecular repertoire (N_G). The number of bits encoded by a sequence of length N from a repertoire of size N_G is $N \log_2(N_G)$. The number of bits encoded by an assembly of size N from a repertoire of size N_G is $\log_2\binom{N_G+N-1}{N}$.

prebiotic scenario, when monomer selection has not yet taken place, large monomer repertoires (even in the millions) are possible.

The sequence information of biopolymers is related to the compositional complexity of assemblies. A living cell has very few members of any relevant combinatorial repertoire of sequences. When considering a cell's protein inventory, for example, one would tend to focus on the information contained within each of the individual polypeptides and its relationship to function. But what is significant in terms of information content is that cells are biased to contain only a very small subset of all the possible strings of amino acids. For every protein present, an astronomically large number of amino acid strings are not. Thus, an important aspect of information content in living cells (and by analogy, in protocells) is their idiosyncratic chemical composition.

The idiosyncratic composition within amphiphile-based assemblies may underlie emergent physical properties not explicitly considered in our model. For example, the lipid composition of the membrane of a simple bacterium (the micoplasma *Acholeplasma laidlawii*) determines the membrane curvature and curvature stress, which control viability (Gruner, 1989; Osterberg et al., 1995). It is possible that compositional information is the basis for collective emergent properties in protocells, which in turn may exert a key influence on mutual catalysis.

11.2.3 Propagating Compositional Information: The GARD Model

Templating is perfectly suited for propagating sequential information. In contrast, in a hypothetical protocell in which coding mechanisms have not yet evolved, and where function is governed by composition, we have demonstrated that a metabolism-like network of mutually catalytic interactions is an appropriate mechanism for compositional information propagation (see also Kauffman, 1993). This mechanism was shown to be robust, that is, to often withstand small perturbations in composition (Shenhav, Solomon, et al., 2005). One can gain relevant insight from observing that present-day living cells are largely compositional entities. For example, transcriptome analysis basically asks about the tally of every messenger RNA (mRNA) type. Indeed, the first step in generating progeny is the biosynthetic doubling of the compositional counts of all molecules (mRNAs, proteins, etc.), as well as organelles, within the cell.

The Graded Autocatalysis Replication Domain (GARD) model (Segré, Ben-Eli, and Lancet, 2000; Segré and Lancet, 2000; Segré et al., 1998; Shenhav et al., 2003; Shenhav, Kafri, and Lancet, 2004) describes the dynamic behavior of a molecular assembly whose constituents manifest mutual catalysis, resulting in a global self-propagating behavior of the entire assembly. Under certain constraints, this behavior resembles replication or reproduction (Segré et al., 2000; Segré, Shenhav, et al., 2001). In the simplest embodiment, GARD computer simulations depict the behavior of molecules that join and leave the assembly, manifesting only noncovalent interactions with each other (figure 11.3). In this embodiment, no biosynthesis takes place, and the only compounds present within the assembly are those supplied from the outside (complete heterotrophy). Mutual catalysis events occur on thermodynamically favorable (downhill) reactions, fueled by a spontaneous tendency of molecules to join the assembly. This GARD description is best suited for lipidlike amphiphilic assemblies (see section 11.2.4), yet it is also generally applicable to other modes of molecular enclosures.

The essence of information transfer and propagation in GARD relates to the capacity of a "daughter" molecular assembly to inherit a rather faithful replica of the compositional information contained within a "mother" assembly. In GARD dynamics, a molecular assembly derived by a physical split from its ancestor, withstands the "trauma" of fission (e.g., the loss of scarce but essential molecular species) based on prior homeostatic growth (figure 11.4; see Segré et al., 2000; Segré, Shenhav, et al., 2001).

In present-day cells, information transfer between generations depends on protein synthesis, which in turn depends on information inscribed in RNA and DNA. This central dogma pathway is so complex that it is unlikely to have been present in very early protocells (Morowitz et al., 2000; Shapiro, 2000). The alternative proposed in the framework of GARD claims that early on, none of the three components of the

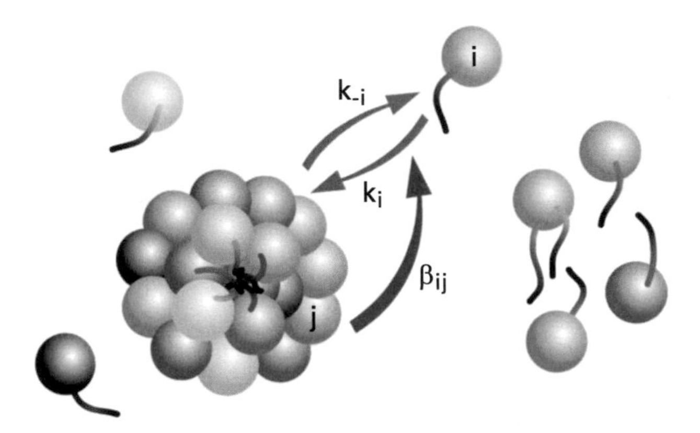

Figure 11.3
GARD: the Graded Autocatalysis Replication Domain model. The main reaction step in the simplest amphiphile-joining GARD formulation, is the reversible exchange of an amphiphilic molecule A_i between the environment and an assembly (black arrows, representing k_i and k_{-i}, respectively the forward and reverse basal rate constants). A key aspect in reaching a kinetic homeostasis is the dependence of the reaction rates on the current composition of the assembly, through mutual catalysis. The matrix β_{ij} signifies the mutual rate enhancement parameters for the catalysis exerted by species A_j on the joining and leaving reactions of A_i (bottom arrow). The β matrix elements are drawn from a probability distribution generated through the Receptor Affinity Distribution model (Lancet et al., 1993; Rosenwald, Kafri, and Lancet, 2002; Segré and Lancet, 1999).

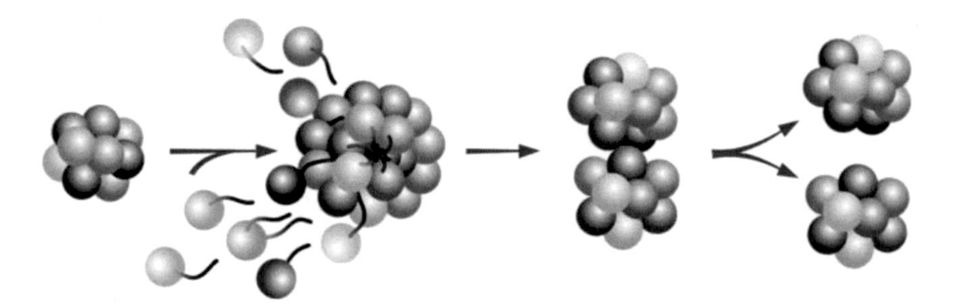

Figure 11.4
The life cycle of a GARD assembly. A GARD assembly undergoes growth through selective joining of molecules from the external environment. As the assembly accretes more and more molecules, it may lose structural stability and "split" into two progenies, as described in Tanaka, Yamashita, and Yamazaki (2004). Some assemblies with distinct compositions manifest homeostatic growth, that is, there is no change in the ratio of the molecular counts within the assembly. Their daughter assemblies are similes of the original assembly with comparable composition (Segré et al., 2000; Segré, Shenhav, et al., 2001), and are likely to undergo homeostatic growth. Thus, a series of "faithful" transmissions of the composition from parent assembly to daughter assembly is generated. Such self-replicating compositions are termed *composomes*.

central dogma existed, and the only resemblance of early replicating and evolving entities to present-day cells was in matters of general design.

Importantly, modern cells undergo a very similar process of homeostatic growth and compositional inheritance during reproduction. According to this view, the molecular templating of DNA is just one of a large set of mechanisms through which it is assured that, after a physical split, the two cellular halves would be capable of replenishing their molecular repertoires en route to the next split. Parallel epigenetic mechanisms would be DNA methylation, the protein-mediated copying of the centrioles and the biosynthetic production of organelles such as ribosomes, mitochondria, and the Golgi apparatus. Such premeiotic processes ensure that, on fission, a mother cell will be endowed with sufficient copies of every necessary component. Epigenetic transmission (e.g., of maternal mRNA to the developing embryo; see King, Messitt, and Mowry, 2005) is no less crucial for the propagation of information than DNA replication. The recent demonstration of potential information flow from such maternal RNA back to DNA (Lolle et al., 2005) further underlines the importance of nongenomic information.

11.2.4 GARD's Lipid World

The basic GARD model assumes a network of mutually catalytic, relatively simple organic molecules, with amphiphilic characteristics, forming small assemblies. The constituent molecules are conceived as "generalized amphiphiles," some resembling present cellular lipids, but others possessing a large variety of head group sizes and functional groups, as well as a diversity of hydrophobic tails. Thus, the term "lipid world" absolutely does not imply the exclusive involvement of the lipids present in contemporary cells. Furthermore, though in present-day cells lipid-type compounds are chiefly involved in constructing enclosures or containers, the lipidlike molecules in the lipid world scenario for life's origin are proposed to have been much more broadly disposed in terms of function, affording rudimentary catalysis and compositional information storage. Notably, present-day lipids within bilayers still preserve some of these unorthodox functions, although in a very rudimentary fashion, as exemplified in the proposed catalysis of ligand-receptor interactions by components of the lipid phase of the cell membranes (Sargent and Schwyzer, 1986; figure 11.5).

The simplest assemblies may be in the form of micelles without internal volume, but vesicles with lipid bilayer boundaries and an aqueous core are equally legitimate GARD embodiments. The "business end" of GARD is in the lipid phase, when the most rudimentary form of rate enhancement is assumed to occur upon amphiphile entry and exit (figure 11.3). The enclosed aqueous lumen may harbor some of the exchanged molecules, but GARD's dynamics and viability do not depend on luminal content in ways invoked for lipid-enclosed protocells that include polynucleotides within their aqueous interior.

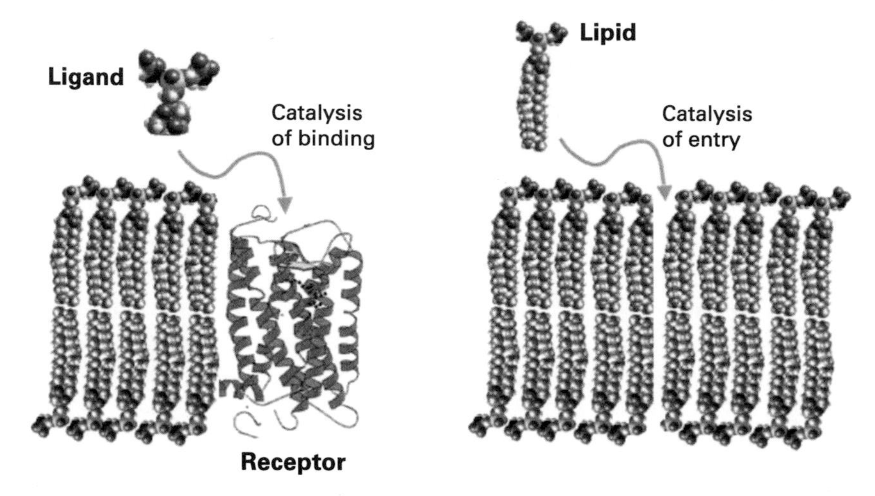

Figure 11.5
(Left) membrane lipid phase as catalyst for peptide-receptor interactions (Sargent and Schwyzer, 1986). (Right) In analogy, neighboring lipids in a micelle or a bilayer may catalyze the entry/exit of another molecule into the amphiphilic assembly. Such a relatively simple notion serves as a core of the GARD model as described in the text and elsewhere (Segré et al., 2000; Segré and Lancet, 2000; Shenhav et al., 2003, 2004). In this context, catalysis implies rate enhancement as described for lipid systems (Fendler and Fendler, 1975), in contrast to enzyme catalysis, which is endowed with a much higher fidelity and specificity.

The basic GARD, which is based simply on amphiphile entry to and exit from the lipidlike phase, still displays a complex dynamic behavior that resembles cellular homeostatic growth (Segré et al., 2000; Segré, 2000; Segré, Ben-Eli, et al., 2001; Ben-Eli and Lancet, 1999). Computer simulations that embody a set of rigorously defined kinetic and thermodynamic constraints demonstrate that such assemblies can grow homeostatically while preserving their compositional information (Segré et al., 2000). GARD assemblies appear to alleviate the chicken-and-egg problem by harboring within them both replicable information and a metabolism-like network.

11.2.5 Evolving GARDs

The most critical aspect of prebiotic evolution is represented in a seeming paradox. In attempting to envision a bridge between nonliving and living matter, there is general agreement that living entities could not have emerged without the involvement of selection and evolution. But most researchers consider evolution an attribute of life, and therefore the term *prebiotic evolution* harbors an intrinsic inconsistency. In other words, crossing the bridge is necessary to construct it.

In an attempt to resolve this paradox, we have proposed (Shenhav et al., 2003) regarding life's emergence as a graded series of steps rather than an abrupt transition. Accordingly, *prebiotic* or *nonliving* would relate solely to organic molecules, even with high covalent complexity, generated by abiotic chemical processes. The terms *biotic* and *living* should be reserved for cells that already contain a broad gamut of elements of the standard life machinery, that is, RNA-like and proteinlike molecules, templating and coding, elaborate metabolism, and self-generated enclosure. To deal with the graded transition between these two dichotomous states of matter, we have employed a new term, *mesobiotic*, to describe the intermediate stages between nonliving and living (Shenhav et al., 2003). It is implied that mesobiotic entities may have harbored a considerably reduced set of organic molecules, and biopolymer components were short and rudimentary in their functions. It is further argued that early protocells were mesobiotic, and attempts to construct them in the laboratory should take this into account. This term, *mesobiotic*, is also appropriate because it conveys the notion that the early molecular assembly-based protocells belong to the realm of mesoscopic entities, on the border between the microscopic and macroscopic.

Since protocells based on the GARD/lipid world conceptual framework have a capacity to store compositional information and propagate it from one generation to another, selection and evolution become possible. We have coined the term *compositional mutations* (Segré et al., 2000) to indicate changes in composition that occur as a result of imperfections in the catalytic network or statistical variations between progeny assemblies on fission. Consequently, during a computational GARD simulation, one homeostatic assembly occasionally gives way to anther, which is homeostatic as well. This is akin to multiple attractors, or fixed points, observed in other dynamic systems. Since these states are characterized by different molecular compositions, we coined the term *composomes* to describe them (Segré et al., 2000). The transitions among composomes are quite abrupt, and they are shown to stem from rapid accumulation of compositional mutations. There seems to be a remote analogy here to genomic mutations that lead to speciation.

Because compositional information in a GARD assembly plays an analogous role to that of DNA/RNA sequence information in present-day genomes, we regard composomes as equivalent to a very primitive "compositional genome." We have shown that composome dynamics (Segré et al., 2000) and GARD dynamics in general (Segré, Pilpel, and Lancet, 1998) are akin to a primitive evolutionary process. It may be argued that division of labor between genome molecules, catalysts, and compartments—the hallmark of full-fledged (biotic) cellular life—was less distinct in early mesobiotic protocells. The question of how such division has emerged would require considerable further research. One potential avenue is the exploration of biopolymer emergence within GARD assemblies, as detailed in the next section.

11.2.6 GARD-Crafted Biopolymers

More advanced versions of our model, called Polymer-GARD or P-GARD (see Shenhav, Bar-Even, et al., 2005), also include events of catalyzed covalent bond formation that lead to oligomerization and the emergence of a more realistic connected metabolism. In this view, simple organic catalysts (enzyme mimetics) underlie the emergence of extended catalytic networks, which govern not only molecular entry and exit, but also the covalent threading of monomers into higher oligomers. It is possible that only after a long progression of evolutionary events did such oligomers lead to the eventual appearance of RNA and folded polypeptides catalysts.

Thus, Polymer-GARD includes reactions of the form

$$A_i + A_j \rightleftarrows A_i - A_j,$$

where A_i and A_j are molecular species present in the assembly and $A_i - A_j$ is their covalent dimer. One chemically realistic embodiment would be two amphiphiles, each with a polar head and a lipophilic tail, joined in a head-to-head fashion. But the model could also represent other possibilities such as the involvement of the tail groups in the formation of a covalent bond, or the linking of an isolated polar head group to an amphiphile head group, generating a more complex head group. The actual chemistry could be diverse, including not only that which is widespread in present-day life (amide and phosphodiester bonds) but also ester, thioester, aldol, and carboxyanhydride condensations. It is also assumed that monomers are either absorbed in an active form from the surroundings or are activated in situ to allow energetically downhill formation of covalent bonds. This implicit assumption is shared with other models for early evolution, including the RNA world.

In the simpler monomer GARD, values of the catalytic enhancements exerted by monomers on the join/leave reactions of other monomers are obtained with a statistical chemistry approach (Lancet, Kedem, and Pilpel, 1994; Lancet, Sadovsky, and Seidmann, 1993; Segré and Lancet, 1999). In P-GARD, in order to avoid an oversized parameter lookup matrix and to represent correlations between oligomer catalyses and those exerted by the constituent monomers, we invoke a formula rate enhancement parameter (Shenhav, Bar-Even, et al., 2005). Considering the special case of dimers-only, for N_G monomer types, a total of $N_G{}^2$ covalent bond formation reactions are added to the basic N_G join/leave reactions of the basic GARD model. The number of rate enhancement factors is thus $(N_G{}^2 + N_G)^2$, compared with $N_G{}^2$ for monomer GARD. This considerable enhancement of complexity is a computing impediment but at the same time harbors a promise for more elaborate, lifelike dynamics.

More recently, we have begun to explore a defined version of polymer GARD, limited to oligomers up to length 3 (trimer GARD or T-GARD). It is computationally more feasible than the model with unlimited polymer length, but at the same time captures much of the complexity and dynamic significance of the more general model. Two approaches have been considered. In the first, the external environment

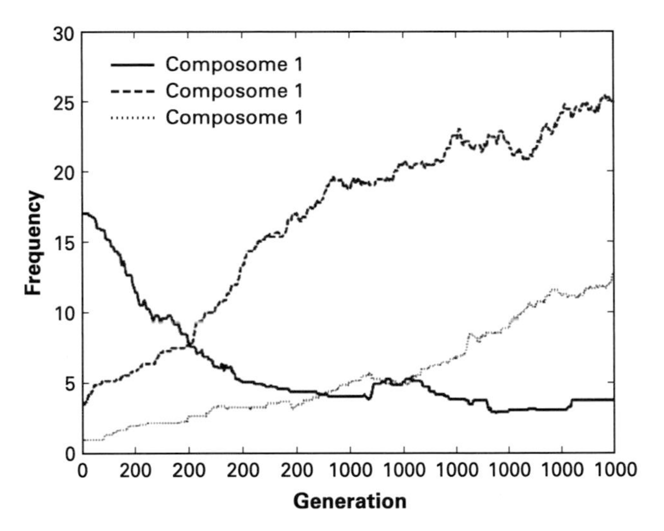

Figure 11.6
Evolution-like behavior in "oligomers exchange" GARD model: the frequency of observing three different composomes in a compositional trace of an assembly for 2,000 consecutive generations (using a window of 100 generations). The frequency of composome 1 decreases from about 16% to less than 4%, whereas the frequencies of composomes 2 and 3 increase. The change in the frequencies reflects an evolution-like process in which composomes 2 and 3 become fitter as the environment changes due to the process of "oligomer exchange." In this analysis, as in earlier GARD analyzes, composomes represent dynamic steady states away from equilibrium, which may prevail for certain periods of time, spanning several cycles of growth and splitting. At a higher level, the system displays "open-ended" evolutionary behavior, whereby the system may still undergo much slower changes in which one composome gives way to another.

consists only of monomers, as in monomer GARD. The second approach allows environmental exchange of molecules of all sizes. The latter—"oligomers exchange" T-GARD model—displays a novel evolutionlike behavior. The simulations show the emergence of novel composomes, whereas previously prevailing composomes become rare (figure 11.6). This dynamic behavior seems to mimic an evolutionary process, likely the combined result of the enhanced complexity due to oligomer formation, and the more realistic assumption of environmental oligomer exchange. It would be important to see whether this trend is accentuated when higher oligomer sizes are allowed.

11.3 Conclusion: Looking into the Future

The question of how life appeared is one of the last true frontiers of science. Most students of this topic realize that it is rewarding to ask the question, but at the same time, we may never know the answer. Our view on this issue is that we tend to limit ourselves to research that is feasible now or within the next decade. Could it be that, to address the riddle of life's inception, we will have to wait much longer? In the perspective of the history of science, waiting a few centuries is not out of question. This last section addresses what the more remote future might bring to origin of life studies in general, and to protocell research in particular.

11.3.1 Futuristic Experimental Technologies

The gist of our laboratory's research is that early protocellular life may have utilized chemistries very different from those of their present-day counterparts. An extreme proposition is that life began with inorganic clays (Cairns-Smith, 1982), although a somewhat more conservative view centered on organic compounds seems more appropriate. Many of the treatises of early life adhere to the notion that a protocell would strongly resemble present-day cells in terms of its chief components: lipids for enclosure, proteins for catalysis, and nucleic acids for information storage. The foregoing arguments should not be taken as an indication that such assumptions are wrong, only that an alternative approach is needed.

The unorthodox view we preach here could be promising, but at the same time risky. The notions of compositional inheritance and enzyme-mimetic organic catalysis allow us to consider highly unusual protocellular chemistries. It is obvious that a primitive protocell in the form of an assembly of arbitrary organic molecules cannot be expected to undergo reproduction-like dynamic changes at rates that are measurable by standard devices. It is not impossible that the time constants in question would be months, years, or even decades. This means setting up an experiment and handing it to the next generation to complete and decipher. It is interesting to draw an analogy to space travel. For a given spaceship speed, a galactic target's accessibility is often judged by the yardstick of a human being's life span. But if we free ourselves from such constraints, new approaches to both galactic exploration and origin of life may be legitimately considered. It should be pointed out that for a process such as life's origin, which most people agree must have lasted millions of years, considering extremely long-duration experiments is not an untenable suggestion. Still, shorter timescale experiments may provide support for some of the components of the GARD model. An example is a study currently underway to quantify mutual rate enhancement in amphiphiles joining to micelles with fluorescent label-based technology as described by Hanczyc, Fujikawa, and Szostak (2003).

Another confounding relevant parameter is volume. We are used to doing experiments in test tubes or Erlenmeyer flasks. Yet, the original "experiment" was performed on an entire planet. We have pointed out in the past that the "planetary probability" of life's emergence is related to the ratio between the total hydrosphere volume and that of an individual protocellular assembly. A conservative estimate comes up with a rough number of 1,000 km^3 divided by 1 μm^3. This is a ratio of 10^{12} m^3 to 10^{-18} m^3, that is, 10^{30}. Even if the events that seeded life had a probability of "only" 10^{-25}, then the volume required for any hope of reproducing such chemical events would be 0.01 km^3, or 100,000 tons of water.

The bottom line of these arguments is that, in parallel to standard laboratory experiments with protocells made up of components extracted from living cells, we should be prepared to consider experiments on very large time and volume scales. At the same time, we could become prepared with modes of analysis for what is ahead. Obviously, if the GARD scenario is correct, then at any time it would be necessary to perform microanalyses of assemblies with idiosyncratic compositions. This would require methodologies and instrumentation that can perform high throughput analysis of extremely small samples, of the order of individual micelles or vesicles. It will require considerable resilience to go through billions of assemblies showing no sign of lifelike activity before reaching those few assemblies endowed with just the right composition to display slow but perceptible reproductive capacity.

11.3.2 Computed Protocells of the Twenty-Second Century

Given the limitations of test-tube approaches, an alternative mode of analysis needs to be explored. We have argued (Shenhav and Lancet, 2004) that as computational tools improve, it may eventually become possible to perform faithful computer-based simulations of prebiotic evolution. This was in the framework of what we have named "computational origins of life endeavor" (CORE; see Shenhav and Lancet, 2004).

One point of view in early evolution research is that computer analyses and simulations should be regarded mainly as a supporting tool aimed at predictions or post-analyses of experimental results. Computing tools have been used for simulation of theoretical models by others (Alves et al., 2001; Bagley et al., 1991; de Oliveira and Fontanari, 2001; Eigen, 1971; Eigen and Schuster, 1979; Farmer, Kauffman, and Packard, 1986; Fontana and Buss, 1994; Kauffman, 1993; Nir and Lahav, 1997) as well as by our own group (Naveh et al., 2004; Segré, Philpel, and Lancet, 1998, 2000; Shenhav, Bar-Even, et al., 2005). Computer-aided scrutiny also serves as the basis for artificial life research (Adami, 1998; Sipper, 2002; Wilke and Adami, 2002). This led some to regard certain analyses of the genre as "abstract and mathematical without consideration of the properties of real molecules" (Anet, 2004, p. 656). It is necessary to ask whether such criticism is justified in view of a slow but persistent revolution in the merger of computing and experimental sciences. In the life sciences, this happens within the realms of bioinformatics or computational biology, which have profound impact, particularly in the wake of the genome revolution (Lander et al., 2001; Venter et al., 2001). Examples are simulations of transcription control networks (Banerjee and Zhang, 2002) and metabolic pathways (Christensen and Nielsen, 2000), and in the field of systems biology (Ideker, Galitski, and Hood, 2001). A new term, *in silico* biology, comprises computer models for entire cells (Loew and Schaff, 2001; Tomita, 2001) or even organisms (Kam et al., 2003).

Ever since the advent of digital computers half a century ago, computing power has been used to simulate chemical reactions and molecular structure. Quantum chemical calculations (Sponer, Leszczynski, and Hobza, 1996) are routinely used to predict the lengths, angles, and energies of covalent bonds within molecules, as well as the rate of their formation. One of the glaring paradoxes of biological chemistry, however, is the difficulty encountered when attempting to predict how sets of weaker, noncovalent interactions govern processes such as protein folding and ligand-protein recognition (Bures and Martin, 1998; Schneider and Bohm, 2002). Similar difficulties are encountered when attempting to perform *ab initio* calculations of enzyme catalysis. Thus, when it comes to predicting the detailed behavior of a molecular network, it is practically impossible to do so based on first principles of quantum chemistry.

Despite these limitations, computer tools can be found in almost any chemical area (Dessy, 1996). For simulations of molecular systems, too complex for detailed description based on quantum principles, computational chemists use other simulation approaches, such as molecular dynamics, which is based on Newtonian dynamics. As an example, various molecular dynamics simulations of chemical systems with different lipid molecules have shown the time course description of the formation of bilayers and micelles (Goetz and Lipowsky, 1998; Lindahl and Edholm, 2000; Marrink and Mark, 2002; Pohorille and Benjamin, 1993; Tieleman, Marrink, and Berendsen, 1997). Chemistry-based simulations improve as a result of developments in the underlying algorithms such as in molecular dynamics (Tuckerman and Martyna, 2000) or in stochastic chemistry (Cao, Gillespie, and Petzold, 2005; Gibson and Bruck, 2000; Gillespie, 2001). Constant improvement is gained also from the continuous progress of hardware. Thus, contemporary simulation tools are capable of rather exact simulation, albeit of relatively small chemical systems, of up to thousands of molecules, and for relatively short periods of time.

What will happen 100 years from now? Our prediction is that in the beginning of the twenty-second century, computed chemistry will merge with true bench chemistry. It will be possible to compute accurately anything from molecular structure to protein folding to enzyme kinetics. In such an era, accurate and fast computing of the dynamics of an amphiphile assembly would become rather easy. With enough computing power it will perhaps become possible to simulate the fate of billions of GARD assemblies over periods of hundreds of years. This may bring true hope to answer the current open questions of the very early steps of life's emergence and evolution.

References

Adami, C. (1998). *Introduction to artificial life*. New York: Springer Verlag.

Alves, D., Campos, P. R. A., Silva, A. T. C., & Fontanari, J. F. (2001). Group selection models in prebiotic evolution. *Physical Review E, 63* (1), 011911.

Anet, F. A. (2004). The place of metabolism in the origin of life. *Current Opinion in Chemical Biology, 8* (6), 654–659.

Bagley, R. J., Farmer, D. J., & Fontana, W. (1991). Evolution of a metabolism. In C. G. Langton, C. Taylor, J. D. Farmer, & S. Rasmussen (Eds.), *Artificial life II* (pp. 141–158). Redwood City, CA: Addison-Wesley.

Banerjee, N., & Zhang, M. Q. (2002). Functional genomics as applied to mapping transcription regulatory networks. *Current Opinion in Microbiology, 5* (3), 313–317.

Bartel, D. P., & Unrau, P. J. (1999). Constructing an RNA world. *Trends in Biochemical Sciences, 24* (12), M9–M13.

Basiuk, V. A., & Navarro-Gonzalez, R. (1996). Possible role of volcanic ash-gas clouds in the Earth's prebiotic chemistry. *Origin of Life and Evolution of the Biosphere, 26* (2), 173–194.

Bures, M. G., & Martin, Y. C. (1998). Computational methods in molecular diversity and combinatorial chemistry. *Current Opinion in Chemical Biology, 2* (3), 376–380.

Cairns-Smith, G. (1982). *Genetic takeover and the mineral origins of life*. Cambridge, UK: Cambridge University Press.

Cao, Y., Gillespie, D., & Petzold, L. (2005). Multiscale stochastic simulation algorithm with stochastic partial equilibrium assumption for chemically reacting systems. *Journal of Computational Physics, 206* (2), 395–411.

Christensen, B., & Nielsen, J. (2000). Metabolic network analysis: A powerful tool in metabolic engineering. *Advances in Biochemical Engineering and Biotechnology, 66*, 209–231.

Chyba, C. F., Thomas, P. J., Brookshaw, L., & Sagan, C. (1990). Cometary delivery of organic molecules to the early Earth. *Science, 249*, 366–373.

Clark, B. C. (1988). Primeval procreative comet pond. *Origin of Life and Evolution of the Biosphere, 18* (3), 209–238.

Cody, G. D., Boctor, N. Z., Filley, T. R., Hazen, R. M., Scott, J. H., Sharma, A., & Yoder, H. S., Jr. (2000). Primordial carbonylated iron-sulfur compounds and the synthesis of pyruvate. *Science, 289* (5483), 1337–1340.

de Oliveira, V. M., & Fontanari, J. F. (2001). Extinctions in the random replicator model. *Physical Review E, 64* (5), 051911.

Dessy, R. E. (1996). New reagents, new reactions: Computers in chemistry. *Technology in Society, 18* (2), 137–149.

Dyson, F. (1999). *Origins of life*. Cambridge: Cambridge University Press.

Eigen, M. (1971). Self-organization of matter and the evolution of biological macromolecules. *Naturwissenschaften, 58* (10), 465–523.

Eigen, M., & Schoster, P. (1979). *The hypercycle*. Berlin: Springer.

Ertem, G., & Ferris, J. P. (1996). Synthesis of RNA oligomers on heterogeneous templates. *Nature, 379* (6562), 238–240.

Farmer, J., Kauffman, S., & Packard, N. (1986). Autocatalytic replication of polymers. *Physica D, 22*, 50–67.

Fedor, M. J., & Williamson, J. R. (2005). The catalytic diversity of RNAs. *Nature Reviews Molecular Cell Biology, 6* (5), 399–412.

Fendler, H. J., & Fendler, E. J. (1975). *Catalysis in micellar and macromolecular systems*. New York: Academic Press.

Fontana, W., & Buss, L. W. (1994). What would be conserved if the tape were played twice. *Proceedings of the National Academy of Sciences of the United States of America, 91* (2), 757–761.

Gibson, M. A., & Bruck, J. (2000). Efficient exact stochastic simulation of chemical systems with many species and many channels. *Journal of Physical Chemistry A, 104* (9), 1876–1889.

Gillespie, D. T. (2001). Approximate accelerated stochastic simulation of chemically reacting systems. *Journal of Chemical Physics, 115* (4), 1716–1733.

Goetz, R., & Lipowsky, R. (1998). Computer simulations of bilayer membranes: Self-assembly and interfacial tension. *Journal of Chemical Physics, 108* (17), 7397–7409.

Greenberg, J. M., & Mendoza-Gomez, C. X. (1992). The seeding of life by comets. *Advances in Space Research, 12* (4), 169–180.

Gruner, S. M. (1989). Stability of lyotropic phases with curved interfaces. *Journal of Physical Chemistry, 93* (22), 7562–7570.

Hanczyc, M. M., Fujikawa, S. M., & Szostak, J. W. (2003). Experimental models of primitive cellular compartments: Encapsulation, growth, and division. *Science, 302* (5645), 618–622.

Ideker, T., Galitski, T., & Hood, L. (2001). A new approach to decoding life: Systems biology. *Annual Reviews in Genomics and Human Genetics, 2*, 343–372.

Jain, S., & Krishna, S. (2001). A model for the emergence of cooperation, interdependence, and structure in evolving networks. *Proceedings of the National Academy of Sciences of the United States of America, 98* (2), 543–547.

Kam, N., Marelly, R., Kugler, H., Pnueli, A., Harel, D., Hubbard, E. J., & Stern, M. J. (2003). Formal modeling of *C. elegans* development: A scenario-based approach. *Proceedings of the International Workshop on Computational Methods in Systems Biology, Rovereto, Italy*. New York: Springer.

Kauffman, S. A. (1993). *The origins of order*. New York: Oxford University Press.

Keefe, A. D., & Miller, S. L. (1995). Are polyphosphates or phosphate esters prebiotic reagents? *Journal of Molecular Evolution, 41* (6), 693–702.

King, M. L., Messitt, T. J., & Mowry, K. L. (2005). Putting RNAs in the right place at the right time: RNA localization in the frog oocyte. *Biology of the Cell, 97* (1), 19–33.

Lancet, D., Kedem, O., & Pilpel, Y. (1994). Emergence of order in small autocatalytic sets maintained far from equilibrium: Application of a probabilistic receptor affinity distribution (RAD) model. *Berichte Der Bunsen-Gesellschaft-Physical Chemistry Chemical Physics, 98* (9), 1166–1169.

Lancet, D., Sadovsky, E., & Seidemann, E. (1993). Probability model for molecular recognition in biological receptor repertoires: Significance to the olfactory system. *Proceedings of the National Academy of Sciences of the United States of America, 90* (8), 3715–3719.

Lander, E. S., Linton, L. M., Birren, B., Nusbaum, C., Zody, M. C., Baldwin, J., et al. (2001). Initial sequencing and analysis of the human genome. *Nature, 409* (6822), 860–921.

Leif, R. N., & Simoneit, B. R. (1995). Confined-pyrolysis as an experimental method for hydrothermal organic synthesis. *Origin of Life and Evolution of the Biosphere, 25* (5), 417–429.

Lifson, S. (1997). On the crucial stages in the origin of animate matter. *Journal of Molecular Evolution, 44* (1), 1–8.

Lindahl, E., & Edholm, O. (2000). Mesoscopic undulations and thickness fluctuations in lipid bilayers from molecular dynamics simulations. *Biophysical Journal, 79* (1), 426–433.

Loew, L. M., & Schaff, J. C. (2001). The virtual cell: A software environment for computational cell biology. *Trends in Biotechnology, 19* (10), 401–406.

Lolle, S. J., Victor, J. L., Young, J. M., & Pruitt, R. E. (2005). Genome-wide non-Mendelian inheritance of extra-genomic information in *Arabidopsis. Nature, 434* (7032), 505–509.

Marrink, S. J., & Mark, A. E. (2002). Molecular dynamics simulations of mixed micelles modeling human bile. *Biochemistry, 41* (17), 5375–5382.

Maurette, M. (1998). Carbonaceous micrometeorites and the origin of life. *Origin of Life and Evolution of the Biosphere, 28* (4–6), 385–412.

Miller, S. L. (1953). A production of amino acids under possible primitive earth conditions. *Science*, *117*, 528–529.

Miller, S. L., Schopf, J. W., & Lazcano, A. (1997). Oparin's "origin of life:" Sixty years later. *Journal of Molecular Evolution*, *44* (4), 351–353.

Miller, S. L., & Urey, H. (1959). Organic compound synthesis on the primitive earth. *Science*, *130*, 245–251.

Miyakawa, S., Yamanashi, H., Kobayashi, K., Cleaves, H. J., & Miller, S. L. (2002). Prebiotic synthesis from CO atmospheres: Implications for the origins of life. *Proceedings of the National Academy of Sciences of the United States of America*, *99* (23), 14628–14631.

Morowitz, H. J. (1999). A theory of biochemical organization, metabolic pathways, and evolution. *Complexity*, *4*, 39–53.

Morowitz, H. J. (2002). *The emergence of everything*. Oxford: Oxford University Press.

Morowitz, H. J., Kostelnik, J. D., Yang, J., & Cody, G. D. (2000). The origin of intermediary metabolism. *Proceedings of the National Academy of Sciences of the United States of America*, *97* (14), 7704–7708.

Naveh, B., Sipper, M., Lancet, D., & Shenhav, B. (2004). Lipidia: An artificial chemistry of self-replicating assemblies of lipid-like molecules. In J. Pollack, M. Bedau, and R. A. Watson *Artificial Life IX* P. Husbands, T. Ikegami, (Eds.) Cambridge, MA: MIT Press.

Nir, S., & Lahav, N. (1997). Emergence of template-and-sequence-directed (TSD) synthesis. II: A computer simulation model. *Origins of Life and Evolution of the Biosphere*, *27*, 567–584.

Oparin, A. I. (1924). *Proiskhozhdenie zhizny*. Moscow: Izd. Moskovski Rabochi.

Oparin, A. I. (1936). *Vozniknovenie zhizny na zemle*. Moscow: Izd. AN SSSR.

Oparin, A. I. (1938). *The origin of life*. New York: Macmillan.

Oparin, A. I. (1953). *The origin of life*. New York: Dover.

Oparin, A. I. (1967). The origin of life. In J. D. Bernal (Ed.), *The origin of life* (pp. 199–234). London: Weidenfeld and Nicolson.

Oparin, A. I. (1976). Evolution of the concepts of the origin of life, 1924–1974. *Origin of Life*, *7* (1), 3–8.

Osterberg, F., Rilfors, L., Wieslander, A., Lindblom, G., & Gruner, S. M. (1995). Lipid extracts from membranes of *Acholeplasma laidlawii* A grown with different fatty acids have a nearly constant spontaneous curvature. *Biochimica et Biophysica Acta*, *1257* (1), 18–24.

Pohorille, A., & Benjamin, I. (1993). Structure and energetics of model amphiphilic molecules at the water liquid vapor interface: A molecular-dynamics study. *Journal of Physical Chemistry*, *97* (11), 2664–2670.

Rosenwald, S., Kafri, R., & Lancet, D. (2002). Test of a statistical model for molecular recognition in biological repertoires. *Journal of Theoretical Biology*, *216* (3), 327–336.

Sargent, D. F., & Schwyzer, R. (1986). Membrane lipid phase as catalyst for peptide-receptor interactions. *Proceedings of the National Academy of Sciences of the United States of America*, *83* (16), 5774–5778.

Schneider, G., & Bohm, H. J. (2002). Virtual screening and fast automated docking methods. *Drug Discovery Today*, *7* (1), 64–70.

Schwartz, A. W. (1995). The RNA world and its origins. *Planetary and Space Science*, *43* (1–2), 161–165.

Segré, D., Ben-Eli, D., Deamer, D. W., & Lancet, D. (2001). The lipid world. *Origins of Life and Evolution of the Biosphere*, *31* (1–2), 119–145.

Segré, D., Ben-Eli, D., & Lancet, D. (2000). Compositional genomes: Prebiotic information transfer in mutually catalytic noncovalent assemblies. *Proceedings of the National Academy of Sciences of the United States of America*, *97* (8), 4112–4117.

Segré, D., & Lancet, D. (1999). A statistical chemistry approach to the origin of life. *Chemtracts—Biochemistry and Molecular Biology*, *12* (6), 382–397.

Segré, D., & Lancet, D. (2000). Composing life. *EMBO Reports*, *1* (3), 217–222.

Segré, D., Lancet, D., Kedem, O., & Pilpel, Y. (1998). Graded autocatalysis replication domain (GARD): Kinetic analysis of self-replication in mutually catalytic sets. *Origins of Life and Evolution of the Biosphere*, *28* (4–6), 501–514.

Segré, D., Pilpel, Y., & Lancet, D. (1998). Mutual catalysis in sets of prebiotic organic molecules: Evolution through computer simulated chemical kinetics. *Physica A*, *249* (1–4), 558–564.

Segré, D., Shenhav, B., Kafri, R., & Lancet, D. (2001). The molecular roots of compositional inheritance. *Journal of Theoretical Biology*, *213* (3), 481–491.

Shapiro, R. (1984). The improbability of prebiotic nucleic acid synthesis. *Origin of Life*, *14* (1–4), 565–570.

Shapiro, R. (2000). A replicator was not involved in the origin of life. *Life (IUBMB)*, *49*, 173–176.

Shenhav, B., Bar-Even, A., Kafri, R., & Lancet, D. (2005). Polymer GARD: Computer simulation of covalent bond formation in reproducing molecular assemblies. *Origins of Life and Evolution of the Biosphere*, *35* (2), 111–133.

Shenhav, B., Kafri, R., & Lancet, D. (2004). Graded artificial chemistry in restricted boundaries. *Proceedings of Artificial Life IX, Boston, MA*. Cambridge, MA: MIT press.

Shenhav, B., & Lancet, D. (2004). Prospects of a computational origin of life endeavor. *Origins of Life and Evolution of the Biosphere*, *34* (1–2), 181–194.

Shenhav, B., Segré, D., & Lancet, D. (2003). Mesobiotic emergence: Molecular and ensemble complexity in early evolution. *Advances in Complex Systems*, *6* (1), 15–35.

Shenhav, B., Solomon, A., Lancet, D., & Kafri, R. (2005). Early systems biology and prebiotic networks. *Transactions on Computational Systems Biology*, *1*, 14–27.

Sipper, M. (2002). *Machine nature: The coming age of bio-inspired computing*. New York: McGraw-Hill.

Sponer, J., Leszczynski, J., & Hobza, P. (1996). Hydrogen bonding and stacking of DNA bases: A review of quantum-chemical *ab initio* studies. *Journal of Biomolecular Structure and Dynamics*, *14* (1), 117–135.

Stark, B. C., Kole, R., Bowman, E. J., & Altman, S. (1978). Ribonuclease P: An enzyme with an essential RNA component. *Proceedings of the National Academy of Sciences of the United States of America*, *75* (8), 3717–3721.

Tanaka, T., Sano, R., Yamashita, Y., & Yamazaki, M. (2004). Shape changes and vesicle fission of giant unilamellar vesicles of liquid-ordered phase membrane induced by lysophosphatidylcholine. *Langmuir*, *20* (22), 9526–9534.

Tieleman, D. P., Marrink, S. J., & Berendsen, H. J. C. (1997). A computer perspective of membranes: Molecular dynamics studies of lipid bilayer systems. *Biochimica et Biophysica Acta (BBA)—Reviews on Biomembranes*, *1331* (3), 235–270.

Tomita, M. (2001). Whole-cell simulation: A grand challenge of the 21st century. *Trends in Biotechnology*, *19* (6), 205–210.

Tuckerman, M. E., & Martyna, G. J. (2000). Understanding modern molecular dynamics: Techniques and applications. *Journal of Physical Chemistry B*, *104* (2), 159–178.

Venter, J. C., Adams, M. D., Myers, E. W., Li, P. W., Mural, R. J., Sutton, G. G., et al. (2001). The sequence of the human genome. *Science*, *291* (5507), 1304–1351.

Wächtershäuser, G. (1997). The origin of life and its methodological challenge. *Journal of Theoretical Biology*, *187* (4), 483–494.

Wilke, C. O., & Adami, C. (2002). The biology of digital organisms. *Trends in Ecology and Evolution*, *17* (11), 528–532.

Zaug, A. J., & Cech, T. R. (1986). The intervening sequence RNA of Tetrahymena is an enzyme. *Science*, *231* (4737), 470–475.

12 Evolutionary Microfluidic Complementation Toward Artificial Cells

John S. McCaskill

12.1 Introduction

12.1.1 Artificial Cells as Microsystems

This chapter addresses the physical, chemical, and evolutionary problems faced by the simplest living chemical systems that aspire to be self-contained, self-reproducing, and self-metabolizing, and proposes that the synthesis of such complex chemical systems can best be achieved in concert with a novel physical machine termed the *Omega Machine*. The Omega Machine, so termed because of its goal-oriented function, *complements* missing cell functionality using an intelligent microscale physicochemical system, which through its monitoring and processing of information (for instance, information about individual emerging cells) effectively complements and regulates them. This one-on-one cell support system has become possible only now, because of advances in microsystem technology. With N. Packard, S. Rasmussen, and M. Bedau, the author has also proposed (in the "Programmable Artificial Cell Evolution (PACE)" proposal, see Acknowledgments) that the support can be withdrawn successively, in an evolutionary technology context, to achieve the free-standing complex functionality of living cells.

This chapter argues that electronic microfluidic systems are indeed the right tools to aid us in the combinatorial synthesis, functionalization, and programming of the first artificial cells, and outlines some current steps in this direction. Biological cells are themselves micro- and nanosystems par excellence, with autonomous self-construction (metabolism), self-assembly (containment), and self-reproduction (heredity). Modern electronic systems are optimized with a universal and convenient serial programming interface. When equipped with parallel micro- or nanoscale actuators and sensors, electronic programmability can become an integral part of a complex chemical environment for artificial cells. From the perspective of computer science, these are embedded systems. They are special in allowing a novel form of evolvable hardware in which genetic information is shared between molecules and electronic memory.

Artificial cell research is an integral part of a complete physicochemical understanding of living systems. In the same way that organic chemists have learned much more about the systematic mechanisms of chemical reactions through the process of synthesis than from the resulting molecules themselves, artificial cells can be regarded as a pinnacle of true synthetic biology (intended here not as the experimental wing of systems biology, but as the process of understanding life by attempting to design artificial living systems). The term *living technology* is preferable, as it emphasizes the unique and distinctive feature of engineering systems with the properties of life.

A first appraisal of a cell from a microsystem perspective shows some obvious features: The cell is spatially localized with strong chemical gradients reinforced by phase separations between amphiphilic molecules (lipids) and water-soluble components. Molecules are subjected to strong electric fields (at least across membranes), and the regulation of molecule transport plays a very significant role in the maintenance of cellular processes. Transport of molecules at the microscale can be much more complex than simple diffusion, and directed and modulated by statistical mechanical nanoscale effects induced by both fields and phases. All of these features are also the subject of study in microfluidic systems, in which researchers have begun to explore the nonequilibrium potential of multiphase systems and inhomogeneous fields. It thus should, and does, appear natural to scientists to begin to employ microsystem technologies in dealing with cells (Wolf et al., 1998). The idea to use combinatorially complex microsystems to aid the design of artificial cell systems—the subject of this chapter—is more recent.

12.1.2 The Integration Challenge of Artificial Cells

Artificial cells, however, still present a major challenge for physical and chemical synthesis. Although most of the core processes central to life—like self-replication of genetic molecules, self-assembly of vesicle containers, and elementary synthetic reactions of metabolism—have been recreated separately in vitro, that is, outside of natural cells, their integration into a single artificial cell is still extremely difficult and has not been achieved. There are problems with achieving high enough catalytic rates, reducing side reactions (reaction specificity), and with buffer compatibility between functional subsystems. A particular example is the need to make the conditions for proliferating surfactant chemistry compatible with those for genetic self-replication. In addition, there are serious issues of both dynamical (e.g., cellular regulation) and evolutionary stability (e.g., genetic error correction and protection against parasites), problems that increase with the diversity of integrated functional components. Some of these issues are discussed elsewhere in this volume (chapter 17). The author was an early proponent of addressing how spatial effects keep evolutionary parasitism at bay (McCaskill, 1992) with theory and experimentation.

Self-replication in the absence of protein enzymes has not yet led to very significant amounts of inherited information (see chapter 13). Indeed, the stable coupling of chemical self-replication to the proliferation of amphiphilic structures is still in its infancy, and despite the fact that it is now twenty years since the discoveries of

- self-replicating molecules by von Kiedrowski (1986), and Rebek (Tjivikua, Ballester, and Rebek, 1990);
- catalytic RNA by Cech (1987), providing a potential way to avoid protein translation machinery; and
- self-reproducing micelles by Luisi (Luisi and Varela, 1990),

a purely chemical solution to replication still seems difficult, whether through combinatorial (e.g., Bartel polymerase ribozyme; see Lawrence and Bartel, 2003) or rational design.

12.1.3 Microfluidic Complementation

Although the achievement of chemical autonomy in cellular functions remains a key goal of protocell research, cellular functions can potentially also be complemented by a microsystem of comparable physical size. Complementation means that certain cellular functions are supported or taken over by the microsystem. One simple example is containment, whereby electric fields generated by electrodes can restrain the dilution of genetic material or nutrients to the environment. Other microsystem containment approaches we have employed use phase boundaries, including those of hydrogels and oils, as well as optical, magnetic, and mechanical restraints. Another example is using microflows to create directed nutrient and waste fluxes, which simplify for the cell the task of accumulating resources and removing wastes. As we shall see later, and as table 12.1 hints at, the actual range of possibilities is much larger, including catalytically active complementation and the use of active separation technologies. The richness of microsystem complementation comes in part from the possibilities of creating multicompartmental cells, with dedicated chemistries in separate locations, and directed transport mediating the exchange of material between the compartments. We are especially interested in microflow systems (Squires and Quake, 2005; Stone, Stroock, and Ajdari, 2004) rather than open-surface microstructures, despite exciting recent developments (Bishop, Gray, Fialkowski, and Grzybowski, 2006), because of their ability to support long-term homeostasis in chemical systems. Complementation brings the additional advantage of providing a means of programming artificial cells: The choice of complementation can direct the spatial structure, the resource structure, the metabolism, and the replication cycle in specific directions.

Microfluidic systems provide a complete repertoire of solutions for physical complementation (see table 12.1). They are on the right scale for complementing artificial

Table 12.1
Collection of physical complementation strategies for protocells

Supply of buffer, nutrients and/or energy
Removal of waste
Local concentration cycling of molecules
External provision of (some) catalysts
External provision of informational molecules
Surface immobilization
Regulated temperature and thermocycling
Physical containment and isolation
Selective separation of molecules (artificial membranes)
Phase control, e.g., hydrophobic patterning
Support system for structured multistep chemistry
Support system for parallel screening: channel networks
Material transport between cells and regulated cell division

Note: Though the list is not intended to be exhaustive, it serves to illustrate the broad range of options open to an integrated complementation system.

cells on a one-on-one individual basis; they have electronic interfaces to digital computers; they allow a high density, quasi-planar optical interface; they are evolvable via reconfiguration; and they have natural extension to nanofluidics.

12.1.4 Evolutionary Complementation

The complementation of protocells can be regarded as analogous to the complementation by "music minus one" electronic media that provides a complete orchestral accompaniment without the soloist's part. A programmable complementation system can incrementally reduce the level of support, increasing the degree of autonomous orchestration of the chemical system. In principle, a smooth succession of environmental challenges can be provided to ease the combinatorial search for viable artificial cells into manageable steps, starting from a fully supported system in which all aspects of the protocell are under microfluidic control. Of course, the difficulty of the individual steps depends not only on the innate evolutionary potential of the chemical system at each stage, but also on the aptness of engineered changes to its composition. In addition, and perhaps less obviously, a combinatorial succession or array of environments can be used to enhance the effective evolvability of the chemical system, especially at early stages.

This chapter first reviews the developments in evolution research using microsystems that set the stage for the current initiative toward artificial cell evolution. This work highlights the importance of spatial effects such as isolation by distance for molecular evolution, and demonstrates that chemical systems that evolve are, indeed, available. Based on this motivation, section 12.3 focuses on the issue of artificial compartmentation in microsystems, exploring geometric, hydrogel, and (briefly, for comparison's sake) also electric field–induced compartmentation systems, which all complement the self-assembly of lipid membranes. The electric field–induced compartmentation deserves a more extended treatment, which is provided in section 12.5 after the major developments in reconfigurable and programmable microsystems are portrayed, allowing us to envisage single-cell scale feedback control for whole populations. Section 12.6 extends this work from containment to a discussion of the complementation of the three major subsystems of the cell, as well as their integration. Finally, section 12.7 provides more details of the Omega Machine and how it can be programmed, before drawing conclusions in the final section.

12.2 Replication and Selection in Microsystems

12.2.1 Early Spatially Resolved Evolution Experiments

The correlation among genetic traits of organisms with spatial proximity is one of the hallmarks of the evolutionary process, and in fact one that prompted Charles Darwin's discoveries (Darwin, 1859). When Eigen proposed his physicochemical analysis of what is required for molecular systems to evolve genetic information (Eigen, McCaskill, and Schuster, 1989), space was implicated as a necessary cofactor in hypercyclic organization, but limited theoretical work was available on the influence of spatial effects on evolutionary self-organization. The experimental program on cell-free (in vitro) molecular evolution in Eigen's lab (Biebricher, Eigen, and Gardiner, 1983, 1984, 1985) and elsewhere (e.g., Johns and Joyce, 2005; Wright and Joyce, 1997) concentrated on homogeneously mixed populations in serial transfer experiments (from test tube to test tube). Chemical spatial self-organization, as investigated with reaction-diffusion systems in particular since Turing's seminal paper (1952), had never been combined experimentally with evolutionary self-organization. This synthesis of spatial and evolutionary self-organization, also necessary for the evolution of artificial cells, was a key motivator for the development of experimental techniques for spatially resolved laboratory evolution, starting with the capillary traveling wave (Bauer, McCaskill, and Otten, 1989). Simulations of general models of evolving automata (McCaskill, 1988) were already revealing the essential role of spatial structure in facilitating complex evolution of functionally differentiated multimolecular reaction systems.

Another factor prompting examination of spatial effects in molecular evolution was the remarkable phenomenon of "de novo" synthesis of RNA templates in the original Qß system for in vitro evolution (Biebricher, Eigen, and McCaskill, 1993), which seemed to defy Pasteur's famous experiment, and was later established to rely on rapid evolution of at least the vast majority of the genetic information in the templates (Biebricher and Luce, 1993). If major evolutionary events occur on laboratory timescales, then the spatial correlation of emergent molecular diversity becomes an exciting object of inquiry. The significance of finding major evolutionary change on laboratory timescales for the current projects in artificial cells is obvious. The first experiments designed to see molecular proliferation as a spatial phenomena, inoculating a thin-film replication mix with RNA template molecules at defined locations, proved inconclusive because of containment difficulties in an open film and the time delays before detection thresholds could be reached. The breakthrough came in 1989 (Bauer et al., 1989), with the use of long submillimeter diameter polyethylene tubes, which provided both a secure seal and a one-dimensional medium connected by diffusive molecular transport. Nonlinear traveling waves of template concentration were found, analogous to the Fisher waves (Fisher, 1937), but involving multiple molecular intermediates in the amplification process, which propagated with precise velocities (constant better than 1:1,000) measurable by fluorescence of RNA intercalating dyes such as ethidium bromide and thiazole orange.

The traveling waves could be shown to initiate from single introduced RNA template molecules, or, in the case of the delayed de novo process mentioned earlier, in the absence of added template. This led to a diversity of replication rates corresponding to the macroscopic diversity in migration behavior observed in gel analysis on macroscopic samples of de novo synthesis by Biebricher and coworkers (1993). Other experiments (McCaskill and Bauer, 1993) with scarce supply of one of the four ribose nucleotides (U), confirmed evolution to be common inside the traveling wavefronts, with sharp changes in front velocities as the observable phenotype,

$$v = 2\sqrt{\kappa D}, \tag{12.1}$$

where κ is the overall exponential phase amplification rate and D is a time-weighted diffusion coefficient for the various template complex intermediates. Figure 12.1 (color plate 6) summarizes these experiments. Typical front velocities are in the range of 0.5 μm/s. The main significance for later work on artificial cells was the realization that spatial phenomena could be harnessed to both simplify and enable the analysis of evolving systems, and that traveling waves provide idealized local populations of evolving molecules with a readily measurable phenotype. This principle was extended in later work with John Yin (Yin and McCaskill, 1992) to the analysis of bacterial virus evolution in traveling waves. Additional experiments that we performed in Eigen's lab, to modulate the front velocities in capillaries using electric fields, showed that traveling waves could be accelerated or stopped. These were early forerunners of the programmable electronic environments currently being developed for artificial cells.

12.2.2 Spatially Resolved Flow Reactors: The Line Reactor

The major limitation of the capillary traveling wave reactor was the inability to replenish chemical resources and remove waste after a traveling wave passes. Replen-

(a)

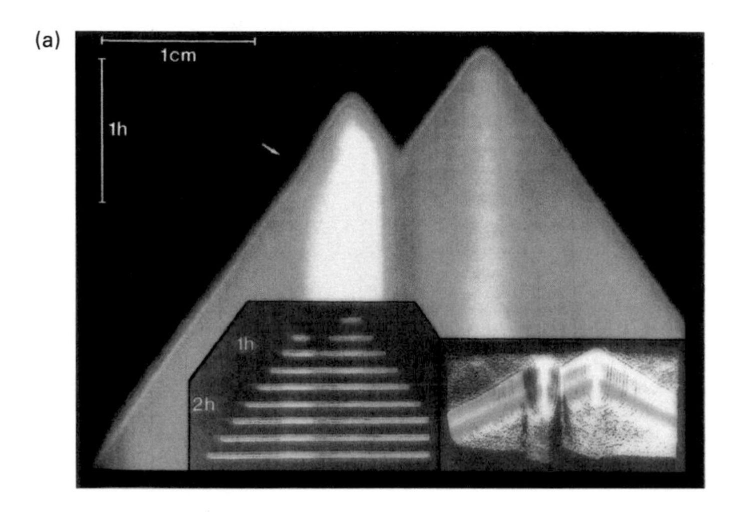

(b)

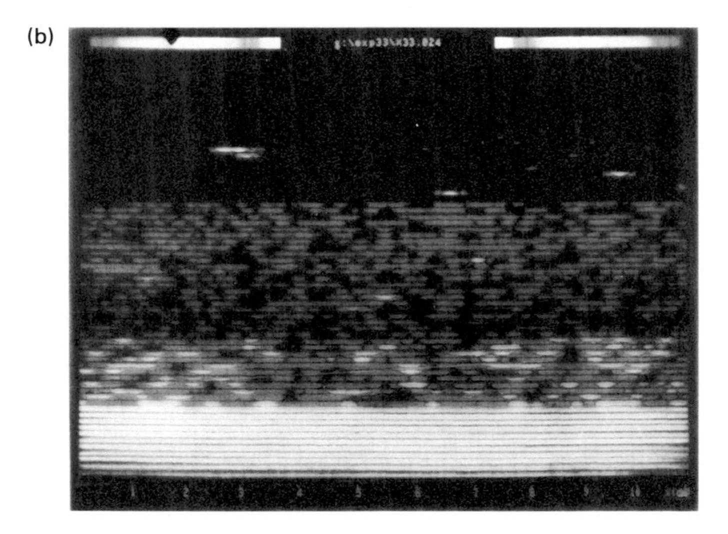

Figure 12.1 (color plate 6)
Traveling wave experiments with Qß enzyme and minivariant RNA. (a) Fluorescence image of capillary wound back and forth into parallel 12-cm segments on a frame (up to 72 horizontal segments in total) showing RNA concentrations resulting from single templates, with increasing dilution of RNA introduced in the upper segments. (b) Space-time fluorescence image of single capillary segment showing two constant velocity wavefront pairs (with 3D data inset) and a single evolutionary event causing a change in front velocity at the arrow, manifested by a change in slope of the growth triangle on the space-time image. Image reproduced with permission from Bauer and McCaskill, 1993.

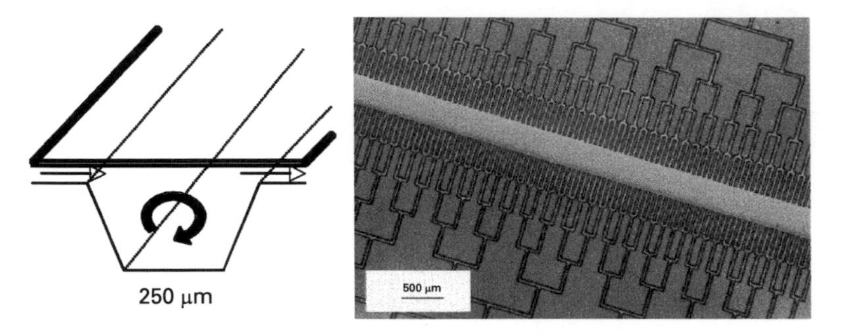

Figure 12.2
Line flow reactor. The reactor is built from a microstructured silicon wafer anodically bonded between two Pyrex glass wafers. The laminar flow at low Reynolds numbers means that the cross flow causes a local dilution flux of the reactor without inducing mixing along the reactor line, so that a 1D reaction-diffusion system with continuous replenishment of resources can be achieved (McCaskill, 1997). The thick circular arrow is intended to portray relatively rapid diffusive mixing, and not circular flow, which is absent in this low-Reynolds-number regime. Bifurcation cascades are employed to ensure uniformity of flow in multiple channels from a single source channel.

ishing resources by passive diffusion from a reservoir, as in gel slice reactors, developed during the 1990s for other reaction diffusion systems (Castets, Dulos, Boissonade, and De Kepper, 1990), proved not to be robust toward the contamination by single molecules that characterizes template evolving systems such as Qß. In fact, we had already extended the traveling wave experiments to other enzymatic proliferation schemes such as 3SR and SDA during the early 1990s (Dapprich, McCaskill, Voelker, and Krause, 1994), so we knew this was not a problem just with the Qß system.

A solution was sought from the then emerging field of microfluidics technology, in which lithography could be used together with anodic bonding to create closed micrometer-scale channel networks of predefined geometry, sandwiched between anodically bonded silicon and Pyrex glass wafers. One such reactor, the line reactor (McCaskill, 1997), is shown in figure 12.2; it provided a chemically open, contamination-free, but diffusion-limited one-dimensional media, in which amplification and evolution processes could, in principle, be sustained indefinitely by a cross flow of nutrient-rich solution that also removed waste products. In such a system, self-organizing pattern formation, more complex than traveling waves, is theoretically possible depending on the nonlinear reaction kinetic mechanisms. In particular, the author realized that self-replicating spot patterns (Pearson, 1993), in which localized concentration spots grew and divided under diffusive resource limitation, could potentially be designed into evolving systems in such a reactor, providing a membrane-free analogy to a protocell, if the appropriate nonlinear kinetic mechanism could be found (see below). The phenomenon is not peculiar to the line reactor, but the latter provides a well-defined experimental prototype for natural extended media such as porous channels or films. In fact, open two-dimensional reactors with cross flow were also designed in my lab in the 1990s (Schmidt et al., 1998), but these proved difficult to control with constant flows.

12.2.3 Replication, Cooperation, and Inhibition

Not only are nonlinear reaction kinetics decisive in allowing chemical reaction systems to create spatial structures by reaction and diffusion, they are also implicated in the origin of life and the transition to cellular structures. To see the significance of this for protocells, we must review some of the basic physicochemical theory of evolving systems. The information threshold for the origin of life is caused by limited copying fidelity in the absence of a complex set of enzyme catalysts (Eigen et al., 1989). Limited copying fidelity restricts the complexity of catalysts that can be passed on from one generation to the next, and hence also the set of molecules responsible for high copying fidelity. Either all catalysts are encoded on a single genetic molecule, so that the error threshold limits the total information available to the system, or the genetic information must be split up on different molecules and maintained dynamically as a population or aggregate. Eigen and Schuster's hypercycle theory addressed the dynamical stability of maintaining a set of separate information carriers, each responsible for only a part of the information necessary for an early precellular or cellular organism (Eigen and Shuster, 1979).

A related but more basic problem limits the self-organization of hereditary physicochemical systems, even in the case when all genetic information is located on a single molecule, regardless of whether the information encodes to enhance containment, metabolism, or templated self-replication. Consider the simplest scheme, $X \rightarrow X + R$, where a genetic molecule X influences other molecules in the system, causing them to form molecules with the property R. This could be a catalytic function associated with self-replication, an activated molecule associated with metabolism, or a molecule with a surfactant property associated with containment. The general feature of molecules with the property R is that they are not self-replicating: The information required to create them resides in X, not R. The positive impact on the self-replication of X may be written $X + R \rightarrow 2X + R$. In cases where R is consumed in the self-replication process of X, as with metabolism when R may stand for, for example, an activated monomer, we have $X + R \rightarrow 2X$. The difficulty for an emerging system stabilizing the presence of both R and X, is that another species, Y, that does not contain the genetic information necessary to create R may also benefit from R, $Y + R \rightarrow 2Y + R$. This evolutionary instability is the same as that which affects the cooperative replication scheme $2X \rightarrow 3X$, obtained in the special case when R is replaced by X. Table 12.2 summarizes the kinetic impact of template-induced properties of protocells. The evolutionary stability of spatially resolved

Table 12.2
Common kinetic impact of template-induced properties of protocells

	Self-Replication	Mutation to Y	Exploitation by Y	
	$X - R \rightarrow 2X + R$	$X + R \rightarrow Y + X + R$		Self-replication
	$X \rightarrow 2X$	$X \rightarrow Y + X$		
	$X \rightarrow R_X$			Self-replication
	$X + R_X \rightarrow 2X + R_X$	$X + R_X \rightarrow Y + X + R_X$	$Y + R_X \rightarrow 2X + R_X$	catalysis
	$2X \rightarrow 3X$	$2X \rightarrow Y + 2X$	$Y + X \rightarrow 2Y + X$	
Resource	$X + A \rightarrow Y + B$			Resource
catalysis	$X + B \rightarrow 2X$	$X + B \rightarrow Y + X$	$Y + B \rightarrow 2Y$	catalysis
	$2X \rightarrow 3X$	$2X \rightarrow Y + 2X$	$Y + X \rightarrow 2Y + X$	
Container	$X + pL \rightarrow X + L$			Container
catalysis	$L + L + \cdots + L \rightarrow L_n$			catalysis
	$X + L_n \rightarrow 2X$	$X + L_n \rightarrow X + Y$	$Y + L_n \rightarrow 2Y$	
	$2X \rightarrow 3X$	$2X \rightarrow Y + 2X$	$Y + X \rightarrow 2Y + X$	

Note: The general problem of cooperation, posed by the integration of protocells, is common to the three central cell functionalities: self-replication, resource, and container catalysis. All three of these functions are open to the same kind of second-order exploitation by variants Y, which do not contribute to the functionality.

systems of this type depends on stochastic effects, and has been investigated in the PRESS framework (McCaskill, Fuchslin, and Altmeyer, 2001).

An experimental system, CATCH (Ehricht, Ellinger, and McCaskill, 1997) was designed and constructed to provide a test object for this class of generic molecular cooperation. In partly theoretical and partly experimental papers, it was established that CATCH should show the self-replicating spot spatial patterns (Kirner, Ackermann, Ehricht, and McCaskill, 1999) and was subject to parasitic exploitation, but resistance to parasites could evolve in a temporally size-varying system or spatially resolved system (Ellinger, Ehricht, and McCaskill, 1998).

12.2.4 The Standing Wave Fan Reactor

As a next step toward artificial cells, we investigated whether the nice features of traveling concentration wavefronts could be transferred to a stable stationary replication wave. This could form the basis of a self-maintaining nonequilibrium replication process. The more obvious way to do this, by balancing the traveling wave velocity along a channel with an opposing fluid flow, had already been examined. Because of the parabolic profile of velocities across a capillary, the concentration wavefront no longer remains perpendicular to the channel walls, and lateral diffusion plays an important role in determining the front velocity under flow. More seriously, an exact balance of the modified wave velocity with flow rate is required, and proves experimentally unstable. A solution is to use a fan-shaped reactor, as shown in figure 12.3 (Ehses et al., 2005). The radial increase in cross section in the fan reactor causes an inversely proportional decrease in local flow velocity (actually of the form $a/(a + r)$,

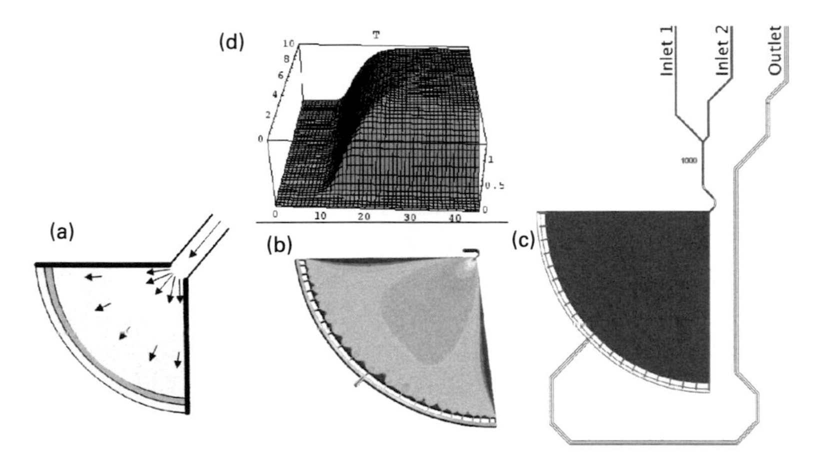

Figure 12.3
Fan reactor achieving stationary replication wave: (a) conceptual design showing radial flow at decreasing flow rates after entering the quadrant, (b) simulated velocity profile (by Thomas Palutke), and (c) mask design of reactor (by Patrick Wagler and Thomas Palutke). Results with this reactor have verified the existence of a stable stationary wavefront arc for both enzyme-dependent (SDA) and enzyme-free replication systems (Ehses et al., 2005). (d) Shows the radial versus time template concentration for the SDA reaction from kinetic simulation.

where a is the entry channel diameter). The upshot is that a traveling concentration wave will propagate toward the center of the fan until its velocity is matched by the increasing local flow. This provides a self-regulating mechanism in which replication is confined to a specific location in the microreactor. There is still a vertical parabolic profile, but the shallow depth of the fan allows relatively rapid vertical mixing, and so the perturbation on ideal (plug flow) behavior is small. Of course, a better plug flow could be achieved with an electroosmotic drive (Clifton, 1996), but this would have to be applied uniformly throughout the fan.

12.3 Artificial Compartmentation in Microsystems

Compartmentation supports evolutionary persistence of artificial cells in multiple ways. One can distinguish two key aspects: (a) containment of beneficial local impact of genetic material, and (b) protection against exploitation by foreign genetic material. The former is more critical to cellular functionality, allowing critical concentrations of nutrients and building blocks to be accrued, and appropriate chemical conditions (e.g., pH, buffer, and ion concentrations) to be maintained independently of those in the environment. There are two immediate negative consequences of containment: The free uptake of resources and the removal of waste products are likely to be inhibited. Either active transport mechanisms across the container for selective substances must be achieved, or the disadvantages of slowed accrual and increased levels of waste must be offset by the previously listed benefits of containment. In artificial environments, it is possible to supply high concentrations of nutrients and ideal buffer conditions, so that containment for its beneficial local impact may be unnecessary.

Protection against exploitation by foreign genetic material takes two forms: In the first case, the foreign material arises by mutation from the organism itself; in the second, it comes from other organisms. In the former case, the threat is from within, and the ongoing process of cell division, creating new boundaries each generation, can isolate the foreign material stochastically, as analyzed in many population genetic models (Kimura, 1983), including the stochastic corrector model (Szathmáry and Demeter, 1987). It is not clear that boundaries improve the situation in this case, compared with a reaction-diffusion mechanism such as replicating spots (as above). If the exploitation comes from foreign material, strong containment would appear to be inevitably advantageous, allowing limited passage of foreign genes. However, then the negative impacts of strong containment on the cell's resource accumulation, waste removal, and the ease of self-reproduction need to be considered. Finally, a limited amount of foreign genetic material may be useful in providing stronger innovative potential in times of changing environments.

On the other hand, it is clear from the analysis in section 12.2, that some defined degree of spatial genetic isolation is essential, and for many environments resource isolation is necessary for achieving an effective metabolism. So if containment can be modulated and tuned by multiphase systems such as membranes, then an optimal situation for the cell can be achieved. This section investigates the complementation of compartmentation using microsystems.

12.3.1 Isolation by Channel Distance

As with the original capillary reactor described previously, replication systems can be compartmentalized in a single-phase fluid simply by means of the scaling law of diffusion with distance: Templates diffuse across the gap between two replication domains in mean times proportional to the square of the separation distance. With typical diffusion coefficients on the order of less than 10^{-10} m^2s^{-1}, traversing distances of 100 μm requires a minute, 1 mm requires an hour, and 1 cm requires several days. As a result, in the absence of replication, the mixing and exploitation of information along a channel is slowed to these timescales. In contrast, nonlinear replication waves, supported by template proliferation, propagate at constant velocity given by equation (12.1) and hence much faster over longer distances. Typical replication times are on the order of a minute.

In open reactors such as the line reactor, both stable and self-replication spot patterns can occur for appropriate proliferation mechanisms. Since many of the artificial cell hereditary mechanisms are based on the Kiedrowski nonenzymatic template scheme (von Kiedrowski, 1993), which shows a square root dependence on concentration and subexponential parabolic growth law at accessible concentrations, the superlinear increase in replication rate concentration, which is necessary for stable pattern formation in replication systems, appears to be absent. However, saturable inhibition has been proposed (Böddeker, 1995) and shown to be an alternate mechanism for attaining such effectively higher-order kinetics. Actually, template growth at low concentrations, according to von Kiedrowski's mechanism (von Kiedrowski, 1993), is in fact exponential, but this cannot normally be observed experimentally because it is masked by the background reaction of nontemplated ligation. Saturable inhibition can be attained for these replicator systems by introducing a small quantity of competing binders such as PNA. In principle, flow rates in an open line reactor can be adjusted so that the early templates formed by the background process are captured by complementary inhibitors, and only when the inhibitor concentration is exhausted (at higher template concentrations) can a net growth of templates occur, which is then predominantly by self-replication. This has, however, not yet been experimentally verified, but could lead to self-replicating spot patterns for nonenzymatic replicators.

Another mechanism for increased isolation of template replication centers in microfluidic systems is shown in figure 12.4 (color plate 7). The figure presents a variant of the line reactor in which slow exchange domains alternate with higher-flow domains. Because of the increased template dilution in the high-flow domains, the natural separation by diffusion is enhanced, and this enhancement persists in the presence of normal template proliferation.

12.3.2 Separation in Two-Phase Systems

Hydrocarbon-water systems with or without added surfactant exhibit a remarkable range of phases, which are extensively exploited by biological cells. The complex phase diagram maps concentrations and physical conditions (such as temperature) to particular phases. The oil-water-surfactant equilibrium phase diagram has been the object of intense study in recent years and is quite well understood (Elliott, Szleifer, and Schick, 2006). Microfluidic systems are capable of generating phase changes (both equilibrium and nonequilibrium) induced by changes in geometry or constitution in channel networks. A classic example is the creation of alternating oil and water droplets at a channel junction of the pure fluids: Pressure builds up in the entry channel not delivering the current droplet, creating the force necessary to make additional interfacial surface between the droplets (Thorsen et al., 2001). Fully enclosed droplets can be created at specially designed nozzles (Cristini and Tan, 2004; Tan, Cristini, and Lee, 2006), and more complex droplet-droplet fusion protocols can be achieved in appropriately designed chambers. In addition to droplet creation,

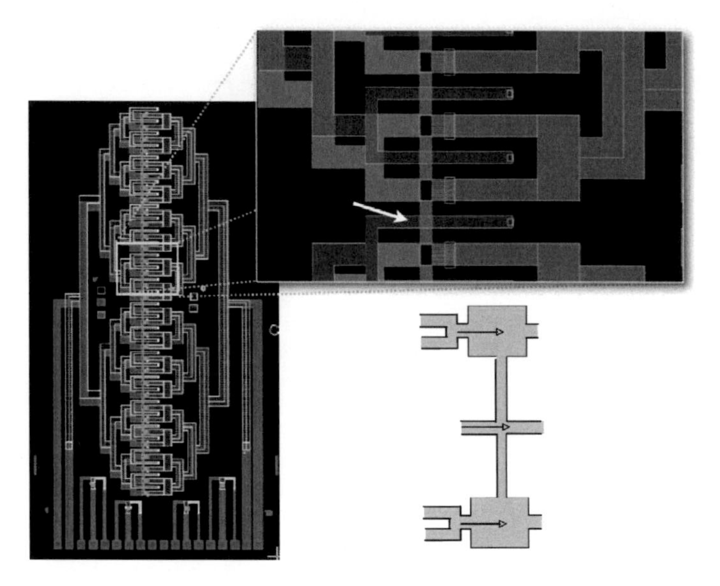

Figure 12.4 (color plate 7)
Design of an extended open line reactor. The blue and green bifurcation trees on the right-hand side (on top of each other) deliver two different solutions from the lower and upper chip layers. Products are collected in the bifurcation cascade on the left. The lower right diagram shows a simplified schematic blowup of the section of the horizontal line reactor between two inflows, emphasizing the separation between top and bottom chambers. The separating flow in the center creates a tunable diffusive connection between the two outer reactors: If the flow here is fast, no interchange occurs on the line. The line reactor is shown in more detail vertically in the mask design on the left, and in the blowup (upper right); an arrow indicates the separation control channel. This design allows mixing of two fluids immediately before entry to the line reactor. Multilayer design is implemented by T. Palutke and P. Wagler, BioMIP.

lamellar structures with alternating water and oil phases perpendicular to the direction of flow can be generated at appropriately designed junctions (Kenis, Ismagilov, and Whitesides, 1999), in the same way as alternating fluids of different composition have been created for kinetic studies (Knight et al., 1998). Such striated two-phase flow could serve to enhance the separation by distance described in the previous section.

It is clear that aqueous droplets in oil, whether or not the water is also in contact with channel walls, can provide an artificial container, and indeed such artificial containers have already been exploited for sample separation in microfluidic systems. Oil or lipid droplets in water provide an alternative basis for constructing "hydrophobic" artificial cells in an aqueous medium. Nutrients could be delivered to such droplets by droplet-droplet fusion, or in the form of (inverse) micelles. Droplet division (fission) can be induced by a similar mechanism to the spontaneous generation of droplets described in the previous paragraph and, for example, in Stone and coworkers (2004), acting to split double-length single-phase sections of fluid along a channel, or by the related phenomena of extrusion-induced fission (Hanczyc and

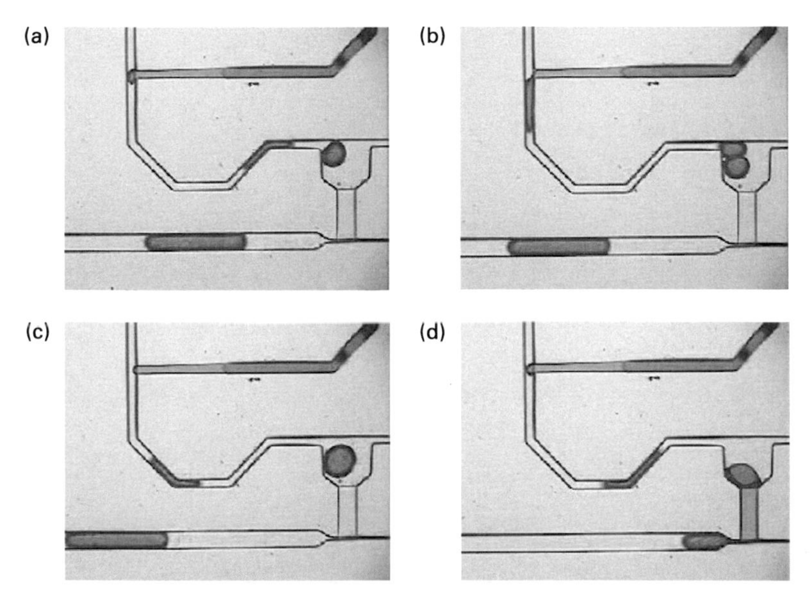

Figure 12.5
Droplet manipulation in microreactors. A sequence of four successive images (a through d), showing droplet formation (at T-junction at upper left) and fusion (chamber at lower right). The red dye–labeled aqueous droplets spontaneously separate from the colorless paraffin separation fluid. Actually, in this reactor, three droplets must fuse before they depart from the fusion chamber. Pictures courtesy of Patrick Wagler and Steffen Chemnitz, BioMIP.

Szostak, 2004). Reactors are currently being designed in our group to create autonomous cycles of droplet container growth and division exploiting these effects. A microreactor manipulating droplets is shown in figure 12.5.

In principle, lamellae created with surfactant molecules can be thinned to the point where they form membranes. Static black lipid membranes have been created across holes in microstructures (Bloom, Peterman, and Ziebarth, 2005) but are relatively unstable toward pressure fluctuations and rupture. Self-replenishing membranes under flow in appropriately designed chambers with pressure balancing could provide more stable membranes to serve as molecularly modulated interfaces. Membranes have also been created, with the use of specially designed nozzles, allowing the spontaneous creation of lipid vesicles under flow (Lee and Tan, 2004). Clearly, this research is opening up a large field of microfluidic engineering for cellular surrogates.

12.3.3 Hydrogel Barriers Created by Laminar Flow

Gels play an important role in biotechnological separations; in particular, the differential mobility of biopolymers driven by electric fields through gels has been in the development of genomics. Though gels are already well established in capillary electrophoresis applications, in which the entire capillary or microstructure can be filled

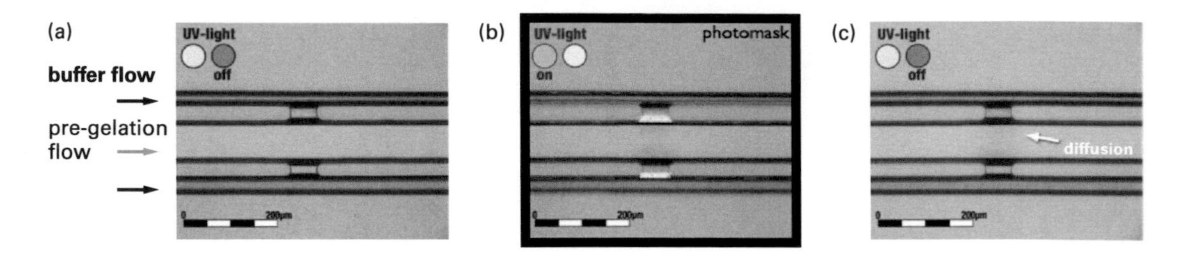

Figure 12.6 (color plate 8)
Dynamical generation of diffusion barriers between microfluidic channels/chambers based on photopolymerization of hydrogels. (a) The central horizontal channel is initially filled with a flowing photopolymerizable pregelation mix (PEG-DA (poly-(ethyleneglycol)-diacrylate)), a cross-linking agent (methylenebisacrylamide) and the photoinitiator, HMPP (2-hydroxy-2-methyl-propiophenone), in TAE buffer solution, whereas the upper and lower channels contain sample solution (labeled buffer flow) dyed red in this experiment. UV illumination through a rectangular mask in (b) actual image during illumination, results in polymerization of the central channel, which is in contact with the buffer flows via shallow ledges. The resulting in-flow gelled barriers are shown in (c), where slow diffusion of dye through the gel from the buffer channels can be observed. They may be used for selectively reduced molecular transport of material to a microcell, functioning like a membrane as semipermeable molecular filters. Experiments by Patrick Wagler, Steffen Chemnitz, and Farsaneh Sadeghar, BioMIP.

with gel material, producing localized gel plugs at specific locations in a microfluidic system is more involved. One configuration that is especially attractive is creating a gel plug between two channels, which could then allow differential mobility (possibly driven by electric fields; see section 12.5). Although photosensitive gel materials are available, the creation of sharp gel boundaries in closed channel networks is problematic. One technique is to use the low Reynolds–number, highly laminar flow to sandwich a flow of gel material between two flows of aqueous buffer. Such an arrangement can be supported by differential depth of channels as shown in figure 12.6 (color plate 8). Photopolymerization of specific plugs, using an appropriate mask, can be achieved at geometry-supported locations without the gel material tearing on gelation.

Apart from differential mobility, gels also impart an ability to create selective molecular filters. Complementary structured guest molecules can be immobilized in the gel material, and a form of affinity chromatography over short distances can be used to induce differential transfer of specific molecules across the gel plugs. Coupled with pulsed electric fields as described later, a high level of control of differential transport to and from "artificial cells" separated from supply channels by gel plugs is conceivable.

12.3.4 Compartmentation by Electric Fields

Electric fields can be used to support compartmentation together with other cell functionalities; this subject is important enough to warrant a separate section (see section 12.5). Before discussing this further, however, we wish to expand on the general relationship between degree of reconfiguration, programmability, and evolvability.

12.4 Reconfigurable and Programmable Microsystem Technology

12.4.1 Reconfiguration Enables Evolvability of Structures

Evolution is understood to depend on the selection of progeny redesigned as variants of their parents. In the biological context, this happens in populations with natural selection depending typically on differential survival and proliferation rates (in contests with other individuals or in interactions with the environment). In artificial contexts, population sizes may be held constant at values as low as 1, and evolutionary processes can be as simple as making a series of variations based on the best design sampled so far. Although organisms are not reconfigured but reproduced, the fundamental process of evolution can occur with a single piece of reconfigurable hardware.

Reconfigurable hardware (Sanchez and Tomassini, 1991) involves essentially hardware whose function depends on the combined state of a set of bi-stable (or more generally multistable) components. A key example is an electronic circuit whose logic gates are determined by one or more bi-stable flip-flops, which are themselves (non-reconfigurable) gates. Changing the state of a flip-flop, for instance, may change an AND gate into an OR gate, which changes the electronic processing of the circuit. Another example is a mechanical robot whose joints can be configured bi-stably as either stiff or flexible. Reconfigurable hardware allows different individual designs to be tested without having to rebuild devices, and hence allows real-world (as opposed to simulated) evolution without the need to redo all the difficult processes of construction for each individual design test.

The conceptual relationship of reconfigurable to living systems pursued here views the organism as based on pieces of reconfigurable machinery (i.e., raw materials, laws of physics, and the environment) that can be reconfigured by genetic information. Minimally, evolution with reconfigurable systems involves one piece of reconfigurable hardware and a current string of genetic information (encoding the configuration) together with a (randomly) generated variation of this configuration. If the behavioral/functional result of operating the hardware with the original configuration is inferior to that of the modified configuration, then the new genome replaces the old one as the basis of variation. With two pieces of reconfigurable hardware, evolution can proceed through contests between the two individuals (placed in a common environment), and so on. Multiple reconfigurable units, if available, can be used to speed evolution (McCaskill, Tangen, and Ackermann, 1997), and in principle thousands of reconfigurable cell microenvironments can be investigated in parallel in an integrated microsystem.

12.4.2 Reconfigurable Microfluidics

The concept of reconfigurability has been extended to microfluidic systems (Ikuta, Hirowatari, and Ogata, 1994; McCaskill and Wagler, 2000), where in the simplest case, in analogy to electronics (channels replacing wires and valves replacing gates), sets of bi-stable valves can be configured to produce any of a large number of microchannel networks. In addition, reconfigurable separation devices and reaction control devices are required for a general reconfigurable microreaction technology, but much can already be done with reconfigurable mixing devices. Unfortunately, the construction and control of the large number of valves that would be necessary to support reconfigurability has proved prohibitive, despite attempts to induce switching only with the combined effect of X- and Y-signals acting on whole rows or columns (e.g., through pressure; see Thorsen, Maerkl, and Quake, 2002), which reduces the number of control elements for an n by n array from n^2 to $2n$. In fact, three distinct degrees of reconfigurability can be distinguished: irreversible (customizable by user only once), repeated global reconfiguration by means of an external global reset, and individually resettable components. An example of the second category is the photopolymerization of gel plugs, by which the entire reactor can be returned to a fluid state by global heating or washing with appropriate solvents. A design for building up general multiarm fluid junctions with a minimal set of simple plug elements has been proposed (McCaskill, Wagler, and Maeke, 2003).

Reconfigurable separation devices can be achieved with magnetic beads, as we have shown in the context of integrated DNA computing (McCaskill, 2001). An example of such a microreactor is shown in figure 12.7. The principle is to immobilize chosen structurally complementary probe molecules on the surface of magnetic beads at particular locations in the microreaction network. This can be achieved under programmable optical control with devices such as digital mirror devices. Chosen beads can be moved between adjacent laminar flows to transfer complementary binding molecules from one fluid to another, thereby achieving separation. A methodology for repeated application of such separations for DNA has been described (McCaskill, 2001) and experimentally tested (van Noort, Wagler, and McCaskill, 2002).

12.4.3 Electrically Reconfigurable Microfluidics

Digital electronic control would provide a direct interface to the integrated electronics of current computers, and allow a significant increase in the level of reconfigurability and parallel control. Reconfigurable electronic chips (field-programmable gate arrays or FPGAs) have captured a significant fraction of the total digital electronics market, and their customizability provides advantages despite the up to tenfold overhead in gate resources compared with dedicated silicon. The following section describes the reaction control mechanisms we are exploiting, which make

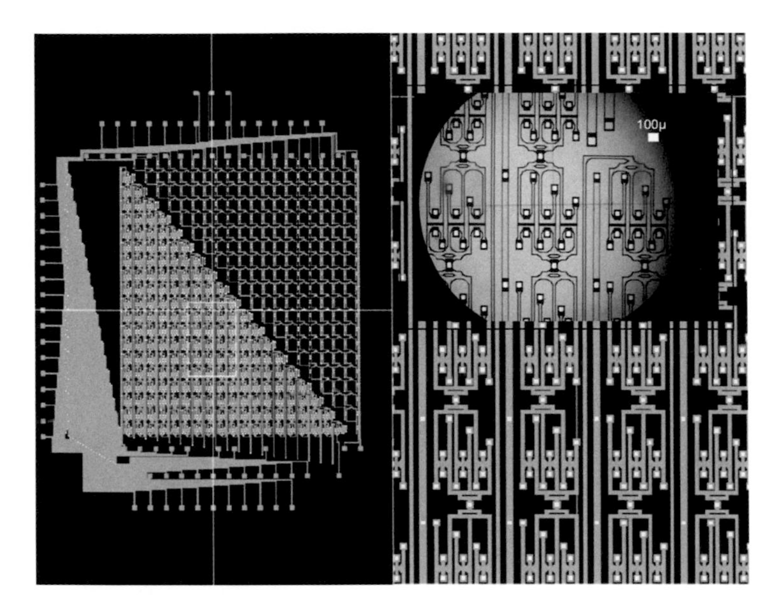

Figure 12.7
DNA computing in microfluidic reactors; optically programmable DNA computer that is designed to solve any maximal clique problem with up to a million variants ($n = 20$). In contrast with other methods, a single set of DNA is required and no problem-specific pipetting. The background shows the overall design (left) and a close up (right) of the regular array of ternary selection modules. The inset shows a microscope view of a portion of the multilayer array, with the squares showing interlayer connections that allow fluidic crossovers. Such designs, involving up to a thousand reaction chambers (van Noort, Wagler, and McCaskill, 2002) in continuous flow, laid the groundwork for a microfluidic approach to artificial cells.

use of electrodes operating at conventional fixed digital voltages (e.g., 3.3 V) in direct contact with fluids in microsystems. If such electric fields can be used to strongly influence the local flow, separation, or reaction properties in a microsystem, then these elements can be combined in a flow system to create a reconfigurable microfluidic device. The high frequencies available with digital electronics can be used together with the natural (capacitive) response times of electrolyte systems to allow control multiplexing, reducing the number of electronic signals needed to control a given number of microfluidic sites (Tangen, Maeke, and McCaskill, 2002; Tangen et al., 2003).

In other projects, noncontact dielectric control, through kilohertz to megahertz AC fields at higher voltages, is being developed (Fiedler et al., 1998). For open microfluidic systems, electrowetting can also be employed to manipulate fluid droplets individually on surfaces (Cho et al., 2002; Gascoyne et al., 2004). This digital droplet manipulation technology contrasts with the continuous flow closed microsystems approach described in this chapter, but it could be employed in the artificial cell context together with a purely digital droplet manipulation approach. Digital manipulation of oil-water droplets has been described recently in a programmable context by Urbanski and coworkers (2006), using pneumatically activated elastomeric valves. Here *digital* refers to the fluid handling in constant volumes, rather than to digital electronic signals, and there are corresponding limitations on ultimate integration.

Programmability of experimentation has received additional attention since the pioneering work of King and coworkers established that a robot scientist could be competitive in genomics research experimentation (2004).

12.5 Electronic Compartmentation and Control in Microsystems

12.5.1 The Use of Electric Fields in Microsystems

This section quickly introduces only those parts of this rich and diverse field of research that are immediately relevant. Further information can be found in general reviews (e.g., Squires and Quake, 2005). Many further electric field effects can be exploited (e.g., the dielectrophoretic effects and electrowetting; see previous section). Static electric fields can support separation, concentration, and compartmentation in microsystems through two main effects: electrophoresis and electroosmosis. *Electrophoresis* can be used to restrain and concentrate charged molecules against a diffusion or flow field, and *electroosmosis* can induce a bulk flow for all molecules in appropriate ionic solutions by accelerating ion surpluses near the walls of microchannels. Although electrophoretic effects depend on both the electrical properties (such as charge z, or more generally also the polarizability) of molecules and their diffusive mobility (primarily determined hydrodynamically by size $r =$ Stokes radius)

$$v_{ep} = \frac{z}{6\pi\eta r}, \tag{12.2}$$

electroosmotic drift effects depend on the electrolytic properties of channel walls (zeta potential ζ), with the electroosmotic mobility given by the Helmholtz-Schmolchowski equation

$$v_{osm} = \frac{\varepsilon\varsigma}{4\pi\eta}, \tag{12.3}$$

where η is the solvent viscosity and ε its dielectric constant. Both equations (12.2) and (12.3) give the proportionality constant between applied electric field parallel to the channel and the induced velocities. Both effects are exploited in capillary electrophoresis to separate biomolecules as they flow down a long (typically 1 m) capillary with a lateral dimension (diameter) of only 100 μ under the influence of an external field in the range of several kilovolts (typical field strengths of 3 V/mm). In contrast, we shall be making use of local and dynamically switchable electric fields imposed by electrodes with typical separations of only 100 μm and digital voltages (e.g., 3.3 V). The field strengths can then be as large as 30 V/mm. Note that electric fields can also be used to influence the migration of micelles and vesicles (Khaledi, 1997), composed of either charged or neutral amphiphiles.

12.5.2 Digital Pulsed Electric Fields and Biomolecular Transport Control

Electric potentials at the surface of electrodes, which exceed the redox potential for electrolysis in saline water (2.19 eV, or just 0.83 eV in acidic solutions), will induce gas bubble creation. Whereas digital voltages (e.g., 3.3 V) are significantly above this level and do indeed cause bubble formation if applied in direct current, a pulsed field duty cycle in which the field is applied intermittently can be effectively employed to limit electrolysis and emulate the effects of lower electric fields. Here, it is not so much the finite capacitance of the electrolyte system that prevents electrolysis by integrating the field, although this plays a major role with pulse times below 10 ms, but more the rapid diffusion and resolvation of hydrogen and chlorine or oxygen during the field-free intervals that avoids macroscopic bubble formation.

This has enabled us to develop a general digital field-programmable control of microsystems, referred to as Field Programmable Fluid Arrays (Wagler et al., 2004). An example of an FPFA is shown in figure 12.8.

Figure 12.8
FPFA (field-programmable fluid array) module with integrated microfluidics and electronics (left) together with reconfigurable FPGA chip and interface (right). Fluidic connections in parallel are shown leaving the module on the far left. Background: pre-diced electrode circuit on silicon wafer. Design and construction are from Wagler et al. (2004).

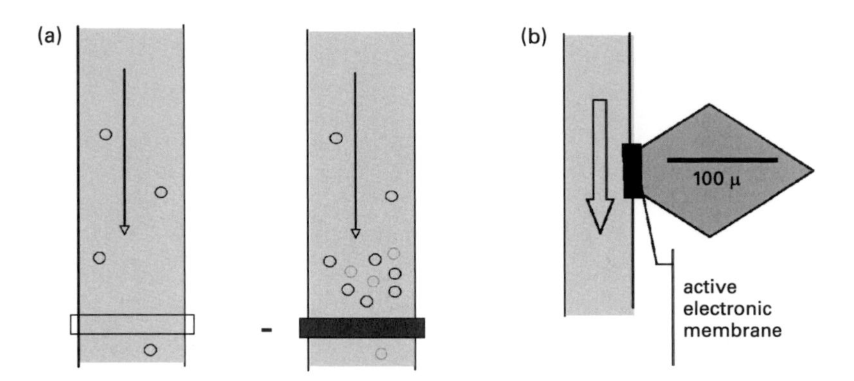

Figure 12.9
Electronic barriers (a) in-line and (b) an electronic "billabong." In (a), the electrode at the base of the flattened channel, when active, is sufficient to repel the charged molecules in the fluid passing over it, causing an increase in concentration that is ultimately limited by diffusion. In (b), diffusive interchange without flow between a side-chamber and a main channel is modulated by the voltage on electrodes across the shallow connecting channel. These designs are implicated in the further development to H reactors shown in figure 12.10.

12.5.3 Electronic Barriers to Diffusion

One configuration, which highlights the potential of electric fields to control compartmentation and concentration of material, is shown in figure 12.9a. Charged molecules are retarded under flow, causing a buildup of concentration upstream of the electrodes (straddling the channel like a weir). Electric fields have been used like this in nanochannels to concentrate ions for detection (Stern, Geis, and Curtin, 1997). The use of such electronic concentration to control reactions is discussed in the next subsection. A greater degree of control over the concentration effects of an electric field can be attained in a second configuration (called the e-billabong; figure 12.9b), whereby molecules can be concentrated in a side chamber separated via a connecting channel from a main flow channel. The diffusive interchange can be limited by the length and relative cross section of the connecting channel, and controlled by electric fields across the entry point. The restraint of molecules by electric fields in microfluidic systems is, of course, a major focus for chip studies in the lab, and more sophisticated geometries and surface effects are being exploited.

12.5.4 Concentration of Biomolecules and Reaction Control

A rather general module for manipulating molecules in microfluidic systems is the H-structure, an example of which is shown in figure 12.10. It can be regarded as an extension of the e-billabong configuration figure 12.9b, in which the side chamber is connected at the rear to an additional flow. This has two advantages. First, it allows for reliable filling of the side chamber by cross flow, and second, it enables pressure-

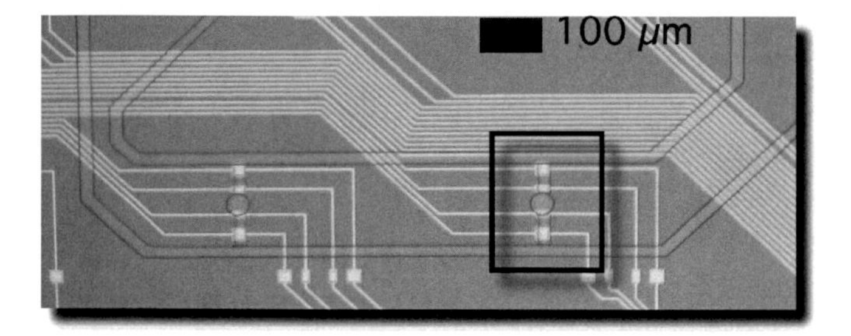

Figure 12.10
Two H-shaped reactors, with central circular chamber for microcell support. One can be identified, framed in black, and both "H"s lie on their side. Manufacturing uses PDMS microchannels (rose colored broader channels) of two or more different depths on a silicon substrate, supporting deposited metal electrodes with connecting wires that are insulated from the channel everywhere except at the rectangular electrodes. For function, see text and Tangen et al. (2006).

driven flow to act in opposition to electroosmotic or electrophoretic flow, thereby providing additional flexibility in the control of molecular concentrations. In the configuration shown in figure 12.10, which we have implemented (Tangen et al., 2006), the shallow channel depths of the silicon ledges ("sand bars") that separate the side chamber from the main channels suppress fluctuating pressure-driven flow between the two main flows. They also serve to increase the control of both fluid and molecular mobilities through suitably placed electrodes. The concentration of molecules from one of the two main channel flows occurs linearly with time, up to values 10 to 1,000 times higher than in the inflow. Activation of appropriately poled electrodes on the opposite side from the molecular source induces electroosmotic flow into the chamber, and further serves to repel negatively charged species such as the DNA oligomers used in these experiments. More details, including computation of the magnitude of competing flows, may be found in Chemnitz and coworkers (2006).

Replication reactions are intrinsically autocatalytic, and can (but usually do not) exhibit higher-order concentration dependence on the replicated template. (A linear concentration dependence leads to exponential growth, whereas most ligation-based schemes exhibit lower-order parabolic growth, equivalent to a square root dependence on concentration. Some cooperative replication schemes such as CATCH (Kirner et al., 1999) involve multiple templates and have a 3/2 power dependence on template concentration.) There are two critical reference reactions for self-replication: the loss or degradation of templates and the nonspecific production rate of alternative products from the same building blocks (background reaction). In a homogeneous system, the loss process is conventionally unimolecular, related to the chemical stability or residence time of the templates in the reaction chamber, and proportional to template concentration, whereas the background reaction is typically independent of template concentration. Replication reactions also require multiple building blocks to be assembled before covalent bonding, so that they typically exhibit higher-order (nonlinear) substrate dependences. Changing the concentrations of templates or building blocks can have a major impact on the outcome of replication processes, even determining whether the replication process has critical gain compared with loss or background processes. For this reason, the electronic concentration control of replication processes looks promising. An additional opportunity is provided by replication inhibitors (e.g., strong template binders), when their depletion in one place by electrodes can turn replication on, and their accumulation in another can turn replication off.

Of course, electric fields will generally simultaneously affect the concentrations of multiple species in a mixture. Depending on the field strengths, charges, and mobility of the chemical species involved, different concentration gradients are achievable for different species. In one scenario, it would be sufficient to attract charged templates, building blocks, or both to an array of microscale electrodes to achieve a critical concentration for net replication, thereby achieving electronically regulated replication. In another scenario, the replication of neutral templates and building blocks (e.g., PNA) could be inhibited by a charged strongly binding species that, when concentrated at a similar array of electrodes, allows replication with net gain in the intervening spaces. Currently, work is in progress on an electronically regulated replication process, with the properties of a chemical transistor, in which a bias voltage gates the level of signal amplification.

Another strong concentration dependence that can open a pathway to introduce electronic gating control into artificial cellular functions is the lipid self-assembly process itself. For the micellar version of the artificial cell, there is a well-understood phenomenon of critical micelle concentration (CMC; see Heerklotz and Seelig, 2000): a (strongly nonlinear) threshold concentration of amphiphiles for the formation of micellar phases. If replication takes place only in connection with micelles, then there is a very strong concentration dependence on the amphiphile concentration, and at least for charged amphiphiles; the electric field modulation of replication can proceed indirectly and sensitively by regulating the local accumulation of amphiphiles. Experimental investigation of these phenomena is commencing in PACE, with the first step being the electronically regulated reversible local formation of micelles. Similar effects are also possible with vesicles and other self-assembling structures.

12.6 Microfluidic Complementation Routes to Artificial Cells

Current bottom-up overall routes to artificial cells rely on the integration of modified versions of existing subsystems for the three basic functionalities of cells: self-replicating template molecules, self-reproducing containers, and catalytic metabolism. This section discusses the ways in which microfluidic compartmentation can assist this integration by supporting missing individual functionalities or aiding their integration with one another. Table 12.1 summarizes a list of particular artificial cell facets that can be complemented in microfluidic systems. We now turn to a structured consideration of complementing the three main subsystems of artificial cells, and then consider complementation for their integration.

12.6.1 Complementation for Replication

Replication for simple nanoscale artificial cells must avoid using biomolecules such as protein enzymes as catalysts, since their construction, necessary for cellular autonomy, would require complex biochemical machinery (protein translation and regulation apparatus). Many groups are using complex proteins for synthetic biology, but this is simply not the goal of a truly artificial cell. Of course, the simplest complementation schemes will provide these complex components as a part of the external environment (as in emerging endosymbionts such as the single-celled organism *Hatena*; see Okamoto and Inouye, 2005). In fact, external enzyme-supported replication systems such as Qß were the first noncellular biochemical systems to show laboratory evolution (Spiegelmann, 1971), and efficient 3SR (Guatelli et al., 1990) and SDA (Walker et al., 1992) and more complex replication systems such as CATCH (Ehricht, Ellinger, and McCaskill, 1997) and its more recent ribozyme cousin (Kim and Joyce, 2004) have been designed as discussed in section 12.2.

A further step in the direction of catalyst relegation to the external environment is to employ catalytic surfaces, thereby separating the phases of the environment from the catalytic action. Continuous-flow replication (Breaker, Banerji, and Joyce, 1994) and evolution processes using immobilized enzymes have been described (Bauer, 1990). Problems with such reactors are the inactivation and loss of catalysts from the solid support and contamination of the catalysts with self-replicating template species (usually DNA or RNA) exhibiting preferential binding.

The systems described here require special enzymatic catalysts because the double-stranded template formed during replication must be separated to allow the next round of replication, and this presents a significant obstacle to efficient replication. The Qß enzyme complex, which has been applied in connection with vesicles (Oberholzer et al., 1995; see also chapters 2 and 3), resolves this problem through a subunit with helicase activity, which unwinds the ds-RNA during new RNA polymerization. The SDA reaction uses a restriction enzyme to nick one side of the double-stranded DNA (ds-DNA) to allow a polymerase to displace one strand during the synthesis of a new one. The 3SR reaction uses the strand displacement polymerization of RNA transcription from ds-DNA, together with reverse transcription from RNA to DNA, to avoid the dead end of double-stranded template. In contrast, the polymerase chain reaction (PCR) technique uses temperature cycles together with thermostable polymerases to effect synchronous rounds of primed template synthesis (creating ds-DNA) followed by strand separation (Saiki et al., 1985). Temperature cycling is of course a complementation strategy, simplifying the task of replication. Temperature cycling has been implicated in Blum's model (Blum, 1962) for the origin of life and could be employed in artificial cells. PCR has been implemented in microfluidic systems (see Zhang et al., 2006, for a recent review).

As detailed in the work of Orgel's lab, particularly in von Kiedrowski's successful solution of the nonenzymatic replication problem (von Kiedrowski, 1986), the double-stranded template cul de sac plays an especially critical role in nonenzymatic replication. Temperature cycling might also be employed here to assist strand separation, but only if the rate-limiting step is not the ligation.

Strand separation can also be achieved in so-called chemical PCR, by a cyclic change of buffer conditions for the templates undergoing replication. In principle, since ds-DNA dissociates in 50 to 100 mM sodium hydroxide (NaOH) solution, repetitive acid-base titrations could be used to cycle the solution between two pH values. However, the accumulation of salt during such a process limits the number of cycles achievable to a handful, and the polymerase enzymes normally require stably buffered pH for efficient operation. Magnetic beads can be used to transfer templates from one buffer solution to the next, avoiding the problem of salt accumulation. In a microfluidic context, laminar flows of two different buffer solutions in direct contact can be created, and magnetic beads can be used to transfer templates between the flows (Penchovsky and McCaskill, 2002). Care must be taken to ensure that templates are retained for the next round of amplification, and doing this in a continuous-flow system requires a special reactor cascade of modules involving transfer, release, and retrieval or a cyclical transfer of templates in a crossed flow reactor (Rücker, 2005). These reactors have not yet proved functional for chemical PCR, for which the individual processes and hydrodynamic stability of the reactors still needs to be optimized.

Instead of using two buffer solutions, replication in enzymatic and nonenzymatic systems can be assisted by transferring templates between two catalytic surfaces, each one catalyzing the synthesis of the opposite strand. The Spread amplification technique introduced by von Kiedrowski (Luther, Brandsch, and von Kiedrowski, 1998) implements this using two glass plates, and electric fields are used to implement the transfer (by electrophoresis) from one surface to the next. This is, of course, a strong form of environmental complementation, but one close to the spirit of the PACE project. Actually, replication catalyzed at an appropriate single surface in a flowing system can also avoid the product inhibition issue and achieve exponential growth, as von Kiedrowski and Szathmáry's (2000) model calculation indicates.

With electric fields, additional forms of complementation become possible. Our digital programmable microsystems have achieved cyclic rounds of concentration and dilution using microelectrodes. Since many replication processes are strongly concentration dependent, this provides a powerful and general alternative approach to environmental complemented replication.

Experiments in this direction are now under way. Note that the forces generated by electric fields, although considerable, are not usually sufficient to allow a direct field-driven dissociation of a double strand (with one side immobilized to a surface or supporting gel), despite the fact that commercial electronic chips such as Nanogen's workstation can improve the specificity of binding by "electrostringency." The field-driven dissociation of complementary matching DNA proved problematic in experiments to utilize this effect for programmable DNA computing (data not shown).

A simple kinetic example of concentration cycling can be seen for the Kiedrowski replicator mechanism

$$X + A + B \underset{k_{-1}}{\overset{k_1}{\rightleftharpoons}} XAB \overset{k}{\rightarrow} X_2 \underset{k_2}{\overset{k_{-2}}{\rightleftharpoons}} 2X. \tag{12.4}$$

For longer templates, the dissociation rate of the duplex X_2 can be rate limiting; for short oligonucleotides the ligation reaction (with rate coefficient k) is rate limiting. Rather generally, and primarily for entropic reasons, the trimolecular complex XAB is less stable than the duplex X_2. Von Kiedrowski estimates for the hexanucleotide replicator, for instance (von Kiedrowski, 1993), that the free energy difference amounts to 3.75 kcal/mol at 25°C (−12.6 entropy units) at experimentally accessible concentrations of A and B. (Increasing A and B too much causes the nonspecific, nontemplating background reaction to dominate the kinetics of ligation.) For short oligonucleotides, concentration cycling (e.g., between two chambers) will not provide any significant kinetic advantage; for longer templates and ligation reactions in the range of 0.01 s^{-1}, however, concentration cycling can provide major increases in the replication rate. This is analogous to PCR, in which the rate-limiting step of duplex separation is avoided by temperature cycling (Rabinow, 1996). Note that the positive effect of concentration cycling holds even if there is no selective concentration. Concentration cycling of templates only would suffice for this effect.

Template replication in amphiphilic structures like micelles should be much more susceptible to concentration effects. The strongly nonlinear concentration dependence of collective self-assembly (akin to a phase transition) is evidenced, for example, in the critical concentration of surfactants for micelle formation (Heerklotz and Seelig, 2000). If we follow the path proposed by Rasmussen, Chen, Nilsson, and Abe (2003), with replication of PNA templates in a hydrophobic phase, then cyclic rounds of replication can be initiated by simple bulk flow that is opposed electrophoretic concentration of lipid amphiphiles at microelectrodes (see electroelution). In some significant sense, the individual microelectrodes become focal points for individual multimicellar artificial cells.

12.6.2 Complementation for Compartmentation

The use of microfluidic systems and electric fields to support the isolation of genetic amplification centers in single- and multiphase systems was discussed in sections 12.3 and 12.5. This section discusses additional opportunities for complementing compartmentation in connection with replication for multiphase amphiphilic systems. Of course, the most immediate and complete complementation system involves a cyclic process of compartment growth and division, as discussed in section 12.3. Such processes can indeed be complemented completely by an appropriate microfluidic physical system: Nozzles can be used to create compartments at defined locations, the supply of surfactant molecules (either dissolved in solvent or as micelles) from a separate channel or juxtaposed laminar flow facilitates compartment growth (micelles or vesicles), and extrusion through a small pore in a microfluidic system can initiate compartment division. Such systems must either be well timed or synchronized with internal replication of templates. Theoretical models of cooperation in compartmentalized systems indicate that it is possible, in principle, to achieve cooperative functionality of genes in a proliferation scheme involving local amplification followed by global mixing and settlement in new sites (Matessi and Jayaker, 1976).

Molecular selective compartmentation can also be supported by microfluidic systems, as discussed in section 12.3. Sections 12.4 and 12.5 outlined how such effects can be electronically programmed. Different electrophoretic mobilities, for example, can be exploited. In addition, if gels are employed, chromatographic separation based on differential binding to immobilized targets in the gel (e.g., immobilized DNA) has been shown to allow differential transmission of genetic sequences across gels (Baba, 1999). Free-flow electrophoresis (Clifton, 1996) provides a further technique for online separation of molecules for use in connection with artificial cells, and microfluidic chips for free-flow electrophoresis have been designed (Raymond, Manz, and Widmer, 1994).

12.6.3 Complementation for Metabolism

Metabolic complementation, in the extreme case, involves waste removal and the delivery of suitably high concentrations of all nutrients required for the cell (including utilizable energy-rich molecules and appropriate building blocks for the synthesis of template, compartments, and any catalysts or regulatory inhibitors of replication and self-reproduction). The first step toward autonomous metabolism for an artificial cell requires the synthesis of at least one building block for self-reproducing compartmentation or self-replicating templates. All of the other energy-rich materials would still be delivered to the protocell. Actually, the flow scenarios described in section 12.5 and figures 12.9 and 12.10 can be used to buffer the concentrations of nutrients and provide programmable control of their delivery to sites for artificial cells. Such limited metabolisms can also be complemented by means of separate microreaction chambers, setting the conditions in these chambers such that synthesis is initiated only when the appropriate templates are detected.

To ensure that replication occurs only locally ("inside the cell"), the concentration of monomers can be maintained externally well below the Michaeli's constant (typically in the 0.1- to 1.0-mM range for NTP monomers, lower for ligation reactions). The replication rate is often a strongly nonlinear function of metabolic building blocks with threshold concentrations required (Biebricher, Eigen, and Gardiner, 1983, 1984, 1985). This is an ideal situation for concentration control. The main difficulty with this approach is simply ensuring that the concentration of all required species (including low-molecular-weight ions) is appropriate. In fact, as mentioned previously, the main technical difficulty is that high ionic strength can limit the efficacy of electric field manipulation.

Concerning the processes of activation of building blocks (energy metabolism), complementation can take a variety of forms: from electronically controlled redox catalysis to photocomplementation using a light source. Caged molecules (e.g., caged ATP) provide another chemical variant for (spatially and temporally) controlled release. Of these methods, it appears that the electronic control of surface catalysis will provide the most flexible and powerful metabolic control. There are many known surface catalytic effects in prebiotic chemistry, and silicates have also been implicated, which could be exploited here.

12.6.4 Complementation for Integration

A scheme for integration was proposed in 2003 (see chapter 4), and is reproduced in figure 12.11 (color plate 9). Microfluidic complementation support of this route breaks down into the three primary component strands, as discussed earlier, and their integration. Various steps of partial integration have already been discussed. Three main strategies support integration: to separate reactions and then bringing together compatible products for joint action, provide in situ support for the compatibility of reactions through regulation, and support spatial self-organization of component processes.

The first strategy includes but goes beyond the microfluidic equivalents of pipetting robot functionality to a chemical microplant methodology. The basic concept is that of continuous flow and triggered release reactors, where resources are converted

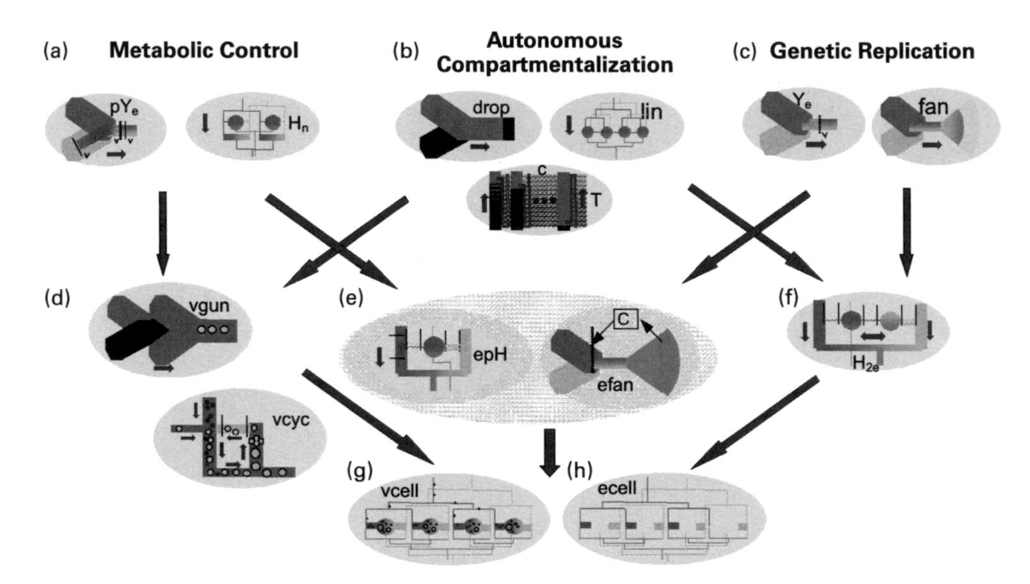

Figure 12.11 (color plate 9)
Microfluidic complementation routes to the protocell. This scheme follows the general subsystem buildup as proposed in figure 4.3 (chapter 4). Electronic microfluidics can complement these routes by providing (a) regulated spatial control and (b) a further range of specific complementation, as described in table 12.2 and the figures. Top row illustrates microfluidic complementation of the separate subsystems with (from left to right) (a) *metabolism* (electronically regulated pH and concentration inflows, and gradient chemical control), (b) *containment* (autonomous formation of channel emulsion droplets, phase space screening with crossed gradient reactor, and geometric control of isolation in linear cross-flow reactor), and (c) *replication* (inflow electronic concentration gate for genetic material, and self-regulation fan reactor for localized homeostatic replication). Second row illustrates complementation support for pairwise combination of subsystems: (d) *metabolism + container* (self-regulated vesicle gun and vesicle replication cycle complemented by regulated micellar feed), (e) *metabolism + replication* (electronically regulated pH gradient controlling replication, and electronically regulated mixing ratio for metabolites controlling replication in fan reactor), and (f) *container + replication* (alternating chamber regulation and enabling of rapid replication). Third row illustrates two targets for complemented protocells integrating all three subsystems: (g) complemented vesicle-nucleic acid protocells, and (h) fully electronically complemented protocells with support for all three subsystems and electronic genomes.

to intermediate products in separate processes upstream and the products are then delivered and mixed to allow new reactions (possibly in multiple stages). The microscale allows cell-individual processing streams, compatible with low genetic copy numbers and evolutionary stability. The second strategy uses microfluidic control (in particular electronic programmable control) to regulate chemical reactions as they occur, for instance, to prevent certain reactions from consuming all the available resources. Many variants are possible.

The third and perhaps most interesting strategy involves regulating the formation of multiple local multiple phases formation separation (for example) to allow a balance of phase-specific reactions to occur. An example is maintaining a local amphiphile membrane where certain reactions can occur (as in von Kiedrowski's catalipid and Rasmussen's protocells; see Rasmussen et al., 2003) and at the same time maintaining a local volume that has controlled exchange with both media channels and the membrane.

12.7 The Omega Machine and Evolving Artificial Cells

12.7.1 Evolvable Route to Artificial Cells

Is there an incremental route to artificial cells through successive withdrawal of complementation support and solving chemical subtasks? The solution of each subtask could then be tackled either by rational design or by combinatorial search (or even fully iterative evolution). The question divides into two parts: the existence of effective complementation strategies to complete an artificial cell, and the ability to remove these in a stepwise process toward an autonomous artificial cell. We have addressed pieces of the complementation process, including integration, in the previous section. Here we concentrate on the removal problem. Of course, the actual route to protocells will most likely differ in many respects. Here, the conceptual work is simply to demonstrate that a route does exist in principle.

For the sake of concreteness, then, let us take the following scenario for (1) *an initial microfluidics-supported artificial cell*, with the following features: (a) the artificial cell is bounded by electronic fields and geometric restriction, so that key molecules are concentrated within the cell; (b) enzyme-free replication of potentially amphiphilic nucleic acids such as PNA or catalipids occurs only within the confines of the artificial cell; and (c) active chemical precursors are supplied externally at a concentration too low to support autonomous proliferation of genetic material in the extracellular regions.

Now, consider whether it is possible to withdraw the complementation system to reach (2) *an autonomous artificial cell in a microfluidics environment*, with the following features: (a) the artificial cell is bounded by an amphiphile membrane synthesized in part by the artificial cell itself; (b) enzyme-free replication of amphiphilic nucleic acids occurs only within the confines of the artificial cell; and (c) chemical precursors supplied externally are converted to active components through the presence of the artificial cell (catalyzed either by the membrane or the constituent molecules).

A withdrawal scenario might involve a sequence of small changes in the duration of barrier electronic fields, confining geometry and flow rates for supply channels, and quantities of active and inactive chemical precursors supplied externally, accompanied by combinatorial search for PNA or catalipids, which can fulfill the artificial cell's three functions. Although the step to membrane formation, for instance, may be difficult, requiring very specific genetic amphiphiles, at least for compatibility with self-replication, it could be broken up into separate optimization steps for the

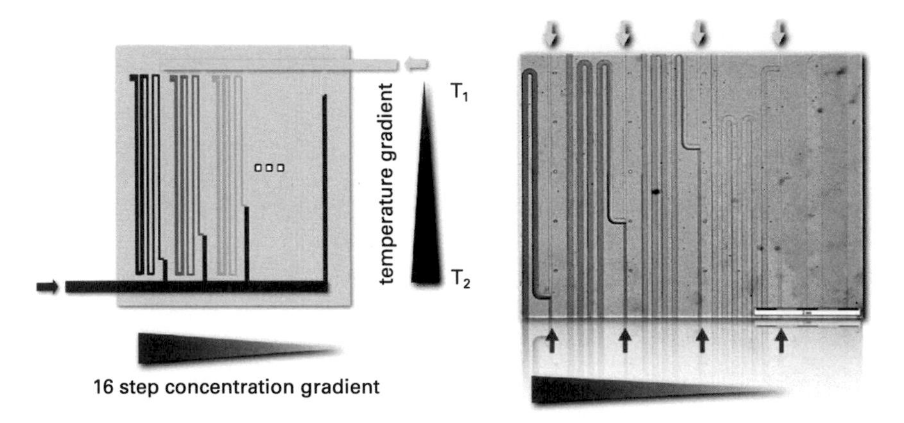

Figure 12.12 (color plate 10)
Gradient microreactor for phase space screening of amphiphile systems. Gradients in temperature (vertical) and concentration ratio of two fluids (horizontal) are shown in the schematic (left) and in a blow up of the actual microreactor (right). The microscope image shows hydrodynamic study in microfluidic gradient system BioGrad^2 with the mixing of two fluids at a sequence of T-junctions. The different channel lengths give different resistivities to hydrodynamic flow, causing a different mixing ratio at the junction. Note that meanders, raising and lowering the mixed fluid temperature, allow one to distinguish kinetic from equilibrium phase change effects. Revised design by Steffen Chemnitz; experimental image by Patrick Wagler and Steffen Chemnitz, BioMIP.

cell's subtasks (a) and (b), or intermediate steps by micelle-forming amphiphiles, starting with the fully supported artificial cell (1).

Note that the support system is not a single point in parameter space. Instead, the microfluidic technology allows one to investigate in parallel a whole combinatorial set of support systems, to screen for capabilities of heightened autonomous functionality. The simplest systems of this type are the gradient reactors being developed in my group (BioMIP), which allow concentration and temperature gradients to be established across a microfluidic chip that enables a kind of spatial phase diagram of behaviors of candidate artificial cells to be evaluated in a single experiment.

Such a prototype gradient search reactor, which does not yet have localized microcell support, is shown in figure 12.12 (color plate 10). Of course, microfluidic gradients have been produced before (Wang, Mukherjee, and Lin, 2006); the emphasis here is on their use for phase space screening.

12.7.2 Phases of the Omega Machine

The Omega Machine is a programmable microfluidic control environment that can "see" the state of the chemical system and appropriately adjust a combinatorially complex set of control signals. It differs from the control center of a chemical plant both in the degree of monitoring and control, with significant combinatorial complexity, and in presence of spatial control elements and spatially resolved sensors on the same microscopic scale as the individual cells formed by the system. This qualitative jump in integration allows one to smoothly transfer functionality and information between the chemical system and the computerized control system: Specific information in the control system can be associated with individual cells. Following the reconfigurable hardware approach to evolvable systems described in section 12.4, the Omega Machine then allows part of the heritable information of a cell to be complemented by information in the computer. The Omega Machine is named to reflect the end-goal ("omega point") oriented evolution that such a programmable environment device makes possible.

Block diagrams describing the basic components and functions of the Omega Machine are shown in figure 12.13. These diagrams highlight the essential online monitoring-control feedback loops at the level of single cells describing Omega Machine components and functionality. The machine incorporates reconfigurable electronics together with interface electrodes to implement combinatorially programmable actuator networks. It incorporates a three-dimensional (3D) microscopy interface, which takes advantage of high-bandwidth CCD computer interfaces and sensitive detection with laser-induced parallel (optionally confocal) fluorescence to provide parallel monitoring of thousands of potential artificial cell sites in 2D or 3D. The microfluidics system processes chemicals from external reservoirs and outputs products (from molecular concentrations or transportable mesoscale structures with specific content, such as vesicles, up to complete artificial cells). The delivery and product selection is made reconfigurable (and thus combinatorially optimizable) through electrode arrangements, which impinge on the fluid. The result is a functional feedback loop integrating local self-organization with information processing and evolution. The functionality of candidate protocells is evaluated and more successful variants selected (spatially in situ through electronic control), and the system is then reconfigured to allow new variant protocells to be formed at interior locations under modified support.

Although the generic Omega Machine described here and in figure 12.13 has already been designed and built in our lab, the functionality is still in its infancy en route to artificial cells. One can identify eight potential development phases in the functionality of microfluidic systems toward a full Omega Machine for artificial cells:

1. Single-phase localized replication, with self-replication confined to a localized region by restraining fields and flows.

2. Dual-phase dynamic containers, making use of phase boundaries (such as membranes or interfaces between hydrocarbons and water) for containment. The two-phase containers should be able to be formed locally from raw material flows within

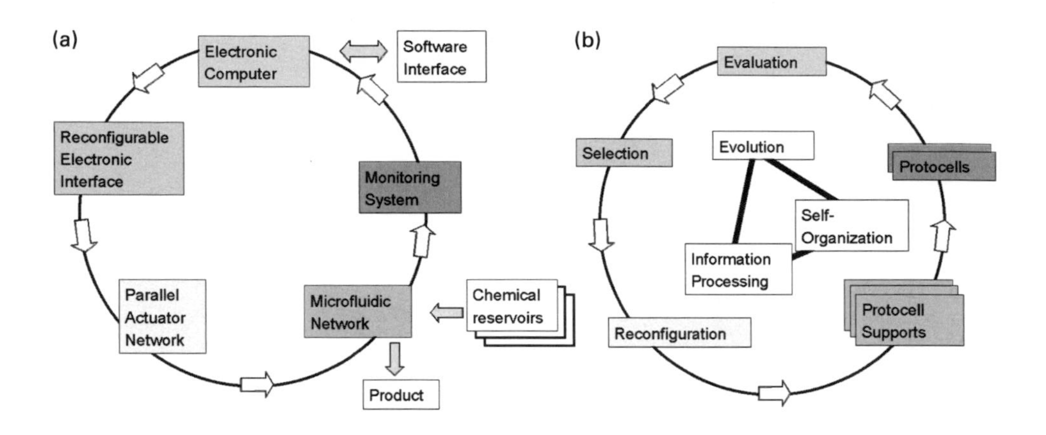

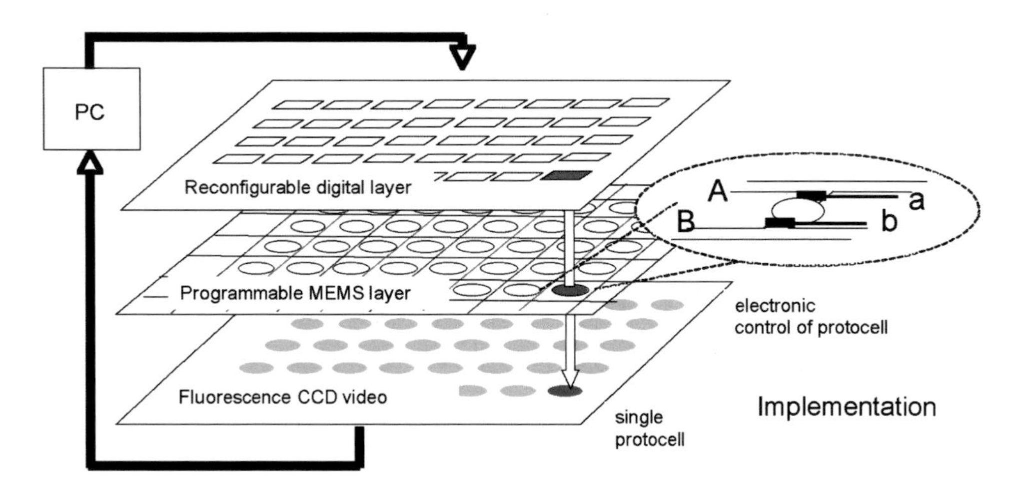

Figure 12.13
The Omega Machine: (a) component subsystems with complementation support and information at the single protocell level; (b) functional flow in the Omega Machine with evolution in a sequence of protocell support systems; (c) implementation principle of feedback cycle for single protocell complementation in the current implementation of the Omega Machine. The blow up shows a simplified single protocell chamber example, placed between two chemical flows (A and B), with electronic control of transport (a, b) between the flows and the cells.

the microfluidic system, and be transported within the system. Container self-reproduction may involve container growth and division, or more simply, transfer of old container contents to two newly formed containers.

3. Regulated chemical self-replication, with the Omega Machine providing gain-control. This is the equivalent of the Q-factor for laser operation, or neutron decelerating material for nuclear chain reactions. The localized regulated replication centers can then deliver replicated genetic material on demand to create the microfluidic transfer system to form new replication centers.

4. A combinatorial niche search machine, capable of generating (under programmable control) a combinatorial repertoire of niches (building on the gradient reactors) that allow joint combinatorial search of molecular libraries and local control patterns. This should be used to search for both genetic molecules with catalytic functionality (e.g., modification of precursors to active building blocks) and collective amphiphilic properties.

5. The integration and coevolution of self-replication and metabolic catalysis.

6. The integration and coevolution of container components and self-replication.

7. The Omega Machine–supported cell, integrating metabolism, self-replication, and containment.

8. Evolutionary transfer of functions from the machine to the artificial cell.

12.7.3 Programming the Omega Machine

From the computer science perspective, the Omega Machine may be regarded as an embedded system. As in the field of robotics, the concept of natural and unnatural divisions of labor existing between a physical system and a computerized control system is important, because the Omega Machine has real-time functionality requirements, to fulfill its role as a programmable local combinatorially complex complementation system. In particular, the concept of embodiment (Brooks, 1999; Pfeiffer and Scheier, 2001) is central to achieving this performance. Physical embodiment is essential for allowing real-time distributed functionality to make strong use of the integrated real-time computational problem-solving power physical systems in direct contact with the environment of interest offer. An example of a key embodiment in the Omega Machine is the electrical capacitance of the microfluidic cell, which can be used to allow multiplcxed sequential addressing of many electrodes with fewer control signals. Another example is the innate regulatory nature of chemical reactions. Delicate tuning of a concentration level of a chemical in the microreactor is often unnecessary, because it can be regulated naturally in a gradient field.

Realizations of the electronic Omega Machine deal with video frequency fluorescence data in a million or more pixel channels, which must be processed in real time and used to control the local environment of up to one thousand candidate artificial cells (30 by 30 pixels per cell) by on the order of ten thousand electrodes (of the order of 10 per cell). Future developments may allow these numbers to be increased to a million cells or more. Of course, experimentation must first deal with a few cells (1 to 10) involving several hundred electrodes before scaling up to millions of cells. The computational requirements of this first prototype are then much reduced compared with the later versions required for sufficient combinatorial search capability. Multichip reconfigurable electronics boards with a standard computer interface, such as that developed in my lab (e.g., Ngen [see McCaskill, Tangen, and Ackerman, 1997] and Meregen [see Tangen, Maeke, and McCaskill, 2002]) which use multiple Xilinx field-programmable gate arrays (FPGAs), each connected to multiple SRAM chips, have been shown to provide the necessary computational capability to deal with the full data processing, feedback, and control tasks. Note that FPGAs allow modular control electrode group responses, involving complex coordinated Hertz to megahertz pulse trains on multiple electrodes, to be set up in digital electronics programmably. They also can be used to do feature extraction from the local portions of images, which have already been made available to a single board reconfigurable card at the full rates described earlier. The role of the standard Intel processors and operating system is then to provide a framework software environment for reconfiguration and non–time-critical software-intensive monitoring and recording tasks.

Because the computer system can deal simultaneously with the individual cells, it can also build up and retain information specific to each of them. This information can then be treated as extended genetic hereditary information for the cell, that is, an electronic genome. Because the information can be in the form of either data (e.g., control parameters) or electronic configurational information (defining the local FPGA electronic processing for feature extraction, control loops, or pulsed electrode actuation), this genetic information has considerable functional flexibility in interaction with the cell. Online reconfiguration can further enhance the computational capabilities as variants of control modules are downloaded to the FPGA chips. The information can thus be proliferated both within the FPGA board and in the microfluidics system (where electrode configurations at different sites can be connected to copies of the same control logic), allowing daughter cells to inherit the same electronic information as their parent cells. In an extreme case, this may be the only significant genetic information in the early artificial cells. It is also clear, then, that this electronic genome can be used to allow the Omega Machine to support an evolutionary process in artificial cell design and functionality. To avoid misunderstanding with "in silico" artificial cell ideas such as the VCell virtual cell simulation (see http://vcell.org), it is important to stress that we are describing true "wet" artificial cells here, with real chemistry, but in online interaction with an electronic control system.

12.8 Conclusions

The main conclusion of this chapter is that an evolutionary approach to artificial cells exists, beyond standard chemical experimentation, using individual-based microfluidic complementation. The first novel artificial cells can arise as hybrid digital electronic and chemical entities bearing both genetic and electronic information. We have shown how current microsystem technology has been and can be further developed to support the chemical and physical functionalities necessary to integrate an artificial cell from its subsystem chemistries.

The EU Programmable Artificial Cell Evolution project (PACE) is actively pursuing these avenues of research, bringing together leading chemists and physicists with experts from the fields of microsystem technology and information technology. The subsystems for container chemistry, self-replicating molecules, and simplest energy metabolisms have been established in the labs of project partners, and the current major focus for integration is the coupling of container chemistry and self-replicating molecule chemistry. The first gradient-based phase space search microreactors are being employed, and localized replication and electronically programmable containment have been experimentally demonstrated. The multiphase container chemistry in the microfluidic systems is already partly functional, with the electronic manipulation of vesicles and emulsion droplets having been achieved. More work is required to complete the repertoire of functional manipulation necessary to support cellular self-reproduction of the containers, but the route to this, using either in situ vesicle formation or microfluidic-controlled extrusion, appears comparatively straightforward.

The work aims to bring about, eventually, not only a truly synthetic artificial cell, but an individual-based evolutionary manufacturing technology suitable for a wide range of tasks. We are also investigating the potential of both artificial cells and the hybrid systems needed to attain them as novel IT technology in their own right.

Acknowledgments

First, I would like to thank all members of the BioMIP research group who have worked consistently on the various microfluidic and electronic components I have used to illustrate this conceptual development, especially P. Wagler, U. Tangen, S. Chemnitz, and T. Maeke. Second, I wish to thank my colleagues S. Rasmussen, N. Packard, and M. A. Bedau for their early conviction of the central importance of the Omega Machine concept, which led to seminal discussions on the subject in 2003, supported in part by the Santa Fe Institute. Third, I appreciate the many helpful and constructive recommendations made by two anonymous referees. The initial enabling technology arose from my work on microfluidic DNA computing at the German National Research Center for Information Technology (GMD), even though the current technological platform is now quite different. Many of the ideas build on understanding developed in the fields of molecular evolution research, in which the author is indebted to Manfred Eigen, and evolutionary biotechnology, in which the support of the IMB Jena and its initial director Peter Schuster is appreciated. The work has been directly and bravely supported by the European Union in the sixth framework program of Future Emerging Technologies in IST (Information Society Technologies). The Fraunhofer Gesellschaft provided initial transitional support for our group but then both aided and delayed our research by making it necessary for us to redesign and reequip in the middle of a project. The author wishes to thank all project partners in the PACE project for their enthusiastic support of this research, and in particular G. von Kiedrowski for both recognizing the impact this research can have on chemistry and consenting to host my group at the Ruhr University of Bochum. Final thanks go to the Friedrich Schiller University in Jena, which has granted me leave of absence to complete this project.

References

Baba, Y. (1999). Capillary affinity gel electrophoresis: New technique for specific recognition of DNA sequence and the mutation detection on DNA. *Journal of Biochemical and Biophysical Methods*, *41*, 91–101.

Bauer, G. J. (1990). *Biochemische Verwirklichung und Analyse von kontrollierten Evolutionsexperimenten mit RNA-Quasispezies in vitro*. Dissertation, Technische Universität Braunschweig.

Bauer, G. J., McCaskill, J. S., & Otten, H. (1989). Traveling waves of in vitro evolving RNA. *Proceedings of the National Academy of Sciences of the United States of America*, *86* (20), 7937–7941.

Biebricher, C. K., Eigen, M., and Gardiner, W. C. Jr. (1983). Kinetics of RNA replication. *Biochemistry*, *22* (1), 2544–2559.

Biebricher, C. K., Eigen, M., and Gardiner, W. C. Jr. (1984). Kinetics of RNA replication: Plus-minus asymmetry and double-strand formation. *Biochemistry*, *23* (14), 3186–3194.

Biebricher, C. K., Eigen, M., and Gardiner, W. C. Jr. (1985). Kinetics of RNA replication: Competition and selection among self-replicating RNA species. *Biochemistry*, *24*, 6550–6560.

Biebricher, C. K., Eigen, M., & McCaskill, J. S. (1993). Template-directed and template-free RNA synthesis by Q beta replicase. *Journal of Molecular Biology*, *231* (2), 175–179.

Biebricher, C. K., & Luce, R. (1993). Sequence analysis of RNA species synthesized without template. *Biochemistry*, *32*, 4848–4854.

Bishop, K. J. M., Gray, T. P., Fialkowski, M., & Grzybowski, B. A. (2006). Microchameleons: Nonlinear chemical microsystems for amplification and sensing. *Chaos*, *16* (3), 037102/1.

Bloom, D. M., Peterman, M. C., & Ziebarth, J. M. (2005). *Microfabricated apertures for supporting bilayer lipid membranes*, U.S. Patent 6,863,833.

Blum, H. F. (1962). On the origin and evolution of living machines. *American Scientist*, *49*, 474–501.

Böddeker, B. (1995). *Physical model for coevolution and simulation on a hardware programmable processor* (Supervisor: J. S. McCaskill, FSU, Jena). Diploma Thesis in Physics, University of Göttingen, Germany.

Breaker, R. R., Banerji, A., & Joyce, G. F. (1994). Continuous *in vitro* evolution of bacteriophage RNA polymerase promoters. *Biochemistry*, *33*, 11980–11986.

Brooks, R. A. (1999). *Cambrian intelligence.* Cambridge, MA: MIT Press.

Castets, V., Dulos, E., Boissonade, J., & De Kepper, P. (1990). Experimental evidence of a sustained standing Turing-type nonequilibrium chemical pattern. *Physical Review Letters*, *64*, 2953–2956.

Cech, T. E. (1987). The chemistry of self-splicing RNA and RNA enzymes. *Science*, *236*, 1532–1539.

Chemnitz, S., Tangen, U., Wagler, P. F., Maeke, T., & McCaskill, J. S. (2006). Electronically programmable membranes for improved biomolecule handling in micro-compartments on chip. Accepted in the 9th International Conference on Microreaction Technology, Potsdam, Germany.

Cho, S. K., Fan, S. K., Moon, H., & Kim, C. J. (2002). Towards digital microfluidic circuits: Creating, transporting, cutting and merging liquid droplets by electrowetting-based actuation. *The Fifteenth IEEE International Conference*, 32–35.

Clifton, M. J. (1996). Continuous-flow electrophoresis in the Taylor regime: A new possibility for preparative electrophoresis. *Journal of Chromatography A*, *757*, 193–202.

Cristini, V., & Tan, Y.-C. (2004). Theory and numerical simulation of droplet dynamics in complex flows—A review. *Lab Chip*, *4*, 257.

Dapprich, J., McCaskill, J. S., Voelker, S., & Krause, F. (1994). Fluorescence imaging of evolving RNA in capillaries. *Berichte-Bunsengesellschaft für Physikalische Chemie*, *98*, 1202–1203.

Darwin, C. (1859). *On the origin of species*. London: John Murray.

Ehricht, R., Ellinger, T., & McCaskill, J. S. (1997). Cooperative amplification of templates by cross-hybridization (CATCH). *European Journal of Biochemistry*, *243* (1–2), 358–364.

Ehses, S., Chemnitz, S., Jünger, M., Maeke, T., Palutke, T., Tangen, U., et al. (2005). Spatially self-regulating molecular replication in a fan microreactor—En route to artificial cells. Preprint.

Eigen, M., McCaskill, J. S., & Schuster, P. (1989). The molecular quasispecies. *Advances in Chemical Physics*, *75*, 149–263.

Eigen, M., & Schuster, P. (1979). *The hypercycle—A principle of natural self-organization*. Berlin: Springer-Verlag. (A combined reprint of *Naturwissenschaften*, *64*, 541–565 (1977), and *65*, 7–41, 341–369 (1978).)

Ellinger, T., Ehricht, R., & McCaskill, J. S. (1998). In vitro evolution of molecular cooperation in CATCH, a cooperatively coupled amplification system. *Chemistry and Biology*, *5* (12), 729–741.

Elliott, R., Szleifer, I., & Schick, M. (2006). Phase diagram of a ternary mixture of cholesterol and saturated and unsaturated lipids calculated from a microscopic model. *Physical Review Letters*, *96*, 098101.

Fiedler, S., Shirley, S. G., Schnelle, T., & Fuhr, G. (1998). Dielectrophoretic sorting of particles and cells in a microsystem. *Analytical Chemistry*, *70*, 1909–1915.

Fisher, R. A. (1937). Wave of advance of an advantageous allele. *Annual Eugenics*, *7*, 355–369.

Gascoyne, P. R. C., Vykoukal, J. V., Schwartz, J. A., Anderson, T. J., Vykoukal, D. M., Current, K. W., et al. (2004). Dielectrophoresis-based programmable fluidic processors. *Lab Chip*, *4*, 299–309.

Guatelli, J. C., Whitfield, K. M., Kwoh, D. Y., Barringer, K. J., Richman, D. D., & Gingeras, T. R. (1990). Isothermal *in vitro* amplification of nucleic acids by a multienzyme reaction modeled after retroviral replication. *Proceedings of the National Academy of Sciences of the United States of America*, *87*, 1874–1878.

Hanczyc, M. M., & Szostak, J. W. (2004). Replicating vesicles as models of primitive cell growth and division. *Current Opinion in Chemical Biology*, *8*, 660–664.

Heerklotz, H., & Seelig, J. (2000). Correlation of membrane/water partition coefficients of detergents with the critical micelle concentration. *Biophysical Journal*, *78* (5), 2435–2440.

Ikuta, K., Hirowatari, K., & Ogata, T. (1994). Three dimensional micro integrated fluid system (MIFS) fabricated by stereo lithography. *Proceedings of IEEE International Workshop on Micro Electro Mechanical Systems (MEMS '94)*, 1–9.

Johns, G. C., & Joyce, G. F. (2005). The promise and peril of continuous in vitro evolution. *Journal of Molecular Evolution*, *61*, 253.

Kenis, P. J. A., Ismagilov, R. F., & Whitesides, G. M. (1999). Microfabrication inside capillaries using multiphase laminar flow patterning. *Science*, *285*, 83–85.

Khaledi, M. G. (1997). Micelles as separation media in high-performance liquid chromatography and high-performance capillary electrophoresis: Overview and perspective. *Journal of Chromatography A*, *780*, 3–40.

Kim, D.-E., & Joyce, G. F. (2004). Cross-catalytic replication of an RNA ligase ribozyme. *Chemistry and Biology*, *11*, 1505–1512.

Kimura, M. (1983). Diffusion model of intergroup selection, with special reference to evolution of an altruistic character. *Proceedings of the National Academy of Sciences of the United States of America*, *80*, 6317–6321.

King, R. D., Whelan, K. E., Jones, F. M., Reiser, P. G. K., Bryant, C. H., Muggleton, S. H., et al. (2004). Functional genomic hypothesis generation and experimentation by a robot scientist. *Nature*, *427*, 247–252.

Kirner, T., Ackermann, J., Ehricht, R., & McCaskill, J. (1999). Complex patterns predicted in an in vitro experimental model system for the evolution of molecular cooperation. *Biophysical Chemistry*, *79*, 163–186.

Knight, J. B., Vishwanath, A., Brody, J. P., & Austin, R. H. (1998). Hydrodynamic focusing on a silicon chip: Mixing nanoliters in microseconds. *Physical Review Letters*, *80*, 3863–3866.

Lawrence, M. S., & Bartel, D. P. (2003). Processivity of ribozyme-catalyzed RNA polymerization. *Biochemistry*, *42* (29), 8748–8755.

Lee, A., & Tan, Y.-C. (2004). *Microfluidic devices for controlled viscous shearing and formation of amphiphilic vesicles*, PCT world patent application, WO 2004/071638.

Luisi, P. L., & Varela, F. J. (1990). Self-replicating micelles—A chemical version of minimal autopoeitic systems. *Origin of Life and Evolution of the Biosphere*, *19*, 633–644.

Luther, A., Brandsch, R., & von Kiedrowski, G. (1998). Surface-promoted replication and exponential amplification of DNA analogues. *Nature (Letters to Nature)*, *396*, 245–248.

Matessi, C., & Jayakar, S. D. (1976). Conditions for the evolution of altruism under Darwinian selection. *Theoretical Population Biology*, *9*, 360–387.

McCaskill, J. S. (1988). *Polymer chemistry on tape: A computational model for emergent genetics.* Internal report of the Max-Planck-Institute for Biophysical Chemistry, Göttingen, Germany.

McCaskill, J. S. (1992). How is genetic information generated? In A. Andersson, S. Andersson, & U. Ottoson (Eds.), *Theory and control of dynamical systems: Applications to systems in biology, Huddinge, Stockholm, 4–10 August 1991*. Singapore: World Scientific Press.

McCaskill, J. S. (1997). Spatially resolved in vitro molecular ecology. *Biophysical Chemistry*, *66*, 145–158.

McCaskill, J. S. (2001). Optically programming DNA computing in microflow reactors. *Biosystems*, *59* (2), 125–138.

McCaskill, J. S., & Bauer, G. J. (1993). Images of evolution: Origin of spontaneous RNA replication waves. *Proceedings of the National Academy of Sciences of the United States of America*, *90* (9), 4191–4195.

McCaskill, J. S., Füchslin, R. M., & Altmeyer, S. (2001). The stochastic evolution of catalysts in spatially resolved molecular systems. *Journal of Biological Chemistry*, *382*, 1343–1363.

McCaskill, J. S., Tangen, U., & Ackermann, J. (1997). VLSE, very large scale evolution in hardware. In P. Husbands & I. Harvey (Eds.), *Proceedings of the fourth European conference on artificial life* (pp. 398–406). Cambridge, MA: MIT Press.

McCaskill, J. S., & Wagler, P. (2000). From reconfigurability to evolution in construction systems: Spanning the electronic, microfluidic and biomolecular domains. *Lecture Notes in Computer Science*, *1896*, 286–299.

McCaskill, J. S., Wagler, P., & Maeke, T. (2003). *Configurable microreactor network.* U.S. Patent 6,599,736.

Oberholzer, T., Wick, R., Luisi, P. L., & Biebricher, C. K. (1995). Enzymatic RNA replication in self-reproducing vesicles—An approach to a minimal cell. *Biochemical and Biophysical Research Communities*, *207*, 250–257.

Okamoto, N., & Inouye, I. (2005). A secondary symbiosis in progress? *Science*, *310* (5746), 287.

Pearson, J. E. (1993). Complex patterns in a simple system. *Science*, *261*, 189–192.

Penchovsky, R., & McCaskill, J. S. (2002). Cascadable hybridization transfer of specific DNA between microreactor selection modules. *Lecture Notes in Computer Science*, *2340*, 46–56.

Pfeiffer, R., & Scheier, C. (2001). *Understanding intelligence*. Cambridge, MA: MIT Press.

Rabinow, P. (1996). *Making PCR: A story of biotechnology*. Chicago: University of Chicago Press.

Rasmussen, S., Chen, L., Nilsson, M., & Abe, S. (2003). Bridging nonliving and living matter. *Artificial Life*, *9* (3), 269–316.

Raymond, D. E., Manz, A., & Widmer, M. (1994). Continuous sample pretreatment using a free-flow electrophoresis device integrated onto a silicon chip. *Analytical Chemistry*, *66*, 2858–2865.

Rücker, T. (2005). Biomolecular information processing in networked microflow reactors (Supervisors, J. S. McCaskill & M. Famulok). Dissertation, Rheinische Friedrich-Wilhelms-Universität Bonn.

Saiki, R. K., Scharf, S., Faloona, F., Mullis, K. B., Horn, G. T., Erlich, H. A., & Arnheim, N. (1985). Enzymatic amplification of beta-globin genomic sequences and restriction site analysis for diagnosis of sickle cell anemia. *Science*, *230*, 1350–1354.

Sanchez, E., & Tomassini, M. (Eds.) (1991). *Towards evolvable hardware: The evolutionary engineering approach. Lecture Notes in Computer Science.* Berlin: Springer.

Schmidt, K., Foerster, P., Bochmann, A., & McCaskill, J. (1998). A microflow reactor for two dimensional investigations of *in vitro* amplification systems. In W. Ehrfeld (Ed.), *Microreaction technology: Proceedings of the first international conference on microreaction technology*. Berlin: Springer-Verlag.

Spiegelmann, S. (1971). An approach to the experimental analysis of precellular evolution. *Quarterly Review of Biophysics*, *4*, 213.

Squires, T. M., & Quake, S. R. (2005). Microfluidics: Fluid physics at the nanoliter scale. *Review of Modern Physics*, *7* (3), 977–1026.

Stern, M. B., Geis, M. W., & Curtin, J. E. (1997). Nanochannel fabrication for chemical sensors. *Journal of Vacuum Science and Technology B*, *15* (6), 2887–2891.

Stone, H. A., Stroock, A. D., & Ajdari, A. (2004). Engineering flows in small devices. *Annual Review of Fluid Mechanics*, *36*, 381–411.

Szathmáry, E., & Demeter, L. (1987). Group selection of early replicators and the origin of life. *Journal of Theoretical Biology*, *128*, 463–486.

Tan, Y.-C., Cristini, V., & Lee, A. P. (2006). Monodispersed microfluidic droplet generation by shear focusing microfluidic device. *Sensors and Actuators B*, *114*, 350–356.

Tangen, U., Maeke, T., & McCaskill, J. S. (2002). Advanced simulation in the configurable massively parallel hardware MereGen. In K. H. Hoffmann (Ed.), *Coupling of biological and electronic systems: Proceedings of the 2nd Caesarium, Bonn, November 1–3, 2000* (pp. 107–118). Berlin: Springer.

Tangen, U., McCaskill, J. S., Maeke, T., Mathis, H., & Füchslin, R. M. (2003). *Electric microfluidics multiplex system and use thereof.* World PCT Patent Application PCT/EP03/05418.

Tangen, U., Wagler, P. F., Chemnitz, S., Goranović, G., Maeke, T., & McCaskill, J. S. (2006). An electronically controlled microfluidics approach towards artificial cells. *ComPlexUs*, *3*, 48–57.

Thorsen, T., Maerkl, S. J., & Quake, S. R. (2002). Microfluidic large-scale integration. *Science*, *298*, 580–584.

Thorsen, T., Roberts, R. W., Arnold, F. H., & Quake, S. R. (2001). Dynamic pattern formation in a vesicle-generating microfluidic device. *Physical Review Letters*, *86*, 4163–4166.

Tjivikua, T., Ballester, P., & Rebek, J., Jr. (1990). A self-replicating system. *Journal of the American Chemical Society*, *112*, 1249–1250.

Turing, A. M. (1952). Chemical morphogenesis. *Philosophical Transactions of the Royal Society of London*, *237*, 37.

Urbanski, J. P., Thies, W., Rhodes, C., Amarasinghe, S., & Thorsen, T. (2006). Digital microfluidics using soft lithography. *Lab on a Chip*, *6* (1), 96–104.

van Noort, D., Wagler, P., & McCaskill, J. S. (2002). Hybrid poly(dimethylsiloxane)-silicon microreactors used for molecular computing. *Smart Materials and Structures*, *11*, 756–760.

von Kiedrowski, G. (1986). A self-replicating hexadeoxynucleotide. *Angewandte Chemie International Edition, 25*, 932–935.

von Kiedrowski, G. (1993). Minimal replicator theory I: Parabolic vs. exponential growth. *Bioorganic Chemistry Frontiers, 3*, 115–146.

von Kiedrowski, G., & Szathmáry, E. (2000). Selection versus coexistence of parabolic replicators spreading on surfaces. *Selection, 1* (1–3), 173–179.

Wagler, P. F., Tangen, U., Maeke, T., Chemnitz, S., Jünger, M., & McCaskill, J. S. (2004). Molecular systems on-chip (MSoC): Steps forward for programmable biosystems. In V. K. Varadan (Ed.), *Smart structures and materials 2004: Smart electronics, MEMS, BioMEMS, and nanotechnology: Proceedings of SPIE—volume 5389* (pp. 298–305).

Walker, G. T., Little, M. C., Nadeau, J. G., & Shank, D. D. (1992). Isothermal *in vitro* amplification of DNA by a restriction enzyme/DNA polymerase system. *Proceedings of the National Academy of Sciences of the United States of America, 89*, 392–396.

Wang, Y., Mukherjee, T., & Lin, Q. (2006). Systematic modeling of microfluidic concentration gradient generators. *Journal of Micromechanics and Microengineering, 16*, 2128–2137.

Wolf, B., Brischwein, M., Baumann, W., Ehret, R., Henning, T., Lehmann, M., & Schwinde, A. (1998). Microsensor-aided measurements of cellular signaling and metabolism on tumor cells: The cell monitoring system (CMS®). *Tumor Biology, 19*, 374–383.

Wright, M. C., & Joyce, G. F. (1997). Continuous in vitro evolution of catalytic function. *Science, 276*, 614–617.

Yin, J., & McCaskill, J. S. (1992). Replication of viruses in a growing plaque: A reaction-diffusion model. *Biophysical Journal, 61* (6), 1540–1549.

Zhang, C., Xu, J., Ma, W., & Zheng, W. (2006). PCR microfluidic devices for DNA amplification. *Biotechnology Advances, 24* (3), 243–284.

III COMPONENTS

The ability to create a protocell depends on chemically integrating a protocell's three functional components: its metabolism, container, and genes. The ability to achieve this integration depends on a deep understanding of the properties of the three individual components. This part focuses on those components. It treats a variety of issues associated with the fundamentals of genetic systems, container systems, and metabolic and bioenergetic questions.

Since protocell systems are so simple, many questions concerning their genetic components concern nonenzymatic replication. It is still an open question under which physicochemical conditions chemical (nonenzymatic) genetic self-replication is possible. The decades-long hunt for a self-replicating RNA molecule has met with only limited success, and it is still unclear how and under what conditions gene monomers can polymerize along a gene template to form a complementary copy. If this is possible only with preactivated monomers, as current experimental work indicates, it leaves unanswered the key question how gene monomers can be activated without the presence of sophisticated metabolic molecular machinery. Even how to identify operational conditions for the simpler template-directed ligation process is still an open question.

A fully integrated protocell can reproduce itself only if its container can grow and divide. Since most protocell containers are lipid aggregates, the central issue about the container component is to identify the conditions under which lipid aggregates autonomously divide into smaller aggregates. This is still an open question, and several ideas currently under investigation are discussed in parts II and III. Further questions include how well the container contents can be transferred to the daughter containers, and whether the reagents needed to drive the growth and division process can be continually resupplied.

Protocell component functionality typically depends on a supply of free energy and material precursors. Although several suggested protometabolic processes are under investigation, these processes are based on breaking rather than forming bonds, thus require complex molecules as resources. So a central question about protocell metabolisms is how to channel external free energy into chemical reaction networks that generate biopolymers.

Part III consists of nine chapters addressing experimental and theoretical issues including minimal gene self-replication, the feasibility of inheritable control in minimal modern cells, parasitism and evolutionary drive for containers, self-assembly and stability of lipid aggregates, the emergence of metabolism, energetics and scaling laws for cell components, and simulations of these components.

In 1986 von Kiedrowski reported parabolic growth resulting from product inhibition as the generic growth pattern in chemical replicator systems. Chapter 13 discusses the issue of parabolic growth versus exponential growth in a variety of simple experimental replicator systems. Because of their importance for Darwinian selection, scientists have been exploring ways to induce exponential growth of simple chemical replicator systems. This chapter reviews the experimental history of this quest, treating issues concerning nucleic acid replicators, chemical ligation to self-replication, competition and cooperation in nonenzymatic self-replicating systems, artificial chemical replicators, and peptide replication. The chapter concludes with an outlook toward exponential protocellular replicators.

Chapter 14 reviews molecular replicators in terms of enzyme catalyzed replication, chemical replication, minimal molecular replicators, ribozymes, and higher-order autocatalysis. It summarizes kinetics theory and the quasispecies model, and expands these into a pre-protocell setting involving coupled gene-membrane growth. The chapter concludes with a discussion of protocell genome kinetics in the context of Darwinian evolution.

Some protocell research teams are exploring the use of nonnatural nucleic acids as genetic material. Chapter 15 specifically discusses the potential of peptide nucleic acid (PNA) for this purpose. After presenting basic PNA chemistry, the chapter reviews investigations of PNA solvability and melting curves in nonpolar solvents and reports how to decorate the backbone of PNA with hydrophobic groups to make it lipophilic. Preliminary PNA replication kinetics in water and the potential catalytic function of PNA for photo-driven metabolic processes through a linking to photosensitizers are also discussed. The chapter ends by considering PNA's potential cooperation with both the container and the metabolism.

Chapter 16 reviews a top-down approach of simplifying existing (modern) small genomes to create an artificial cell. The history and the current efforts in both experimental and computational comparative genomics approaches are discussed. A minimal gene set to maintain the essential metabolic reactions to sustain life is discussed from the simulation point of view; finally, the chapter outlines possible implications for the assembly of top-down protocells.

The nature of parasitism and its relationship with cellularization are the topic of chapter 17. Parasitism is treated in both theoretical and experimental contexts and is shown to be a natural consequence of replicator systems. The influence of space on parasitism in replicator systems is discussed in detail, and it is shown how compartments, under certain circumstances, can eliminate parasites. The authors conclude that if naked replicators are assumed to be the primary component of life, replicator parasitism could be the driving force for the selection of efficient cellularization, and this could eventually lead to the evolution of metabolism and, ultimately, protocells.

Chapter 18 provides an extensive review of the fundamental principles of lipid self-assembly. Thermodynamics, formation, and driving force for the aggregation processes are effectively simulated by both dissipative particle dynamics (DPD) and coarse-grained particle simulation methods. Micellar and tubular filament formation, bilayer formation and dynamics, and vesicle formation and division are all discussed.

Chapter 19 argues that a universal metabolic cycle is to be expected from the details of simple component (geo)chemistry and thermodynamic considerations. Sources of unrelaxed free energy spontaneously induce ordered dynamical states that create channels for their relaxation. Prebiotic sources of free energy and certain properties of the prebiotic chemical species produce ordered chemical cycles that resemble metabolic cycles in modern biological systems. Metabolic cycles are conjectured to be the natural outgrowth of geochemistry, and as such are seen to be both universal and robust. The chapter discusses what all this implies for protocell design.

Chapter 20 discusses the basic energetics of the living state. A protocell has to synthesize its essential constituents, overcome unfavorable entropy to assemble and preserve its structure, replicate itself, and optimize its functions through evolution. All of these processes require an energy flux that may be termed a protometabolism. This chapter examines some universal biological scaling relationships and other energetic considerations, and the resulting constraints on modern living systems, and ends by discussing how these may apply to protocells.

In chapter 21 a rich variety of computational methods simulating protocellular components and subsystems are presented. The methods span multiple spatial scales, and include detailed *ab initio* calculations, semi-empirical quantum calculations, molecular mechanics, molecular dynamics, Brownian dynamics, DPD, lattice gas molecular dynamics, lattice Boltzmann methods, reaction kinetics methods, and Ginzburg-Landau methods. The illustrations of these methods include gene-membrane interactions, phase separation of lipids in membranes, and photo-driven metabolic systems. This chapter concludes by considering how to couple simulation methods at multiple spatial and temporal scales.

13 Self-Replication and Autocatalysis

Volker Patzke and Günter von Kiedrowski

13.1 Introduction

The theory of Darwinian evolution describes the origin of biological information. As Oparin stated in 1924, an evolving system (i.e., an information-gaining system) is generally able to metabolize, to self-replicate, and to undergo mutations. Thus, self-replication is one of the three criteria that enable us to distinguish nonliving from living systems. Since nucleic acids have the inherent ability of complementary base-pairing (and replication), they are very likely candidates for the first reproducing molecules. Kühn and others (Crick, 1968; Eigen and Schuster, 1979; Kühn and Waser, 1981; Orgel, 1968; for a review, see Joyce, 1989) have drawn the picture of an RNA world that might have existed before translation was invented. This picture was supported by the finding that RNA (and also DNA) can act as an enzymelike catalyst (Breaker and Joyce, 1994; Cech, 1986; Joyce, 1989; Sharp, 1985), or even as a "self-replicating ligase ribozyme" (Paul and Joyce, 2002).

A simple three-step model can be used to conceptualize the process of molecular template–directed replication (figure 13.1). In this model, the template molecule, T, is self-complementary and thus able to autocatalytically augment itself. In the first step, the template T reversibly binds its constituents A and B to yield a termolecular complex, M. Within this complex, the reactive ends of the precursors are held in close proximity, which facilitates the formation of a covalent bond between them.

In the following step, the termolecular complex M is irreversibly transformed into the duplex D. Reversible dissociation of D gives two template molecules, each of which can initiate a new replication cycle. The minimal representation given in figure 13.1 has served as a successful aid for the development of nonenzymatic self-replicating systems based on nucleotidic and nonnucleotidic precursors, as will be shown.

Chemical self-replicating systems have been designed in order to identify the minimal requirements for molecular replication, to translate the principle into synthetic supramolecular systems and derive a better understanding of the scope and limitation

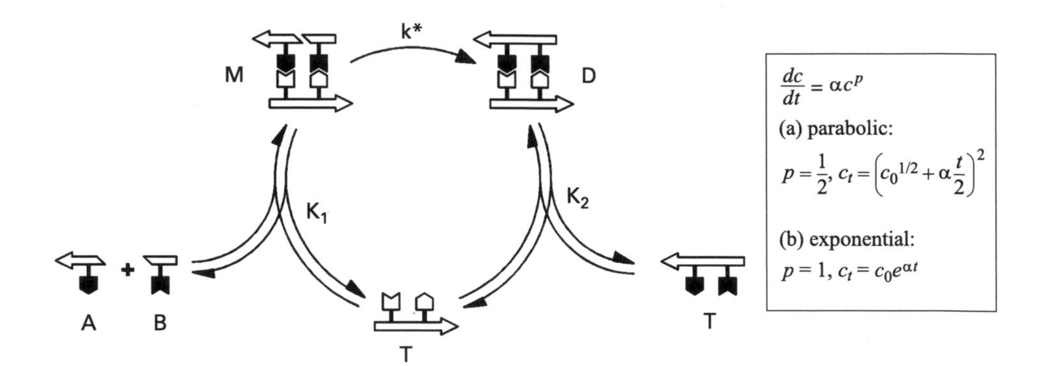

Figure 13.1
Schematic representation of a self-replicating system; rate equations for parabolic and exponential growth.

of self-organization processes believed to be relevant to the origin of life on Earth. Current implementations make use of oligonucleotide analogs, peptides, and other molecules as templates and are based on either autocatalytic, cross-catalytic, or collectively catalytic pathways for template formation. A common problem of these systems is product inhibition, leading to parabolic instead of exponential amplification. The latter is the dynamic prerequisite for selection in the Darwinian sense (Szathmáry and Gladkih, 1989).

13.2 Nucleic Acid Replicators

The earliest replicating molecular systems developed in bioorganic chemistry were nucleic acids, but the field of non-nucleotidic replicators involving molecules such as peptides has also arisen recently. This section reviews the main results concerning nucleic acid replicators, and the following section covers non-nucleotide replicators.

13.2.1 Polycondensations of Activated Mononucleotides Directed by Templates

The template-directed synthesis of oligonucleotides from activated mononucleotide precursors such as monoribonucleoside-5′-phosphorimidazolides has been studied extensively by Orgel and coworkers since 1968 (Inoue and Orgel, 1983; Orgel and Lohrmann, 1974). These studies revealed two basic principles: (1) The polycondensation yields two regioisomeric products, one with a 2′-5′- and one with the natural 3′-5′-linkage. The regioselectivity depends on various parameters (e.g., the template, nature of the leaving group, presence of metal ions, etc.). (2) The polycondensations were found to be rather efficient with a pyrimidine-rich template. In contrast, a purine-rich oligo- or polynucleotide acts as a poor template.

The findings concerning the templates were envisioned as a major obstacle to further attempts to realize self-replicating systems based on mononucleotide precursors. Despite these difficulties, the Orgel group succeeded in demonstrating the nonenzymatic template-directed synthesis of fully complementary products. Templates as long as 14-mers were successfully transcribed (Orgel, 1992). However, no complete replication cycle could be achieved when mononucleotides were used as precursors.

13.2.2 From Chemical Ligation to Self-Replication

The work of Orgel and coworkers clearly demonstrated that transcription with information transfer can occur in the absence of enzymes. As a result of the problems previously described, it seemed worthwhile to use oligonucleotides as building material and chemical ligations instead of polycondensations as coupling reactions. The first example of a chemical ligation, namely, a template-directed condensation of activated oligonucleotides, was reported by Naylor and Gilham (1966). They demonstrated that the condensation of pentathymidylic acid and hexathymidylic acid molecules could be catalyzed by a poly(A) template in the presence of the water-soluble carbodiimide EDC (l-ethyl-3-(3-dimethylaminopropyl)-carbodiimide). Further examples of chemical ligation reactions have been reported by Shabarova's group in Moscow (Dolinnaya et al., 1988, 1991). A successful demonstration of enzyme-free nucleic acid replication based on an autocatalytic chemical system was reported in 1986 (von Kiedrowski, 1986). A 5′-terminally protected trideoxynucleotide 3′-phosphate d(Me-CCG-p) (*1*) and a complementary 3′-protected trideoxynucleotide d(CGG-p′) (*2*) were reacted in the presence of EDC to yield the self-complementary hexadeoxynucleotide (Me-CCG-CGG-p′) (*3*) with natural phosphodiester linkage as well as the 3′-3′-linked pyrophosphate of *1*. The sequences chosen were such that the product (*3*) could act as a template for its own production. The hexamer formation proceeds through a termolecular complex *C* formed from *1*, *2*, and *3*, in which reactive ends are in close spatial proximity and thus ready to be ligated. During the course of the reaction, the activated 3′-phosphate of *1* is attacked by the adjacent 5′-hydroxyl group of *2*, forming a 3′-5′-internucleotide bond between the trimers (see figure 13.2).

The resulting template duplex can then dissociate to yield two free template molecules, each of which can initiate a new replication cycle. Two parallel pathways for the formation of hexameric templates exist, as kinetic studies using high-pressure liquid chromatography (HPLC) show. Both pathways—the template-dependent, autocatalytic and the template-independent, nonautocatalytic—contribute to product formation. The latter pathway has been found to be predominant. Moreover, the experiments revealed that the addition of template did not increase the rate of autocatalytic template formation in a linear sense. Instead, the initial rate of autocatalytic

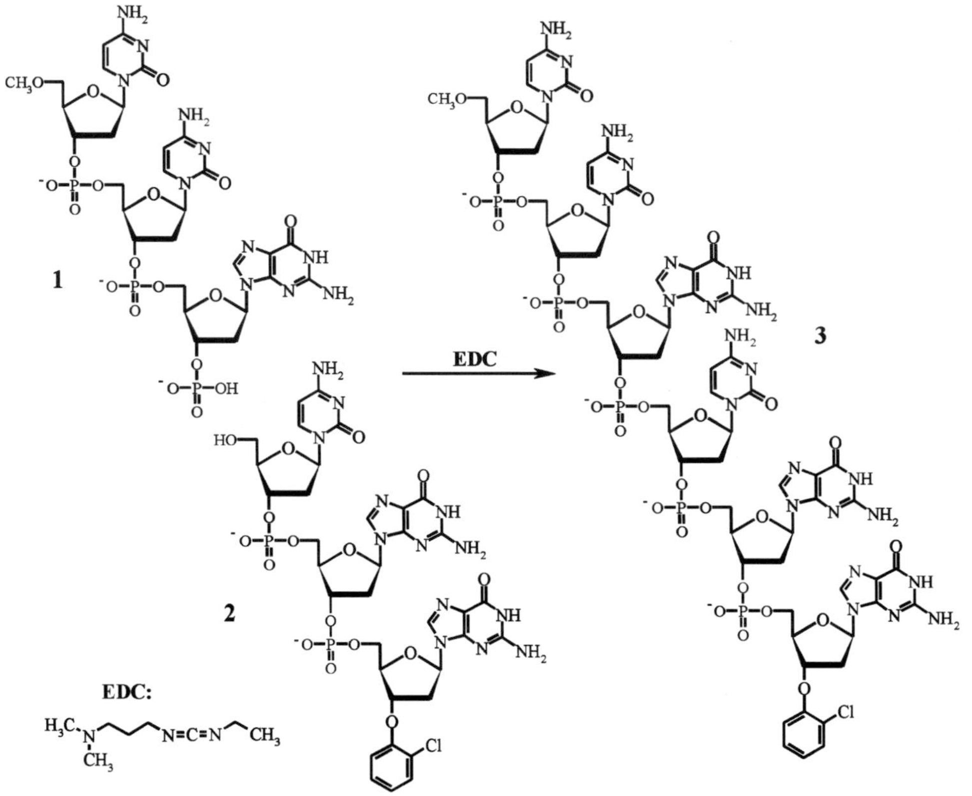

Figure 13.2
The first nonenzymatic self-replicating system.

synthesis was found to be proportional to the square root of the template concentration (a finding that was termed the square root law of autocatalysis). Thus, the reaction order in this autocatalytic self-replicating system was found to be 1/2 rather than 1, in contrast with most autocatalytic reactions known so far. According to theory, a square root law is expected in the previously described cases, in which most of the template molecules remain in their double-helical (duplex) form, leaving them in an "inactive" state. In other words, a square root reflects the influence of both autocatalysis and product inhibition.

Another example of an autocatalytic system following the square root law was published by Zielinsky and Orgel (1987). The diribonucleotide analogs $(G_{NHp}C_{NH2})$ (*4*) and $({}_{p}G_{NHp}C_{N3})$ (*5*) were ligated in the presence of water-soluble carbodiimide (EDC) and self-complementary tetraribonucleotide triphosphoramidate $(G_{NHp}C_{NHp}G_{NHp}C_{N3})$ (*6*), serving as a template (see figure 13.3).

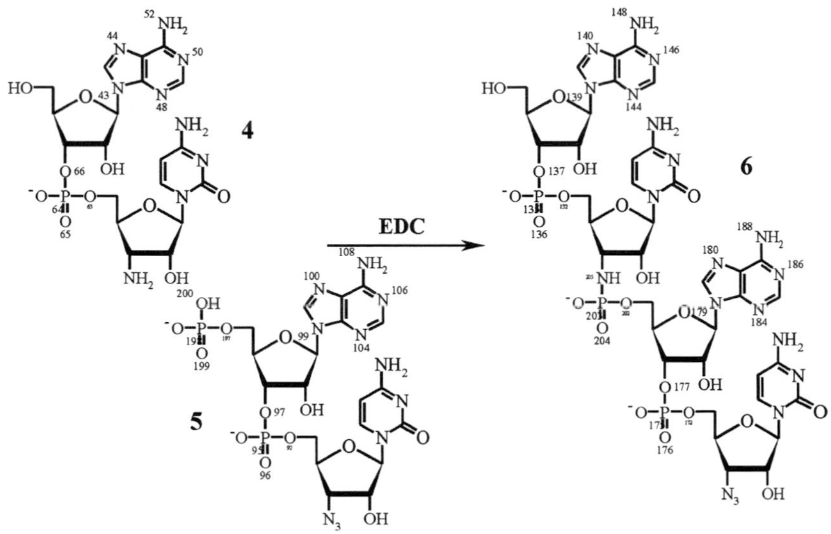

Figure 13.3
The self-replicating tetraribonucleotide, according to Zielinski and Orgel (1987).

This was the first demonstration of self-replication of nucleic acid–like oligomers bearing an artificial backbone structure. In kinetic studies, the autocatalytic nature of template synthesis was obvious from the square root dependence of the initial reaction rate on the template concentration. In theory, every true autocatalytic system should show a sigmoidal concentration-time profile (von Kiedrowski, 1993). Because of the predominance of the nonautocatalytic pathway, this was not true for either system just described. The autocatalytic nature of these systems was ascertained indirectly by observing the increase in the initial reaction rate when the reaction mixtures were seeded with increasing amounts of template.

In the years of research following 1987, a major goal was to enhance template-instructed autocatalytic synthesis while keeping the reaction rate of the noninstructed synthesis as low as possible. It can be shown that autocatalytic synthesis particularly benefits from an increased nucleophilicity of the attacking 5′-group. When trimer *2* was used in its 5′-phosphorylated form instead of the 5′-hydroxyl form, the carbodiimide-dependent condensation with *1* yielded hexamers bearing a central 3′-5′-pyrophosphate linkage. As a result of this modification, the rate of the template-induced hexamer formation was increased by roughly 2 orders of magnitude (von Kiedrowski, Wlotzka, and Helbing, 1989). Replacing the 5′-phosphate by a 5′-amino group led to the formation of a 3′-5′-phosphoramidate bond and resulted in a rate enhancement of almost 4 orders of magnitude compared to the phosphodiester system.

In addition, the autocatalytic synthesis of template molecules was found to be more selective in the case of the faster replicators. The quantity e (a measure of the ratio of autocatalytic over background synthesis) could be increased from 16 $M^{-1/2}$ in the phosphodiester system to 430 $M^{-1/2}$ in the 3′-5′-phosphoramidate system. The first direct evidence for a sigmoidal increase in template concentration was found in the latter system (von Kiedrowski, 1990; von Kiedrowski et al., 1991). Shortly after these observations, a second chemical self-replicating system was reported to exhibit sigmoidal growth (Rotello, Hong, and Rebek, 1991). A sigmoidal shape for template formation gives direct evidence of autocatalytic growth, since this type of growth is a direct consequence of the square root law of autocatalysis. Following the square root law, the increase of template concentration at early reaction times is parabolic rather than exponential. For these early points in time, the integrated form of $dc/dt = \alpha c^{1/2}$ (α is an empirical constant) can be approximated using a second-order polynomial of time whose graph shows a parabola. Parabolic growth is a direct consequence of the square root law in cases where nonautocatalytic synthesis is negligible. A detailed analytical treatment revealed three types of autocatalytic growth as borderline cases (von Kiedrowski, 1993). From theoretical considerations, it was also concluded that the autocatalytic growth order depends solely on the thermodynamics of the coupled equilibria, rather than on the energy of the transition state.

Other studies were devoted to the subject of sequence selectivity in the self-replication of hexadeoxynucleotides, giving further evidence for the autocatalytic nature of this reaction (Wlotzka, 1992). A homologous set of trimer-3′-phosphates bearing the general sequence ^{PG}XYZp (PG = protective group) was synthesized, where X, Y, and Z could represent either C or G monomers. Each of the trimer 3′-phosphates was reacted with the aminotrimer ^{H2N}CGGp^{PG} in the presence of EDC. The trimer 3′-phosphate bearing the sequence CCG was formed significantly faster than all the other trimers. Moreover, addition of template CCGCGG, complementary to the ligated trimers, stimulated only the synthesis of the proper phosphoramidate though it had a negligible influence on the variant sequences. These studies demonstrated that autocatalysis can occur only if the sequences of both trimers match the sequence of the resulting hexamer according to the Watson-Crick base-pairing rules. They also showed that the condensation reactions are controlled predominantly by the stacking of nucleobases flanking the newly formed internucleotide linkage. Hexamers bearing a central G-C subsequence, for example, are formed 1 order of magnitude faster than hexamers with a central C-G subsequence. In general, the following reactivity order was found: G-G > G-C > C-G > C-C, a finding that is in good correspondence with experiments with nonautocatalytic chemical ligations.

Further studies revealed a remarkable temperature dependence in hexadeoxynucleotide self-replication. Each parabolic replicator shows a rate optimum at a certain temperature, which was found to be close to the measurable melting temperature of the respective hexamer duplex (Wotzka, 1992). Again, this is in good correspondence with minimal replicator theory. Generally, the autocatalytic reaction rate is given by k[C], where k equals the rate constant for irreversible internucleotide bond formation and [C] is the equilibrium concentration of the termolecular complex. Because the rate constant k increases with temperature (according to Arrhenius's law) and [C] decreases as a result of the melting of the termolecular complex, the reaction rate as a function of temperature is expected to pass a maximum at the temperature t_{opt}; t_{opt} itself depends on the concentrations of both the template and its precursors as well as the thermodynamic stabilities of the termolecular complex and the template duplex.

13.2.3 Competition and Cooperation in Nonenzymatic Self-Replicating Systems

Following studies in our lab were devoted to the question of information transfer in a more complex system, in which a number of alternative templates can be produced from a set of common precursors. Such a system was realized in an experiment in which the sequence CCGCGG was synthesized from three fragments (Achilles and von Kiedrowski, 1993). The trimer-3′-phosphate ^{PG}CCGp (A), the 5′-aminodimer-3′-phosphate ^{H2N}CGp (B), and the 5′-aminomonomer ^{H2N}G (C) were allowed to react in the presence of 1-methylimidazole (MeIm) and EDC. The five products AB, AC, BC, BB, and ABC, all bearing central 3′-5′-phosphoramidate linkages, could be identified. To monitor the whole system using HPLC kinetic analysis, it was necessary to reduce its complexity. In order to detect possible catalytic, cross-catalytic, or autocatalytic pathways (couplings) induced by the different products, the whole reaction system was divided into less complex subsystems. For example, to analyze the formation of the pentamer AB separately, we employed a 5′-aminodimer B', which was protected at its 3′-phosphate.

In a series of experiments, each subsystem was studied with respect to the effect of each reaction product. Standard oligodeoxynucleotides were employed as model templates. These experiments allowed us to decipher the dynamic structure of the whole system, which can be understood as a catalytic network, namely, an autocatalytic set with a total of six feedback couplings (see figure 13.4).

However, only those couplings with sufficient efficiency—denoted (+)(+) and (+)(+)(+)—exert a notable influence. These strong couplings affect only the synthesis of the hexamer ABC and its pentameric precursor AB. Both products are coupled autocatalysts: They behave as autocatalytic "egoists" and, at the same time, as mutually catalytic altruists (see chapter 17). On the other hand, the tetramer AC, which is formed as the main product through the nonautocatalytic channel (G-G-stack leads to fast condensation), can be described as an isolated autocatalyst. The molecules AB and ABC compete with the main product AC for the incorporation of their common precursors, A and C. This competition was indeed observed: On the seeding

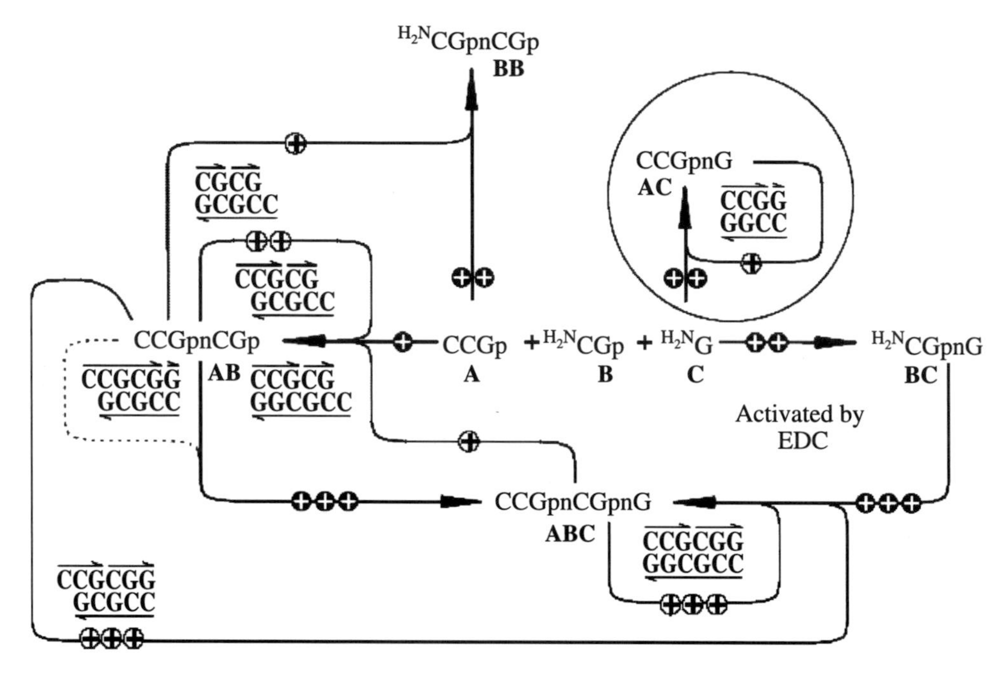

Figure 13.4
A self-replicating system from three starting materials.

of the reaction mixture with the hexameric template CCGCGG (*ABC*), the autocatalysts *AB* and *ABC* were formed more quickly, whereas the formation of the egoistic tetramer, *AC*, decreased. Competition of replicators for common resources is the prerequisite for selection. "Selection," in the biological sense, usually means the takeover of resources by a species that reproduces more efficiently than its competitors (survival of the fittest). However, because selection also depends on the population level of a species (its concentration), a less efficient species may win if it starts at a higher population level. In any case, the population size of a species directs the flow of resource consumption. Our experimental findings resemble selection, but only a rudimentary form; "true" Darwinian selection necessitates exponential, and not parabolic, growth.

All synthetic replicators described so far are based on the simplification of an autocatalytic, self-complementary system. However, the natural prototype of nucleic acid replication uses complementary, rather than self-complementary, strands (that is, replication via (+)- and (−)-strands). The underlying principle of this type of replication is a cross-catalytic reaction in which one strand acts as a catalyst for the formation of the other strand and vice versa. It seemed worthwhile to test whether or not a replication of complementary hexadeoxynucleotides in the absence of any

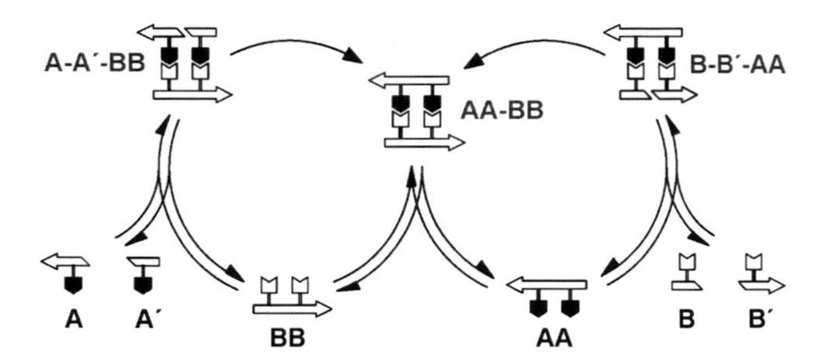

Figure 13.5
Minimal representation of a cross-catalytic self-replicating system.

enzyme could be achieved. A minimal implementation of such a cross-catalytic self-replicating system can be represented by a simple reaction scheme (see figure 13.5), where *AA* and *BB* denote templates, and *A*, *A*′, *B*, and *B*′ denote fragment molecules complementary to the templates.

In this scheme, two parallel pathways exist, both leading to the same template duplex. Since *AA* catalyzes the formation of *BB* and vice versa, one can speak of cross-catalysis. As long as both templates are formed with the same type of chemistry, the same conditions that enable cross-catalytic formation of complementary molecules *AA* and *BB* will also allow for the autocatalytic self-replication of both self-complementary products *AB* and *BA*. For best observation of the cross-catalytically coupled reactions, all four possible condensations between the precursor molecules must be equally efficient. Earlier experiments in our lab have shown that the efficiency of template-directed condensation reactions between suitably protected trideoxynucleotides was predominantly determined by the stacking of nucleobases flanking the reaction site within the termolecular complex (Wlotzka, 1992). Accordingly, the cross-catalytic system could be realized experimentally by Sievers and von Kiedrowski (Sievers et al., 1994; Sievers and von Kiedrowski, 1994) using the trimer precursors *A* (for CCG) and *B* (for CGG) to form a central GC subsequence in all four possible hexamers (*AB*, *BB*, *AB*, *BA*). The formation of 3′-5′-phosphoramidate linkages was used to join the trimers. Kinetic analysis revealed that the two complementary sequences *AA* and *BB* are formed with similar efficiency, despite the difference in their pyrimidine content. When the experiment was performed in such a way that all products could form simultaneously, both the complementary and the self-complementary hexadeoxynucleotides exhibited the same time course of formation. In other words, cross-catalysis was as efficient as autocatalysis. On the other hand, in single experiments in which only one hexamer could form, the autocatalysts were formed much faster than the complementary oligonucleotides. This is the expected result, since by definition only the autocatalyst is able to accelerate its own synthesis.

Cross-catalysis, again by definition, necessitates the simultaneous formation of both complementary products and thus cannot be observed in single experiments in which only one product is formed. As expected, the cross-catalytic self-replication of complementary hexamers revealed parabolic growth characteristics due to product inhibition.

The general scheme of a minimal self-replicating system (see figure 13.1) has served as a successful aid for the design of various new replicators. In these systems, the rate of autocatalytic synthesis typically depends on the square root of the template concentration, and thus the template growth is parabolic rather than exponential. Szathmáry and Gladkih (1989) showed theoretically that parabolic growth leads to the coexistence of self-replicating templates, which compete for common resources under stationary conditions. Coexistence in this context means that the faster replicator is not able to take over the common resources completely. If two non–self-replicating molecules, denoted C_1 and C_2, are competing for common precursors, the ratio of their concentrations is determined solely by their reactivity: $[C_1]/[C_2] = k_1/k_2$. The quotient $[C_1]/[C_2]$ describes selectivity, which is a metric for coexistence. For two parabolic replicators with the autocatalytic rate constants k_1 and k_2, it follows from Szathmáry's treatment that $[C_1]/[C_2] = (k_1/k_2)^2$. Hence, small differences in reactivity lead to a higher selectivity in parabolic replicators compared to non–self-replicating molecules. This enhancement of selectivity is partly implicit in the results reported.

Szathmáry pointed out that Darwinian selection (survival of the fittest) necessitates exponential growth of competing replicators. For equilibrated self-replicating systems (in which the template-directed condensation is slow compared to internal equilibration), the autocatalytic reaction order (which is 1 in the case of exponential growth and 1/2 in the case of parabolic growth) is determined solely by the population of the complexes involved. A minimal self-replicating system, as presented in figure 13.1, is expected to exhibit exponential growth if the termolecular complex is thermodynamically more stable than the template duplex (for entropic reasons, the usual situation is just opposite to that).

Since exponential growth has been described as the prerequisite for selection in the Darwinian sense, experimental approaches aimed at the realization of molecular evolution through different strategies. One strategy is the introduction of a catalytically active covalently bound leaving group, a minimal replicase, which increases the thermodynamic stability of the termolecular complex. As long as the replicase is part of the termolecular complex, it stabilizes the latter, for example, by wrapping itself around the complex. Stabilization occurs in an intramolecular sense.

Using the technique of directed molecular evolution, Burmeister (Burmeister, von Kiedrowski, and Ellington, 1997) and Azzawi (2001) screened pools of random oligodeoxynucleotides to isolate molecules that are able to catalyze the EDC-driven self-replication of a self-complementary hexa- or decadeoxynucleotide. The goal was to find an oligodeoxynucleotide that would act as a nucleophilic catalyst of the phosphoryl transfer step occurring in the termolecular complex (covalent catalysis). The selection started with a synthetic pool of random oligodeoxynucleotides bearing 5′- and 3′-constant regions as primer-binding sites for PCR amplification. A modified primer containing a 5′-terminal biotin and a 3′-5′-phosphoramidate linkage following the leader sequence CCG (primer-trimer-conjugates) was synthesized. PCR amplification of the random pool using the modified 5′-primer, as well as a conventional 3′-primer, gave a pool of double-stranded DNA. Immobilization of the double-stranded pool on a streptavidin column and subsequent denaturation results in a single-stranded DNA pool containing one 3′-5′-phosphoramidate linkage. The in vitro selection was performed on the column in the presence of the attacking trimer (CGG) and the hexameric template (CCGCGG). The 5′-amino oligomers that was released during selection was eluted from the column. PCR amplification of the column eluate yielded the pool for the next round of selection. Successive rounds of kinetic selection and PCR amplification were expected to result in selective enrichment of active sequences. From a DNA-pool containing 72 randomized positions embedded in two constant sequences, two dominating sequences were isolated. Unfortunately, an examination of these sequences revealed that the oligonucleotides did not catalyze the desired ligation reaction but the hydrolysis of the internal phosphoramidate bond in the presence of the trinucleotide (CGG) as a cofactor.

An exponential nonenzymatic amplification of oligonucleotides was realized in our laboratory on the surface of a solid support in a stepwise replication procedure called SPREAD (Surface-Promoted Replication and Exponential Amplification of DNA analogs). In the procedure described by Luther, Brandsch, and von Kiedrowski (1998) two template molecules were separately immobilized on solid support and hybridized with complementary oligonucleotide building blocks, which were ligated to the corresponding template molecules by EDC (figure 13.6).

The following separation and immobilization of the product molecules yields further material for another replication cycle. The oligonucleotides connected to the solid-phase through a disulfide-bond were removed by reduction and analyzed by RP-HPLC. In this approach, the duplex formation of the template was prevented by immobilization, and therefore no product inhibition could occur. This iterative procedure is the first nonenzymatic exponential replication in a nonautonomous system.

All experiments on self-replication of oligonucleotides have been analyzed by RP-HPLC so far. Since this method is very laborious and provides only limited data, it was necessary to develop rapid techniques for monitoring the chemical kinetics in a parallel fashion. From our group, Schöneborn, Bülle, and von Kiedrowski (2001)

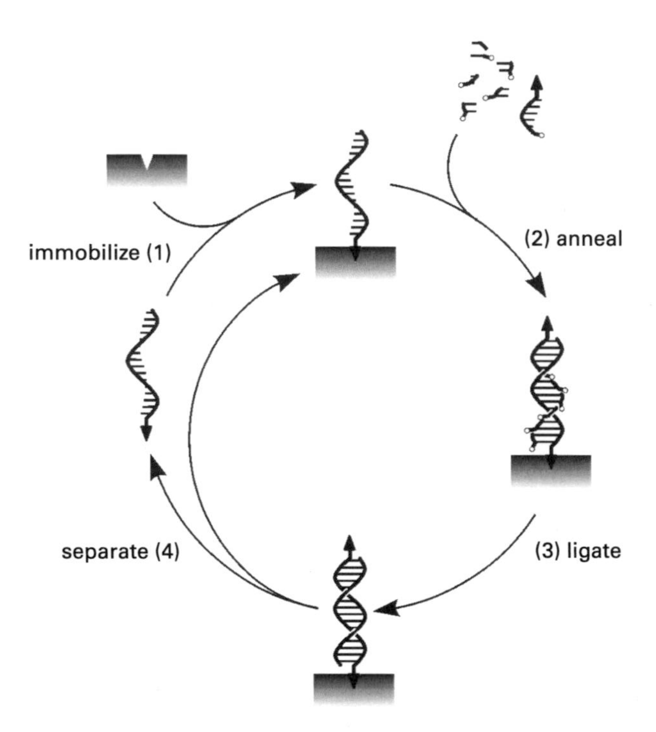

Figure 13.6
General scheme of the SPREAD procedure.

reported on the online monitoring of chemical replication of self-complementary oligonucleotides by means of FRET (fluorescence resonance energy transfer). The chosen oligonucleotide sequences were based on the hexameric replication system described before, which was extended to octameric (TCCGCGGA), decameric (TTCCGCGGAA), and dodecameric (TTTCCGCGGAAA) systems by elongating with dT and dA. The two fluorescent dyes, donor Cy3 and acceptor Cy5, were introduced at the 5′-end to give a labeled educt *A* and a labeled template molecule *C* (see figure 13.7). Since the population of the termolecular complex is small, the product formation in the template duplex C_2 gives the desired FRET signal.

The chemical ligation in this system is achieved through EDC activation to yield a phosphoramidate product, as described earlier. The FRET-based method allows fast examination for a number of self-replicating systems. Further, because the experiments can also be carried out on microtiter plates, it is a suitable technique for the rapid screening of factors that exhibit an influence on the dynamics of autocatalytic growth, like a "minimal replicase," for instance.

Further developments of oligonucleotide self-replicating systems aim at the development of new ligation methods to avoid hydrolysable condensing agents such as

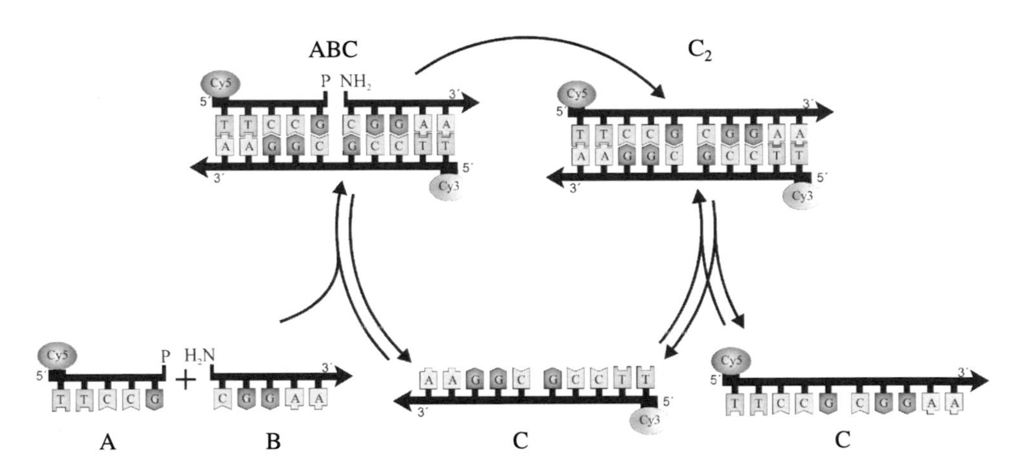

Figure 13.7
Schema of a self-replicating system analyzed by FRET.

EDC. Especially at high temperatures, the formation of side products increases in the EDC-driven ligation.

A number of alternative ligation reactions proceeding without activating reagents were described in the literature. Examples include the reaction of 3′-phosphorthioate or 3′-phosphorselenoate compounds with 5′-iodine–substituted compounds, the formation of monophosphoryl disulfides from 3′-phosphorthioates with 5′-mercaptooligonucleotides or the reaction of 3′- or 5′-hydrazides with 3′- or 5′-aldehydes.

In our group, the formation of a 3′-5′-disulfide bond through disulfide-exchange reactions was recently examined in an oligonucleotide self-replicating system. The reaction was monitored by ultraviolet (UV) spectroscopy as characteristically UV-active leaving groups like Ellman's ion were used for thiol activation (Patzke, 2005). Autocatalysis was proven for this system, but because of the very fast exchange reaction, the noncatalytic reaction was predominant under the chosen conditions. Nevertheless, the disulfide ligation has some advantages over phosphoramidate ligation, because no side products are formed and the reaction rate can be influenced by pH value. Also, a variation or modification of the leaving group can be used to affect the reaction rate and form reactive and stable conjugates for the concept of "minimal replicase."

13.3 Replication in Non-Nucleotidic Model Systems

A related field in bioorganic chemistry is the development of non-nucleotidic replicators. Since molecular recognition and catalysis are common features among various

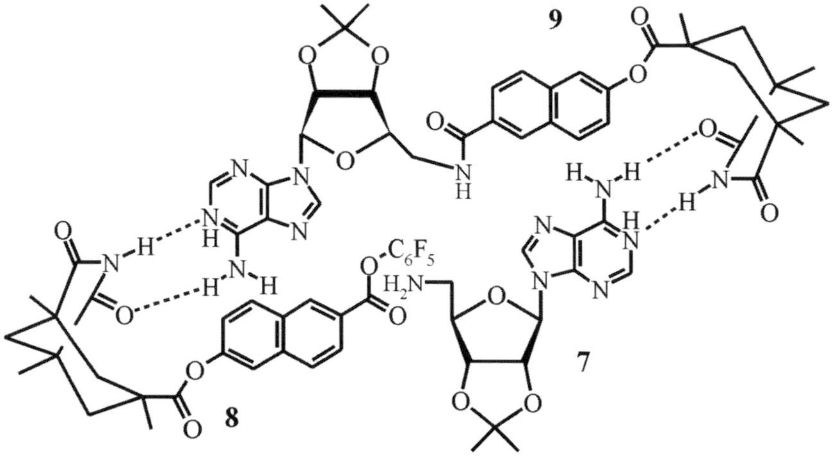

Figure 13.8
The synthetic self-replicating system according to Rebek (Tjivikua, Baluster, and Rebek, 1990).

organic molecules, it seemed feasible to develop new replicators based on truly artificial precursors. Rebek and coworkers designed a replicator consisting of an adenosine derivative as the natural component and a derivative of Kemp's acid as the artificial part (Nowick et al., 1991; Tjivikua, Baluster, and Rebek, 1990). As shown in figure 13.8, the reactive ends of the reactants *7* and *8* come into spatial proximity when the reactants interact with the template, *9*. Nucleophilic attack of the primary amine of *7* on the activated carboxyl ester of *8* leads to amide bond formation, giving a new template molecule of *9*. Dissociation of the self-complementary template duplex closes the replication cycle.

Rebek's replicator challenged us to think about an even simpler self-replicating system. Terfort and von Kiedrowski used the condensation of 3-aminobenzamidine (*10*) and (2-formylphenoxy)-acetic acid (*11*) to develop an artificial self-replicating system based on simple organic molecules (Terfort and von Kiedrowski, 1992). As illustrated in figure 13.9, autocatalytic condensation of *10* and *11* giving the anil *12* was followed by ^{1}H NMR spectroscopy in dimethylsulfoxide. As expected, the autocatalytic contribution of the condensation reaction shows a square root law.

Sutherland presented another artificial self-replicating system in 1997. Analysis revealed that the autocatalytic reaction order in this system lies between parabolic and exponential growth (reaction order of 0.8). The chemical ligation in this system is based on Diels-Alder reaction between a cyclohexadien-derivative and an *N*-substituted maleimid as dienophile. A further increase of the autocatalytic reaction order was achieved by Kindermann, Stahl, and colleagues (2005) in our group (reaction order of 0.89) with another Diels-Alder replicator. In these systems the product

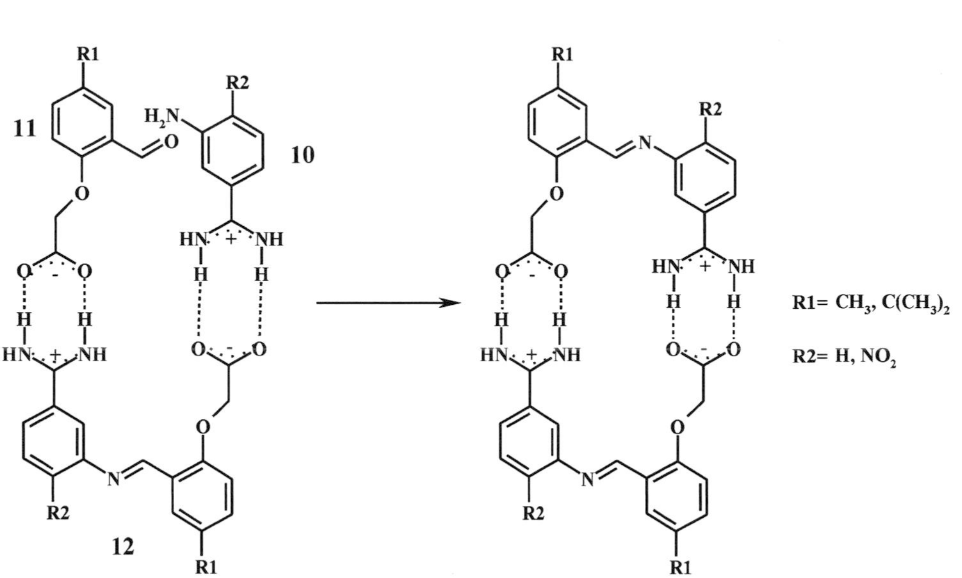

Figure 13.9
Self-replication of amidinium-carboxylate templates.

inhibition seems to be reduced because the association of the template molecules is sterically hindered, whereas the close spatial proximity in the termolecular complex is suitable for ligation. By this stabilization of the termolecular complex compared to the template duplex, the observed reaction order follows the theoretical requirements for exponential growth.

In 1996, Lee and coworkers demonstrated that nonenzymatic self-replication can also be realized in a system based on peptides. The described system is based on a 32 amino acid peptide with an á-helical coiled-coil structure of repetitive 7mer units $(abcdefg)_n$. The reacting fragments were a 17mer as electrophile and a 15mer as the nucleophile, which form an amide bond through Kent ligation. The molecular recognition is based on hydrophobic (positions a and d) and electrostatic (positions e and g) interactions in the coiled-coil (see figure 13.10). The kinetic data from the system was gained by RP-HPLC analysis, and examination of the system revealed parabolic growth as expected. Interestingly, at higher starting concentrations of peptides an increase in autocatalytic growth was observed (reaction order of 0.63). As the reaction order increases with the number of templates involved in the complex, these results can be explained by the formation of a quarternary complex *ABTT*, which leads to a termolecular product complex *TTT*.

A further development of the self-replicating peptides toward exponential growth was also achieved by a destabilization of the template duplex. Chmielewski and coworkers (Issac and Chmielewski, 2002; Li and Chmielewski, 2003) reported on two

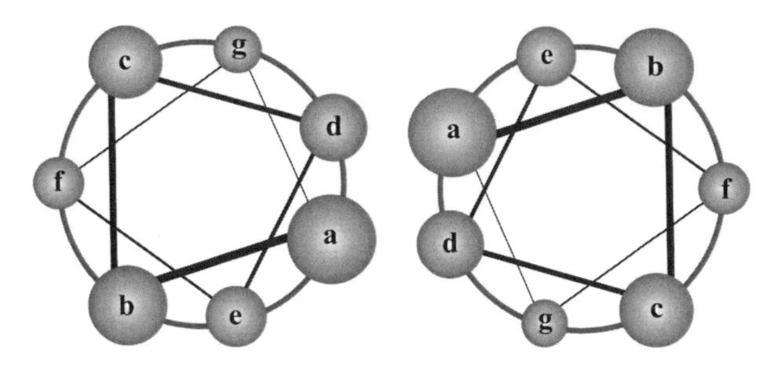

Figure 13.10
Schematic view of a self-replicating peptide system.

replicators with a very high reaction order of 0.91. In one case, this was achieved by shortening the coiled coil to its minimum length of 26 amino acids (2001). In the other case, the insertion of one proline amino acid in the building blocks of the replicating peptides leads to a nick, which destabilizes the template duplex (2003).

13.4 Outlook Toward Exponential Replication

Theoretical considerations about exponential growth (discussed in section 13.2.3) have to some extent been proven in synthetic replicators, where high reaction orders were observed for adequately designed systems with destabilized product duplexes. To overcome product inhibition in minimal self-replicating systems of oligonucleotides, the destabilization of template duplexes might be practicable by changing the ligation chemistry or performing the template replication at surfaces. Using the new techniques for analyzing and ligating oligonucleotides, the search for a "minimal replicase" is still promising for overcoming product inhibition. This concept can even be extended to an artificial cell, where the leaving group is a stabilizing lipid that can transport the oligonucleotide through cell membranes and form new membranes after cleavage.

Acknowledgment

This chapter was previously published in *Arkivoc*, volume 5, 2007.

References

Achilles, T., & von Kiedrowski, G. (1993). A self-replicating system from three starting materials. *Angewandte Chemie International Edition*, *32*, 1198–1201.

Azzawi, A. (2001). *Ligationsreaktionen mit Phosphoramidaten in Modellreaktionen, SELEX-Experimenten und bei Oligonucleotid-Peptid-Konjugaten: Beiträge zum Konzept einer minimalen Replikase.* Dissertation, Ruhr-Universität–Bochum.

Breaker, R., & Joyce, G. F. (1994). A DNA enzyme that cleaves RNA. *Chemistry and Biology*, *1*, 223–229.

Burmeister, J., von Kiedrowski, G., & Ellington, A. (1997). Cofactor-assisted self-cleavage in DNA libraries with a 3′-5′-phosphoramidate bond. *Angewandte Chemie International Edition*, *36*, 1321–1323.

Cech, T. R. (1986). A model for the RNA-catalyzed replication of RNA. *Proceedings of the National Academy of Sciences of the United States of America*, *83*, 4360–4363.

Crick, F. H. C. (1968). The origin of the genetic code. *Journal of Molecular Biology*, *38*, 367–379.

Dolinnaya, N. G., Sokolova, N. I., Gryaznova, O. L., & Shabarova, Z. A. (1988). Site-directed modification of DNA duplexes by chemical ligation. *Nucleic Acids Research*, *16*, 3721–3738.

Dolinnaya, N. G., Tsytovich, A. V., Sergeev, V. N., Oretskaya, T. S., & Shabarova, Z. A. (1991). Structural and kinetic aspects of chemical reactions in DNA duplexes: Information of DNA local structure obtained from chemical ligation data. *Nucleic Acids Research*, *19*, 3073–3080.

Eigen, M., & Schuster, P. (1979). *The hypercycle: A principle of natural self-organization*. Berlin: Springer.

Inoue, T., & Orgel, L. E. (1983). A nonenzymatic RNA polymerase model. *Science*, *219*, 859–862.

Joyce, G. F. (1989). RNA evolution and the origins of life. *Nature*, *338*, 217–224.

Issac, R., & Chmielewski, J. (2002). Approaching exponential growth with a self-replicating peptide. *Journal of the American Chemical Society*, *124*, 6808–6809.

Kindermann, M., Stahl, I., Reimold, M., Pankau, W. M., & von Kiedrowski, G. (2005). Systems chemistry: Kinetic and computational analysis of a nearly exponential organic replicator. *Angewandte Chemie International Edition*, *44* (41), 6750–6755.

Kühn, H., & Waser, J. (1981). Molekulare Selbstorganisation und Ursprung des Lebens. *Angewandte Chemie*, *93*, 495–515.

Lee, D. H., Granja, J. R., Martinez, J. A., Severin, K., & Ghadiri, M. R. (1996). A self-replicating peptide. *Nature*, *382*, 525–528.

Li, X., & Chmielewski, J. (2003). Peptide self-replication enhanced by a proline kink. *Journal of the American Chemical Society*, *125*, 11820–11821.

Naylor, R., & Gilham, P. T. (1966). Studies on some interactions and reactions of oligonucleotides in aqueous solution. *Biochemistry*, *5*, 2722–2728.

Nowick, J. S., Feng, Q., Tjivikua, T., Ballester, P., & Rebek, J. (1991). Kinetic studies and modeling of a self-replicating system. *Journal of the American Chemical Society*, *113*, 8831–8832.

Oparin, A. I. (1924). *The Origin of Life*. Moscow: Pabochii.

Orgel, L. E. (1968). Evolution of the genetic apparatus. *Journal of Molecular Biology*, *38*, 381–393.

Orgel, L. E. (1992). Molecular replication. *Nature*, *358*, 203–209.

Orgel, L. E. (1995). Unnatural selection in chemical systems. *Accounts of Chemical Research*, *28*, 109–119.

Orgel, L. E., & Lohrmann, R. (1974). Prebiotic chemistry and nucleic acid replication. *Accounts of Chemical Research*, *7*, 368–377.

Patzke, V. (2005). *Selbstreplizierende 3′-5′-Thiol-Oligodesoxynucleotidderivate: Zum Einfluss peptidischer Abgangssgruppen auf die Oligonucleotidreplikation*. Dissertation, Ruhr-Universität–Bochum.

Paul, N., & Joyce, F. G. (2002). A self-replicating ligase ribozyme. *Proceedings of the National Academy of Sciences of the United States of America*, *99*, 12733.

Rotello, V., Hong, J. L., & Rebek, J. (1991). Sigmoidal growth in a self-replicating system. *Journal of the American Chemical Society*, *113*, 9422–9423.

Schöneborn, H., Bülle, J., & von Kiedrowski, G. (2001). Kinetic monitoring of self-replicating systems through measurement of fluorescence resonance energy transfer. *ChemBioChem*, *12*, 922–927.

Sharp, P. A. (1985). On the origin of RNA splicing and introns. *Cell*, *42*, 397–400.

Sievers, D., Achilles, T., Burmeister, J., Jordan, S., Terfort, A., & von Kiedrowski, G. (1994). Molecular replication: From minimal to complex systems. In G. R. Fleischaker, S. Colonna, & P. L. Luisi (Eds.), *Self-Production of Supramolecular Structures* (pp. 45–64). Dordrecht: Kluwer Publishers.

Sievers, D., & von Kiedrowski, G. (1994). Self-replication of complementary nucleotide-based oligomers. *Nature*, *369*, 221–224.

Szathmáry, E., & Gladkih, I. (1989). Subexponential growth and coexistence of nonenzymatically replicating templates. *Journal of Theoretical Biology*, *138*, 55–58.

Terfort, A., & von Kiedrowski, G. (1992). Self-replication by condensation of 3-amino-benzamidines and 2-formylphenoxyacetic acids. *Angewandte Chemie International Edition*, *31*, 654–656.

Tjivikua, T., Baluster, R., & Rebek, J. (1990). A self-replicating system. *Journal of the American Chemical Society*, *112*, 1249–1250.

von Kiedrowski, G. (1986). A self-replicating hexadeoxynucleotide. *Angewandte Chemie International Edition*, *25*, 932–935.

von Kiedrowski, G. (1990). Selbstreplikation in chemischen Minimalsystemen. *40 Jahre Fonds der Chemischen Industrie 1950–1990*, 197–218.

von Kiedrowski, G. (1993). Minimal replicator theory I: Parabolic versus exponential growth. *Bioorganic Chemistry Frontiers*, *3*, 113–146.

von Kiedrowski, G., Wlotzka, B., & Helbing, J. (1989). Sequence dependence of template-directed syntheses of hexadeoxynucleotide derivatives with 3′-5′-pyrophosphate linkage. *Angewandte Chemie International Edition*, *28*, 1235–1237.

von Kiedrowski, G., Wlotzka, B., Helbing, J., Matzen, M., & Jordan, S. (1991). Parabolic growth of a self-replicating hexadeoxynucleotide bearing a 3′-5′-phosphoamidate linkage. *Angewandte Chemie International Edition*, *30*, 423–426, and corrigendum 892.

Wlotzka, B. (1992). *Selbstreplizierende Oligonucleotide. Untersuchungen zur Sequenz- und Temperaturabhängigkeit bei der Synthese hexamerer 3′-5′-Phosphoamidate*. Dissertation thesis, University of Göttingen.

Zielinski, W. S., & Orgel, L. E. (1987). Autocatalytic synthesis of a tetranucleotide analogue. *Nature*, *327*, 346–347.

14 Replicator Dynamics in Protocells

Peter F. Stadler and Bärbel M. R. Stadler

14.1 Introduction

In recent years, substantial progress has been made in understanding the requirements for minimal cell-like structures. Several proposals for artificial minimal cells have been put forward (Luisi, Walde, and Oberholzer, 1994; Pohorille and Deamer, 2002; Rasmussen et al., 2003; Szostak, Bartel, and Luisi, 2001). Some of them call for a sophisticated molecular machinery to be enclosed in a lipid vesicle. The model of Szostak and coworkers (Szostak, Bartel, and Luisi, 2001) consists of a vesicle containing an RNA genome with an RNA-replicase ribozyme (e.g., an advanced version of the molecule described in Johnston et al., 2001; Lawrence and Bartel, 2005; Paul and Joyce, 2003), and a functionality that influences the fitness of the vesicle. The construct of Pohorille and Deamer (Pohorille and Deamer, 2002; see also chapter 25), which is even closer to a modern cell, includes transcription and translation functionalities. In contrast, the LANL Bug (Rasmussen et al., 2003) envisions a very simple genetic material in lipid aggregates that actively facilitates an autocatalytic reproduction of lipids as well as the genetic material itself (see chapter 6). It is designed as a minimalistic, thermodynamic coupling between the three functional structures' container, metabolism, and genes.

The integration of these fundamental building blocks requires a detailed knowledge of dynamic properties of each of the subsystems and their interactions. While advances in numerical mathematics make it feasible to simulate such systems, a structural analysis of the kinetic equations is a prerequisite for understanding the principles on which lifelike physicochemical structures operate.

Mathematically, the best-studied subsystem is autocatalytic replication. Template-dependent replication at the molecular level is the basis of reproduction in nature. Indeed, a plausible way of characterizing the origin of life is the emergence of heritable information that, through the interplay of selection and variation, leads to Darwinian evolution (Joyce, 2002). A detailed understanding of the peculiarities of the chemical reaction kinetics associated with replication processes is therefore an indispensable prerequisite for any understanding of evolution at the molecular level.

The notion of a *replicator*—originally invented by Richard Dawkins (Dawkins, 1976, pp. 13–21)—is now used in biology for "an entity that passes on its structure largely intact in successive replications" (Vrba, 1989). Before we turn to the mathematics of replication processes, however, we briefly summarize some of the experimental evidence for replication at the molecular level.

14.2 Molecular Replicators

14.2.1 Enzyme-Catalyzed Replication

Enzyme-catalyzed replication of nucleic acids is a ubiquitous technique in molecular biology today. The most prominent example is the polymerase chain reaction (PCR). However, the first successful attempts to study RNA evolution in vitro were already carried out in the late 1960s (Mills, Peterman, and Spiegelman, 1967; Spiegelman, 1971) with the replicase enzyme of the bacteriophage $Q\beta$. Extensive studies on the reaction kinetics of RNA replication in the $Q\beta$ system revealed kinetic data consistent with a multistep reaction mechanism (Biebricher and Eigen, 1988; Biebricher, Eigen, and Gardiner, 1983). Depending on the concentration of template molecules, [C], one can distinguish three phases of the replication process: (1) at low concentrations, all free template molecules are instantaneously bound by the replicase, E, which is present in excess, and therefore the template concentration grows exponentially; (2) excess of template molecules leads to saturation of enzyme molecules, then the rate of RNA synthesis becomes constant and the concentration of the template grows linearly; and (3) very high template concentrations impede dissociation of the complexes between template and replicase, and the template concentration approaches a constant. This effect is known as product inhibition. We neglect plus-minus complementarity in replication by assuming stationarity in relative concentrations of plus and minus strands (Eigen, 1971) and consider the plus-minus ensemble as a single species. Then, RNA replication in the $Q\beta$ system may be described by the overall mechanism:

$$\mathrm{A} + \mathrm{C} + \mathrm{E} \underset{\bar{k}}{\overset{k}{\rightleftharpoons}} \mathrm{A} + \mathrm{C}\cdot\mathrm{E} \overset{a}{\rightarrow} \mathrm{C}\cdot\mathrm{E}\cdot\mathrm{C} \underset{\bar{k}'}{\overset{k'}{\rightleftharpoons}} \mathrm{C}\cdot\mathrm{E} + \mathrm{C}. \tag{14.1}$$

Here A represents the building blocks, the dot indicates a noncovalently bond complex, and as before, C and E, are template and replicase, respectively. Lowercase letters above or below the reaction arrows represent the reaction rate constants. This simplified reaction scheme reproduces all three characteristic phases of the detailed mechanism and can be readily extended to replication and mutation.

14.2.2 Minimal Molecular Replicators

Minimal molecular replicators typically consist of a template and two substrate molecules that become joined to form a copy of the template. A number of experimental examples of such systems have been described so far, based on nucleic acids (Paul and Joyce, 2003; von Kiedrowski, 1986; Zielinski and Orgel, 1987), peptides (Ashkenazy et al., 2004; Isaac and Chmieleswski, 2002; Lee et al., 1996; Lee, Severin, and Ghadiri, 1997; Yao et al., 1998) and small organic molecules (Tijvikua, Ballester, and Rebek, 1990; Wintner, Conn, and Rebek, 1994); see (Paul and Joyce, 2004) for a recent review.

The ligation-based mechanism of all these experimental systems is encapsulated by a common chemical reaction scheme. Here C is the template, A and B are the building blocks, and ABC denotes the complex in which A and B are properly aligned to the template C. The irreversible step is the ligation reaction, which converts ABC into C_2. The complete system of chemical reactions reads

$$\begin{array}{lll} A + C \underset{\bar{a}}{\overset{a}{\rightleftharpoons}} AC & AC + B \underset{\bar{h}}{\overset{h}{\rightleftharpoons}} ABC & ABC \overset{r}{\rightarrow} C_2 \\ B + C \underset{\bar{b}}{\overset{b}{\rightleftharpoons}} BC & BC + A \underset{\bar{g}}{\overset{g}{\rightleftharpoons}} ABC & C_2 \underset{\bar{d}}{\overset{d}{\rightleftharpoons}} 2C \end{array} \tag{14.2}$$

Note the difference between 2C (two isolated copies of the molecule C) and C_2 (the complex formed from two hybridized copies of C).

A quite different mechanism of replication proceeds via DNA triple helices (Li and Nicolaou, 1994): A DNA duplex $C \cdot C$ is replicated by first forming an adduct $C \cdot C'DE$ with triple helix geometry, where the template strand forms standard Watson-Crick pairs, whereas the building blocks D and E are attached via Hoogsteen pairs. The fragments are ligated and then the resulting $C \cdot C'C$ complex dissociates along the weaker Hoogsteen pairs. Finally, the single-stranded template sequence is ligated with fragments of its complements and forms a copy of the original duplex DNA. The reaction mechanism can be summarized as follows

$$\begin{array}{lll} C \cdot C + D + E \underset{\bar{b}}{\overset{b}{\rightleftharpoons}} C \cdot C'DE & C \cdot C'DE \overset{r}{\rightarrow} C \cdot C'C & C \cdot C'C \underset{\bar{d}}{\overset{d}{\rightleftharpoons}} C \cdot C + C \\ C + A + B \underset{\bar{a}}{\overset{a}{\rightleftharpoons}} C \cdot AB & C.AB \overset{s}{\rightarrow} C \cdot C & \end{array} \tag{14.3}$$

The proposed LANL Bug (Rasmussen et al., 2003; see also chapter 6) envisions simpler molecules, such as peptide nucleic acids (PNA) (Nielsen, 1993; see also chapter 15) as the genetic material. PNAs should be much easier to couple with a lipid layer than traditional nucleic acids because of their hydrophobic backbone. Note,

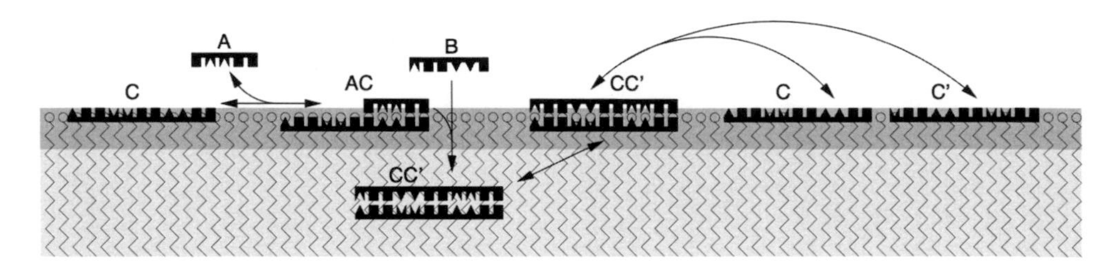

Figure 14.1
Ligase-based replication reaction anchored in a lipid aggregate corresponding to equation (4). Adapted from Rasmussen et al., 2004.

however, that the standard PNA backbone will need to be modified with hydrophobic amino acids for this purpose. As in the other protocell proposals, it utilizes the lipid to keep the cooperative structure together. In contrast to other proposals, the protogenes directly interact with the lipid; this requires a less sophisticated spatial organization than vesicles (see, e.g., Apel, Deamer, and Mautner, 2002), making micelles (Whitten et al., 1998) or even less organized lipid aggregates plausible. A scheme of the replication mechanism is shown in figure 14.1. The overall reaction mechanism for this model can be summarized as follows:

$$\begin{array}{lll} \mathrm{A}+\mathrm{C} \underset{\bar{k}_A}{\overset{k_A}{\rightleftharpoons}} \mathrm{AC} & \mathrm{AC}+\mathrm{B} \overset{a'}{\rightarrow} \mathrm{C}_2^* & \mathrm{C}_2^* \underset{\bar{f}}{\overset{f}{\rightleftharpoons}} \mathrm{C}_2 \\ \mathrm{B}+\mathrm{C} \underset{\bar{k}_B}{\overset{k_B}{\rightleftharpoons}} \mathrm{BC} & \mathrm{BC}+\mathrm{A} \overset{a''}{\rightarrow} \mathrm{C}_2^* & \mathrm{C}_2 \underset{\bar{k}_d}{\overset{k_d}{\rightleftharpoons}} 2\mathrm{C} \end{array} \tag{14.4}$$

Here C_2^* denotes the duplex buried in the lipid phase, whereas C_2 denotes the duplex exposed on the surface where dissociation is thermodynamically feasible. The mechanism envisaged here is only one of several possibilities. Alternatively, one could assume that the CC' duplex dissociates already in the hydrophobic phase. In this case we have to consider the phase equilibrium of the template molecules rather than of the duplexes:

$$\mathrm{C}_2^* \underset{\bar{k}_d^*}{\overset{k_d^*}{\rightleftharpoons}} 2\mathrm{C}^* \quad \text{and} \quad \mathrm{C}^* \underset{\bar{f}'}{\overset{f'}{\rightleftharpoons}} \mathrm{C}. \tag{14.5}$$

14.2.3 Replicase Ribozymes and Higher-Order Autocatalysis

Significant progress has be been made in recent years toward the construction of artificial replicase ribozymes (Ekland and Bartel, 1996; Johnston et al., 2001; Lawrence and Bartel, 2005; McGinness and Joyce, 2003). Though to date, no ribozyme is known that could faithfully replicate a copy of itself, this goal seems to be within

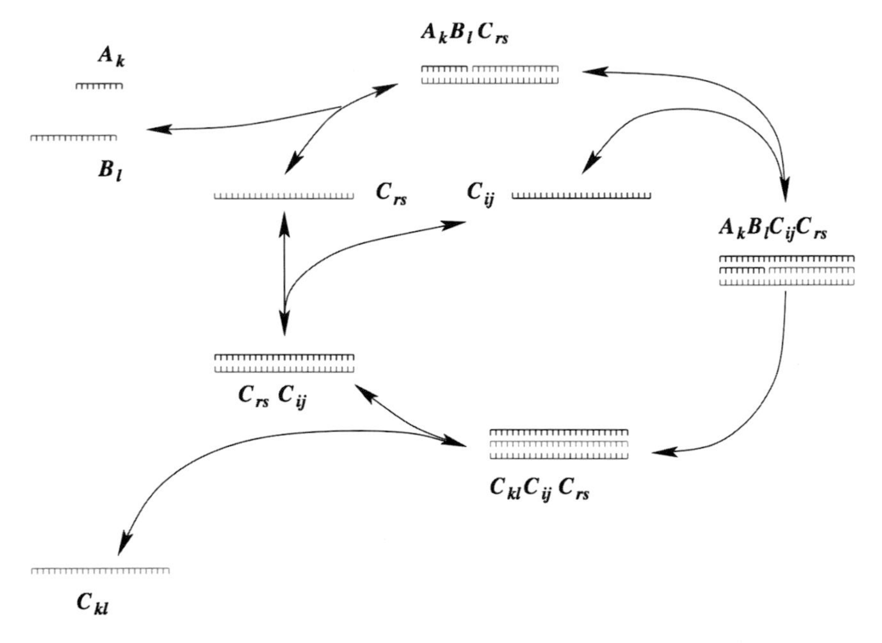

Figure 14.2
A hypothetical mechanism for actively catalyzed ligationlike replication reactions whose dynamics were studied in detail. Adapted from Stadler, Stadler, and Schuster, 2000. Note that in this scheme we have tacitly assumed that template instruction is direct rather than complementary. This amounts to assuming that all involved sequences are palindromic. Alternatively, one could complete the reaction mechanism by including a corresponding cycle for the production of offspring from the complementary templates. It is argued (e.g., in Stadler, 1991a) that as far as the dynamics is concerned, one may view a complementary pair of replicators as a single species.

experimental reach. If successful, such a riboreplicase, C, would be capable of performing template-directed, actively catalyzed replication, following a replication mechanism of the form

$$\mathrm{C}+\mathrm{C}+\mathrm{A} \underset{\bar{k}}{\overset{k}{\rightleftharpoons}} \mathrm{CC}+\mathrm{A} \underset{\bar{a}}{\overset{a}{\rightleftharpoons}} \mathrm{C}\cdot\mathrm{CA} \overset{r}{\rightarrow} \mathrm{C}\cdot\mathrm{CC} \underset{\bar{d}'}{\overset{d'}{\rightleftharpoons}} \mathrm{C}+\mathrm{CC} \quad \text{and} \quad \mathrm{CC} \underset{\bar{d}''}{\overset{d''}{\rightleftharpoons}} 2\mathrm{C}. \tag{14.6}$$

Theoretical models for actively catalyzed ligation-like replication are investigated in (Stadler, Stadler, and Schuster, 2000). Examples are shown in figure 14.2. *Molecular ecologies* of strongly interacting molecular replicators have also been investigated experimentally (McCaskill, 1997; Wlotzka and McCaskill, 1997).

14.3 Replicator Dynamics

The mathematical analysis of the reaction schemes described in the previous section starts by translating the reaction mechanism into kinetic differential equations using the law of mass action (see, e.g., Stadler, Stadler, and Wills, 2001; von Kiedrowski, 1993; Wills et al., 1998). As an example, consider equation (1). We obtain

$$\frac{d[C]}{dt} = -k[A][E][C] + \bar{k}[A][C \cdot E] + k'[C \cdot E \cdot C] - \bar{k}'[C \cdot E][C]$$

$$\frac{d[C \cdot E]}{dt} = k[A][E][C] - \bar{k}[A][C \cdot E] - a[A][C \cdot E] + k'[C \cdot E \cdot C] - \bar{k}'[C \cdot E][C] \quad (14.7)$$

$$\frac{d[C \cdot E \cdot C]}{dt} = a[A][C \cdot E] - k'[C \cdot E \cdot C] + \bar{k}'[C \cdot E][C]$$

Numerical integration can now be used to gain a very detailed understanding of particular model systems, provided the microscopic rate constants can be either measured directly or at least estimated. Examples include the $Q\beta$ replicase system (Biebricher and Eigen, 1988), self-replicating peptides (Islas et al., 2003), and the RNA ligase ribozyme (Bergman, Johnston, and Bartel, 2000). In this contribution, however, we are interested in the qualitative and structural properties of the kinetic differential equations.

We are most interested in the total concentration c of the replicator, which is the sum of free replicator concentrations $[C]$ and the concentrations of the intermediate species that contain the replicator: $c = [C] + [CE] + 2[CEC]$. One observes, by adding up the differential equations for the individual contributions, that the net production of the replicator, $\dot{c}$, is determined by the single irreversible step. In the previous example, this yields

$$\dot{c} = a[A][C \cdot E] \quad (14.8)$$

Under a wide variety of circumstances, one can assume that the concentrations of the reaction intermediates are stationary. This is known as the *quasi-stationary state approximation* (QSSA) (Borghans, de Boer, and Segel, 1996; Segel and Slemrod, 1989). This leads to a set of algebraic equations for the concentrations of the intermediates, which can then be substituted into the growth law for $\dot{c}$. Usually, one makes additional assumptions, for example, that the total concentration of the enyzme E is constant, $[E] + [C \cdot E] + [C \cdot E \cdot C] = E_0$, and that building material A is "buffered," $[A] = a_0$ in our example.

For example, the variants of minimal replicators discussed in the previous section all lead to the same effective dynamics of the form

$$\dot{c} = \alpha c \psi(\beta c) \quad \text{where} \quad \psi(u) = \frac{2}{u}(\sqrt{1+u} - 1) \quad (14.9)$$

where α and β can be expressed in terms of the microscopic reaction rate constants. Of course, one obtains different (and usually very complicated) expressions for α and

β for different models. Since we will not need the explicit equations, we refer to the literature for further details (Rasmussen et al., 2004; Stadler and Stadler, 2003; Wills et al., 1998).

The function ψ appears through the solution of a quadratic equation for [C] in terms of c. Similarly, [C] and c are related by a cubic equation in the model of enzyme-catalyzed replication, in equations (1) and (7). Higher-order algebraic equations also arise in the case of higher-order autocatalytic systems, such as the mechanism in figure 14.2, leading to much more complex functional dependencies.

This approach readily translates to systems with different competing replicators, C_k. In the most general case, we obtain vector fields of the form

$$\dot{c}_k = c_k F_k(\vec{c}) \tag{14.10}$$

where F_k is a continuous function of the concentrations of the different replicator species. In general, it is hard or impossible to obtain a closed form for the vector field $F_k(\vec{c})$.

Most of the work on such coupled chemical reaction systems has been considered either a *continuously stirred tank reactor* (CSTR), which amounts to an additional unspecific degradation term $-rc_k$ or *constant organization.* The latter constraint fixes the total concentration $c = \sum_k c_k$ at a constant level c_0. This is equivalent to a regulated outflux, $-c_k\Phi(\vec{c})$, which is determined by the net production of replicators:

$$\Phi(\vec{c}) = \sum_j \frac{c_j}{c} F_j(\vec{c}). \tag{14.11}$$

In the case of homogeneous interaction functions, $F_k(\lambda\vec{c}) = h(\lambda)F_k(\vec{c})$, one can show that the CSTR and the constant organization model are the same up to a rescaling of the time axis (Schuster and Sigmund, 1985). An analogous result can be shown for the limit of small flux rates r in the CSTR and arbitrary interaction functions $F_k(\vec{c})$ (Happel and Stadler, 1999). It is thus useful to rewrite the dynamics in terms of relative concentrations $x_k = c_k/c$. From equation (10), we obtain

$$\dot{x}_k = x_k \left[F_k(c \cdot \vec{x}) - \sum_j x_j F_j(c \cdot \vec{x}) \right]. \tag{14.12}$$

Again, as demonstrated in Schuster and Sigmund (1985), the total concentration c amounts to only a rescaling of the time axis in the case of homogeneous interaction functions $F_k(.)$. Equation (12) is the general form of a *replicator* equation (Schuster and Sigmund, 1983). This class of dynamical systems has been the subject of a large number of research papers as well as of Hofbauer and Sigmund's book (1988).

A few cases have been studied in great detail:

• $F_k = a_k$ is a constant fitness value.
In this case, we have strong selection (*survival of the fittest*), i.e., only the sequence with the largest value of k can survive.

• $F_k(c \cdot \vec{x}) = c \sum_j A_{kj} x_j$.
These *second-order replicator equations* also describe the dynamics of strategies in evolutionary games (Taylor and Jonker, 1978). Hofbauer showed that they are topologically equivalent to the Lotka-Volterra equations (Hofbauer, 1981). Their equivalence to the Price equation is demonstrated by Page and Novak (2002). A famous special case of a second-order replicator equation is the *hypercycle model* of cooperative replicators (Eigen and Schuster, 1979). Here sequence $k-1$ catalyzes the replication k in a cyclic arrangement. The most important property of hypercycles is permanent coexistence, that is, the fact that, independently of initial conditions, the relative concentrations x_k are bounded from below by a fixed constant after a transient initial time (Schuster, Sigmund, and Wolf, 1979). Such cooperative behavior, however, is very rare in second-order replicator equations (Happel and Stadler, 1998; Stadler and Happel, 1993).

The dynamics of second-order replicator equations can be extremely complicated despite the rather simple form of the differential equation (figure 14.3). In the case of two independent variables ($n = 3$, the state space is an equilateral triangle), there are 35 generic-phase portraits (Bomze, 1983; Stadler and Schuster, 1990). In the case of three independent variables, that is, $n = 4$ species, there are heteroclinic orbits (Brannath, 1994; Stadler, 1996), multiple limit cycles (Hofbauer and So, 1994), and strange attractors (Arneodo, Coullet, and Tresser, 1980; Forst, 1996; Gilpin, 1978; Schnabl et al., 1991; Vance, 1978; see figure 14.4).

• $F_k(c \cdot \vec{x}) = a_k \psi(c x_k)$, where ψ is a monotonically decreasing function.
Such systems were investigated in detail in (Hofbauer, 1981). The minimal replicators described in the previous section are examples of this class of dynamical systems (Stadler, Stadler, and Wills, 2001; Wills et al., 1998). There is a unique fixed point, $\hat{x}$, that is eventually reached by all trajectories that start in the interior of the state space, that is, for which all initial concentrations are nonzero. There is a *survival threshold*, a^*, such that all species with a fitness $a_k \geq a^*$ can coexist, whereas those with $a_k < a^*$ eventually die out. Models with parabolic growth (Szathmáry and Gladkih, 1989; Varga and Szathmáry, 1997; von Kiedrowski, 1993) can be regarded as a limiting case in which the survival threshold is low enough to allow permanent coexistence (Wills et al., 1998).

• $F_k(c \cdot \vec{x}) = \vartheta(c \cdot [\mathrm{A}x]_k)$, where ϑ is a monotonically increasing function. The dynamics of this system are very similar to the second-order replicator equation with the same interaction matrix A (Stadler and Stadler, 1991).

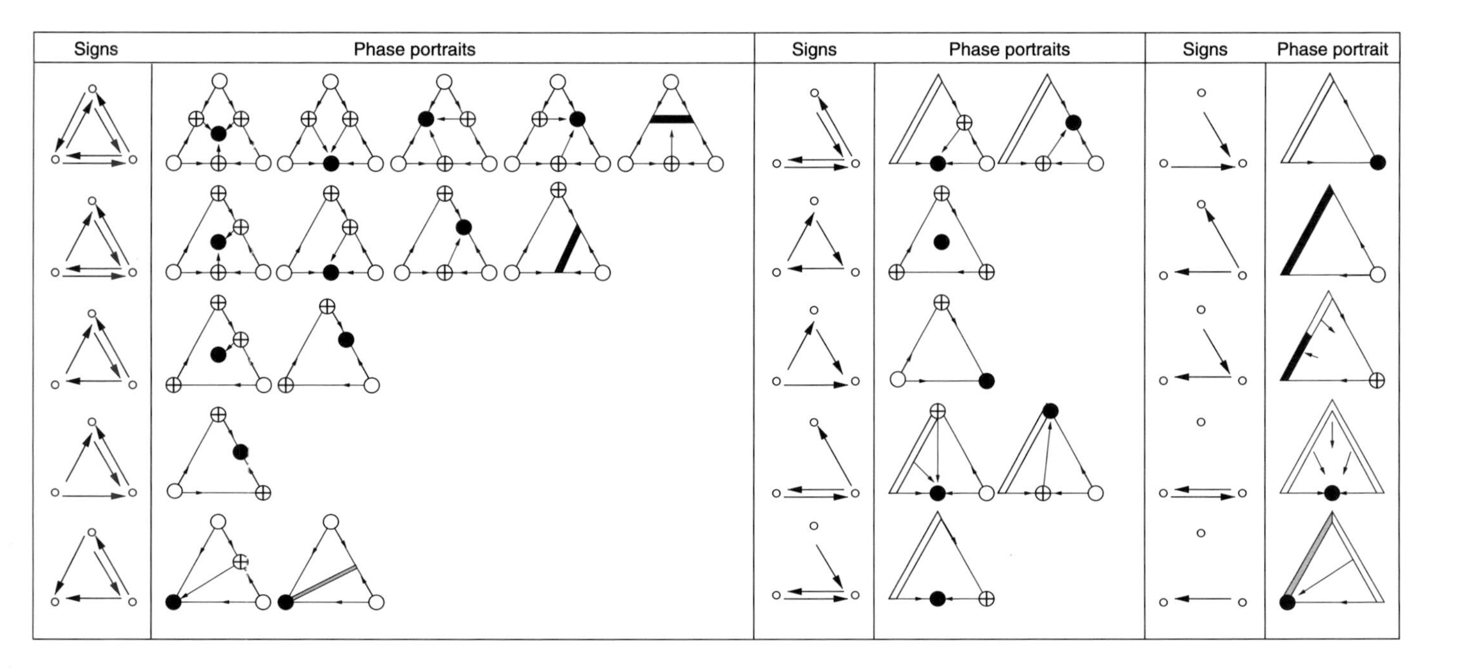

Figure 14.3
Autocatalytic networks are the subclass of second-order replicator equations in which there is no self-catalysis ($A_{ii} = 0$) and all other interactions are cooperative $A_{ij} \geq 0$. Their structure is readily represented by a graph with an arrow, $i \rightarrow j$ iff i catalyzes the replication of j, that is, iff $a_{ji} > 0$. The diagrams here summarize the diversity of qualitative dynamical behavior of three-species autocatalytic networks. Symbols in the phase portraits: • stable fixed point (sink), ○ unstable fixed point (source), ⊕ saddle point; thick lines indicate lines consisting entirely of fixed points. Adapted from Schuster and Stadler, 2002.

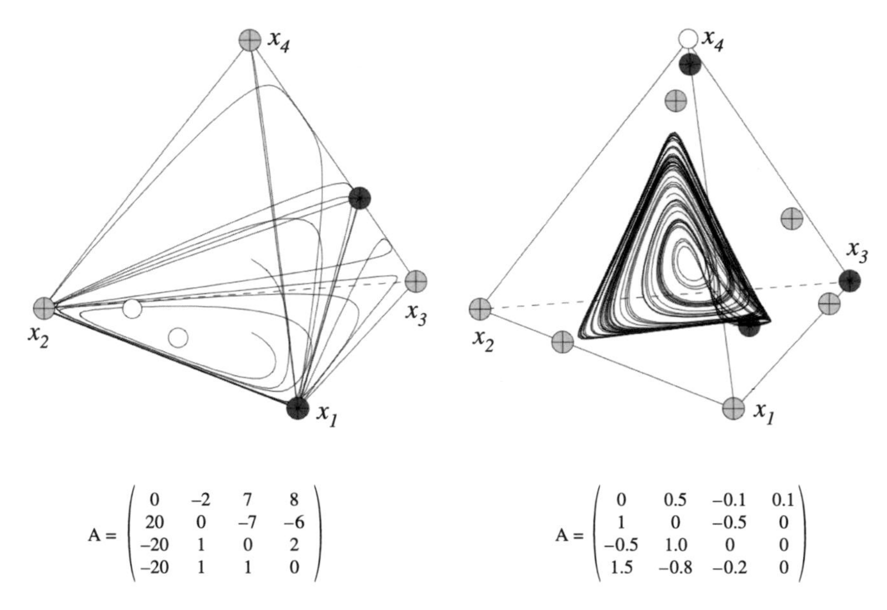

Figure 14.4
Complex dynamics in second-order replicator equations. (Left) An attracting heteroclinic orbit in the central plane (Stadler, 1996). (Right) A chaotic attractor of class described in Arneodo, Coullet, and Tresser, 1980. Fixed points are distinguished by the number of stable directions: ○ sources without stable direction, ⊕ with one stable direction, and ⊛ with two stable directions.

Beyond a few general results for arbitrary F_k, which are discussed in detail in (Hofbauer and Sigmund, 1988, 1998), and the functional forms listed earlier, very little is known about replicator equations with nonlinear response functions. A few special cases are discussed, for example, in Bomze (1983), Forst (1996), Hofbauer, Schuster, and Sigmund (1982), and Stadler and coworkers (2000; Stadler, Schuster, and Perelson, 1994).

14.4 Replicator-Mutator Equations

Mutation can be included in a straightforward way. Denote by Q_{kj} the probability to produce an offspring of type k from a type j template. The dynamics of such a system are then described by the *replicator-mutator* equation

$$\dot{x}_k = \sum_j Q_{kj} x_j F_j(\vec{x}) - x_k \Phi. \tag{14.13}$$

This expression has been used in population genetics (Hadeler, 1981), autocatalytic reaction networks (Stadler and Schuster, 1992), game theory (Bomze and Bürger, 1995), and language evolution (Nowak, Komarova, and Niyogi, 2001).

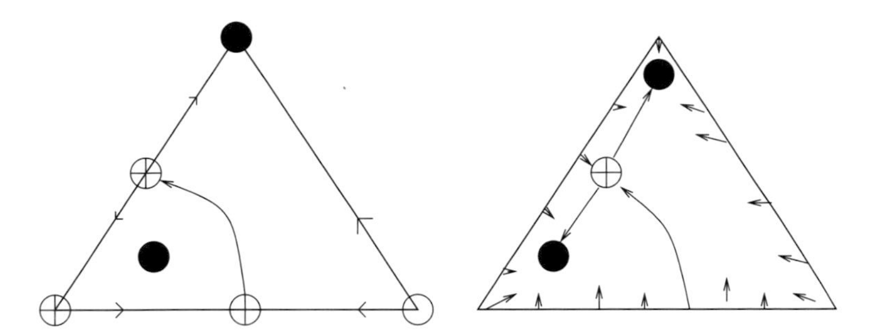

Figure 14.5
Rest point migration theorem. (Left) Phase portrait of selection-only system, that is, a replicator equation. (Right) The same selection part is superimposed with a mutation vector field that points inwards at the boundary. As a result, saturated fixed points are driven into the interior of the state space, while nonsaturated fixed points move into the physically inaccessible exterior.

Mutation in general can be interpreted as an additional contribution to the vector field in (relative) concentration space that points inward at the boundary of the state space, that is, it generates additional species that are not present in a given initial condition. Under certain conditions, namely, that the off-diagonal elements in the mutation rate matrix Q are small enough and some (mild) technical conditions on the vector field $\vec{F}$ (described in detail in Stadler and Schuster, 1992) hold, mutation can be treated as a perturbation. Its qualitative effects on the selection dynamics are then captured by the *rest point migration theorem*.

A fixed point is saturated if it is stable against invasion, that is, if the boundary of the concentration simplex is attracting in its vicinity. Small mutation rates deform the vector field of a replicator equation in such a way that saturated boundary equilibria move into the interior of the state space, while nonsaturated boundary equilibria move into the (nonphysical) outside (figure 14.5). Small amounts of mutations therefore simplify the phase portrait of the selection dynamics and do not change stable fixed points (and limit cycles).

For constant F_k, equation (13) specializes to the *quasispecies* model (Eigen, 1971; Eigen, McCaster, and Schuster, 1989). The most salient feature of this model is the existence of an *error threshold*, which restricts the amount of information that can be sustained under error-prone replication. It is plausible that genetic inheritance is also limited in the general case of frequency-dependent selection, albeit no formal proof for this claim exists. Numerical studies for the hypercycles model are reported in Forst (2000).

14.5 Dynamics of a Pre-Protocell

In Cavalier-Smith (2001), a scenario is considered in which membranes initially functioned as supramolecular structures to which different replicators attached and were

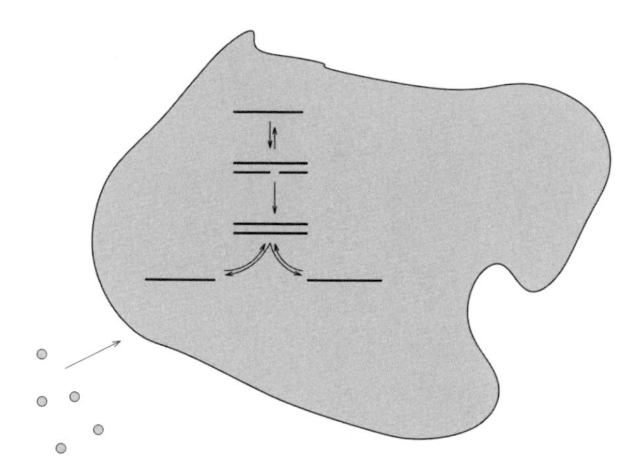

Figure 14.6
Model of a pre-protocell: Replicating polymers are attached to the surface of a lipid aggregate, which can grow by incorporating amphiphilic molecules from the environment.

selected as a higher-level reproductive unit. This picture is conceptually simpler than micellar or vesicular protocells, since it avoids the difficulties of modeling the regulation of both growth and fission. Our pre-protocell, (figure 14.6) consists of a lipid aggregate that can grow by inclusion of amphiphilic molecules that are present in the environment. Attached to its surface is a suitable nucleic-acid analog, maybe some variant of a PNA, that undergoes uncatalyzed replication in the spirit of the membrane-linked replication cycle of the *Los Alamos Bug* (Rasmussen et al., 2003; see also chapter 6). The type and property of a membrane fragment are determined by its inventory of genetic material.

Denote by Ω_a the total surface area of type-a membranes and let n_{ka} be the number of macromolecules with sequence k embedded in it. Then

$$\dot{n}_{ka} = n_{ka} F_k(\vec{c}_a) \tag{14.14}$$

where $\vec{c}_a$ is the vector of concentrations of the different PNA sequences and F is a growth law, for example, one of those described in the previous sections. We have $c_{ka} = n_{ka}/\Omega_a$ and hence

$$\dot{c}_{ka} = c_{ka} F_k(\vec{c}_a) - c_{ka} \frac{\dot{\Omega}_a}{\Omega_a}. \tag{14.15}$$

In terms of the total concentration of replicating polymers in type-a membranes, $c_a = \sum_k c_{ka}$, and relative polymer concentrations $x_{ka} = c_{ka}/c_a$ (i.e., $\vec{c}_a = c_a \vec{x}_a$) we obtain an internal dynamics of the genetic material governed by the replicator equation

$$\dot{x}_{ka} = x_{ka}\left[F_k(c_a \cdot \vec{x}_a) - \sum_j x_j F_j(c_a \cdot \vec{x}_a)\right] = x_{ka}[F_k(c_a \cdot \vec{x}_a) - \Phi_a] \tag{14.16}$$

and a growth law for the total concentration of polymers that is linked to the membrane growth

$$\dot{c}_a = c_a \sum_j x_j F_j(c_a \cdot \vec{x}_a) - c_a \frac{\dot{\Omega}_a}{\Omega_a} = c_a\left[\Phi_a - \frac{\dot{\Omega}_a}{\Omega_a}\right]. \tag{14.17}$$

As expected, the concentration of the genetic material is determined by the balance of replication and membrane growth. It follows directly from this equation that a feedback is needed between the net production of genetic material, Φ_a, in the type-a membrane and its growth rate, $\dot{\Omega}_a$. If this were not the case, either the replicating material would be diluted out of the system or it would completely pack the membrane. At the latter limit, any realistic model will, of course, show feedback.

To complete this model, we need to specify how the membrane growth depends on the concentrations of the attached replicators. Assuming that they have a certain catalytic activity that can increase or inhibit the incorporation of monomers into the membrane (or catalyze their formation from precursors), we expect a growth law of the general form

$$\dot{\Omega}_a = \Omega_a[g + c_a G(\vec{x}, c_a)]. \tag{14.18}$$

We see that, as one would expect, the membrane will become asymptotically devoid of genetic material if

$$\Phi_a < g \tag{14.19}$$

that is, if the replication rate of the polymer is smaller than the autonomous growth rate of the membrane.

On the other hand, if the polymer concentration c_a approaches a steady state, then membrane growth is determined by the net production of replicators: $\dot{\Omega}_a \to \Phi_a \cdot \Omega_a$. In the case of vesicles that enclose replicating RNA, it was demonstrated that the osmotic pressure exerted on the membrane drives the uptake of membrane components from the environment (Chen, Roberts, and Szostak, 2004). The growth rate of a vesicle is thus directly determined by the net production, Φ_a, of replicators in its interior.

One would expect that equations similar to those discussed here would hold for vesicular and micellar systems, with the added complication that fission of the protocells needs to be modeled. However, such cell models incur an additional complication compared to the simple membrane model discussed previously: Due to the small size of protocells, a certain fraction of their daughter cells will not inherit the complete set of genomic molecules, and thus will not be viable. This effect further reduces the effective cellular replication rate.

In modern cells, the feedback between genomic replication and cellular growth is mediated indirectly by a complex regulatory cascade of gene expression and metabolic control. Recent advances in DNA chemistry demonstrate the possibility of "translating" the information stored in a nucleic acid sequence directly into nonpolymeric compounds without the help of sophisticated enzymes (Calderone and Liu, 2004; Gartner and Liu, 2001; Halpin and Harbury, 2004). Such mechanisms might provide a physicochemical basis for the direct influence of the genome on the lipid aggregate, which is implicitly postulated by the function $G(\vec{x}, c_a)$.

A particularly interesting facet of this model is the approximate dynamic independence of the individual components. In the case of homogeneous interaction functions F_k, the eventual compositions, $\vec{x}_a$, of the genomes (mathematically speaking, the ω-limits of the replicator equations) are independent of the concentrations c_a. This follows immediately from the arguments in Schuster and Sigmund (1985). In the limiting case of slowly varying c_a, a kind of "adiabatic" approximation allows us to predict the dynamic outcomes (see Stadler, 1991b, for some specific examples).

In this section we have presented only a cursory analysis of the dynamics of replicators interacting with a growing lipid aggregate. A more detailed investigation will be necessary to understand, for example, the conditions under which a steady state is reached. Another research topic is the elucidation of the relationship of the present model with simpler group-selection models that have been proposed in a prebiotic context, such as the *stochastic corrector* (Szathmáry and Demeter, 1987). Clearly, the system has the potential for open-ended evolution. Mutations of the genetic material may lead to an increase in net flux Φ_a, thus giving rise to a new, more competitive species. Collisions of aggregates with different genetic contents essentially play the role of "recombination."

14.6 Concluding Remarks: Evolution of a Protocell Genome

The dynamics of self-replicating heteropolymers set the stage for the evolution of the information that is encoded by these protogenomes. Indeed, a central issue in models of prebiotic evolution is the integration of information necessary to bridge the gap between a simple system of replicating molecules and the complexity of a modern cell (Eigen and Schuster, 1979; Kauffman, 1993). The template length is limited by the accuracy of the replication mechanism, which is necessarily error-prone because of mutations. As an order of magnitude estimate, the length of directly replicated genome n is limited by the inverse $1/p$ of the per-digit mutation rate p (Eigen, 1971). In principle, the error threshold can be circumvented by evolving more accurate replicases that could be encoded by longer sequences (Poole, Jeffares, and

Penny, 1999; Scheuring et al., 2003; Szabó et al., 2002). Such a bootstrapping mechanism, however, requires a functional replicase-ribozyme to start with.

The error threshold, however, could be drastically relaxed in the simple model outlined here, since the dynamics of selection on this level are only weakly dependent on the dynamic details of replication. The latter could be organized cooperatively, for example, in the case of a hypercycle (Eigen and Schuster, 1979), thereby substantially increasing the genetic storage capacity. Such cooperative models are notoriously plagued by the *parasite problem*: Mutants without suitable catalytic activity can exploit, and eventually destroy, the entire system. Starting with the work of Boerlijst and Hogeweg (1991), it has been demonstrated, however, that the problem of parasite invasion can be alleviated by considering spatially organized systems (Altmeyer and McCaskill, 2001; Cronhjort and Blomberg, 1994; Streissler, 1992; Tereshko, 1999; Zintzaras, Santos, and Szathmáry, 2002). Replication kinetics that include product inhibition can have a similar effect in some parameter ranges (Stadler, Stadler, and Schuster, 2000; Stadler et al., 2001). It is conceivable that with the simple coupling of replication to a container, growth under "genetic" control is already sufficient to bridge the *information gap* between uncatalyzed self-replication of nucleic acids with at most 20nt, and plausible replicase ribozymes, which could have a length of 100 to 200nt, based on a comparison with known ribozymes.

The shape of the fitness function, and, in particular, the accessibility of mutants from a given population, crucially influences the dynamics of evolution (Fontana and Schuster, 1998; Schuster et al., 1994; Stadler et al., 2001). In the case of RNA, it has been demonstrated that the sequence-structure relation is dominated by neutral mutations: Single point mutations often leave structure, and thus also function, intact. This implies that functionally equivalent sequences form so-called neutral networks that percolate through sequence space. With selection acting on structure or function rather than directly on sequence, neutrality implies a significant redundancy at the sequence level and replaces the genotypic error threshold by a relaxed *phenotypic error threshold* (Forst, Reidys, and Weber, 1995; Huynen, Stadler, and Fontana, 1996). It has been argued that this could be sufficient to bridge the information gap (Kun, Santos, and Szathmáry, 2005).

From a dynamical systems point of view, neutrality implies that the interplay of selection and mutation can efficiently explore sequence space by means of neutral drift confined to the neutral networks (Huynen, 1996; Huynen, Stadler, and Fontana, 1996; Schuster et al., 1994). Recently, it was shown that a similar mechanism allows a population of autocatalytic self-replicators to explore sequence space in a diffusion-like manner (Stadler, 2002; Stephan-Otto Attolini and Stadler, 2004).

Our simplistic pre-protocells from the previous section can therefore be expected to show all hallmarks of Darwinian evolution. They are, of course, extreme heterotrophs: We have not discussed at all where the energy-rich building material the protocells need to replicate their genomes and grow their membranes comes from. That, of course, is another story.

Acknowledgments

This work is supported in part by the DFG bioinformatics initiative, grant no. BIZ-6/1-2, and the COST action D27.

References

Altmeyer, S., & McCaskill, J. S. (2001). Error threshold for spatially resolved evolution in the quasispecies model. *Physical Review Letters, 86*, 5819–5822.

Apel, C. L., Deamer, D. W., & Mautner, M. N. (2002). Self-assembled vesicles of monocarboxylic acids and alcohols: Conditions for stability and for the encapsulation of biopolymers. *Biochimica Biophysica Acta, 1559*, 1–9.

Arneodo, A., Coullet, P., & Tresser, C. (1980). Occurrence of strange attractors in three-dimensional volterra equations. *Physics Letters A, 79*, 259–263.

Ashkenazy, G., Jagasia, R., Yadav, M., & Ghadiri, M. R. (2004). Design of a directed molecular network. *Proceedings of the National Academy of Sciences of the United States of America, 101*, 10872–10877.

Bergman, N. H., Johnston, W. K., & Bartel, D. P. (2000). Kinetic framework for ligation by an efficient RNA ligase ribozyme. *Biochemistry, 39*, 3115–3123.

Biebricher, C. K., & Eigen, M. (1988). Kinetics of RNA replication by Qβ replicase. In E. Domingo, J. J. Holland, & P. Ahlquist (Eds.). *RNA genetics. Vol. I: RNA directed virus replication* (pp. 1–21). Boca Raton, FL: CRC Press.

Biebricher, C. K., Eigen, M., & Gardiner Jr., W. C. (1983). Kinetics of RNA replication. *Biochemistry, 22*, 2544–2559.

Boerlijst, M. C., & Hogeweg, P. (1991). Spiral wave structure in pre-biotic evolution: Hypercycles stable against parasites. *Physica D, 48*, 17–28.

Bomze, I. (1983). Lotka-Volterra equations and replicator dynamics: A two-dimensional classification. *Biological Cybernetics, 48*, 201–211.

Bomze, I., & Bürger, R. (1995). Stability by mutation in evolutionary games. *Games and Economic Behavior, 11*, 146–172.

Borghans, J. A. M., de Boer, R. J., & Segel, L. A. (1996). Extending the quasi-steady state approximation by changing variables. *Bulletin of Mathematical Biology, 58*, 43–63.

Brannath, W. (1994). Heteroclinic networks on the tetrahedron. *Nonlinearity, 7*, 1367–1384.

Calderone, C. T., & Liu, D. R. (2004). Nucleic-acid-templated synthesis as a model system for ancient translation. *Current Opinion in Chemical Biology, 8*, 645–653.

Cavalier-Smith, T. (2001). Obcells as proto-organisms: Membrane heredity, lithophosphorylation, and the origins of the genetic code, the first cells, and photosynthesis. *Journal of Molecular Evolution, 53*, 555–595.

Chen, I. A., Roberts, R. W., & Szostak, J. W. (2004). The emergence of competition between model protocells. *Science, 305*, 1474–1476.

Cronhjort, M. B., & Blomberg, C. (1994). Hypercycles versus parasites in a two dimensional partial differential equations model. *Journal of Theoretical Biology, 169*, 31–49.

Dawkins, R. (1976). *The selfish gene*. Oxford: Oxford University Press.

Eigen, M. (1971). Self-organization of matter and the evolution of biological macromolecules. *Naturwissenschaften, 58*, 465–523.

Eigen, M., McCaskill, J. S., & Schuster, P. (1989). The molecular quasi-species. *Advances in Chemical Physics*, *75*, 149–263.

Eigen, M., & Schuster, P. (1979). *The hypercycle*. New York, Berlin: Springer-Verlag.

Ekland, E. H., & Bartel, D. P. (1996). RNA-catalysed RNA polymerization using nucleoside triphosphates. *Nature*, *382*, 373–376.

Fontana, W., & Schuster, P. (1998). Continuity in evolution: On the nature of transitions. *Science*, *280*, 1451–1455.

Forst, C. V. (1996). Chaotic interactions of self-replicating RNA. *Computers and Chemistry*, *20*, 69–83.

Forst, C. V. (2000). Molecular evolution of catalysis. *Journal of Theoretical Biology*, *205*, 409–431.

Forst, C. V., Reidys, C. M., & Weber, J. (1995). Evolutionary dynamics and optimization: Neutral networks as model-landscape for RNA secondary structure folding-landscapes. In F. Morán, A. Moreno, J. Merelo, & P. Chacón (Eds.), *Advances in artificial life*, vol. 929 of *Lecture notes in artificial intelligence* (*ECAL '95*) (pp. 128–147). Berlin, Heidelberg, New York: Springer.

Gartner, Z. J., & Liu, D. R. (2001). The generality of DNA-templated synthesis as a basis for evolving nonnatural small molecules. *Journal of the American Chemical Society*, *123*, 6961–6963.

Gilpin, M. E. (1978). Spiral chaos in a predator prey system. *The American Naturalist*, *133*, 306–308.

Hadeler, K. P. (1981). Stable polymorphisms in a selection model with mutation. *SIAM Journal of Applied Mathematics*, *41*, 1–7.

Halpin, D. R., & Harbury, P. B. (2004). DNA display I. Sequence-encoded routing of DNA populations. *PLoS Biology*, *2*, e173.

Happel, R., & Stadler, P. F. (1998). The evolution of diversity in replicator networks. *Journal of Theoretical Biology*, *195*, 329–338.

Happel, R., & Stadler, P. F. (1999). Autocatalytic replication in a CSTR and constant organization. *Journal of Mathematical Biology*, *38*, 422–434.

Hofbauer, J. (1981). On the occurrence of limit cycles in Volterra-Lotka equations. *Nonlinear Analysis*, *5*, 1003–1007.

Hofbauer, J., Schuster, P., & Sigmund, K. (1982). Game dynamics in Mendelian populations. *Biological Cybernetics*, *43*, 51–57.

Hofbauer, J., & Sigmund, K. (1988). *Dynamical systems and the theory of evolution*. Cambridge: Cambridge University Press.

Hofbauer, J., & Sigmund, K. (1998). *Evolutionary games and population dynamics*. Cambridge: Cambridge University Press.

Hofbauer, J., & So, J. (1994). Multiple limit cycles for three-dimensional Lotka-Volterra equations. *Applied Mathematics Letters*, *7*, 65–70.

Huynen, M. A. (1996). Exploring phenotype space through neutral evolution. *Journal of Molecular Evolution*, *43*, 165–169.

Huynen, M. A., Stadler, P. F., & Fontana, W. (1996). Smoothness within ruggedness: The role of neutrality in adaptation. *Proceedings of the National Academy of Sciences of the United States of America*, *93*, 397–401.

Isaac, R., & Chmieleswski, J. (2002). Approaching exponential growth with a self-replicating peptide. *Journal of the American Chemical Society*, *124*, 6808–6809.

Islas, J. M., Pimienta, V., Micheau, J.-C., & Buhse, T. (2003). Kinetic analysis of artificial peptide self-replication: Part I: The homochiral case. *Biophysical Chemistry*, *103*, 191–200.

Johnston, W. K., Unrau, P. J., Lawrence, M. J., Glasner, M. E., & Bartel, D. P. (2001). RNA-Catalyzed RNA polymerization: Accurate and general RNA-templated primer extension. *Science*, *292*, 1319–1325.

Joyce, G. F. (2002). The antiquity of RNA-based evolution. *Nature*, *418*, 214–221.

Kauffman, S. A. (1993). *The origin of order*. New York, Oxford: Oxford University Press.

Kun, A., Santos, M., & Szathmáry, E. (2005). Real ribozymes suggest a relaxed error threshold. *Nature Genetics*, *37*, 1008–1011.

Lawrence, M. S., & Bartel, D. P. (2005). New ligase-derived RNA polymerase ribozymes. *RNA*, *11*, 1173–1180.

Lee, D. H., Granja, J. R., Martinez, J. A., Severin, K., & Ghadiri, M. R. (1996). A self-replicating peptide. *Nature*, *382*, 525–528.

Lee, D. H., Severin, K., & Ghadiri, M. R. (1997). Autocatalytic networks: The transition from molecular self-replication to ecosystems. *Current Opinion in Chemical Biology*, *1*, 491–496.

Li, T., & Nicolaou, K. C. (1994). Chemical self-replication of palindromic duplex DNA. *Nature*, *369*, 218–221.

Luisi, P. L., Walde, P., & Oberholzer, T. (1994). Enzymatic RNA synthesis in self-reproducing vesicles: An approach to the construction of a minimal synthetic cell. *Berichte der Bunsen-Gesellschaft Physical Chemistry*, *98*, 1160–1165.

McCaskill, J. S. (1997). Spatially resolved *in vitro* molecular ecology. *Biophysical Chemistry*, *66*, 145–158.

McGinness, K. E., & Joyce, G. F. (2003). In search of an RNA replicase ribozyme. *Chemical Biology*, *10*, 5–14.

Mills, D. R., Peterson, R. L., & Spiegelman, S. (1967). An extracellular Darwinian experiment with a self-duplicating nucleic acid molecule. *Proceedings of the National Academy of Sciences of the United States of America*, *58*, 217–224.

Nielsen, P. E. (1993). Peptide nucleic acid (PNA): A model structure for the primordial genetic material? *Origins of Life and Evolution of the Biosphere*, *23*, 323–327.

Nowak, M. A., Komarova, N. L., & Niyogi, P. (2001). Evolution of universal grammar. *Science*, *291*, 114–118.

Page, K. M., & Novak, M. A. (2002). Unifying evolutionary dynamics. *Journal of Theoretical Biology*, *219*, 93–98.

Paul, N., & Joyce, G. F. (2003). A self-replicating ligase ribozyme. *Proceedings of the National Academy of Sciences of the United States of America*, *99*, 12733–12740.

Paul, N., & Joyce, G. F. (2004). Minimal self-replicating systems. *Current Opinion in Chemical Biology*, *8*, 634–639.

Pohorille, A., & Deamer, D. (2002). Artificial cells: Prospects for biotechnology. *Trends in Biotechnology*, *20*, 123–128.

Poole, A., Jeffares, D., & Penny, D. (1999). Early evolution: Prokaryotes, the new kids on the block. *Bioessays*, *21*, 880–889.

Rasmussen, S., Chen, L., Nilsson, M., & Abe, S. (2003). Bridging nonliving and living matter. *Artificial Life*, *9*, 269–316.

Rasmussen, S., Chen, L., Stadler, B. M. R., & Stadler, P. F. (2004). Proto-organism kinetics: Evolutionary dynamics of lipid aggregates with genes and metabolism. *Origins of Life and Evolution of the Biosphere*, *34*, 171–180.

Scheuring, I., Czaran, T., Szabo, P., Karyoli, G., & Toroczkai, Z. (2003). Spatial models of prebiotic evolution: Soup before pizza? *Origins of Life*, *33* (4–5), 329–355.

Schnabl, W., Stadler, P. F., Forst, C., & Schuster, P. (1991). Full characterization of a strange attractor. *Physica D*, *48*, 65–90.

Schuster, P., Fontana, W., Stadler, P. F., & Hofacker, I. L. (1994). From sequences to shapes and back: A case study in RNA secondary structures. *Proceedings of the Royal Society of London*, *B225*, 279–284.

Schuster, P., & Sigmund, K. (1983). Replicator dynamics. *Journal of Theoretical Biology*, *100*, 533–538.

Schuster, P., & Sigmund, K. (1985). Dynamics of evolutionary optimization. *Berichte der Bunsen-Gesellschaft Physical Chemistry*, *89*, 668–682.

Schuster, P., Sigmund, K., & Wolff, R. (1979). Dynamical systems under constant organization III: Cooperative and competitive behaviour of hypercycles. *Journal of Differential Equations*, *32*, 357–368.

Schuster, P., & Stadler, P. F. (2002). Networks in molecular evolutions. *Complexity*, *8*, 34–42.

Segel, L. A., & Slemrod, M. (1989). The quasi-steady state assumption: A case study in perturbation. *SIAM Review*, *31*, 446–477.

Spiegelman, S. (1971). An approach to the experimental analysis of precellular evolution. *Quarterly Review of Biophysics*, *4*, 213–253.

Stadler, B. M. R. (1996). Segregation distortion and heteroclinic cycles. *Journal of Theoretical Biology*, *183*, 363–379.

Stadler, B. M. R. (2002). Diffusion of a population of interacting replicators in sequence space. *Advances in Complex Systems*, *5* (4), 457–461.

Stadler, B. M. R., & Stadler, P. F. (1991). Dynamics of small autocatalytic reaction networks III: Monotonous growth functions. *Bulletin of Mathematical Biology*, *53*, 469–485.

Stadler, B. M. R., & Stadler, P. F. (2003). Molecular replicator dynamics. *Advances in Complex Systems*, *6*, 47–77.

Stadler, B. M. R., Stadler, P. F., & Schuster, P. (2000). Dynamics of autocatalytic replicator networks based on higher order ligation reactions. *Bulletin of Mathematical Biology*, *62*, 1061–1086.

Stadler, B. M. R., Stadler, P. F., Wagner, G., & Fontana, W. (2001). The topology of the possible: Formal spaces underlying patterns of evolutionary change. *Journal of Theoretical Biology*, *213*, 241–274.

Stadler, B. M. R., Stadler, P. F., & Wills, P. R. (2001). Evolution in systems of ligation-based replicators. *Zeitschrift fur Physikalische Chemie*, *216*, 21–33.

Stadler, P. F. (1991a). Complementary replication. *Mathematical Biosciences*, *107*, 83–109.

Stadler, P. F. (1991b). Dynamics of small autocatalytic reaction network IV: Inhomogeneous replicator equations. *BioSystems*, *26*, 1–19.

Stadler, P. F., & Happel, R. (1993). The probability for permanence. *Mathematical Biosciences*, *113*, 25–50.

Stadler, P. F., & Schuster, P. (1990). Dynamics of small autocatalytic reaction networks I: Bifurcations, permanence and exclusion. *Bulletin of Mathematical Biology*, *52*, 485–508.

Stadler, P. F., & Schuster, P. (1992). Mutation in autocatalytic networks—An analysis based on perturbation theory. *Journal of Mathematical Biology*, *30*, 597–631.

Stadler, P. F., Schuster, P., & Perelson, A. S. (1994). Immune networks modelled by replicator equations. *Journal of Mathematical Biology*, *33*, 111–137.

Stephan-Otto Attolini, C., & Stadler, P. F. (2004). Evolving towards the hypercycle: A spatial model of molecular evolution. *Physica D: Nonlinear Phenomena*, *217*, 134–141.

Streissler, C. (1992). *Autocatalytic networks under diffusion*. Ph.D. thesis, University of Vienna.

Szabó, P., Scheuring, I., Czaran, T., & Szathmáry, E. (2002). In silico simulations reveal that replicators with limited dispersal evolve towards higher efficiency and fidelity. *Nature*, *420*, 278–279.

Szathmáry, E., & Demeter, L. (1987). Group selection of early replicators and the orgin of life. *Journal of Theoretical Biology*, *128*, 463–486.

Szathmáry, E., & Gladkih, I. (1989). Sub-exponential growth and coexistence of non-enzymatically replicating templates. *Journal of Theoretical Biology*, *138*, 55–58.

Szostak, J., Bartel, D., & Luisi, P. L. (2001). Synthesizing life. *Nature*, *409* Supplemental, 387–390.

Taylor, P. D., & Jonker, L. B. (1978). Evolutionary stable strategies and game dynamics. *Mathematical Biosciences*, *40*, 145–156.

Tereshko, V. (1999). Selection and coexistence by reaction-diffusion dynamics in fitness landscapes. *Physics Letters A*, *260*, 522–527.

Tijvikua, T., Ballester, P., & Rebek Jr., J. (1990). A self-replicating system. *Journal of the American Chemical Society*, *112*, 1249–1250.

Vance, R. R. (1978). Predation and resource partitioning in one predator—Two prey model communities. *The American Naturalist*, *112*, 797–813.

Varga, S., & Szathmáry, E. (1997). An extremum principle for parabolic competition. *Bulletin of Mathematical Biology*, *59*, 1145–1154.

von Kiedrowski, G. (1986). A self-replicating hexadeoxynucleotide. *Angewandte Chemie International Edition English*, *25*, 932–935.

von Kiedrowski, G. (1993). Minimal replicator theory I: Parabolic versus exponential growth. In *Bioorganic chemistry frontiers* (Vol. 3, pp. 115–146). Berlin, Heidelberg: Springer-Verlag.

Vrba, E. S. (1989). Levels of selection and sorting with special reference to the species level. In P. H. Harvey & L. Partridge (Eds.), *Oxford surveys in evolutionary biology* (Vol. 6, pp. 114–115). Oxford: Oxford University Press.

Whitten, D. G., Chen, L., Geiger, H. C., Perlstein, J., & Song, X. (1998). Self-assembly of aromatic-functionalized amphiphiles: The role and consequences of aromatic-aromatic noncovalent interactions in building supramolecular aggregates and novel assemblies. *Journal of Physical Chemistry B*, *102*, 10098–10111.

Wills, P. R., Kauffman, S. A., Stadler, B. M., & Stadler, P. F. (1998). Selection dynamics in autocatalytic systems: Templates replicating through binary ligation. *Bulletin of Mathematical Biology*, *60*, 1073–1098.

Wintner, E. A., Conn, M. M., & Rebek Jr., J. (1994). Self-replicating molecules: A second generation. *Journal of the American Chemical Society*, *116*, 8877–8884.

Wlotzka, B., & McCaskill, J. S. (1997). A molecular predator and its prey: Coupled isothermal amplification of nucleic acids. *Chemistry & Biology*, *4*, 25–33.

Yao, S., Ghosh, I., Zutshi, R., & Chmielewski, J. (1998). Selective amplification by auto- and cross-catalysis in a replicating peptide system. *Nature*, *396*, 447–450.

Zielinski, W. S., & Orgel, L. E. (1987). Autocatalytic synthesis of a tetranucleotide analogue. *Nature*, *327*, 346–347.

Zintzaras, E., Santos, M., & Szathmáry, E. (2002). "Living" under the challenge of information decay: The stochastic corrector model vs. hypercycles. *Journal of Theoretical Biology*, *217*, 167–181.

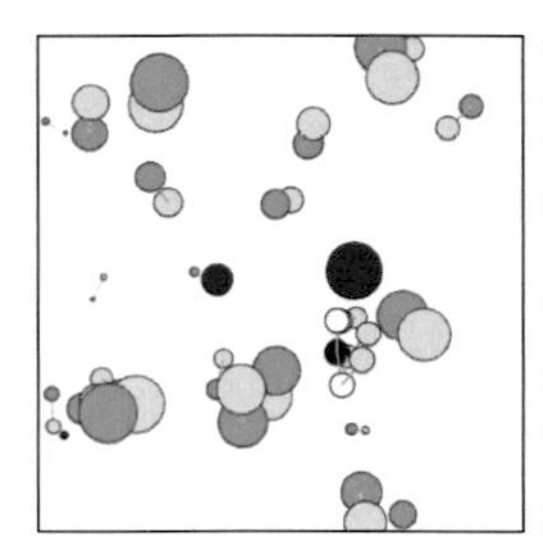
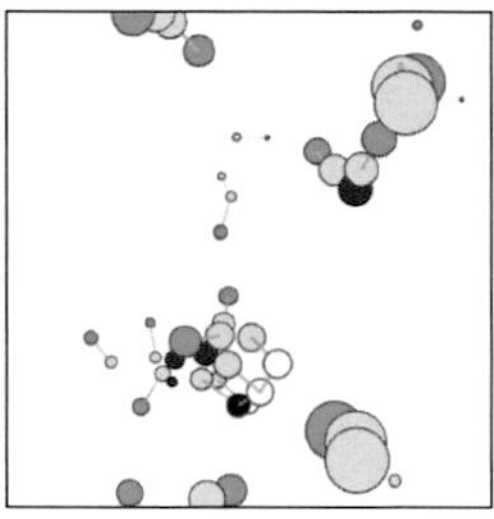
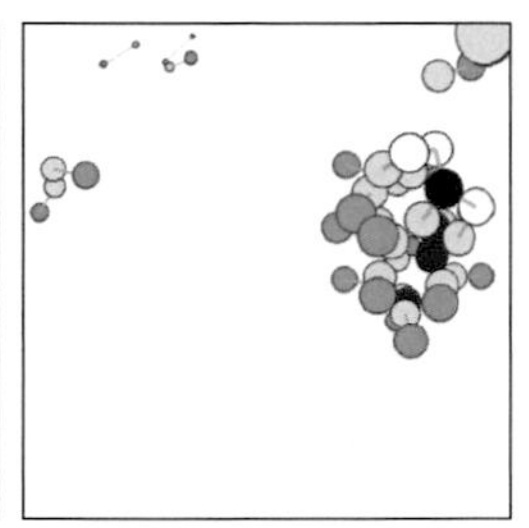

Plate 1 (figure 6.2)
Simplified dissipative particle dynamics (DPD) simulation (Fellermann et al. 2007) of protocellular component self-assembly in water (water not shown). Lipid molecules (L) are represented as amphiphilic dimers with a hydrophilic head (green) and a hydrophobic tail (yellow), the (hydrophobic) sensitizer molecules (Z) are represented as red particles, and the gene (a 4-mer template, T) is represented by a polymer with black and white monomers, each representing different bases, and yellow hydrophobic anchors on each monomer. The self-assembly dynamics is shown starting from random initial conditions (left), initial assembly (middle), and the fully assembled protocell (right). Note how the yellow lipid hydrocarbon tails define the micellar interior while the green lipid head groups orient themselves toward the water. The hydrophobic sensitizer molecules are located in the nonpolar micellar interior, while the amphiphilic gene is located at the water lipid interface. A hydrophobic (or amphiphilic) sensitizer could also be covalently linked to the backbone of the gene (not shown in this simulation).

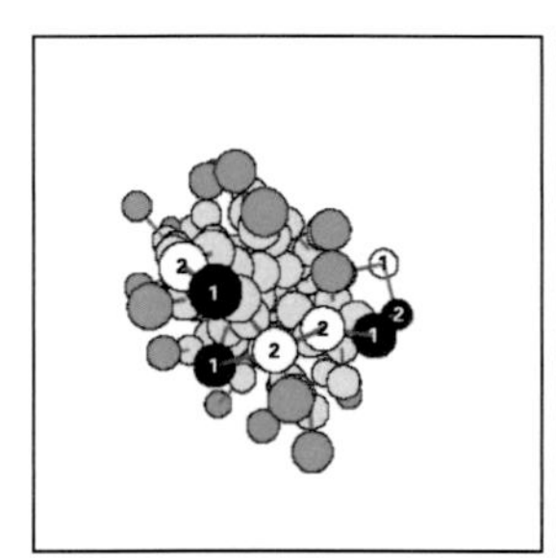
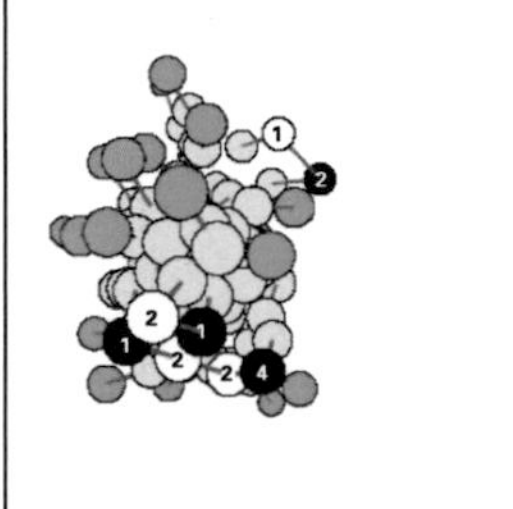
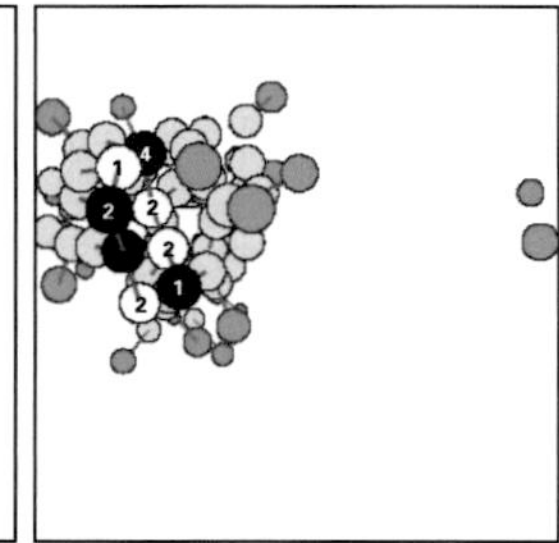

Plate 2 (figure 6.11)
DPD simulation of the three steps of template-directed (gene) ligation at a micellar interface (water not shown), where the micelle is loaded with oil-like (lipid precursor) molecules. The amphiphilic fatty acids are represented as green and yellow dimers (green=hydrophilic, and yellow=hydrophobic), and the oil-like dimers consist of two yellow beads. The simplified lipophilic PNA gene template consists of four monomers, ABBA. The A base component monomer is white, whereas the B base component monomer is black. This template can hybridize with two small PNA dimers consisting of BA and AB sequences (monomer A base pair with monomer B). The hydrophobic component of the PNA backbone is modeled by a (yellow) hydrocarbon tail attached to each PNA base component monomer. For details, see Fellermann et al. (2007). In the left panel, the three pieces of PNA are associated to the lipid aggregate interface. In the middle panel, one of the dimers is hybridized, and in the right panel, both dimers are hybridized and the ligation process occurs.

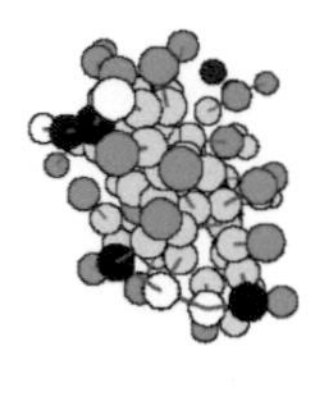
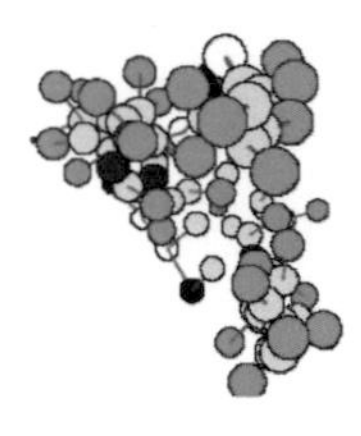
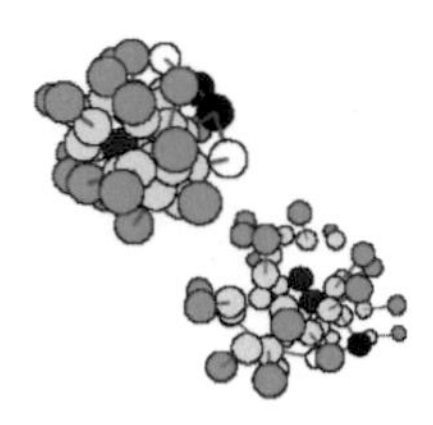

Plate 3 (figure 6.12)
DPD simulation of the protocellular division starting from preloaded micelle, free photosensitizers, and two dehybridize, single-stranded genes (left panel). For details, see Fellermann et al. (2007). Water molecules are not shown, the lipids are represented as green-yellow (head-tail) dimers, the photosensitizers are red, and the oily precursor lipids are yellow-yellow dimers. As the oily precursors are "digested" and transformed into fatty acids driven by the sensitizers and catalyzed by the genes, the micellar size exceeds its equilibrium size and becomes unstable (middle), and it eventually splits into two micelles (right). Note the nice partition of sensitizers and gene molecules in this particular simulation, which does not always occur. Further it should be noted that reloading ("feeding") the micelles with the oil-like precursor lipids is nontrivial. Simulation studies show that a delicate balance between feeding and diffusion rates is necessary to prevent the formation of large oil droplets that can absorb multiple micelles and thus destroy their individuality.

(a)

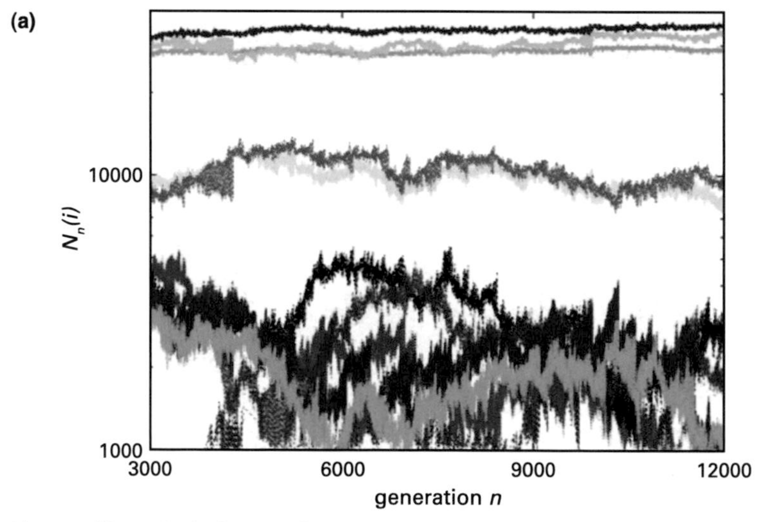

Plate 4 (figure 8.4) above and opposite
The number of molecules $N_n(i)$ for the species i is plotted as a function of the generation, n (after n division events). In (a), a random network with $k=500$ and $p=0.2$, and in (b) a random network with $k=200$ and $p=0.2$, was adopted, with $N=64000$ and $\mu=0.01$. Only some species (species whose population becomes large at some generation) are plotted. In (a), dominant species change successively by generation, whereas in (b) three quasi-recursive states are observed. Reproduced from Kaneko (2003a).

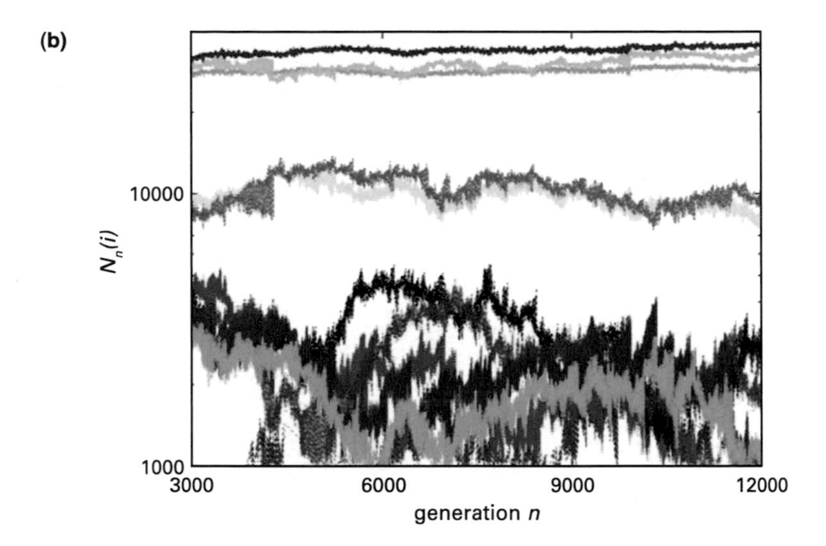
(b)
10000
1000
$N_n(i)$
3000
6000
9000
12000
generation n

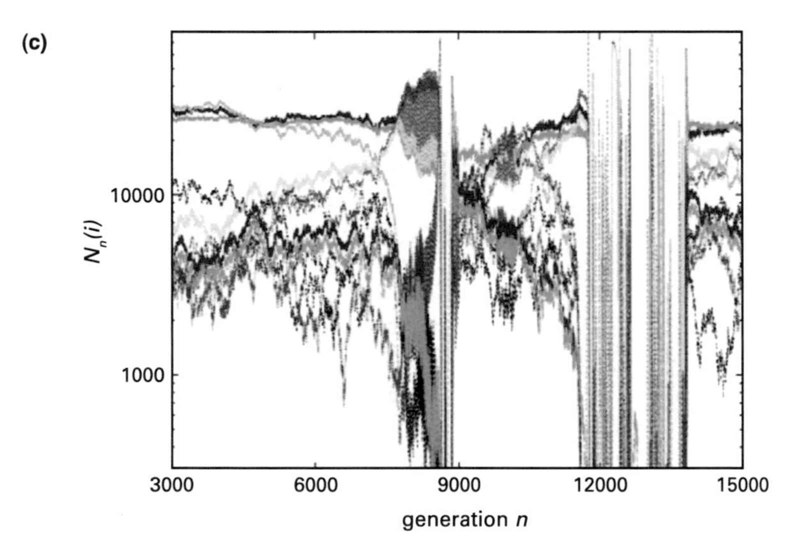
(c)
10000
1000
$N_n(i)$
3000
6000
9000
12000
15000
generation n

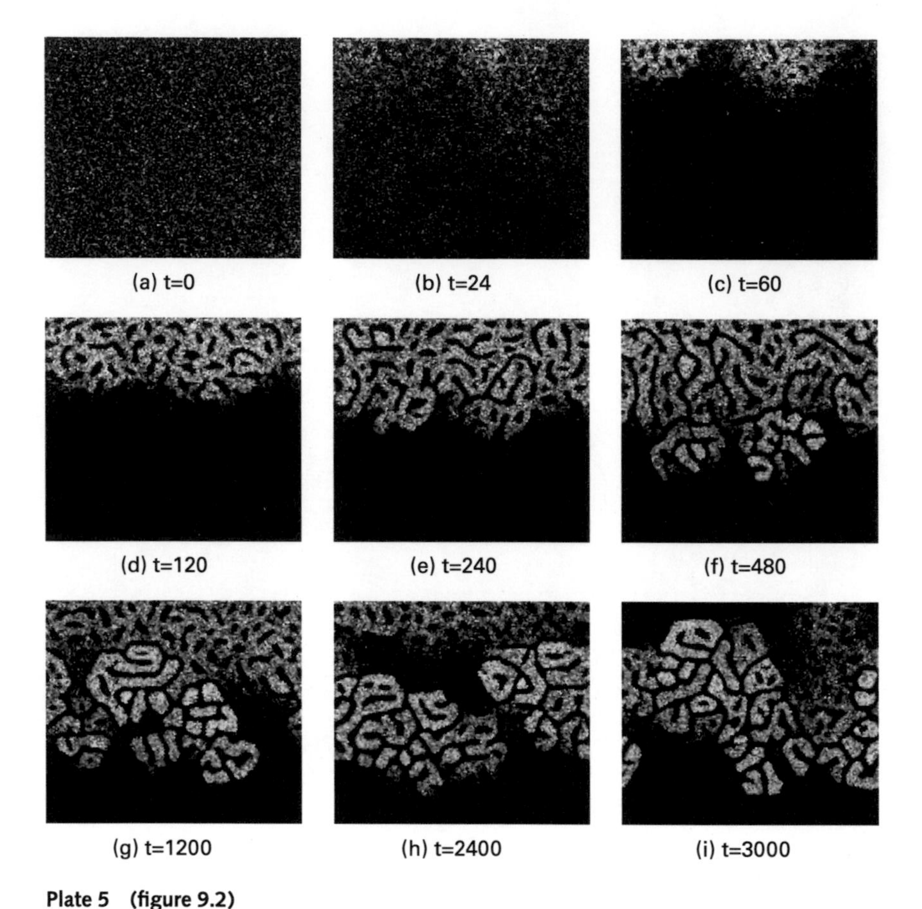

Plate 5 (figure 9.2)
Evolution from precellular metabolism to protocells. Resource supply decreases gradually from top to bottom. In the beginning, autocatalysts produce few membrane particles and they survive only in the richest area (a, b, c). Due to the selection which occurs, more efficient catalysts multiply and the total amount of membrane grows. Once a protocell is organized, it begins to reproduce itself by growing and then dividing. Protocells become more stable through evolution until they finally replace the precellular autocatalysts.

(a)

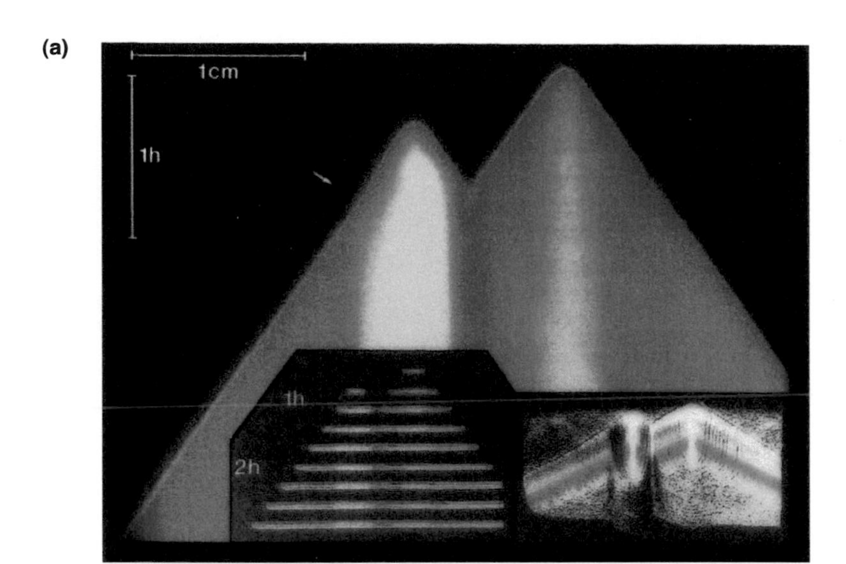

(b)

Plate 6 (figure 12.1)
Traveling wave experiments with Qß enzyme and minivariant RNA. (a) Fluorescence image of capillary wound back and forth into parallel 12-cm segments on a frame (up to 72 horizontal segments in total) showing RNA concentrations resulting from single templates, with increasing dilution of RNA introduced in the upper segments. (b) Space-time fluorescence image of single capillary segment showing two constant velocity wavefront pairs (with 3D data inset) and a single evolutionary event causing a change in front velocity at the arrow, manifested by a change in slope of the growth triangle on the space-time image. Image reproduced with permission from Bauer and McCaskill (1993).

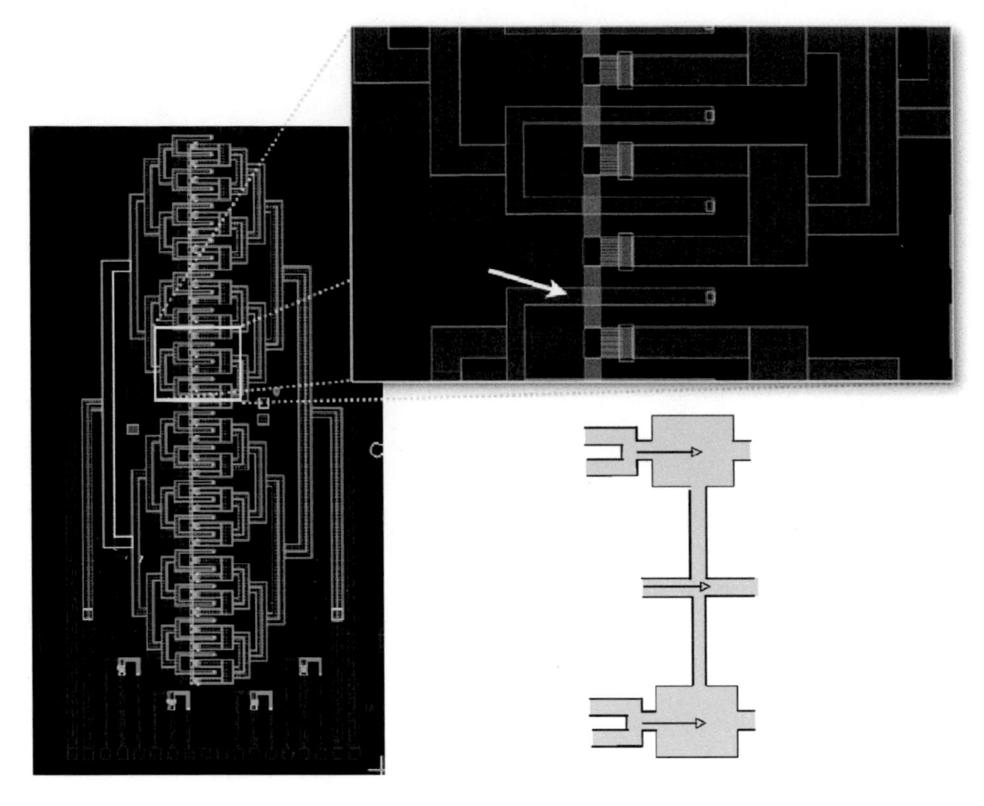

Plate 7 (figure 12.4)
Design of an extended open line reactor. The blue and green bifurcation trees on the right-hand side (on top of each other) deliver two different solutions from the lower and upper chip layers. Products are collected in the bifurcation cascade on the left. The lower right diagram shows a simplified schematic blowup of the section of the horizontal line reactor between two inflows, emphasizing the separation between top and bottom chambers. The separating flow in the center creates a tunable diffusive connection between the two outer reactors: If the flow here is fast, no interchange occurs on the line. The line reactor is shown in more detail vertically in the mask design on the left, and in the blowup (upper right); an arrow indicates the separation control channel. This design allows mixing of two fluids immediately before entry to the line reactor. Multilayer design is implemented by T. Palutke and P. Wagler, BioMIP.

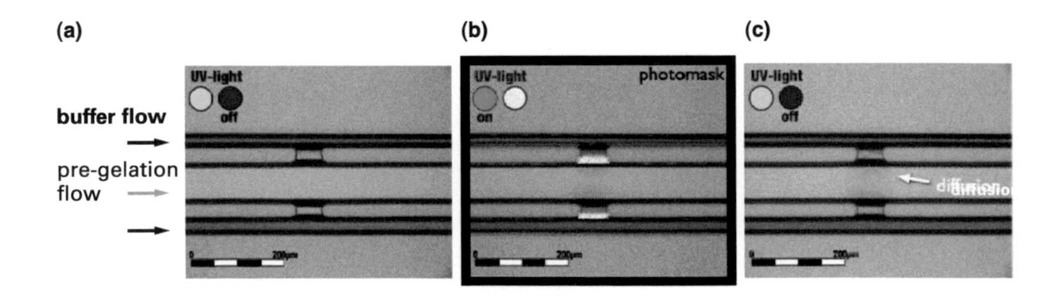

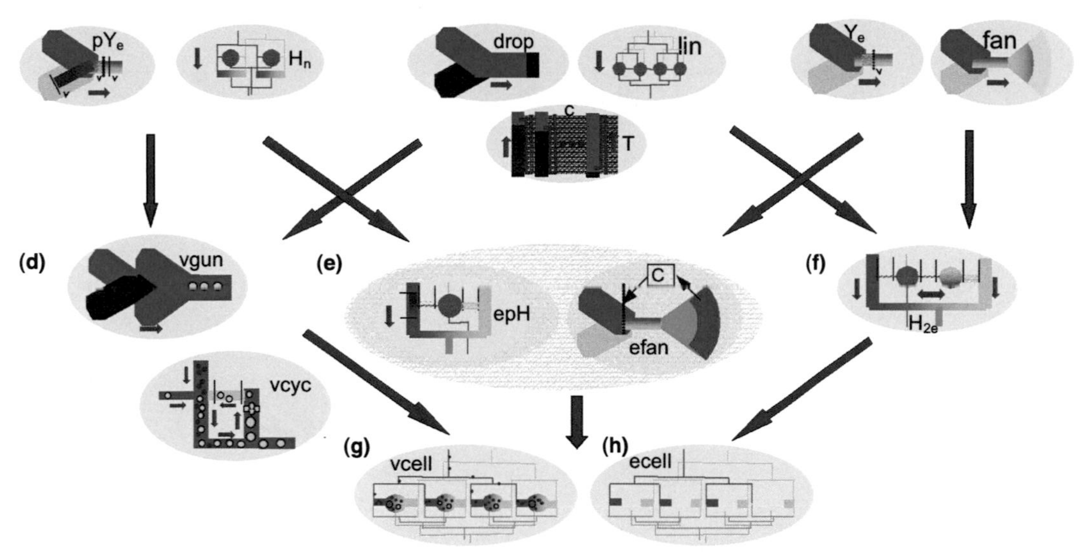

Plate 9 (figure 12.11)
Microfluidic complementation routes to the protocell. This scheme follows the general subsystem buildup as proposed in figure 4.3 (chapter 4). Electronic microfluidics can complement these routes by providing (a) regulated spatial control and (b) a further range of specific complementation, as described in table 12.2 and the figures. Top row illustrates microfluidic complementation of the separate subsystems with (from left to right) (a) *metabolism* (electronically regulated pH and concentration inflows, and gradient chemical control), (b) *containment* (autonomous formation of channel emulsion droplets, phase space screening with crossed gradient reactor, and geometric control of isolation in linear cross-flow reactor), and (c) *replication* (inflow electronic concentration gate for genetic material, and self-regulation fan reactor for localized homeostatic replication). Second row illustrates complementation support for pairwise combination of subsystems: (d) *metabolism+container* (self-regulated vesicle gun and vesicle replication cycle complemented by regulated micellar feed), (e) *metabolism+replication* (electronically regulated pH gradient controlling replication, and electronically regulated mixing ratio for metabolites controlling replication in fan reactor), and (f) *container+replication* (alternating chamber regulation and enabling of rapid replication). Third row illustrates two targets for complemented protocells integrating all three subsystems: (g) complemented vesicle-nucleic acid protocells, and (h) fully electronically complemented protocells with support for all three subsystems and electronic genomes.

◄ **Plate 8 (figure 12.6)**
Dynamical generation of diffusion barriers between microfluidic channels/chambers based on photopolymerization of hydrogels. (a) The central horizontal channel is initially filled with a flowing photopolymerizable pregelation mix (PEG-DA (poly-(ethyleneglycol)-diacrylate)), a cross-linking agent (methylenebisacrylamide) and the photoinitiator, HMPP (2-hydroxy-2-methyl-propiophenone), in TAE buffer solution, whereas the upper and lower channels contain sample solution (labeled buffer flow) dyed red in this experiment. UV illumination through a rectangular mask in (b) actual image during illumination, results in polymerization of the central channel, which is in contact with the buffer flows via shallow ledges. The resulting in-flow gelled barriers are shown in (c), where slow diffusion of dye through the gel from the buffer channels can be observed. They may be used for selectively reduced molecular transport of material to a microcell, functioning like a membrane as semipermeable molecular filters. Experiments by Patrick Wagler, Steffen Chemnitz, and Farsaneh Sadeghar, BioMIP.

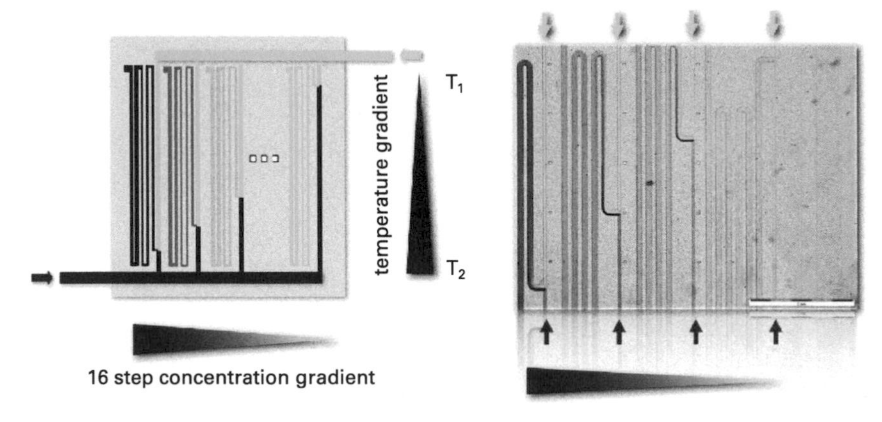

Plate 10 (figure 12.12)
Gradient microreactor for phase space screening of amphiphile systems. Gradients in temperature (vertical) and concentration ratio of two fluids (horizontal) are shown in the schematic (left) and in a blowup of the actual microreactor (right). The microscope image shows hydrodynamic study in microfluidic gradient system BioGrad² with the mixing of two fluids at a sequence of T-junctions. The different channel lengths give different resistivities to hydrodynamic flow, causing a different mixing ratio at the junction. Note that meanders, raising and lowering the mixed fluid temperature, allow one to distinguish kinetic from equilibrium phase change effects. Revised design by Steffen Chemnitz; experimental image by Patrick Wagler and Steffen Chemnitz, BioMIP.

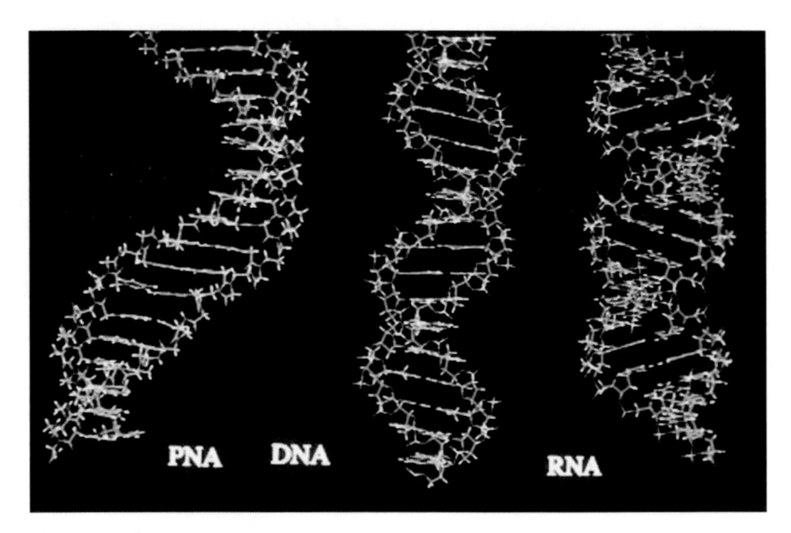

Plate 11 (figure 15.3)
Helical structures of homoduplexes of PNA, DNA, and RNA.

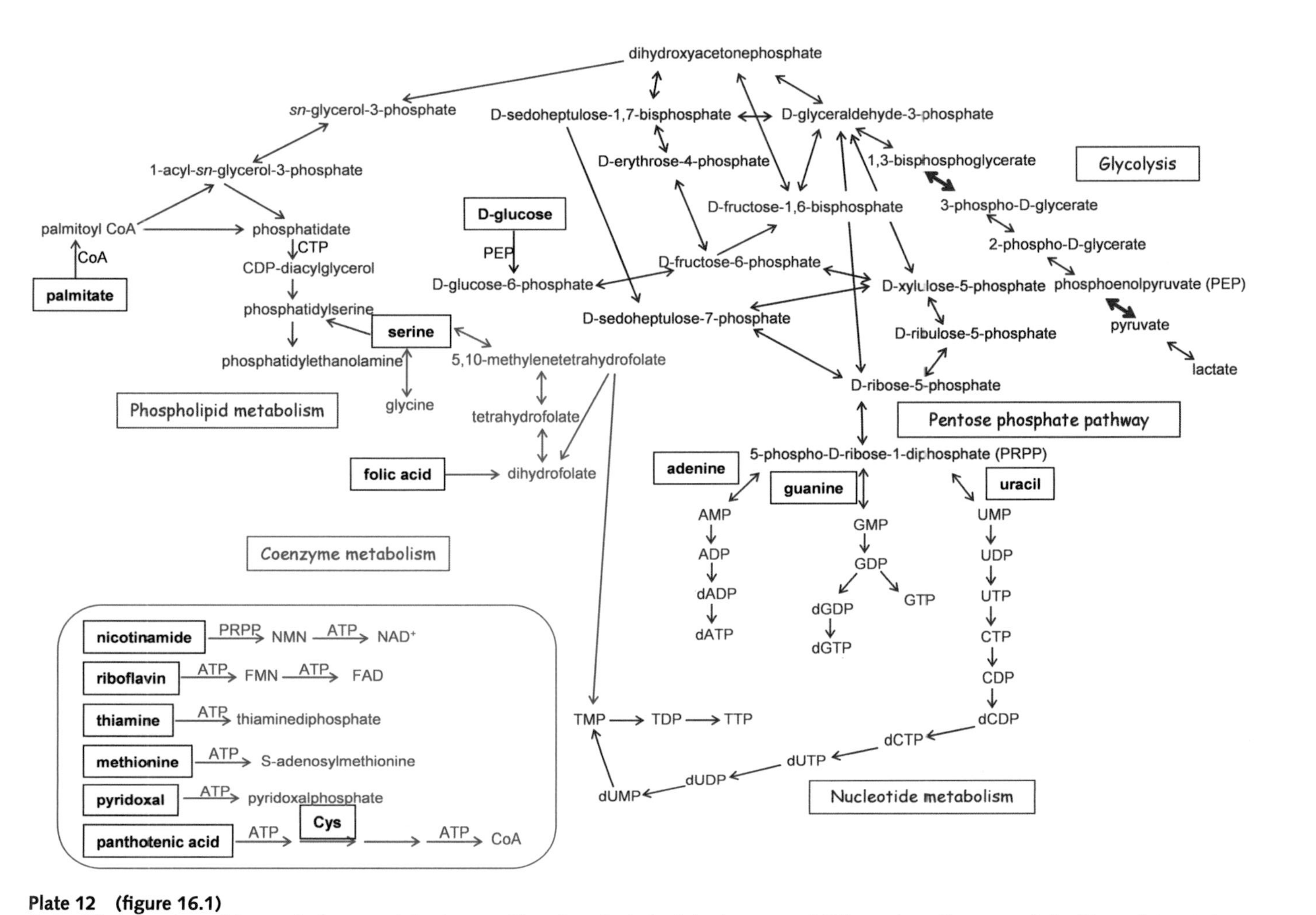

Plate 12 (figure 16.1)
A simplified overview of the metabolic network implemented by a hypothetical minimal genome of 206 protein-coding genes derived by an integrated approach taking into account genomewide computational, experimental and metabolic studies on completely sequenced bacterial genomes. Names of substrates freely available for the hypothetical minimal cell are represented in boldface characters and inside a frame. Coenzyme metabolism (except the folate metabolism linked to nucleotide metabolism) is shown in the inset and was not considered in the network analysis. Wider arrows in the glycolytic pathway indicate the only two steps in which ATP is synthesized by substrate-level phosphorylation.

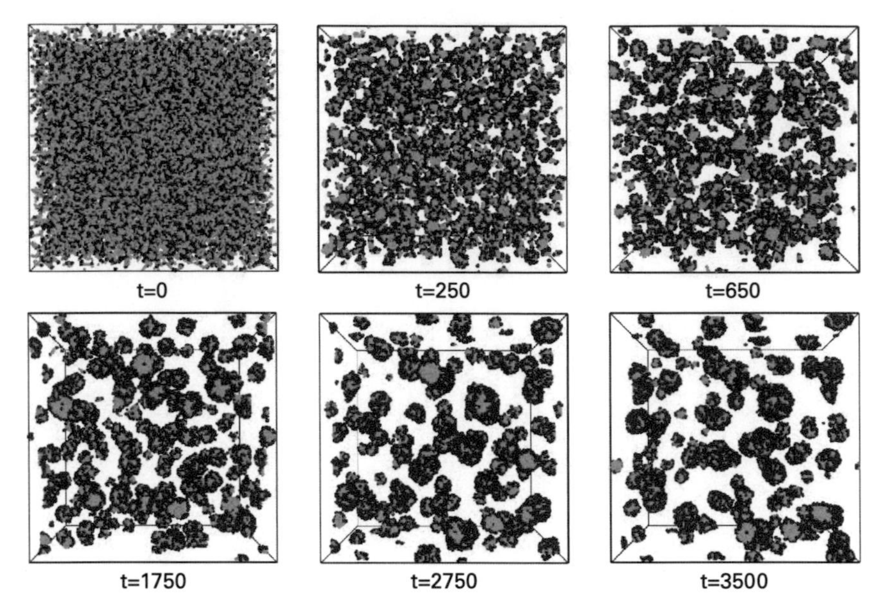

Plate 13 (figure 18.9)
Snapshots of micelle formation from a random mixture of HT_6 lipids simulated by dissipative particle dynamics. The dimension of the cubic box is $L=96r_0$, and the number density is $\rho r_0^3 = 3$. The model parameters are adapted from Shillcock and Lipowsky (2002). Head-group beads are red and tail beads are green. For the sake of clarity, the water beads are not shown. The DPD timescale, t, is in units of $t_0 = \sqrt{m_0 r_0^2 / k_B T}$. The individual beads are drawn with a radius corresponding to $0.5r_0$.

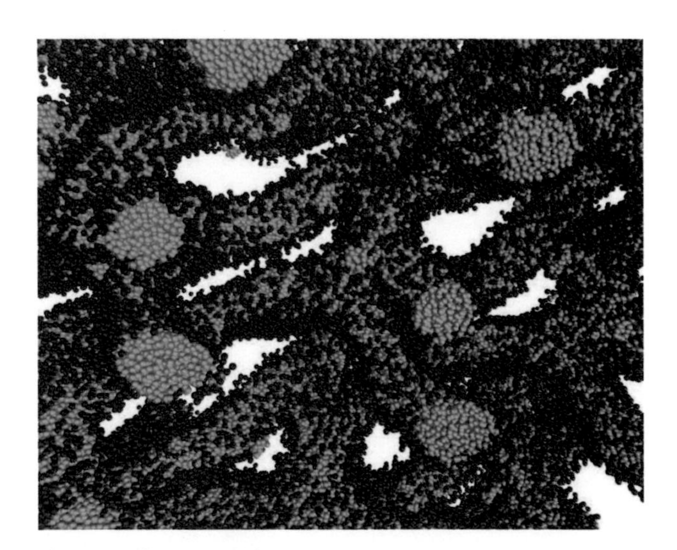

Plate 14 (figure 18.10)
Snapshot of a configuration of lipid aggregates in the form of a filament of rodlike micelles formed by $H_3(T_5)_2$ lipids and simulated by dissipative particle dynamics. The dimension of the cubic box is $L=32r_0$, and the number density is $\rho r_0^3 = 3$. The model parameters are adapted from Groot and Rabone (2001). The red beads represent hydrophilic head groups (H) and the green beads represent the tail beads (T). For the sake of clarity, the water beads are not shown.

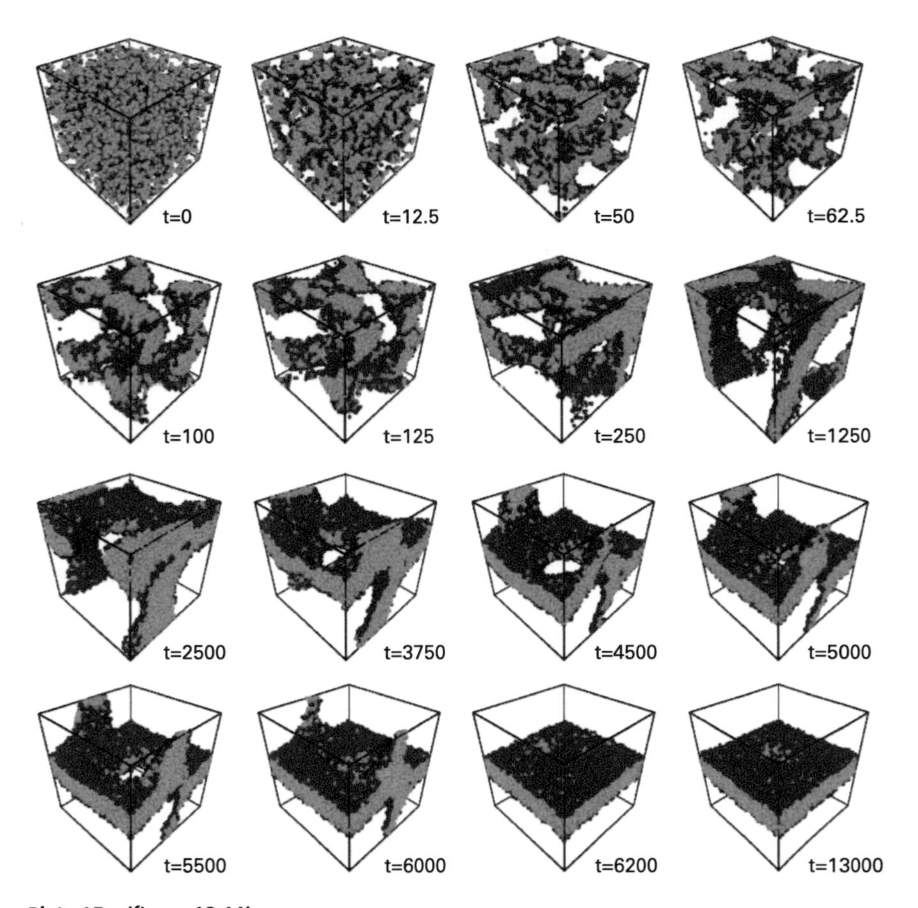

Plate 15 (figure 18.11)

Snapshots of bilayer formation from a random mixture of $H_3(T_6)_2$ lipids simulated by dissipative particle dynamics. The dimension of the cubic box is $L = 32r_0$ and the number density is $\rho r_0^3 = 3$. The model parameters are from Shillcock and Lipowsky (2002). Head-group beads are red and tail beads are green. For the sake of clarity, the water beads are not shown. The DPD timescale, t, is in units of $t_0 = \sqrt{m_0 r_0^2 / k_B T}$.

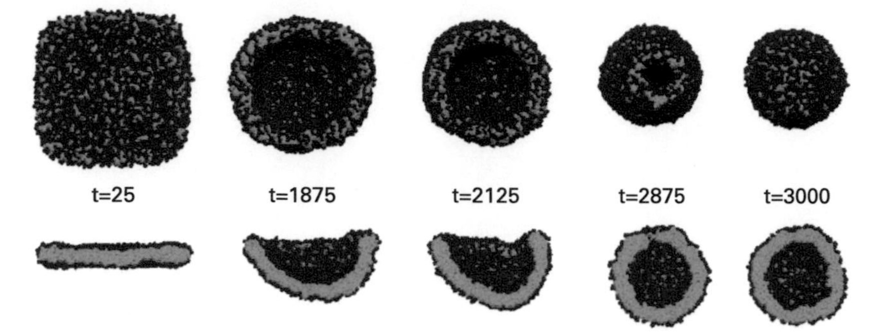

Plate 16 (figure 18.12)
Snapshots of vesicle formation from a preassembled bilayer sheet of HT3 lipid molecules simulated by dissipative particle dynamics. The dimension of the cubic box is L=60r_0 and the number density is $\rho r_0^3 = 3$. Parameters are adapted from Yamamoto, Maruyama, and Hyodo (2002). These parameters produce a totally interdigitated bilayer. The bottom row of snapshots show a cut through the bilayer. Head-group beads are red and tail beads are green. Water beads are not shown. The DPD timescale, t, is in units of $t_0 = \sqrt{m_0 r_0^2 / k_B T}$.

Plate 17 (figure 18.14)
Brownian dynamics simulation of the fusion of two vesicles revealing the formation of a fusion intermediate in the form of a membrane stalk that connects only the outer monolayers of the two fusing vesicles. Time proceeds from left to right, and only cross-sections of the vesicles are shown. The total number of molecules in the simulations was 1,000. Adapted from Noguchi and Takasu (2001).

(a)

(b)

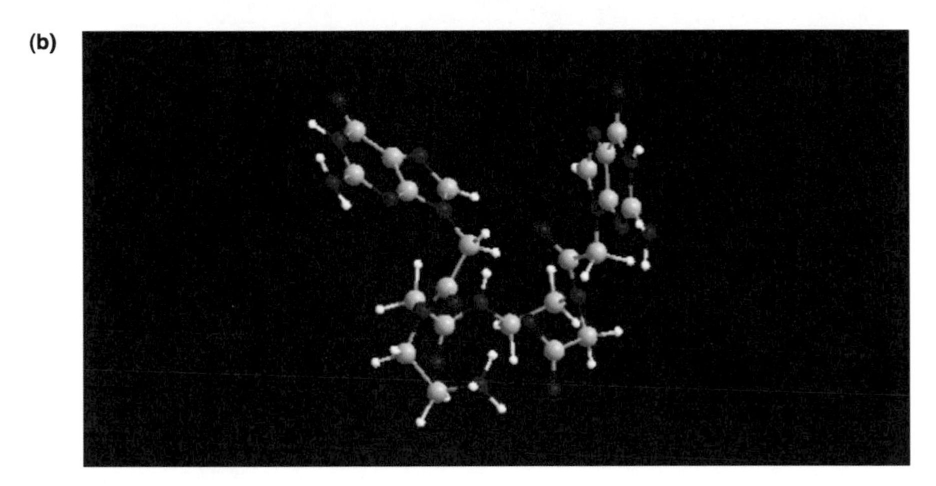

Plate 18 (figure 19.2)
Semiempirical calculation of small peptide nucleic acid (PNA) segment, PNA-g-g, in water. Two guanine bases point upward with the backbone on the bottom. (a) The initial stacked configuration of the guanine pair. (b) After energy minimization using MOPAC (see appendix I) with AM1 method, the guanine pair becomes less parallel. Red=oxygen, blue=nitrogen, gray=carbon.

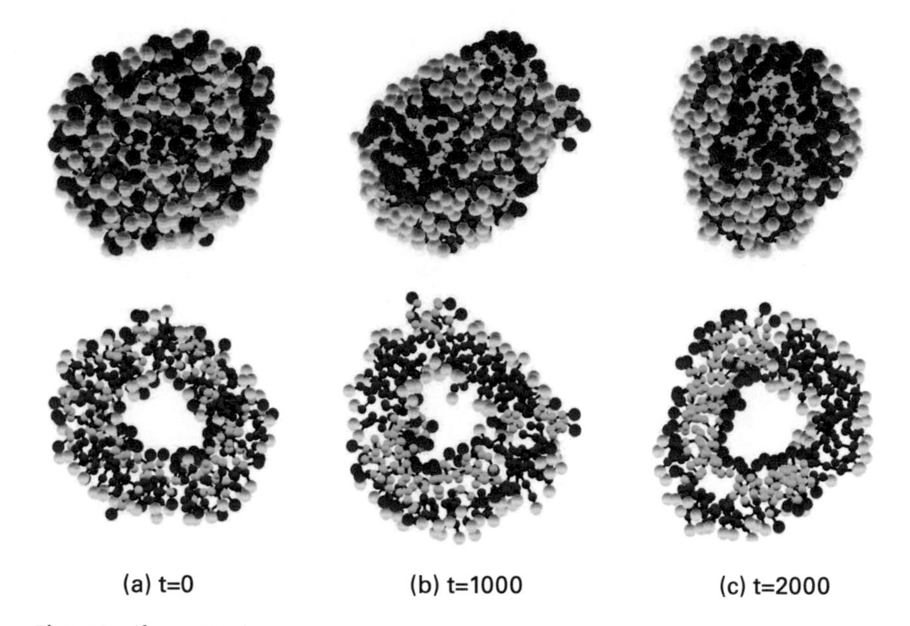

Plate 19 (figure 19.4)
DPD simulation of phase separation of a two-component lipid vesicle. The interaction parameters are adapted from Yamamoto and Hyodo (2003). Two types of lipids are indicated as yellow head, blue tail, and red head, cyan tail, respectively. The top row shows the surface view and the bottom row shows the cross-sectional view.

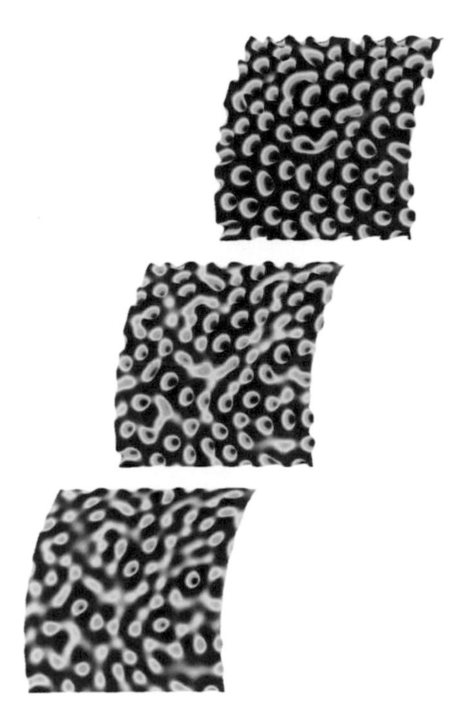

Plate 20 (figure 19.8)
Ginzburg-Landau simulation of binary lipid phase separation coupled to local deformation (equations 19.9 and 19.10). Color encodes the lipid concentration field. Red and blue are two lipid components and the initial condition is a curved membrane surface with one constant curvature and homogeneous lipid mixture.

Plate 21 (figure 24.4)
A cell made from a semipermeable calcium carbonate membrane surrounded by a purple solution resulting from the formation of a complex of iodine and starch. The green-blue color inside the cell is caused by the copper catalyst.

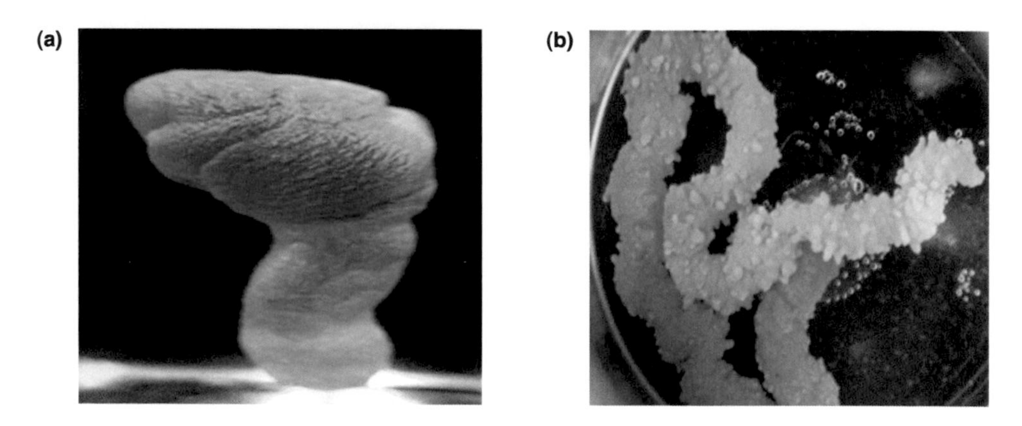

Plate 22 (figure 24.9)
Influence of an additional compound on cell formation. In (a) the cell developed in a Cu^{2+}-silicate system, whereas in (b) the structure developed in a Cu^{2+}-silicate-glycine system. The additional structures seen on this cell's surface results because glycine is both a complexing agent and a surfactant able to form its own cells.

15 Peptide Nucleic Acids as Prebiotic and Abiotic Genetic Material

Peter E. Nielsen

15.1 Introduction

Peptide nucleic acids (PNA) are structural mimics of nucleic acids based on a pseudopeptide backbone composed of N-(aminoethyl)glycine (AEG) units (Nielsen et al., 1991; see figure 15.1). PNA oligomers are hydrophilic, noncharged, and achiral. They exhibit extraordinarily high chemical and biological stability (Demidov et al., 1994), and they are easy to synthesize and chemically modify (e.g., Christensen et al., 1995; Ganesh and Nielsen, 2000; Thomson et al., 1995).

Because of these properties, combined with their ability to hybridize in a sequence-specific way to complementary strands of DNA, RNA, or another PNA (Egholm et al., 1993; Jensen et al., 1997; Wittung et al., 1994), these molecules have attracted widespread attention in science ranging from drug discovery and development to molecular biology and genetic diagnostics to chemistry and nanotechnology (e.g., Nielsen, 1999, 2004; Stender, 2003).

It came as a great surprise that a molecule that is essentially a peptide could mimic, and in principle functionally replace, the genetic material of modern life, that is, DNA in prebiotic or abiotic life forms. Thus, the creation of PNA added a new perspective to discussions of the origin of life on Earth (Nielsen, 1993). In particular, it has been shown that chemical flow of genetic information between PNA oligomers, as well as from a PNA oligomer to an RNA oligomer, is in principle possible with use of homooligomer templates and chemically activated precursor substrates (figure 15.2; Böhler, Nielsen, and Orgel, 1995; Schmidt, Nielsen, and Orgel, 1997). However, efficient *prebiotic replication* has not yet been demonstrated. Furthermore, aminoethyl glycine (AEG) PNA backbone as well as nucleobase acetic acid PNA building blocks have been identified in "prebiotic soup" experiments (Nelson, Levy, and Miller, 2000), and precursors for aminobutyric acid- or ornithine-based PNA oligomers (Nielsen, 1993) have been identified in meteorites (Meierhenrich et al., 2004). Therefore, PNA-like oligomers should be considered as

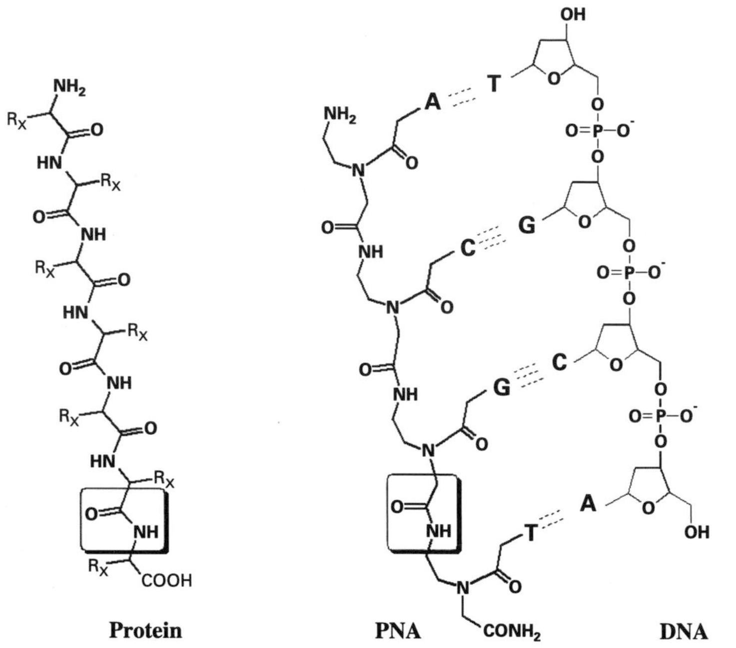

Figure 15.1
Chemical structures of protein, PNA, and DNA illustrating the close chemical similarity between PNA and protein.

possible prebiotic genetic material that may have played a role in the early stages of the origin of life on Earth, in the universe, or both. Of course, PNA is also an intriguing candidate as the genetic material for the construction of artificial life forms that in some ways mimic the fundamental properties of biological life but in other ways are distinct.

15.2 PNA Chemistry

PNA oligomers are conveniently available through conventional solid-phase peptide synthesis, and they can be conjugated to peptides and a variety of ligands on the solid support. Furthermore, a variety of methods employed in peptide chemistry for ligations in solution are also available for PNA. Although single-stranded PNA oligomers are very flexible without any distinct structure, PNA-PNA duplexes adopt a P-form helix conformation reminiscent of the well-known B-form DNA helix but with a larger pitch (18 bp instead of 10.5 bp) and a larger diameter (20 Å rather than 16 Å) (Rasmussen et al., 1997; figure 15.3, color plate 11). Because the PNA

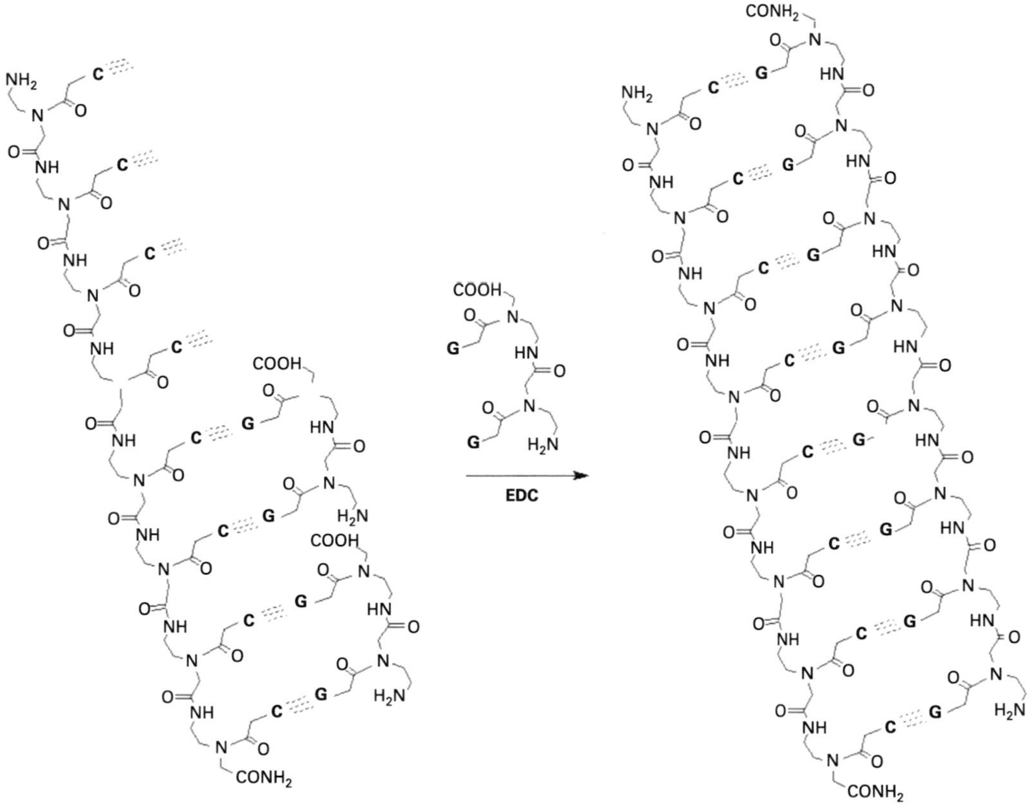

Figure 15.2
Schematic drawing of the oligomirization of PNA-G_2 on a PNA C_{10} template.

molecule is inherently achiral, PNA double helices exist in an equilibrium between a right- and a left-handed form (Wittung et al., 1994). Chirality can, however, be induced in the helix by appended chiral ligands such as amino acids (Wittung et al., 1994, 1995) or nucleosides (Kozlov, Orgel, and Nielsen, 2000). Furthermore, PNA oligomers of appropriate sequence may form secondary and tertiary structures, including hairpins (Armitage et al., 1998), triplexes (Petersson et al. 2005; Wittung, Nielsen, and Nordén, 1997), and quadruplexes (Datta et al., 2005; Krishnan-Ghosh, Stephens, and Balasubramanian, 2004) analogous to those formed by DNA. Thus, one can imagine an achiral PNA world containing a variety of PNA structures and catalytic PNA oligomers that, through a selection process, became a chiral PNA world, for example, through interactions with chiral RNA, along the way to converting to an RNA world.

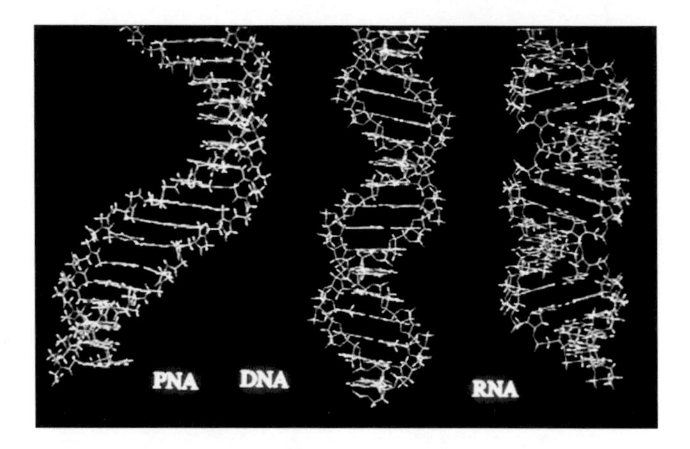

Figure 15.3 (color plate 11)
Helical structures of homoduplexes of PNA, DNA, and RNA.

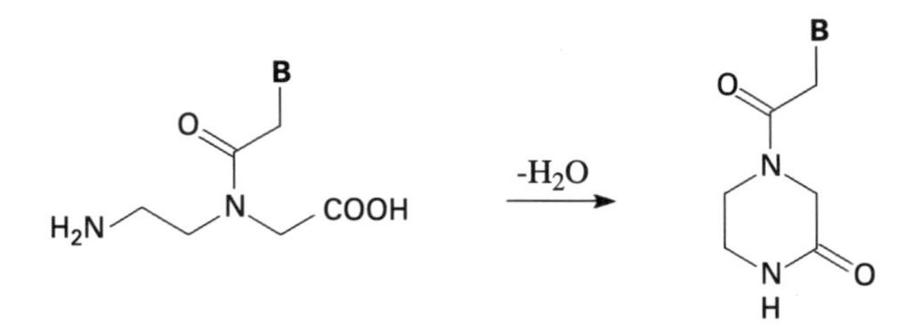

Figure 15.4
Formation of piperazinone from a PNA monomer.

15.3 PNA Replication

No enzymes that recognize or use PNA as a substrate have yet been discovered or developed. PNA- (or RNA-) directed PNA replication must therefore rely on chemical activation and catalysis. It has been demonstrated that a PNA G-oligomer can be synthesized on a PNA homocytosine decamer with the use of PNA G-dimers as substrates and EDC as condensing/activating agent (Böhler, Nielsen, and Orgel, 1995; figure 15.2). It is necessary to use G-dimers rather than G-monomers as precursors because of the ease with which the monomer cyclizes to the piperazinone (figure 15.4). Longer PNA oligomers may also be assembled by PNA-directed PNA-PNA ligation (Mattes and Seitz, 2001; figure 15.5). For neither of these processes has the efficiency and the fidelity (specificity) of the "replication" been studied in any detail. Nonetheless, in a prebiotic or abiotic context, this system suffers from the same type of product inhibition that prohibits exponential replication (Sievers and von Kie-

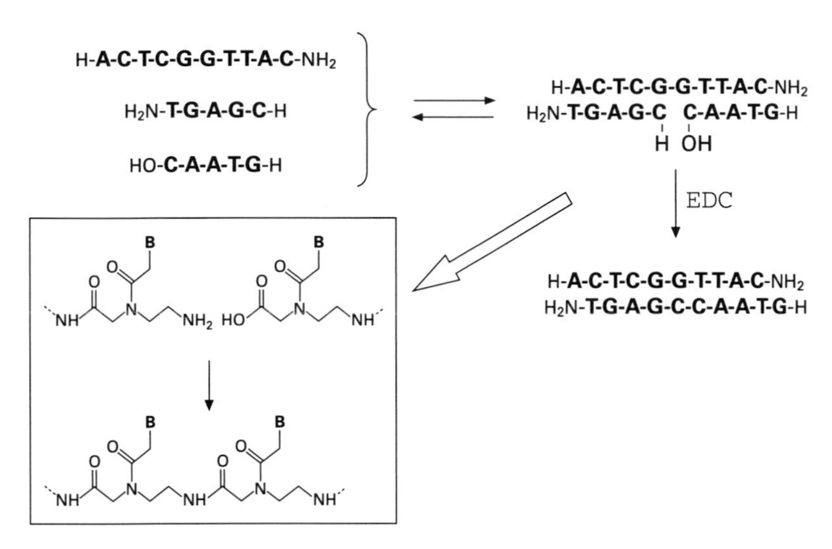

Figure 15.5
Schematic drawing of PNA-directed PNA-PNA ligation by EDC activation (arbitrary sequence). The box shows the chemical structure at the ligation point.

drowski, 1994), and for longer templates only a stoichiometric amount of product is formed (figure 15.6).

15.4 PNA Catalysis

A prebiotic scenario based solely on PNA would require that different PNA oligomers/polymers have varying advantages for evolutionary selection. In analogy to our present-day protein/RNA/DNA world and the hypothesized early RNA World (see chapter 17), this evolutionary advantage would most likely be catalytic activity. PNA oligomers with specific ligand affinities (PNA aptamers) or catalytic PNA oligomers (PNazymes) have not yet been identified or discovered. However, it is now well established that PNA oligomers of appropriate sequences can form well-defined three-dimensional folded structures (Petersson et al., 2005), and there is no doubt that future research will show that some of these behave as aptamers and others possess catalytic activity. Whether such properties are sufficient to sustain a primitive form of life is, of course, still an open question.

Furthermore, the nucleobases of PNA should be able to function as a charge diffusion (capture) relay in analogy to the behavior of double-stranded DNA (Giese and Biland, 2002; Wagenknecht, 2003). This effect may be exploited for direct chemical (phenotype-genotype) coupling of a photoelectronic metabolic process to the sequence of the genetic material.

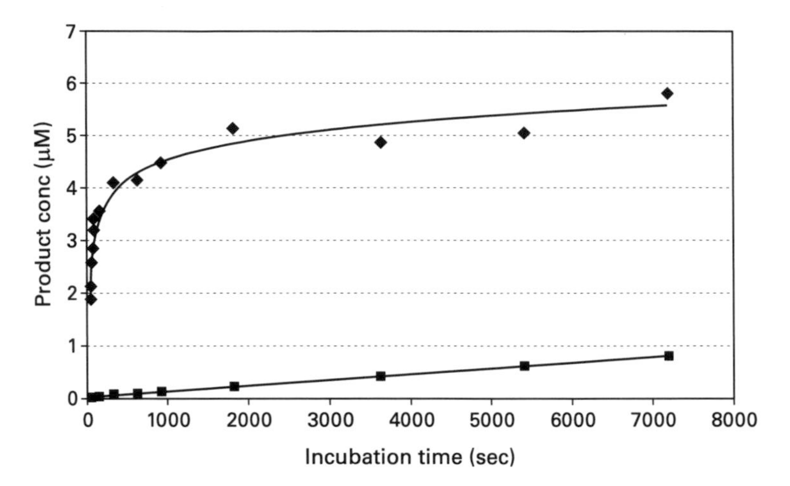

Figure 15.6
Kinetics of PNA-template-directed PNA ligation using two pentamer PNAs on a decamer template (see figure 15.5) analyzed by HPLC. Diamonds and squares show ligation in the presence and absence of template, respectively (Sen and Nielsen, unpublished).

15.5 PNA in Protocells

One protocell model is composed of three interconnected functional components that can be optimized and analyzed independently (see chapter 6). Ultimately, the model requires that the genetic material be stably associated with a lipid container (relative to the reproduction time of the protocell), such that a strong connection between genotype and phenotype is maintained. Furthermore, a coupling between the genetic template and the metabolic chemistry is envisaged, as this would allow selection and evolution on the genetic material on the basis of metabolic performance.

It should be evident from the preceding information about PNA that it is of eminent interest for the synthesis of protocells that are not closely based on present-day biological chemistry. Self-sustained protocells should consist of physically or otherwise confined compartments containing hereditable (genetic) material coupled to a metabolic system. Furthermore, the heritable material in protocells must have some chemical function (genotype-phenotype coupling) and be capable of mutating in order to support evolutionary selection.

Since PNA is inherently an uncharged molecule, it might well have a distinct advantage over RNA and other phosphodiester-based material for confinement within compartments. Efforts for physical incorporation (association) of the template with the lipid phase involve chemically modifying PNAs either within the backbone (for

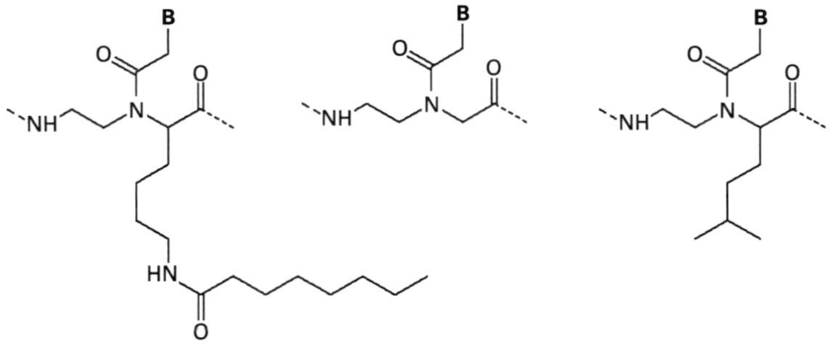

Figure 15.7
Chemical structures of octanoyl-lysine PNA (left), normal PNA (middle), and lecine PNA (right). B is a nucleobase.

embedding and dissolution in the lipid phase) or by chemical conjugation of lipophilic groups. Chemical modification such as lipid conjugation (Ljungstrøm, Knudsen, and Nielsen, 1999) and lipophilic backbone modification (figure 15.7; see Püschl et al., 1998) permit one to construct PNA molecules with affinity for micelles or liposomes (Ljungstrøm, Knudsen, and Nielsen, 1999; Vernille, Kovell, and Schneider, 2004). Thus, one might envision a system in which the PNA is physically integrated into the lipid compartment instead of being merely entrapped by it, as are DNA and RNA in modern life. Indeed, it should be possible to construct PNA oligomers that are freely soluble in organic solvents and thus in micelles. Encouraging recent results have indicated that base-paired PNA duplexes, in contrast with DNA duplexes, are stable under such conditions and will retain sequence-specific base-pair recognition (Sen and Nielsen 2006, 2007). Specifically, it is found that PNA duplexes retain stability in the presence of aprotic organic solvents such as dimethylformamide or dioxane (Sen and Nielsen, 2006), and an extrapolation to 100% organic solvent indicates only a limited decrease in thermal stability of such PNA double helices (figure 15.8; Sen and Nielsen, 2007).

Finally, along the lines of direct coupling between the genetic material and the metabolism of a protocell (Rasmussen et al., 2003; see also chapter 6), conjugating "metabolically active" ligands, such as polyaromatic ruthenium complexes that may exploit the PNA nucleobase as electron relay, to PNA oligomers is relatively straighforward (figure 15.9). Consequently, the chemicophysical and chemical properties of PNA open novel avenues for construction of protocells that are not readily available with the use of biological DNA- or RNA-based systems. The fact that PNA is an artificial molecule, however, creates limitations because PNA chemistry cannot take advantage of the complex enzymatic reactions that have evolved in the history of life.

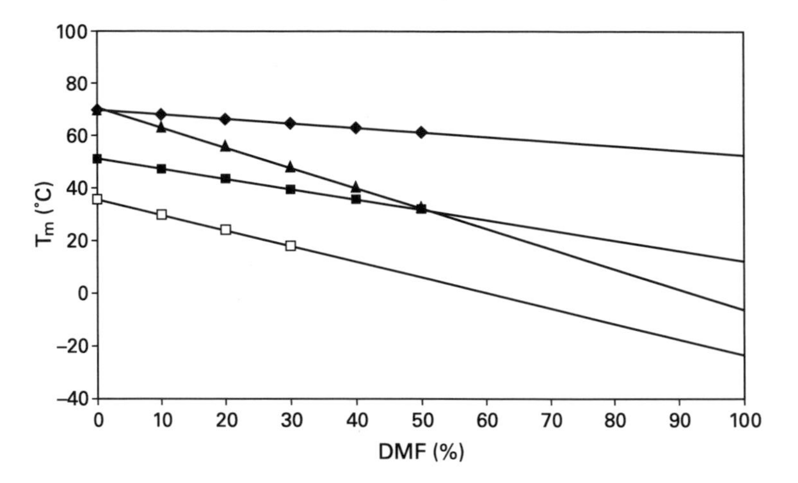

Figure 15.8
Plots of T_m of a PNA duplex (—◆—), a PNA-DNA duplex (—■—), and DNA duplexes (—□—, —▲—) as a function of the amount of DMF in the solvent.

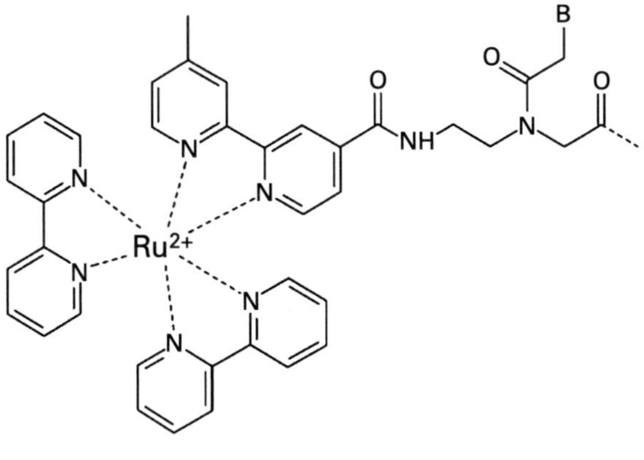

Figure 15.9
Example of the structure of a PNA-Ru-tri-bipyridyl conjugate.

References

Armitage, B., Ly, D., Koch, T., Frydenlund, H., Ørum, H., & Schuster, G. B. (1998). Hairpin-forming peptide nucleic acid oligomers. *Biochemistry, 37*, 9417–9425.

Böhler, C., Nielsen, P. E., & Orgel, L. E. (1995). Template switching between PNA and RNA oligonucleotides. *Nature, 376*, 578–581.

Christensen, L., Fitzpatrick, R., Gildea, B., Petersen, K. H., Hansen, H. F., Koch, T., et al. (1995). Solid-phase synthesis of peptide nucleic acids. *Journal of Peptide Science, 1*, 185–183.

Datta, B., Bier, M. E., Roy, S., & Armitage, B. A. (2005). Quadruplex formation by a guanine-rich PNA oligomer. *Journal of the American Chemical Society, 127*, 4199–4207.

Demidov, V. V., Potaman, V. N., Frank-Kamenetskii, M. D., Egholm, M., Buchardt, O., Sönnichsen, S. H., & Nielsen, P. E. (1994). Stability of peptide nucleic acids in human serum and cellular extracts. *Biochemical Pharmacology, 48*, 1310–1313.

Egholm, M., Buchardt, O., Christensen, L., Behrens, C., Freier, S. M., Driver, D. A., et al. (1993). PNA hybridizes to complementary oligonucleotides obeying the Watson-Crick hydrogen-bonding rules. *Nature, 365*, 566–568.

Ganesh, K. N., & Nielsen, P. E. (2000). Peptide nucleic acids: Analogs and derivatives. *Current Organic Chemistry, 4*, 931–943.

Giese, B., & Biland, A. (2002). Recent developments of charge injection and charge transfer in DNA. *Chemical Communications*, 667–672.

Jensen, K. K., Ørum, H., Nielsen, P. E., & Nordén, B. (1997). Kinetics for hybridization of peptide nucleic acids (PNA) with DNA and RNA studied with the BIAcore technique. *Biochemistry, 36*, 5072–5077.

Kozlov, I. A., Orgel, L. E., & Nielsen, P. E. (2000). Remote enantioselection transmitted by an achiral peptide nucleic acid backbone. *Angewandte Chemie, 39*, 4292–4295.

Krishnan-Ghosh, Y., Stephens, E., & Balasubramanian, S. (2004). A PNA(4) quadruplex. *Journal of the American Chemical Society, 126*, 5944–5945.

Ljungstrøm, T., Knudsen, H., & Nielsen, P. E. (1999). Cellular uptake of adamantyl conjugated peptide nucleic acids. *Bioconjugate Chemistry, 10*, 965–972.

Mattes, A., & Seitz, O. (2001). Sequence fidelity of a template-directed PNA-ligation reaction. *Chemical Communications*, 2050–2051.

Meierhenrich, U. J., Caro, G. M. M., Bredehöft, J. H., Jessberger, E. K., & Thiemann, W. H. P. (2004). Identification of diamino acids in the Murchison meteorite. *Proceedings of the National Academy of Sciences of the United States of America, 101*, 9182–9186.

Nelson, K. E., Levy, M., & Miller, S. L. (2000). Peptide nucleic acids rather than RNA may have been the first genetic molecule. *Proceedings of the National Academy of Sciences of the United States of America, 97*, 3868–3871.

Nielsen, P. E. (1993). Peptide nucleic acid (PNA): A model structure for the primordial genetic material. *Origins of Life and Evolution of the Biosphere, 23*, 323–327.

Nielsen, P. E. (1999). Peptide nucleic acid: A molecule with two identities. *Accounts of Chemical Research, 32*, 624–630.

Nielsen, P. E. (2004). The many faces of PNA. *Letters in Peptide Science, 10*, 135–147.

Nielsen, P. E., Egholm, M., Berg, R. H., & Buchardt, O. (1991). Sequence-selective recognition of DNA by strand displacement with a thymine-substituted polyamide. *Science, 254*, 1497–1500.

Petersson, B., Nielsen, B. B., Rasmussen, H., Larsen, I. K., Gajhede, M., Nielsen, P. E., & Kastrup, J. S. (2005). Crystal structure of a partly self-complementary peptide nucleic acid (PNA) oligomer showing a duplex-triplex network. *Journal of the American Chemical Society, 127*, 1424–1430.

Püschl, A., Sforza, S., Haaima, G., Dahl, O., Nielsen, P. E. (1998). Peptide nucleic acids (PNAs) with a functional backbone. *Tetrahedron Letters, 39*, 4707–4710.

Rasmussen, S., Chen, L., Nilsson, M., & Abe, S. (2003). Bridging non-living and living matter. *Artificial Life, 9*, 269–316.

Rasmussen, H., Kastrup, J. S., Nielsen, J. N., Nielsen, J. M., & Nielsen, P. E. (1997). Crystal structure of a peptide nucleic acid (PNA) duplex at 1.7 Å resolution. *Nature Structural and Molecular Biology*, *4*, 98–101.

Schmidt, J. G., Nielsen, P. E., & Orgel, L. E. (1997). Information transfer from peptide nucleic acids to RNA by template-directed syntheses. *Nucleic Acids Research*, *25*, 4797–4802.

Sen, A., & Nielsen, P. E. (2006). Unique properties of purine/pyrimidine asymmetric PNA-DNA duplexes: Differential stabilization of PNA-DNA duplexes by purines in the PNA strand. *Biophysical Journal*, *90*, 1329–1337.

Sen, A., & Nielsen, P. E. (2007). On the stability of peptide nucleic acid duplexes in the presence of organic solvents. *Nucleic Acids Research*, *35*, 3367–3374.

Sievers, D., & von Kiedrowski, G. (1994). Self-replication of complementary nucleotide-based oligomers. *Nature*, *369*, 221–214.

Stender, H. (2003). PNA FISH: An intelligent stain for rapid diagnosis of infectious diseases. *Expert Review of Molecular Diagnostics*, *3*, 649–655.

Thomson, S. A., Josey, J. A., Cadilla, R., Gaul, M. D., Hassman, C. F., Luzzio, M. J., et al. (1995). Fmoc mediated synthesis of peptide nucleic acids. *Tetrahedron*, *51*, 6179–6194.

Vernille, J. P., Kovell, L. C., & Schneider, J. W. (2004). Peptide nucleic acid (PNA) amphiphiles: Synthesis, self-assembly, and duplex stability. *Bioconjugate Chemistry*, *15*, 1314–1321.

Wagenknecht, H. A. (2003). Reductive electron transfer and transport of excess electrons in DNA. *Angewandte Chemie International Edition*, *42*, 2454–2460.

Wittung, P., Eriksson, M., Lyng, R., Nielsen, P. E., & Nordén, B. (1995). Induced chirality in PNA-PNA duplexes. *Journal of the American Chemical Society*, *117*, 10167–10173.

Wittung, P., Nielsen, P. E., Buchardt, O., Egholm, M., & Nordén, B. (1994). DNA-like double helix formed by peptide nucleic acid. *Nature*, *368*, 561–563.

Wittung, P., Nielsen, P., & Nordén, B. (1997). Observation of a PNA-PNA-PNA triplex. *Journal of the American Chemical Society*, *119*, 3189–3190.

16 The Core of a Minimal Gene Set: Insights from Natural Reduced Genomes

Toni Gabaldón, Rosario Gil, Juli Peretó, Amparo Latorre, and Andrés Moya

16.1 Introduction

Analyses of minimal genomes aim to define the repertoire of genes that is necessary and sufficient to support cellular life. Besides shedding light on which functions are essential for modern cells, this information might be applied to other purposes, including synthesizing a living cell. This represents an important scientific challenge that a century ago was already considered the "ideal goal" of biology (Loeb, 1906, p. 23). Nevertheless, a large technological gap still exists between what has been achieved by genetic engineering to date and the ability to actually create life. While working on the development of the appropriate technology, scientists are moving toward the design of artificial minimal life forms in two opposite but complementary ways, defined as the bottom-up and the top-down approaches (Luisi, 2002; Szathmáry, 2005; see also chapter 3).

The top-down approach aims at simplifying existing small genomes, starting with the information about minimal genomes already obtained from computational and experimental studies, to construct a living cell. Following this approach, Craig Venter, in 2002, announced his intention to build a synthetic chromosome as a first step toward engineering an organism with the desired biotechnological features. He predicted that, in about 3 years, his research group will be able to construct a synthetic genome, and insert it into a cell to generate what can be called a semisynthetic minimal cell (Zimmer, 2003). However, several challenges must be overcome before such a goal can be achieved. First, it is not possible to accurately synthesize the long stretch of DNA necessary to make a minimal genome with present DNA synthesis methodology, although its application to exploit whole-genome sequence information is guiding scientists toward steady and rapid progress in this field. The largest genome synthesized to date is that of the 7,500-nucleotide poliovirus (Cello, Paul, and Wimmer, 2002), far from the size of a minimal genome that, according to current estimates, should include around 200 genes. This means that it will be more than 200 kb long, increasing enormously the risk of introducing mistakes into the sequence. Recent methodological advances in this field, which improve both accuracy and pace, have been reported. Smith, Hutchison, Pfannkoch, and Venter (2003) were able to synthesize fully infectious *f*X174 virions, with a 5386-bp genome, in only 2 weeks; a procedure for precise assembly of linear DNA constructs as long as 20 kb, using long PCR-based fusion of several fragments, has been designed (Shevchuk et al., 2004); and a contiguous 32-kb polyketide synthase gene cluster was successfully synthesized and introduced in *E. coli* (Kodumal et al., 2004). But aside from the technical problems, synthesizing a complete bacterial genome, however small, is far more complicated, since it is not obvious which genes should be included, in which order, or which regulatory sequences it must include. Furthermore, once the newly synthesized genome is introduced into a cell, it is unknown whether the biochemical machinery of the host cell will recognize it or not (Zimmer, 2003). A profound challenge remains, and scientists are learning as they go in working toward creating a semisynthetic minimal cell.

There is general agreement that a top-down approach will not achieve the construction of the minimal possible cell in chemical terms. Doubts arise from the fact that extant cells, however small, have very complex transcription and translation systems, and it seems unrealistic that the simplest living chemical system would require such components. The bottom-up approach aspires to construct such artificial simplest chemical supersystems or protocells by adding the basic nonliving components that confer on a system the properties of living matter. Considering life as a property that emerges from the union of three subsystems—a metabolic network, an informational genome, and a boundary (Gánti, 2003)—it is possible to conceive, with the current knowledge which would be the minimal requisites that the simplest protocell should possess to be considered living (Szathmáry, Santos, and Fernando, 2005). To be able to synthesize a living cell from its basic components, all components must work in concert while at the same time forming a stable system. Although no such experimental system exists yet, the recent advances in genomic technology and membrane biophysics make the possibility of synthesizing protocells an imaginable goal (Pohorille and Deamer, 2002; Rasmussen et al., 2004; Szathmáry, 2005; Szostak, Bartel, and Luisi, 2001), which will provide fascinating insights into the essence of cellular life and may offer some clues of how life first evolved on Earth.

16.2 Approaching the Minimal Genome

Even the simplest unicellular organisms on Earth display an amazing degree of complexity, but such complexity does not seem to be a necessary attribute of cellular life. Although we still have a long way to go before we understand the specific functions that are essential for different organisms in their natural environments, we are already aware of most essential cellular functions and, clearly, modern cells possess many functions that would be dispensable in an ideally controlled environment.

Metabolism and genetics can be considered as the two central pillars that sustain life (Peretó, 2005), since every living being has some form of metabolism and genetic replication from a template, both taking place within a boundary that, in modern cells, is a phospholipidic membrane. Considering that the functional parts of a living cell are proteins and RNA molecules, and that the instructions for making these parts are encoded by genes, we can define the necessary elements to keep a minimal cell alive by knowing its complete gene set, which has been called a minimal genome (Mushegian, 1999). Such a minimal genome must contain the smallest number of genetic elements sufficient to allow the cell to maintain a minimal metabolic network within a boundary, reproduce, and evolve, three main properties of living cells (Islas et al., 2004; Luisi, Oberholzer, and Lazcano, 2002; Ruiz-Mirazo, Peretó, and Moreno, 2004). The concept of a minimal genome needs to be associated with a defined set of environmental conditions and, therefore, the absolute minimal genome will contain the smallest possible group of genes that would be sufficient to sustain cellular life in the most favorable conditions, that is, in a rich environment in which all essential nutrients are provided, and in the absence of any adverse factors (Koonin, 2000).

Models to define a minimal genome have to be tied to particular levels of biological organization. The increasing knowledge about complete genomes from bacteria makes these prokaryotes a suitable model for defining what a modern minimal genome should be like. In recent years, several theoretical and experimental studies have attempted to outline the minimal gene-set for bacterial life, as an important step toward creating simple organisms.

16.2.1 Comparative and Evolutionary Genomics

Comparative genomic analyses have an evolutionary basis, since they are primarily based on alignment of DNA or protein sequences for the purpose of identifying orthologous genes in genomes of distantly related species. Although the definition of a minimal genome based on the comparative analysis of known genomes depends on the number of genomes used to define it, and the genomic core size decreases as the number of phylogenetically distant species included in the comparison increases (Charlebois and Doolittle, 2004), this kind of study has proven to be very useful in understanding which are the essential functions that define a living cell. In any case, nonorthologous gene displacements (i.e., genes that perform the same function but do not derive from the same ancestral gene) must be taken into account.

The first attempt to define a minimal genome based on comparative genomics (Mushegian and Koonin, 1996) was made soon after the first two bacterial genomes, from *Haemophilus influenzae* (Fleischmann et al., 1995) and *Mycoplasma genitalium* (Fraser et al., 1995), were completely sequenced. Because of their parasitic lifestyle, these two bacteria present reduced genomes compared to other phylogenetically related free-living species. In addition, *M. genitalium* and *H. influenzae* are gram-positive and gram-negative bacteria, respectively, so genes conserved across the large phylogenetic distance between these two species are good candidates to be considered essential. The analysis led to the reconstruction of a first minimal gene set composed of only 256 genes. This hypothetical minimal genome is rich in universally conserved proteins (71%), with more than half of the genes involved in genetic information storage and processing, including more or less complete systems for replication, translation, and transcription, and a surprisingly large set of molecular chaperones. The rest of the genes encode proteins necessary to sustain a simplified metabolism, a limited repertoire for transport systems and protein export, and 18 apparently essential genes of uncharacterized function.

Later on, the computational comparative analysis of 21 complete genomes of bacteria, archaea, and eukaryotes (Koonin, 2000) suggested that a set of about 150 genes would be sufficient to maintain a living cell that possesses basal systems for replication, transcription, and translation, a reduced repair machinery, a small set of molecular chaperones, an intermediate metabolism reduced to glycolysis, a primitive transport system, and no cell wall.

A more recent approach to the minimal gene-set has been obtained by including the reduced genomes of insect endosymbionts in these computational comparisons. Mutualistic and obligate insect-bacteria symbioses have been intensively studied in the last years, particularly those involving the g-proteobacteria *B. aphidicola* (three different strains), *Wigglesworthia glossinidia*, and two species of *Blochmannia*, primary endosymbionts of aphids, tsetse flies, and carpenter ants, respectively, whose genomes have been fully sequenced (Akman et al., 2002; Degnan, Lazarus, and Wernegreen, 2005; Gil et al., 2003; Shigenobu et al., 2000; Tamas et al., 2002; van Ham et al., 2003). All these genomes have experienced a massive genome reduction after the establishment of the respective symbioses, are relatively similar in size, and encode a quite similar number of genes in each functional COG (cluster of orthologous genes) category (Tatustov et al., 2001). A comparative analysis performed among the first five sequenced genomes of endosymbionts showed that they share only 277 protein-coding genes (281 if nonorthologous gene displacement is taken into account), corresponding to half of their coding capacity (Gil et al., 2003). Interestingly, as was already observed in the first computational approach to the minimal genome, about one-third of the genes in all five genomes are devoted to information storage and processing. This is a result of preservation of most information storage and processing genes shared by all of them, and of a considerable number of molecular chaperones. Some of these shared genes must be involved in endosymbiotic processes, whereas the rest should be essential for any kind of cellular life. To identify the latter subset of genes, the complete set of genes shared by all five endosymbionts was compared with the reduced genome of the epicellular parasite *M. genitalium*. The study showed that these six genomes share only 180 housekeeping protein-coding genes (Gil et al., 2003). Once more, the analysis unveiled the essentiality of the genes involved in informational processes, since about half of these genes belong to this category. The number of shared genes among endosymbiotic and parasitic bacteria was further reduced to 156 when the intracellular parasites *Rickettsia prowazekii* and *Chlamydia trachomatis* were added to the comparison (Klasson and Andersson, 2004).

16.2.2 Three Main Experimental Approaches

Several genomewide analyses to identify genes that are essential under particular growth conditions have been performed using three different experimental approaches: massive transposon mutagenesis (the most widely used approach, reviewed in Judson and Mekalanos, 2000), antisense RNA to inhibit gene expression (Forsyth et al., 2002; Ji et al., 2001), and systematic inactivation of each individual gene present in a genome (Gerdes et al., 2003; Kang et al., 2004; Kobayashi et al., 2003; Mori et al., 2000). All these approaches yielded minimal gene-sets that are compatible with the comparative genomics inferences. Very few genes included in the computationally derived minimal genomes were found to be dispensable, and among these, it remains to be determined whether their dispensability reflects experimental artifacts or unexpected functional redundancy (Koonin, 2003).

16.2.3 A Combined Approach

All the preceding experimental and computational approaches to the minimal genome provided sets of essential genes with similar functional features, which are distinct from those of the general population of conserved bacterial genes (the latter represented in the database of protein COGs). They are substantially rich in genes encoding components of genetic-information processing systems, mainly the transcriptional apparatus, and contain relatively few genes for metabolic enzymes plus a very small fraction of functionally uncharacterized genes. However, the maintenance of metabolic homeostasis is one of the essential functions that define life, and therefore the minimal genome must include the necessary genes to maintain a minimalist metabolism.

Taking this consideration into account, a combined investigation of all previously used computational and experimental strategies for addressing this issue was performed, checking that all genes involved in essential pathways needed to maintain

Table 16.1
Classification of genes included in the minimal genome for bacterial life proposed by Gil et al. (2004)

Functional Category	Number of Genes
DNA metabolism	16
Basic replication machinery	13
DNA repair, restriction and modification	3
RNA metabolism	106
Basic transcription machinery	8
Translation: aminoacyl-tRNA synthesis	21
Translation: tRNA maturation and modification	6
Translation: ribosomal proteins	50
Translation: ribosome function, maturation, and modification	7
Translation factors	12
RNA degradation	2
Protein processing, folding, and secretion	15
Protein posttranslational modification	2
Protein folding	5
Protein translocation and secretion	5
Protein turnover	3
Cellular processes	5
Cell division	1
Transport	4
Energetic and intermediary metabolism	56
Glycolysis	10
Proton motive force generation	9
Pentose phosphate pathway	3
Lipid metabolism	7
Biosynthesis of nucleotides	15
Biosynthesis of cofactors	12
Poorly characterized	8
TOTAL	206

a reasonable metabolic homeostasis were included (Gil et al., 2004). The analysis rendered a minimal genome containing 206 protein-coding genes (table 16.1; figure 16.1, color plate 12). Nevertheless, it still remains questionable whether such a minimalist cell (or any of those previously mentioned) could survive under any realistic conditions. It should also be mentioned that there is no conceptual or experimental support for the existence of *one* form of minimal bacterial cell, at least from a metabolic point of view, since different essential functions can be defined depending on the environmental conditions, and numerous versions of minimal genomes can be conceived to fulfill such functions even for the same set of conditions. There is no doubt that future studies will highlight a diversity of minimal ecologically dependent metabolic charts supporting a universal genetic machinery.

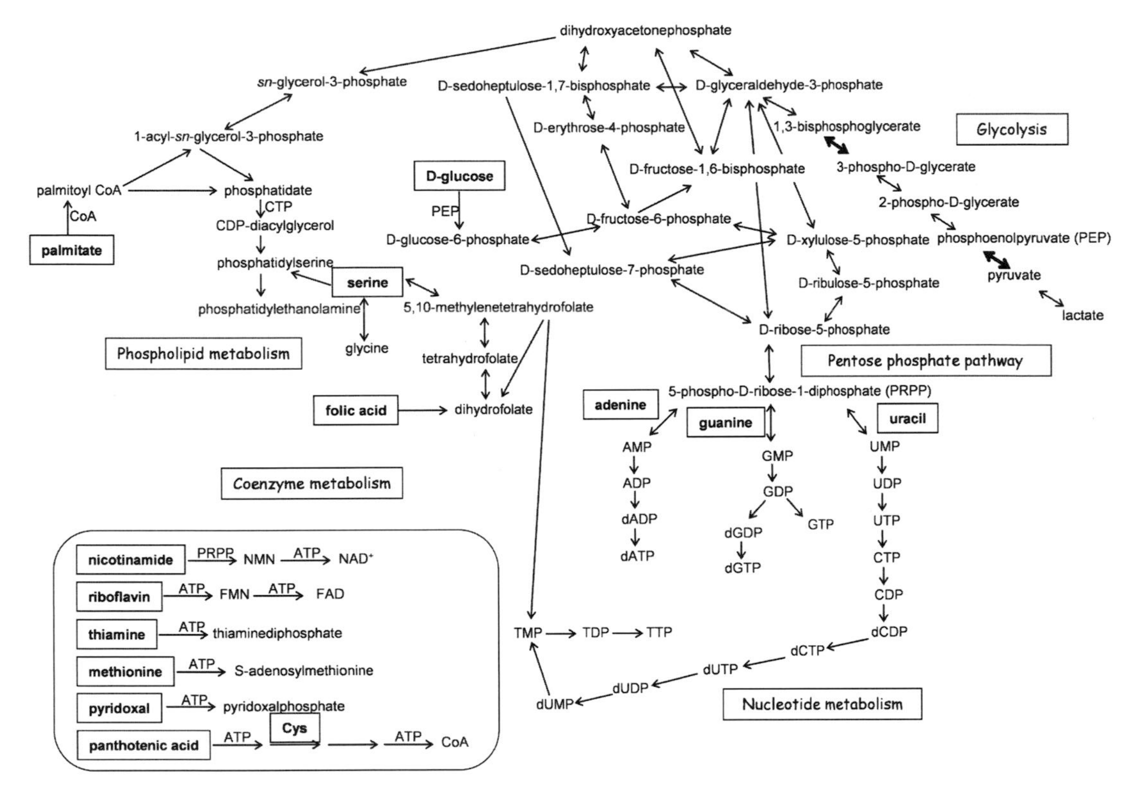

Figure 16.1 (color plate 12)
A simplified overview of the metabolic network implemented by a hypothetical minimal genome of 206 protein-coding genes derived by an integrated approach taking into account genomewide computational, experimental and metabolic studies on completely sequenced bacterial genomes. Names of substrates freely available for the hypothetical minimal cell are represented in boldface characters and inside a frame. Coenzyme metabolism (except the folate metabolism linked to nucleotide metabolism) is shown in the inset and was not considered in the network analysis. Wider arrows in the glycolytic pathway indicate the only two steps in which ATP is synthesized by substrate-level phosphorylation.

16.3 Genome Minimization in Nature

All described attempts to approximate a minimal genome resemble those already performed by nature (see table 16.2 for genomes recently sequenced that are not included in such attempts), where genome reduction is mainly associated with a transition from free-living to a host-dependent lifestyle. The genome of a bacterium that lives in such close relationship with a eukaryotic host is referred to in the literature as resident genome (Anderson and Kurland, 1998). Among the completely sequenced resident genomes, many correspond to obligate pathogens of the classes Mollicutes, Chlamydiae, Spirochetes, Actinobacteria, and a-proteobacteria. In addition,

Table 16.2
Gene content in last sequenced naturally reduced genomes and their comparison with the hypothetical minimal genome containing 206 protein-coding genes

Species	Genome Size (Mb)	Protein-Coding Genes	Genes Shared with the Minimal Genome	Reference
Buchnera aphidicola BCc	0.42	363	161	Our group (unpublished)
Blochmannia pennsylvanicus	0.79	658	193	Degnan et al., 2005
Wolbachia wBm	1.08	806	170	Foster et al., 2005
Prochlorococcus marinus MED4	1.66	1716	180	Rocap et al., 2003
Prochlorococcus marinus SS120	1.75	1882	183	Dufresne et al., 2003

several resident genomes of mutualistic endosymbionts have been sequenced, all corresponding to two classes of the phylum proteobacteria: the six g-proteobacteria endosymbionts of insects already mentioned, and two species of *Wolbachia* (a-proteobacteria), endosymbionts of the fruit fly and nematode, respectively. In bacteria, gene content is basically correlated with genome size (Casjens, 1998) and, therefore, the reduced size of resident genomes reflects the presence of a smaller number of genes, compared with their free-living relatives, because of a relaxed selection on the maintenance of genes that are rendered unnecessary in the protected environment provided by the host.

Recently, it has been shown that natural genome reduction can also take place in free-living bacteria. Three species of *Prochlorococcus*, free-living cyanobacteria that represent the smallest and most abundant photosynthetic organism in the ocean, have been sequenced (Dufresne et al., 2003; Rocap et al., 2003; see also table 16.2). Two of these species present genomes that have been downsized about 30% in comparison with other close relatives and can be considered nearly minimal oxyphototrophic genomes.

16.4 The Core of a Minimal Metabolic Set

As mentioned previously, a common feature of all life forms is their ability to maintain homeostasis in a given environment. Moreover, to accomplish cellular growth and division, a minimal cell would also require the ability to transform and assemble its building blocks using the energy provided by the environment. It seems, therefore, that a minimal cell would require a minimal metabolism to fulfill both essential aspects. A first approximation to this core metabolism is provided by the analysis of the enzymatic functions encoded by the theoretically inferred minimal gene set from the abovementioned combined approach. Figure 16.1 (color plate 12) provides representation of the metabolic network encoded by the theoretically inferred minimal gene set, which is thought to comprise the minimal set of metabolic reactions to sustain a bacterial cell under ideal nutrient supply conditions (i.e., glucose, fatty acids, amino acids, nucleobases, and vitamins).

The comparison of this theoretically inferred minimal metabolism, in terms of metabolic capacities, with naturally reduced genomes reveals many parallels, since the procedure to determine this minimal set includes genes that are shared by most endosymbiotic bacteria. In the minimal gene set, the intermediary metabolism is mainly reduced to ATP synthesis by substrate-level phosphorylation during glycolysis and the nonoxidative pentose phosphate pathway, whereas amino acid biosynthesis is virtually absent. So it is with de novo biosynthesis of nucleotides, although the complete salvage pathways for most of them can be found. Lipid biosynthesis is limited to condensation of fatty acids with glycerol phosphate, and there are no pathways for biosynthesis of fatty acids. Altogether the minimal metabolic core seems devoted to the production of energy from glucose and the interconversion, rather than the net biosynthesis, of essential cellular building blocks, most of which would be readily provided by a rich environment. However, adding some complexity to this heterotrophic metabolism, one could envisage a hypothetical autotrophic minimal metabolism, like the one conjectured by Benner (1999).

16.4.1 Metabolic Networks from Minimal and Natural Reduced Genomes

As we have seen, the study of the metabolic capacities of minimal and reduced genomes provides important insights into the minimal metabolic core necessary to sustain life, as well as into the required properties of its surrounding environment. However, to gain insight on the organizational and evolutionary principles of the metabolism of living organisms, it is necessary to go beyond the analyses of its components and explore its emerging system properties. Several groups have recently addressed this issue by performing topological analyses of complex networks derived from cellular metabolisms (Arita, 2004; Jeong et al., 2000; Ma and Zeng, 2003; Wagner and Fell, 2001). They all found that metabolic networks can be best described as scale-free networks in which most metabolites have few connections while a few highly connected metabolites act as hubs of the network. Differences among the various studies arise from the use of different definitions of connections between metabolites. For instance, Jeong and coworkers (2000) considered connections between all substrates and products of a reaction and described the *Escherichia coli* metabolic network as a "small world" with a short average path length of only 3.2 steps. At the other extreme, Arita (2004) considered connections between metabolites only when a carbon atom is transferred between them; he computed a long average path length of 8.4 steps for the same metabolism and therefore discarded the small-world attribute. To perform comparative analyses on metabolic networks, it is thus important to strictly use the same network reconstruction method for all organisms. The reconstruction procedure might also be important in revealing structural differences between metabolisms of different organisms. It appears that, since some highly connected metabolites such as coenzymes, inorganic phosphate, or water tend to dominate the overall network topology, reconstruction methods that remove the most frequent metabolites are better for detecting differences between different groups of organisms (Ma and Zeng, 2003).

In the context of the study of protocells, it is important to address whether metabolisms from reduced genomes share specific properties and how these compare with the theoretically inferred minimal gene set. From the abovementioned surveys, only that of Ma and Zeng (2003) reported some deviations for metabolic networks from organisms with reduced genomes, although these were not considered in detail. Here, we perform a more detailed analysis of the topological properties of metabolisms from a representative sample of species with reduced genomes, and compare them with those of the minimal gene set and those from related species covering a wide range of genome sizes (table 16.3). The metabolic reconstruction method used here is similar to that described in Gabaldón and Huynen (2003), which automatically maps the annotated gene functions onto KEGG metabolic pathways (Kanehisa and Goto, 2000). The reaction database was derived from the 35.0 release of the LIGAND database after removing polymerization reactions and reactions involving macromolecules. The direction of the reaction was taken into account. To eliminate connections through frequent metabolites, we considered connections only through metabolites represented in the pathway maps. When more than one substrate or more than one product was represented in the map, we considered connections only through pairs of compounds that have at least one carbon atom in common on the two sides of the reaction, according to their atomic mappings in the RPAIR database (Hattori et al., 2003). Errors in the directionality of the reaction detected by others (Ma and Zeng, 2003) and other obvious errors were automatically corrected. To reconstruct the metabolic network of a genome, the annotated functions of all encoded genes were mapped onto the reaction database and the corresponding network connections were generated. It must be noted that the functional annotation of many genes in all genomes is incomplete and error prone and, therefore, so are their reconstructed networks. For the convenience of their mathematical analysis, the reconstructed metabolic networks are represented as a directed graph in which nodes and edges correspond to metabolites and the enzymatic reactions connecting them, respectively (figure 16.2). In our set, the number of nodes in the metabolic networks inferred from genomes with more than about 2,000 protein-coding genes varies within a wide range of 800 to 1,300 nodes. For smaller genomes, amid a high degree of variation, we observe a tendency for the number of nodes in the network to decrease

Table 16.3
Topological parameters of the inferred metabolic networks from the minimal gene set and natural genomes of various sizes

Bacterial Division	Species	p-c genes	n	L	D
g-proteobacteria	*Buchnera aphidicola*	504	443	7.76	25
	Wigglesworthia glossinidia	617	561	11.4	35
	Blochmannia floridanus	583	634	8.47	26
	Escherichia coli (K-12)	4237	1215	10.3	35
	Escherichia coli (CFT073)	5379	1120	10.2	34
	Haemophilus influenzae (d)	1657	775	10	30
a-proteobacteria	*Rickettsia prowazekii*	886	517	8.41	24
	Wolbachia (Bma)	1195	516	8.76	28
	Agrobacterium tumefaciens (w)	5402	1147	9.45	33
	Bradyrhizobium japonicum	8317	1282	10.2	35
	Brucella melitensis	3198	1197	8.54	31
	Meshorizobium loti	7272	1209	9.71	33
Firmicutes	*Mycoplasma genitalium*	484	207	7.49	23
	Clostridium acetobutylicum	3848	784	9.56	25
	Lactobacillus plantarum	3059	864	9.64	26
Actinobacteria	*Tropheryma whipplei*	839	426	11.6	43
	Mycobacterium tuberculosis	3991	1139	9.98	31
	Streptomyces coelicolor	8154	1119	10.1	29
	Nocardia farcinica	5936	1089	9.79	30
Cyanobacteria	*Prochlorococcus marinus*	1760	844	10.5	30
	Anabaena sp.	6131	970	9.76	29
	Synechocystis sp.	3264	918	10.5	30
	Minimal gene set	206	165	5.34	18

Note: The size of the genome (p-c genes) is expressed by the number of protein-coding genes; the following columns indicate: n, number of nodes, L, average path length, D, network diameter. Power-law distribution is observed in all connectivity degree distributions of the inferred networks.

with the genome size (figure 16.3). This reduction in the number of metabolites involved in the cellular metabolism might be related to the fact that species with reduced genomes are usually in a rich environment in which many required metabolites are provided, and therefore their biosynthetic pathways can be omitted. The minimal gene set displays a number of nodes, which is consistent with its size and falls within the observed trend in natural genomes.

16.4.2 Topological Analyses of Metabolic Networks

Mechanical statistics and graph theory provide us with several quantitative parameters to describe the global topology of complex networks (Albert and Barabási, 2002). Due to space limitations we focus here on three of the most relevant topological properties: connection degree distribution, average path-length, and network diameter.

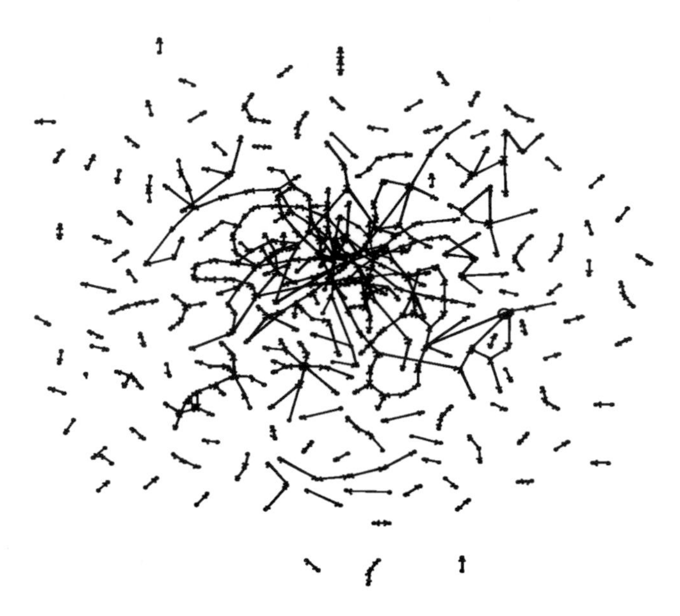

Figure 16.2
Example of an inferred metabolic network. Each of the 426 metabolites present in the *Tropheryma whipplei* metabolic network are depicted as black dots (nodes); arrows (edges) connecting the different nodes represent enzymatic reactions.

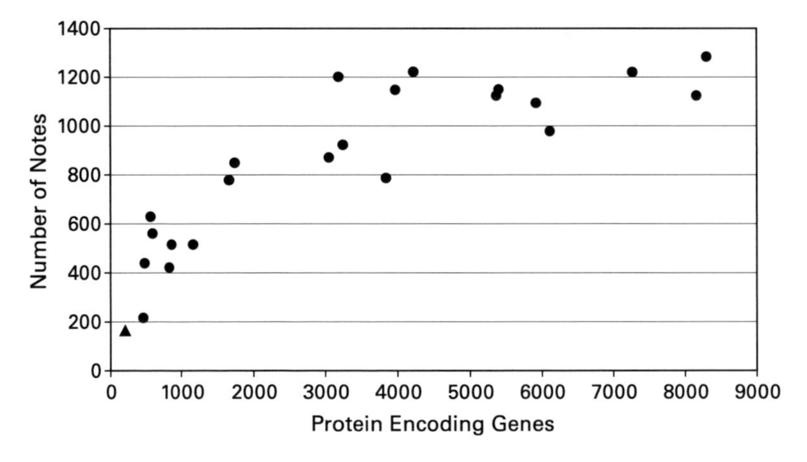

Figure 16.3
Number of nodes in the network versus number of protein-coding genes for all analyzed genomes. The minimal gene set is represented as a gray triangle.

The connection degree (k_i) of a certain metabolite, *i*, in a metabolic network is defined as the number of metabolites that are directly connected to it. When considering directionality of the reaction, k_i can be decomposed into output and input connection degrees (number of connections starting from or ending at that metabolite, respectively). When the connection degree distribution of a network follows a power law ($P(x) = Kx^{-a}$; where x is the variable and K and a are constants), the network is said to be scale free. As mentioned earlier, all metabolic networks investigated to date, regardless of the network reconstruction method used, are scale free. This is also the case for all natural genomes analyzed here. It seems therefore that scale-freeness is a universal property of natural metabolic networks, regardless of their size or the species lifestyle. Interestingly, the theoretically inferred minimal genome set also displays such behavior.

Another interesting property of complex networks is their average path length (L). The path length between two metabolites is defined as the number of edges in the shortest pathway between them. Note that because directionality of the reaction is considered, the path from metabolite *i* to *j* is not necessarily the same as the path from *j* to *i*. The average path length of the network is computed by averaging path lengths over all pairs of connected metabolites in the network. For this purpose, we identified all reachable metabolites in the network by the Breath First Searching algorithm (Broder et al., 2000). Our results (table 16.3) show this parameter to vary from 7.5 to 11 steps in most networks. With the exception of *Tropheryma whipplei*, we observe a slight tendency of smaller networks to have shorter average path lengths. If this trend is assumed, the average path length corresponding to the minimal gene set would fall within the expected range for a metabolic network of its size.

We also computed the network diameter (D), which is formally defined as the length of the longest pathway among all shortest pathways in the network. Our results (figure 16.4) show a linear relationship of this parameter to the size of the network. In general, bigger networks display longer diameters and, again, the minimal gene set seems to have a network diameter within the expected range. Also in this parameter, *T. whipplei* shows an unexpectedly high value. Its longest pathway consists of 43 steps going from dihydrofolate to TTP. The conservation of this long pathway in this human pathogen, despite an extensive genome reduction, might reflect adaptation and could provide useful information for elucidating the pathogenic mechanisms of this bacterium, highlighting the potential applicability of metabolic network analyses. Nevertheless, the possibility should not be ignored that errors or incompleteness in the annotation of this genome may have resulted in this anomalous value.

Altogether, our results show that some topological properties scale down with the size of the network in naturally reduced genomes, although a significant degree of variation does exist. Most important, the metabolic network from the theoretically

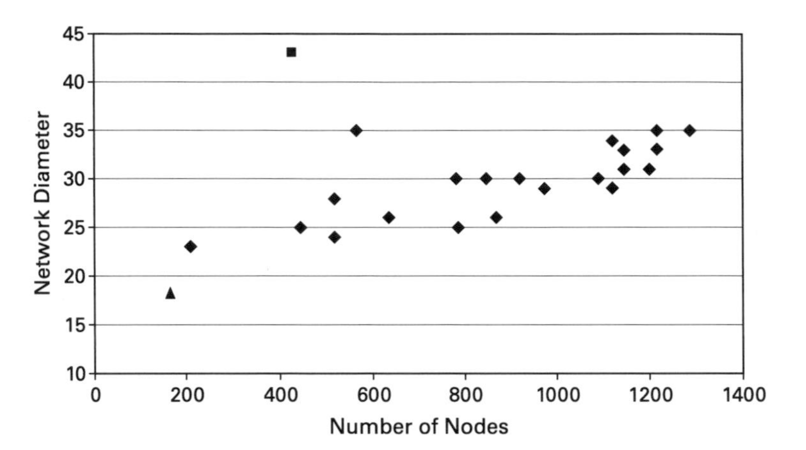

Figure 16.4
Network diameter versus number of nodes in the network for all analyzed genomes. The minimal gene set is represented as a gray triangle. *Tropheryma whipplei* (gray square) shows an unexpectedly very long diameter.

inferred minimal gene set appears to behave as would be expected for a naturally reduced genome of its size, underscoring its consistency.

16.4.3 Robustness Analysis of the Minimal Metabolic Network

As a consequence of the power law connectivity distribution, scale-free networks are sensitive to sequential removal of the most connected nodes, which causes sharp increases in topological parameters such as average path length. However, this type of network shows a significant robustness against random mutations. These tendencies have also been observed in natural metabolic networks, which were shown to be robust and error-tolerant (Jeong et al., 2000).

To investigate the robustness of the minimal metabolic network against random mutations, we simulated a random mutation attack. As opposed to what has usually been done (Jeong et al., 2000), and to perform a more biologically meaningful simulation, we chose to sequentially remove enzymatic activities instead of nodes. This is closer to the real situation, in which natural metabolisms can evolve by the loss of enzymatic functions encoded in the genes. Every simulation involved the sequential removal of up to 20 of the 101 enzymatic activities encoded in the minimal genome. The process was repeated 100 times, and every mutated network was reconstructed to measure its average path length and diameter (figure 16.5).

In our simulation, the removal of an enzymatic activity from the network can be seen as the removal of the edges and nodes that correspond to the reactions catalyzed exclusively by that enzyme. Since a degree of overlap in substrates and reactions exists among enzymes, earlier mutations and mutations affecting enzymatic activities

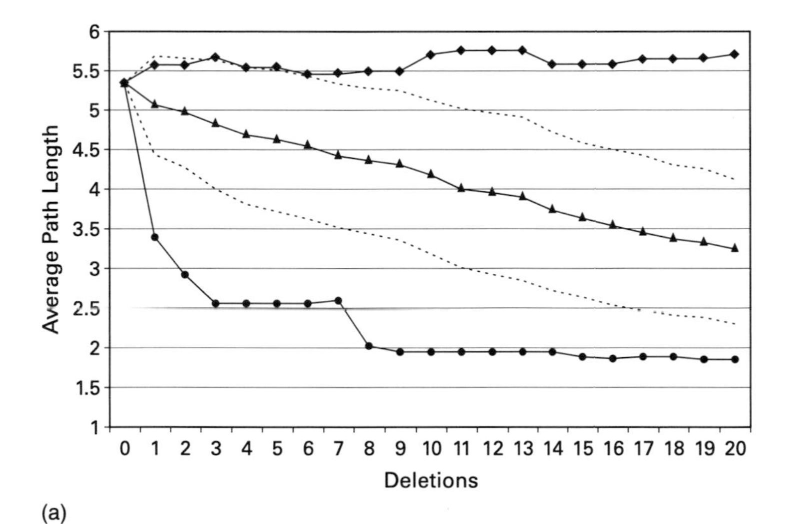

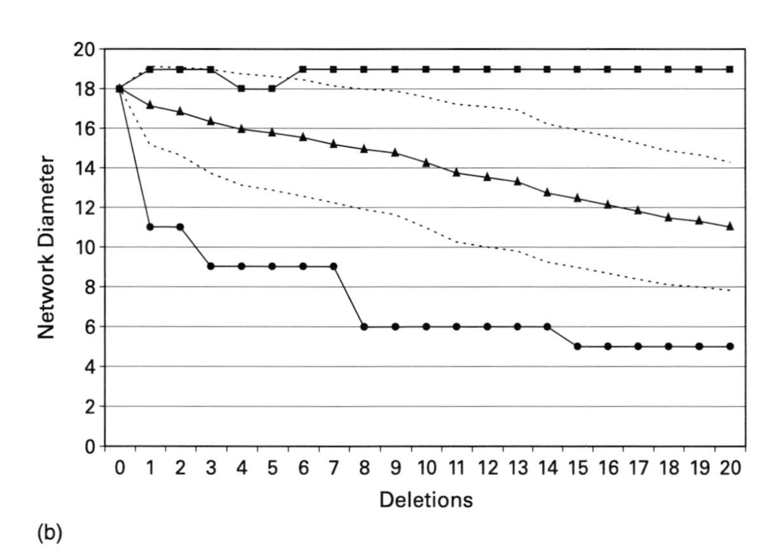

Figure 16.5
Robustness analysis of the minimal metabolic network. The results from 100 samples of sequential random removal of up to 20 enzymatic activities are summarized, indicating for each parameter (average path length, A, and network diameter, B): the maximal value (squares), the average value for all 100 simulations (triangles) and the minimal value (circles). Dashed lines indicate the values of the average, ±1 standard deviation (top and bottom dashed lines, respectively).

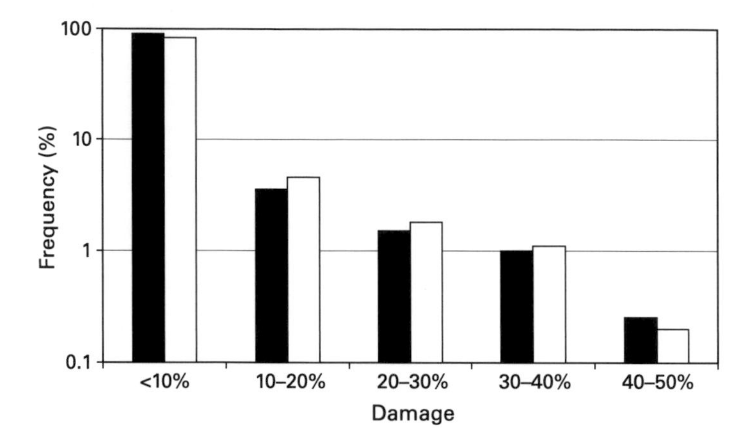

Figure 16.6
Frequency of deletions causing a certain topological damage, in terms of average path length (black bars) and network diameter (white bars). Note the logarithmic scale of the frequencies. Topological damage is measured as the percentage of variation in a given parameter caused by the deletion. Levels of damage were grouped into categories: <10%, 10–20%, 20–30%, 30–40% and 40–50% damage.

with higher overlaps are likely to have smaller effects on the network topology. In contrast, removal of enzymes that are exclusive for certain reactions and act in longer pathways has more drastic effects in the network topology. We observed a great variability among the different simulations. Despite the overall tendency of reducing both the average path-length and the network diameter, a few mutations seem to have a drastic effect while most mutations have a rather small effect; 82% and 89% of the mutations had a limited effect in the overall topology, varying in less than 10% the average path length and network diameter, respectively (figure 16.6). In contrast, the removal of few key enzymatic activities triggered abrupt reductions of up to 50% in both parameters (figure 16.6). For instance, the removal of the enzyme 3-glycerophosphate acyltransferase (EC 2.3.1.15) from the complete minimal network has a drastic effect, reducing the average path length from 5.35 to 3.39 and the network diameter from 18 to 12. This behavior, in which most random mutations have minor effects while mutations in a few key elements may cause the overall topology to change abruptly, is typical of robust, error-tolerant networks (Jeong et al., 2000). It seems, therefore, that the minimal metabolism also complies with this general property of natural metabolisms. However, the small size of the network makes it sensitive to sustained random attacks, and most simulations produced a collapsed network after 20 random mutations. Nevertheless, a significant fraction of the simulations (8–12%) displayed a high conservation of their average path length and network diameter, respectively, with variations lower than 10% after removal of 20 (~20%) of the enzymatic activities. Whether these reduced versions, derived from the mutagenesis of the minimal gene set while maintaining the overall network topology, are viable remains an open question. The topological organization of a metabolic network is just one of the aspects that might be related to viability. Metabolic coherence and viability necessarily depend on the specific components of the network, enzymes, and the metabolites that can support the homeostasis and multiplication of the cell. The most likely scenario here is that a certain overall topology of the metabolic network is necessary but not sufficient to sustain life.

16.5 Conclusion

The notion of a minimal genome is, at this point, mostly a conceptual tool for discussion of the minimal requisites for cellular life and the possible experimental design and construction of simpler cellular models. A minimal genome cannot be sharply described, given that the essential cellular functions can be differently defined depending on the environmental conditions. Nevertheless, it is possible to try to delineate which functions should be performed in any modern living cell, and list the genes that would be necessary to maintain such functions, although it should be noticed that numerous alternative minimal genomes can be conceived to fulfill such functions even for the same set of conditions (see discussion in chapter 3). Once a minimal genome is conceived, its system properties, such as the encoded metabolic network, can be analyzed from different perspectives to compare it with those of natural genomes and ascertain its consistency and expected properties. The development of new sophisticated techniques for genomic engineering, together with continued efforts in defining the minimal genome, will help us achieve the exciting goal of experimentally constructing a modern-type minimal living cell in the perhaps not so distant future.

Acknowledgments

This work was supported by grants BMC2003-00305 from the Ministerio de Educación y Ciencia (Spain) and Grupos03/04 from Conselleria d'Empresa, Universitat i Ciència (Generalitat Valenciana, Spain). T.G. was supported by a fellowship from the European Molecular Biology Organization (EMBO LTF 402-2005). R.G. is a recipient of a Ramón y Cajal contract from the Ministerio de Eduación y Ciencia (Spain).

References

Akman, L., Yamashita, A., Watanabe, H., Oshima, K., Shiba, T., Hattori, M., & Aksoy, S. (2002). Genome sequence of the endocellular obligate symbiont of tsetse flies, *Wigglesworthia glossinidia*. *Nature Genetics*, *32* (3), 402–407.

Albert, R., & Barabási, A. L. (2002). Statistical mechanics of complex networks. *Reviews of Modern Physics, 74* (1), 47–97.

Anderson, S. G. E., & Kurland, C. G. (1998). Reductive evolution of resident genomes. *Trends in Microbiology, 6* (7), 263–268.

Arita, M. (2004). The metabolic world of *Escherichia coli* is not small. *Proceedings of the National Academy of Sciences of the United States of America, 101*, 1543–1547.

Benner, S. A. (1999). How small can a microorganism be? In *Size limits of very small microorganisms* (pp. 126–135). Washington, DC: National Academy Press.

Broder, A., Kumar, R., Maghoul, F., Raghavan, P., Rajagopalan, S., Stata, R., et al. (2000). Graph structure in the web. *Computer Networks, 33*, 309–320.

Casjens, S. (1998). The diverse and dynamic structure of bacterial genomes. *Annual Review of Genetics, 32*, 339–377.

Cello, J., Paul, A. V., & Wimmer, E. (2002). Chemical synthesis of poliovirus cDNA: Generation of infectious virus in the absence of natural template. *Science, 297* (5583), 1016–1018.

Charlebois, R. L., & Doolitle, W. F. (2004). Computing prokaryotic gene ubiquity: Rescuing the core from extinction. *Genome Research, 14* (12), 2469–2477.

Degnan, P. H., Lazarus, A. B., & Wernegreen, J. J. (2005). Genome sequence of *Blochmannia pennsylvanicus* indicates parallel evolutionary trends among bacterial mutualists of insects. *Genome Research, 15* (8), 1023–1033.

Dufresne, A., Salanoubat, M., Partensky, F., Artiguenave, F. I., Axmann, M., Barbe, V., et al. (2003). Genome sequence of the cyanobacterium *Prochlorococcus marinus* SS120, a nearly minimal oxyphototrophic genome. *Proceedings of the National Academy of Sciences of the United States of America, 100* (17), 10020–10025.

Fleischmann, R. D., Adams, M. D., White, O., Clayton, R. A., Kirkness, E. F., Kerlavage, A. R., et al. (1995). Whole-genome random sequencing and assembly of *Haemophilus influenzae* Rd. *Science, 269* (5223), 496–512.

Forsyth, R. A., Haselbeck, R. J., Ohlsen, K. L., Yamamoto, R. T., Xu, H., Trawick, J. D., et al. (2002). A genome-wide strategy for the identification of essential genes in *Staphylococcus aureus. Molecular Microbiology, 43* (6), 1387–1400.

Foster, J., Ganatra, M., Kamal, I., Ware, J., Makarova, K., Ivanova, N., et al. (2005). The *Wolbachia* genome of *Brugia malayi*: Endosymbiont evolution within a human pathogenic nematode. *PLoS Biology, 3* (4), e121.

Fraser, C. M., Gocayne, J. D., White, O., Adams, M. D., Clayton, R. A., Fleischmann, R. D., et al. (1995). The minimal gene complement of *Mycoplasma genitalium. Science, 270* (5235), 397–403.

Gabaldón, T., & Huynen, M. A. (2003). Reconstruction of the proto-mitochondrial metabolism. *Science, 301* (5633), 609.

Gánti, T. (2003). *The principles of life.* Oxford: Oxford University Press.

Gerdes, S. Y., Scholle, M. D., Campbell, J. W., Balazsi, G., Ravasz, E., Daugherty, M. D., et al. (2003). Experimental determination and system level analysis of essential genes in *Escherichia coli* MG1655. *Journal of Bacteriology, 185* (19), 5673–5684.

Gil, R., Silva, F. J., Peretó, J., & Moya, A. (2004). Determination of the core of a minimal bacterial gene set. *Microbiology and Molecular Biology Reviews, 68* (3), 518–537.

Gil, R., Silva, F. J., Zientz, E., Delmotte, F., González-Candelas, F., Latorre, A., et al. (2003). The genome sequence of *Blochmannia floridanus*: Comparative analysis of reduced genomes. *Proceedings of the National Academy of Sciences of the United States of America, 100* (16), 9388–9393.

Hattori, M., Okuno, Y., Goto, S., & Kanehisa, M. (2003). Development of a chemical structure comparison method for integrated analysis of chemical and genomic information in the metabolic pathways. *Journal of the American Chemical Society, 125* (39), 11853–11865.

Islas, S., Becerra, A., Luisi, P. L., & Lazcano, A. (2004). Comparative genomics and the gene complement of a minimal cell. *Origins of Life and Evolution of Biospheres, 34* (1–2), 243–256.

Jeong, H., Tombor, B., Albert, R., Oltvai, Z. N., & Barabási, A. L. (2000). The large-scale organization of metabolic networks. *Nature*, *407*, 651–654.

Ji, Y., Zhang, B., van Horn, S. F., Warren, P., Woodnutt, G., Burnham, M. K. R., & Rosemberg, M. (2001). Identification of critical staphylococcal genes using conditional phenotypes generated by antisense RNA. *Science*, *293* (5538), 2266–2269.

Judson, N., & Mekalanos, J. J. (2000). Transposon-based approaches to identify essential bacterial genes. *Trends in Microbiology*, *8* (11), 521–526.

Kanehisa, M., & Goto, S. (2000). KEGG: Kyoto encyclopaedia of genes and genomes. *Nucleic Acid Research*, *28*, 27–30.

Kang, Y., Durfee, T., Glasner, J. D., Qiu, Y., Frisch, D., Winterberg, K. M., & Blattner, F. R. (2004). Systematic mutagenesis of the *Escherichia coli* genome. *Journal of Bacteriology*, *186* (15), 4921–4930.

Klasson, L., & Andersson, S. G. (2004). Evolution of minimal-gene-sets in host-dependent bacteria. *Trends in Microbiology*, *12* (1), 37–43.

Kobayashi, K., Ehrlich, S. D., Albertini, A., Amati, G., Andersen, K. K., Arnaud, M., et al. (2003). Essential *Bacillus subtilis* genes. *Proceedings of the National Academy of Sciences of the United States of America*, *100* (8), 4678–4683.

Kodumal, S. J., Patel, K. G., Reid, R., Menzella, H. G., Welch, M., & Santi, D. V. (2004). Total synthesis of long DNA sequences: Synthesis of a contiguous 32-kb polyketide synthase gene cluster. *Proceedings of the National Academy of Sciences of the United States of America*, *101* (44), 15573–15578.

Koonin, E. V. (2000). How many genes can make a cell: The minimal-gene-set concept. *Annual Review of Genomics and Human Genetics*, *1*, 99–116.

Koonin, E. V. (2003). Comparative genomics, minimal gene-sets and the last universal common ancestor. *Nature Reviews Microbiology*, *1* (2), 127–136.

Loeb, J. (1906). *The dynamics of living matter*. New York: Macmillan.

Luisi, P. L. (2002). Toward the engineering of minimal living cells. *The Anatomical Record*, *268* (3), 208–214.

Luisi, P. L., Oberholzer, T., & Lazcano, A. (2002). The notion of a DNA minimal cell: A general discourse and some guidelines for an experimental approach. *Helvetica Chimica Acta*, *85* (6), 1759–1777.

Ma, H., & Zeng, A. P. (2001). Reconstruction of metabolic networks from genome data and analysis of their global structure for various organisms. *Bioinformatics*, *19* (2), 270–277.

Mori, H., Isono, K., Horiuchi, T., & Miki, T. (2000). Functional genomics of *Escherichia coli* in Japan. *Research in Microbiology*, *151* (2), 121–128.

Mushegian, A. (1999). The minimal genome concept. *Current Opinion in Genetics and Development*, *9* (6), 709–714.

Mushegian, A. R., & Koonin, E. V. (1996). A minimal gene set for cellular life derived by comparison of complete bacterial genomes. *Proceedings of the National Academy of Sciences of the United States of America*, *93* (19), 10268–10273.

Pohorille, A., & Deamer, D. (2002). Artificial cells: Prospects for biotechnology. *Trends in Biotechnology*, *20* (3), 123–128.

Peretó, J. (2005). Controversies on the origin of life. *International Microbiology*, *8* (1), 23–31.

Rasmussen, S., Chen, L., Deamer, D., Krakauer, D. C., Packard, N. H., Stadler, P. F., & Bedau, M. A. (2004). Transitions from nonliving to living matter. *Science*, *303* (5660), 963–965.

Rocap, G., Larimer, F. W., Lamerdin, J., Malfatti, S., Chain, P., Ahlgren, N. A., et al. (2003). Genome divergence in two *Prochlorococcus* ecotypes reflects oceanic niche differentiation. *Nature*, *424* (6952), 1042–1047.

Ruiz-Mirazo, K., Peretó, J., & Moreno, A. (2004). A universal definition of life: Autonomy and open-ended evolution. *Origins of Life and Evolution of the Biospheres*, *34* (3), 323–346.

Shevchuk, N. A., Bryksin, A. V., Nusinovich, Y. A., Cabello, F. C., Sutherland, M., & Ladisch, S. (2004). Construction of long DNA molecules using long PCR-based fusion of several fragments simultaneously. *Nucleic Acids Research*, *32* (2), e19.

Shigenobu, S., Watanabe, H., Hattori, M., Sakaki, Y., & Ishikawa, H. (2000). Genome sequence of the endocellular bacterial symbiont of aphids *Buchnera* sp. APS. *Nature, 407* (6800), 81–86.

Smith, H. O., Hutchison, C. A. III, Pfannkoch, C., & Venter, J. C. (2003). Generating a synthetic genome by whole genome assembly: ϕX174 bacteriophage from synthetic oligonucleotides. *Proceedings of the National Academy of Sciences of the United States of America, 100* (26), 15440–15445.

Szathmáry, E. (2005). In search of the simplest cell. *Nature, 433* (7025), 469–470.

Szathmáry, E., Santos, M., & Fernando, C. (2005). Evolutionary potential and requirements for minimal protocells. *Topics in Current Chemistry, 259*, 167–211.

Szostak, J. W., Bartel, D. P., & Luisi, P. L. (2001). Synthesizing life. *Nature, 409* (6818), 387–390.

Tamas, I., Klasson, L., Canback, B., Naslund, A. K., Eriksson, A. S., Wernegreen, J. J., et al. (2002). 50 million years of genomic stasis in endosymbiotic bacteria. *Science, 296* (5577), 2376–2379.

Tatusov, R. L., Natale, D. A., Garkavtsev, I. V., Tatusova, T. A., Shankavaram, U. T., Rao, B. S., et al. (2001). The COG database: New developments in phylogenetic classification of protein from complete genomes. *Nucleic Acids Research, 29* (1), 22–28.

van Ham, R. C. H. J., Kamerbeek, J., Palacios, C., Rausell, C., Abascal, F., Bastolla, U., et al. (2003). Reductive genome evolution in *Buchnera aphidicola. Proceedings of the National Academy of Sciences of the United States of America, 100* (2), 581–586.

Wagner, A., & Fell, D. A. (2001). The small world inside large metabolic networks. *Proceedings of the Royal Society: Biological Sciences, 268*, 1803–1810.

Zimmer, C. (2003). Tinker, tailor: Can Venter stitch together a genome from scratch? *Science, 299* (5609), 1006–1007.

17 Parasitism and Protocells: Tragedy of the Molecular Commons

Jeffrey J. Tabor, Matthew Levy, Zachary Booth Simpson, and Andrew D. Ellington

17.1 Introduction

It seems likely to many people that life arose by means of a self-replicating molecule or assembly of self-replicating molecules. Because of the almost unique complementarity inherent in nucleobases, it has been suggested that nucleic acids or nucleic acid–like molecules were the first self-replicators. Although some factions have held that self-replicating peptides or lipid amalgams may have preceded or arisen in parallel with nucleic acids, an early nucleic acid replicator is also consistent with the prevalence of molecular fossils in modern cells that are related to nucleic acids, such as ATP and other cofactors (White, 1976). Indeed, given the fact that the ribosome is at some level a ribozyme, it seems highly likely that a complex RNA world preceded the modern world of protein catalysts. This RNA world may have, in turn, descended from an earlier nucleic acid replicator, by recruitment, duplication, parasitism, and diversification.

A full review of ribozyme catalysis and the RNA world is beyond the scope of this chapter, but we will present a brief overview that encompasses one plausible scenario for the evolution of the earliest living systems. Following the prebiotic synthesis of nucleobases, sugars, nucleotides, and short oligonucleotides (reviewed in Orgel, 2004), the first replicases may have been similar to those described by von Kiedrowski and Orgel (von Kiedrowski, 1986; Zielinski and Orgel, 1987; see also chapter 13). In these systems, short oligonucleotides would have served as catalysts for the template-directed ligation of other oligonucleotides. Going forward, it seems reasonable that the first function acquired by these nascent self-replicators would have been the ability to catalyze self-assembly through ligation. Indeed, this idea has been experimentally buttressed by the discovery of numerous, short ribozyme ligases (for examples see Ekland, Szostak, and Bartel, 1995; Robertson and Ellington, 1999; Rogers and Joyce, 2001). In fact, Paul and Joyce (2002) have implemented a self-replication system based on a small RNA ligase. In this system, the ribozyme is divided into two pieces, much like the simpler oligonucleotide replicases of von Kiedrowski, and the pieces are joined by (and forming) the parental ligase. It is possible that such "simple" ribozyme ligases and replicases may have, in turn, served as the precursors of more complex ribozyme polymerases, similar to the Bartel polymerase (Johnston et al., 2001) but capable of more lengthy nucleotide polymerization and ultimately self-replication (Levy and Ellington, 2001). Indeed, we have previously suggested that the evolution of the earliest oligonucleotide ligases would have been driven by increasing needs for fidelity (James and Ellington, 1999). In this view, even if life began as a series of simple but dedicated self-replicators, it would inevitably have transitioned to a system in which polymerases were generally responsible for the propagation of all genetic information in a cell.

Once a complex, replicating template had arisen, duplication and divergence would have resulted in the evolution of a variety of supporting RNAs such as kinases (Lorsch and Szostak, 1994), nucleotide transferases (Huang and Yarus, 1997), and even nucleotide synthetases (Unrau and Bartel, 1998), and ultimately in the evolution of the first genome. It is relatively easy to imagine how the activities of these ribozymes may have been coordinated in higher-order replicative systems such as hypercycles (Eigen and Schuster, 1977, 1978a, 1978b).

In its most complex form, the RNA world would have possessed a complexity quite similar to modern metabolism. To this end, ribozymes capable of carbon-carbon bond formation (Seelig and Jaschke, 1999; Tarasow, Tarasow, and Eaton, 1997), Michael additions (Sengle et al., 2001), amide bond formation (Sun et al., 2002; Wiegand, Janssen, and Eaton, 1997), sulfur and nitrogen alkylation (Wecker, Smith, and Gold, 1996; Wilson and Szostak, 1995), and even redox chemistry (Tsukiji, Pattnaik, and Suga, 2004) would have provided much of the same catalytic infrastructure that now comes from protein enzymes. Moreover, RNA would have served as an excellent signal transduction and regulatory molecule. Allosteric ribozymes that are activated by small molecules have been selected (Robertson and Ellington, 1999, 2000; Soukup and Breaker, 1999; Tang and Breaker, 1997), and *riboswitches* that undergo conformational changes and control the regulation of gene expression have been found in modern organisms (Barrick et al., 2004; Mandal and Breaker, 2004; Winkler, Nahvi, and Breaker, 2002).

While the early evolution of an RNA world from an RNA replicase is supported by both natural history and experimental data, fundamental, theoretical problems still surround the emergence of the first replicators, RNA or otherwise. In particular, the sequence space surrounding a replicator will contain numerous molecules that lack the ability to replicate themselves, but that can still be copied by active replicases. Indeed, it is likely that there are many more mutations that inhibit the catalytic functionality of a replicator than inhibit its ability to serve as a template. Thus, early molecular replicators would have quickly accumulated parasites, and under many circumstances should have been outcompeted by these parasites. By more closely examining the problems posed by parasitism, we can hopefully come to a more accurate understanding of the forces that would have shaped early evolution, and thereby more accurately depict the earliest replicators. These insights will also illuminate attempts to synthesize protocells, for synthetic protocells also will likely face various analogous forms of molecular parasitism.

17.2 Theoretical Treatments of Parasitism

Experimental demonstrations of the plausibility of the RNA world hypothesis have often been anticipated by theoretical treatments. For example, Eigen originally suggested that in order to successfully propagate information, a replicase enzyme of significant length would eventually have been required (Eigen, 1971). However, early genetic information would have been limited to short nucleic acids as a result of the error-prone nature of replication in the absence of sophisticated mutation correction systems. If a nucleic acid sequence became longer than 50 to 100 nucleobases, the accumulation of errors during replication (from mismatches in base-pairing) would likely have produced a significant loss of genetic information and in the catalytic ability of the replicator (Eigen, 1971). The first proposed solution to this paradox was the hypercycle (Eigen and Schuster, 1977, 1978a, 1978b), a cooperative series of nucleic acids each of which was responsible for the replication of the next member of the cycle. For a hypercycle composed of four replicators, 1 would replicate 2, 2 would replicate 3, 3 would replicate 4 and 4 would replicate 1. Each member of the hypercycle could be short enough to avoid catastrophic information loss resulting from mutation, and the sum total of information stored in the cycle could be large. The practical implementation of Eigen's hypercycle model with short oligonucleotide replicases can be imagined, and as information accumulated, high-fidelity polymerases or mutation-correction systems would have evolved and ultimately led to the more modern organization of information in genomes.

Nonetheless, a major problem facing an early hypercycle, or any other self-replicating RNA system, would have been the emergence of parasitic mutants derived from members of the cycle or from other, unrelated replicators in solution. Given the error-prone nature of uncorrected replication, any early RNA species would have existed as a quasispecies composed of the original, master sequence, and all mutated versions of that sequence that arose through replication events (Eigen et al., 1981). In the case of the hypercycle, two major classes of parasites could emerge: selfish parasites, which are replicated by a member of the cycle but do not contribute to the replication of any other member, and "shortcut" parasites, which reduce the information maintained in the cycle by replicating a species downstream of their original target (Bresch, Niesert, and Harnasch, 1980).

One of the most parsimonious solutions to the problem of parasitism is to segregate replicators from a larger population of intermixing parasites (Altmeyer and McCaskill, 2001; McCaskill, 1984). Two strategies of segregation have been discussed at length in the theoretical literature: spatial segregation of replicating species on two-dimensional surfaces, and compartmentalization of replicators in vesicles, or protocells. In the case of hypercyclic replication, if members of the cycle are immobilized onto a regular surface and then localized, mutualistic interactions that exclude selfish parasites can arise (Boerlijst and Hogeweg, 1991). This strategy, however, is still not robust against the emergence of shortcut parasites.

Szathmáry and coworkers have recently demonstrated that nonhypercyclic replicators immobilized onto a two-dimensional surface also tend to evolve toward increased complexity (length), fidelity, and speed while remaining stable against the detrimental effects of parasitism (Szabo et al., 2002). This phenomenon arises mostly as a result of the benefits of spatially mandated mutualism. When a good replicase emerges and replicates a neighboring template that also encodes a good replicase, both species thrive. Any parasite that emerges in this environment will depend on the activity of the local replicators to survive. If the parasite is too active, it will extinguish itself locally. However, if a less virulent parasite emerges in a particular location, efficient local replicator populations can subsist. These benefits are not retained when the replicators are allowed to diffuse more freely, however, as localized mutualism is eliminated and parasites can more efficiently compete for the enzymatic activity of best replicators.

Another well-studied strategy for parasite resistance is compartmentalization. Early theoretical models suggested that compartmentalization of autonomously replicating systems would provide a mechanism for parasite resistance by allowing natural selection to act directly on a vesicle or compartment (Bresch, Niesert, and Harnasch 1980; Eigen, Gardiner, and Schuster, 1980; Eigen et al., 1981; Niesert, Harnasch, and Bresch, 1981). If replicators and their parasites are physically isolated from other replicators, and the fitness of a specific replicator/parasite population is subject to selection, then vesicles containing deleterious parasites will be promptly outcompeted by vesicles containing less harmful parasites.

Szathmáry and Demeter have described another fortuitous property of compartmentalization—that random partitioning of the molecules during vesicle duplication can provide a natural mechanism for sampling optimal stoichiometries of molecules in multicomponent replicative systems (Szathmáry and Demeter, 1987). These authors envision a *metabolic hypercycle* in which multiple independent self-replicating molecules also have catalytic function, contributing to a metabolism that benefits all of the replicators. For example, each of four RNA self-replicators could be responsible for the biosynthesis of one nucleotide monomer. In this way, all four species must be present and contribute to the common metabolism for any of the species to replicate. If hyperactive or parasitic replicator variants arise and harm the stability of the hypercycle, random partitioning during vesicle division could counteract their negative effects. That is, since different numbers of each replicator will be randomly partitioned to daughter vesicles, those vesicles that randomly acquire distributions of the replicators that are optimal for efficient community replication can be selected for and propagated. A similar selection of optimal molecular stoichiometries would be difficult if not impossible in bulk solution.

The evolutionary drive to avoid parasitism is analogous to the economic argument for the "tragedy of the commons" first proposed by Lloyd in 1833 and popularized by Hardin (1968) to explain how the "invisible hand" of self-interest envisioned by Adam Smith can have dire social consequences. Hardin's explanation is that for any given common and limited resource, each user of this resource gains one unit of utility by exploiting it but loses only a small fraction of utility from their share of the destruction. Thus, it is in every user's immediate interest to consume as much of the common resource as possible until the resource is depleted—for example, individual farmers each increasing the size of their herds and in turn overgrazing common pastureland.

In the case of parasitic replicators, the analogous common resource is "replicative capacity," that is, the fractional share per unit time of all replicase capacity. Individual parasites that manage to have themselves copied but do not produce replicases gain one copy of themselves but reduce the common replicative capacity in proportion to their genome length divided by the total genome length of all replicases. This consumption may be small at first, but it will grow until the common replication capacity is inevitably overwhelmed by the parasites. In 1999, Turner and Chao demonstrated experimentally that when multiple biological replicators (bacteriophages) compete in a common environment (a single bacterial cell), mutants capable of parasitizing the common replication machinery evolve and subsist (Turner and Chao, 1999). The presence of the parasites lowered the rate of replication of the population as a whole, but the parasites were replicated faster than the true replicators, presumably because they had streamlined their genomes. The parasites had directly gained by drawing more heavily from a common resource pool, and the result was the compromised replication rate of the population as a whole. The tragedy of the commons has many economic regulatory solutions, but the simplest is privatization whereby the commons are divided into discrete parcels and exclusion of nonowners is afforded through either law or violence. Compartmentalization or cellularization of early RNA replication systems may have been a similar response. Indeed, Turner and Chao also demonstrated that when bacteriophages were isolated from one another by cell membranes (i.e., one phage infecting one cell), the parasites did not persist (Turner and Chao, 1998).

From this vantage, parasitism may have been *the* selective pressure driving the transition from a homogenous solution-based RNA world to a cellularized world. Moreover, it is apparent that once compartmentalized, individual parasites might eventually become the fodder for selection and serve as protogenes in a nascent genome. In a bulk-solution RNA world, mutated replicators may acquire nascent catalytic functionality, but there is very little selective pressure to maintain the mutant in the absence of a direct benefit to the replicator. Compartmentalization in protocells would not only allow negative selection against deleterious parasites, but also positive selection for mutants with beneficial functionality (e.g., cell membrane biosynthesis), paving the way for the dawn of genes and the evolution of modern genomes.

Of course, it should be pointed out that evading parasites would not have been the only driving force for compartmentalization. The formation of primitive membranes also provides a mechanism for concentrating reactants, which would have been essential in the development of a complex RNA world. In addition, segregation of ions by membranes is a common means of generating chemical gradients that can, in turn, be used for the production of energy-rich compounds.

17.3 Experimental Examples of Parasitism of Molecular Replicators

Whereas work with bacteriophage bears out theoretical models for parasitism, unfortunately, very little experimental work on molecular replicators and parasites applies directly to theoretical models. Nonetheless, the general theoretical conclusions are bolstered by numerous experimental examples of molecular parasitism. In some of the earliest in vitro replication and selection experiments with purified Qβ replicase, viral templates were quickly displaced by deletion variants, which contained significantly shorter nucleotide sequences (Biebricher, Eigen, and Luce, 1986; Biebricher and Luce, 1993; Chetverin, Chetverina, and Munishkin, 1991; Kacian et al., 1972; Mills et al., 1975). These results were consistent with the earlier assertion that many more molecules can act as templates rather than as catalysts in the sequence space surrounding a given replicase. In fact, many of the parasite RNAs contained relatively small amounts of information (terminal sequence tags and some conserved secondary structures) that allowed them to be used as templates, relative to other encoded information. Once a shorter parasite had arisen, it was recognized just as well by the replicase, could be more quickly replicated, and therefore outcompeted its longer parent. To the extent that the amount of information required to act as a template will almost always be smaller than the amount of information required to act as or encode a catalyst, molecular parasites that can more efficiently replicate and thereby usurp resources are likely an inherent feature of any replicating system.

In a different experiment, Konarska and Sharp identified highly amplifiable RNA parasites of the DNA-dependent RNA polymerase of bacteriophage T7 (Konarska and Sharp 1989, 1990). These RNAs (termed RNA X and Y) were present within the original T7 RNAP enzyme preparation either in undetectably small amounts or not at all, but were capable of completely dominating the polymerization capacity of an in vitro transcription reaction. These selfish replicators were highly self-structured; similar results were later obtained by Biebricher and Luce (1996). Most extant DNA-dependent RNA polymerases possess RNA-dependent RNA polymerization activities that are not obviously important for the fitness of the organism or virus. Parasitic RNAs invariably take advantage of these activities, leading to the interesting question of whether they do indeed contribute to the fitness of the organism, or whether it is just too mechanistically difficult for many RNA polymerases to exclude RNA as a template.

Multienzyme systems that use several template-copying steps have also been shown to be overrun by parasitic molecular species. For example, Breaker and Joyce (1994) developed a directed evolution scheme for ribozymes that used a continuous isothermal nucleic acid amplification system known as 3SR (self-sustained sequence replication) or NASBA (nucleic acid sequence-based amplification) (figure 17.1a). NASBA makes use of two protein enzymes, reverse transcriptase (RT) and T7 RNA polymerase (RNAP), to continuously amplify RNA molecules present in solution. Breaker and Joyce modified this strategy to select for a ribozyme catalyst by limiting the association of the second DNA primer to RNA variants capable of ligating a promoter bearing tag sequence to its own 5′ end (figure 17.1b). However, instead of selecting only for the desired property, catalysis, this scheme quickly led to the evolution of parasites that incorporated molecular sequence "tags" (promoters) and foldback structures that allowed them to take advantage of both polymerase enzymes while shortcutting the catalytic requirement (figure 17.1c). This class of parasites, termed RNA Z, arose by accumulating minimal amounts of information required to act as a template, while avoiding the information required for catalysis. It is again noteworthy that even for an extremely simple ribozyme ligase (as opposed to a more complex ribozyme or protein polymerase) the information requirements for templating were much smaller than those for catalysis.

The inherent problem of molecular parasitism cannot be solved by simply increasing the complexity of the replication system (i.e., adding more enzymes), or by building in feedback loops between the components of the replication machinery. Ehricht and coworkers (1997) devised a coupled NASBA system that contained two RNA oligonucleotides (we will call them A and B) along with primers specific to each RNA (P_A and P_B). P_A and P_B could anneal to A and B and be extended by the enzyme RT into cDNAs (dA and dB). The copies dA and dB were partially complementary to one another at their 3′ ends, and could anneal and be extended by RT to form a fully double-stranded DNA template that contained promoters for the transcription (amplification) of A and B by a DNA-dependent RNA polymerase (RNAP) (figure 17.2). This process could be continuously iterated, and in this way

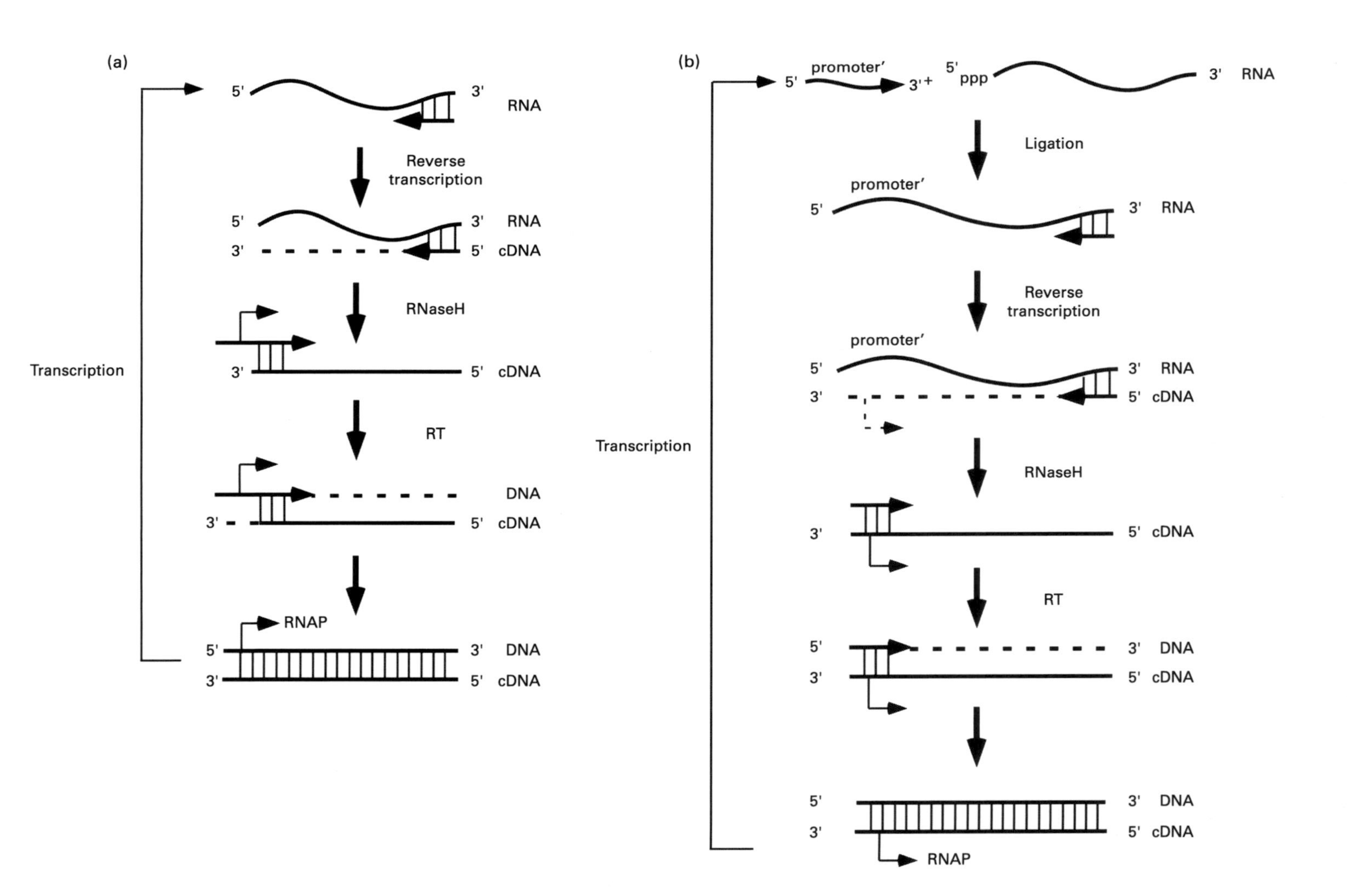
(a)
5'
3'
RNA
Reverse transcription
5'
3'
RNA
3'
5'
cDNA
RNaseH
Transcription
3'
5'
cDNA
RT
DNA
3'
5'
cDNA
RNAP
5'
3'
DNA
3'
5'
cDNA
(b)
5'
promoter'
3'+
5' ppp
3'
RNA
Ligation
promoter'
5'
3'
RNA
Reverse transcription
promoter'
5'
3'
RNA
3'
5'
cDNA
Transcription
RNaseH
3'
5'
cDNA
RT
5'
3'
DNA
3'
5'
cDNA
5'
3'
DNA
3'
5'
cDNA
RNAP

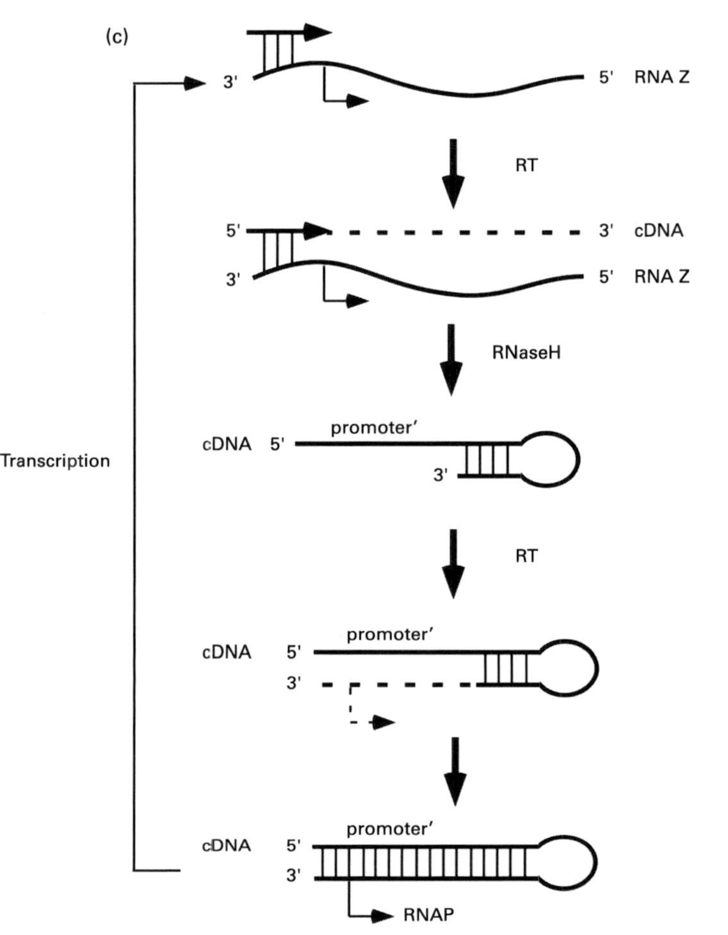

Figure 17.1
(a) NASBA (nucleic acid sequence-based amplification) scheme. A template RNA is amplified in a single, isothermal batch reaction. A sequence-specific DNA primer (black arrow) anneals to the 3′ end of the template RNA and is extended by reverse transcriptase (RT) into a cDNA. The template RNA strand is then specifically digested by RNaseH, leaving only the cDNA. The 3′ end of the cDNA then anneals to a second primer with an embedded promoter sequence that is extended by RT into a full-length double-stranded template for transcription of the original RNA by RNA polymerase (RNAP). (b) The Breaker and Joyce NASBA-based scheme for continuous in vitro selection of catalytic RNAs. A transcribed pool of RNAs is incubated with an RNA substrate bearing the reverse complement of a promoter sequence (promoter'). Those variants capable of catalyzing the ligation of the substrate to their 5′-end can template the RT of a promoter bearing cDNA (dashed arrow, dashed line). A second primer then specifically recognizes the promoter-bearing cDNA and the complement strand is synthesized by RT. This provides a full-length template for the transcription of the original RNA sequence by RNAP (adapted from Breaker and Joyce, 1994). (c) Mechanism of replication of RNA Z. RNA Z contains an internal promoter tag sequence that can anneal to the second primer from (b) and be directly extended to form a cDNA bearing a promoter' sequence. This cDNA has internal complementarity, allowing the formation of a partial hairpin structure that can prime its own extension into a full-length hairpin by RT. The full-length hairpin bears a proper double-stranded promoter that is used to transcribe around the hairpin, producing more copies of RNA Z (adapted from Breaker and Joyce, 1994).

A and B could cooperate in their mutual replication. Parasites capable of selfishly exploiting one half of the system (e.g., A and P_A alone) invariably arose despite thorough attempts at condition optimization. The predominance of these shortcut parasites was likely a result of their smaller information content and size, which led to more rapid replication.

The amount of information required for templating is typically less than the amount required for encoding additional functionality. However, it is not necessarily true that less information means more efficient templating. For example, if a particular nucleotide sequence is required for primer- or polymerase-binding, shortening this sequence may lead to less efficient replication. Similarly, it is possible that a parasite can increase its fecundity not only by reducing the amount of information that is not involved in templating, but also by increasing the amount of information to increase the efficiency of templating. In this regard, it is instructive that Marshall and Ellington (1999) used NASBA to evolve fast-replicating molecular parasites that were actually longer (150 nt) than their parental templates (80 nt). On sequencing, the parasites appeared to be extensions or aggregates of their parents. Although the mechanism of replicative advantage was not clear, the parasites were capable of dominating the replication machinery unless they were outcompeted by an initial excess of the unevolved parental template in the reaction.

The amount of information required for templating can increase in other ways, as well. As the complexity of a replication system evolves, the capacity for parasitism by longer and more complex species may also evolve. For example, even though RNA Z (100 nt) was shorter than its parent (800 nt), it contained several sequence and structural features that allowed its serial interaction with different polymerases during NASBA amplification. Given that the Marshall replicators also evolved during NASBA amplification, we may eventually find that interactions with multiple enzymes could best be encoded by augmenting or melding parental templates that otherwise contained too few signals for efficient templating.

The overall conclusions of both the theoretical and experimental treatments of parasitism of molecular replicators are roughly congruent with one another. However, neither treatment can yet answer key questions, such as what the relative ratio of information required for templating versus catalysis or other functionalities may be. This ignorance is, for the most part, a result of different replicators having very idiosyncratic requirements for both information and catalysis: Promoters or other sequences that ensure templating can be longer or shorter, and replicases come in various lengths. The only truism that continues to hold is that the information required for templating is less than the information required for additional functionality. This is a genuine truism, since a replicase must contain both types of information, whereas a parasite need contain only one. In this regard, the path forward in

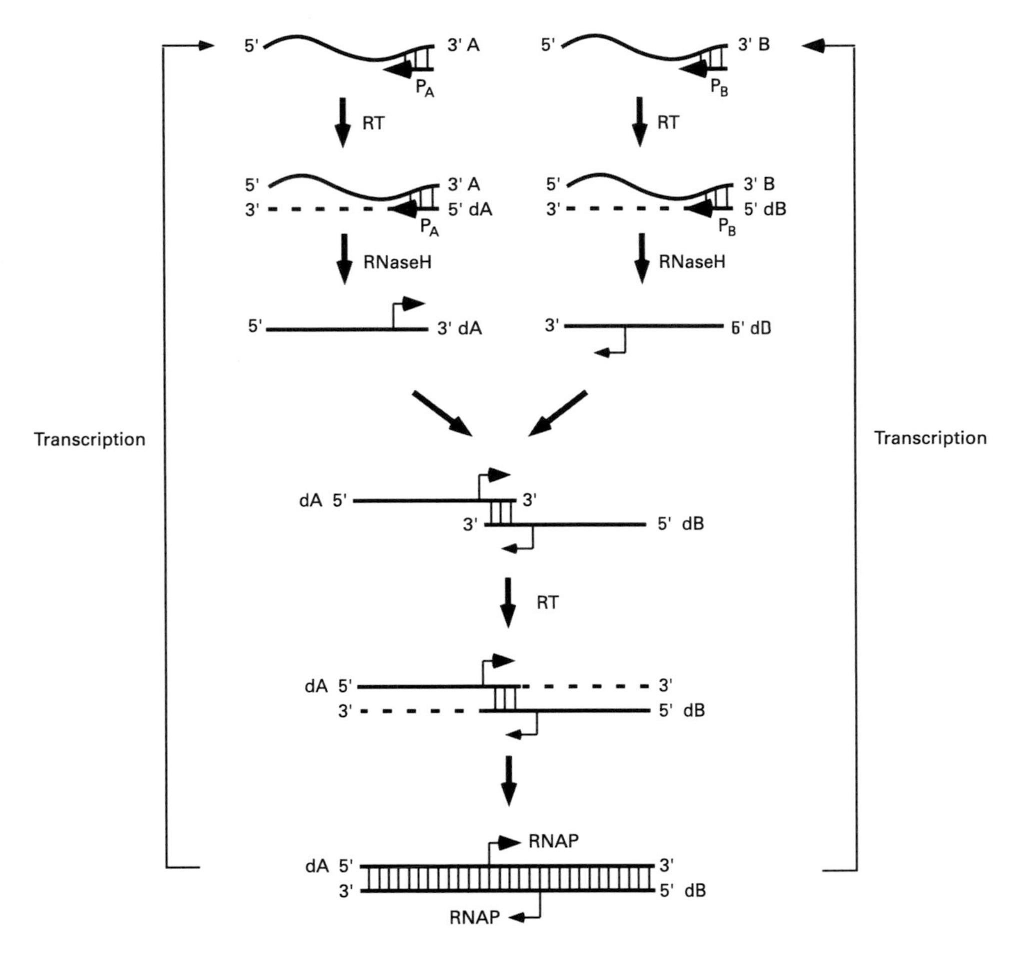

Figure 17.2
Mechanism of cooperative amplification of templates by cross-hybridization (CATCH). CATCH is a NASBA scheme modified to be dependent on the hybridization of two independently derived cDNAs, which are then extended to form a full-length template for the amplification of two starting RNA molecules. In the reaction are two RNAs (A and B), each with a specific 3′ primer (P_A and P_B), which primes the synthesis of a corresponding cDNA (dA and dB) with an embedded single-stranded promoter sequence (black arrows). dA and dB are complementary to each other at their 3′ ends, and on base-pairing can be extended into a full-length double-stranded DNA bearing functional promoters for the transcription of A and B by RNAP (adapted from Ehricht, Ellinger, and McCaskill, 1997).

understanding molecular parasitism might be blazed largely by experimentalists. As experiments are devised to generate systems that are increasingly capable of self-replication, experimentalists should be able to increasingly provide existence proofs about how parasites can arise. We further suggest that the analyses of these experiments will be best guided by economic models, which tend to take into account the cost-benefit analyses of individual molecular actors in idiosyncratic situations, as opposed to identifying unifying principles that will be applicable in any replication regime.

17.4 Cellularization as an Experimental Solution to Parasitism

The transition from a noncompartmentalized or solution-phase RNA world to a cellularized world in which individual replicators were isolated from one another should have vastly changed the dynamics of parasitism. In solution, replicative capacity would have been the most important component of fitness, and freely diffusing parasites would therefore have disproportionately slowed the replication of the most fit members of the population. With the advent of a cellularized RNA world, however, any RNA species that spawned a less functional parasite would have been independently subjected to Darwinian selection. If the parasite bestowed a negative effect on either cellular or replicator phenotype, then that cell or replicator would have been selected against independently of all other cellularized replicators. As previously suggested (Bresch, Niesert, and Harnasch, 1980; Eigen, Gardiner, and Schuster, 1980; Eigen et al., 1981; Niesert, Harnasch, and Bresch, 1981), this realization suggests that cellularization may have been an optimal mechanism for the evasion of parasites. It is, of course, difficult to recapitulate any aspect of origins; however, our and others' experiences with in vitro selection experiments demonstrate just how easy it is to generate parasites, even though selection schemes typically go to great lengths to avoid them (e.g., the discontinuous purification of molecular species of a "correct" size). To the extent that all molecular replicators must have produced parasites at some point, the competition between replicators would have essentially been a competition to determine who could first be cellularized.

If cellularization or compartmentalization was essential to the evolution of replicators, it should be possible to demonstrate this experimentally. It has now been shown that systems of molecular replicators can be ensconced within protocells (Oberholzer, Albrizio, and Luisi, 1995a; Oberholzer et al., 1995b; see also chapters 2 and 3). However, these systems in general involve protein-based replication of nucleic acid templates. In an attempt to better understand how cellularization might influence the evolution of nucleic acid catalysts similar to those that would have been present at origins, we and others have recently developed techniques for compartmentalized selection of a ribozyme using an oil and water emulsion (Agresti et al., 2005; Levy,

Griswold, and Ellington, 2005). Whereas the directed evolution of ribozymes typically involves *cis* reactions in which a substrate is added to or detached from a given ribozyme; these systems have no obvious relevance in a prebiotic milieu. In contrast, compartmentalized systems allow reactions in *trans*, and hence allow the evolution of multiple turnover catalysts similar to those that would almost certainly have been present at or near origins. These selection schemes are clearly not prebiotic, but these nascent experiments begin to move us in a direction that should ultimately lead to a greater understanding of the effects of cellularization on evolution. Indeed, recent work from the Szostak lab has demonstrated that protocells composed of myristoleic acid and its glycerol monoester are sufficiently stable to encapsulate functional ribozymes, and are also capable of vesicle growth and division with the addition of exogenous amphiphiles (see chapter 5). Perhaps more important, the vesicles were also shown to be permeable to small molecules (mononucleotides) and ions (Mg^{2+}; see Chen et al., 2005), paving the way for a more prebiotically plausible ribozyme selection strategy.

In the case of our selection strategy, we used a water-in-oil emulsion system previously used for the directed evolution of proteins (Bernath, Magdassi, and Tawfik, 2005; Levy, Griswold, and Ellington, 2005). In this selection scheme (figure 17.3), a microbead served as a means to link genotype and phenotype. To do this, a double-stranded DNA pool was immobilized on a streptavidin-coated microbead such that there was, on average, one dsDNA sequence per bead. In addition, each bead was

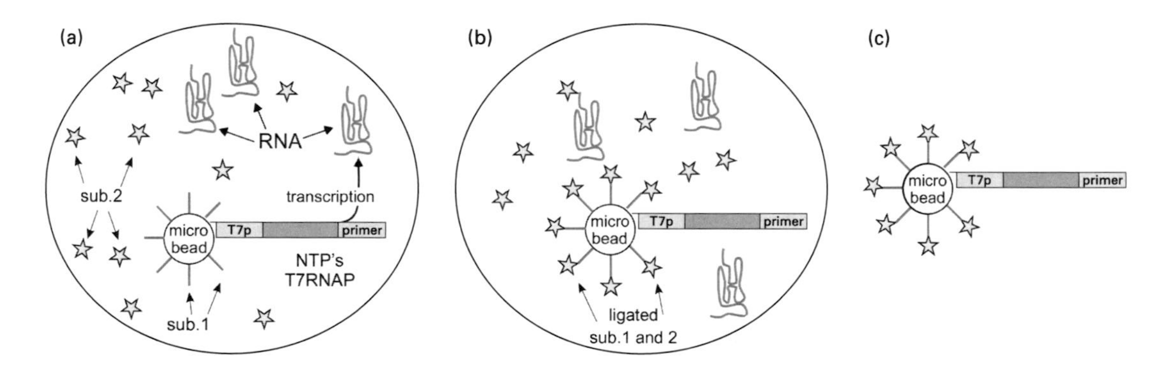

Figure 17.3
Compartmentalized selection scheme for the isolation of multiple turnover ligase ribozymes. (a) Microbeads bearing a dsDNA pool as well as one half of the substrate molecule (sub. 1) are emulsified with the components of an in vitro transcription reaction and the second half of the RNA substrate (sub. 2; star). Transcription from the dsDNA results in the production of RNA. (b) The production of functional RNA ligases results in the tagging of the microbead with the second half of the substrate molecule (star), linking genotype and phenotype. (c) After the emulsification is broken, the labeled beads can be isolated by fluorescent-activated cell sorting (FACS), and the attached dsDNA encoding the functional RNA amplified by polymerase chain reaction (PCR).

loaded with multiple copies (~100,000) of one half of an RNA ligation substrate (figure 17.3, sub. 1). The beads were then used as the input for an in vitro transcription reaction. In addition to the standard components of transcription (T7 RNA polymerase and NTPs), the reaction also contained the second half of the RNA ligation substrate (figure 17.3, sub. 2). The transcription reaction and microbeads were encapsulated within a water-in-oil emulsion by adding the aqueous reaction to an iced mixture of mineral oil and nonionic detergents followed by rapid mixing. This process produces $\sim 10^{10}$ discrete aqueous compartments per milliliter of oil. Compartments are not capable of exchanging materials with neighboring compartments, and each contained either one or no microbead. Following encapsulation, the emulsion was warmed to room temperature, allowing transcription to occur. During this time, functional RNA ligases produced through transcription could ligate the two substrate halves within their compartment and thus tag the microbead bearing their own gene. When the emulsification was broken, the tagged beads could be isolated by FACS and the individual dsDNA amplified by PCR. As with most directed evolution schemes, repetition by emulsification, ligation, and sorting would result in further enrichment of functional ribozyme ligases in the population.

To test this method, we generated a doped sequence library based on the highly efficient *trans*-acting variant of the class I Bartel ligase (Bartel and Szostak, 1993). Ninety positions within the pool were doped at a level of 30% such that on average each member of the pool contained 27 mutations. We then carried out four rounds of selection in which we decreased the reaction time allowed for each round. In addition, the selection of false positives was reduced by decreasing the gene-to-bead ratio during each round so that, during the final round of selection, only one in 10 microbeads had a gene on it. This stricture may have a prebiotic counterpart, in that early vesicles may have randomly divided and been widely dispersed.

Considering that any active ribozymes would have assorted randomly with substrates and templates in a nonemulsified system, the selection was brutally effective. After only two rounds of selection, the pool displayed *trans*-ligation activity. Sequence analysis of 34 clones isolated from the round-three population yielded only a single active clone. By round four, this clone comprised more than 50% of the population (14 out of 25). Surprisingly, the dominant clone contained only five mutations relative to the parental ribozyme. Statistically this variant would have been expected to occur less than once in the starting population.

Even though selection was extremely stringent, compartmentalization did not eliminate parasitism. The 11 remaining clones in the round-four population contained numerous mutations relative to the parental ribzoyme and showed no appreciable ligation activity under the conditions employed in the selection. These sequences may have resulted from the equilibration of dsDNA genes between the microbeads; this explanation is supported by the fact that excess biotin blocks equilibration, presumably by blocking free-binding sites following the dissociation of templates. The ultimate breaching of compartmentalized defenses should not have been too surprising. Even following cellularization, molecular parasitism readily occurs. Viroids (hepatitis delta virus) frequently arise that confer no fitness advantage but can slow the replication of their parental viruses (hepatitis B virus; Lai, 1995).

The presence of parasites in a compartmentalized selection was even more impressive given that replication was not continuous, as in the isothermal amplification systems examined earlier, but was discontinuous in that in each round a discrete amplification step (PCR) was employed that was spatially and temporally separate from the selection for function (ligation). It has been suggested that such discontinuous replication systems should limit the production and spread of parasites that can usurp the replicative machinery (Bull and Pease, 1995). To the extent that these results and their interpretation are valid, they immediately suggest a testable hypothesis. It seems likely that, if the continuous selection methodology developed by Wright and Joyce (1997) were used to select for compartmentalized ligation function, then even more nonfunctional parasitic sequences would arise and be fixed. Overall, one might expect that parasitism should affect experimental systems in the following order: solution phase > compartmentalized with continuous amplification > compartmentalized with discontinuous amplification. Should such a correlation be shown to exist, it would provide excellent evidence that modern experiments can not only recapitulate the types of reactions that may have occurred prebiotically, but can also mimic the evolutionary pressures the early replicators would have experienced.

Ultimately, these results speak not only to how parasitism might have led to cellularization, but also to the early development of cellular defenses capable of repelling parasites or even other replicators. Such defenses would inevitably have led to the diversification of genetic information and function in the nascent protocell.

17.5 Conclusion

It can be argued that no replicator can evade parasites, and the evolution of cellularization will therefore be driven by parasitism. Given the idiosyncratic nature of most replicating systems, it remains the province of experimentalists to demonstrate the evolutionary benefits of compartmentalization and how such benefits may have been played out at or near life's origins. In creating a private economy that avoids the tragedy of the molecular commons, cellularization also, of course, provides substantive benefits for the evolution of metabolism (concentration of metabolites, development of chemical gradients). Parasitism and cellularization therefore may have also been the driving forces for the advent of genes and ultimately genomes.

References

Agresti, J. J., Kelly, B. T., Jaschke, A., & Griffiths, A. D. (2005). Selection of ribozymes that catalyze multiple-turnover Diels-Alder cycloadditions by using in vitro compartmentalization. *Proceedings of the National Academy of Sciences of the United States of America, 102* (45), 16170–16175.

Altmeyer, S., & McCaskill, J. S. (2001). Error threshold for spatially resolved evolution in the quasispecies model. *Physical Review Letters, 86*, 5819–5822.

Barrick, J. E., Corbino, K. A., Winkler, W. C., Nahvi, A., Mandal, M., Collins, J., et al. (2004). New RNA motifs suggest an expanded scope for riboswitches in bacterial genetic control. *Proceedings of the National Academy of Sciences of the United States of America, 101* (17), 6421–6426.

Bartel, D., & Szostak, J. (1993). Isolation of new ribozymes from a large pool of random sequences. *Science, 261* (5127), 1411–1418.

Bernath, K., Magdassi, S., & Tawfik, D. S. (2005). Directed evolution of protein inhibitors of DNA-nucleases by in vitro compartmentalization (IVC) and nano-droplet delivery. *Journal of Molecular Biology, 345* (5), 1015–1026.

Biebricher, C., Eigen, M., & Luce, R. (1986). Template-free RNA synthesis by Q beta replicase. *Nature, 321* (6065), 89–91.

Biebricher, C., & Luce, R. (1993). Sequence analysis of RNA species synthesized by Q beta replicase without template. *Biochemistry, 32* (18), 4848–4854.

Biebricher, C., & Luce, R. (1996). Template-free generation of RNA species that replicate with bacteriophage T7 RNA polymerase. *European Molecular Biology Organization, 15* (13), 3458–3465.

Boerlijst, M. C., & Hogeweg, P. (1991). Spiral wave structures in prebiotic evolution: Hypercycles stable against parasites. *Physica D, 48*, 17–28.

Breaker, R. R., & Joyce, G. F. (1994). Emergence of a replicating species from an in vitro RNA evolution reaction. *Proceedings of the National Academy of Sciences of the United States of America, 91*, 6093–6097.

Bresch, C., Niesert, U., & Harnasch, D. (1980). Hypercycles, parasites and packages. *Journal of Theoretical Biology, 85* (3), 399–405.

Bull, J., & Pease, C. (1995). Why is the polymerase chain reaction resistant to in vitro evolution? *Journal of Molecular Evolution, 41* (6), 1160–1164.

Chen, I. A., Salehi-Ashtiani, K., & Szostak, J. W. (2005). RNA catalysis in model protocell vesicles. *Journal of the American Chemical Society, 127* (38), 13213–13219.

Chetverin, A., Chetverina, H., & Munishkin, A. (1991). On the nature of spontaneous RNA synthesis by Q beta replicase. *Journal of Molecular Biology, 222* (1), 3–9.

Ehricht, R., Ellinger, T., & McCaskill, J. S. (1997). Cooperative amplification of templates by cross-hybridization (CATCH). *European Journal of Biochemistry, 243*, 358–364.

Eigen, M. (1971). Self organization of matter and the evolution of biological macromolecules. *Naturwissenschaften, 58* (10), 465–523.

Eigen, M., Gardiner, W. C. Jr., & Schuster, P. (1980). Hypercycles and compartments. Compartments assist—but do not replace—hypercyclic organization of early genetic information. *Journal of Theoretical Biology, 85* (3), 407–411.

Eigen, M., Gardiner, W., Schuster, P., & Winkler-Oswatitsch, R. (1981). The origin of genetic information. *Scientific American, 244* (4), 88–92, 96, et passim.

Eigen, M., & Schuster, P. (1977). The hypercycle: A principle of natural self-organization. Part A: Emergence of the hypercycle. *Naturwissenschaften, 64* (11), 541–565.

Eigen, M., & Schuster, P. (1978a). The hypercycle: A principle of natural self-organization. Part B: The abstract hypercycle. *Naturwissenschaften, 65*, 7–41.

Eigen, M., & Schuster, P. (1978b). The hypercycle: A principle of natural self-organization. Part C: The realistic hypercycle. *Naturwissenschaften, 65*, 341–369.

Ekland, E. H., Szostak, J. W., & Bartel, D. P. (1995). Structurally complex and highly active RNA ligases derived from random RNA sequences. *Science, 269* (5222), 364–370.

Hardin, G. (1968). The tragedy of the commons. *Science*, *162*, 1243–1248.

Huang, F., & Yarus, M. (1997). Versatile 5′ phosphoryl coupling of small and large molecules to an RNA. *Proceedings of the National Academy of Sciences of the United States of America*, *94* (17), 8965–8969.

James, K. D., & Ellington, A. D. (1999). The fidelity of template-directed oligonucleotide ligation and the inevitability of polymerase function. *Origins of Life and Evolution of the Biosphere*, *29* (4), 375–390.

Johnston, W. K., Unrau, P. J., Lawrence, M. S., Glasner, M. E., & Bartel, D. P. (2001). RNA-catalyzed RNA polymerization: Accurate and general RNA-templated primer extension. *Science*, *292* (5520), 1319–1325.

Kacian, D., Mills, D., Kramer, F., & Spiegelman, S. (1972). A replicating RNA molecule suitable for a detailed analysis of extracellular evolution and replication. *Proceedings of the National Academy of Sciences of the United States of America*, *69* (10), 3038–3042.

Konarska, M., & Sharp, P. (1989). Replication of RNA by the DNA-dependent RNA polymerase of phage T7. *Cell*, *57* (3), 423–431.

Konarska, M., & Sharp, P. (1990). Structure of RNAs replicated by the DNA-dependent T7 RNA polymerase. *Cell*, *63* (3), 609–618.

Lai, M. M. (1995). The molecular biology of hepatitis delta virus. *Annual Review of Biochemistry*, *64*, 259–286.

Levy, M., & Ellington, A. (2001). The descent of polymerization. *Nature Structural Biology*, *8* (7), 580–582.

Levy, M., Griswold, K. E., & Ellington, A. D. (2005). Direct selection of trans-acting ligase ribozymes by in vitro compartmentalization. *RNA*, *11* (10), 1555–1562.

Lorsch, J. R., & Szostak, J. W. (1994). In vitro evolution of new ribozymes with polynucleotide kinase activity. *Nature*, *371* (6492), 31–36.

Mandal, M., & Breaker, R. R. (2004). Adenine riboswitches and gene activation by disruption of a transcription terminator. *Nature Structural and Molecular Biology*, *11* (1), 29–35.

Marshall, K., & Ellington, A. (1999). Molecular parasites that evolve longer genomes. *Journal of Molecular Evolution*, *49* (5), 656–663.

McCaskill, J. S. (1984). A localization threshold for macromolecular quasispecies from continuously distributed replication rates. *The Journal of Chemical Physics*, *80*, 5194–5202.

Mills, D., Kramer, F., Dobkin, C., Nishihara, T., & Speigelman, S. (1975). Nucleotide sequence of microvariant RNA: Another small replicating molecule. *Proceedings of the National Academy of Sciences of the United States of America*, *72* (11), 4252–4256.

Niesert, U., Harnasch, D., & Bresch, C. (1981). Origin of life between Scylla and Charybdis. *Journal of Molecular Evolution*, *17* (6), 348–353.

Oberholzer, T., Albrizio, M., & Luisi, P. (1995a). Polymerase chain reaction in liposomes. *Chemistry and Biology*, *2* (10), 677–682.

Oberholzer, T., Wick, R., Luisi, P., & Biebricher, C. (1995b). Enzymatic RNA replication in self-reproducing vesicles: An approach to a minimal cell. *Biochemical and Biophysical Research Communications*, *207* (1), 250–257.

Orgel, L. E. (2004). Prebiotic chemistry and the origin of the RNA world. *Critical Reviews in Biochemistry and Molecular Biology*, *39* (2), 99–123.

Paul, N., & Joyce, G. F. (2002). Inaugural article: A self-replicating ligase ribozyme. *Proceedings of the National Academy of Sciences of the United States of America*, *99* (20), 12733–12740.

Robertson, M. P., & Ellington, A. D. (1999). In vitro selection of an allosteric ribozyme that transduces analytes to amplicons. *Nature Biotechnology*, *17* (1), 62–66.

Robertson, M. P., & Ellington, A. D. (2000). Design and optimization of effector-activated ribozyme ligases. *Nucleic Acids Research*, *28* (8), 1751–1759.

Rogers, J., & Joyce, G. F. (2001). The effect of cytidine on the structure and function of an RNA ligase ribozyme. *RNA*, *7* (3), 395–404.

Seelig, B., & Jaschke, A. (1999). A small catalytic RNA motif with Diels-Alderase activity. *Chemistry and Biology*, *6* (3), 167–176.

Sengle, G., Eisenfuhr, A., Arora, P. S., Nowick, J. S., & Famulok, M. (2001). Novel RNA catalysts for the Michael reaction. *Chemistry and Biology*, *8* (5), 459–473.

Soukup, G. A., & Breaker, R. R. (1999). Engineering precision RNA molecular switches. *Proceedings of the National Academy of Sciences of the United States of America*, *96* (7), 3584–3589.

Sun, L., Cui, Z., Gottlieb, R. L., & Zhang, B. (2002). A selected ribozyme catalyzing diverse dipeptide synthesis. *Chemistry and Biology*, *9* (5), 619–628.

Szabo, P., Scheuring, I., Czaran, T., & Szathmáry, E. (2002). In silico simulations reveal that replicators with limited dispersal evolve towards higher efficiency and fidelity. *Nature*, *420* (6913), 340–343.

Szathmáry, E., & Demeter, L. (1987). Group selection of early replicators and the origin of life. *Journal of Theoretical Biology*, *128* (4), 463–486.

Tang, J., & Breaker, R. R. (1997). Rational design of allosteric ribozymes. *Chemistry and Biology*, *4* (6), 453–459.

Tarasow, T. M., Tarasow, S. L., & Eaton, B. E. (1997). RNA-catalyzed carbon-carbon bond formation. *Nature*, *389* (6646), 54–57.

Tsukiji, S., Pattnaik, S. B., & Suga, H. (2004). Reduction of an aldehyde by a NADH/Zn2+-dependent redox active ribozyme. *Journal of the American Chemical Society*, *126* (16), 5044–5045.

Turner, P. E., & Chao, L. (1998). Sex and the evolution of intrahost competition in RNA virus phi6. *Genetics*, *150* (2), 523–532.

Turner, P. E., & Chao, L. (1999). Prisoner's dilemma in an RNA virus. *Nature*, *398* (6726), 441–443.

Unrau, P. J., & Bartel, D. P. (1998). RNA-catalyzed nucleotide synthesis. *Nature*, *395* (6699), 260–263.

von Kiedrowski, G. (1986). A self replicating hexadeoxynucleotide. *Angewandte Chemie International Edition*, *25*, 932–935.

Wecker, M., Smith, D., & Gold, L. (1996). In vitro selection of a novel catalytic RNA: Characterization of a sulfur alkylation reaction and interaction with a small peptide. *RNA*, *2* (10), 982–994.

White, H. B. (1976). Coenzymes as fossils of an earlier metabolic state. *Journal of Molecular Evolution*, *7* (2), 101–104.

Wiegand, T. W., Janssen, R. C., & Eaton, B. E. (1997). Selection of RNA amide synthases. *Chemistry and Biology*, *4* (9), 675–683.

Wilson, C., & Szostak, J. W. (1995). In vitro evolution of a self-alkylating ribozyme. *Nature*, *374* (6525), 777–782.

Winkler, W., Nahvi, A., & Breaker, R. R. (2002). Thiamine derivatives bind messenger RNAs directly to regulate bacterial gene expression. *Nature*, *419* (6910), 952–956.

Wright, M. C., & Joyce, G. F. (1997). Continuous in vitro evolution of catalytic function. *Science*, *276*, 614–617.

Zielinski, W. S., & Orgel, L. E. (1987). Autocatalytic synthesis of a tetranucleotide analogue. *Nature*, *327* (6120), 346–347.

18 Forming the Essential Template for Life: The Physics of Lipid Self-Assembly

Ole G. Mouritsen and Ask F. Jakobsen

18.1 Introduction: Life from Molecules

All forms of life as we know it are made from the same four classes of small organic molecules: the amino acids, the nucleotides, the carbohydrates, and the fatty acids. Molecules from the same class or from different classes combine chemically to form larger macromolecular entities, specifically the polysaccharides (sugars), the proteins (polypeptides), the nucleic acids (polynucleotides like DNA and RNA), and the fats (lipids). Whereas polysaccharides, nucleic acids, and proteins are all polymers, that is, long-chain and possibly branched (in the case of sugars) molecules bound by strong covalent forces, the fats do not normally polymerize. There are no polylipids in nature. In contrast, the lipids organize into different types of macromolecular assemblies such as micelles and membranes bound by weaker physical forces. The assembly process takes place spontaneously in the presence of the biological solvent, water. It is the physics of such self-assembly processes that is the focus of the present chapter.

It is likely that the first forms of life on Earth required for their formation and evolution certain templates made of lipidlike molecules that provided the appropriate embedding or encapsulation medium for (1) the information-storing molecules capable of reproduction, (2) the enzymelike catalysts encoded by that information and capable of enhancing reproduction rates, and (3) the molecules capable of storing energy and using this energy to convert molecules into organized assemblies of biologically active molecules (Deamer, 1986). By forming dynamic structures characterized by specific interfaces and compartments, the lipids were able to provide both the proper reaction conditions for, and the necessary protection of, the molecules described here in order to permit self-reproduction, growth, and evolution.

We have just witnessed a scientific revolution at the end of the 20th century that took us through the genomics era with a focus on sequencing genes and mapping gene products. An increasing number of almost complete genomes of organisms from worms to man is being determined. We are now in the middle of the proteomics era in which the multitude of proteins encoded in the genes are being identified and their functions unveiled. Whereas proteins and polynucleotides (including DNA, RNA, and complete genomes) attracted enormous attention among scientists throughout the 20th century, the lipids have been somewhat overlooked molecules (Mouritsen, 2005). The reason for this is not only the great success of molecular and structural biology focused on genomics and proteomics, but also the perception that lipids are somewhat dull molecules. Whereas the genes encode the information for constructing the proteins, and the proteins perform most of the functions of living systems, the lipids are foremost structure builders whose properties are characterized by fuzzy terms like adaptability, diversity, plasticity, and softness. This characterization implies that lipids and lipid assemblies have considerable elements of disorder whereas proteins and polynucleotides are characterized by well-defined molecular structure and a high degree of order. The relationships between structure and function for lipids are therefore far more subtle than for proteins and genes. As a consequence, lipid systems have obtained the doubtful status as the grease within which all the beautiful machinery of life controlled by genes and proteins are imbedded.

The fact that lipids in living systems engage in structures of considerable disorder of the type characteristic of fluids makes them difficult to study and describe quantitatively. The hidden elements of order, which are subtle and difficult to measure, are strongly influenced by thermal agitation. Hence, the stability of systems involving lipids is to a large extent controlled by entropy and colloidal forces. Furthermore and importantly, the properties of lipid assemblies are inextricably related to the properties of the solvent, in particular the peculiar hydrogen-bonding dynamics of water.

In recent years, lipids have received increasing attention because of the recognition that lipids are key players in the various signaling pathways in the cell. Moreover, it is becoming clear that lipids modulate protein and enzyme function and they are targets for many types of drugs. In front of us we have a new scientific era recognizing the role of lipids. The science of lipidomics is emerging (Mouritsen, 2005; Rilfors and Lindblom, 2002). Lipidomics involves a quantitative experimental and theoretical study of, for example, lipid and membrane self-assembly, lipid-protein interactions, lipid-gene interactions, and the biophysical properties of lipid structure, function, and dynamics. All these aspects are relevant to consider when elucidating the necessary conditions for forming protocells.

In the present chapter, we describe the phenomenology of lipid self-assembly in water and how structures like lipid micelles, lipid bilayers, and lipid vesicles arise. We draw on recent results obtained from both experiments and computer-simulation calculations. The information provided should be considered as a necessary supplement to the information encoded in the genome for a living organism. Knowing the genome of an organism implies information about which proteins the cells of this organism can produce. However, knowing the genome of an organism does not imply that one necessarily knows which function a given protein can carry out. Furthermore, it is unlikely that it is written in the genome how a cell and its various parts are assembled from the molecular building blocks. The information contained in the genome is, in this sense, not complete, and additional principles have to be invoked to describe and understand the complex organization of the molecules of life. This is the point at which the physics of self-assembly and complex systems come in. Physics provides us with the tools to predict and describe the emergent and often disordered properties resulting from many molecules interacting with each other.

In this chapter, we first describe the phenomenology of self-assembly processes of amphiphilic molecules in water. The various types of structures and morphologies are described, including micelles, planar bilayers, and closed vesicles, and we consider the transitional behavior between the different structures. The propensity to form curved interfaces is described in terms of the effective shape of the molecules. Next, we review the trans-bilayer and lateral structure of lipid bilayers and discuss them in relation to protein and enzyme function. Then we introduce a specific numerical technique called dissipative particle dynamics (DPD) and show it to be capable of simulating the entire dynamic process of lipid self-assembly in water, leading to micelles, tubular filaments, planar lipid bilayers, and closed compartments like vesicles (liposomes). We demonstrate the spontaneous process of closing planar lipid bilayers into closed vesicles or liposomes. Finally, we describe some recent work on the morphological transitions of the vesicular state into buds, blebs, tubes, and fully developed fission and vesiculation processes.

18.2 Phenomenology of Lipid Self-Assembly

Lipids are amphiphilic molecules that have a hydrophilic head and a hydrophobic tail. The tail usually consists of one or two acyl chains of fatty acids. The head group comes in different sizes and may be uncharged, zwitter-ionic, or positively/negatively charged. The fatty acid chains vary in length and degree of saturation. When mixed with water, lipid molecules organize spontaneously into supramolecular aggregates of different sizes and symmetries. The main thermodynamic driving force is often referred to as the hydrophobic effect which, to substantial extent, is entropic and related to the hydrogen-bonding dynamics of liquid water. The phenomenology of self-assembly is well understood (Cates and Safran, 1997; Chen and Rajagopalan, 1990; Evans and Wennerström, 1999; Gelbart, Ben-Shaul, and Roux, 1994; Jönsson et al., 1998). Some examples of lipid aggregates in water are illustrated in figure 18.1.

Extended two-dimensional lipid monolayers as shown in figure 18.1a are easily formed by spreading a lipid solution at an air/water interface. Micelles as in figure

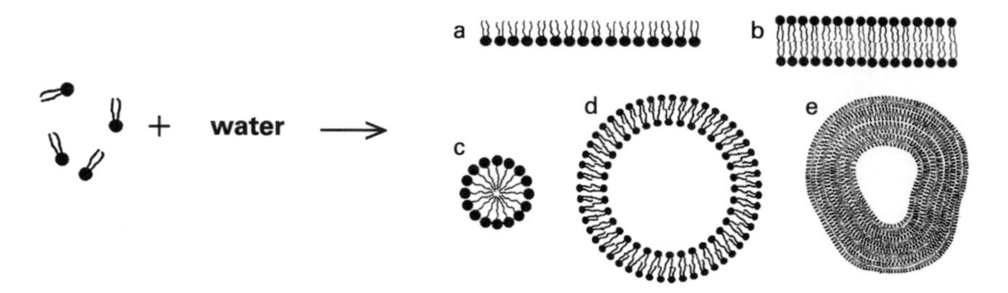

Figure 18.1
Schematic illustration of the self-assembly process of lipids into supramolecular aggregates in association with water: (a) Lipid monolayer; (b) lipid bilayer; (c) spherical micelle of amphiphiles with a single hydrocarbon chain; (d) vesicle (liposome), which is a closed lipid bilayer; (e) multilamellar vesicle (liposome).

18.1c, well-known from soap solutions, form when the monomer concentration in water is larger than the critical micelle concentration. The aggregation number and the size of micelles depend on the nature of the head group and the length of the fatty acid chains. Some lipids form bilayers, as shown in figure 18.1b. Lipid bilayers close onto themselves and form either unilamellar vesicles (liposomes), as in figure 18.1d, or multilamellar vesicles like those in figure 18.1e. Vesicles can be formed with diameters ranging from as low as 30 nm to hundreds of micrometers. Lipid bilayers are lyotropic liquid crystals (smectics) (Petrov, 1999). Experimental realizations of large and small unilamellar and multilamellar vesicles are shown in figure 18.2.

The relative stability of the different types of lipid aggregates is controlled by the associated chemical potentials (see e.g., Israelachvili, 1992). If we consider the set of aggregation equilibria

$$nL_1 \rightleftharpoons L_n, \tag{18.1}$$

where L_1 is a monomer lipid and n is the aggregation number, the condition of thermodynamic equilibrium in a dispersion of aggregates of different size n can be expressed in terms of the chemical potentials as $\mu_1 = \mu_2 = \mu_3, \ldots,$ where

$$\mu_n = \mu_n^\circ + \frac{\mathrm{k_B}T}{n} \ln\left(\frac{f_n x_n}{n}\right). \tag{18.2}$$

x_n is the molar concentration and f_n is the activity coefficient of aggregates with aggregation number n. For simplicity, we put $f_n = 1$ for all n in the following; μ_n° is the standard chemical potential for the aggregate type in question. Using monomers as a reference, we can solve equation (18.2) for the molar concentration of n-aggregates as

$$x_n = n\left(x_1 \exp\left[\frac{\mu_1^\circ - \mu_n^\circ}{\mathrm{k_B}T}\right]\right)^n. \tag{18.3}$$

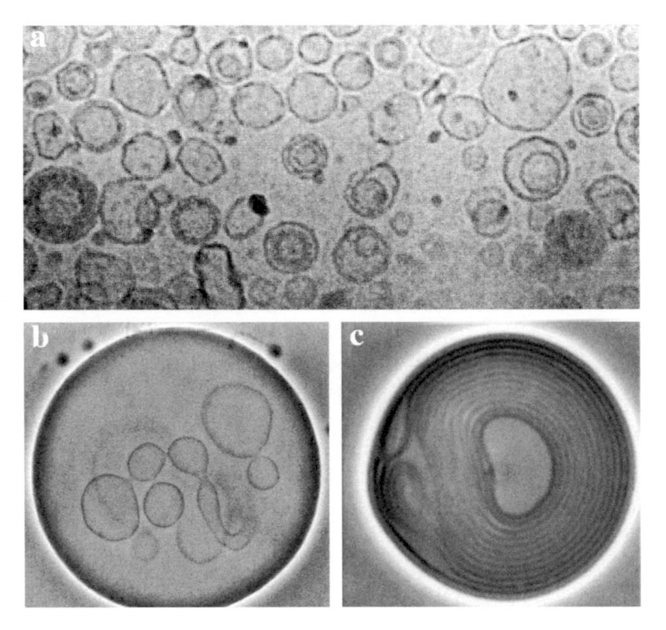

Figure 18.2
Experimental realizations of vesicles (liposomes) as obtained by microscopy techniques. (a) Unilamellar vesicles of approximately 100-nm diameter formed by extrusion through porous filters. Some vesicles have smaller vesicles trapped inside. Adapted from Callisen and Talmon (1998). (b) A large unilamellar vesicle with a diameter around 70 μm. Many smaller unilamellar vesicles are trapped inside. (c) A large multilamellar vesicle with an outer diameter of 40 μm. The bilayers are arranged in the form of a onionlike structure. Courtesy of Dr. Jonas Henriksen.

A necessary condition for forming an aggregate is that the standard chemical potential of the aggregate is less than the monomer standard chemical potential, $\mu_n^\circ < \mu_1^\circ$. The actual form of the standard chemical potential, $\mu_{n,\mathrm{A}}^\circ$, depends on the symmetry of the actual type of aggregate (A) in question and it may have a nonmonotonous dependence on n such that the aggregate is stabilized over monomers for a range of values of the aggregation number n. Hence, an equilibrium situation for a one-phase system may be described by a distribution of aggregate sizes corresponding to a concentration distribution x_n. In the case of micelles, A = micelle, where $\mu_{n,\mathrm{micelle}}^\circ$ has a minimum at an aggregate size m, the standard chemical potential can be approximated by a parabolic form (Israelachvili, 1992), $\mu_{n,\mathrm{micelle}}^\circ = \mu_{m,\mathrm{micelle}}^\circ + \Lambda(n-m)^2$, where Λ is an energy constant. From equation 18.3 it then follows that the concentration distribution is

$$x_{n,\mathrm{micelle}} = n x_1^n \exp\left(\frac{\mu_1^\circ - \mu_{m,\mathrm{micelle}}^\circ + \Lambda(n-m)^2}{\mathrm{k_B}T}\right)^n. \tag{18.4}$$

Equation 18.4 has a near-Gaussian shape with the approximate width $\sqrt{k_B T/2n\Lambda}$. This distribution is also expected to be approximately valid for unilamellar spherical vesicles.

The dependence of the standard chemical potential, $\mu^\circ_{n,A}$, on the symmetry of the aggregate can, in the simplest case, be expressed in terms of the strength of the monomer-monomer cohesion energy, α (in units of the thermal energy), as

$$\mu^\circ_{n,A} = \mu^\circ_\infty + \alpha k_B T n^{-1/d_A}, \tag{18.5}$$

where d_A is the effective dimensionality of the aggregate; $d_{rod} = 1$ for linear aggregates (e.g., long cylindrical micelles), $d_{plate} = 2$ for two-dimensional sheets (e.g., bicelles), and $d_{sphere} = 3$ for three-dimensional objects (e.g., micelles and vesicles); μ°_∞ is the bulk-free energy per molecule in an infinite aggregate. Since $\alpha > 0$, equation 18.5 shows that the standard chemical potential for an aggregate of a given type decreases strongly with the aggregation number, hence shifting the equilibrium from a dispersion of monomers toward aggregation.

In the following section we apply this formalism to lipid aggregates and specifically take into account the possibility of curvature.

18.3 Molecular Packing in Lipid Aggregates

Whether a lipid solution at excess water conditions forms one or the other type of aggregate and with which symmetry depend on a number of factors that have to enter the standard chemical potential $\mu^\circ_{n,A}$. We would expect these factors, which are buried in the coherence energy, α in equation 18.5, to include a bulk term, a surface term, a curvature term, and a packing term. These terms would compete with $\mu^\circ_{1,A}$ and the entropy of a dispersion of monomers. Formation of aggregates is favored by cohesion energy and gain of entropy in the solvent, but disfavored by a decreasing mixing entropy and penalties incurred by ineffective molecular packing in the aggregate, surface tension, and bending energy of possible curved structures. The characterization of these latter contributions is complicated by the fact that the corresponding thermodynamic forces act in different parts of the aggregates. An effective repulsion (steric, electrostatic) acts between the head groups at the surface of the aggregate, a tensile force (interfacial tension) acts at the interface between the hydrophilic and the hydrophobic parts of the aggregate, and an effective steric (entropic) repulsion acts between the hydrocarbon chains in competition with the van der Waals attractive forces. In the simplest formulation, $\mu^\circ_{n,A}$ may be expressed in terms of the interfacial energy per lipid molecule, γ, and the effective surface area, A, per molecule as

$$\mu^\circ_{n,A} \simeq \gamma A + \frac{C}{A}, \tag{18.6}$$

where the materials constant C can be related to the equilibrium molecular area, A_0, as $A_0 = \sqrt{C/\gamma}$. Substituting back into equation 18.6 this leads to

$$\mu^{\circ}_{n,\mathrm{A}} \simeq 2\gamma A_0 + \gamma \frac{(A - A_0)^2}{A}. \tag{18.7}$$

We shall return to equation 18.7 in section 8.4 in the context of lipid bilayer aggregates.

The stable shape of a lipid aggregate may be gauged from the effective shape of the lipid molecules. Aggregates of planar and lamellar symmetry are most easily formed by lipid molecules of effectively cylindrical shape, which can pack nicely into a condensed bilayer. In contrast, lipid molecules with an effective incompatibility between the average cross-section of the polar head and the effective cross-section of the hydrophobic tail tend to stabilize curved structures, like micelles.

This phenomenology is conveniently described by Israelachvili's molecular packing parameter (Israelachvili, 1992)

$$P = \frac{V}{Al}, \tag{18.8}$$

where V is the molecular volume, A is the cross-sectional area of the polar head group, and l is the length of the molecule, as illustrated in figure 18.3.

Lipid molecules with $P \simeq 1$ are good bilayer formers, whereas molecules with $P < 1$ and $P > 1$ tend to engage in curved structures as illustrated in the figure.

Variations in the effective shape of lipid molecules may be caused by a large number of procedures, as illustrated in figure 18.4.

In this context it is interesting to note that the lipid contents of many biological membranes, when depleted of membrane proteins and dissolved in water, do not form bilayer membranes but curved structures (Epand, 1996), such as the hexagonal and cubic phases shown in figure 18.3. Obviously, noncylindrical molecules incorporated into lipid bilayers and membranes introduce a curvature stress field, which we shall discuss more elaborately in section 18.4.

It should be remarked that the very simple considerations presented here, particularly the notion of a molecular packing parameter, are at best useful guides for developing intuitions about lipid aggregate shapes and morphologies. A much more extensive treatment would be required to develop a quantitatively reliable theory.

18.4 Curvature Stress in Lipid Bilayers

In the case of a lipid bilayer, the elastic energy of area compression/expansion is given by $\frac{1}{2}\kappa_A(A - A_0)^2/A$, where κ_A is the elastic area compressibility modulus. It

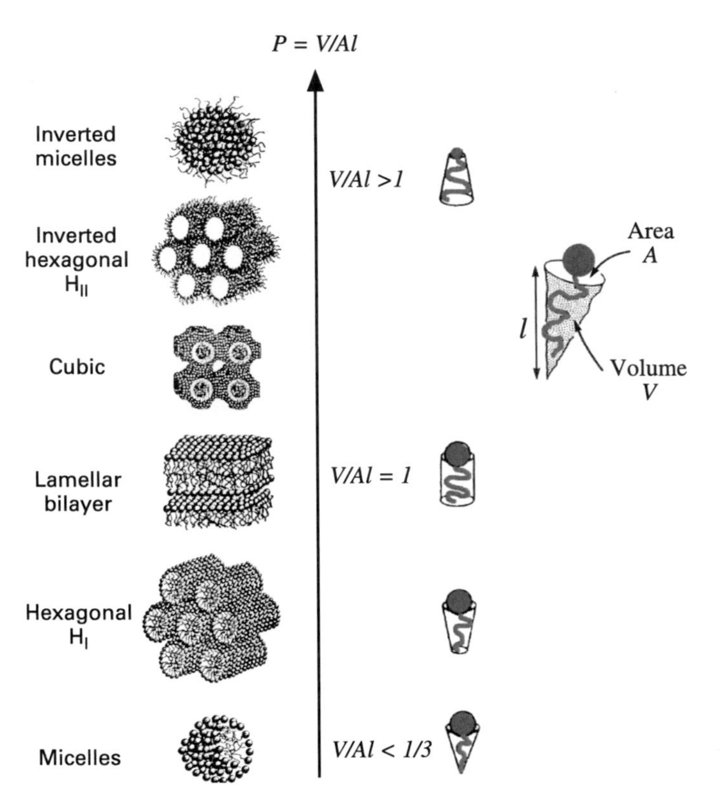

Figure 18.3
Schematic illustration of lamellar and nonlamellar lipid aggregates formed in water. The different structures have different curvatures and are arranged in accordance with the value of packing parameter $P = V/Al$. Adapted from Jönsson et al. (1998).

therefore follows from equation 18.7 that γ for a bilayer (with two interfaces) is related to the elastic area compressibility modulus as $\kappa_A = 4\gamma$.

The two monolayers of a bilayer may have an intrinsic tendency to curve, a so-called spontaneous curvature. When forming a bilayer of identical monolayers with spontaneous curvature, the resulting bilayer, if stable, will by itself not possess a spontaneous curvature. If the two monolayers are not identical, an asymmetric bilayer with spontaneous curvature results. Most cell membranes are asymmetric. If a bilayer is made of monolayers with intrinsic curvature, the bilayer suffers from a curvature stress field. This stress field may be enhanced or suppressed by incorporating lipid molecules with an appropriate shape and therefore a propensity for forming curved aggregates. This curvature stress field is illustrated in figure 18.5 in the cases of negative, zero, and positive spontaneous curvature of the involved monolayers.

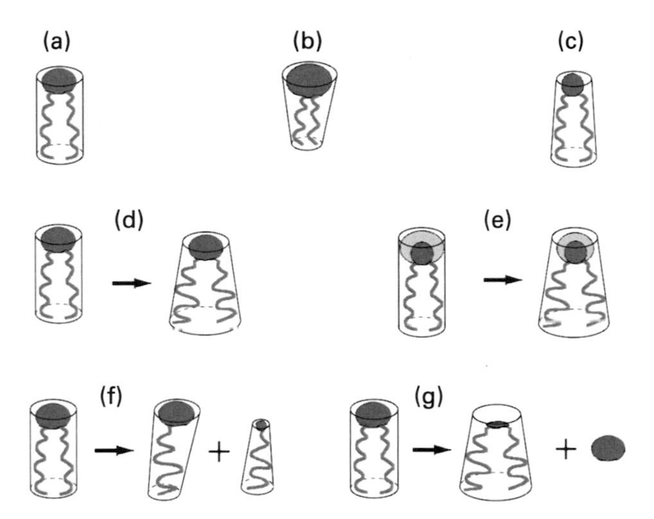

Figure 18.4
Effective shapes of lipid molecules. (a) Cylindrical: similar cross-sectional area of head and tail; (b) cone: big head and skinny tail; (c) inverted cone: small head and bulky tail (e.g., with unsaturated fatty-acid chains); (d) going conical by increasing temperature; (e) going conical by changing the effective size of the head group, for example, by changing the degree of hydration or the effective charge of an ionic head group; (f) going conical by removing fatty-acid chain, for example, by the action of phospholipase A_2, which forms a lysolipid molecule and a free fatty acid; (g) going conical by chopping off the entire polar head group, for example, by the action of phospholipase C.

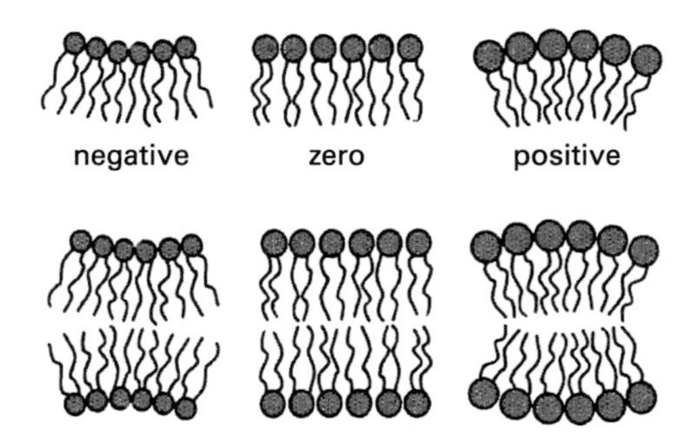

Figure 18.5
Illustration of the destabilization of a lipid bilayer composed of lipids with conical shapes that promote a tendency for the two monolayers to curve. Bilayers made of monolayers with a nonzero curvature have a built in curvature stress. Courtesy of Dr. Olaf Sparre Andersen.

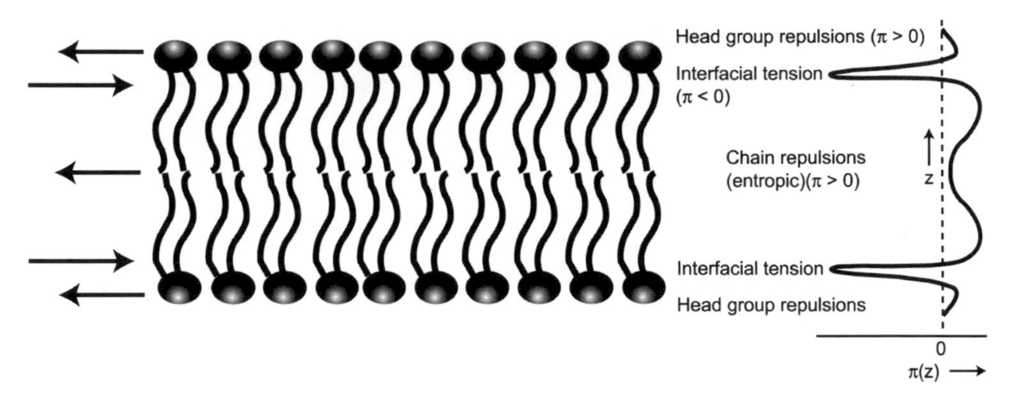

Figure 18.6
Lateral pressure profile of a lipid bilayer. The left-hand side is a schematic illustration of a cross-section through a symmetric lipid bilayer with indication of the forces that act within the layer (arrows indicate direction and relative magnitude of the force). To the right is the resulting pressure or stress profile, $\pi(z)$.

The forces stabilizing and acting within lipid bilayers lead to the peculiar profile of lateral stresses illustrated in figure 18.6. This lateral stress profile (or lateral pressure profile) is a consequence of the stabilizing forces acting in different planes of the bilayer, as indicated in the figure. The corresponding variations in local pressure density are huge, of the order of $4\gamma/d$, where $d \simeq 5$ nm is the bilayer thickness. Since $\gamma \sim 50$ mN/m for an oil-water interface, this may amount to hundreds of atmospheres (Cantor, 1999). These stresses imply that the bilayer is a very "hostile" environment in which huge forces are at work. These forces have been shown to be large enough to induce conformational changes in proteins and peptides that are imbedded in the bilayer (Cantor, 1999).

18.5 Lateral Structure of Lipid Bilayers

To complete the description of the structure of lipid-bilayer aggregates, it should be mentioned that the bilayer not only has a highly stratified trans-bilayer structure over its 5 nm thickness, but it is also structured laterally on scales from nanometers to the size of the object. This lateral structure is controlled by a number of in-plane phase equilibria and phase-separation phenomena (Keller et al., 2005; Mouritsen, 1998). In biological membranes, this lateral structure, in the form of lipid domains, or so-called rafts, is known to play a significant role in cell function (Edidin, 2003).

Figure 18.7 shows an example of this kind of lateral structure for lipid bilayer membranes composed of the lipids and proteins that constitute the pulmonary surfactants in the lung lining (Bernardino de la Serna et al., 2004).

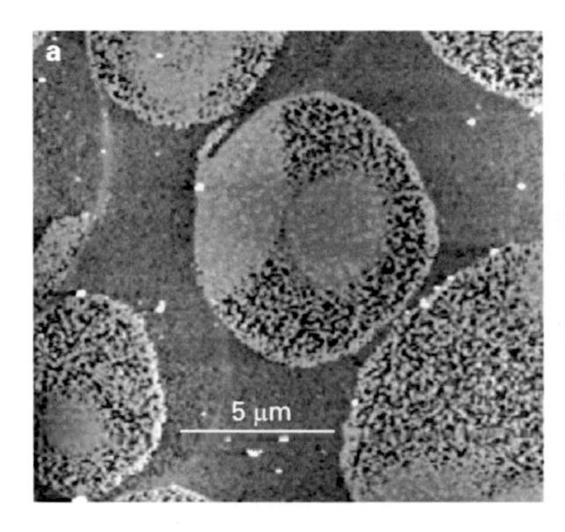

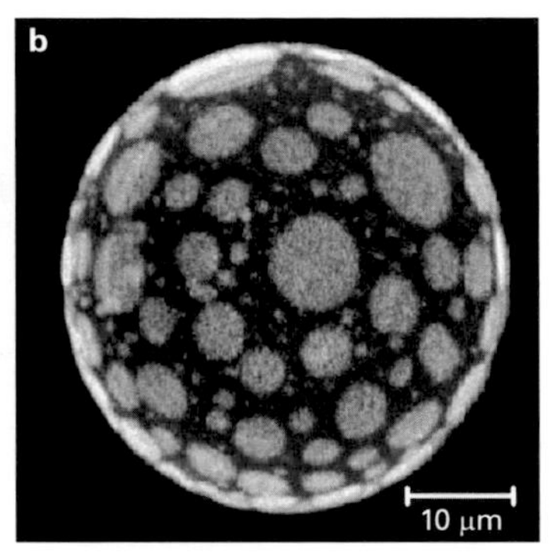

Figure 18.7
Lipid domains in native pulmonary surfactant membranes consisting of lipids and proteins. (a) Atomic force microscopy image of a supported bilayer on mica. The individual spikes in this granular structure are presumably single lung surfactant proteins. (b) Fluorescence microscopy image of a giant unilamellar liposome. The round domains are seen to have a granular structure. On the micron scale, the individual proteins cannot be discerned. Adapted from Bernardino de la Serna et al. (2004). Courtesy of Drs. Adam C. Simonsen and Luis Bagatolli.

18.6 Computer Modeling of Lipid Self-Assembly

Lipid aggregates in water have a very complicated phase space and exhibit, as we have described phenomenologically earlier, a rich spectrum of phenomena occurring on timescales from picoseconds to minutes and on length scales from 0.1 nm to micrometers. It is computationally very demanding and challenging to design models and simulation algorithms that can span all these scales. This is particularly troublesome since many important biological phenomena take place over a range of time and length scales, such as vesicle formation, budding, fusion, and raft formation. Questions at mesoscopic scales cannot be addressed by atomic-scale molecular dynamics (MD) simulation because of the computational demands. The number of degrees of freedom are simply too large. As an example, current large-scale supercomputer simulations, at best, cover hundreds of nanoseconds and deal with tens of nanometer membrane patches consisting of the order of a thousand lipids (Lindahl and Edholm, 2000; Marrink and Mark, 2003) or preassembled lipid bilayers of a few hundred lipid molecules with single peptides or trans-membrane proteins (Jensen and Mouritsen, 2004; Jensen, Mouritsen, and Peters, 2004). One of the major problems derives from the fact that self-assembled lipid systems exist only in water, thus the simulations have to deal with a large number of water molecules.

The usual strategy to circumvent some of these problems is to turn to various coarse-grained models and simulation techniques that span large time and length scales at the expense of molecular detail and a well-defined time parameter (Attig et al., 2004; Karttunen, Vattulainen, and Lukkarinen, 2004). For example, simplified discrete lattice models or random surface triangulation, together with stochastic Monte Carlo methods, cellular automata, and Lattice-Boltzmann techniques, can be used to study lipid self-assembly and the formation of equilibrium lipid aggregates (Nilsson and Rasmussen, 2003). The dynamics of such simulations are often rather artificial, however, and since they are not always fully momentum conserving, they neglect the hydrodynamic modes.

In recent years, a new type of hybrid coarse-grained model treated with stochastic momentum-conserving dynamics has been introduced—so-called dissipative particle dynamics (DPD) (Español and Warren, 1995; Groot and Warren, 1997). This approach has pushed the limits of dynamic simulations for soft-matter systems toward larger systems and longer times (for a recent list of references to the DPD literature, see Jakobsen, Mouritsen, and Besold, 2005). Among other things, the method has been used to study self-assembly of lipids into micelles and bilayers, membrane fission, fusion, budding and rupture, as well as specific effects such as lipid-protein interactions and lateral stress distributions in bilayers.

18.6.1 Dissipative Particle Dynamics

Dissipative particle dynamics can be thought of as molecular dynamics simulation with a stochastic local momentum-conserving thermostat. Dissipative particle dynamics is typically combined with some interactions that are very different from those used in molecular dynamics. In molecular dynamics, the interactions are "hard" in the sense that the potential energy diverges rapidly if two atoms/molecules overlap (Lennard-Jones interaction), and the forces (and thereby the accelerations) are calculated as minus the derivative of the potential energy. Since the bond frequency in a covalent bond is typically of the order of 10 femtoseconds, hard potentials require the use of very small time steps to avoid rapid error build-up in the integration scheme. Furthermore, the electrostatic interactions are computationally expensive to evaluate for a system with periodic boundary conditions. When the potentials are hard, it takes a long time for molecules to "pass each other," since they tend to "rattle in a box" until they have explored enough of the local phase space to actually diffuse. This explains the large timescale difference between, for example, hydrogen bond dynamics in water and the actual timescale characterizing the diffusion of a water molecule in bulk water.

In DPD, the number of degrees of freedom is reduced by lumping large parts of the molecules into larger entities or "beads." The idea is to simplify the underlying system as much as possible, while retaining the key properties that are expected to govern the processes of interest. In the case of a symmetric lipid molecule, a coarse-grained model is denoted $H_n(T_m)_k$, where n is the number of hydrophilic head group beads (H), k is the number of hydrophobic chains in the tail, and m is the number of

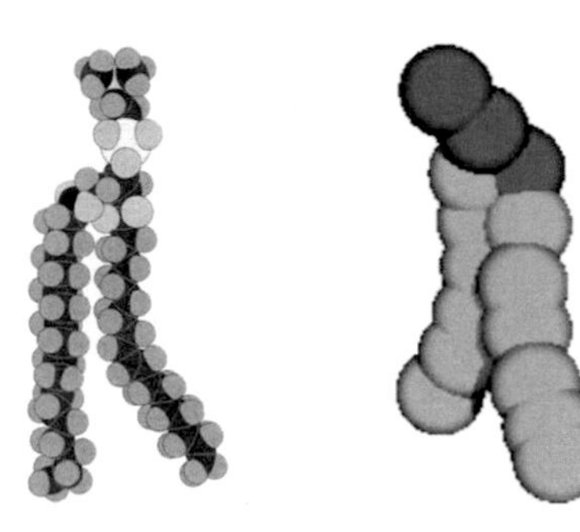

Figure 18.8
Schematic structure of a coarse-grained phospholipid molecule. To the left is an all-atom space-filling model of a phospholipid molecule. To the right is a coarse-grained $H_3(T_6)_2$ lipid molecule. The dark beads represent the hydrophilic head group (H) and the light beads represent the tail beads (T).

hydrophilic beads (T) in each tail chain. This is illustrated in figure 18.8 in the case of a lipid molecule with a hydrophobic tail consisting of two hydrocarbon chains. Effective potentials are then constructed as coarse-grained averages over time *and* space. These effective interactions are soft (Klapp, Diester, and Schoen, 2004), that is, the potential energy remains finite even when the particles overlap.

In DPD the total force on particle i at position $\mathbf{r}_i$ and with momentum $\mathbf{p}_i$ is given by

$$\dot{\mathbf{p}}_i = \mathbf{F}_i^C + \sum_{j \neq i} \mathbf{F}_{ij}^D + \sum_{j \neq i} \mathbf{F}_{ij}^R, \tag{18.9}$$

where $\mathbf{F}_i^C$ is a conservative force exerted on particle i from particle j. The two forces, $\mathbf{F}_{ij}^D$ and $\mathbf{F}_{ij}^R$, which are not present in MD, make up the thermostat. To be momentum conserving and provide the correct temperature, the form of the forces has to fulfill certain constraints.

The dissipative force $\mathbf{F}_{ij}^D$ and the random force $\mathbf{F}_{ij}^R$ are chosen to have the forms

$$\mathbf{F}_{ij}^D = -\Gamma \omega_D(r_{ij})(\mathbf{e}_{ij} \cdot \mathbf{v}_{ij})\mathbf{e}_{ij} \tag{18.10}$$

$$\mathbf{F}_{ij}^R = \sigma \omega_R(r_{ij})\xi_{ij}\mathbf{e}_{ij}, \tag{18.11}$$

where $\mathbf{r}_{ij} = \mathbf{r}_i - \mathbf{r}_j$, $r_{ij} = |\mathbf{r}_{ij}|$, $\mathbf{e}_{ij} = \mathbf{r}_{ij}/r_{ij}$, $\mathbf{v}_{ij} = \mathbf{v}_i - \mathbf{v}_j$, $\mathbf{v}_i$ is the velocity of particle i, and ω_D and ω_R are arbitrary weight functions. The variable ξ_{ij} represents Gaussian white noise with $\xi_{ij} = \xi_{ji}$ and the following stochastic properties

$$\langle \xi_{ij}(t) \rangle = 0$$

$$\langle \xi_{ij}(t)\xi_{i'j'}(t') \rangle = (\delta_{ii'}\delta_{jj'} + \delta_{ij'}\delta_{ji'})\delta(t - t'). \tag{18.12}$$

The parameter Γ controls the strength of the dissipation and the parameter σ the strength of the noise. The thermostat consisting of the random and dissipative forces in equation 18.9 conserves momentum pairwise, so DPD is a momentum-conserving thermostat. An intuitive way of thinking about the DPD forces is that beads are considered to have unobserved internal degrees of freedom that give rise to the dissipative force, and the beads are coupled to the local temperature of the fluid environment, which in turn is the source of the random forces.

It has been proved that the corresponding Fokker-Planck equation of equation 18.9 has the canonical equilibrium distribution as a solution given the following constraints (Español and Warren, 1995):

$$[\omega_R(r)]^2 = \omega_D(r) \tag{18.13}$$

$$\sigma^2 = 2\Gamma \mathrm{k_B} T. \tag{18.14}$$

The preceding is thus a fluctuation-dissipation theorem for the DPD system. Therefore, DPD naturally samples the canonical ensemble (NVT). In other methods (Jakobsen, 2005), DPD is coupled to a barostat so it is possible to sample various constant-pressure ensembles. For example, constant surface tension and constant normal pressure are relevant for bilayer simulations.

In DPD simulations, $\omega_D(r)$ is often chosen to be

$$\omega_D(r) = \begin{cases} 1 - r/r_0 & r < r_0 \\ 0 & r \geq r_0 \end{cases}, \tag{18.15}$$

where r_0 is the cutoff distance.

The nonbonded interactions that provide the beads with their particular identity are chosen very differently from what is the case in conventional MD. A conservative soft-core repulsion is usually modeled simply by

$$\mathbf{F}_{ij}^S = \mathscr{A}_{ij}\omega_D(r_{ij})\mathbf{e}_{ij}, \tag{18.16}$$

where ω_D is chosen as in equation 18.15. The repulsion parameter $\mathscr{A}_{ij}$ gives the different particle/bead species their identity. To create model polymers or lipids, beads are tied together by harmonic springs, and to add rigidity to the structure, angle-bending potentials are applied, all of which are conservative interactions. The fact that all nonbonded interactions in DPD are repulsive are in line with the fact that entropic interactions (which are repulsive) tend to dominate the properties of soft-matter systems.

All beads have the same mass, m_0, and cutoff, r_0 (effective size). The energy scale is set by $\mathrm{k_B}T$. We will use units where all these quantities are unity. From m_0, r_0, and

k_BT, a timescale $t_0 = \sqrt{m_0 r_0^2 / k_B T}$ can be extracted. This timescale cannot be applied directly in interpreting the time-evolution of the results.

Water molecules are modeled by single beads, which usually are taken to be the same as the lipid head-group beads (H). The link to experimental data is the isothermal compressibility of water at room temperature. The interaction between the tail beads and water is determined by comparing solubility of dissimilar DPD fluids with Flory-Huggins theory of immiscible polymers (Groot and Warren, 1997). The parameters found with this crude method give a more or less faithful representation of the mechanical properties of a real bilayer.

It is difficult to extract physical timescales of dynamical phenomena from DPD, but the method might give an idea of relative timescales (or size of barriers). Since the parameters are chosen to fit only a few physical quantities of real systems, one cannot expect model calculations based on DPD to get all system properties correct. For example, parameters determined for modeling bilayers might not be appropriate for modeling micelles, and so on. This is somewhat similar to the problems that occur in molecular dynamics. As an example, biologically relevant parameters that have been "optimized" for aggregates solvated in water would probably not give good results if they were used to model aggregate formation in nonpolar solvents.

DPD simulations of self-assembly processes are now feasible on fast single-processor standard computers. As an example, a typical simulation on a 100,000-bead system takes on the order of a few days' CPU time.

In the following sections, we illustrate the use of DPD and coarse-grained models to simulate lipid self-assembly, and we demonstrate that this approach is capable of producing different kinds of supramolecular structures, such as micelles, tubular filaments, bilayers, and closed vesicles (liposomes).

18.6.2 Formation of Micelles and Tubular Filaments

Figure 18.9 (color plate 13) shows a time series of snapshots derived from extensive DPD simulations for a system composed of 12,500 lipids of HT_6 (7 beads). The total number of beads in the simulation is 2,654,208. The coarse-grained HT_6 lipids are of the type we would expect could form micelles because their packing fraction, P in equation 18.8 is small (cf. figure 18.3) since the head group is relatively large compared to the size of the single, rather short tail.

The lipids in this model are indeed observed to aggregate into a large number of micellelike structures at early times. As time lapses, the small aggregates grow by incorporating more monomers, and some aggregates merge to form larger micelles that are close to having an average spherical shape. At the end of the simulation, the system is believed to be close to equilibrium, although it cannot be ruled out that even longer simulations would lead to a size distribution of a somewhat different

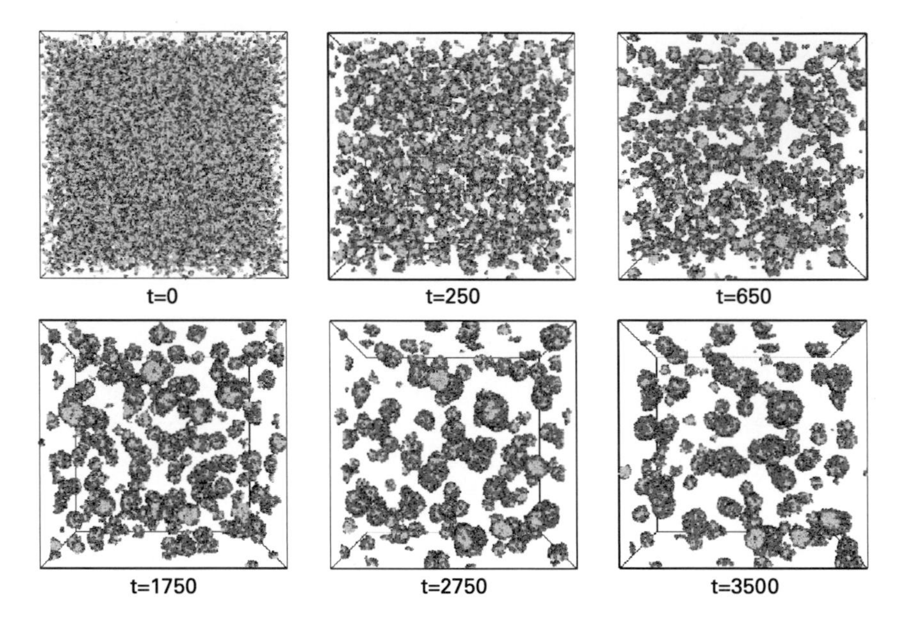

Figure 18.9 (color plate 13)
Snapshots of micelle formation from a random mixture of HT_6 lipids simulated by dissipative particle dynamics. The dimension of the cubic box is $L = 96r_0$, and the number density is $\rho r_0^3 = 3$. The model parameters are adapted from Shillcock and Lipowsky (2002). Head-group beads are red and tail beads are green. For the sake of clarity, the water beads are not shown. The DPD timescale, t, is in units of $t_0 = \sqrt{m_0 r_0^2 / k_B T}$. The individual beads are drawn with a radius corresponding to $0.5r_0$.

shape. The actual average micellar size and the size distribution depends on the effective shape of the lipid molecules, that is, the packing parameter, P.

Figure 18.10 (color plate 14) provides a snapshot of the lipid configuration obtained from a DPD simulation for a system composed of 1,200 lipids of the type $H_3(T_5)_2$ (13 beads). The total number of beads in the simulation is 98,304. The coarse-grained $H_3(T_5)_2$ molecules are of the type we would expect to form long, rod-like micelles because their packing fraction, P in equation 18.8, falls between the values characterizing bilayers and spherical micelles (cf. figure 18.3). The structure is seen to be a tubular filament consisting of long, flexible rods with an inner core of hydrophobic lipid tails.

18.6.3 Formation of Bilayers

Figure 18.11 (color plate 15) shows a time series of snapshots derived from DPD simulations for a system composed of 1,685 lipids of the type $H_3(T_6)_2$ (15 beads) in a simulation consisting of 98,304 beads. The coarse-grained $H_3(T_6)_2$ lipid molecule is the one depicted in figure 18.8. This molecule has a packing parameter close to 1 ($P \simeq 1$) and is therefore expected to be a good bilayer former.

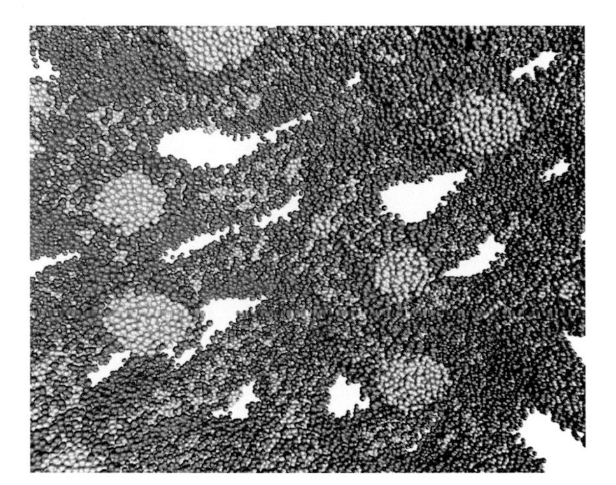

Figure 18.10 (color plate 14)
Snapshot of a configuration of lipid aggregates in the form of a filament of rodlike micelles formed by $H_3(T_5)_2$ lipids and simulated by dissipative particle dynamics. The dimension of the cubic box is $L = 32r_0$, and the number density is $\rho r_0^3 = 3$. The model parameters are adapted from Groot and Rabone (2001). The red beads represent hydrophilic head groups (H) and the green beads represent the tail beads (T). For the sake of clarity, the water beads are not shown.

The snapshots in figure 18.11 (color plate 15) show that the lipids very quickly organize spontaneously into filamentlike structures that rapidly merge into one large, connected structure. This structure is very stable. The stability is partly controlled by the periodic boundary conditions.[1] The line tension is then finally reduced by quickly closing and forming the bilayer. A short while after the aggregate has closed into a bilayer, a lipid molecule was observed (not shown) to be expelled from the upper leaflet. This lipid molecule diffuses in the aqueous phase for a while before it is reabsorbed into the same leaflet. This suggests that the self-assembly process has resulted in slightly too many lipids in the upper leaflet. One might eventually see this asymmetry evened out by diffusion from the upper to the lower leaflet, through either the water phase or very slow lipid flip-flop processes between the two monolayer leaflets.

18.6.4 Formation of Closed Vesicles

Figure 18.12 (color plate 16) shows a time series of snapshots derived from DPD simulations of 4,092 HT_3 (4 beads) lipids in a simulation consisting of 648,000 beads. The HT_3 molecules have both small heads and small tails and will be expected to be candidates for forming bilayers.

The lipid molecules are preassembled in a flat bilayer sheet with hydrophobic edges exposed to the water phase. Initially, the lipid sheet wobbles back and forth. This state of the system is unstable because of the line tension at the edges. After

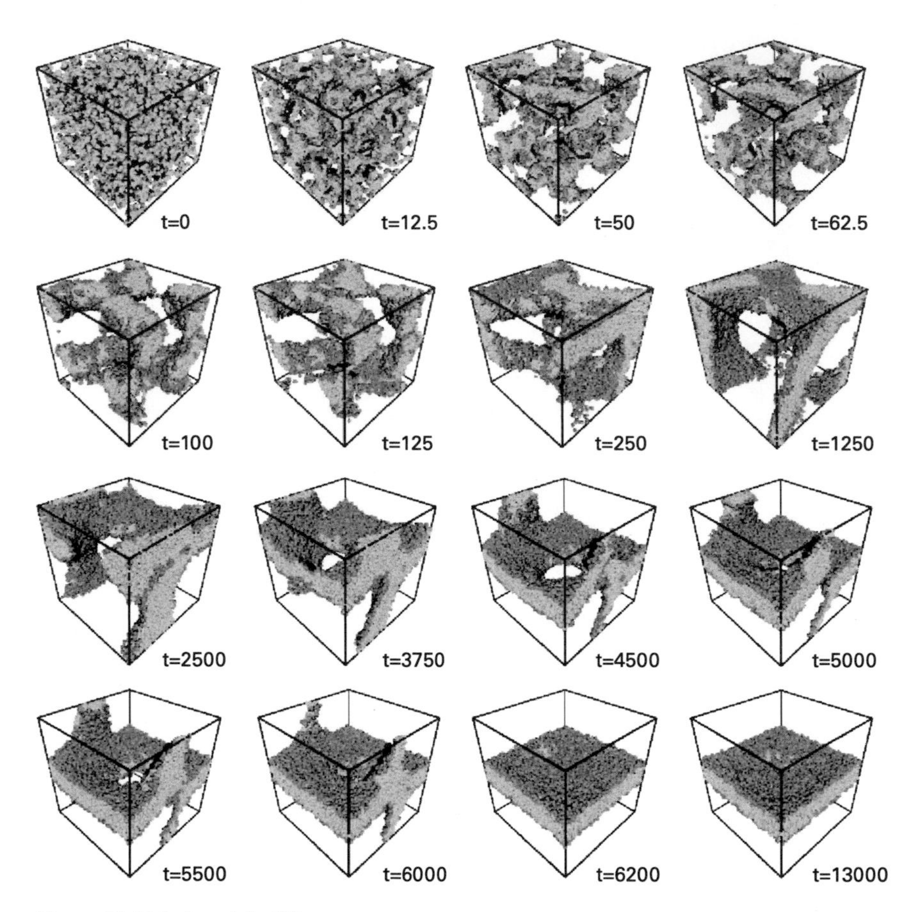

Figure 18.11 (color plate 15)
Snapshots of bilayer formation from a random mixture of $H_3(T_6)_2$ lipids simulated by dissipative particle dynamics. The dimension of the cubic box is $L = 32r_0$ and the number density is $\rho r_0^3 = 3$. The model parameters are from Shillcock and Lipowsky (2002). Head-group beads are red and tail beads are green. For the sake of clarity, the water beads are not shown. The DPD timescale, t, is in units of $t_0 = \sqrt{m_0 r_0^2 / k_B T}$.

some time, the sheet minimizes its open-edge line tension by "choosing" a direction in which to curl up. Subsequently, the line tension is slowly minimized by reducing the length of the sheet's water-exposed edge at the expense of an increasing bending energy. The final closing of the hole is fast.

18.7 Morphological Transitions of Lipid Vesicles

Various computer simulation methods have been applied to a range of models of lipid bilayers to investigate morphological transitions, such as vesiculation (cytosis),

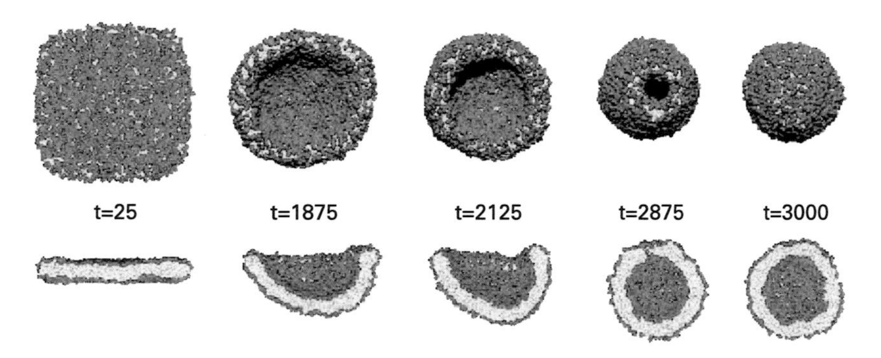

Figure 18.12 (color plate 16)
Snapshots of vesicle formation from a preassembled bilayer sheet of HT_3 lipid molecules simulated by dissipative particle dynamics. The dimension of the cubic box is $L = 60r_0$ and the number density is $\rho r_0^3 = 3$. Parameters are adapted from Yamamoto, Maruyama, and Hyodo (2002). These parameters produce a totally interdigitated bilayer. The bottom row of snapshots show a cut through the bilayer. Head-group beads are red and tail beads are green. Water beads are not shown. The DPD timescale, t, is in units of $t_0 = \sqrt{m_0 r_0^2 / k_B T}$.

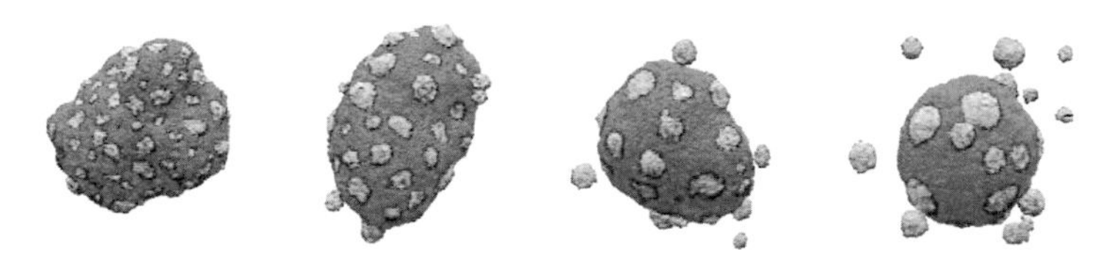

Figure 18.13
Dissipative particle dynamics simulation of budding and vesiculation process based on a unilamellar lipid vesicle consisting of two lipid species indicated by blue and yellow. The water beads are not shown. The total number of beads in the simulation was 1,536,000. Adapted from Laradji and Kumar (2004).

fusion, and fission processes. All these processes are important in the context of biological membrane function. We make no attempt to review the vast literature on this topic, but only mention the results of two recent studies based on coarse-grained lipid models.

The first study used DPD simulation methods, as described previously, to investigate the dynamics of phase separation and vesiculation of a fluid two-component vesicle, taking full account of the water solvent (Laradji and Kumar, 2004). Hence, this model is faithful to the hydrodynamics of the problem. Furthermore, the constraint of a fixed surface-area-to-volume ratio was imposed by rendering the bilayer impermeable to the water beads. The coarse-grained lipid model was of the very simple type HT_3. Two types of lipids were distinguished by a different head-group character. The simulations showed different dynamical regimes corresponding to coalescence of flat patches, capping, budding and vesiculation, as well as coalescence of caps as illustrated in figure 18.13.

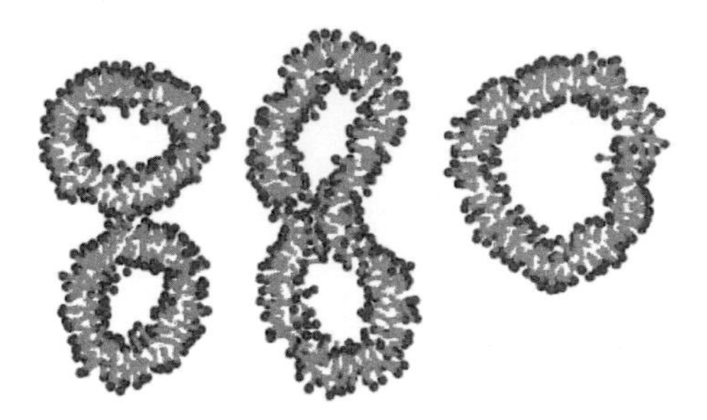

Figure 18.14 (color plate 17)
Brownian dynamics simulation of the fusion of two vesicles revealing the formation of a fusion intermediate in the form of a membrane stalk that connects only the outer monolayers of the two fusing vesicles. Time proceeds from left to right, and only cross-sections of the vesicles are shown. The total number of molecules in the simulations was 1,000. Adapted from Noguchi and Takasu (2001).

In another study, Noguchi and Takasu (2001) used a coarse-grained lipid model consisting of three beads, one hydrophilic head and two hydrophobic tail beads, confined as a rigid rod on a line. The solvent was not taken into account explicitly. The simulations, in this case, were performed by means of Brownian dynamics. Figure 18.14 (color plate 17) shows the time evolution of a setup of two self-assembled, almost spherical vesicles that undergo a fusion process. Simulations of this type allow the intermediate states of the fusion process to be examined.

18.8 Conclusion

We have, in this brief topical review, given an account of the fundamental principles of lipid self-assembly processes in water. We have described the thermodynamics and the elementary driving forces for aggregation and characterized the different types of aggregates and their symmetry, such as micelles, bilayers, vesicles, and various structures with curved interfaces. In the case of bilayers, we have described both the transbilayer and the lateral bilayer structure. We have pointed out that dissipative particle dynamics methods, used together with appropriate coarse-grained lipid models, can effectively simulate lipid self-assembly processes, the formation and dynamics of different types of lipid aggregates, as well as the morphological properties of vesicles. It is to be expected that dissipative particle dynamics simulations and other coarse-grained particle simulation approaches will gain further relevance in the future since they offer a way of exploring the self-organization processes and principles in future proposed models of protocells.

The current state of the art with regard to both simulational and experimental self-assembly and its theoretical underpinnings holds promise for exploiting this insight to model, simulate, and experimentally realize the kind of nano-scale structures and encapsulation technologies that may form the essential templates for biological function and hence help bridge the gap between nonliving and living matter.

Acknowledgments

This work was funded in part by PACE (Programmable Artificial Cell Evolution), a European Integrated Project in the EU FP6-IST-FET Complex Systems Initiative. The simulations were carried out in part at the Danish Center for Scientific Computing (DCSC). MEMPHYS–Center for Biomembrane Physics is supported by the Danish National Research Foundation.

Note

1. It should be noted that the simulation results may be affected by interaction of the lipid aggregate with the periodic images caused by the applied periodic boundary conditions. This makes it difficult to unambiguously relate the aggregate symmetry to the packing parameter. The periodic boundary conditions are likely to influence physical properties calculated for the bilayer, and systematic finite-size analyses would have to be performed to accurately calculate, for example, thermomechanical properties.

References

Attig, N., Binder, K., Grudmüller, H., & Kremer, K. (Eds.) (2004). *Computational soft matter: From synthetic polymers to proteins.* Julich, Germany: John von Neumann Institute of Computing (NIC) Research Centre, 23.

Bernardino de la Serna, J., Perez-Gil, J., Simonsen, A. C., & Bagatolli, L. A. (2004). Cholesterol rules: Direct observation of the coexistence of two fluid phases in native pulmonary surfactant membranes at physiological temperatures. *Journal of Biological Chemistry, 279*, 40715–40722.

Callisen, T. H., & Talmon, Y. (1998). Direct imaging by cryo-TEM shows membrane break-up by phospholipase A_2 enzymatic activity. *Biochemistry, 37*, 10987–10993.

Cantor, R. S. (1999). The influence of membrane lateral pressures on simple geometric models of protein conformational equilibria. *Chemistry and Physics of Lipids, 101*, 45–56.

Cates, M. E., & Safran, S. A. (1997). Theory of self-assembly. *Current Opinion in Colloid and Interface Science, 2*, 359–387.

Chen, S.-H., & Rajagopalan, R. (1990). *Micellar solutions and microemulsions.* New York: Springer-Verlag.

Deamer, D. W. (1986). Role of amphiphilic compounds in the evolution of membrane structure on the early Earth. *Origins of Life, 17*, 3–25.

Edidin, M. (2003). The state of lipid rafts: From model membranes to cells. *Annual Review of Biophysics and Biomolecular Structure, 32*, 257–283.

Epand, R. (1996). Functional roles of nonlamellar farming lipids. *Chemistry and Physics of Lipids, 81*, 101–104.

Español, P., & Warren, P. (1995). Statistical mechanics of dissipative particle dynamics. *Europhysics Letters, 30*, 191–194.

Evans, D. F., & Wennerström, H. (1999). *The colloidal domain: Where physics, chemistry, biology, and technology meet*, 2nd Ed. New York: VCH Publishers, Inc.

Gelbart, W. M., Ben-Shaul, A., & Roux, D. (Eds.) (1994). *Micelles, membranes, microemulsions, and monolayers.* Berlin: Springer-Verlag.

Groot, R. D., & Rabone, K. L. (2001). Mesoscopic simulation of cell membrane damage, morphology change and rupture by non-ionic surfactants. *Biophysical Journal, 81*, 725–736.

Groot, R. D., & Warren, P. B. (1997). Dissipative particle dynamics: Bridging the gap between atomistic and mesoscopic simulation. *Journal of Chemical Physics, 107*, 4423–4435.

Israelachvili, I. (1992). *Intermolecular and surface forces*, 2nd Ed. London: Academic Press.

Jakobsen, A. F. (2005). Constant-pressure and constant-surface tension simulations in dissipative particle dynamics. *Journal of Chemical Physics*, 124901–124908.

Jakobsen, A. F., Mouritsen, O. G., & Besold, G. (2005). Artifacts in dynamical simulations of coarse-grained model lipid bilayers. *Journal of Chemical Physics, 122*, 204901–204911.

Jensen, M. Ø., & Mouritsen, O. G. (2004). Lipids do influence protein function: The hydrophobic matching hypothesis revisited. *Biochimica et Biophysica Acta, 1666*, 205–226.

Jensen, M. Ø., Mouritsen, O. G., & Peters, G. H. (2004). Molecular dynamics study of the interactions between an acylated C14-peptide and DPPC bilayers. *Biophysical Journal, 86*, 3556–3575.

Jönsson, B., Lindman, B., Holmberg, K., & Kronberg, B. (1998). *Surfactants and polymers in aqueous solution*. New York: John Wiley & Sons.

Karttunen, M., Vattulainen, I., & Lukkarinen, A. (Eds.) (2004). *Novel methods in soft matter simulations*. Lecture notes in physics (p. 640). New York: Springer-Verlag.

Keller, D., Larsen, N. B., Møller, I. M., & Mouritsen, O. G. (2005). Decoupled phase transitions and grain-boundary melting in supported phospholipid bilayers. *Physical Review Letters, 94*, 025701–025704.

Klapp, H. L., Diestler, D. J., & Schoen, M. (2004). Why are effective potentials 'soft?' *Journal of Physics: Condensed Matter, 16*, 7331–7352.

Laradji, M., & Kumar, P. B. S. (2004). Dynamics of domain growth in self-assembled fluid vesicles. *Physical Review Letters, 93*, 198105.

Lindahl, E., & Edholm, O. (2000). Mesoscopic undulations and thickness fluctuations in lipid bilayers from molecular dynamics simulations. *Biophysical Journal, 79*, 426–433.

Marrink, S. J., & Mark, A. E. (2003). Molecular dynamics simulation of the formation, structure, and dynamics of small phospholipid vesicles. *Journal of the American Chemical Society, 125*, 15233–15242.

Mouritsen, O. G. (1998). Self-assembly and organization of lipid-protein membranes. *Current Opinion in Colloid and Interface Science, 3*, 78–87.

Mouritsen, O. G. (2005). *Life—As a matter of fat. The emerging science of lipidomics*. Heidelberg: Springer-Verlag.

Nilsson, M., & Rasmussen, S. (2003). Cellular automata for simulating molecular self-assembly. *Discrete Mathematics and Theoretical Computer Science, AB* (DMCS), 31–42.

Noguchi, H., & Takasu, M. (2001). Fusion pathways of vesicles: A Brownian dynamics simulation. *Journal of Chemical Physics, 115*, 9547–9551.

Petrov, A. G. (1999). *The lyotropic state of matter: Molecular physics and living matter physics*. Amsterdam: Gordon and Breach.

Rilfors, L., & Lindblom, G. (2002). Regulation of lipid composition in biological membranes—Biophysical studies of lipids and lipid synthesizing enzymes. *Colloidal Surfaces B: Biointerfaces, 26*, 112–124.

Shillcock, J. C., & Lipowsky, R. (2002). Equilibrium structure and lateral stress distribution of amphiphilic bilayers from dissipative particle dynamics simulation. *Journal of Chemical Physics, 117*, 5048–5061.

Yamamoto, S., Maruyama, Y., & Hyodo, S. (2002). Dissipative particle dynamics study of spontaneous vesicle formation of amphiphilic molecules. *Journal of Chemical Physics, 116*, 5842–5849.

19 Numerical Methods for Protocell Simulations

Yi Jiang, Bryan Travis, Chad Knutson, Jinsuo Zhang, and Pawel Weronski

19.1 Introduction

One could take as the defining characteristics of a protocell or protolife the ability to ingest resources and convert them into building blocks, the ability to grow and self-reproduce, and the ability to evolve. Other chapters in this book describe a rapidly expanding family of experiments whose collective goal is to make possible the advent of artificially created, minimal living structures. A scheme proposed by Rasmussen and coworkers (Rasmussen et al., 2003), which we will refer to as the Los Alamos (LANL) protocell (details discussed in chapter 6), is our prime example. In simple terms, these experiments involve self-assembly of a container (lipid and/or surfactants), loading containers with other molecules (sensitizers, precursors), photo/electrochemical interactions to produce building blocks from precursors, RNA (DNA or PNA) replication and polymerization, and a whole system working in synchrony (Rasmussen et al., 2003). Figure 19.1 illustrates the length and timescales covered by the processes involved in the protocell assembly process.

There are vast, uncharted areas for each group of experiments. These are complex systems wherein the large number of degrees of freedom makes experimental resolution difficult. It would be prohibitively expensive if not impossible for experiments to explore the complete parameter space. An alternative (indeed adjunct) to experimentation is modeling. Modeling can provide guidance by predicting experimental results where the underlying mechanisms are known, testing various hypotheses for the underlying mechanisms, exploring parameter territories, helping to make design choices and define what is worth measuring, and estimating sensitivity of processes to various parameters involved.

This chapter reviews the state of the art of molecular modeling approaches at different scales, from *ab initio* and atomic scale to continuous description. We will strive to emphasize their relevance to protocell experiments and provide examples where possible. The review is divided according to the scales of the models, increasing in length scale from micro to macro in sequence as *ab initio*, semiempirical

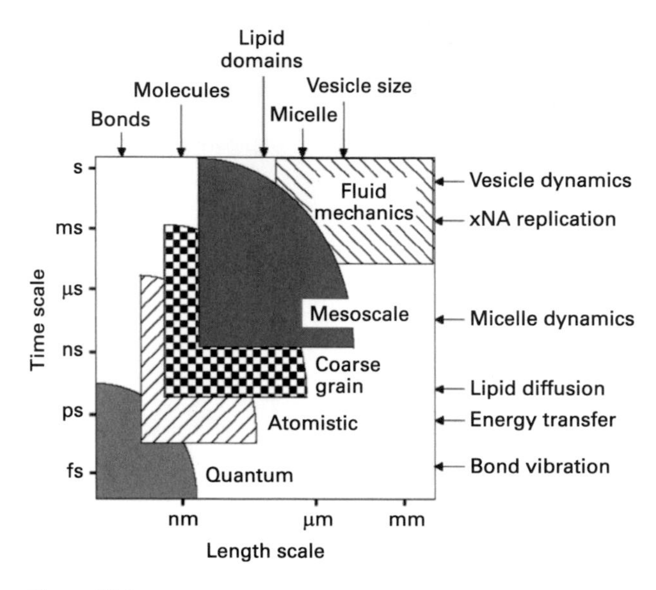

Figure 19.1
Schematic of the length and timescales covered by the processes involved in protocell assembly.

methods, molecular dynamics, coarse-grained molecular dynamics (including Brownian dynamics, dissipative particle dynamics, and lattice gas molecular dynamics), lattice Boltzmann method, chemical kinetics, and the Ginzburg-Landau theoretic method. In each of these sections, we first introduce the method, then demonstrate with an example, and finish with comments on their capabilities and limitations. The last section reviews current efforts in coupling of different scales and our vision of a multiscale model for the protocell project.

19.2 Numerical Methods Tied to Scale

19.2.1 *Ab Initio* Methods

Ab initio electronic structure (or molecular orbital) methods are based on quantum mechanics, mathematically rigorously solving the Schrödinger equation, and therefore provide the most accurate and consistent predictions for chemical systems. However, *ab initio* methods are extremely computationally expensive. These methods are suitable for small systems of tens or hundreds of atoms, and are the best tool for studying electronic transition with high accuracy (e.g., Levine, 2000; Szabo and Ostlund, 1982).

Strictly speaking, the *ab initio* method can rigorously solve only the two-body problem. For systems of more than two interacting particles, the Schrödinger equation cannot be solved exactly. Therefore, all *ab initio* calculations for many-body systems (e.g., a regular molecule) involve some level of approximation and empirical parameterization, though they all must satisfy a set of stringent criteria (Szabo and Ostlund, 1982). Common approximations include the Born Oppenheimer approximation, which assumes that nuclear positions are fixed because electrons are much lighter than nuclei, the single-particle approximation, whose most popular example is the Hartree-Fock theory that describes each electron as moving in the average electric field generated by other electrons and nuclei, and the density functional theory (DFT) that considers the electronic density instead of the many-body electronic wave function and thus allows modeling electron-electron correlations.

The approach of all *ab initio* techniques is to build the total wave function from a basis set of mathematical functions capable of reproducing critical properties of the system. Minimal basis sets contain the minimum number of basis functions needed to describe each atom (e.g., 1s orbitals for H and He; 1s, 2s, 2px, 2py, 2pz orbitals for Li to Ne). Most common *ab initio* calculations are based on a linear combination of atomic orbitals (LCAO), also known as the tight-binding approximation. Standard *ab initio* software packages like Gaussian (see appendix I) offer a choice of basis sets containing contracted Gaussians, linear combinations of "primitive" Gaussian functions, optimized to reproduce the chemistry of a large range of molecular systems. Molecular properties can be assessed from a user-specified input (single-point energy calculation), or the molecule can be allowed to relax to a minimal energy configuration (geometry optimization).

These calculations are capable of high accuracy predictions over a wide range of systems. Rapid advances in computer technology are making *ab initio* methods increasingly more practical for use with realistic chemical systems. Similarly, computationally cheaper techniques such as density functional theory calculations are continually being refined. These methods show promise of providing consistent and accurate chemical predictions for complicated systems requiring explicit treatment of electronic structure, such as energy transfers in excited molecules. The literature in this area has focused on studying the formation and hydrolysis of peptide bonds and ester bonds, their reaction condition, reaction constant, transition states, reaction mechanisms, and catalysis or environmental dependence. For the Los Alamos project, we expect to have a detailed understanding of the charge transfer processes for the metabolic photofragmentation process that produces the protocell components based on quantum mechanics time-dependent density functional theory.

19.2.2 Semiempirical Methods

Semiempirical methods are founded on quantum mechanics, but speed up computation by replacing some explicit calculations with approximations (e.g., by limiting choices of molecular orbitals or considering only valence electrons) based on experimental data. This method is less computationally demanding than *ab initio* calculations, but it requires *ab initio* or experimental data for its parameters. It is suitable for medium system sizes of hundreds to thousands of atoms. It calculates transition states and excited states and electronic transition as do *ab initio* methods, but with reduced accuracy. In recent years, semiempirical methods have been calibrated to typical organic or biological systems, but they tend to be inaccurate for problems involving hydrogen-bonding and chemical transitions (Clark, 1985; Pilar, 2001).

Several semiempirical methods are available and appear in commercial computational chemical software packages such as Gaussian, Gamess, and Chem3D (see appendix I). The semiempirical methods can be grouped according to their treatment of electron-electron interactions: the extended Hückel method (which neglect all electron-electron interactions), NDO (which neglect of differential overlap, or neglect some but not all e-e interactions), Austin method version 1 (AM1; Dewar et al., 1985), and parameterization model version 3 (PM3; Stewart, 1989). Figure 19.2 (color plate 18 gives an example of AM1 energy minimization of a two-base segment of a PNA molecule (PNA-GG). In the Los Alamos protocell design, the PNA molecule serves as both information and metabolic molecule. For the latter, we need to understand the capability of guanines as an electron donor in the charge transfer reaction (for details see chapter 6). The stacking of guanine pairs would be an indication of whether the spacing between the guanine bases at equilibrium is ideal for electron transfer. The AM1 semiempirical calculation shows that the guanines become less "stacked" after energy minimization in water.

19.2.3 Molecular Mechanics or Molecular Dynamics (MM or MD)

Molecular mechanics, MM, often referred to as molecular dynamics or MD (we will hereafter use MM and MD as synonyms) techniques, consider a system of N particles, whose positions and momenta follow the equations of motion as dictated by Newton's law of motion. (The *ab initio* MD literature almost exclusively uses the Lagrangian formalism instead of directly integrating the equation of motion.) The forces on the particles are derived from the potential. Table 19.1 lists the typical achievable simulation timescale and simulation size for various methods.

Molecular dynamics simulations allow realistic simulation of equilibrium and transport properties. Ensemble averages of MD simulations can be used for statistical mechanics calculations. MD simulations can offer time evolution of chemical reactions and phase transitions. They are also useful in searching for reaction paths and exploring phase space.

Normally, however, MD refers to a purely empirical method based on the principles of classical physics by using a classical potential, which completely neglects electronic structure, and is therefore severely limited in scope; MD, though, often

(a)

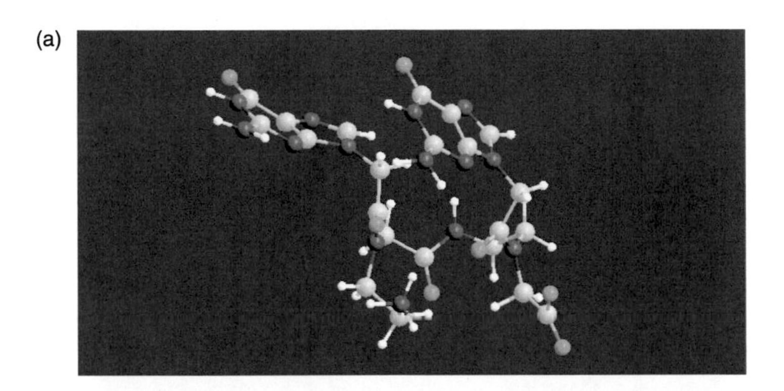

(b)

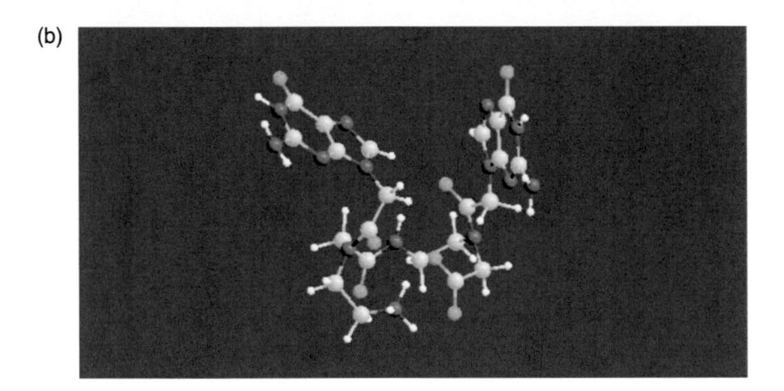

Figure 19.2 (color plate 18)
Semiempirical calculation of small peptide nucleic acid (PNA) segment, PNA-g-g, in water. Two guanine bases point upward with the backbone on the bottom. (a) The initial stacked configuration of the guanine pair. (b) After energy minimization using MOPAC (see appendix I) with AM1 method, the guanine pair becomes less parallel. Red = oxygen, blue = nitrogen, gray = carbon.

Table 19.1
Typical timescale and size achievable by molecular dynamics simulations with different methods

Correlated Methods	Hartree-Fock	DFT	Semiempirical	Classical Potential
1 ps	10 ps	10 ps	100 ps	10,000 ps
10 atoms	10 atoms	100 atoms	1,000 atoms	1,000,000 atoms

provides the only means to study large chemical systems (e.g., amphiphile aggregates in solutions) or nonhomogeneous mixtures (e.g., multiphase lipid aggregates or loading of amphiphile aggregates with other molecules) with some accuracy.

The MD method calculates a molecular system's time evolution by numerically solving Newton's equations of motion based on the system's potential function. The empirical, classical potential function is

$$E = \sum_{bonds} \frac{1}{2} k_{ij}^{b} (r_{ij} - r_{ij}^{0})^2 + \sum_{angles} \frac{1}{2} k_{ijk}^{\theta} (\theta_{ijk} - \theta_{ijk}^{0})^2 + \sum_{dihedrals} k_{ijkl}^{\phi} \{1 + \cos[n(\phi_{ijkl} - \phi_{ijkl}^{0})]\} + \sum_{i<j} \frac{q_i q_j}{\varepsilon r_{ij}} + \sum_{i<j} \left(\frac{A_{ij}}{r_{ij}^{12}} - \frac{B_{ij}}{r_{ij}^{6}} \right) \quad (19.1)$$

where r_{ij}, θ_{ijk}, and ϕ_{ijkl} are, respectively, the length of a chemical bond *ij* or the interatomic distance between atoms *i* and *j*, the angle between chemical bonds *ij* and *jk*, and the dihedral angle formed by four consecutive atoms *i*, *j*, *k*, and *l*. The first two terms describe chemical bonds and angles, using a harmonic oscillator approximation of the real potentials. The equilibrium parameters θ_{ijk}^{0} and ϕ_{ijkl}^{0} as well as force constants k_{ij}^{b} and k_{ijk}^{θ} are obtained from either *ab initio* calculations or vibrational spectroscopy. The third term describes the dihedral interaction with a simplified cosine potential, where k_{ijkl}^{ϕ} are the heights of the potential barriers obtained either from experimental data or *ab initio* calculations, and ϕ_{ijkl}^{0} are the equilibrium values of the dihedral angles. The fourth term, where q_i and q_j are fixed partial atomic charges fit to best reproduce the electrostatic potential of the molecule as calculated *ab initio*, and $\varepsilon = 1$ is a macroscopic parameter, describes the electrostatic interactions between charged atoms, assuming pointlike atoms and neglecting electronic polarizability. The last term uses the semiempirical Lennard-Jones potential for the Born repulsion and van der Waals attraction, where parameters A_{ij} and B_{ij} are difficult to determine theoretically. These nonbonded interactions are assumed to be pair interactions; interactions involving three or more atoms are neglected. Also, because of the simplified potentials, the force parameters obtained from *ab initio* calculations should be treated as a starting guess and may need adjustment to ensure reasonable thermodynamics in agreement with experimental data.

The most commonly used MD force field software packages are CHARMM, AMBER, NAMD, GROMACS, LAMMPS, NWCHEM, DL_POLY, and Tinker (see appendix I). CHARMM, AMBER, NAMD, GROMACS, NWCHEM, and Tinker are designed primarily for modeling biological molecules. CHARMM and AMBER use atom-decomposition (replicated-data) strategies for parallelism; NAMD, LAMMPS and NWCHEM use spatial-decomposition approaches. Tinker is a serial code. DL_POLY and GROMACS include potentials for a variety of biological and nonbiological materials. In addition, MMx (MM2, MM3, etc., see appendix I) are optimized for structural and thermodynamic studies of small nonpolar molecules. The various MMx versions differ primarily in their parameterizations; the higher versions are more recent and address deficiencies in their predecessors. In addition, CPMD (see appendix I) is an implementation of DFT specially designed for *ab initio* MD.

Some packages allow more sophisticated simulations in addition to the classical MD approach. For example, NWChem and VASP (see appendix I) can combine classical and quantum descriptions (QM/MM, see below). Tinker and MOIL contain long-time dynamics algorithms based on stochastic difference equations, thus extending significantly the simulation time step up to the order of milliseconds. The main drawback of the last method is the reduction of the entropy effect on a system's trajectory. Also, the method needs both a starting and a final system conformation as the input, and the latter can be difficult to predict.

The MD method has a number of limitations. For all the simulations, the time step must be short enough to assume constant values of the forces during the time step (typically a few femtoseconds if vibrational modes can be neglected). Consequently, the typical accessible timescale of the method is on the order of 10 ns at the performance of present-day computers. The computational performance has also limited the simulation length scale, currently on the order of 10 nm for the condensed phase, and the system size is on the order of 10^6 atoms. Related to the system size limitation is a minimum concentration of about 10^{-4} M. One of the consequences of this limitation is the range of pH we can correctly simulate. The limits on the allowable concentrations of H^+ and OH^- demand that only the upper and lower ends of the pH spectrum can be solved (that is, $pH < 4$ or $pH > 10$) in addition to $pH = 7$. Most force field parameters found in the software packages mentioned earlier correspond to the neutral pH. Therefore, systems at different pH values can be modeled only indirectly with the use of different protonation states if the corresponding parameters are available. The system size limitation also prevents us from simulating larger-scale phenomena, for example, the undulation of a lipid bilayer, and calculating such parameters as the large or small partitioning coefficient. In addition, finite domain size also requires that the MD simulations use periodic boundary conditions; this in turn demands the use of particle mesh Ewald algorithms to minimize the cutoff error resulting from the electrostatic force scaling and the electroneutral condition. Finally, there is another difficulty in applying MD: Frequently, we must use macroscopic parameters in the microscale (constant V, A, e), which to some degree suppresses thermodynamic fluctuations. Despite these various limitations, MD is still capable of examining important questions related to, for example, lipid dynamics. Figure 19.3 shows an example of an atomic scale MD simulation of a

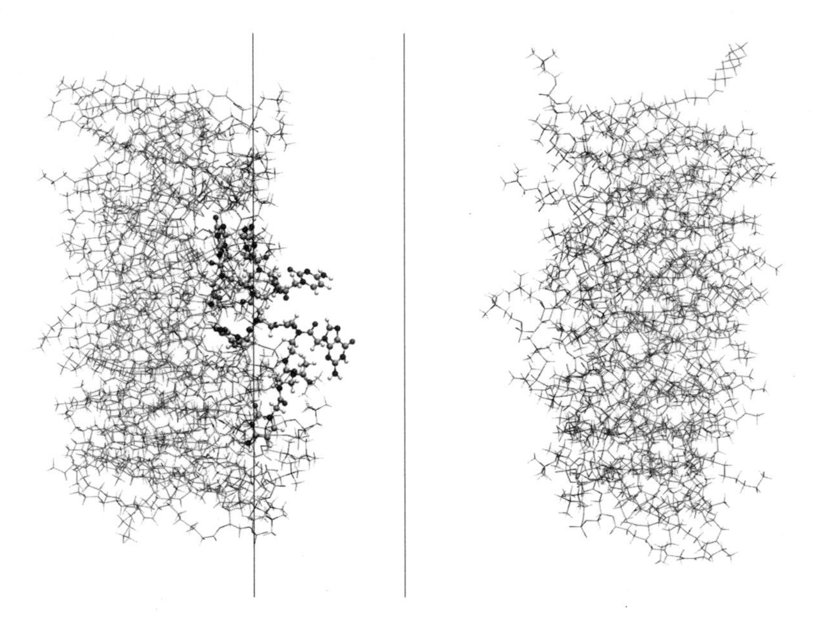

Figure 19.3
PNA molecule (protein data bank identifier 1PUP) at a lipid-water interface (water molecules are invisible). The lipid bilayer is composed of two palmitoyl oleoyl phosphatidyl choline (POPC) lipid layers. The PNA molecule is attached to the upper interface. The picture presents the system at equilibrium. Simulations conducted using NAMD (Kale et al., 1999) and VMD (Humphrey, Dalke, and Schulten, 1996) packages. Reproduced from Weronski, Jiang, and Rasmussen (2007), with permission.

system containing water, lipid clusters, and one peptide nucleic acid (PNA) molecule; the PNA molecule becomes trapped at a lipid-water interface (Weronski, Jiang, and Rasmussen, 2007).

19.3 Quantum Mechanics and Molecular Mechanics

One key process of the LANL protocell assembly is the photo-activated catalytic reaction involving PNA and surfactant precursors and sensitizers. As we have discussed, *ab initio* methods are capable of modeling catalytic reactions but only for a very small system; molecular mechanics methods, on the other hand, cannot study the reactions, but can handle other classical dynamics of large numbers of atoms.

Over the past couple of decades, methods have emerged that are capable of overcoming the practical problems associated with modeling catalytic reactions. With the development of modern density functional methods (Parr and Yang, 1989), *ab initio* quantum chemistry (QM) is now capable of describing reactive chemistry for systems involving hundreds of atoms with a very respectable level of accuracy. Molecular mechanics force fields (e.g., CHARMM, AMBER, NAMD) have been constructed that provide a remarkably good description of conformational energetics and non-bonded interactions for proteins, nucleic acids, and lipids. Finally, mixed quantum mechanics/molecular mechanics (QM/MM) approaches have been created (Field, Bash, and Karplus, 1990; Sherwood et al., 2003) that can seamlessly join QM and MM representations for different sectors of a complex condensed-phase system. The conjunction of these technologies contains the elements necessary to properly describe the potential energy surfaces relevant to enzymatic chemistry, at least to a first-order approximation. QM/MM has been applied successfully to model protein enzymatic chemistry (Friesner and Guallar, 2005). The reactive region of the active site can be treated with a robust *ab initio* QM methodology, employing a QM region of sufficient size to encompass any important electronic structure effects. The remainder of the protein can be modeled at the MM level, providing the appropriate structural constraints, and electrostatic and van der Waals interactions with the core reactive region. A suitably parameterized QM/MM interface technology ensures that large errors are not made in coupling the two regions together.

19.4 Coarse-Grained Molecular Dynamics

The heterogeneous nature of biological surfactant assemblies poses many challenges to theorists. They typically reside in aqueous environments that bring all of the difficulties associated with modeling water to the forefront, and require that full hydrodynamic information be maintained to truly capture the dynamics of any assembly processes. The constituent species range from relatively small lipids to large polymeric proteins while the assemblies themselves display some cooperative, structure-wide dynamics. Thus, important timescales range from picoseconds to minutes while the germane length scales can span Angstroms to microns. Moreover, any fully atomistic study must include the computationally expensive calculation of the long-range charged interactions. This is a daunting set of barriers to answering the many fundamental questions prerequisite to a quantitative understanding of these basic processes. Indeed, the state of the art for atomistic MD simulations of lipids simulates only 100 ns for 1,024 surfactant molecules (Bogusz, Venable, and Pastor, 2001). For longer time dynamics, for example in micellar stability or vesicle deformation, or diffusion of amphiphiles within the lipid aggregates, we will have to resort to coarse-grained molecular dynamics. A number of coarse-grained lipid MD models have been developed (Goetz, Gompper, and Lipowski, 1999; Goetz and Lipowski, 1998; De Vries, Mark, and Marrink, 2004; Shelley et al., 2001) based on lumping atoms into particles and approximating their interactions with Lennard-Jones potentials. However, in these methods the truncation of electrostatics and severe simplifications of the lipid molecular structure can potentially alter the membrane's macroscopic material properties, for example, the bulk modulus. We highlight three coarse-grained MD methods below, which seem promising in modeling processes relevant to protocell assembling.

19.4.1 Brownian Dynamics

In Brownian dynamics, the motion of the particles is described by the Langevin equation,

$$\ddot{r}_i = -\nabla U_i - \Gamma \dot{r}_i - W_i(t), \tag{19.2}$$

which consists of inertial, force field, frictional drag, and noise terms, respectively. The potential U typically includes two-, three-, and four-body interactions. This method simulates the effect of individual solvent molecules through the noise W, which is drawn from a Maxwell-Boltzmann distribution. The friction coefficient Γ is related to the autocorrelation function of W through the fluctuation-dissipation theorem,

$$\langle W_i(t) \cdot W_j(t') \rangle = \delta_{ij}\delta(t - t')6kT\Gamma \tag{19.3}$$

where δ_{ij} is the Kronecker delta function. Since in BD the effect of the solvent is implicit, one can only estimate the relationship between simulation time and real time. Because of the coarse-grained nature of the united-atom models often used in BD, the time unit is expected to be longer than that of an atomistic model, on the order of the monomeric relaxation time rather than the atomic relaxation time. Based on the example of liquid argon, the relaxation time for a typical solvent molecule is on the order of 10^{-12} s. We therefore expect the relaxation time of the united atom models to be also on the order of 10^{-12} s. Since in each iteration of a simulation the noise is uncorrelated and the time step is typically on the order of 0.001, one time unit in a simulation is on the order of 10^{-10} s (Welch and Muthukumar, 1998, 2000). Because the BD method treats electrostatics explicitly, pH dependence can be simulated either directly by including the appropriate number of ions or by the appropriate corresponding protonation states of molecules.

19.4.2 Dissipative Particle Dynamics

Dissipative particle dynamics (DPD) is based on the simulation of soft sphere particles, whose motion is governed by certain collision rules and Newton's equations (Groot and Warren, 1997). Instead of representing an individual atom as in MD, one DPD particle represents a collection of molecules or molecular groups. Therefore, a DPD simulation is able to incorporate much microscopic physics but still keep the computational costs at a reasonable level, and can cover larger time and length scales to study such processes as phase transitions. Several solution molecules, such as water molecules, can be modeled as one DPD particle. Using bead and

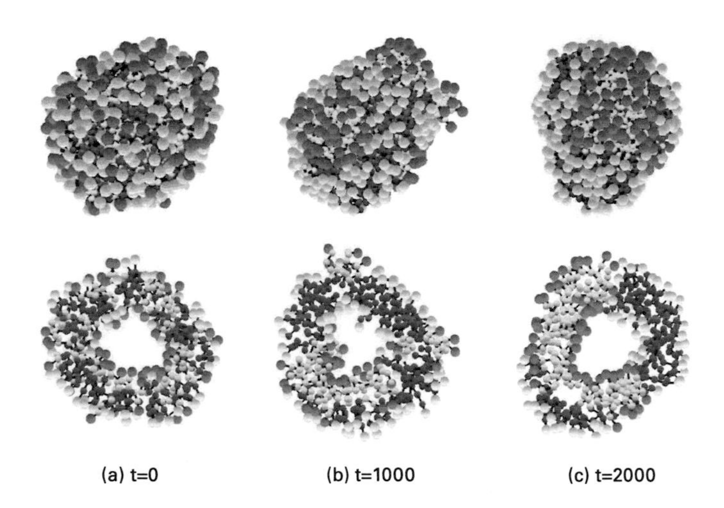

Figure 19.4 (color plate 19)
DPD simulation of phase separation of a two-component lipid vesicle. The interaction parameters are adapted from Yamamoto and Hyodo (2003). Two types of lipids are indicated as yellow head, blue tail, and red head, cyan tail, respectively. The top row shows the surface view and the bottom row shows the cross-sectional view.

spring-type particle chains, one can simulate a lipid as a simple flexible amphiphilic chain with hydrophilic and hydrophobic regions (Yamamoto and Hyodo, 2003).

Each DPD particle follows the equation of motion, which is typically integrated with a modified velocity-Verlet algorithm. The total force on the particle consists of three parts—the dissipative, the conservative, and the random forces—the sum of which is essentially repulsive. By changing the interaction parameters, we can adjust the nature of the solution or the type of the lipid. Within one amphiphile tail, the chemical bond between two bonded particles is described as an elastic spring with an equilibrium bond distance and a spring constant.

Figure 19.4 (color plate 19) shows the phase separation of a two-component vesicle. Each lipid molecule has one head particle and three tail particles. Two types of molecules (type A [yellow head, blue tail] and type B [red head, cyan tail]) reorganize at both inner and outer surfaces of the vesicle, and form domains of a single lipid type. If we change the interaction strengths between the two types of lipids, a domain may bud off from the parent vesicle to form smaller vesicles or micelles, depending on the number of lipids in the domain.

The main advantage of DPD is that the effective interaction is "soft core," that is, does not diverge at small distances. This feature allows much longer time steps, compared to traditional MD, when integrating the equations of motion. One shortcoming, however, is that the interaction is essentially only repulsive. Only hydrophobic effects of different degrees, that is, strong and less-strong repulsions, can be consistently modeled with DPD. Any attractive forces, such as electrostatic force, cannot be treated in the present framework. It is also not yet clear whether DPD actually faithfully captures the thermodynamic behavior of polymers in solution. In addition, deriving DPD parameters from microscopic molecular details is nontrivial.

19.4.3 Lattice Gas Molecular Dynamics

The lattice gas molecular dynamics (LGMD) model is based on classical lattice gas models but allows for more complex interactions between particles (Nilsson et al., 2003). It fills a gap between highly detailed and computationally expensive MD, and lattice models too oversimplified to capture many of the interesting physicochemical and thermodynamic properties of self-assembly processes. The LGMD allows detailed electromagnetic interactions between particles. The lattice spacing is assumed to be equal to the length of a hydrogen bond (3 angstroms), and the time steps are on the order of picoseconds. The length and time scales allow for sufficient numbers of molecules and time steps so that large-scale molecular ordering may be observed. The steps of the algorithm are propagation of the interactions between molecules, molecular rotation, collisions, and molecular movement. Interactions among molecules are determined from simplified quantum calculations. Potential energy comes from hydrogen bonds, dipole-induced dipole interactions, induced dipole-induced dipole interactions, and cooperative effects. Cooperative effects allow the strength of individual dipole interactions to increase as the number of interactions for a molecule increases. For each node, the potential energy is calculated from the sum of each of these interactions,

$$V_{total} = \sum_{i=1}^{n}\sum_{j=1}^{q} V^{i,j}_{H-bond} + \sum_{i=1}^{n}\sum_{j=1}^{q} V^{i,j}_{dip.-ind.dip.} + \sum_{i=1}^{n}\sum_{j=1}^{q} V^{i,j}_{ind.dip.-ind.dip.} + \sum_{i=1}^{n}\sum_{j=1}^{q} V^{i,j}_{coop.}, \tag{19.4}$$

where n is the number of different types of molecules, and q is the number of nearest neighbors for each node. Molecules are rotated in order to minimize the free energy of interactions with nearest neighbors. For a water molecule, the orientation determines the three neighboring nodes with which hydrogen bonds may be formed. For simulations involving water and lipid, the relative strengths of hydrogen bonds to all other bonds is approximately 5 to 1. During the collision step, molecules that are located at the same node after the movement step exchange momentum according to a determined set of collision rules. During the movement step, each molecule moves a unit step in the direction with a probability that is proportional to its velocity.

The LGMD model has been used to simulate clustering of amphiphilic molecules dissolved in water (Mayer and Rasmussen, 2000). A critical result from these simulations is that the hydrophobic effect is emergent from the model: No ad hoc forces are necessary to achieve phase separation. Phase separation dynamics in simulations agree with experimentally observed dynamics (Nilsson et al., 2003).

19.5 Lattice Boltzmann

The lattice Boltzmann (LB) method is a kinetic model that uses simple collision and streaming processes that may capture macroscale fluid behavior. A broad range of applications is made possible by suitable choice of the collision process. These include single-phase, single-component fluids, multicomponent fluids, and multiphase fluids (Chen and Doolen, 1998). Recently, the LB method has been extended to simulation of amphiphilic fluids (Chen et al., 2000; Nekovee et al., 2000).

The ternary LB model (Nekovee et al., 2000) employs two interacting fluids (e.g., oil and water) and a surfactant. The oil and water fluids are modeled as collections of particles, and the surfactant is modeled as a dipole particle with two equal and opposite affinities for water and oil, respectively. In this way, the hydrophobic effect can be parameterized. The fluid physics are contained in the distribution functions that describe the probability of finding particles at a particular location with a particular velocity. Multiple fluid components may be simulated by including sets of distribution functions for each component. Amphiphilic fluids are modeled as molecules with two bonded particles, one of which is polar and the other nonpolar, separated by a distance much less than the grid spacing. A dipole vector, d, is used to measure the degree of alignment among the amphiphilic particles, and undergoes collision and streaming as well. The interaction forces of the amphiphilic fluids depend on the separation distance in addition to the density of each component. Figure 19.5 shows results from a sample simulation. Initially, the densities of fluid A and an amphiphile are randomly distributed throughout a two-dimensional periodic system so that the average densities are 0.5 and 0.1, respectively. The system is allowed to equilibrate for 2,000 time steps, and form a lamellar structure that is two nodes wide.

LB algorithms can, in principle, be used to model the dynamics of micelles and vesicles. However, limitations in the surfactant model of the LB algorithm are hampering its application to protocell assembly studies.

19.6 Chemical Kinetics

Chemical kinetics focuses on characterizing chemical reactions in a system and lumps atomistic-level details into rate coefficients. Chemical kinetics captures the dynamics of a system through a set of coupled, frequently nonlinear, differential equations. The

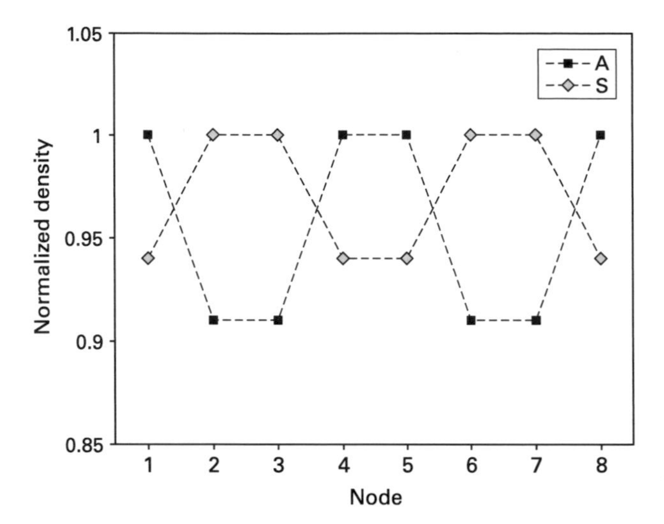

Figure 19.5
Normalized equilibrium density of fluid A and amphiphile S for initially random density distributions.

spatial distribution of components may be ignored as well, leaving the equation set as ordinary differential equations (ODEs), or spatial coupling may be included through mechanisms such as diffusion and advection, changing the equation set to partial differential equations (PDEs). There is a rich body of mathematics for dealing with ODEs and PDEs. Considering protocell assembly from a chemical stoichiometric and kinetic viewpoint rather than an atomic and molecular dynamics viewpoint provides a new tool for investigating the requirements for protocell stability.

Successful replication of protocells requires production of lipids and oligomers from precursor materials. In this section, we consider the rate of formation of lipids from the metabolic scheme. The energy required to generate lipids from precursors is obtained by photoexcitation of a sensitizer molecule.

To fix ideas, we use the metabolic scheme proposed by Rasmussen, as shown in figure 19.6. The first step in the reaction scheme is the excitation of the sensitizer (Z^*) from the ground state (Z) by a photon. The excited electron is transferred to the lipid precursor (pL), producing a charge-separated state (Z^+/pL^-). At this step, either the electron may be transferred back to the sensitizer or an electron from PNA may neutralize the sensitizer, returning it to the ground state. The lipid (L) is produced by breaking an ester bond in the negatively charged lipid precursor. The differential equations for these compounds in the proposed scheme are given by

$$\frac{d[\mathrm{pL}]}{dt} = -k_{ip}[\mathrm{Z}^*][\mathrm{pL}] + k_{back}[\mathrm{Z}^+\mathrm{pL}^-]$$

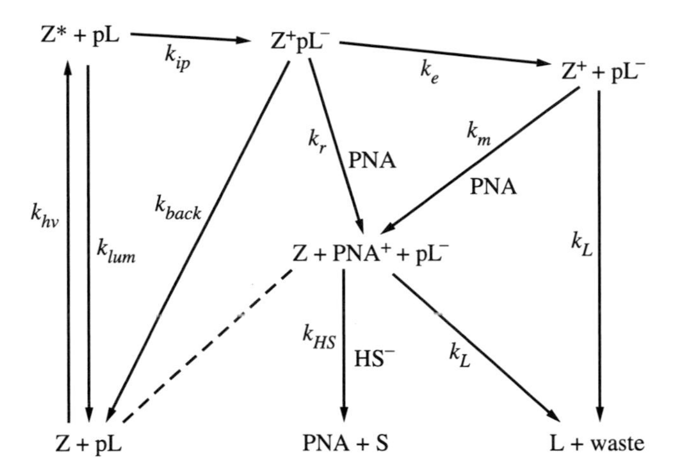

Figure 19.6
Reaction diagram for the metabolic scheme to produce lipids for the protocell membrane proposed by Rasmussen et al. (2003). Z is the sensitizer molecule, pL is the lipid precursor, PNA is used as an electron donor, and sulfide is the sacrificial electron donor.

$$\frac{d[\mathrm{Z}]}{dt} = -k_{hv}[\mathrm{Z}] + k_{lum}[\mathrm{Z}^*] + k_{back}[\mathrm{Z}^+\mathrm{pL}^-] + (k_r[\mathrm{Z}^+\mathrm{pL}^-] + k_m[\mathrm{Z}^+])[\mathrm{PNA}]$$

$$\frac{d[\mathrm{Z}^*]}{dt} = k_{hv}[\mathrm{Z}] - k_{lum}[\mathrm{Z}^*] - k_{ip}[\mathrm{Z}^*][\mathrm{pL}]$$

$$\frac{d[\mathrm{Z}^+\mathrm{pL}^-]}{dt} = k_{ip}[\mathrm{Z}^*][\mathrm{pL}] - k_{back}[\mathrm{Z}^+\mathrm{pL}^-] - k_e[\mathrm{Z}^+\mathrm{pL}^-] + k_r[\mathrm{Z}^+\mathrm{pL}^-][\mathrm{PNA}]$$

$$\frac{d[\mathrm{Z}^+]}{dt} = k_e[\mathrm{Z}^+\mathrm{pL}^-] + k_r[\mathrm{Z}^+\mathrm{pL}^-][\mathrm{PNA}] - k_L[\mathrm{pL}^-]$$

$$\frac{d[\mathrm{pL}^-]}{dt} = k_e[\mathrm{Z}^+\mathrm{pL}^-] + k_r[\mathrm{Z}^+\mathrm{pL}^-][\mathrm{PNA}] - k_L[\mathrm{pL}^-]$$

$$\frac{d[\mathrm{L}]}{dt} = k_L[\mathrm{pL}^-], \tag{19.5}$$

where brackets designate concentrations of the species. This scheme has yet to be tested in the laboratory, so reaction rates and even orders of reactions are considered estimates.

Several assumptions may be made to simplify this set of equations. We assume that the concentrations of all forms of the sensitizer (Z, Z^*, Z^+, Z^+pL^-) quickly

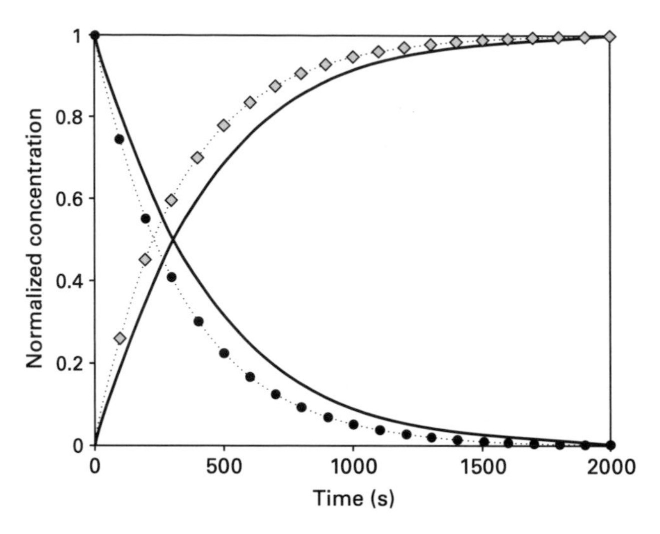

Figure 19.7
Concentrations of lipids and precursor lipids for slow photoexcitation. Numerical simulations of the full reaction system (solid lines) are compared to the analytical solution of the simplified equations (symbols). Curves starting at concentration equal to 1 correspond to pL, and curves starting at concentration equal to zero correspond to L. Values for the reaction rate constants are the following: pL0 = 0.01 M, Z0 = 0.0001 M, [PNA] = 0.0001 M, khv = 1 1/s, kback = 104 1/s, kr = km = 109 (M s)-1, klum = 108 1/s, kip = 3 × 109 (M s)-1, and kL = 105 1/s.

reach steady state. The slowest step in the reaction scheme is assumed to be the photoexcitation of Z. Then, the production rate of lipids can be estimated by

$$\frac{d[\mathrm{L}]}{dt} \approx \frac{k_{h\upsilon}}{k_{lum}} k_{ip} \mathrm{Z}_0 [\mathrm{pL}], \tag{19.6}$$

where Z_0 is the initial concentration of sensitizer in the solution (Knutson et al., 2007).

In figure 19.7, results from the approximate solution are compared to the numerical solutions of the full scheme. There is some difference in the results from the approximate solution at intermediate times for the set of values for rate constants used here. However, currently not enough experimental data exist to obtain good estimates for most of the rate constants in the scheme. Laboratory experiments are testing the correctness of the model reactions and attempting to determine values for their rate constants.

19.7 Ginzburg-Landau

In Ginzburg-Landau (GL) models, the free energy is described by a spatially localized scalar parameter (or a set of such parameters). Global minimization of the free energy is then transformed into a local partial differential equation, which becomes the fundamental mathematical model for the system in equilibrium. This method has been successful in reproducing qualitatively correct phase diagrams.

At a macroscopic level, since the thickness of the lipid membrane (~5 nm) is orders of magnitude smaller than the size of the membrane (~μm), a lipid membrane can be treated as a two-dimensional fluid with a bending elastic energy (Helfrich, 1973):

$$F\alpha = \int d^2r\left[\frac{\kappa_1}{2}(H - H_0)^2 + \kappa_2 K\right], \tag{19.7}$$

where H and K are the mean curvature and Gaussian curvature, and κ_1 and κ_2 are the corresponding elastic moduli, respectively. This bending elastic energy, under area and volume constraints, can produce many shape changes in vesicles (Miao et al., 1994).

Because molecules are free to move in the plane of the membrane, a multicomponent membrane can exhibit phase separation of different components. In real cell membranes, phase separation plays an important role in the stabilization of vesicles by forming lipid rafts and in the fission of small vesicles after budding. We can model the phase separation kinetics with the elastic properties of a lipid membrane by a linear coupling term in the total free energy of the system (Jiang, Lookman, and Saxena, 2000):

$$F = \int dS\left[\frac{\kappa_1}{2}(H - H_0)^2 + \kappa_2 K + \frac{\xi}{2}(\nabla\varphi)^2 + \frac{1}{4}\alpha\varphi^4 - \frac{1}{2}\beta\varphi^2 + \Lambda\varphi H\right], \tag{19.8}$$

where φ is the relative concentration of one lipid component or the order parameter field. The first two terms are the bending elastic energy, the third term describes the energy associated with creating domain boundaries; the fourth and fifth terms together form the double well potential energy that favors phase separation of mixtures; the last term is the linear coupling between the concentration field and the mean curvature.

If the lipid molecules do not change (mass conservation), the evolution equation found by minimizing the free energy (8) with respect to φ is

$$\frac{\partial\varphi}{\partial t} = M_\varphi\nabla^2\frac{\delta F}{\delta\varphi} + \eta, \tag{19.9}$$

where M is mobility and η is Gaussian noise. We model the phase separation and the accompanying shape deformation of the membrane. On the other hand, if we let the order parameter φ be the concentration of the lipid aggregate (aggregate numbers can change), the evolution equation becomes

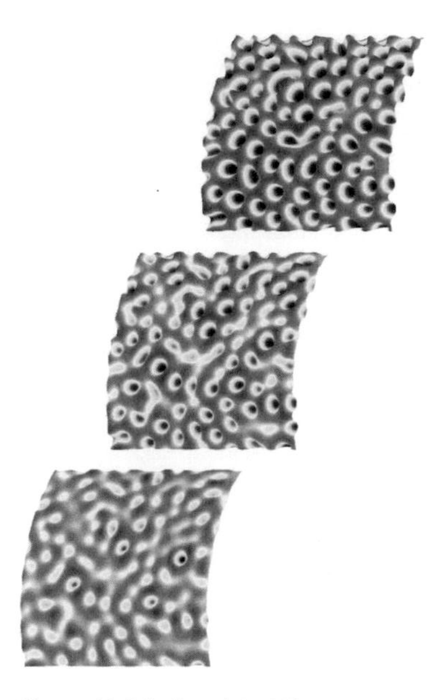

Figure 19.8 (color plate 20)
Ginzburg-Landau simulation of binary lipid phase separation coupled to local deformation (equations 19.9 and 19.10). Color encodes the lipid concentration field φ. Red and blue are two lipid components and the initial condition is a curved membrane surface with one constant curvature and homogeneous lipid mixture.

$$\frac{\partial \varphi}{\partial t} = -\Gamma \frac{\delta F}{\delta \varphi} + \eta, \tag{19.10}$$

where Γ is a rate constant that sets the timescale for the phase evolution. We can model the nucleation of lipid aggregates and the time evolution of the aggregates. Figure 19.8 (color plate 20) illustrates results from a coupled field GL simulation for deformation and lipid phase separation in membranes with conserved dynamics. In addition, we can couple the dynamics of phase and shape into a flow field; for example, in convection-diffusion equations instead of simple diffusion equations, we can model the dynamics of a phase-separating vesicle in a flow environment. This model will shed light on vesicles in microfluidic devices. However, the parameterization of the underlying microdynamics limits the connection to microscopic details in realistic systems.

19.8 Multiscale Methods

Clearly, the optimal strategy for modeling complex systems such as protocell assembly must not rely solely on a single simulation technique best suited to a given size or timescale. A random assortment of independent methods, however, will not fully capture the relevant physics. Lest modelers end up in the situation of the blind men describing the elephant, the modeling community needs to find a path forward that connects the various time and length scales and the differing viewpoints (particle versus continuum) into a coherent approach.

Some range of time and space scales can be covered by more than one method. An example is vesicle formation: MD (De Vries, Mark, and Marrink, 2004), coarse-grained MD (Marrink and Mark, 2003), BD (Nogochi and Takasu, 2001), and DPD (Yamamoto, Maruyama, and Hyodo, 2002) have all been used to model vesicle formation. When discrepancies arise between these models, which one should we trust most?

Circumventing the limitations of each method by mapping from one scale to another scale, in particular from fine scale to coarser scales, might not only provide new insight into the relationship between, for example, molecular structure and free energy, but also revolutionize our capability to model real chemistry and design new materials. Thus motivated, many researchers have attempted to stitch together the patchwork of particle and continuum descriptions into a cohesive framework.

A particle-based method can approximate the molecular-level dynamics as well as easily permit major structural rearrangements; a field-theoretical model (e.g., GL) can capture the large-scale phase separation dynamics coupled to the structural elasticity. Coupling the GL model to a particle method will provide key insight into the relationship between molecular-level details and the macroscopic phenomenology. Recently, it was shown that GL models can be directly parameterized from particle-based simulations of the simpler case of binary blend phase separation (Welch et al., 2005). The key to this technique lies in extracting the composition fields as a function of time from the particle-based simulation method and fitting that data to the GL equation of motion.

Voth's group (Ayton et al., 2002a, 2002b; Ayton and Voth, 2002) has coupled atomic-scale molecular dynamics and a macroscopic mechanics model (the materials point method, or MPM), to model the large-scale undulation of a lipid vesicle surface (Ayton et al., 2002a), lateral diffusion and composition-dependent (lipid/cholesterol mixture) phenomena (Ayton et al., 2002b; Ayton and Voth, 2002). Treating the membrane as a viscoelastic material, the MPM discretizes the membrane surface into a grid, which is used to evaluate continuum-level strain and strain rates. At each grid point, the parameters are calculated using nonequilibrium molecular dynamics. In other words, they used the MD simulation cell as a "material property" template and then coarse-grained the membrane into domains with a length scale approximately the size of the MD cell (grid). The grid points are then parameterized to respond to local plane strain, as predicted by atomic-scale MD simulations. This model employs the microscopically calculated bulk moduli, thickness, and density as key parameters to model bulk elastic response of the membrane (Ayton et al., 2002a).

Kevrekidis and coworkers, over the last few years, have developed an "equation-free" approach for extracting coarse-grained information from microscopic models, which provides an alternative to direct, long-term atomistic/stochastic simulations (Theodoropoulos, Qian, and Kevrekidis, 2000). This approach establishes a systematic connection between traditional, continuum analysis and microscopic, atomistic, or particles-based and stochastic simulations. The procedure relies on the expected evolution of a few slow, coarse-grained observables, and on the timescale separation between these slow and the remaining fast variables. Short bursts of appropriately initialized fine-scale simulations are used to estimate the local dynamics of the evolution of the slow observables. These estimates are then used to accelerate the evolution to computational stationarity through traditional continuum algorithms. This method circumvents the derivation of explicit, closed equations and allows microscopic simulators (e.g., MD, DPD) to perform system-level tasks directly. Applications of this method have ranged from liquid crystal rheology (Sietto et al., 2003) to micelle formation (Kopelevich, Panagiotopoulos, and Kevrekidis, 2005a, 2005b), surface reactions (Makeev et al., 2004), and chemotaxis of cells (Setayeshgar et al., 2005). It relies, however, on the assumption of smooth behavior for the slow variable; so, as with virtually every algorithm or conceptual idea, there are limitations.

Our vision of a multiscale simulation model couples dynamics at the molecular scale all the way up to macroscale behavior, but relies on experimentation as well as on computation. For example, loading of PNA into a lipid container is a vital step in the LANL protocell assembly (see chapter 6). Experiments such as light-scattering, absorption, or fluorescence spectroscopy, and solute trapping permeability evaluation using light and electron microscopy, will reveal stoichiometry and kinetics of partitioning of PNA into micelles or vesicles, and conformation of PNA. This data set will test MD fine-scale atomistic models, improving model predictions, for example, for free energy change resulting from PNA association. Then, based on both experiments and simulations, effective interaction potentials, force-field parameters, and thermodynamic observables will be derived to drive coarse-grained MD (e.g., MDLG, BD, DPD) simulations of PNA association with lipid aggregates. The addition of spatial variability to chemical kinetic packages will be necessary to capture the dynamics of a real protocell.

As a first step toward a multiscale modeling tool, we have developed a general hyperdynamics method to accelerate atomistic MD for entropic systems (Zhou et al., 2006). Dynamics of most systems can be characterized by spending most of the time in metastable regions whose dynamics can be described by thermodynamic relationships, and fast transitions between metastable states. The hyperdynamics method, first proposed by Voter (1997), takes advantage of these characteristics and uses bias potentials to accelerate the transition between metastable regions. This method has been very successful in simulating small solid systems, but cannot work for fluids or macromolecular systems, because entropic effects are important in these systems such that the potential energy alone cannot determine the metastable regions. Our general hyperdynamics method coarse-grains space and time in metastable regions by biasing the system's potential with a few collective variables, while remaining faithful to the transition regions. We have demonstrated the slow dynamics of gas–to-liquid transition in argon in a biased simulation that is 6 orders of magnitude faster than normal MD. Further development in this direction can be very promising for simulating the slow dynamics in macromolecular systems with atomistic details.

The protocell world is not complete without the presence of an information-preserving component, the genes of a protocell, bringing more complexity to computational models as well as experiments. At each scale, the questions that need to be answered change. Our multiscale model will need to compile these questions and explicitly state how level n questions depend on level n-1 answers. We envision an iterative process, in which experimentation validates or improves simulation, and simulation inspires additional experiments. Parameters needed for larger-scale modeling will be derived from appropriate averaging or burst-sampling (Armaou, Kevrekidis, and Theodoropoulos, 2005) of smaller-scale dynamics. Analysis techniques such as two-timing (Cole, 1968) are also available, with a novel aspect of coupling models containing both particle and continuum representations. The availability of massively parallel processing computers makes multiscale coupling feasible. We anticipate that protocell research will not only lead to a new suite of ground-breaking experiments, but will also oversee the development of truly multiscale simulation capability.

19.9 Conclusion

Two areas of modeling will surely be the focus of great activity in the coming decade. The first, multiscale modeling, makes use of large parallel computing systems to bridge across islands in the time-length spectra. The second, experiment-simulation iteration, uses statistical and inverse methods to couple forward predictive models and experimental data sets to rigorously bound model parameters, and possibly suggest new experiments as well as new modeling concepts in an iterative process.

This chapter has described a number of numerical simulation tools, each focusing on a particular part of the time and length spectrum, which researchers are using to help understand the requirements for a stable protocell. Regardless of the specificity of individual examples given here, these algorithms are all widely applicable. Computational methods have progressed greatly in the last three decades, but, despite the optimistic tone conveyed here, there are still significant hurdles to overcome. A major concern is that a full theory does not exist for protocells, nor are proposed experiments needed to provide parameter values in models (let alone clarify the essential conceptual ideas) fully carried through or understood yet. Models, then, are either forced to use poorly constrained parameters or are restricted to aspects of protocell dynamics for which relevant parameters are known. In this situation, modeling can nevertheless be useful for exploring "what if" scenarios, considering consequences of ranges of parameter values. To be fair, not all the "parameters" of the process explored in laboratory experiments at the forefront of a discipline are known. Just as laboratory experiments are designed to test ideas and reveal relationships, numerical/computational experimentation may also lead to advances in conceptual thinking.

Will we need so many experiments to fix the parameters that in the end we will not need models? That is a criticism sometimes heard from nonmodelers. Although it may be effectively true for the first successful protocell project, the capabilities acquired will then be useful for a growing body of other applications. The current situation may be much like the state of physics and engineering a hundred years ago. The basic governing equations were known for important areas of physics, but measurements of material properties, equations of state, and constitutive relationships were still needed to actually simulate real systems. Eventually most of that data became available, and now models are used for building our cities, factories, water and energy systems, roads, airplanes, automobiles, computers, and medicines. A similar evolution is expected in the world of protocells.

Acknowledgments

This work is supported by US DOE under contract DE-AC52-06NA25396.

References

Armaou, A., Kevrekidis, I. G., & Theodoropoulos, C. (2005). Equation-free gaptooth-based controller design for distributed complex/multiscale processes. *Computers and Chemical Engineering*, *29*, 731–740.

Ayton, G., Smondyrev, A. M., Bardenhagen, S., McMurtry, P., & Voth, G. A. (2002a). Calculating the bulk modulus for a lipid bilayer with nonequilibrium molecular dynamics simulation. *Biophysical Journal*, *82*, 1226–1238.

Ayton, G., Smondyrev, A. M., Bardenhagen, S., McMurtry, P., & Voth, G. A. (2002b). Interfacing molecular dynamics and macro-scale simulations for lipid bilayer vesicles. *Biophysical Journal*, *83*, 1026–1038.

Ayton, G., & Voth, G. A. (2002). Bridging microscopic and mesoscopic simulations of lipid bilayers. *Biophysical Journal*, *83*, 3357–3370.

Bogusz, S., Venable, R. M., & Pastor, R. W. (2001). Molecular dynamics simulation of octyl glucoside micelles: Dynamic properties. *Journal of Physical Chemistry*, *105*, 8312–8321.

Chen, H., Boghosian, B. M., Coveney, P. V., & Nekovee, M. (2000). A ternary lattice Boltzmann model for amphiphilic fluids. *Proceedings of Royal Society of London A*, *456*, 2043–2057.

Chen, S., & Doolen, G. D. (1998). Lattice Boltzmann method for fluid flows. *Annual Reviews of Fluid Mechanics, 30*, 329–364.

Clark, T. (1985). *A handbook of computational chemistry*. New York: John Wiley & Sons.

Cole, J. D. (1968). *Perturbation methods in applied mathematics*. New York: Blaisdell Publishing.

De Vries, A. H., Mark, A. E., & Marrink, S. J. (2004). Molecular dynamics simulation of the spontaneous formation of a small DPPC vesicle in water in atomistic detail. *Journal of the American Chemical Society, 126*, 4488–4489.

Dewar, M. J. S., Zoebisch, E. G., Healy, E. F., & Stewart, J. J. P. (1985). AM1: A new general-purpose quantum-mechanical molecular-model. *Journal of the American Chemical Society, 107* (13), 3902–3909.

Field, M. J., Bash, P. A., & Karplus, M. (1990). A combined quantum mechanical and molecular mechanical potential for molecular dynamics simulations. *Journal of Computational Chemistry, 11*, 700–733.

Friesner, R. A., & Guallar, V. (2005). Quantum mechanics/molecular mechanics (QM/MM) methods for studying enzymatic catalysis. *Annual Review of Physical Chemistry, 56*, 389–427.

Goetz, R., Gompper, G., & Lipowky, R. (1999). Mobility and elasticity of self-assembled membranes. *Physical Review Letters, 82*, 221–224.

Goetz, R., & Lipowsky, R. (1998). Computer simulations of bilayer membranes: Self-assembly and interfacial tension. *Journal of Chemical Physics, 108*, 7397–7409.

Groot, R. D., & Warren, P. B. (1997). Dissipative particle dynamics: Bridging the gap between atomistic and mesoscopic simulation. *Journal of Chemical Physics, 107*, 4423–4436.

Helfrich, W. (1973). Elastic properties of lipid bilayers: Theory and possible experiments. *Z Naturforsch, 28c*, 693–703.

Humphrey, W., Dalke, A., & Schulten, K. (1996). VMD: Visual molecular dynamics. *Journal of Molecular Graphics, 14*, 33–38.

Jiang, Y., Lookman, T., & Saxena, A. (2000). Phase separation and shape deformation on a two-phase membrane. *Physical Review E, 61*, R57–R60.

Kale, L., Skeel, R., Bhandarkar, M., Brunner, R., Gursoy, A., Krawetz, N., et al. (1999). NAMD2: Greater scalability for parallel molecular dynamics. *Journal of Computational Physics, 151*, 283–312.

Knutson, C. E., Benko, G., Rocheleau, T., Mouffouk, F., Maselko, J., Chen, L., et al. (2007). Metabolic photo-fragmentation kinetics for a minimal protocell: Rate limiting factors, efficiency, and implications for evolution. *Journal of Artificial Life*, in press.

Kopelevich, D. I., Panagiotopoulos, A. Z., & Kevrekidis, I. G. (2005b). Coarse-grained computations for a micellar system. *Journal of Chemical Physics, 122*, 044907.

Kopelevich, D. I., Panagiotopoulos, A. Z., & Kevrekidis, I. G. (2005b). Coarse-grained kinetic computations for rare events: Application to micellar formation. *Journal of Chemical Physics, 122*, 044908.

Levine, I. N. (2000). *Quantum chemistry* (5th ed.). Upper Saddle River, NJ: Prentice Hall.

Makeev, A. G., Maroudas, D., Panagiotopoulos, A. E., & Kevrekidis, I. G. (2004). Coarse bifurcation analysis of kinetic Monte Carlo simulations: A lattice-gas model with lateral interactions. *Journal of Chemical Physics, 117*, 8229–8240.

Marrink, S. J., & Mark, A. E. (2003). Molecular dynamics simulation of the formation, structure and dynamics of small phospholipids vesicles. *Journal of the American Chemical Society, 125*, 15233–15242.

Mayer, B., & Rasmussen, S. (2000). Dynamics and simulation of micellar self-reproduction. *International Journal of Modern Physics C, 11*, 809–826.

Miao, L., Seifert, U., Wortis, M., & Döbereiner, H. G. (1994). Budding transitions of fluid-bilayer vesicles: The effect of area-difference elasticity. *Physical Review E, 49*, 5389–5407.

Nekovee, M., Coveney, P. V., Chen, H., & Boghosian, B. M. (2000). A lattice-Boltzmann model for interacting amphiphilic fluids. *Physical Review E, 62*, 8282–8294.

Nilsson, M., Rasmussen, S., Mayer, B., & Whitten, D. (2003). Molecular dynamics (MD) lattice gas: 3-D molecular self-assembly. In C. Griffeath & C. Moore (Eds.), *New constructions in cellular automata* (p. 183). New York: Oxford University Press.

Noguchi, H., & Takasu, M. (2001). Self-assembly of amphiphiles into vesicles: A Brownian dynamics study. *Physical Review E, 64*, 041913.

Parr, R. G., & Yang, W. (1989). *Density functional theory of atoms and molecules*. New York: Oxford University Press.

Pilar, F. L. (2001). *Elementary quantum chemistry* (2nd Ed.). New York: Dover Publications.

Rasmussen, S., Chen, L., Nilsson, M., & Abe, S. (2003). Bridging nonliving and living matter. *Artificial Life, 9*, 269–316.

Setayeshgar, S., Gear, C. W., Othmer, H. G., & Kevrekidis, I. G. (2005). Application of coarse integration to bacterial chemotaxis. *Multiscale Modeling and Simulation, 4*, 307–327.

Shelley, J. C., Shelley, M. Y., Reeder, R. C., Bandyopadhyay, S., Moore, P. B., & Klein, M. L. (2001). Simulations of phospholipids using a coarse grain model. *Journal of Physical Chemistry B, 105*, 9785–9792.

Sherwood, P., de Vries, A. H., Guest, M. F., Schreckenbach, G., Catlow, C. R. A., et al. (2003). QUASI: A general purpose implementation of the QM/MM approach and its application to problems in catalysis. *Journal of Molecular Structure: THEOCHEM, 632*, 1–28.

Sietto, C. I., Graham, M. D., & Kevrekidis, I. G. (2003). Coarse Brownian dynamics for nematic liquid crystals: Bifurcation, projective integration, and control via stochastic simulations. *Journal of Chemical Physics, 118*, 10149–10156.

Stewart, J. J. P. (1989). Optimization of parameters for semiempirical methods: 1. methods & 2. applications. *Journal of Computational Chemistry, 10*, 209–220; 221–264.

Szabo, A., & Ostlund, N. S. (1982). *Modern quantum chemistry: Introduction to advanced electronic structure theory*. New York: Macmillan.

Theodoropoulos, C., Qian, Y., & Kevrekidis, I. G. (2000). "Coarse" stability and bifurcation analysis using time-steppers: A reaction-diffusion example. *Proceedings of the National Academy of Sciences of the United States of America, 97*, 9840–9843.

Voter, A. F. (1997). Hyperdynamics: Accelerated molecular dynamics of infrequent events. *Physical Review Letters, 78*, 3908–3911.

Welch, P., & Muthukumar, M. (1998). Tuning the density profile of dendritic polyelectrolytes. *Macromolecules, 31*, 5892–2897.

Welch, P., & Muthukumar, M. (2000). Dendrimer-polyelectrolyte complexation: A model guest-host system. *Macromolecules, 23*, 6159–6167.

Welch, P., Rasmussen, K. O., Shitanvis, S. M., Lookman, T., & Sewell, T. D. (2005). Direct determination of free energy functionals for phase separating systems. Preprint.

Weronski, P., Jiang, Y., & Rasmussen, S. (2007). Molecular dynamics study of small PNA molecules in lipid-water system. *Biophysical Journal, 92*, 3081–3091.

Yamamoto, S., Maruyama, Y., & Hyodo, S. (2002). Dissipative particle dynamics study of spontaneous vesicle formation of amphiphilic molecules. *Journal of Chemical Physics, 116*, 5842–5849.

Yamamoto, S., & Hyodo, S. (2003). Budding and fusion dynamics of two-component vesicles. *Journal of Chemical Physics, 118*, 7937–7943.

Zhou, X., Jiang, Y., Kramer, K., Rasmussen, S., & Ziock, H. (2006). Hydrodynamics for entropic systems: Space-time compression and pair correlation function approximation. *Physical Review E, 74*, 03571 (R).

Appendix I

Web sites for software packages (all web sites accessed 15 March, 2007):

AMBER	http://amber.scripps.edu
CHARMM	http://www.charmm.org

Chem3D	http://products.cambridgesoft.com
CPMD	http://www.cpmd.org
DL_POLY	http://www.cse.clrc.ac.uk/msi/software/DL_POLY
Gamess	http://www.msg.chem.iastate.edu/gamess/
Gaussian	http://www.gaussian.com
GROMACS	http://www.gromacs.org
LAMMPS	http://lammps.sandia.gov/
MMx	http://europa.chem.uga.edu
MOIL	http://cbsu.tc.cornell.edu
MOPAC	http://www.cachesoftware.com/mopac
NAMD	http://www.ks.uiuc.edu/Research/namd
NWCHEM	http://www.emsl.pnl.gov/docs/nwchem/nwchem.html
Tinker	http://dasher.wustl.edu/tinker
VASP	http://cms.mpi.univie.ac.at/vasp/

20 Core Metabolism as a Self-Organized System

Eric Smith, Harold J. Morowitz, and Shelley D. Copley

20.1 Introduction

Life exists in a nonequilibrium flux of energy flowing both from the sun and from geochemical sources, through the biosphere, to heat that is radiated into space. Here, we develop the view that the basic chemical reaction networks that underlie life have emerged as a form of structure that is a direct result of the prebiotic version of this energy flow. The emergence of these structures during the origin of life is a particular instance of more general thermodynamic phenomena in which sources of unrelaxed free energy can spontaneously induce ordered dynamical states that create channels for their relaxation by means of energy flow through the channels.

In this chapter we first review the origin of life, giving particular attention to transition from ergodicity to contingency. We then consider a key example of dynamical self-organization in reversible thermal systems that serves as a model for the more complicated case of emergent chemical reaction networks. Next, we discuss the mechanism by which ordered chemical reaction networks could have emerged on the primordial Earth, and how those initial ordered structures affected the subsequent emergence of life. Finally, we comment on implications of our view for the problem of artificial protocell design.

20.2 The Continuum Between Contingency and Necessity

Fully cellular life with sophisticated metabolic and genetic machinery appears to have emerged on Earth as little as 0.2 billion years after condensation of the oceans (Fenchel, 2002). We argue that this origin of life was a natural outcome of chemical processes occurring on the early Earth rather than an outcome contingent on a fortuitous set of circumstances. This issue is important in a philosophical sense, as it affects our perception of the place of life within the larger set of understood physical and chemical processes in the universe, and in a practical sense, as it affects the approaches that might be considered for engineering artificial forms of life.

Our knowledge of extant life, in which chance plays an important role and natural selection operates on a set of characteristics that never includes all possible variations, has profoundly influenced thinking about the origin of life. The existence of shared traits in biological organisms is generally a result of descent from a common ancestor, although examples of convergent evolution and horizontal gene transfer are widely recognized. Thus, biological systems today are highly contingent on both previous events and current conditions. This is, indeed, expected for systems with such large configuration spaces that they can never be sampled exhaustively or redundantly. In contrast, physical and chemical systems with large numbers of small and simple components can sample their more limited configuration spaces more thoroughly, and their responses to experimental boundary conditions are essentially deterministic. However, such systems can give rise to ordered, recurring, and persistent structures. Currents, convective storms, and cyclones are examples in oceanic and atmospheric systems.

We posit that the conditions under which life originated are most properly considered at the necessity end of the continuum, between contingency and necessity. The collection of small organic molecules from which biomass is built (which we refer to as the metabolic substrate) is a domain in which most energetically accessible molecular configurations and most kinetically accessible transformations can be sampled over geological timescales on the order of millions of years.

For a macroscopic system consisting of many small molecules, the space of microscopic molecular configurations (the "configuration space") is explored as the system evolves dynamically. Statistical mechanics describes the behavior of macroscopic observables (e.g., temperature, pressure, reaction rates) in terms of the microscopic constituents, and most statistical mechanical descriptions rely on the notion of *ergodicity*, which holds that as a system explores its microscopic configurations, it repeatedly visits configurations and transitions between configurations that are representative in the sense that measurements correspond to averages over the representative configurations and transitions.

Contingency arises when the microscopic dynamics of a system cause it to sample one set of configurations and transitions instead of another, because of a fluctuation or the presence of new microscopic elements or both, breaking the ergodic sampling of configurations. Ergodicity, when it applies, allows the observations of the system to reflect physical and chemical boundary conditions, rather than the results of sampling bias.

Although the development of the underlying network of chemical reactions preceding the emergence of life was likely ergodic, further developments necessary for the emergence of life, and finally life itself, likely became progressively less ergodic and more contingent, in the following sense. The sequence space of modern genes and proteins is so large that it could not have been explored by all organisms that have ever lived, so that the most likely way for an organism to arrive at any particular trait is through inheritance, either directly or through horizontal gene transfer. Thus, the particular form of every organism is in some sense improbable, and all known living organisms have recent common ancestry, compared to the time that would be required to exhaustively sample the sequence spaces for their components. Thus, observations of the system become contingent on particular molecules (both configurations and transitions) that arise, and do not arise from ergodic sampling of configurations.

The proposition that a self-organized protometabolism emerged deterministically as the first step toward life is also a proposition that the dynamics of life are continuous with the dynamics of geochemistry, particularly in the realm of metabolism. If mechanisms existed to create a self-organized protometabolism within the domain of ergodic sampling, the most natural role for replicating macromolecules, once they arose, was to "freeze in" the order in that core, refining but generally not altering its main pathways. Thus, the metabolic core likely served as an early arbiter of usefulness for the macromolecules. Although the basic structure of core metabolic pathways likely derived from primordial chemical reaction networks that emerged by ergodic sampling, the emergence of replicating macromolecules gradually led to a loss of ergodicity. Once an acceptable macromolecular structure was found to serve a given function, further exploration of sequence space became less important. At this stage, contingency became the primary driver of biological evolution. Biological dynamics contingent on improbable events, which requires truly new principles to understand, arose only as complexity accreted around more inevitable properties of small-molecule networks.

20.3 Useful Physical Insights from Dynamical Self-Organization in Reversible Thermal Systems

The characteristics of simple physical systems provide a useful conceptual framework for thinking about the emergence of order in a primordial chemical network. The most familiar paradigms of self-organization in simple physical systems are *dissipative structures* (Glansdorff and Prigogine, 1971) or Turing mechanisms. These mechanisms are useful for understanding morphogenesis and other problems of high-level assembly (Muller and Newman, 2003), but they are poor models for the origin of metabolism. Order forms by these mechanisms only when reaction and diffusion, or some other transport/dissipation pair, occur at certain balanced rates. The need for balance of a transport and a dissipation process in all these mechanisms is reflected by the intrinsic irreversibility built into their fundamental dynamical equations. Metabolism is notably more robust, orderly, and rich than the conventionally recognized dissipative structures noted previously. Dissipation is certainly not excluded from metabolism; the important observation is that in most cases it is not *essential* to the maintenance of metabolic order.

A better model for metabolism is found in the classical concepts of engine and refrigeration cycles, which can be applied in the chemical as well as the more familiar thermal domain. Engines convert energy supplied by some type of thermodynamic reservoir into work. To understand thermodynamics of engines and refrigerators, we will briefly discuss the relevant concepts of thermodynamics, which is properly understood as the application of the theory of statistical inference to problems in physical (including chemical) dynamics (Jaynes, 1983). Thermodynamics applies to systems that may have many degrees of freedom (such as energies and momenta of a large number of individual particles), but in which only a small subset of those degrees of freedom (such as the whole-system energy or particle number) are given values determined by the boundary conditions. The degrees of freedom controlled by the boundary conditions are equivalent to ordinary "mechanical" properties, whereas those left unspecified by the boundary constraints, which often evolve in very complex ways, are treated stochastically. In thermodynamic usage, *work* refers to energy in degrees of freedom that are explicitly controlled, and *heat* to energy in the stochastic degrees of freedom. The Kelvin statement of the second law of thermodynamics is that energy cannot be transferred from uncontrolled to controlled degrees of freedom (converting heat to work) with no other effect. Energy distributed among uncontrolled degrees of freedom in the system is accessible to an external probe only at the cost of increasing entropy and possibly the number of particles in the probe, which increases and must then be expelled to some reservoir if the probe's interaction with the system is to be performed cyclically. Though individually uncontrolled, some statistical properties of the stochastic degrees of freedom do have regular relations to the controlled degrees of freedom, which define the *state variables* of the system and the constraint on their joint changes under transformation, known as the *equation of state*. For instance, the change in entropy (the logarithm of the accessible phase space volume) per change in internal energy is the inverse of the temperature, whereas the change in internal energy per change in the number of particles is the chemical potential. Engines and refrigerators operate by exchanging energy between the uncontrolled and controlled degrees of freedom, under the constraints implied by the associated exchanges of entropy.

Various biochemical engines extract energy from chemical potentials in their environments. That energy is stored either in the form of electrochemical gradients across cell membranes or in high-energy molecules such as ATP. In the sense that such energy is conducted through mechanical motions (tightly controlled degrees of freedom) such as the machinery of the ATP synthase, energy put into or retrieved from these stores qualifies as "work." Refrigerators (of the familiar thermal type) use work to create order by moving heat from a colder reservoir to a hotter one. The concept of refrigeration applies in the chemical domain when molecules or functional groups are moved from reservoirs with lower chemical potentials to reservoirs with higher chemical potentials. Living systems implement chemical refrigeration cycles when they use the work from precisely controlled degrees of freedom to construct or change the concentrations of molecules, for example, by removing monomers from the cytosol to construct polymers or by concentrating useful molecules in the cytoplasm while removing waste products to the exterior.

The coupling of engines to refrigerators is common in engineered systems, where the thermal gradient driving the engine can be coupled, through the transfer of work from the engine to the refrigerator, to the construction of a different thermal gradient in the refrigerated system. In the chemical domain, this coupling of engines to refrigerators results in extraction of energy from chemical (or photochemical) sources and its use to create the nonequilibrium distributions of other chemical species that constitute biomass. In this sense, chemical refrigerators can do something seen infrequently in the thermal domain: Their process of refrigeration can build the molecular machinery of which both the engine and refrigerator are made. We will discuss later a particular thermal system in which the exact analog is seen.

Traveling-wave thermoacoustic engines (see figure 20.1) provide a collection of useful, exactly solvable physical models of engine-driven self-assembly (Atchley, 1994; Atchley, Bass, Hofler, and Lin, 1992; Ceperley, 1979, 1982, 1985; Swift, 1988) that provide an intellectual framework for thinking about the emergence of order in chemical systems. Thermoacoustic engines use thermal gradients in a tube filled with pressurized gas to produce oscillations in pressure (i.e., soundwaves, in the form of either standing waves or traveling waves); these oscillations in pressure can be used to do work, for example, by driving a piston or an alternator to generate electricity. Some thermoacoustic engine cycles (those based on standing waves) require the diffusion of heat through a thermal boundary layer, and its associated time lag relative to compression of the gas, for the formation of coherent sound. These systems are

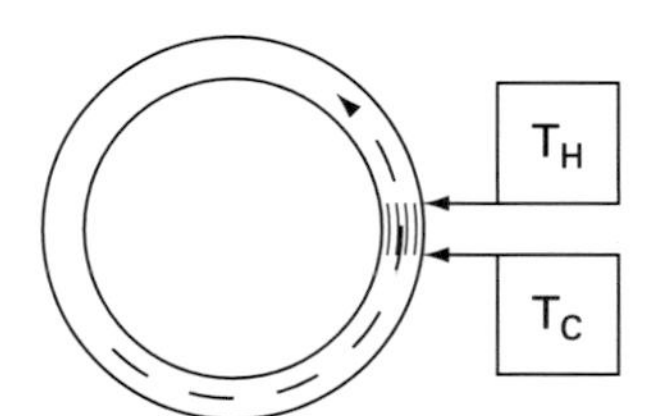

Figure 20.1
Schematic diagram of a reversible thermoacoustic engine. The annulus represents a resonator filled with the pressurized gas that is the working fluid of the engine. Squares represent thermal reservoirs at temperatures T_H and T_C, coupled to either end of a stack of plates in the flowstream of the gas. Dashed arrow represents the traveling soundwave direction that is spontaneously induced when $T_H > T_C$ and the temperature difference exceeds the critical threshold T_{Crit}.

thus intrinsically irreversible, and could therefore be categorized as dissipative structures. The other thermoacoustic engine cycles (those based on traveling waves), of primary interest in this discussion, do not require dissipation for formation of structure, and thus preserve their interesting features if their dynamics are approximated under the idealization of thermodynamic reversibility.

The formation of order in thermoacoustic engines occurs by a form of stimulated emission of soundwaves similar in some ways to optical Dicke superradiance or laser onset. A stochastic soundwave in a resonator coupled to a heat source and cold sink leads to flow of entropy from the source to the sink, accompanied by in-phase amplification of the soundwave, if it is of the correct wavelength and direction. When the temperature difference between the source and sink ($T_H - T_C$) lies below a critical threshold (T_{crit}), the gas remains quiescent. In other words, its distribution of sound is that of an ideal gas in thermal equilibrium. When $T_H - T_C > T_{crit}$, in-phase amplification causes a random fluctuation, of a particular frequency but arbitrary phase, to grow exponentially into a classically identifiable soundwave that accounts for a finite fraction of the energy in the gas. In the idealized limit of a completely reversible system, $T_{crit} = 0$ and the amplitude of the soundwave grows exponentially (see equation 20.1). That the formation of order should begin at any nonzero driving-temperature difference can be directly understood as a consequence of reversibility. If there were a quiescent phase with nonzero driving-temperature difference, the resulting flow would be intrinsically dissipative, which is ruled out by restriction to the reversible limit. With $T_{crit} = 0$, any flow through the system occurs in a phase where it results in work output that amplifies the sound energy in the engine, thus adding to the ordered component.

In real systems, dissipation, though not relevant to the mechanism creating order, does bring an end to the initial exponential growth phase, and the sound in the engine approaches a finite steady-state amplitude.

$$\frac{dA}{dt} = \alpha \frac{T_H - T_C - T_{crit}}{T} A \tag{20.1}$$

Equation (20.1) describes the exponential growth or decay of sound amplitude A with time, when the driving temperature difference is in the neighborhood of the critical value. This equation applies smoothly through the critical threshold for spontaneous onset of sound, and also describes the so-called resonant quality factor for decay of induced perturbations when the driving strength is below the threshold for onset; α is a constant of proportionality that depends on details of engine geometry and composition.

An important property of the transition from disorder to order in thermoacoustic engines is the impedance that the system presents to the reservoirs. When the engine is quiet, its impedance is high, so it is a barrier to heat transport between the reservoirs. When the engine enters the ordered state, its impedance is in linear proportion to the energy in the soundwave (Swift, 1988). Thus, the ordered state forms a channel for energy conduction from the high-temperature reservoir to the low-temperature reservoir.

Thermoacoustic engines are special physical models because they require no external moving machinery to guide the gas (also called the *working fluid*) through the sequence of states that constitute a running engine. All transformations in the thermoacoustic engine associated with either engine or refrigeration cycles are performed on the working fluid by its own coherent motion and thermal conduction. The gas itself receives the energy from those transformations, in the form of amplified coherent motion (soundwaves). Within the gas, different excitations play fundamentally different roles, either as working fluid or as "machinery." Vibrations in a gas at nonzero temperature may be described with a dense spectrum of high-frequency modes, which give it its character as a fluid, as well as a sparser spectrum of low-frequency modes, generally determined by the shape of its container. The latter can be individually manipulated and are therefore more mechanical in nature. In a quiescent gas, all modes are populated incoherently in proportion to the Boltzmann factor for the modal energy at the ambient temperature. No single mode accounts for more than a vanishingly small part of the total energy in the gas. Classical sound represents an ordered state of the gas, in which a single mode accounts for a substantial fraction of the whole system energy. Such a coherent excitation constitutes the machinery of the engine. Since a coherent excitation is a nonequilibrium configuration for the gas, its creation and maintenance depend on the "refrigeration" process performed by the interactions of the soundwave with the environment capable of providing work and conducting away entropy.

For thermoacoustic engines, it is important to recognize the existence of *two* relevant pairs of reservoirs. The obvious pair is the thermal hot/cold pair driving the engine dynamics, which is external to the engine, and for which the detailed structure can be ignored. In biological systems, the counterpart would be a redox couple that can be used to drive biochemical engines. The other reservoir pair in thermoacoustic engines comes from the possible states of order of the gas itself. The nature of the gas as working fluid comes from the large background of high-frequency, thermally populated modes, which could be regarded as a background reservoir of incoherent sound energy. The one or more special modes excited to a macroscopic energy content by the dynamics of the engine are a second reservoir for sound energy. To the extent that self-organization resulting in production of a soundwave happens by transfer of thermal energy from the background of incoherent excitations to the special mode, it is an entropy-reducing transfer of energy from a low-potential reservoir to a high-potential reservoir. (In the ideal of thermodynamic reversibility, such transfer obeys Carnot's theorem, that entropy flux out of one reservoir equals that into the other; hence the self-organizing dynamic is properly described as refrigeration; see Smith, 1998, 1999, 2003, 2005.)

In biological systems, self-organization occurs by removal of organic compounds from a low-potential reservoir (e.g., small molecules in solution) to states in which they have higher chemical potential, for example, polymers such as nucleic acids and proteins. Thus, the ordered states of the gas and their dynamics in thermoacoustic engines are sensible mathematical analogs to the ordered states of chemical species involved in biological engine and refrigeration cycles. Furthermore, in both the thermoacoustic engines and their chemical analogs, sound or chemical order (respectively) is not merely an interesting pattern to observe, but a kind of machinery that performs work.

As mentioned previously, biological systems have the special property that their refrigeration cycle builds the machinery of the engine that drives the refrigeration cycle. Thermoacoustic engines also have this property. As shown in equation (20.1), the rate of growth of the amplitude of the soundwave is proportional to the soundwave's magnitude, yielding exponential growth in the absence of dissipation. The emergence of order in a prebiotic reaction network might be expected to follow similar dynamics when the network topology allows production of molecules in a cycle at a rate that depends directly on the concentrations of cycle intermediates. A case in point—the reductive tricarboxylic acid (TCA) cycle (rTCA)—is discussed in detail below.

The self-organizing thermoacoustic engines demonstrate two important properties that may be useful in understanding the emergence of metabolism. The first is that stress on a thermodynamic system from its boundary conditions can make an ordered state with rich dynamics statistically favored over the equilibrium state. In the context of life, sources of free energy available on the early Earth may have required the emergence of nonequilibrium ensembles (of which life is one) to serve as channels to relax the free energy stress. To the extent that this causation applies, the biotic state on Earth can be regarded as a result of its ability to provide channels for the relaxation of chemical and photochemical energy not provided by weather or any other abiotic process.

The second point that holds, at least for these models, is that the ordered state spontaneously forms because there is a second-order phase transition, away from the equilibrium state. In other words, the ordered phase is reached by a continuous deformation of the statistical distribution from that of the disordered phase, without an intervening kinetic barrier. Thus, the emergence of order is rapid. Furthermore, it does not require a rare or improbable event.

We suggest that the emergence of order in the form of a primordial chemical reaction network may have important parallels to the emergence of order in thermoacoustic engines. Thus, the ordered chemical reaction network that provided the metabolic underpinnings of life was likely a natural consequence of the presence of free energy stresses, and contingent on conditions only insofar as the environment provided particular types of free energy sources and catalysts capable of promoting certain pathways for relief of those free energy stresses. Further, the emergence of order may have occurred rapidly and spontaneously without a need to cross a high kinetic barrier.

20.4 An Ordered Biotic State Is a Dynamic Attractor on Earth

Life is an incredibly robust feature in the geosphere. Severe shocks, both exogenous (e.g., the heavy bombardment by meteorites early in Earth's history) and endogenous (e.g., habitat destruction and anthropogenic pollution), have led to numerous mass extinctions. Yet life regenerates rapidly and robustly. Although the suite of organisms that recolonizes an empty ecological niche may be different from its predecessor, the core of biochemical processes constituting the net autotrophic metabolism of the ecosystem is preserved. The rapid, avalanche-like reorganization of ecosystems after perturbations suggests that the underlying metabolic network is more robust than local optima of cooperating metabolic specialists.

It thus seems that an Earth with some fraction of biomass with our core biochemistry is a dynamical attractor, to which the geosphere returns after perturbations. This observation suggests that an Earth with a living state is more stable than an abiotic Earth. The perturbations need not eliminate all life to attest to this stability. The observation that recolonization regenerates biomass in "emptied niches," while death does not appear ever to have propagated out from an extinction event to end all life on Earth, identifies the biotic state as locally stable. We conclude that an Earth filled with life is not unstable with respect to collapse to an entirely abiotic state. Rather, abiotic Earth was once a metastable state, and the emergence of life represents a transition to a state of greater stability. Stability of living ecologies does not by itself imply that the transition from an abiotic to a biotic state was facile, and a rapid emergence of life remains logically consistent with improbable crossing of a large barrier, though it is uninformative about the barrier if that was the actual case. However, it is possible that the barrier was not high, and that life emerged rapidly because the early levels of metabolic order formed essentially deterministically in response to geochemical conditions. The challenge in describing this transition is to relate such large-scale statistical arguments to the particular chemistry of life and the transitional stages through which it was reached.

20.5 The Emergence of Order in a Primordial Chemical Reaction Network

20.5.1 The Origin of the Organic Material Required for the Emergence of Life

Identification of the source of the organic material from which metabolism emerged is critical for consideration of the emergence of an ordered chemical reaction network because it defines both the types of molecules available to participate in the network and the conditions under which these molecules can interact.

Enormous quantities of organic material were delivered to the early Earth by comets and meteors. Immediately after the primary collisional period (about 4 billion years ago), approximately 10^{12} kg of organic material were delivered over a period of a million years (Pizzarello, 2004). However, this material may have been unimportant as the source of the building blocks of early macromolecules. It was delivered over a very long period of time into the bulk ocean, where it would have been diluted to tiny concentrations. Furthermore, the suite of molecules found in extraplanetary materials overlaps, but does not coincide with the molecules used by living systems. For example, about 55 amino acids detected in carbonaceous chondrites are not found in living organisms, while the biotic amino acids phenylalanine, lysine, histidine, and arginine are not found in carbonaceous chondrites (Engel, Andrus, and Macko, 2005). It is not obvious why only certain components of the mixture of chemicals delivered from this source would have been incorporated into living systems.

Considerable attention has been given to the possibility of atmospheric processes as the source of amino acids and purines. The demonstration that amino acids can be produced by electrical discharge in a simulated atmosphere consisting of CH_4, NH_3, H_2, and H_2O was particularly influential (Miller, 1953). However, it is now recognized that the atmosphere of the early Earth was more oxidized, consisting primarily of CO_2 and N_2 (Fenchel, 2002), and amino acids do not form readily under such conditions. Furthermore, the levels of reactive compounds such as HCN that could be achieved by atmospheric processes are not high enough to allow synthesis of biomolecules. For example, polymerization of HCN has been reported to yield amino acids, purines, and pyrimidines (Ferris, Joshi, and Lawless, 1977; Oro and Kamat, 1961; Oro and Kimball, 1961). However, such reactions require concentrations of HCN greater than 0.01 M and pH values of 9.2 for condensation reactions to compete favorably with hydrolysis of HCN to formamide and formic acid (Ferris et al., 1978). Based on estimates of the rate of atmospheric formation of HCN, Stribling and Miller (1987) estimate that the steady-state concentration of HCN in the primitive ocean was only in the μM range.

A more attractive source of organic molecules is geochemical processes occurring in environments such as hydrothermal vents in which organic molecules can be sequestered and concentrated. Hydrothermal vents occur at or near midoceanic spreading centers. Seawater that is convected through the seafloor is superheated and loaded with minerals dissolved from hot rock layers overlying magma chambers or hot crystalline rocks. The composition of the hydrothermal fluid depends on the temperature and composition of the rocks through which the fluid percolates before being vented back to the seafloor. Typical fluids are highly enriched in H_2, H_2S, Mn,

Fe, Cu, Zn, Co, CO_2, and CH_4 (Kelley, Baross, and Delaney, 2002). (The compositions of these fluids undoubtedly include contributions from a robust sub-seafloor microbial community, and thus are not necessarily representative of prebiotic hydrothermal fluids.) As the hydrothermal fluids are extruded and mix with cold seawater, various minerals precipitate and form chimney structures. Some vents are composed of carbonates (e.g., the Lost City field), whereas others are composed primarily of sulfide minerals (e.g., the Trans-Atlantic Geotraverse, the East Pacific Rise, and Endeavor fields) The iron-sulfide vents are particularly interesting because of the catalytic abilities of iron sulfide minerals. The walls of the vent structures are porous, and a substantial amount of hydrothermal fluid percolates through the porous walls. Thus, these structures might have served as primordial chemical reactors, fed by a stream of water containing small molecules such H_2, sulfide, and CO_2. The porous chambers could have provided both compartmentalization, allowing accumulation of organic material, and catalysis conferred by the reactive iron sulfide wall. Furthermore, the fluid cools as it traverses the walls. Thus, certain molecules such as pyruvate might be formed at high temperatures, and more fragile molecules such as sugars, nucleotides, and peptides might be formed in the cooler regions near the exterior of the structure. As mentioned earlier, the temperature and composition of the vent fluids varies considerably among sites. Consequently, different types of reaction networks might have developed at different sites. Several authors have written about the possibility of prebiotic chemistry in vent structures (Baross and Hoffman, 1985; Corliss, 1990; Corliss et al., 1979; Koonin and Martin, 2005; Martin and Russell, 2003; Russell and Hall, 1997). We find this hypothesis compelling based on the ability of vent structures to compartmentalize organic molecules within a structure lined with catalytic sites and fed by a steady stream of inorganic precursors.

A further consideration is the complexity of the organic material that provided the initial building blocks for macromolecules and later, true cellular organisms. Many attempts to generate simple biomolecules have resulted in complex mixtures. The Miller-Urey experiment used gas-phase free-radical reactions to produce a complex array of organic products, in which biomolecules were minor constituents. The formose reaction, which produces sugars from condensation of formaldehyde and glycolaldehyde in alkaline solutions containing calcium hydroxide, famously produces a complex mixture, of which ribose, the most relevant sugar, represents less than 1% of the product (Ricardo et al., 2004). The very complexity of the products formed in such reactions may suggest that they are not the means by which the building blocks of life were synthesized. A major challenge is to define the mechanisms by which relatively sparse networks in which biologically relevant molecules dominate could have emerged. Free-radical reactions were unlikely to be important because the high reactivity of free radicals leads inevitably to complex mixtures. Formation of new bonds by processes involving attack of nucleophiles on electrophiles restricts the positions at which reactions can occur, and thus simplifies the spectrum of products. Further, mineral surfaces or dissolved metal ions may play an important role in either controlling reactivity or stabilizing certain products, or both. For example, borate minerals stabilize ribose because the 1,2-diol coordinates nicely to borate. Addition of borate minerals to the formose reaction results in high yields of the pentoses arabinose, lyxose, xylose, and ribose (Ricardo et al., 2004).

20.5.2 The Importance of Small-Molecule Catalysis in Primordial Chemical Reaction Networks

Metabolic reactions allow relaxation of environmental stresses, such as redox couples, only when a kinetically accessible reaction pathway is coupled to the source of energy. Thus, the availability of catalysts for particular reactions will be a critical factor in determining what pathways are used, and thus what reaction intermediates are formed. On the primordial Earth, the earliest catalysts were undoubtedly mineral surfaces (Cairns-Smith, Hall, and Russell, 1992; Wächtershäuser, 1990, 1992). Mineral surfaces can catalyze reactions by concentrating and orienting reactants on a surface and increasing the reactivity of molecules because of interactions with specific atoms at the surface. For example, divalent metal cations can polarize carbonyl groups or stabilize negative charges on leaving groups in substitution or elimination reactions. The iron-sulfide and chalcopyrite mineral surfaces formed at hydrothermal vents promote carbonylation reactions that allow synthesis of long-chain hydrocarbons (Cody, 2004; Cody et al., 2000). Certain types of surfaces might also have promoted redox reactions. NiMo and CoMo sulfides deposited on ceramics like aluminum oxide are used for reducing alkenes and aromatics during petrochemical refining (Gosselink, 1998). Plausible mechanisms by which transition metal catalysts might promote reduction of carbonyls and alkenes using reducing equivalents provided by sulfides are discussed further later.

An important step in the emergence of more effective catalysts would have been the incorporation of amino groups into organic molecules, because amino groups can catalyze proton transfer reactions and act as nucleophiles to generate intermediates that are more reactive than the initial reactant. Such small molecules could have begun to catalyze certain reactions in the chemical reaction network, leading to predominance of certain pathways and products. Although small-molecule catalysts are likely to be rather nonspecific and to provide only modest rate enhancements, even modest rate enhancements in the context of a network of chemical reactions would be expected to result, over time, in the predominance of certain pathways and components. Examples of small-molecule catalysis include the catalysis of aldol and Michael reactions by proline (List, 2002) and the catalysis of the hydrolysis of DNA, proteins, and esters by the dipeptide Ser-His (Li et al., 2000). The cofactor pyridoxal phosphate is particularly impressive; it catalyzes the decarboxylation of amino acids by 10 orders of magnitude (Zabinski and Toney, 2001).

Small-molecule catalysis provides the potential for nonlinear positive feedback. For example, we have proposed that dinucleotides might have catalyzed the synthesis of amino acids (Copley, Smith, and Morowitz, 2005). Since some amino acids (aspartate and glycine) are precursors for the formation of the pyrimidine and purine bases, such catalysis allows a feedback loop in which catalysis favors synthesis of more molecules of the catalyst. Feedback is the essential kinetic feature capable of elevating a few pathways to dominance within the large graph of small-molecule reactions, and thus producing order in both protometabolism and its modern descendent. Simulations carried out by Segré and Lancet have shown that mutual catalysis within a network subject to dilution can lead to predominance of certain products even when all the possible products have equivalent stabilities (Segré, Ben-Eli, and Lancet, 2000; Segré and Lancet, 1999; Segré, et al., 1998; Segré, Pilpel, and Lancet, 1998; Segré et al., 2001).

A stage at which catalysis was provided by small molecules such as peptides, oligonucleotides, and cofactors is a plausible intermediate between an initial stage of catalysis provided by mineral surfaces and the more powerful catalysis provided by macromolecules such as RNA and proteins. Remnants of these early catalysts are likely preserved in the many extant enzymes that use metal ions and cofactors as integral parts of their catalytic machinery. Indeed, 52% of enzymes use organic cofactors (White, 1976) and 30% use metal ions or clusters (Voet, Voet, and Pratt, 2006).

20.5.3 The Importance of Network Topology in Primordial Chemical Reaction Networks: The Reductive TCA as an Engine of Organosynthesis

An alternative type of catalysis can be provided by cycles with a topology that allows a cycle of reactions to produce a product at a higher rate than alternative direct syntheses, and in such a way that the cycle is self-sustaining or self-enhancing. A network can be *catalytic* when a molecule within a cyclic pathway acts as a "seed" to enable production of a product through a series of reactions that regenerates the seed molecule at the completion of the cycle. It is the cycle, rather than a particular molecule, that provides the catalytic effect. Network catalysis can be elevated to network *autocatalysis* when the concentration of the seed molecule is not only maintained, but doubled on completion of the cycle, so that even under perturbations or parasitic losses, the cycle can be self-sustaining. The simplest biochemical cycle that is autocatalytic in this sense is the reductive tricarboxylic acid (the reductive TCA or rTCA) cycle (see figure 20.2). (The TCA cycle in the oxidative direction is also known as the *citric acid cycle* or the *Krebs cycle*). Notably, this cycle has many similarities to the self-organizing engine system mentioned earlier. The seed molecule is counterpart to the soundwave in the engine, and it couples to the inorganic carbon source CO_2 and enables the reduction of carbon (an energy-yielding process), much as the soundwave couples to the reservoirs and enables transport of energy from hot to cold. The branching topology of the rTCA reaction graph (shown in detail in

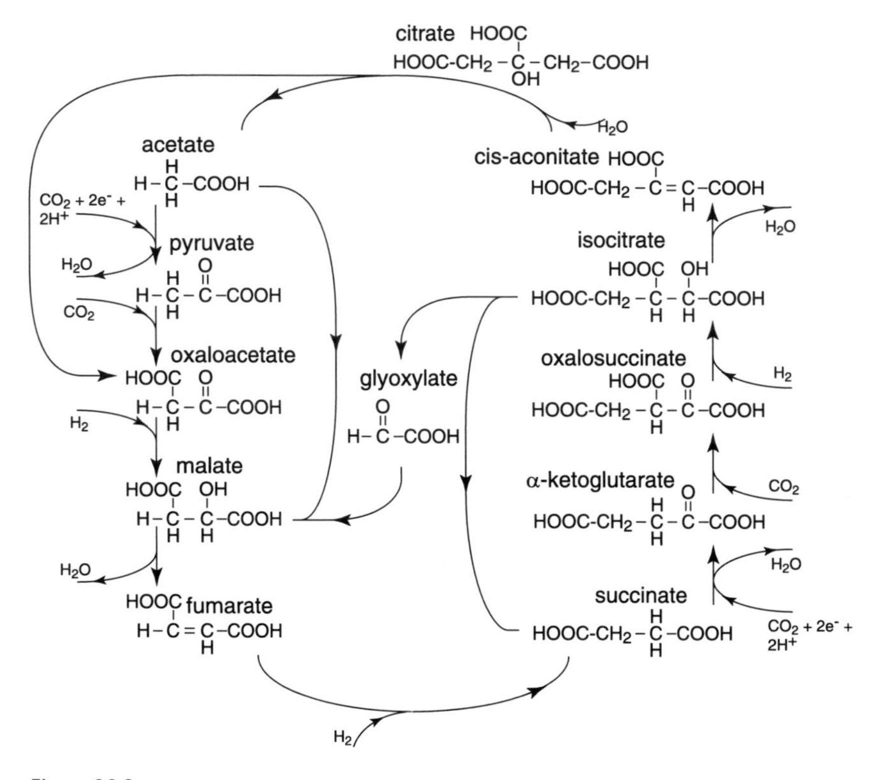

Figure 20.2
The reactions and primary reagents in the reductive tricarboxylic acid (rTCA) cycle, and the closely related reactions that encompass the glyoxylate bypass. All arrows in the outer ring indicate the reaction directions in the rTCA cycle. If the glyoxylate bypass is used, the arrows from succinate to pyruvate are reversed. The glyoxylate bypass is thus an oxido-reductive pathway, reflecting the range of oxidation states in its carbon source acetate.

figure 20.2) allows production of two seed molecules from one as part of the process of CO_2 reduction, much as in-phase amplification of soundwaves increases the transport capacity of the engine in the course of extracting energy from the reservoirs.

In autocatalysis by a biosynthetic network, we see the simplest form of reproduction. When multiple such networks couple to common, limited sources of atomic species and energy, differences in autocatalytic efficiency result in competitive exclusion. Darwinian reproduction and selection are not so much novel concepts, as recapitulations of these simple chemical processes in the more complex context in which heritable variations *not* contained in the basic chemical substrate become possible.

The TCA cycle appears to be a fundamental core of metabolism for living systems because it provides precursors for all major categories of biomolecules. Acetate initiates fatty acid synthesis, pyruvate initiates sugar synthesis, and succinate initiates synthesis of pyrroles in cofactors such as heme and chlorophyll. Pyruvate, oxaloacetate, and α-ketoglutarate are precursors of the simplest amino acids. Aspartate (synthesized from oxaloacetate) and glycine (which can be made from glyoxylate obtained by cleavage of malate or isocitrate) are the precursors of pyrimidines and purines, respectively (Metzler, 2003). Most extant organisms use the TCA cycle in the oxidative direction to oxidize pyruvate to carbon dioxide. The reverse sequence of the same reactions, called the reductive TCA (rTCA) cycle, is used by organisms in reducing environments for de novo organosynthesis (Lengeler, Drews, and Schlegel, 1999).

Carbon flows through the rTCA reactions from CO_2 to the more reduced cycle intermediates or biomolecules and wastes synthesized from them. This carbon flow energetically resembles the flow through the thermoacoustic soundwave from hot to cold reservoirs. In reductive ecosystems, the rTCA cycle forms a biotic channel for carbon reduction, effectively lowering the impedance to carbon flow relative to a network with only abiotic pathways. Through syntrophy, the rTCA cycle and a few alternative pathways (e.g., the Wood-Ljungdahl pathway) account for essentially all carbon flow through biological channels in reducing ecologies, much as the soundwave accounts for a large fraction of energy in the ordered state of the engine, and all of its supradiffusive flow between the reservoirs. In this sense, carbon flow through the rTCA cycle and the Wood-Ljungdahl pathway defines an ordered state of the geochemical network, with alternative channels populated only at the level of near-equilibrium backgrounds.

A more complex framework is required to understand ecosystems based on oxidation, probably all of which are ultimately connected to photosynthesis on the modern earth. Whereas anabolism remains essentially reductive, when the energy to produce reductant is included in the synthetic reactions, they are endergonic, thus requiring the linkage of energy-capture mechanisms to anabolism. In this domain the Calvin-Benson cycle is the main channel for primary organosynthesis from CO_2 and photosynthetically generated reductant. The complexity of photosynthetic metabolism suggests, however, that it arose within the context of well-developed complex (reducing) cellular life, and perhaps that it could have arisen *only* from such a background.

We have argued elsewhere (Smith and Morowitz, 2004) that a chemical pathway is eligible to form a "winner-take-all" flux channel for relaxation of a chemical stress (e.g., by carbon reduction), if it has three properties, all of which are exhibited by the rTCA cycle:

- Stability: The chemical species in the channel should be reasonably stable;
- Facility: The reactions within the cycle should be facile relative to alternative reactions that compete for the same inputs;

• Positive feedback: The reactions in the cycle should be subject to nonlinear positive feedback, so that the rate of increase of the concentrations of cycle intermediates is enhanced by increases in the flux around the cycle.

Regarding the first property, stability, the intermediates in the rTCA cycle are all reasonably stable, with the exceptions of oxaloacetate and oxalosuccinate. These β-keto acids undergo facile decarboxylation. Decarboxylation of oxaloacetate produces pyruvate, the precursor of alanine and sugars. Although this reaction will deplete the concentration of oxaloacetate, at equilibrium or in a driven steady state, some nonzero concentration of oxaloacetate will be available for reduction to malate. A comparable situation exists for oxalosuccinate. Decarboxylation of oxalosuccinate simply produces the preceding intermediate in the cycle (a-ketoglutarate). Again, a nonzero concentration of oxalosuccinate should be available for conversion to isocitrate. In both cases, the position of the equilibrium will be determined by the pH and the concentration of CO_2.

Regarding the second property, facility, the rTCA pathway involves four types of chemical reactions that depend on inputs of CO_2, a reductant, or both: (1) the carbonylation reactions that form pyruvate from acetyl CoA and a-ketoglutarate from succinyl CoA; (2) the carboxylation of the a-keto acids pyruvate and a-ketoglutarate to form oxaloacetate and oxalosuccinate, respectively; (3) the reduction of the carbonyl groups of oxaloacetate and oxalosuccinate; and (4) the reduction of the double bond of fumarate to form succinate. The feasibility of the carbonylation reactions has been demonstrated by Cody and coworkers (2000), who observed synthesis of pyruvate under simulated hydrothermal conditions, apparently by carbonylation of acetate. The rate of carboxylation of a carbon alpha to a carbonyl is enormously enhanced compared to that of a typical methyl group because of the stabilization of the carbanion by delocalizing the negative charge into the adjacent carbonyl (see figure 20.3), leading to a higher concentration of the carbanionic intermediate that reacts with CO_2. Thus, the a-keto acids in the cycle should compete well with other types of molecules for the available CO_2.

The feasibility of the remaining reduction reactions depends on the types and the concentrations of reductants and the types of catalysts that are available. In extant enzymes, the carbonyl groups of oxaloacetate and oxalosuccinate are reduced by hydride transfer from NADH, and the double bond of fumarate is reduced by hydride transfer from a reduced flavin cofactor. Before these cofactors were available, these reactions might have been mediated by sulfide or alkyl thiols (Heinen and Lauwers, 1996) in combination with a surface containing metal ions that acted as Lewis acid catalysts (see figures 20.4 and 20.5). Alternatively, transition metal sulfides might have transferred electrons derived from H_2 or sulfide to the carbonyl or alkene acceptors in these molecules. Experimental investigations to assess the feasibility of these reactions are being initiated.

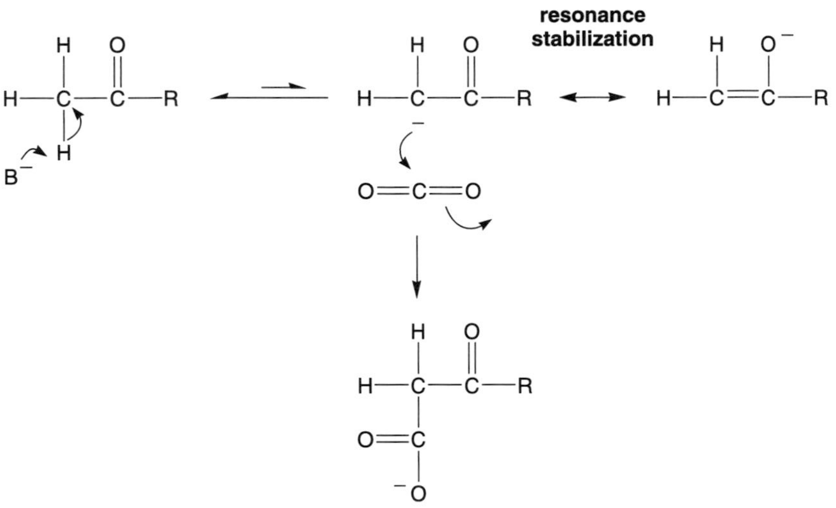

Figure 20.3
Resonance stabilization of a carbanion facilitates carboxylation reactions by increasing the concentration of the reactive carbanion.

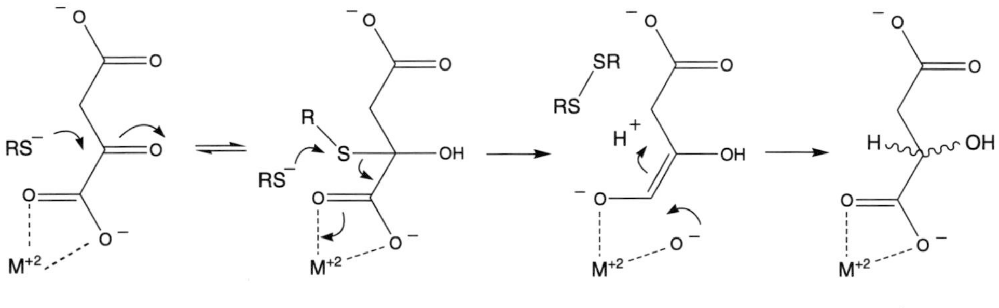

Figure 20.4
Possible mechanism for reduction of oxaloacetate mediated by sulfide or alkyl sulfides. Coordination to a metal surface would accelerate the reaction by stabilizing the aci-anion intermediates (in which a carbon bears two negatively charged oxygens). The wavy line indicates undefined stereochemistry. Coordination of oxaloacetate to a surface would likely favor formation of one stereoisomer.

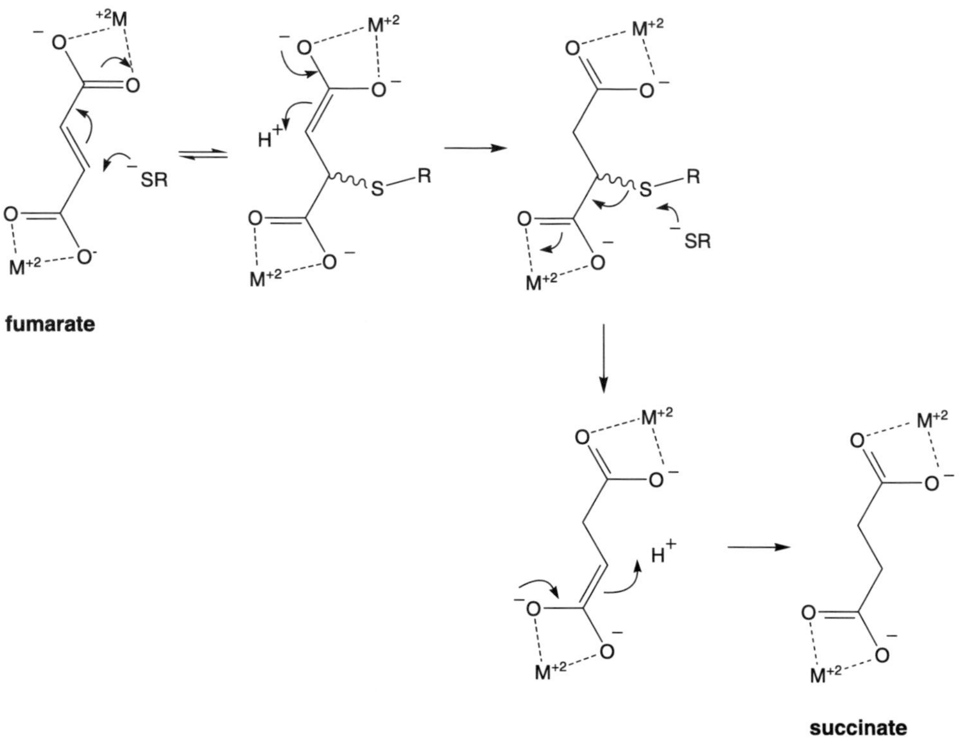

fumarate

succinate

Figure 20.5
Possible mechanism for reduction of fumarate mediated by sulfide or alkyl sulfides. Coordination to a metal surface would accelerate the reaction by stabilizing the aci-anion intermediates.

Regarding the third property, positive feedback, recall that such feedback characterizes self-constructing engines, whose rate of work (self-building) is proportional to the capacity of the engine (the amount already built) at any moment (Smith, 1998, 1999). In the rTCA cycle, this feedback is accomplished by cleavage of citrate to acetate and oxaloacetate, in combination with the pathway for conversion of acetate into a second molecule of oxaloacetate. This combination of reactions elevates the network-catalytic flow of oxaloacetate into network-autocatalytic flow. Thus, each turn of the cycle doubles the concentrations of the cycle intermediates in the absence of parasitic side reactions. However, the cycle can also continue to operate while generating precursors for various synthetic pathways as long as intermediates are drained off at a rate that does not deplete the cycle intermediates.

On the basis of these three properties, the rTCA cycle, excited as a bulk process above the threshold for network autocatalysis, may have been a key departure from inorganic geochemistry toward the emergence of life. Like the ordered state of the thermoacoustic engine, we expect that flow of a finite fraction of all available carbon through rTCA was an attractor in the network of chemical reactions, continuously accessible from the equilibrium background and favored when the chemical potential of CO_2 and H_2 exceeded a critical threshold. The accessibility of the rTCA cycle for flow of carbon will depend on the availability of more chemical species than CO_2 and reductant. Thioesters and phosphate esters (see DeDuve, 1991, 2006) play key roles in at least two steps in the modern cycle (the carbonylation of acetate and succinate), and their availability in abiotic environments may have been an important constraint on the possibility that an ordered rTCA cycle could emerge.

Other pathways for chemoautotrophic organosynthesis exist, such as the reductive acetyl CoA pathway and the glyoxylate cycle. The former uses a complex mechanism to reduce one carbon to a methyl group before combining it with a carbonyl to make acetate. To serve as a foundation for autotrophic carbon synthesis it must be followed by so-called anaplerotic reactions, which produce intermediates in the TCA cycle as biosynthetic precursors (Lengeler, Drews, and Schlegel, 1999, p. 242). The glyoxylate cycle is an acetate-incorporating pathway that includes several of the reactions in the TCA cycle (Lengeler, Drews, and Schlegel, 1999, p. 173). It bypasses the steps leading to a-ketoglutarate, however, thus requiring anaplerotic reactions starting from phosphoenolpyruvate to produce a complete set of biosynthetic precursors. These pathways lack the appealing network-autocatalytic features of the rTCA cycle. However, they might have emerged under circumstances in which the rTCA cycle was relatively inefficient because of a limiting supply of a critical catalyst or reactant.

20.6 Implications for the Emergence of Modern Metabolic Pathways

Oparin and Haldane proposed that emergent life depended on its environment for organic inputs (Haldane, 1967; Oparin, 1967). This conjecture has been contrasted to one in which the earliest life forms synthesized all their organic inputs from inorganic precursors. Along the lines of this dichotomy, metabolism-first scenarios have been characterized as heterotrophic versus autotrophic origin theories (Fry, 2000). In a way, the expectation that origin stories would fall within either of these essentially *ecological* categorizations of modern life is overly simplistic. It is unlikely that a complex organic soup capable of providing all the needs of a heterotrophic microbe existed on the early Earth. However, it is difficult to explain how a completely autotrophic microbe could have emerged, given that it would have had to use macromolecules made of various types of building blocks to synthesize those building blocks—this is the classic chicken-and-egg problem, and its very insolubility suggests that something is missing in this scenario. We propose a metabolism-first view that is better characterized as one in which particular geochemical pathways became ordered before life as an integrated system could form, but were later subsumed and coordinated within that system. Cairns-Smith nicely articulated a similar view of the emergence of evolutionary innovations: "[T]he first organisms would have been made in a way that was particularly accessible, especially in keeping with a current local geochemistry. Then later on initial subsystems would have been replaced by more efficient ones. This is how evolutionary and other forms of technological innovation commonly proceed—starting with what is easiest, ending with what is most efficient" (1992, p. 163).

If the order of life can be understood as an outgrowth of the order of geochemical processes, the small-molecule metabolic substrate would be expected to be the point of connection between prebiotic chemistry and modern metabolism. Examining the metabolic pathways of extant organisms reveals the existence of a common core of metabolic processes consisting of the TCA cycle and synthetic pathways for construction of simple amino acids, ribose, purines, and pyrimidines. These would have been the basic requirements for a primitive organism that needed to synthesize the building blocks of macromolecules from small organic compounds available in the environment. Although many organisms have shed pieces of the metabolic core as a consequence of adaptation to ecological niches in which certain metabolic abilities are dispensable, and there are some differences in a few biosynthetic pathways, the conservation of this metabolic core is striking and consistent with the origin of all known extant life from the LUCA (last universal common ancestor), leading us to refer to the metabolic core as *universal*. Remarkably, this universal metabolic core is made up of fewer than 450 small metabolites (molecules weighing less than 400 Daltons), from which all biomass is synthesized (Metzler, 2003). This universality of core metabolism, together with the possibility that it originated in an ergodic chemical network, suggests that many core metabolic pathways may be predictable from a knowledge of the geochemical conditions on the early Earth (Fenchel, 2002).

We propose that transition metal and small-molecule catalysis resulted in an ordered protometabolic network within the porous walls of hydrothermal vents in regions where organic material was concentrated, in the vicinity of catalytic transition metal surfaces. When RNA, and later proteins, emerged from the monomers provided by protometabolism, they gradually took over catalytic functions from the small molecules, often preserving them as coenzymes. Since improved catalysts were probably "discovered" one at a time, it is unlikely that novel multistep synthetic pathways would have been invented de novo. "Discovery" of a catalyst for, say, the third step in a novel pathway would be of no use if its substrate could not be made and its product further utilized. It is more likely that ever better catalysts were incorporated into the preexisting network structure, thereby preserving the underlying structure of the protometabolic network. Thus, "information" flowed upward from the small-molecule core onto networks of catalysts and genes. This view contrasts with that presented by the RNA world theory (Gesteland, 1999), which supposes that RNA sequences were first selected for their ability to replicate, and only later for their abilities to synthesize the metabolic substrate autonomously. In such a case, the structure of modern metabolic networks would have little or no relation to a primordial reaction network, but would have been determined solely by chance according to the types of catalytic RNA molecules that emerged.

The issue of whether extant metabolic pathways are continuous with prebiotic pathways has been debated for decades, and has been variously supported (Cleaves and Miller, 2001; Lipmann, 1965; Yamagata et al., 1990) or criticized (Lazcano and Miller, 1999; Orgel, 2003) by many authors. We believe that continuity between geochemical and biochemical processes is logical for synthesis of core biomolecules such as the simple amino acids, purines, pyrimidines, and sugars. This point of view provides a conceptual framework for investigations of protometabolism. Efforts to understand prebiotic chemical processes should be guided by knowledge of geochemistry. Little is gained by showing that biologically relevant molecules can be formed under geochemically irrelevant conditions. Furthermore, substantial hints about the nature of abiotic protometabolic processes may be preserved in extant metabolic pathways.

20.7 Lessons from Extant Life for the Design of Artificial Protocells

Our observations about metabolism suggest that it is stable in the geosphere because it is a very good (and perhaps very accessible) solution to the problem of creating relaxation channels for certain stresses generated by geochemistry. Though we have focused on reducing metabolism here, similar arguments should apply to photosynthesis and the stress created by sunlight relative to the thermal microwave background, and the oxidizing metabolism it made possible on a large scale. An important question is whether multiple good and accessible solutions exist. It is difficult at this stage, when we are just beginning to glimpse how life as we know it might have emerged out of a primordial chemical reaction network, to generalize to situations involving different kinds of stresses, feedstocks, reductants, catalysts, and perhaps even solvents. Thus, it is difficult to predict which types of metabolisms might be suited for artificial protocells that differ substantially from the familiar metabolic networks found in extant life. We can, however, suggest some principles based on our understanding of the emergence of order in physical systems that can inform the design of protocells.

Artificial cells will function stochastically. A Newtonian design philosophy, which applies down to micron-scale robotics, will necessarily give way to statistical methods of operation and possibly of assembly in nanorobotics and the design of protocells. Machine components will become single molecules, and if artificial protocells are to interact with (or become equivalent to) natural cells, many of these molecular components will have low mass. As a result their interactions, and possibly which molecular structures arise, will be stochastically determined.

Robust protocells must use catalysts to accelerate the rates of useful reactions above those of competing reactions. The difficulty of evading disordering mechanisms, especially in complex living structures where most local modifications to a functional form yield nonfunctional forms, acutely worried Schrödinger (1992), and has led to the subfield of physics devoted to uncovering principles of *self-organization* (Glansdorff and Prigogine, 1971; Haken, 1983). A factor in the success of living cells is their ability to separate the timescales of catalyzed and uncatalyzed reactions. Separation of timescales permits useful reactions to occur repeatedly and selectively over the lifetime of the catalyst, rather than infrequently among many competing reactions in an ergodically sampled reaction network. For macromolecular catalysts, however, the problem of selecting the catalysts could in principle be more demanding, and require cells to maintain more information, than the original problem of culling a cloud of reactions to a sparse network of useful ones. Statistical physics has no general "closure theorems" explaining how the information specifying a particular reaction network can be protected from decay to disorder by processes that occur entirely within that network, especially when the components that must be specified span multiple timescales and levels of complexity, and depend on one another. Empirically, the long lifetimes of catalysts, and the DNA through which their forms are inherited, seem to be required for the complexity of modern cells, but it would be an error to understand the mere existence of long lifetimes as making cell processes deterministic, and thus overcoming the concerns raised by intuitions gained from statistical physics. Schrödinger called the lack of a theory to explain how apparently mechanical, deterministic behaviors can emerge as a characteristic of small molecular systems the problem of explaining "order from order."

Peter Gács's (2001) error-correcting one-dimensional noisy cellular automaton provides an important illustration of how separation of timescales can make error correction possible in a self-contained driven system. Though it is not a model of a living system, Gacs's construction is an excellent example of how much a separation of timescale can cause a driven stochastic process to differ qualitatively from one in static equilibrium, but also how difficult self-contained error correction can be to achieve.

For dissipative structures, maintenance of a condition of macroscopic order is recognizably a consequence of statistical interactions, usually among small, simple components interacting as peers without direct control from higher-level structures in a hierarchy (Glansdorff and Prigogine, 1971). In cells, the statistical character of maintenance of order among low-level components seems to be removed by the stability and persistence of catalysts, whereas ensuring the fidelity of the inventory of catalysts seems more naturally described as a process of *self-repair*. We emphasize that this change in description does not actually change the essential statistical nature of the process of maintaining the ordered state, but if we use the concept of self-repair to refer to systems with hierarchical division, in which higher-level components must be maintained by the coordinated action of collections of lower-level processes, it follows that *robust protocells must be capable of self-repair.* The wide array of mechanisms for repair of damaged DNA, chaperoning of protein folding, repair or degradation of macromolecules damaged by free radicals, and pumping of ions and other species to maintain the internal cell state, attests that cells must correct constantly for random events, and emphasizes the essential similarity of repair as a statistical process to conventionally recognized statistical mechanisms for maintenance of ordered states in nonhierarchical systems.

Robust protocells must be self-starting, self-organizing engines. It is natural in chemistry to ignore the concept of inertia. All the information about how a chemical reactor or organism responds to its environment is contained in a static description of its chemical state and physical configuration. We recognize that inertia keeps mechanical engines running, and the absence of inertia makes the spontaneous transition from the quiescent to the running state impossible. Chemical systems may have similar barriers in the form of system coordination and aggregate activation energies. As illustrated by the thermoacoustic engines, the distinction between self-starting and non–self-starting arises from a distinction between second-order and first-order phase transitions. It is interesting that even many multicellular organisms are self-starting, like simple chemical reactors. Their metabolism can be suspended indefinitely by freezing (suspension in liquid nitrogen has been demonstrated, at which temperature metabolic reactions have been slowed sufficiently that further cooling to liquid helium temperatures would not change the result), and then restarted to normal function when the organisms are thawed and placed in an appropriate environment. For some organisms the tolerance of mild freezing is an adapted function in the normal life cycle, but at the subcellular component level, the ability to restore function from inactive structure has probably been essential to the robustness of all organisms against environmental shocks. In particular, the ability to self-start may be intrinsic to the ability to self-organize, as any self-maintaining process must correct perturbations of its internal state. Self-starting, rather than collapse into equilibrium, allows a self-maintaining process to take advantage of plentiful resources in a fluctuating environment while surviving periods in which resources are in short supply.

An alternative and its consequences are demonstrated by large multicellular organisms, which cannot spontaneously restart if they are killed by disruption of dynamical coordination (e.g., by electric shocks). While such fragility of individual organisms appears to have been an acceptable price to pay for the greater sophistication of multicellularity, it may create an uncontrollable instability for all cellular processes to be similarly unable to restart after disruptions, and this may be why single cells, including unicellular organisms, are observed to be self-starting if their structure is preserved.

Metabolism of protocells should be based on pathways with the lowest thresholds to autocatalysis. Extremely simple mechanisms for the selective population of reaction pathways in a driven system are created by the dynamics of self-organization, and it is quite plausible that natural selection arose in part from such mechanisms in early stages of life, and may continue to depend partly on them today. Protocellular metabolisms that arise as attractors in ergodic networks should require minimal catalytic enhancement to be further amplified above parasitic processes, and metabolisms based on pathways with the lowest threshold to autocatalysis may automatically use the chemical potential of the inputs as the primary mechanism to exclude competing pathways. Artificial protocellular metabolisms could also be designed to use pathways with low thresholds to autocatalysis. Under such circumstances, if the rate of delivery of resources is limited, the first autocatalytic cycle that reaches its threshold will consume resources and thereby limit the driving potential to a level below that required to drive other cycles above their thresholds.

One of the more satisfying explanations in physics of self-organization to the critical state (Dickman, Vespignani, and Zapperi, 1998) starts with dissipative models that undergo phase transitions between ordered and disordered dynamical states (the so-called absorbing state phase transitions), and shows that, when such systems are embedded in environments that deliver resources at a limited rate, the difference in resource flow permitted by the system in its ordered and disordered phases is sufficient to hold the dynamics on the boundary between these two phases, where long-range structure formation is richest as ordered regions just begin to form. When the system is disordered, it permits energy and material to flow only as a low-rate parasitic loss between the high- and low-potential reservoirs in the environment, and even when resources are delivered slowly, the delivery rate is adequate to compensate for such losses and maintain the driving strength at a high value. In contrast, when the system is ordered, it actively transports energy and material between the high- and low-potential reservoirs, at a rate faster than the delivery mechanism can match, lowering the driving strength indefinitely. Neither the disordered nor the fully ordered state of the system can be maintained as a steady state in such an environment, because the existence of either state creates environmental conditions ultimately favoring the other state. The net result is that the system is balanced on its transition point between order and disorder, and the environment is held at a state where its delivery of resources just supplies the driving strength needed to maintain the critical state.

The mechanism we are proposing for the selection of protocellular metabolism is not as specific as self-organization to a dynamical critical point. It relies, rather, on the use of an ordered phase from one critical transition to deplete resources below the conditions needed to induce *other* phase transitions to ordered phases, a condition that can hold across a range of chemical potentials. The Dickman mechanism for self-organization merely demonstrates, in a tractable model, the way in which coordinated modulation of a system state with its environment can occur.

20.8 Summary

We suggest that the order of life should be understood as a statistical phenomenon, from the smallest to the largest structures, and from the earliest chemical stages to the persistence of modern forms. In particular, we argue that core metabolism has emerged at the boundary of contingency and necessity during the origin of life. We have developed intuition for the emergence of structure for driven systems using an example from physics, the thermoacoustic refrigerator. Then we have described in some detail the emergence of primordial biochemical pathways and their preservation in modern core metabolism. Finally, we suggest questions about later and higher levels of organization that bear on the ability to engineer stable systems that share characteristics of extant life.

Acknowledgments

The authors were supported in this collaboration by grant FIBR: The Emergence of Life: From Geochemistry to the Genetic Code, Award 0526747, from the National Science Foundation. DES thanks Insight Venture Partners for individual support during the course of this work. We also wish to thank N. Packard and S. Rasmussen for comments and editorial improvements to the manuscript.

References

Atchley, A. A. (1994). Analysis of the initial buildup of oscillations in a thermoacoustic prime mover. *Journal of the Acoustical Society of America, 95*, 1661–1664.

Atchley, A. A., Bass, H. E., Hofler, T. J., & Lin, H.-T. (1992). Study of a thermoacoustic prime mover below onset of self-oscillation. *Journal of the Acoustical Society of America, 91*, 734–743.

Baross, J. A., & Hoffman, S. E. (1985). Submarine hydrothermal vents and associated gradient environments as sites for the origin and evolution of life. *Origins of Life and the Evolution of the Biosphere, 15*, 327–345.

Cairns-Smith, A. G., Hall, A. J., & Russell, M. J. (1992). Mineral theories of the origin of life and an iron sulfide example. *Origins of Life and the Evolution of the Biosphere, 22*, 161–180.

Ceperley, P. H. (1979). A pistonless Stirling engine: The traveling wave heat engine. *Journal of the Acoustical Society of America, 66*, 1508–1513.

Ceperley, P. H. (1982). Gain and efficiency of a traveling wave heat engine. *Journal of the Acoustical Society of America, 72*, 1688–1694.

Ceperley, P. H. (1985). Gain and efficiency of a short traveling wave heat engine. *Journal of the Acoustical Society of America, 77*, 1239–1244.

Cleaves, H. J., & Miller, S. L. (2001). The nicotinamide biosynthetic pathway is a by-product of the RNA world. *Journal of Molecular Evolution*, *52*, 73–77.

Cody, G. (2004). Transition metal sulfides and the origin of metabolism. *Annual Review Earth and Planetary Science*, *32*, 569–599.

Cody, G. D., Boctor, N. Z., Filley, T. R., Hazen, R. M., Scott, J. H., Sharma, A., & Yoder, H. S. (2000). Primordial carbonylated iron-sulfur compounds and the synthesis of pyruvate. *Science*, *289*, 1337–1340.

Copley, S. D., Smith, E., & Morowitz, H. J. (2005). A mechanism for the association of amino acids with their codons and the origin of the genetic code. *Proceedings of the National Academy of Sciences of the United States of America*, *102*, 4442–4447.

Corliss, J. B. (1990). Hot springs and the origin of life. *Nature*, *347*, 624.

Corliss, J. B., Dymond, J., Gordon, L. I., Edmond, J. M., von Herzen, R. P., Ballard, R. D., et al. (1979). Submarine thermal springs on the Galapagos Rift. *Science*, *203*, 1073–1083.

DeDuve, C. (1991). *Blueprint for a cell: The nature and origin of life*. Burlington, NC: Patterson.

DeDuve, C. (2005). *Singularities: Landmarks on the pathways of life*. London: Cambridge University Press.

Dickman, R., Vespignani, A., & Zapperi, S. (1998). Self-organized criticality as an absorbing-state phase transition. *Physical Review E*, *57*, 5095–5105.

Engel, M. H., Andrus, V. E., & Macko, S. A. (2005). Amino acids: Probes for life's origin in the solar system. *NATO Science Series, Series I: Life and Behavioral Sciences*, *366*, 25–37.

Fenchel, T. (2002). *Origin and early evolution of life*. New York: Oxford University Press.

Ferris, J. P., Joshi, P. C., Edelson, E. H., & Lawless, J. G. (1978). Hydrogen cyanide: A plausible source of purines, pyrimidines and amino acids on the primitive earth. *Journal of Molecular Evolution*, *11*, 293–311.

Ferris, J. P., Joshi, P. C., & Lawless, J. G. (1977). Pyrimidines from hydrogen cyanide. *BioSystems*, *9*, 81–86.

Fry, I. (2000). *The emergence of life on Earth: A historical and scientific overview*. New Brunswick, NJ: Rutgers University Press.

Gács, P. (2001). Reliable cellular automata with self-organization. *Journal of Statistical Physics*, *103*, 45–267.

Gesteland, R. F. (1999). *The RNA world* (2nd ed.). Cold Spring Harbor, NY: Cold Spring Harbor Laboratory Press.

Glansdorff, P., & Prigogine, I. (1971). *Thermodynamic theory of structure, stability and fluctuations*. New York: John Wiley & Sons.

Gosselink, J. W. (1998). Metal sulphides and refinery processes. In T. Weber, H. Prins, & R. A. van Santen (Eds.), *Transition metal sulphides*. (pp. 311–355) Dordrecht: Kluwer.

Haken, H. (1983). *Advanced synergetics: Instability hierarchies of self-organizing systems and devices*. New York: Springer-Verlag.

Haldane, J. B. S. (1967). The origin of life. In J. D. Bernal (Ed.), *The origin of life* (pp. 242–249). London: Wiedenfeld and Nicolson.

Heinen, W., & Lauwers, A. M. (1996). Organic sulfur compounds resulting from the interaction of iron sulfide, hydrogen sulfide and carbon dioxide in an anaerobic aqueous environment. *Origins of Life and the Evolution of the Biosphere*, *26*, 131–150.

Jaynes, E. T. (1983). E. T. Jaynes: Papers on probability, statistics, and statistical physics. Dordrecht: Kluwer.

Kelley, D. S., Baross, J. A., & Delaney, J. R. (2002). Volcanoes, fluids, and life at mid-ocean ridge spreading centers. *Annual Review Earth and Planetary Science*, *30*, 385–491.

Koonin, E. V., & Martin, W. (2005). On the origin of genomes and cells within inorganic compartments. *Trends in Genetics*, *21*, 647 654.

Lazcano, A., & Miller, S. L. (1999). On the origin of metabolic pathways. *Journal of Molecular Evolution*, *49*, 424–431.

Lengeler, J. W., Drews, G., & Schlegel, H. G. (1999). *Biology of the prokaryotes.* New York: Blackwell Science.

Li, Y., Zhao, Y., Harfield, S., Wan, R., Zhu, Q., Li, X., et al. (2000). Dipeptide seryl-histidine and related oligopeptides cleave DNA, protein, and a carboxyl ester. *Bioorganic and Medicinal Chemistry, 8,* 2675–2680.

Lipmann, F. (1965). Projecting backward from the present stage of evolution of biosynthesis. In S. W. Fox (Ed.), *Origins of prebiololgical systems and their molecular matrices, conference proceedings, Wakulla Springs, Florida* (pp. 259–273). New York: Academic Press.

List, B. (2002). Proline-catalyzed asymmetric reactions. *Tetrahedron, 58,* 5573–5590.

Martin, W., & Russell, M. J. (2003). On the origin of cells: An hypothesis for the evolutionary transitions from abiotic geochemistry to chemoautotrophic prokaryotes, and from prokaryotes to nucleated cells. *Philosophical Transactions of the Royal Society of London, 358B,* 27–85.

Metzler, D. E. (2003). *Biochemistry: The chemical reactions of living cells.* San Diego, CA: Academic Press.

Miller, S. L. (1953). Production of amino acids under possible primitive earth conditions. *Science, 117,* 528–529.

Muller, G. B., & Newman, S. (2003). *Origin of organismal form: Beyond the gene in developmental and evolutionary biology.* Cambridge, MA: MIT Press.

Oparin, A. I. (1967). The origin of life. In J. D. Bernal (Ed.), *The origin of life* (pp. 199–234). London: Wiedenfeld and Nicolson.

Orgel, L. E. (2003). Some consequences of the RNA world hypothesis. *Origins of Life and the Evolution of the Biosphere, 33,* 211–218.

Oro, J., & Kamat, S. S. (1961). Amino-acid synthesis from hydrogen cyanide under possible primitive earth conditions. *Nature, 190,* 442–443.

Oro, J., & Kimball, A. (1961). Synthesis of purines under possible primitive earth conditions. I. Adenine from hydrogen cyanide. *Archives of Biochemistry and Biophysics, 94,* 217–227.

Pizzarello, S. (2004). Chemical evolution and meteorites: An update. *Origins of Life and the Evolution of the Biosphere, 34,* 25–34.

Ricardo, A., Carrigan, M. A., Olcott, A. N., & Benner, S. A. (2004). Borate minerals stabilize ribose. *Science, 303,* 196.

Russell, M. J., & Hall, A. J. (1997). The emergence of life from iron monosulphide bubbles at a submarine hydrothermal redox and pH front. *Journal of the Geological Society of London, 154,* 377–402.

Schrödinger, E. F. (1992). *What is life? The physical aspect of the living cell.* New York: Cambridge University Press.

Segré, D., Ben-Eli, D., & Lancet, D. (2000). Compositional genomes: Prebiotic information transfer in mutually catalytic noncovalent assemblies. *Proceedings of the National Academy of Sciences of the United States of America, 97,* 4112–4117.

Segré, D., & Lancet, D. (1999). A statistical chemistry approach to the origin of life. *Chemtracts-Biochemistry and Molecular Biology, 12,* 382–397.

Segré, D., Lancet, D., Kedem, O., & Pilpel, Y. (1998). Graded autocatalysis replication domain (GARD): Kinetic analysis of self-replication in mutually catalytic sets. *Origins of Life and the Evolution of the Biosphere, 28,* 501–514.

Segré, D., Pilpel, Y., & Lancet, D. (1998). Mutual catalysis in sets of prebiotic organic molecules: Evolution through computer simulated chemical kinetics. *Physica A, 249,* 558–564.

Segré, D., Shenhav, B., Kafri, R., & Lancet, D. (2001). The molecular roots of compositional inheritance. *Journal of Theoretical Biology, 213,* 481–491.

Smith, E. (1998). Carnot's theorem as Noether's theorem for thermoacoustic engines. *Physical Review E, 58,* 2818–2832.

Smith, E. (1999). Statistical mechanics of self-driven Carnot cycles. *Physical Review E, 60,* 3633–3645.

Smith, E. (2003). Self-organization from structural refrigeration. *Physical Review E, 68,* 046114.

Smith, E. (2005). Thermodynamic dual structure of linear-dissipative driven systems. *Physical Review E, 72*, 36130.

Smith, E., & Morowitz, H. J. (2004). Universality in intermediary metabolism. *Proceedings of the National Academy of Sciences of the United States of America, 101*, 13168–13173.

Stribling, R., & Miller, S. L. (1987). Energy yields for hydrogen cyanide and formaldehyde syntheses: The HCN and amino acid concentrations in the primitive ocean. *Origins of Life and the Evolution of the Biosphere, 17*, 261–273.

Swift, G. W. (1988). Thermoacoustic engines. *Journal of the Acoustical Society of America, 84*, 1145–1180.

Voet, D., Voet, J. G., & Pratt, C. W. (2006). *Fundamentals of biochemistry: Life at the molecular level* (2nd Ed.). New York: John Wiley & Sons.

Wächtershäuser, G. (1990). Evolution of the first metabolic cycles. *Proceedings of the National Academy of Sciences of the United States of America, 87*, 200–204.

Wächtershäuser, G. (1992). Groundworks for an evolutionary biochemistry: The iron-sulfur world. *Progress in Biophysics and Molecular Biology, 85*, 85–201.

White, H. B. (1976). Coenzymes as fossils of an earlier metabolic state. *Journal of Molecular Evolution, 7*, 101–104.

Yamagata, Y., Sasaki, K., Takaoka, O., Sano, S., Inomata, K., Kanemitsu, K., et al. (1990). Prebiotic synthesis of orotic acid parallel to the biosynthetic pathway. *Origins of Life and the Evolution of the Biosphere, 20*, 389–399.

Zabinski, R. F., & Toney, M. D. (2001). Metal ion inhibition of non-enzymatic pyridoxal phosphate catalyzed decarboxylation and transamination. *Journal of the American Chemical Society, 123*, 193–198.

21 Energetics, Energy Flow, and Scaling in Life

William H. Woodruff

21.1 Introduction

The processing of energy—its conversion, flow, and utilization—is an essential feature of life (Alberts et al., 1994, 2002; Lengeler, Drews, and Schleger, 1999). Major sources of biological energy include photochemistry (e.g., photosynthesis), thermally activated (nonphotochemical) electron transfer (e.g., aerobic respiration), fermentation (e.g., conversion of sugars to alcohols), and chemosynthesis or lithotrophy (metabolism based on inorganic compounds and carbon dioxide, e.g., in hydrothermal deep-sea vents). These broad categories comprise very diverse arrays of processes, depending on the nature of the organisms (e.g., plants and animals) and, in many cases, the environment of a specific organism. The latter option generally applies only to prokaryotes; for example, the bacterium *Escherichia coli* can employ either aerobic or anaerobic respiration and *Rhodobacter sphaeroides* can grow either photosynthetically or by aerobic respiration.

Schematic diagrams of two major biological energy conversion themes, aerobic respiration and green plant photosynthesis, are given in figure 21.1 (adapted from Alberts, et al., 1994). In this figure, the relative redox and photochemical energies of the protein complexes that compose the energy conversion systems are indicated on the vertical axes of the insets, while the direction of electron flow is shown on the horizontal axes (electrons flow from left to right in both cases). The membrane protein complexes and the modular structures of the two systems are represented above the insets. In addition, these structures are hierarchical, proceeding from the membrane protein complexes shown to mitochondria or chloroplasts (in eukaryotes) to cells and higher orders of organization (multicellular organisms, populations, ecosystems, etc.). Figure 21.1 demonstrates that photosynthesis and respiration are almost exact chemical reverses of one another. Respiration uses electrons from NADH (i.e., reduced nicotinamide adenine dinucleotide, NAD^+, a near-universal biological reducing agent) to reduce O_2 to water, harvesting the associated energy (1.14 eV) in the process and storing it as a transmembrane electrochemical (proton) gradient.

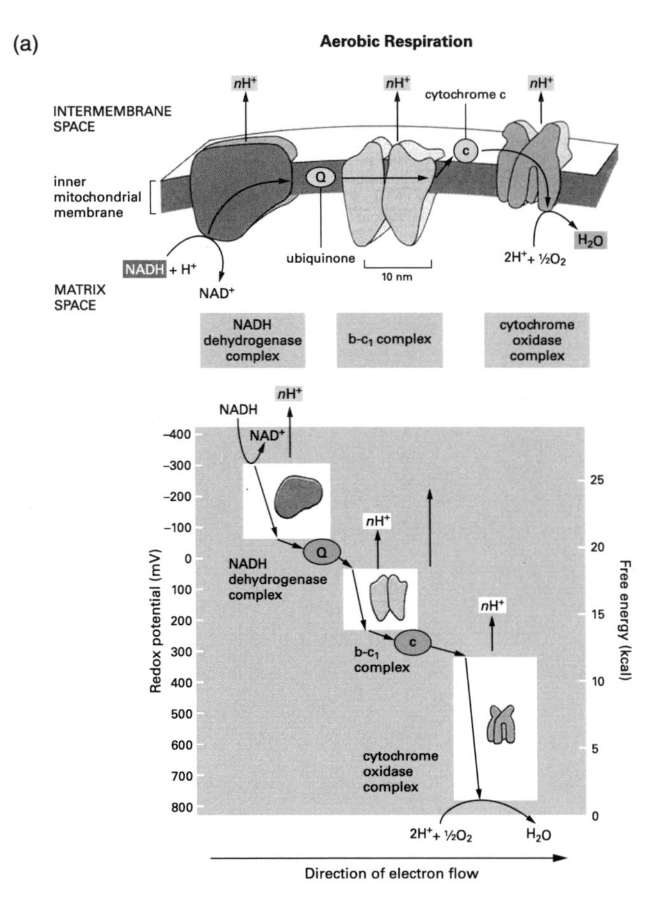

Figure 21.1
Schematic representations and energetics of aerobic respiration (a) and photosynthesis (b); adapted from Alberts et al. (1994). (c) An example of a synthetic nonbiological photosystem (see text).

Photosynthesis drives this process uphill, producing O_2 from water and transferring the electrons to $NADP^+$ ($NADP^+$ is phosphorylated NAD^+) to form NADPH while conserving the excess photochemical energy as a transmembrane proton gradient. As an example of a synthetic photosystem that has similar energetics and is one possible energy conversion system for protocells, the transition metal complex *tris*-2,2′-bipyridineruthenium(II), $[(Ru(II)(bpy)_3]^{2+}$ is shown. As is the case in the bioenergetic systems, the relative energies of the ground electronic state of this complex, the metal-to-ligand charge transfer (MLCT) excited state, and the two possible intermediate redox states are shown along the vertical axis.

The bioenergetic systems use the proton gradients that are generated to drive the synthesis of adenosine triphosphate (ATP, not shown), which in turn serves as the universal energy currency of life. This strategy for energy conversion that nature has

(b)

Photosynthesis

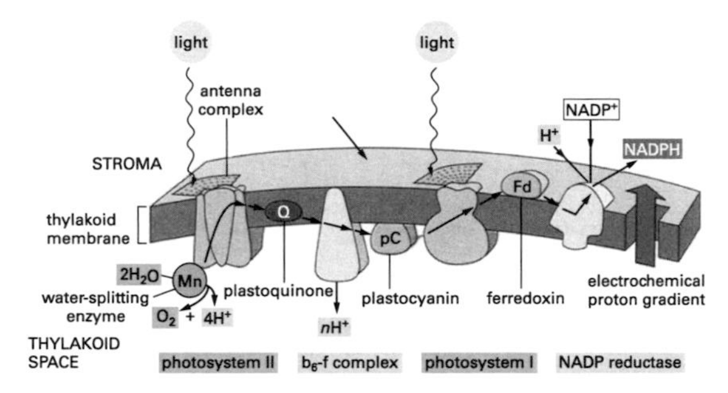

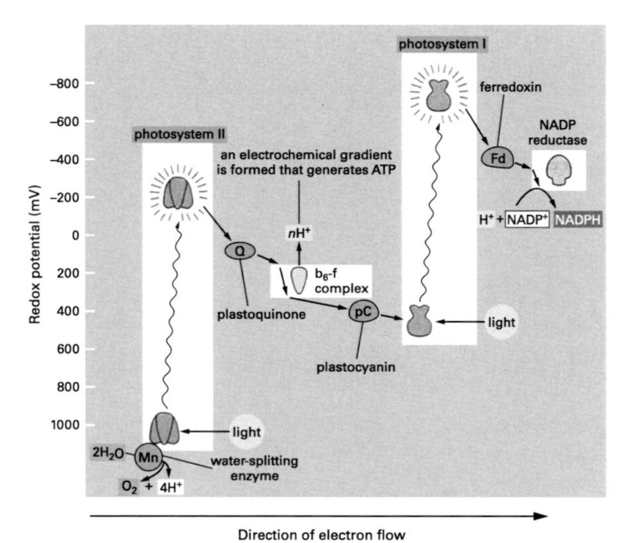

(c)

Synthetic Photosystem

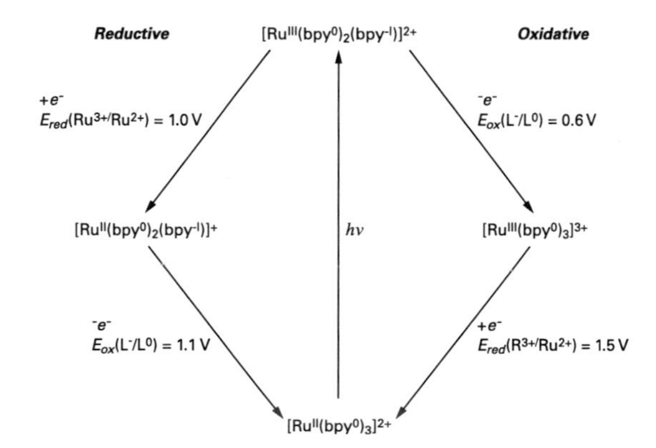

Figure 21.1
(continued)

evolved is probably too complex to be a viable approach to producing systems that mimic life. Rather, the energy of synthetic systems (such as the $[Ru(II)(bpy)_3]^{2+}$ photosystem depicted above) may be harvested by direct electron transfer to or from substrates, analogous to fermentation. In addition, the modular architecture of the bioenergetic systems (figure 21.1) provides strong guidance for the design of the biomimetic systems.

The range of nutrients that organisms need as energy sources or as building blocks for their structural or functional components is very diverse. Autotrophs grow using CO_2 as their sole carbon source (lithoautotrophs grow entirely on inorganic compounds) while heterotrophic bacteria may require very complex organic compounds to survive. Extreme examples of the latter are some mycoplasma that can exist only as parasites within or on more complex cells. Viruses, of course, have no metabolic apparatus of their own and are entirely parasitic, depending solely on a host cell's energy production and biosynthesis to support their replication.

The size and complexity of living cells also spans a vast range. The fundamental categories of cells are prokaryotes (archaebacteria and eubacteria, small cells with relatively simple internal structure) and eukaryotes, which are generally larger and much more complex. The primary distinction between prokaryotes and eukaryotes is the existence in the latter of discrete internal compartments: *organelles* such as nuclei, mitochondria, and chloroplasts.

The simplest known free-living cells are *Mycoplasma*, with cellular masses of a few femtograms (fg, 10^{-15} g), cellular diameters about 200 nm, and genomes encoding only about 500 proteins (smaller cells sometimes termed *nanobacteria* or *nanobes* have been claimed (Folk, 1993; Maniloff et al., 1997; McKay et al., 1996; Miyoshi, Iwatsuki, and Naganuma, 2005; Uwins, Webb, and Taylor, 1998; see also Harvey, 1997). In general cellular size of prokaryotes ranges from 10^{-15} to 10^{-10} g (Shuter et al., 1983), but in rare cases bacteria with cellular masses in the 10^{-6} g range have been observed (Schulz and Jorgensen, 2001). Similarly, cellular sizes in eukaryotes generally range from small protists and yeasts, about 10^{-11} g, to giant amoebae, 10^{-4} g (Hemmingsen, 1950, 1960; Shuter et al., 1983). However, some protists and unicellular plants are larger and some acellular marine algae have "cellular masses" of about 100 g (Raven, 1999).

Accordingly, typical (equivalent spherical) diameters of prokaryotic cells range from 200 nm to about 5 microns (μm, 10^{-6} m), whereas for eukaryotes the typical range is from about 2 μm to nearly 1 mm. By comparison, sizes of model "containers" that have been suggested for protocells—synthetic systems that embody the minimal characteristics of cellular life—range from about 5 to 50 nm or larger for micelles (Bales and Almgren, 1995; Tanford, 1974). Unilamellar vesicles begin at about 30 nm and typically range up to about 1 μm, with *giant unilamellar vesicles* (GUVs) ranging up to 50 μm or larger (Veatch and Keller, 2003). Multilamellar vesicles, though arguably not very suitable as protocell containers, can be much larger. So, vesicles as protocell containers overlap the entire biological size range of prokaryotes and a significant fraction of that of eukaryotes, and both vesicles and micelles offer sizes smaller than those of known living cells.

21.2 Metabolic Power in Living Organisms

Aerobic organisms, including prokaryotes, unicellular eukaryotes, fungi, plants, and metazoa (multicellular animals, including cold-blooded, ectothermic, and warm-blooded, endothermic animals), exhibit metabolic rates that depend on organismal size (Peters, 1983; Schmidt-Nielsen, 1984). For the aerobes, these rates are generally determined by measuring O_2 consumption, inasmuch as there is an energetic equivalency in the respiratory reaction shared by all of these organisms:

$$2NADH + O_2 + 2H^+ = 2NAD^+ + 2H_2O, \quad \Delta G^o = -440 \text{ kJ/mol } O_2. \tag{21.1}$$

Thus O_2 consumption per unit time can be expressed as power (Joules per second = watts). The power-size (equivalently, power-mass) relationships can be expressed as an empirical power law,

$$P = P_o M^b, \tag{21.2}$$

where P is metabolic power, M is organismal mass (both measured) and P_o and b are empirically determined parameters (the mass coefficient and scaling exponent, respectively). Beginning with measurements early in the twentieth century (Kleiber, 1932), the scaling exponent b has been established to be 0.75 by numerous measurements on organisms (Peters, 1983; Savage et al., 2004; Schmidt-Nielsen, 1984), populations, or ecosystems (Enquist, Brown, and West, 1998), up to the total biomass on Earth (Hochochka and Somero, 2002).

The reasons why equation (21.2) is a power law with $b = 3/4$ rather than, for example, a linear relationship ($b = 1$) or a surface-to-volume relationship ($b = 2/3$) have been elusive. Recently, a physical model was proposed, based on transport-limited constraints on energetics and the fractal-like geometry of biological transport systems (West and Brown, 2005; West, Brown, and Enquist, 1997). The constraints imposed by the fractal-like transport networks imply that biological transport cannot occur by simple bulk diffusion, even within unicellular organisms or individual cultured cells. This model accounts for the value of b in plants and animals, and allows quantitative prediction of numerous other biological scaling relationships (e.g., Savage et al., 2004).

Figure 21.2 plots log(whole-organism metabolic power) versus log(organismal mass) for prokaryotes, unicellular eukaryotes, and mammals (West, Woodruff, and

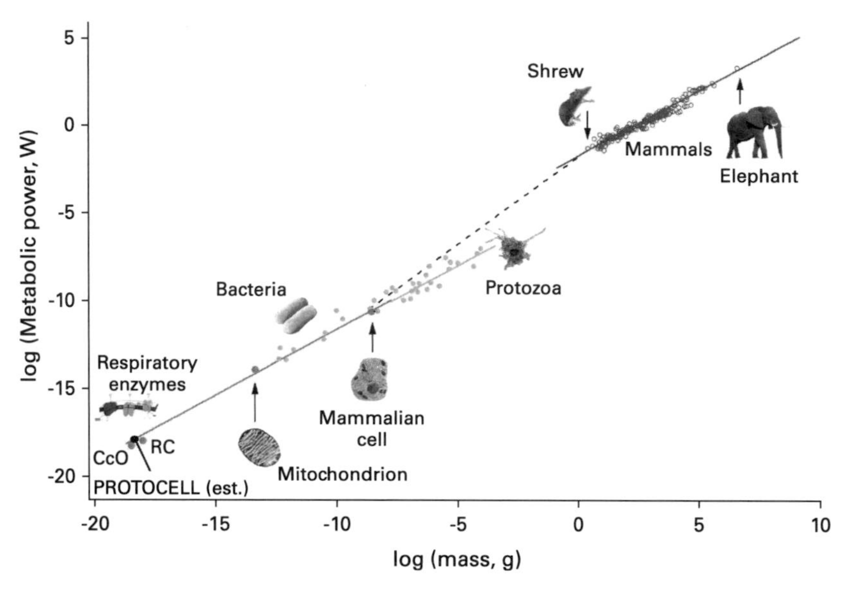

Figure 21.2
Log-log plot of whole-organism metabolic power (Watts, W = J/s) versus mass for subcellular components, prokaryotes, unicellular eukaryotes, and mammals (adapted from West, Woodruff, and Brown, 2002; see text). An estimate of the position of a micelle-based protocell is included.

Brown, 2002). It is noteworthy that subcellular components—namely, mitochondria, the system of cellular respiratory enzyme complexes (RC, NADH dehydrogenase or Complex I, plus cytochrome bc_1 or Complex III, plus cytochrome *c* oxidase or Complex IV), and the terminal O_2 reductant cytochrome *c* oxidase (CcO) itself—all fit the plot and its extrapolation to the molecular masses of the enzymes. For the mitochondria, this probably reflects this organelle's heritage as a free-living bacterium before it was recruited by evolution as the energy-producing endosymbiont for the protoeukaryote. Why the respiratory enzymes also fit seems hard to understand until one realizes that their intracellular concentrations parallel the mass-specific metabolic power of the organisms ([enzymes] μ $P/M = P_o M^{-0.25}$). In other words, metabolic powers are subject to the same contraint by transport of resources to the respiratory system by fractal-like networks, as suggested previously—rather than by simple bulk diffusion, biosynthesis of the respiratory enzymes—because the cells generally will not synthesize more of any component than they can use. The relationship in figure 21.2 covers 27 orders of magnitude in size from the molecular masses of the enzymes (about 10^{-19} g) to the mass of an elephant (about 10^6 g), and undoubtedly applies to the great whales (about 10^8 g) as well. Considering the applications to ecosystems cited earlier and the extrapolation to the total biomass on Earth (about 10^{20} g), this profoundly significant relationship spans some 40 orders of magnitude. The metabolic power of any system based on aerobic metabolism can be estimated from relationships like this, given mass information. Although it is not obvious that protocells will fit the same relationship with living systems, it will be very interesting to see where they reside in mass/power space and how they scale with one another.

21.3 The Energetic Equivalency of Biomass

The amount of energy needed to produce unit biomass obviously depends on what the starting materials are. Lithoautotrophs, which must synthesize their cellular components from simple inorganic compounds including CO_2 as the sole carbon source and water, obviously will need more energy to produce a gram of biomass than heterotrophs, which ingest and metabolize, inter alia, complex organics including the basic building blocks of life: proteins, lipids, and nucleic acids. In general, we expect protocells to be "ultimate heterotrophs" that are provided complex starting materials and perform minimal protobiosynthesis. So, the energy that living heterotrophs require to replicate their biomass might be expected to be an upper bound for the energy that a protocell would need to replicate its protobiomass.

This energy equivalency of biomass can be determined. The metabolic power of organisms is known (see figure 21.2; Hemmingsen, 1950, 1960; Peters, 1983), as are the cell division lifetimes of many unicells and cell cultures (Altman and Katz, 1976). Thus, the energy needed to create a new cell is simply the metabolic power per cell (J/s) times the doubling time of the population. Both the metabolic power and the doubling time may vary with growth conditions, generally in a compensatory manner; that is, conditions that decrease metabolic power tend to increase doubling time (see figure 21.3). The product—the "doubling energy," cellular metabolic power times doubling time—includes a certain amount of "housekeeping energy" that the cell would require to maintain its structures and functional components, even if it were not doubling. Dividing the doubling energy by cellular mass gives the energy required to produce the unit's new biomass. The relevant plot is shown (log-log) in figure 21.3, for unicellular masses over 7 orders of magnitude from small bacteria (about 100 fg) to amoeba (about 1 μg), and including cultured mammalian cells.

For these heterotrophic unicellular organisms (which synthesize their biomass from starting materials that are relatively complex molecular species, and are similar regardless of the identity of the organism), the energy needed to produce biomass is remarkably constant over the size range shown, about 500 J/g. The relationship is essentially linear, as shown by the log-log slope, close to unity. It is worth noting that this quantity is independent of either doubling time or metabolic power, separately. The more active bacteria double in less than an hour, whereas the mammalian cells require about 18 hours and the amoebae require several days. Among the bacteria, this tradeoff between doubling time and metabolic power can be seen as well;

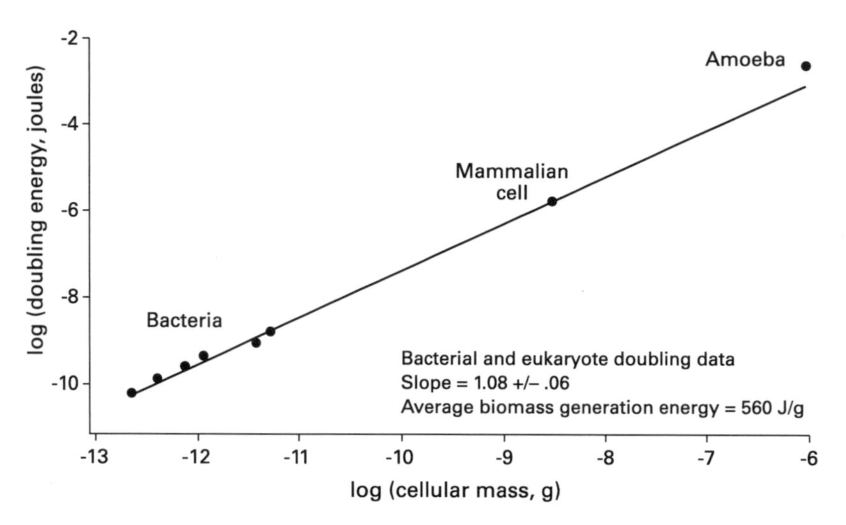

Figure 21.3
Log-log plot of "doubling energy" versus cellular mass for a series of bacteria and eukaryotes (cultured mammalian cells and protozoa).

one bacterium plotted is *Mycobacterium tuberculosis*, which has a very long doubling time (about 9 hours) but also a very low metabolic rate; the other bacteria have more typical doubling times, in the vicinity of 30 minutes. All of the mass-specific energies, however, are similar. Despite the exceptions such as *M. tuberculosis*, the doubling times of unicellular organisms generally increase as mass to the $+1/4$ power, while P/M decreases as mass to the $-1/4$ power; this is responsible for the (approximately) invariant energy equivalence of biomass.

Finally, how do these energies compare to the energies of biomass production for autotrophs? The most common examples are the green plants, which are photoautotrophs that synthesize carbohydrates according to the reaction

$$CO_2 + H_2O = (CH_2O)_n + O_2 \quad \Delta G^o = +460 \text{ kJ/mol carbon}, \tag{21.3}$$

where the large unfavorable energy of the reaction is provided by photons (see Photosystem II in figure 21.1). The reaction requires approximately 4.8 eV. The *bandgap*, the energy available to drive chemistry, for one photon absorbed by the chlorophyll-based Photosystem II of green plants is approximately 1.4 eV, therefore, four successive photochemical events are required per CO_2 molecule to produce O_2. Photosystem II of green plants accomplishes this by storing the photonic energy as electron transfers from a cluster of manganese ions (see, e.g., Spiro and Stigliani, 1996). Thus, from equation (21.3) the energy per unit biomass for this particular autotrophy is approximately 15,000 J/g, 30 times the energy derived above for biomass production by heterotrophs.

21.4 Energy and Information

The information system of living organisms is the genome. The genome has four major functions. It must encode the information needed to synthesize protein sequences (which in turn assemble and perform most cellular functions), including the proteins' amino acid sequences themselves plus the RNA species that do the actual operations of synthesis. It must allow mechanisms for transcribing the protein sequence information onto the RNA systems. It must allow mechanisms for replicating the genome from one generation of organisms (or cells) to the next. And, in the long term, it must allow mechanisms for evolution (Alberts et al., 2002). All of these operations except evolution explicitly require energy (in fact, protein synthesis is one of the primary demands on cellular energy), and it can be argued (e.g., Gillooly et al., 2005) that evolution is a consequence of energy production through radical damage to the genome by reactive oxygen species that result from aerobic respiration. Accordingly, it should be no surprise that some characteristics of the genome, such as the number of different proteins that can be encoded (i.e., the number of genes in the genome and—because the *average* length of a gene is essentially constant for all organisms, about 1,200 DNA base pairs—total coding genome length) are constrained by cellular energy production. Because cellular metabolic power itself is a function of cellular mass (figure 21.2 and equation (21.2), $b = 0.75$), commensurate effects on the number of genes per genome may be expected, at least for unicells.

Living cells, whether prokaryotes or eukaryotes (including plant protoplasts), are generally composed of the same fraction of protein, approximately 17 to 18% of wet weight or about 50% of dry weight. Other ways of stating this are that the *total* concentration of protein (protein molecules per unit volume, without regard to the identity of the molecules) is essentially invariant among organisms, or the number of protein molecules per cell scales linearly with cellular mass. If the concentration of each specific protein (gene product) were also invariant (on the average), then the number of different gene products per cell would have to be independent of cellular mass as well, and genetic and functional diversity would be severely constrained. But we know that is not the case; in general, more complex organisms (for unicells, generally larger cells) have genomes that encode more different gene products (we acknowledge complexities that are not explicitly considered here—for example, differing expression levels of different gene products; our arguments are simplified by assuming a number of average conditions). Moreover, invariant concentrations of enzymes that produce or utilize cellular energetics, or process materials that are equivalent to energy, are inconsistent with the observation that metabolic power scales with cellular mass to the 3/4 power. The *mass-specific* number density (i.e., concentration) of key energy-processing enzymes should scale with the *mass-specific* metabolic power (P/M, see above), namely, as cellular mass to the $-1/4$ power. This is observed in many cases (Darveau et al., 2002; Weibel et al., 2004). Given that the total protein concentration is constant, this generally allows the number of different gene products encoded by the genome to increase as cellular mass to the 1/4 power, to the extent that the energetic constraint applies.

Total genome length in unicellular eukaryotes, including unicellular plants, increases linearly with cellular mass (Shuter et al., 1983). Metazoa follow the same linear relationship for average cell size, notwithstanding cell differentiation. In all of these cases, however, only a fraction of the eukaryotic genome encodes gene product, and that fraction diminishes as cellular mass increases (e.g., about 50% for yeast, 1% for mammals). Accordingly, even given modern genome sequencing, obtaining coding genome length for eukaryotes is not straightforward. In prokaryotes, however, essentially the whole genome encodes gene product. Thus, if the preceding prediction is correct and the number of different gene products a prokaryotic cell can use is constrained by metabolic power, then genome length in prokaryotes should increase as cellular mass to the 1/4 power. Figure 21.4 shows that this is the case (data from Shuter et al., 1983, and the TIGR database, www.tigr.org).

To satisfy most definitions of life, protocells must have information systems that perform functions analogous to those of genes in living cells. It is not obvious that energy and information will be interdependent in protocells, in the manner discussed for living cells. However, such relationships may exist. Investigating this point will be

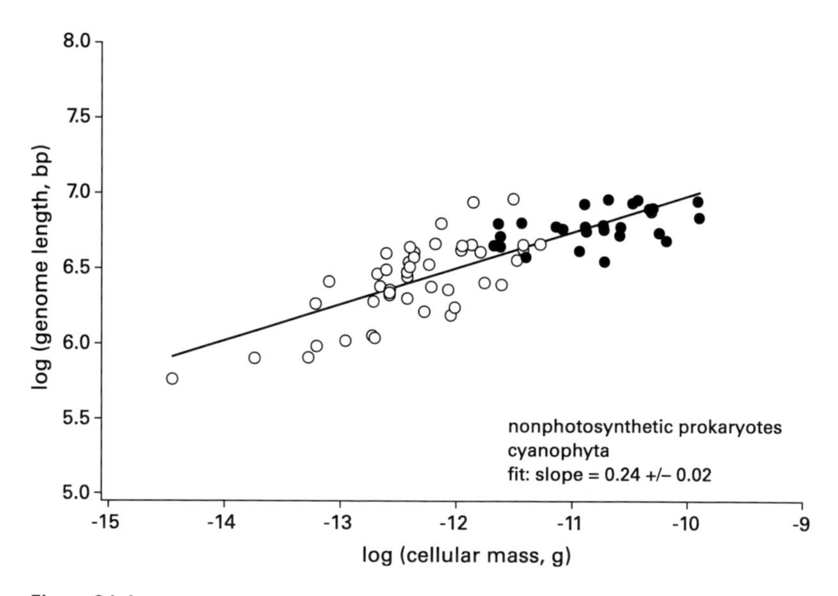

Figure 21.4
Log-log plot of genome length of prokaryotes versus cellular mass. Genome length is expressed as number of DNA base pairs, bp.

a useful aspect of protocell science, and it will be interesting to see whether protoevolution works to establish or strengthen similar interdependencies. In this regard, it may be significant or coincidental that sodium dodecyl sulfate micelles grow linearly as the 1/4 power of the total detergent concentration (Bales and Almgren, 1995; Bezzobotnov et al., 1988).

21.5 Protocell Energetics

The main energy sources used by living systems—photochemistry, thermally activated electron transfer, fermentation, and chemosynthesis—may be adaptable to protocell design. The protocell has to perform functions that are similar to those of living cells: It has to synthesize its own building blocks, grow, reproduce, and evolve. In all probability, the protocell will have to be an "ultimate heterotroph," that is, the precursors for the container, protometabolic system, and information system will have to be supplied to make the protocell operations that produce the final building blocks needed for growth and replication as simple as possible. The energy of replication can be estimated for a hypothetical protocell that is very "simple," consisting of a micelle, one metabolic system (of whatever nature), and one information system. Assume that the micelle constituent is SDS or some amphiphile similar in molecular mass (about 250 g/mol or 4×10^{-22} g/molecule), the information system is an octamer of DNA or something similar (about 10^{-20} g/octamer), and the molecular mass of the metabolic system is approximately 10^{-21} g/molecule. Then (assuming that the micelle comprises about 100 amphiphile molecules), the mass of the assembled protocell is about 5×10^{-20} g. Further, assume that in order to form the amphiphile from its precursor, one chemical bond has to be broken, requiring 1 eV (96.5 kJ/mol or 1.6×10^{-19} J/molecule), that the metabolic system is furnished intact, and that the energy required to replicate the information system needs the formation of only one chemical bond (e.g., a DNA octamer from two tetramers by forming one phosphate ester bond). Further, assume that the energy needed to form this one bond is negligible compared to the energy needed to form the micelle constituents. Then the energy required to replicate this hypothetical protocell (the *doubling energy*; see figure 21.3) is 1.6×10^{-17} J or approximately 300 J/g. This is close to and, satisfyingly, below (as predicted in the foregoing) the approximately 500 J/g "energy equivalency of biomass" observed for living unicellular heterotrophs. A related estimate by others predicted 1,300 J/g energy equivalence for protocells (Rasmussen et al., 2003), which is in very good agreement with the foregoing considering the approximations and differing assumptions involved. Note also that the energy required per turnover of the protometabolic system to produce the protocell building blocks (the precursor bond energy, about 1 eV) is very close to that of the aerobic respiratory system, 1.14 eV (this estimated point for protocells is shown in figure

21.2), and well below that of the photoautotrophs of 4.8 eV per mol CO_2. How these issues will actually play out in real protocells, of course, remains to be determined. For example, it will be interesting to see whether, or at what point in protocell complexity, the hierarchical organization that is a hallmark of biological systems manifests itself or becomes necessary.

Acknowledgments

Support of this work by NIH Grant DK36263 to WHW, and by LANL/LDRD, is gratefully acknowledged.

References

Alberts, B., Bray, D., Lewis, J., Raff, M., Roberts, K., & Watson, J. D. (1994). *The molecular biology of the cell* (3rd ed.). New York: Garland Science.

Alberts, B., Johnson, A., Lewis, J., Raff, M., Roberts, K., & Walter, P. (2002). *The molecular biology of the cell* (4th ed.). New York: Garland Science.

Altman, P. L., & Katz, D. D. (Eds). (1976). *Cell biology*. Bethesda, MD: Federation of American Societies of Experimental Biology.

Bales, B. L., & Almgren, M. (1995). Fluorescence quenching of pyrene by Cu(II) in sodium dodecyl sulfate micelles: Effect of micelle size as determined by surfactant concentration. *Journal of Physical Chemistry, 99*, 15153–15162.

Bezzobotnov, V. Y., Borbely, S., Cser, L., Farago, B., Gladkih, I. A., Ostanevich, Y. M., & Vass, S. (1988). Properties of sodium dodecyl sulfate micelles determined from small angle neutron scattering experiments. *Journal of Physical Chemistry, 92*, 5738–5743.

Darveau, C. A., Suarez, R. K., Andrews, R. D., & Hochochka, P. W. (2002). Allometric cascade as a unifying principle of body mass effects on metabolism. *Nature, 417*, 166–170.

Enquist, B. J., Brown, J. H., & West, G. B. (1998). Allometric scaling of plant energetics and population density. *Nature, 395*, 163–165.

Folk, R. L. (1993). SEM imaging of bacteria and nanobacteria in carbonate sediments and rocks. *Journal of Sedimentary Research, 63*, 990–999.

Gillooly, J. F., Allen, A. P., West, G. B., & Brown, J. H. (2005). The rate of DNA evolution: Effects of body size and temperature on the molecular clock. *Proceedings of the National Academy of Sciences of the United States of America, 102*, 140–145.

Harvey, R. P. (1997). Nanobacteria: What is the evidence? *naturalSCIENCE, 1*, article 7.

Hemmingsen, A. M. (1950). The relationship of standard (basal) energy metabolism to total fresh weight of living organisms. *Reports of the Steno Memorial Hospital and the Nordisk Insulinlaboratorium, Copenhagen, 4*, 1–58.

Hemmingsen, A. M. (1960). Energy metabolism as related to body size and respiratory surfaces, and its evolution. *Reports of the Steno Memorial Hospital and the Nordisk Insulinlaboratorium, Copenhagen, 9*, 1–110.

Hochochka, P. W., & Somero, G. N. (2002). *Biochemical adaptation*. Oxford: Oxford University Press.

Kleiber, M. (1932). Body size and metabolism. *Hilgardia, 6*, 315–363.

Lengeler, J. W., Drews, G., & Schlegel, H. G. (1999). *Biology of the prokaryotes*. New York: Blackwell Science.

Maniloff, J., Nealson, K. H., Psenner, R., Loferer, M., & Folk, R. L. (1997). Nanobacteria: Size limits and evidence. *Science, 276*, 1773–1776.

McKay, D. S., Gibson, E. K. Jr., Thomas-Keprta, K. L., Vali, H., Romanek, C. S., Clemett, S. J., et al. (1996). Search for past life on Mars: Possible relic biogenic activity in martian meteorite ALH84001. *Science*, *273*, 924–930.

Miyoshi, T., Iwatsuki, T., & Naganuma, T. (2005). Phylogenetic characterization of 16S rRNA gene clones from deep-groundwater microorganisms that pass through 0.2-micrometer-pore-size filters. *Applied and Environmental Microbiology*, *71*, 1084–1088.

Peters, R. H. (1983). *The ecological implications of body size*. Cambridge: Cambridge University Press.

Rasmussen, S., Chen, L., Nilsson, M., & Abe, S. (2003). Bridging nonliving and living matter. *Artificial Life*, *9*, 269–316.

Raven, J. A. (1999). The size of cells and organisms in relation to the evolution of embrophytes. *Plant Biology*, *1*, 2–12.

Savage, V. M., Gillooly, J. F., Woodruff, W. H., West, G. B., Allen, A. P., Enquist, B. J., & Brown, J. H. (2004). The predominance of quarter-power scaling in biology. *Functional Ecology*, *18*, 257–270.

Schmidt-Nielsen, K. (1984). *Scaling: Why is animal size so important?* Cambridge: Cambridge University Press.

Schulz, H. N., & Jorgensen, B. B. (2001). Big bacteria. *Annual Review of Microbiology*, *55*, 105–137.

Shuter, B. J., Thomas, J. E., Taylor, W. D., & Zimmerman, A. M. (1983). Phenotypic correlates of genomic DNA content in unicellular eukaryotes and other cells. *The American Naturalist*, *122*, 26–44.

Spiro, T. G., & Stigliani, W. M. (1996). *Chemistry of the environment*. Upper Saddle River, NJ: Prentice Hall.

Tanford, C. (1974). Thermodynamics of micelle formation: Prediction of micelle size and size distribution. *Proceedings of the National Academy of Sciences of the United States of America*, *71*, 1811–1815.

Uwins, P., Webb, R., & Taylor, A. (1998). Novel nano-organisms from Australian sandstones. *American Mineralogist*, *83*, 1541–1550.

Veatch, S. L., & Keller, S. L. (2003). A closer look at the canonical "raft mixture" in model membrane studies. *Biophysical Journal*, *84*, 725–726.

Weibel, E. R., Bacigalupe, L. D., Schmidt, B., & Hoppeler, H. (2004). Allometric scaling of maximal metabolic rate in mammals: Muscle aerobic capacity as determinant factor. *Respiratory Physiology and Neurobiology*, *140*, 115–132.

West, G. B., & Brown, J. H. (2005). The origin of allometric scaling laws in biology from genomes to ecosystems: Towards a quantitative unifying theory of biological structure and organization. *Journal of Experimental Biology*, *208*, 1575–1592.

West, G. B., Brown, J. H., & Enquist, B. J. (1997). A general model for the origin of allometric scaling laws in biology. *Science*, *276*, 122–126.

West, G. B., Woodruff, W. H., & Brown, J. H. (2002). Allometric scaling of metabolic rate from molecules and mitochondria to cells and mammals. *Proceedings of the National Academy of Sciences of the United States of America*, *99*, 2473–2478.

IV BROADER CONTEXT

This part presents diverse perspectives on protocell research. The chapters include discussions of the logical or statistical elements of living systems, reports on the biochemical regularities underpinning the origin of life, treatments of the inorganic, dynamical backbone of adaptive patterns, and introductions to the technological and social issues raised by the prospect of protocells. These discussions help embed protocell research in a broader scientific and social context.

The chapters do not only concern the larger implications of protocell research. In addition, the broader issues they raise can feed back into and inform our view of protocell research itself. Expanding our horizons can increase our understanding of the local environment. In particular, our view of what forms of life could be made in the laboratory could suffer from anthropocentric constraints, derived from our preoccupation with existing, typically multicellular forms of life. Science sometimes must confront counterfactuals and pursue unifying ideas that transcend the physical reality that we observe. Perhaps the deeper principles that unify living systems can be discerned by reflecting on minimal adaptive systems—viruses, polynucleotides, chemical automata, physical pattern formation—which are typically viewed to be outside the scope of the living.

The Hungarian chemical engineer Tibor Gánti was one of the earliest figures to write carefully about how to understand and engineer cellular living systems. Chapter 22 provides an account of Gánti's work, his characterization of living systems by means of life criteria, and his formulation of the chemoton, a model system of stoichiometrically coupled chemical reactions aiming to satisfy his life criteria. Gánti was the first to propose that life involves the integration of a chemical replicator (the genetic material), an autocatalytic chemical cycle (minimal metabolism), and a membrane for compartmentalization of the first two functions—the general organizing principle for this book. The technical challenges associated with the artificial synthesis of each of these three constituents are reviewed, including the origins of replicating templates, catalysis, and mechanisms of group selection. Gánti's work now seems somewhat dated, partly because of its lack of connection to contemporary experimental efforts such as those described in part II, and partly because of the realization that stoichiometric coupling must be augmented by other chemical reactions such as self-assembly processes in realistic protocells. Nevertheless, this review of Gánti's work still offers valuable guidance to contemporary protocell efforts.

Chapter 23 questions the categorical dichotomy between nonliving and living and proposes a more continuous conceptualization based on autonomy and individuality. Individuality is a statistical concept relating to how the sequence of a genome in the future depends on its sequence in the present, and autonomy captures the extent of the dependency on environmental factors for completion of a life cycle. These ideas are illustrated with viruses, because viruses are often dismissed as nonliving because they lack a metabolism. The essential metabolic resources used by viruses are distributed throughout a network of living systems involving hosts; individual viruses represent the propagation of a highly reduced metabolic form of life that can destabilize more complex forms of life.

It is well known that simple chemical systems maintained far from equilibrium can generate complex spatiotemporal patterns. Accordingly, chapter 24 addresses protocell research from the perspective of simple nonlinear chemical dynamical systems, making use of a continuously stirred tank reactor. It is shown how a simple driven system containing only salts can form inorganic cell-like structures surrounded by membranes, and also a diversity of self-repairing morphologies. This chapter thus raises the question whether these processes could form a dynamical backbone to evolved protocell biochemistry.

There is no necessity that the process of creating protocells will resemble in any detail the process by which the first forms of life on Earth arose from their earlier molecular ancestors, but understanding one process will illuminate the other. Chapter 25 examines the principles of physics and chemistry thought to underlie the origin of the earliest forms of life. This chapter emphasizes the widespread distribution of amphiphilic molecules capable of forming spherical vesicles in a wide range of solutions (including those found on meteorites) and their role in channeling reactive molecular species. It also examines the role that random peptide networks could play in generating autocatalytic cycles constituting a primitive form of replication. In this chapter, hereditary molecules are assumed to have evolved subsequently, as a means of stabilizing more complex reaction networks.

Chapter 26 reviews progress in the synthesis of biochemical monomers and polymers under realistic prebiotic conditions. It examines alternatives to RNA that possess the same catalytic and structural properties—a regularly repeating linkage type—that provide information storage capacity. The shift from reducing to oxidizing conditions on the primitive Earth places significant constraints on prebiotic synthesis, and requires reevaluation of the potential role for both extraterrestrial and hydrothermal input of the building blocks of life. This chapter also argues that genes that are chemically simpler than RNA could be the basis for early life.

Once protocells exist, they will likely be found to have many practical applications. Chapter 27 examines the prospect of engineering a cell-like entity designed to solve practical problems that are difficult or prohibitively expensive to address with current technologies. This cell-like entity would be a low-energy microscale building block that could self-assemble into a variety of structures, depending on the functional niche, and would be capable of self-repair and adaptive task differentiation. It might become possible to use such cell-like entities to solve pressing engineering, medical, and environmental problems.

Chapter 28 discusses some of the social and ethical consequences of a successful protocell research program. The very attributes of self-reproduction and evolvability that are deemed desirable goals of scientific protocell research could make protocells a threat to the environment and human health. Evaluating the proper courses of action will significantly stress our current practices of risk analysis and precautionary thinking. In addition, making new forms of life will raise concerns about violating the sanctity of life and playing God. Thinking through these issues forces us to confront the question of what kind of say in determining our future with protocells should be given to protocell scientists, to private industry, to professional ethicists, to nongovernmental organizations, to governmental officials, and to the general public.

The discussions in these chapters indicate the broader context for protocell research, but of course there are further issues in the broader context not covered here. Space limitations, among other considerations, have made it impossible to treat these further issues in any detail in this volume; however, we will briefly mention some highlights of these additional topics that figure in the larger background of protocell research.

Synthetic biology It is useful to consider how protocell research fits into the related emerging fields of synthetic biology and systems biology. Synthetic biology has been defined as the design and construction of new biological parts, devices, and systems. Since fully functional protocells would be biological systems, although perhaps quite unlike familiar forms of life, protocell science and technology should be classified as a form of synthetic biology. Most of the current work in synthetic biology involves genetic or nongenetic manipulation of living cells, which is most akin to top-down protocell research (see the general introduction to this volume). The bottom-up approach to creating protocells from nonliving materials through processes involving self-assembly and emergent engineering (see the general introduction) would then be seen as a bottom-up form of synthetic biology that counterbalances the gene-centric and top-down engineering orientation characteristic of much current work in synthetic biology.

Systems biology Systems biology includes research on virtually any biological system with an emphasis on the networks of dependencies among component parts. It most typically implies quantitative approaches to genomewide dynamical interactions at the subcellular level, and populations of cells forming tissues (see recent issues of *IET Systems Biology*, *BMC Systems Biology*, and *Molecular Systems Biology*). The defining features of living systems (see the general introduction) are themselves systemic properties par excellance. So one may regard protocell research efforts to fabricate living macromolecular systems as a fundamental form of systems biology.

Artificial life The place of protocell research in the larger context of artificial life can best be seen by summarizing artificial life's grand challenges. Bedau and coworkers (2000) classified the main open problems for artificial life into three broad issues: the transition to life, the potentials and limits of living systems, and the relations among life, mind, machines, and culture. This book is mainly concerned with issues in the first category, which concerns how life arises from nonliving matter. This challenge has been divided into several more specific challenges, including (a) generating a molecular protoorganism in vitro, (b) achieving the transition to life in an artificial chemistry in silico, (c) determining whether fundamentally novel living organizations can arise from inanimate matter, and (d) explaining how rules and symbols are generated from physical dynamics in living systems. Challenges (a) and (b) are addressed in previous sections of this book, and chapters in this part address (c) and (d) as well as some further artificial life challenges about the potentials and limitations of living systems and the relations among life, machines, mind, and culture.

Astrobiology Research on the origins of life on the Earth (discussed in chapters 25 and 26) is clearly related to the bottom-up protocell research described in this book, the main difference being the significant additional constraints on the starting materials in origins of life work. Astrobiology, or the study of life elsewhere in the universe, is a new interdisciplinary science that seeks answers to fundamental questions on the origin, evolution, distribution, and destiny of life throughout the universe. Astrobiology encompasses the disciplines of astronomy, chemistry, biology, paleontology, geology, physics, and many of their subdisciplines. Astrobiology and protocell research share a broad scope that covers any possible form of life, even if quite unlike the life forms that now exist on the Earth. The main difference is that astrobiology is concerned with what forms of life might now exist in the universe (see, e.g., Ehrenfreund et al., 2006), whereas protocell research has the additional goal of creating new forms of life that have never yet existed anywhere in the universe.

Novel soft materials and complex fluids Research in soft materials and complex fluids addresses issues that substantially overlap those addressed in protocell research; these issues include self-assembly, functional interfaces, reactive multiphase colloidal flows, and combined soft-hard materials compositions (see de Gennes, 1991, and new journals such as *Soft Matter* and *Soft Materials*). In addition, progress of protocell research in synthesizing self-replicating materials will provide valuable information to the bio-inspired materials research areas. Also, recent advances in assembling molecular scaffolding such as branched DNA structures (Rothemund, 2006) could be helpful for protocell research, as these structures could lead to new self-replicating materials. Being composed of templating strings, they could in principle also undergo evolution, which would give them key properties in common with protocells.

Energy transformations Research on molecular energy transduction and transformation, such as the development of artificial photosynthesis systems, has striking similarities with some of the minimal metabolic systems developed in protocell research. The study of energy transduction mechanisms in industrial processes, for example, how energy-rich electrons can be transported along molecular complexes to produce useful materials such as fuels, are equally important for protocell metabolisms, where the goal of the metabolic processes is to transform resource molecules into building blocks.

Molecular sensors, actuators, and motors The field of molecular sensor development has taken off recently because of a perceived need to develop sensors for toxins and bioweapons. Molecular actuators are systems that cause a particular action based on some signal (e.g., a sensor). An actuator is typically a switch, which initiates some other action, including signal amplification. Molecular motors are, as the term indicates, devices that transform energy, usually chemical energy, into organized molecular motion. Examples include flagella rotor complexes, kinesin fibers that move material along the cellular microtubules, and RNA polymerases that transcribe DNA to RNA. Once protocell self-replication and programmability have been achieved, the incorporation of molecular sensors, actuators, and motors will help enable protocells to achieve more complex and useful properties.

Potential applications Even though commercial applications of protocell research are still years away, the broader context for protocell research includes the practical technologies that protocells could enable in the future. Chapter 27 discusses one set of possible future applications. It is tempting to speculate about a few further practical opportunities. When imagining future applications, one should remember two things: that protocells might be radically unlike existing cells (e.g., much simpler and metabolize unfamiliar raw materials), and that we will have some capacity to program protocells to do things we find useful.

One can imagine using protocells to address global warming by using a novel metabolism to sequester carbon dioxide as a carbonate. Another likely application area is alternative energy. One can imagine protocells that capture light energy from the sun and produce hydrogen (H_2) more efficiently and under less demanding conditions than existing forms of life. Alternatively, protocells could revolutionize environmental remediation. Synthetic bottom-up protocells could be engineered with chemically novel genetics and metabolisms so that they can ingest only the undesirable toxins, and they are less likely than modern genetically modified organisms to interact with the existing biosphere and thus pose less risk to the environment. Protocells could also be used as chemical factories. We already use genetically modified bacteria to synthesize a number of useful proteins, and vastly simpler protocells might accomplish the same tasks much more efficiently with much higher yields.

These examples only begin to enumerate the future possibilities, because protocells will be an enabling technology that will have application in a wide variety of fields.

References

Bedau, M. A., McCaskill, J. S., Packard, N. H., Rasmussen, S., Adami, C., Green, D. G., et al. (2000). Open problems in artificial life. *Artificial Life*, *6*, 363–376.

Ehrenfreund, P., Rasmussen, S., Cleaves, J., & Chen, L. (2006). Experimentally tracing the key steps in the origins of life: The aromatic world. *Astrobiology*, *6*, 490–520.

de Gennes, P.-G. (1991). *Soft matter*, Nobel lecture. Available at http://nobelprize.org/nobel_prizes/physics/laureates/1991/gennes-lecture.pdf (accessed June, 2007).

Rothemund, P. (2006). Folding DNA to create nanoscale shapes and patterns. *Nature*, *440*, 297–302.

22 Gánti's Chemoton Model and Life Criteria

James Griesemer and Eörs Szathmáry

22.1 Introduction

In this chapter, we review the history and promise of chemoton theory as a platform for theoretical and empirical modeling and simulation of infrabiological and protocellular living systems. To the extent that contemporary laboratory constructions satisfy life criteria, these provide insights into the origins of living systems. To the extent that origins of life research provides evidence of historical steps toward satisfying life criteria, these provide hints for significant laboratory constructions.

Tibor Gánti's chemoton theory is composed of a set of life criteria and a heuristic model, the chemoton, of a minimal chemical organization that satisfies its life criteria. Gánti's technical method of cycle stoichiometry offers a mode of presentation and analysis for known and hypothetical autocatalytic chemical systems with the potential for life. We consider the light chemoton theory sheds on (1) problems shared by origins of life research and experimental constructions of artificial living systems, and (2) potential historical pathways from chemical to living systems that represent alternative research programs for the construction of artificial living systems.

The chemoton model has its origin in Gánti's conception of living systems as combinations of two basically different processes: a so-called main cycle, which drives the system through a series of ontogenetic changes, and the equilibrating system, which is responsible for maintaining organization despite changes in the environment and the organism itself. This idea appeared first in Gánti's book on molecular biology in 1966 (the first account of the field in Hungarian). The author, a chemical engineer by profession, then went on to refine his theory about the general characterization of living systems, which led to the publication in 1971 of *The Principle of Life* in Hungarian, in a popular science disguise (the reasons for this camouflage are given in Szathmáry, 2003). From then on one can rightly speak of chemoton theory, which consists of two main parts: a phenomenological characterization of living systems and the presentation of the simplest, minimal living system model. The term *chemoton*, short for "chemical automaton," refers to this minimal model.

The 1971 version of the chemoton consisted of only two subsystems, exactly corresponding to the two components identified in 1966: a genetic material undergoing template replication (the main cycle), consuming as monomers material produced by an autocatalytic chemical cycle, which is effectively a minimal metabolic network producing more of its own components as well as template monomers at the expense of the difference between source and waste materials (the equilibrating system). We emphasize that the chemoton as we refer to it today is different, in that it includes yet another subsystem: an autocatalytically growing bilayer membrane for compartmentation. This form of the chemoton, now understood as the basic form, appeared in 1974. This historical discussion is relevant to the theoretical and empirical problems considered below. The ways in which a modern, three-subsystem chemoton model can be built from either Gánti's original two-subsystem model (genetic subsystem plus metabolism) or other possible two-subsystem models reflect theoretical challenges to explaining the origins of life as well as empirical problems for the construction of artificial living systems. The main challenges are to explain (1) why some pathways from primitive chemical cycles to living autocatalytic supersystems may be more likely than others in the history of life, and (2) whether this apparent chemical modularity of subsystems, making up a space of possible alternative protobiological systems, should play a role in research programs of artificial synthesis.

A chemoton is any autocatalytic chemical supersystem capable of self-maintenance, proliferation, and evolution, and composed of three component autocatalytic subsystems: a genetic material undergoing template replication, an equilibrating metabolic network, and a growing membrane (cf. the account of the minimal protocell by Morowitz, 1992; Morowitz, Heinz, and Deamer, 1988). The subsystems may each be stoichiometric, but the supersystem as a whole may be stoichiometrically "indefinite" in the sense that its overall dynamics are stoichiometric while individual reactions or subsystem couplings need not be (see Gánti 2003b, vol. 1, p. 184).

The conception of the chemoton model had at least two motivations: first, to understand the organization of life in its minimal form, and second, to apply it to the problems of natural and artificial biogenesis (as apparent in the subtitle of Gánti's first monograph published in English in 1979, after lying dormant for about 5 years in the drawers of the Publishing House of the Hungarian Academy). A similar motivation, but with more emphasis on the origin of life, was behind Manfred Eigen's 1971 paper in *Naturwissenschaften*. Both approaches recognize the importance of template replication and autocatalytic processes, although Gánti emphasizes metabolism in the form of a network and compartmentation, which Eigen began to incorporate only since 1981.

The other, phenomenological part of Gánti's theory is a proposition of life criteria, which a living system must obey irrespective of the details of its organization. Gánti accepted the common wisdom that bacteria as well as lions are regarded as alive, but he recognized that it would be next to impossible to draw a common mechanistic model for them, partly because lions consist of units that are also alive (the cells), which is not true for bacteria or for minimal cells. That is why the life criteria remained phenomenological. A further key distinction between absolute and potential life criteria is that absolute criteria must be fulfilled by all living individuals, whereas potential criteria, such as reproduction, are important only for the formation of populations (and the constitution of a living world) and not for the living state of individuals as such. As we discuss later, the distinction between absolute and potential life criteria is important for artificial synthesis of living systems, since the potential for evolution may be considered a criterion of success, yet it is not an absolute criterion for life. The chemoton model was conceived in such a way that the minimal living cell could form a whole biota; thus, the potential life criteria are also satisfied by the system.

Gánti deliberately aimed to formulate his ideas in an exact and quantitative manner. For the dynamic description, members of his group had been using standard chemical kinetics (Békés, 1975). It was also in 1975 that the first English publication of the chemoton model was presented (Gánti, 1975). But Gánti realized a problem of the formal description: Standard chemical stoichiometry cannot properly handle chemical cycles, let alone autocatalytic ones. This observation led to the conception of cycle stoichiometry, which is a remarkable combination of mass balance with overall dynamics. In order to show this, let us take the simple case of an enzyme E transforming substrate S into product P:

$$E + S \rightarrow ES$$

$$ES \rightarrow EP$$

$$EP \rightarrow E + P.$$

The overall balance equation of these coupled reactions in the standard way yields

$$S \rightarrow P,$$

which conveys the false information that the enzyme does not take part in the reactions. To remedy this defect, the custom in biochemistry is to place E above the arrow:

$$S \xrightarrow{E} P,$$

which conveys only qualitative information about the enzyme's role, in contrast to the quantitative nature of stoichiometry. Gánti has therefore introduced the cyclic process sign that enables one to incorporate both qualitative and quantitative information into the balance equation:

$$E + S \xrightarrow{\;E\;(1)\;} E + P,$$

where the "1" in the sign means that the balance is correct after one "turn" of the cycle. In other words, after one reaction of E and S to form products E and P, there is stoichiometric balance between E + S and E + P. Since E appears on both sides, S and P are in stoichiometric balance. This can be called the "turning" of a cycle because *E* is "recycled" in this one-step reaction rather than consumed or transformed: As E turns, S goes in and P comes out. Generalization of the formalism for *u* number of turns is straightforward:

$$E + u\,S \xrightarrow{\;E\;(u)\;} E + u\,P,$$

and it is understood that the materials above the cyclic process sign cannot be canceled from the overall balance.

The idea of a turn is better suited to the more complicated cases of autocatalytic sequences of coupled reactions, in which a product of one reaction is an input of the next so that the sequence forms a closed loop or cycle—the product of the last reaction in the sequence forming an input to the first reaction. In such cases, there need not be enzyme catalysts in the conventional sense at each reaction step; in the following reactions, no reagent is conserved in any given reaction, but the set of A_i is conserved by the set of reactions as a whole by virtue of the set forming a closed loop. In this set of reactions, a turn of the cycle means the production of a given A_i from A_i as input:

$$A_1 + X_1 \rightarrow A_2 + Y_1$$

$$A_2 + X_2 \rightarrow A_3 + Y_2$$

$$A_3 + X_3 \rightarrow A_1 + Y_3$$

The basic idea of this formalism was presented in thc 1971 edition of the *Principles*; its current version is developed in detail in Gánti (2003b, volume 1). The formalism allows one to construct an overall equation of the chemoton as a whole, and it was successfully applied to the stoichiometric description of the so-called prebiotic chemoton, comprising well over 100 reactions, all taken from prebiotic chemical experiments or postulated on the basis of plausibility (Gánti, 2003b). Cycle stoichiometry has also been used to describe the following well-known systems of reactions: the citric acid cycle, the formose cycle, the reductive citric acid cycle, template polycondensation of DNA, the urea cycle, the glyoxylate cycle, industrial synthesis of FDP, and industrial synthesis of ATP (Gánti, 2003b).

The introduction of the membrane subsystem in 1974 was compelled by the observation that something must keep the constituents of the system together, since they function in the fluid phase. This observation has formed a basis for speculations about the origins of living systems since Oparin (1924), Haldane (1929) and Goldacre (1958) (see Deamer and Fleischaker, 1994). Hargreaves, Mulvihill, and Deamer (1977) showed that precursors to membrane lipids could be formed without catalysts and vesicles could be formed in presumptive prebiotic conditions. Lipid bilayer membranes are fluid in two dimensions but rigid in the third dimension, so the membrane represents a rudimentary morphology of the system. Because the chemoton model was designed as a "fluid state automaton," incorporation of a membrane subsystem into the chemoton model was possible only *after* publication of the fluid mosaic model of Singer and Nicolson in 1972.

The current basic chemoton model (figure 22.1) depicts strict stoichiometric coupling between template polycondensation and membrane growth. This is not an absolute necessity of the model, but its stability against changes in concentration and kinetic rate constants is greatly increased. It is instructive to call attention to a systematic distinction between AND coupling and OR coupling of reactions (Gánti, 2003b). In the basic model (figure 22.1), coupling between template replication and membrane growth is of the AND nature, and therefore stoichiometrically determined. Figure 22.2 shows an example in which the coupling is of the OR nature: A_3 is transformed into either A_4 OR V'. Kinetic analysis (Csendes, 1984) reveals that the latter system is sensitive to relative reaction rate constants. For example, if the rate of formation of V' vastly exceeds that of T', then template replication may go on without the possibility of division of the system: Ultimately, it may result in rupture of the membrane. Needless to say, attempts to synthesize minimal cells in vitro must take into account that chemical supersystems may be stoichiometrically stable only for certain ranges of k values.

22.2 Three Challenges for a Chemoton Theory of Protobiological Evolution

The functioning of the chemoton, including its dynamics, spatial proliferation, and genetics, has been analyzed in detail: Gánti's monograph is a goldmine for future considerations. Here we would like to deal explicitly with three issues: (1) the

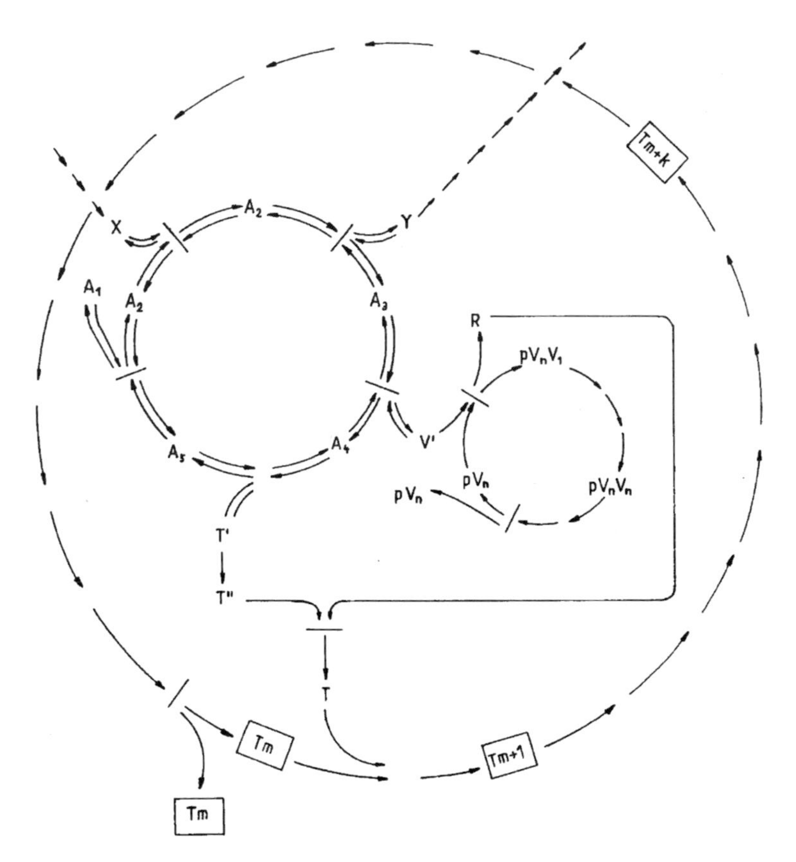

Figure 22.1
Network of an abstract chemoton (after Gánti, 1997, figure 1). This is the most commonly used representation of the system comprising three autocatalytic subsystems. A, V, T, R, X, and Y stand for different abstract kinds of molecular species. Materials A_i are intermediates of the metabolic cycle, pV_n is the template polymer consisting of monomers V, produced by the metabolic subsystem in precursor form V′. T is the molecule that spontaneously builds the membrane, the precursor of which is T′, which has to react the condensation side-product R from template polycondensation. The latter establishes a strict stoichiometric coupling between the subsystems so that growth is synchronized. Theoretical studies indicate that this system can divide in space, following growth at the expense of the difference of source (X) and waste (Y) material. Compare to figure 22.2, showing a different chemoton network without such a strict coupling.

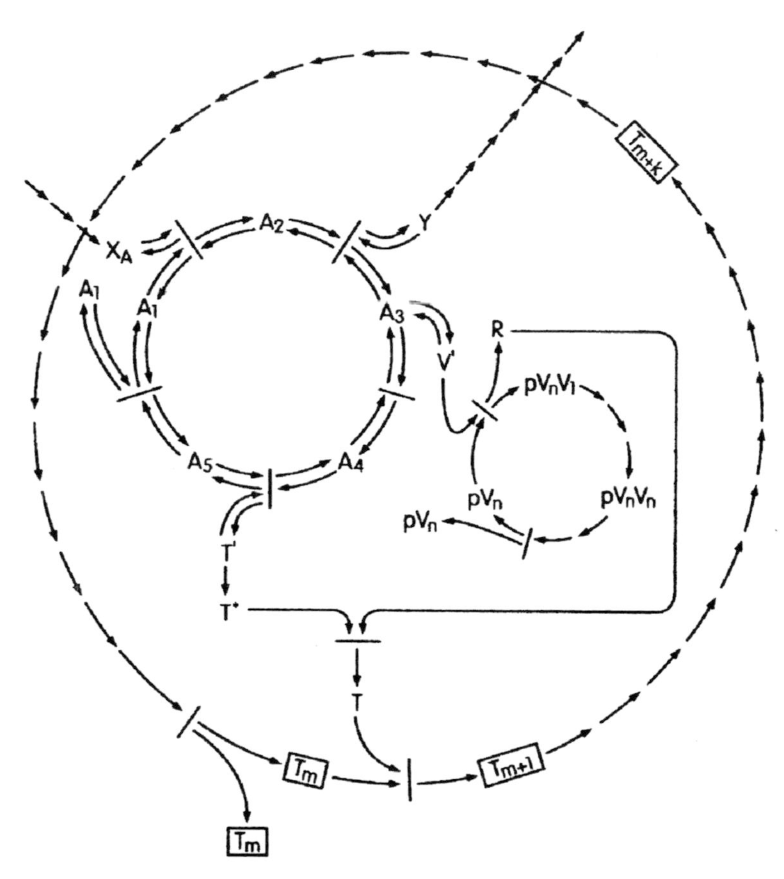

Figure 22.2
An alternative chemoton network with OR as well as AND branches (after Gánti, 2003b, volume 1, figure 4.10). Cycle intermediate A3 may yield either intermediate A4 OR V′. The OR branch is indicated by arrow pairs branching directly from A3, whereas in figure 22.1, the AND branch from A3 was shown by arrow pairs branching from the bar on the path between A3 and A4. Though the membrane still cannot grow without template copying (because R is required to form T from T*), template replication can "parasitize" the system and many rounds of replication can occur without division.

idealization that only reactions good for the chemoton are considered, (2) the genetics of a bag of genes, and (3) the incorporation of enzymatic template (ribozymes) in the model.

22.2.1 The Side Reaction Problem as the Key Problem of the Origin of Life

As we mentioned earlier, OR-type couplings tend to limit the stability of the chemoton even when they link stoichiometrically stable subsystems. It is easy to see that this concern applies even more when we consider reactions that lead *out of* the system to unwanted side reactions. The basic chemoton can either grow and divide or come to a standstill, depending on the concentration of *X* and *Y*. This is, of course, an idealization: Real cells die after a while when nutrients are depleted. They need metabolism not only for *growth* but also for *maintenance*, a well-known fact about microbial cultures. The chemoton model is idealized in that it does not deal with the maintenance problem (Szathmáry, 1989).

The chemoton has the capacity of exponential growth. The trouble is that spontaneous decay also follows exponential kinetics. Faster decay than growth obviously kills the system. We know that, for contemporary biological systems, growth can be much faster than decay, but this is because enzymes catalyze the wanted reactions. It can be said that the role of evolved catalysts is primarily to increase the reaction rates of the wanted reactions *relative to* those of the side reactions. By definition, wanted reactions are a tiny fraction of possible reactions. Despite this, present-day enzymatic systems are efficient enough to overcome this problem. The question is wide open for primitive systems, as is that of the significance of artificially constructed systems using present-day enzymes for inferences about the evolution of primitive, preenzyme systems. One can thus pose the question of under what conditions prebiotic reactions lead to life or to a complex but uninteresting substance such as tar. The answer to this problem is unknown, either experimentally or theoretically. We not only require that subsystems be able to maintain themselves despite spontaneous decay, but also that the appropriate coupling between them should be maintainable. Compounds in the three subsystems can react with other members of the set in countless ways; somehow, again, the wanted reactions should dominate.

These considerations are related to the concern raised by Shapiro (1986), who pointed out that many reactions of the Miller type, although progressing gracefully in isolation from each other, turn out to be *chemically incompatible*, which prohibits coupling and the wanted increase in complexity. The formation of microstructures under Miller-Urey conditions (Fraser and Folsome, 1975) shows that the side reaction burden can be overcome in certain regions of chemical space, and even proliferating structures can form spontaneously, but we know nothing about the internal organization of these structures. Particularly noteworthy in this regard is the finding (Folsome and Brittain, 1981) that ultraviolet (UV) irradiation of a gas mixture of N_2, CH_4, and CO_2 in presence of liquid water results in the spontaneous formation of microspheres, which grow autocatalytically. The structures catalyze the photoreduction of carbonate, mostly to formaldehyde. The nature of this reaction needs to be revisited by modern techniques.

On the theoretical side, the issue is also open. Although a lot of work has been done on "reflexively autocatalytic" systems (such as the formation of an autocatalytic protein network by Kauffman, 1986), here we are dealing with something different—the formation of an autocatalytic network of small intermediates, in which most reactions are not catalytic but stoichiometric. The idea that such networks, in a *nonenzymatic* form, are chemically feasible has been around for some time. Candidates include the formose reaction (Breslow, 1959; Butlerov, 1961; Decker and Speidel, 1972), an archaic form of the reductive citric acid cycle (Wächtershäuser, 1992), and the reductive citric acid cycle itself (Morowitz et al., 2000; Smith and Morowitz, 2004). Gánti (1978) recognized the fact that the Calvin cycle is an autocatalytic cycle at the level of sugar phosphates. The fact is that, apart from the formose reaction, there has been no experimental demonstration of any of these suggestions. The formose reaction has been criticized as a transient phenomenon that converts finally to uninteresting products (Shapiro, 1986, 1988). The reductive citric acid cycle has been defended partly on the grounds that many of the side reactions of the cycle are themselves beginnings of biosynthetic pathways—to lipids, sugars, amino acids, or pyrimidines (Smith and Morowitz, 2004).

The most relevant series of theoretical investigations is a result of the late King. First, he managed to demonstrate that in a sufficiently large recycling chemical system the appearance of autocatalytic subnetworks is more probable than not (King, 1982). But there is a snag: He argued that such subnetworks must necessarily have been simple, that is, consisting of only a few reactions. His reason was that he took the side reactions into account, but his analysis now seems to be incomplete. Surely, some of the side reactions' paths will return to the cycle (in the recycling system considered); the question, then, is where the balance lies between drain and replenishment. We would like to know the length *distribution* of maintainable autocatalytic cycles under such conditions.

A way out of this impasse is to apply some nonenzymatic but measurable channeling effect. There are two options: surface catalysis and compartmentation. We discuss them in turn.

Surface catalysis by minerals from clay (Bernal, 1967) to pyrite (Wächtershäuser, 1992) has been promoted by many researchers. The fact that clay has been shown to be important in the formation of longer oligonucleotides (Ferris, 2002) and ribose phosphate in a modified formose reaction is important. Similarly, NiS and FeS promote the reduction of carbon monoxide to activated acetic acid (Huber and Wächtershäuser, 1997). Demonstration that such surfaces can promote chemically more challenging catalysis, such as channeling the steps of the reductive citric acid pathway, is doubtful, since the same surface would have had to catalyze very different reactions (Orgel, 2000).

One final element of the story needs to be emphasized. With complex chemical systems like this, one must abandon the picture of deterministic chemical kinetics. Many compounds initially will be totally absent, and then later turn into other ones; some will persist for a very long time. Some will gain control over others: These will be the recycled (auto-) catalysts. The arising hysteresis will open the gate to simple forms of inheritance.

As King (1982) emphasized, compartmentation greatly reduces the kinetic complexity of chemical systems: Seen from the external world, most reactions within the compartment would be internal rearrangements. Before, we said that the importance of enzymes lies in speeding up favorable reactions relative to unwanted ones. By the same token we can say that the importance of membranes is at least as much the ability to keep many compounds *out of the system* as it is to keep the systemic compounds inside! Selectivity of the membrane is one of the main mechanisms of metabolite channeling. Whether it is sufficient for the nonenzymatic functioning of chemotons, however, is uncertain.

Whatever the outcome of these investigations, one crucial thesis should not be forgotten. Even if a nonenzymatic chemoton turns out to be unfeasible, the core organization of cellular systems, catalyzed by enzymes, needs to be elucidated. Gánti (1978) has emphasized that the basic unit of life remains a triplex composed of metabolism, template replication, and a membrane, whatever the fate of the particular chemoton models he has proposed.

22.2.2 The Origin of Template Replication

This problem can be regarded as a variant of the side reaction issue. In the kinetic treatise of chemoton theory (Gánti, 2003b), template polycondensation is modeled under simplifying assumptions, namely, that the separation of double-stranded molecules happens only above a critical concentration [V]*, and then elongation happens quickly, relative to the opening of the newly formed double-stranded chains. Strand dissociation is spontaneous and the incoming V' monomers prohibit the reformation of the old double-stranded form. But the dynamics of association and dissociation of V' molecules have been neglected in the kinetic analysis of chemotons. To remedy this defect, a detailed treatment of template polycondensation using stochastic kinetics has been initiated (Fernando, von Kiedrowski, and Szathmáry, 2007). The main obstacles to nonenzymatic replication of *long* templates (as opposed to short oligomers; cf. von Kiedrowski, 1986) are as follows:

- Detachment of hydrogen-bonded monomers competes with covalent-bonded elongation—the former is a faster process.
- Short oligomers readily replicate following the von Kiedrowski mechanism.
- Oligomers readily elongate to long templates but the latter fail to replicate.

Thus, at most there is spontaneous reappearance of longer templates in chemotons, but this is by de novo formation from oligomers rather than replication. It is not clear whether a way out of this impasse in a strictly nonenzymatic fashion will be found. Currently, we are investigating conditions for a minimal replicase that would provide enough channeling (i.e., a sufficient suppression of side reactions) for replication of longer strands to become feasible. It may be that, by the time systems became compartmentalized, a replicase was in place, following its origin on mineral surfaces *before* the advent of chemotons (cf. Szabó et al., 2002).

22.2.3 Genes Become Catalytic

As Gánti (1978) emphasized, information in chemical terms can be carried by three mechanisms: (1) the amount of a signal, (2) the proportion of signals, or (3) the geometrical arrangement (e.g., sequence) of signals. In nonenzymatic chemoton forms, only the first two mechanisms come into play. Even if there is only one type of monomer (V), the length of a polymer template pV_n affects the system as whole.

It did not take too long to propose a way to use the sequence of templates in the chemoton (Gánti, 1979). Gánti followed earlier suggestions by Woese (1967), Crick (1968), and Orgel (1968) proposing that templates could act as ribozyme-like catalysts. White's (1976) suggestion that coenzymes are remnants of an earlier metabolic stage, with RNA acting as catalyst, was carried further (Korányi and Gánti, 1981). It was also demonstrated theoretically that the presumptive substrates present in the chemoton could have channeled the self-assembly and evolution of ribozymes (Gánti, 1983). Thus, by the time of the experimental demonstration of ribozymes in contemporary organisms (Kole and Altman, 1981; Kruger et al., 1982) the RNA World (a phrase introduced by Gilbert, 1986) was in full bloom in Gánti's intellectual landscape. The suggestion that ribozymes could be artificially selected based on their replication and affinity to small ligands (Szathmáry, 1989, 1990) was a direct outcome of this line of thought.

22.2.4 Genome Size and Gene Diversity of Early Cells

In 1971, Eigen identified a major problem in the origin of life, now known as the *error threshold.* Put simply, this means that if replication is inaccurate, then the length of the genome maintainable by natural selection must be short. Estimates based on nucleic acid oligomer properties suggested that early replicators could not have been longer than about 100 nucleotides, a very modest genome size indeed (one tRNA is about 77 nucleotides long). If several different genes were needed, some means of dynamical coexistence must have been effective (Eigen, 1971). Inspired by the chemoton model, the stochastic corrector mechanism (SCM) was suggested as a way to ensure such coexistence (Grey, Hutson, and Szathmáry, 1995; Szathmáry and Demeter, 1987). It is assumed that replicators carry out useful functions for the compartment: For example, they perform enzymatic catalysis on different reactions of the metabolic subsystem (just as ribozymes would do). Coexistence hinges on group selection of replicators (figure 22.3), where group structure is provided by compartmentation. Basic population genetics shows that group selection is especially

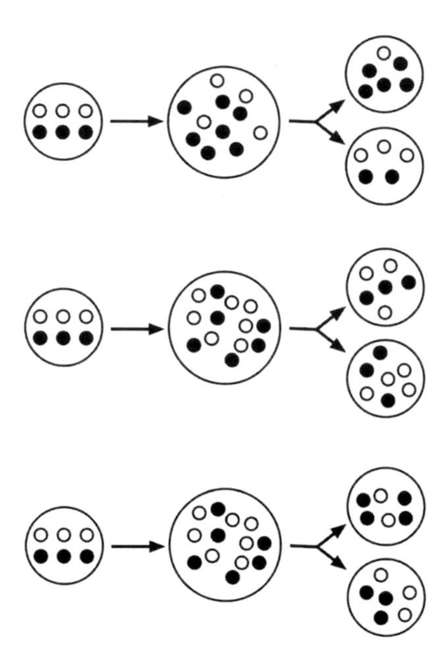

Figure 22.3
The stochastic corrector model (SCM) (after Maynard Smith and Szathmáry, 1995, figure 4.11). Different replicators (black and white) are "sitting in the same boat," that is, reproducing protocells. Replicators are allowed to compete within each compartment. Protocell division rate depends on internal template composition, for example, the templates can act as ribozymes. One of source of stochasticity is clearly indicated: On protocell division, templates sort themselves independently into offspring compartments. There is another important source of stochasticity, not shown on the figure: Since each compartment has a limited number of templates, their growth can be described by stochastic kinetics only. These two sources of stochasticity generate variation among the compartments, on which natural selection (at the compartment level) acts. This is a model for group selection of early replicators.

effective if three conditions are met (Leigh, 1983): (1) Each group has only one parent, (2) the number of groups is much larger than the number of individuals in any one group, and (3) there is no mixing between groups. Clonal reproduction of chemotons eminently satisfies these conditions (Szathmáry, 1989). Compartment fitness depends on how efficiently the genes present run metabolism, and variation is generated by two sources of stochasticity: Within each compartment there is demographic stochasticity, because of low template numbers, and templates randomly resort into offspring compartments on fission.

Yet two questions remain: First, would alternative systems be less or more efficient, and second, what is the maximum genome size maintainable by the stochastic corrector. The first question has been analyzed in some detail with the alternative system being the compartmentalized hypercycle (Zintzaras, Mauro, and Szathmáry, 2002). Now we explain the comparison in some detail (figure 22.4).

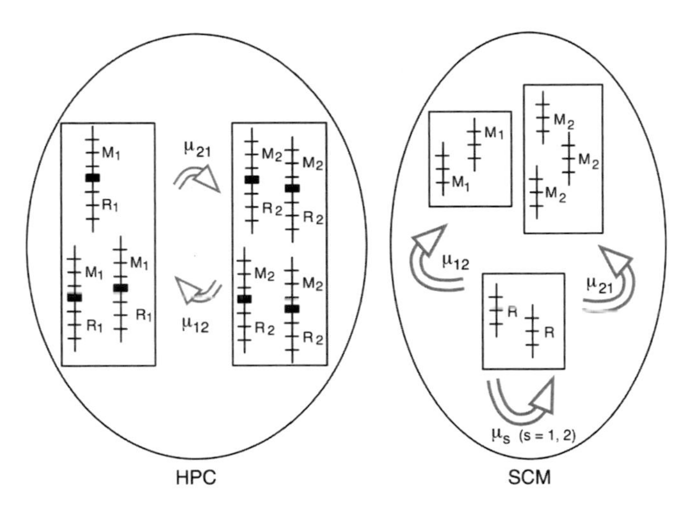

Figure 22.4
A model to compare hypercycles (HPC) with the stochastic corrector model (SCM) (after Zintzaras, Mauro, and Szathmáry, 2002, figure 1). For comparison, both systems are compartmentalized. There are two metabolic genes, M_1 and M_2, each consisting of the same number of mutable sites. The HPC has two replicases, R_1 and R_2, whereas the SCM has only a common replicase R. Linkage relationships are indicated. The distribution of replication rates μ_{ji} critically differs for the two cases. Note the assumption that heterocatalysis is assumed to be stronger that autocatalysis for the HPC, which favors HPC over SCM.

We assume that genes carry out some function useful for metabolism in both systems, but the organization of replication is different. In the SCM, the two metabolic genes are replicated by a common replicase, albeit with different efficiency (there is internal competition); the replicase also replicates itself. In the compartmentalized hypercycle, each replicator has dual functionality: one cistron acts as a replicase, and the linked cistron exerts a metabolic function. To give the hypercycle an advantage, it is assumed that the replicators preferentially replicate the other member of the system. Because of this dual functionality, the length of replicators in the hypercycle is doubled relative to that of the stochastic corrector. It turns out that, for high mutation rates, the fitness of the stochastic corrector exceeds that of the hypercycle, for two reasons. First, longer replicators automatically increase the mutational load. Second, because metabolic genes are physically linked to replicase genes in the hypercycle, linkage disequilibrium is generated by stochastic mutations, which also decreases fitness (bad replicase–good metabolic gene and good replicase–bad metabolic gene combinations cannot be broken).

The latter observation suggests that within-cell recombination may be advantageous and may increase the error threshold, as proposed by Lehman (2003). This proposal was investigated for the SCM (Santos, Zintzaras, and Szathmáry, 2004). Indeed, recombination within a compartment between copies of homologous replicators turns out to be advantageous, but the error threshold is only modestly increased.

22.2.5 The Permeation Problem

As Wächtershäuser (1992) correctly observed, a real danger for early compartments is self-suffocation. If a crucial ingredient of the system cannot be incorporated from the environment, or it cannot be formed de novo within the system, autocatalytic growth cannot be sustained. This is why the experimentally demonstrated replication of RNA (Oberholzer et al., 1995) comes to a halt: The protein replicase is diluted out of the reproducing liposomes.

It is important to realize that the problem of early membrane permeability is tightly linked to the problem of autotrophy versus heterotrophy of early protocells. If one applies the approach taken by different experiments that compartments retain only macromolecules, which can be arranged by appropriate lipid composition of the membrane (Monnard and Deamer, 2001; Oberholzer et al., 1995), then an intermediate metabolism composed of interacting small molecules is ruled out. Conversely, if one assumes a complex metabolism, then the selective permeability of compounds must somehow be solved without protein enzymes, so that raw materials can get in, waste can get out, and intermediates stay inside. The most important barrier is against the transfer of charged molecules like phosphates and amino acids through bilayers that do not allow the spontaneous passage of small molecules in general. One solution is the transient neutralization of such compounds, as demonstrated by Stillwell in a series of investigations. Simple lipid-soluble carriers could perform facilitated diffusion of sugars, cations, and protons (Stillwell, 1980). For example, amino acids can react with aliphatic aldehyde carriers to form a lipid-soluble Schiff base (Stillwell and Rau, 1981). Since facilitated diffusion runs both ways, a crucial assumption is that the molecule carried inside is transformed by reactions within the network into compounds that cannot diffuse back. Also, the facilitating molecule must either be generated internally by metabolism or be spontaneously taken up from the environment. Alternatively, transient membrane defects rather than permeation may explain the movement of some small ions, amino acids, and phosphates across lipid bilayers (Chakrabarti and Deamer, 1992). The constraints on an autocatalytic metabolic network by such transport systems that must be less selective than enzymes need to be investigated further, both experimentally and theoretically.

22.3 A Chemical Perspective on the Units of Life

Gánti's chemical perspective on the units of life and evolution is an integrated engineering philosophy, involving both descriptive and constructive aspects. His absolute and potential life criteria provide a useful general means of describing both the living state and the living world in which units of life can also function as units of evolution. Life criteria of some kind must play a role in judging the success of artificial constructions of living systems as well as in theories of the origins of life. The chemoton model provides a constructive complement to the life criteria, facilitating an iterative approach to theoretical and synthetic investigation. The behavior of the model presents opportunities to test approximate criteria hypothetically and the criteria, which can be checked against well-known cases in the contemporary natural or artificially constructed living world, allow fine-tuning of the model. Together they can provide heuristic guidance to theoretical and empirical research into questions of origins of life and evolutionary transitions (Griesemer, 2003). Three features of this integrated approach are worth noting.

First, descriptive approaches to "defining life" are often constrained by standard ways of describing contemporary, highly evolved living systems. For example, modern organisms all have nucleic acid replicators (DNA or RNA). The properties of genetic organization sufficient to ensure a capacity for hereditary change need not depend on nucleic acid organization, however, so life criteria based on even very general properties of contemporary replicators and reproducers (e.g., Dawkins, 1976; Fleischaker, 1994) may be too narrow to explain how those very properties evolved (see Griesemer, 2000a, b, c). It is important, in other words, not to beg the question of how life evolved by embedding descriptions of modern life into our definitions or by rendering artificial synthesis projects irrelevant to the evolution of life. The engineering approach helps here because no one knows how to build a "modern" living cell from scratch, so constructive projects cannot begin by assuming that synthetic life will resemble modern living things in detail.

Second, constructive approaches to synthesizing life tend to be too little constrained by the full range of criteria by which we recognize living systems and the living world. Since the days of Oparin and Urey-Miller, experimental chemists have explored reaction systems that suggest the contours of synthetic evolutionary biochemistry (e.g., von Kiedrowski, 1986, 1999; Morowitz et al., 2000; Rebek, 1994a, 1994b; Szostak, Bartel, and Luisi, 2001; Wächtershäuser, 1992). Synthesizing constituent parts of living systems, however, is not the same as constructing a living organization. The history of chemistry suggests that "obvious" synthetic routes need not be the most efficient or effective, nor the routes discovered by evolution. The advantage of an iterative engineering approach is the ability to work back and forth between biology and chemistry, description and construction, theory and model, exploration and test. Gánti's insight that life is organized in cycles bound to particular chemical matter, together with his perspicuous cycle stoichiometry, provides a chemical perspective and notation appropriate to biological engineering (Gánti, 1987, 1997, 2003a, 2003b). Cycle stoichiometry bridges the gap between description and construction with a formal language through which to link chemical models to life criteria (Griesemer, 2003).

Third, it is important to point out that Gánti's approach is a heuristic one. The life criteria are *criteria*, not a definition in either a descriptive or essential sense

(Ruiz-Mirazo, Peretó, and Moreno, 2004). Although they do describe life, they function properly only in the context of a specified model. The discovery of counterexamples by application of the criteria to hard cases beyond the model is welcomed because these challenges are the bases for refinements that lend insight and understanding into theoretical models like the chemoton. Similarly, the chemoton model is not intended as an accurate representation that, if implemented exactly, could live. It is, instead, a heuristic guide to the organizational properties of chemical systems that would minimally fulfill the living state. Gánti's goal of an "exact theoretical biology" requires a formal model in which exact results can be achieved, even if these results do not precisely or realistically describe the behavior of actual chemical systems. In other words, the model is bound to falsely represent actual living systems, but this is to be expected, not an undesired side effect. No real living system has a template polymer that pairs a single monomer type like with like, or a membrane with a single type of constituent, or a metabolism consisting of a single catalytic cycle with a single kind of exogenous input. Heuristic theoretical models are tools of construction that can provide opportunities to refine criteria in response to the emergent model behavior, so that the criteria, in turn, can be used as tools of discovery for real systems, whether in the laboratory or in the field. In other words, with an engineering heuristic strategy, false models can lead to "truer" theories (Wimsatt, 1987).

22.3.1 Life Criteria: Absolute and Potential

Gánti's absolute life criteria present a means of characterizing the living state. These are related to, but importantly different from, concepts tracing back to Aristotle. To be alive, an entity must (1) be an *inherent unity* such that its properties are not additive compositions of properties of their parts, (2) perform *metabolism*, that is, chemically transform exogenous energy and matter into its own substance, (3) be dynamically or *inherently stable* in its organization, despite material turnover, (4) have subsystems with "surplus *information*" of potential use by the system as a whole, and (5) have processes that are *regulated and controlled*, ensuring the cyclical maintenance, recurrent functioning (homeostasis, autopoiesis), and directional changes of the system characteristic of development and evolution (Gánti 2003a, pp. 77–78). Whenever an entity exhibits these absolute criteria, it is *alive*. If it formerly exhibited them and is no longer capable of doing so, it is *dead*. If it formerly did and presently does not, but is capable of doing so again, it is *nonliving*. Otherwise, an entity formerly exhibiting the absolute criteria has *ceased to exist*, as when a living bacterium reproduces by fission, moving the parent from the state of being alive to that of nonexistence without passing through death and conversely for the offspring.

In contrast to these absolute criteria, which must be satisfied at every moment by each individual entity qualified as alive, Gánti's potential life criteria must be satisfied by some units at some times for there to be a persistent living world, that is, a world in which living units can serve as units of evolution. The potential life criteria include capacities of living systems for *growth* and *multiplication*, *hereditary change* and *evolution*, and *mortality* (Gánti, 2003a, pp. 78–79). Gánti's motivation for distinguishing between absolute and potential criteria is the observation that contemporary entities can be alive without satisfying all of the classical Aristotelian life functions (reviewed in Feldman, 1992), which mix together some of Gánti's absolute and potential criteria. Frozen organisms are not dead merely because they have ceased metabolizing, but they are nonliving since metabolism is an absolute life criterion. Sterile organisms do not fail to be alive, however, just because they are incapable of reproduction, so reproduction cannot be an absolute life criterion. Nor do nonvarying organisms in a homogeneous population and environment fail to be alive, so evolution (which requires population variation) cannot be an absolute life criterion. It is hard to imagine that sterile modern organisms (such as mules) or cells (such as neurons) could exist, however, if there were not already a living world in which some living systems (e.g., the sterile organism's parents), did have the capacity to multiply and satisfied the other potential life criteria as well. Artificial synthetic systems satisfying Gánti's absolute life criteria thus pose an interesting theoretical problem: Is the construction of such a system likely, thereby, to have evolutionary potential, or can artificial living systems that resemble modern living things lack evolutionary potential? Differently put, to what extent does evolutionary potential function as an operational absolute life criterion, contra Gánti's distinction of absolute and potential life criteria?

22.3.2 Which Came First: Living Things or a Living World?

Whether there can be living systems in the modern world that have no evolutionary potential is one thing, but the possible origin of living systems without a living world containing units of evolution at all is quite another. Gánti's potential life criteria are deeply connected to criteria for units of evolution, so the relationship between absolute and potential life criteria bears on the question of which came first: living systems or a living world. Gánti's chemoton satisfies both absolute and potential life criteria, so it does not directly answer this question. Nevertheless, further detailed examination reveals several clues.

One of Gánti's potential life criteria is a capacity for evolution, although as Szathmáry (2003) pointed out, only a population of varying entities can be a "unit" of evolution. (One could say, equivalently, that to be a unit of evolution, an individual entity must be part of a variable population of such entities.) Multiplication and a capacity for hereditary change are two requirements for units of evolution, along with variation (Maynard Smith, 1986, 1987; Szathmáry, 2002). Maynard Smith (1987) suggests that fitness variation is also a requirement for units of evolution that, therefore, are necessarily units of selection because, according to Maynard

Smith, units of selection are entities having multiplication and fitness variation. The key distinction for Maynard Smith concerns whether the entities in question have heredity. Not all analysts of units agree with this approach, particularly over the issue of what is required for adaptation at a level (see Sober, 1987), but the philosophical disagreement does not bear on the present argument. In addition, multiplication without growth must eventually cease (see Griesemer, 2000a, where this condition is called "general" as opposed to "special," or reproductive "progeneration").

Gánti's final potential life criterion, mortality, seems to us not a strict criterion for a living world, but rather a consequence of a living world in a state of resource limitation. Differently put, mortality is a recycling criterion for the *persistence* of a living world. A living world without mortality could exist so long as resources (including space) are not limiting and therefore need not be recycled to continue. Multiplication without mortality, like multiplication without growth, is ultimately a function of resource limitation.

It follows that the core potential life criteria are multiplication, heredity, and evolution. Furthermore, since multiplication, heredity, and variation are necessary conditions for units of evolution, being a unit of evolution is sufficient for satisfying Gánti's potential life criteria. Moreover, if satisfying the requirement for hereditary change entails satisfying the conditions for variation (without variation, there may be inheritance but not hereditary change), then the potential life criteria are necessary conditions for being a unit of evolution. Finally, if fitness is interpreted as "Malthusian" (Michod, 1999), then hereditary variation in the stoichiometric world of chemotons will almost inevitably result in fitness differences, since the rate of chemoton division will be a function of most variations in its stoichiometrically coupled chemical reactions. In this case, satisfying the potential life criteria will entail satisfaction of Maynard Smith's full requirements for units of evolution: multiplication, heredity, variation, and fitness differences (Maynard Smith, 1987). Thus, entities constitute a living world if and (almost) only if they are units of evolution.

The chemoton in figure 22.1 satisfies the potential life criteria. It grows and multiplies, and has heredity in a limited sense (Maynard Smith and Szathmáry, 1995, 1999; Szathmáry, 1995, 1999, 2002) in that polymers of different length (with a different balance coupling to metabolism and membrane) would be passed to offspring and would affect chemoton reproductive rates differentially. (Later we discuss a chemoton with heredity in a richer sense.) Moreover, a single chemoton will generate a population of varying chemotons according to the SCM (Szathmáry and Demeter, 1987), so the chemoton model represents a kind of organization that can function as a unit of evolution (as it must if it satisfies the potential life criteria). Therefore, a population of such chemotons can evolve.

The chemoton in figure 22.1 also satisfies the absolute life criteria. It is an inherent unity, whose properties are not all additive functions of its chemical properties (more on this later). It performs metabolism. It is inherently stable in that its stoichiometrically correct reactions adjust its internal structure in response to environmental changes (in food and waste concentrations), and its ability to multiply by membrane division serves to maintain its organization. It has a subsystem that carries surplus information: The length of the template polymer carries information about the state of the chemoton as a whole. Indeed, each of the three subsystems carries surplus information: The size of the membrane carries information about the length of the polymer and the concentration of metabolic cycle components; the concentrations of the latter carry information about the length of polymer and size of membrane. The reactions incorporating monomers into the polymer and the membrane both regulate and control the functioning of the metabolism. Indeed, the problems of information-carrying, regulation, and control may be as much a matter of specialized loss of those functions in some subsystems as specialized gain and elaboration in others (cf. division of labor in evolution, Maynard Smith and Szathmáry, 1995).

Gánti's argument that the full chemoton in figure 22.1 is the *minimal* chemical model of the organization of living systems, together with the fact that it satisfies both the absolute and potential life criteria, entails that there cannot be living systems without a living world, except by the convention that the limit case of a single chemoton that has not yet generated a population of variants does not constitute a living world. This does not contradict Gánti's argument that a given living system need not fulfill all the potential life criteria (since some other living systems in the living world must), but simply states that the *origin* of living systems entails the origin of a living world. Gánti's approach to this result is very different from projects of *defining* living systems as satisfying both absolute and potential criteria, so that living systems cannot *logically* precede a living world (Ruiz-Mirazo, Peretó, and Moreno, 2004).

These philosophical considerations are relevant to arguments about the empirical origins of life in the following way: For life to have *evolved* from nonliving chemical systems there must have been chemical units of evolution. Since being a unit of evolution in Maynard Smith's sense entails satisfaction of the potential life criteria in Gánti's sense, a living *world* must have arisen prior to living *systems*. Otherwise, living systems and a living world must have originated by a nonevolutionary process simultaneously, and then living systems subsequently evolved in the coincident living world. Thus, Gánti's life criteria point directly to the problem of whether genuinely chemical *evolution* was involved in the origin of life or not. Szathmáry (2002) argued that units of life and units of evolution are partially overlapping sets. We agree, but here we go further to explore whether life could have originated outside the region of overlap where units of evolution in the chemical world were not yet units of life. Gánti's arguments, together with our previous arguments, show that based on the life criteria assumed here, life could not have originated in living systems prior to the existence of a living world (figure 22.5).

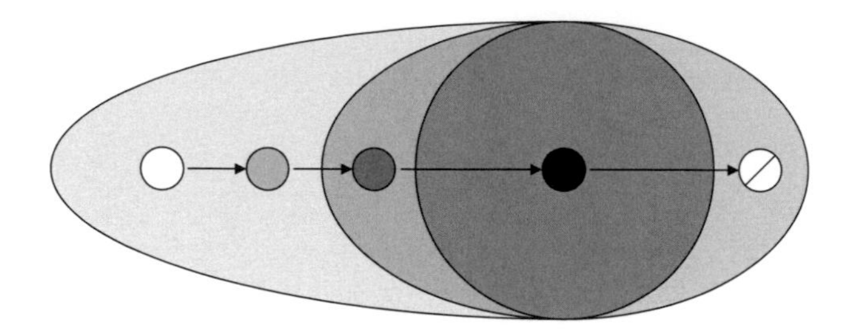

Figure 22.5
Units of living organization (modified from Szathmáry, 2002, figure 1, shading and legends added). Units of evolution (stippled) satisfy Gánti's potential life criteria whereas living systems (gray) satisfy the absolute life criteria (see text for explanation). These constitute overlapping classes. If life evolved from nonliving chemical systems, there must have been a progression of types of systems serving as units of evolution in a chemical world (open circle) through infrabiological systems (light gray circle) to fully living systems (black circle). Whether living systems arose before there was a living world concerns the placement of systems of chemoton grade (dark gray circle) in this progression. The region of overlap between units of evolution and units of living systems includes two subregions: one for systems like the chemoton (dark gray circle) with limited heredity and one for systems like modern cells (black circle) with unlimited heredity. A modern living system lacking reproductive potential would satisfy the absolute criteria for living systems but not the potential ones (circle with slash).

22.3.3 Infrabiology and Protobiology

Assuming Gánti is correct that the chemoton organization in figure 22.1 represents the minimal organization of a living system, we can investigate questions of evolution in the chemical world in terms of conditions under which various kinds of "infrabiological" or "protobiological" systems may fail to satisfy all of the absolute life criteria while satisfying all of the potential ones, thus fulfilling the requirements for chemical units of evolution (Luisi, 1998, calls infrabiological systems "way stations" on the continuous transition to life). Infrabiological systems lack (at least) one essential subsystem of a minimal living system, but exhibit a crucial subset of biological phenomena, for example, a membrane surrounding a self-catalytic ribozyme lacking a proper metabolism (figure 22.6).

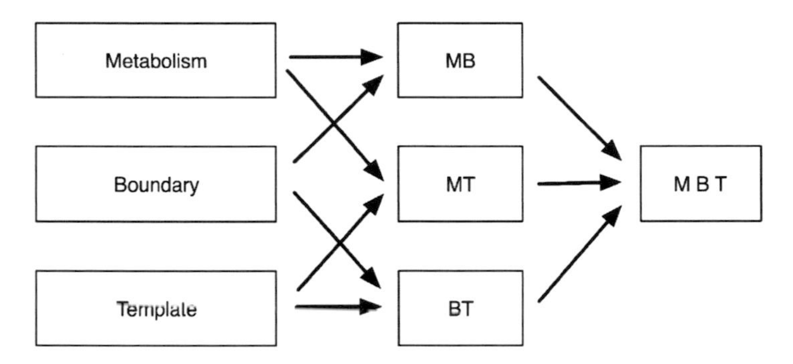

Figure 22.6
Infrabiological systems (after Szathmáry, Santos, and Fernando, 2005, figure 2). An infrabiological system includes some but not all of the three kinds of component subsystems identified as necessary and sufficient for living systems according to Gánti's chemoton model, together with his life criteria. The possibilities include metabolism, a boundary membrane, a template polymer, and all pairwise combinations of these (MB, MT, BT).

A key question is whether any infrabiological systems (metabolism + boundary, metabolism + template, boundary + template) can satisfy all of the potential life criteria—growth and multiplication, hereditary change and evolution, mortality—without *thereby* satisfying the absolute life criteria. Classical RNA world hypotheses fall into this category, but these tend also to assume definitions of life that render such systems living (Luisi, 1998). A related question is which absolute criteria remain unsatisfied in such systems and whether evolutionary paths from them to full chemoton organization are plausible. Infrabiological systems lacking a membrane would presumably have to be rather heterotrophic compared to the other possibilities since valuable products of metabolism would diffuse away, so early Earth conditions would provide additional constraints combining with the theoretical ones regarding satisfaction of criteria for units of evolution. Metabolism-first hypotheses may not be plausible in a hot environment without some means of holding reaction products together, but then hot environments would probably have been hostile to naked RNA ribozymes as well, since these have to fold to do their catalytic work (Penny, 2005). If some type of infrabiological system could satisfy only the potential life criteria, then a living world could emerge before life and life could evolve from chemical systems. The particular form of evolvable infrabiological system would give clues to possible evolutionary routes to the full chemoton grade of organization. If not, then full living systems must have emerged somehow without evolution, which appears implausible given the complexity of even the heuristic chemoton model.

Alternatively, chemical evolution might pass through protochemotons rather than infrabiological systems. That is, prebiotic chemical systems might have involved primitive versions of all three subsystems (boundary, template, and metabolism) without autocatalysis manifested in all of them. Protochemotons would represent a prebiotic grade of organization if the term *protocell* is used to mean minimal *living* system, since, according to Gánti's argument, a full chemoton is the minimal living system and therefore a protochemoton would fail to be a living protocell. Presumably, all such protosystems would have had to be heterotrophic with regard to the subautocatalytic subsystems to drive "multiplication" of the protochemoton as a whole. More research is needed on the relative ease of autocatalysis emerging from already catalytic systems compared to the stoichiometric integration of additional subsystems into already autocatalytic infrabiological systems. This question bears on those artificial synthesis research programs that aim to discover plausible steps in the evolution of life on Earth. It may be possible to construct a variety of infrabiological systems or protochemotons that do not reflect plausible or likely paths of transformation from simple chemical catalytic systems to the full chemoton grade of organization. Gánti's chemoton model thus provides a valuable distinction between infrabiological and protochemoton organization.

A further question concerns whether any infrabiological system or protochemoton could satisfy all of the absolute life criteria without also satisfying the potential life criteria. Because Gánti argued that the chemoton is the minimal organization of a living system and that it also satisfies the potential life criteria, such a situation would violate his chemoton theory (chemoton model plus life criteria): something would have to be alive, yet not of full chemoton organization and also not satisfy the potential life criteria, hence there would be living systems without a living world. In such a situation, life would have arisen "spontaneously," that is, chemically, without an evolutionary process driving it. Since Gánti has argued that satisfying the absolute criteria but not the potential ones must be possible in the context of contemporary living systems in a richly populated living world, his views challenge us to provide answers for the chemical world and the origins of living systems and for artificially synthesized systems. Traditional RNA world scenarios with a ribozymic origin of life (no metabolism and no boundary other than covalent bonding) would fit this conception if we accept life criteria that fit them. It is a conceivable approach, but one that must answer Gánti's arguments that chemoton organization is the minimum for living systems. The engineering approach thus shows the fallacy of defining life to fit the conception of a particular chemical model and the advantages of articulating model and criteria together.

22.3.4 Stoichiometric Freedom of Properties versus Autonomy of Systems

Many of our claims about Gánti's life criteria and their application to the chemoton model are consistent with theories of autopoiesis (Maturana and Varela, 1980; Varela, 1979, 1994; Varela, Maturana, and Uribe, 1974) and biological autonomy (Christensen and Hooker, 2000, Ruiz-Mirazo, Peretó, and Moreno, 2004). All three kinds of theory focus on ways in which living systems are self-producing as well as self-maintaining unities, providing for their own stability, regulation, and control. However, proponents of these various approaches emphasize different aspects and sometimes identify different points in the transition from a chemical world to a biological world as the critical point or marker at which systems count as minimally, fully, or fundamentally living. Ruiz-Mirazo and coworkers (2004) identify life with the transition from a cellular "one-polymer" (RNA) to a "two-polymer" (genotype/phenotype separated) world. Varela's autopoiesis (1994) and Christensen and Hooker's (2000) autonomy theories identify living systems with dynamical individuation of self-generating, reinforcing, and maintaining boundaries, that is, of self-defining unities. Both argue that these have the basic organization of cells, but do not require that cells must include something like Gánti's "genetic polymer" or Ruiz-Mirazo's "informational records." Christensen and Hooker include Rebek's bicyclic molecular catalysts as the simplest autonomous systems, while Varela favors bacterial cells as minimal. Rebek himself classifies his synthetic chemical systems as "extra-biotic," meaning that they may or may not be parts of actual living systems, but are lifelike and yet not living (Rebek, 1994b).

One diagnosis of the differences among these theoretical approaches is that the focus of discussion has been on interpreting how the mechanisms and material organization of chemical systems constitutes *system* autonomy. The main contemporary target of such views is the identification of living systems with those self-sustaining chemical systems that meet criteria for units of evolution, which might include individual ribozyme molecules as living (see Joyce, 1994). From autonomy-theoretic points of view, whether molecules satisfy criteria for units of evolution does not answer the question of their status as living systems, which has more to do with the operation of a self-sustaining metabolism than with its evolvability. That is, they seek an emergent property of the system as a whole that can justify calling it a living organization. They differ on details of what that property is, not so much because they have different ideas about sufficient conditions for the living state of the last universal common ancestor or in an experimenter's test tube, or about concepts of autonomy, but rather about the kinds of chemical matter and organizations that can realize the various necessary component properties of those sufficient conditions. This is a question that is best answered empirically, most likely by synthetic biochemistry: The construction of various chemical systems may show that satisfaction of particular life criteria is easier or harder than anticipated, given the behavior of component molecules on the one hand and theoretical and simulation studies of chemoton models on the other.

Just as it is risky to base an account of life criteria on modern living systems, perhaps the conceptual risk in origins of life research lies in focusing on the emergent character of living systems as opposed to the basis of emergence in the properties of their parts. Characterizing the autopoiesis or autonomy of a system is important to general understanding, but as Ruiz-Mirazo and colleagues (2004) point out, it tends toward a level of abstraction unlikely to yield life definitions that are sufficiently operational to move empirical research forward. To this criticism, we add that remaining in the realm of abstraction also fails to serve in the identification of those properties of system components whose interactions form the basis of those emergent properties of living systems.

John Stuart Mill recognized in the 19th century that chemical reactions have an emergent quality in relation to their physical components, for example, when $NaOH + HCl \rightarrow NaCl + H_2O$ (see O'Connor and Wong, 2005). The empirical problem is not merely to recognize *that* the properties of the products are not physically additive compositions of the properties of the reagents, but to determine *how*. What physical interactions yield chemical properties, and how shall we characterize them? By analogy, how shall we characterize the chemical basis of biologically emergent properties of chemical systems? Life criteria go part of the way, by breaking the problem into subproblems—those features of chemically emergent properties that must be satisfied for a system to be alive, or for a living world to exist. They do not, however, identify the basis of biological emergence in specific properties (or interactions) of chemical components. To do that, aspects of living organization must be bound to specific models of chemical systems with energetic and material constraints specified.

Emergent properties are "Janus faced." One face looks "upward," to the emergent level we wish to characterize and understand. The other face looks "downward," to the level from which the emergent property is composed. Our focus is downward, toward chemical properties that are not by themselves biological, but whose origins in chemical organization create biological potential. Here we focus on the property of *stoichiometric freedom*, a *chemically* emergent property of molecules in infra- or protobiological systems on the way to autonomous or autopoietic living systems. Such chemical properties may constitute intermediate steps or stages to biologically emergent system behavior.

If we consider a slightly more complicated abstract chemoton proposed by Gánti, we can begin to see how to evaluate promising chemical properties of infrabiological systems and protochemotons as chemical systems on the pathway to modern living systems and as potential generators of a modern living world. Let V and W stand for arbitrary monomers, [V] and [W] for their concentrations. Suppose the chemoton's metabolism is elaborated (e.g., by introducing an additional food input) such that it produces two kinds of template monomers, V and W, instead of only one. This generates two new properties in the polymer: a composition property, that is, the relative concentrations of the monomers in the polymer, [V]/[W], and a sequence property, that is, the order of monomers in the polymer, for example, V–W–V–V–W–V– . . . , of which there are 2^{m+n} unique kinds, where $[V] = m$ and $[W] = n$.

The composition property should reflect the concentrations of monomers produced and circulating in the internal milieu of the chemoton. That is, composition is a stoichiometric function of the metabolism that produces the monomers and the polymerization reactions that incorporate them. The order property of monomers, or sequence, however, is a stoichiometrically free property: It does not depend on the stoichiometry of the chemoton, except insofar as possible sequences are constrained to given compositions (and assuming there are no steric constraints among adjacent monomers). An innovation in metabolism that produced a second type of monomer would not *thereby* also constitute a mechanism for the control of the composition of the template polymer because there are no means for feedback of sequence differences to chemical composition of the chemoton. So, for a given composition, each sequence is approximately equiprobable and therefore independent of stoichiometric constraint.

The monomer sequence is not yet a *biological* property because it has no significance for the chemoton, per se. Although the polymer has "surplus information" in its length and composition because of its stoichiometric couplings, the sequence property must be *exploited* for *it* to constitute such a surplus. Since sequence is stoichiometrically free, this exploitation will not be by virtue of new stoichiometric relations, at least not directly. Stoichiometric freedom is the downward-looking face of an emergent property relative to living systems: freedom from constraint, or what political philosophers call "negative freedom" (Berlin, 1958). In order to have positive freedoms—the freedom to do what you will—you must first have negative freedom from constraints that would prevent you from pursuing your wants and desires. We propose that something analogous to negative freedom in the chemical world will be a prominent step toward biologically emergent properties, those available for open-ended evolution, on the way to modern living systems with a genotype/phenotype distinction.

If V and W self- or complement-pair during template autocatalysis (i.e., V with V or V with W), then this elaborated chemoton is a hereditary system with some potential for open-ended evolution. Maynard Smith and Szathmáry (1999) identify this possibility with the presence of a subsystem of "unlimited" heredity. Heredity is unlimited if the number of possible states far exceeds the number realizable in any actual population, such that a selection process cannot quickly discover the global optimum among the variants. Heredity is limited in proportion to how close the number of possible states is to the number of variants in actual populations. These can evolve for a while until the global optimum evolutionarily stable state is discovered.

The simple chemoton of figure 22.1 can pass on changes in polymer length through template reproduction, so it is a hereditary system. However, length is a holistic property of the template—only the transmission of the whole molecule (or its templated complement) transmits the parental length. No parts of a holistic system can be changed (added or deleted) without changing the whole system (Maynard Smith and Szathmáry, 1999). In contrast, a system is modular if a change in one part leaves the others unaltered, so that the causes of a system property change can be decomposed into changes in component parts. Put differently, the effects of changes in parts are local or regional rather than systemic. By the same token, additional features of modular systems must be specified to explain how changes of modules can have systemic effects. In the case of genetic polymers, it is clear in chemoton stoichiometry how a change in polymer length would affect the rest of the system. But a change in polymer sequence would not obviously affect chemoton behavior.

In the elaborated chemoton, the presence of two monomer types results in composition and sequence properties of the polymer. Composition is again a holistic property of the polymer (although if monomer incorporation is random, polymer subsequences may statistically resemble the composition of the whole and thus be "statistically" modular). Interestingly, sequence is a modular property of a holistic hereditary system in the elaborated chemoton. The length property is still holistic, and the only one that feeds back to the regulation of the chemoton, whereas the sequence property is idle in this regard. The degree to which such a polymer can function as an unlimited system of heredity depends on *both* the length of the polymer (relative to the number of monomer-types) and on the further requirement that subsequences have some relevance to the chemical operation of the chemoton, which in the elaborated chemoton they do not have.

This modularity not only means that subsequence properties can be passed on without transmission of the full polymer, but also that the equi-probability of sequences with respect to a given set of concentrations—their stoichiometric freedom—can become the target of new, second-order regulation and control through their interactions with *other* molecules (figure 22.7). Molecules that can bind polymers can, for example, affect their rates of polymerization and template reproduction. Primary interactions in such a system establish chemical relations between monomers in the polymer (both covalent bonds within a strand and hydrogen bonds between strands). They are called primary because they are the product of chemical reactions specified by the basic chemoton organization and stoichiometry. Second-order interactions of molecules binding particular subsequences in the polymer will be the product of two stoichiometry-dependent molecules, the polymer and the binder. Because the order of monomers in the polymer is stoichiometrically free, the interaction is also stoichiometrically free, even if the binding molecule's structure and binding specificity is a stoichiometric function of its components. So, the second-order interactions are stoichiometrically free insofar as they depend on free properties of their interactants.

Thus far, we have made only a preliminary step toward a biologically emergent system property. But now a new possibility arises: interaction between binding mole-

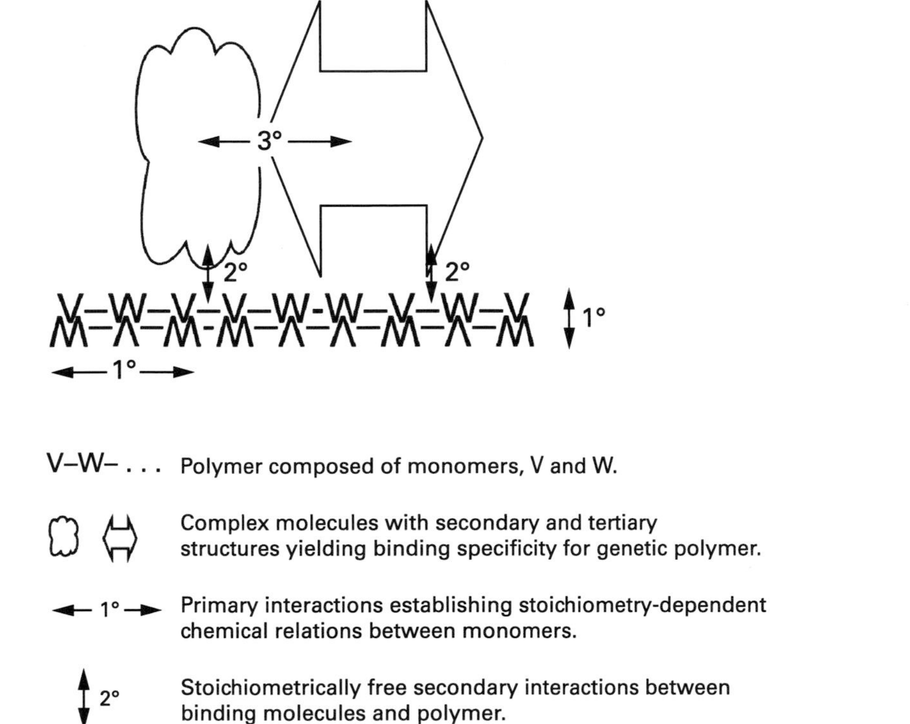

Figure 22.7
Stoichiometrically free interactions. Three types of chemical interaction are depicted. Primary chemical interactions (both covalent and hydrogen bonds) depend on the stoichiometry of the participating monomers produced in the chemoton. Secondary interactions are "stoichiometrically free," in the sense that they occur as a function of the stoichiometry-independent sequence of monomers in the polymer(s) depicted. An interaction is stoichiometrically free if the structure of at least one of its participants is. The tertiary interaction depicted in the figure is also stoichiometrically free because it occurs as a function of stoichiometry-free second-order interactions. (See text for further explanation.)

cules that bind the polymer in a stoichiometrically free interaction. These interactions should be considered third order because they occur as a consequence of second-order interactions between binding molecules and polymer. These third-order interactions have a new property: They can affect the lower-level rates of polymerization and templating of the polymer by virtue of *relations* between subsequences of the polymer. If a molecule binds a particular subsequence by virtue of their stoichiometrically free interaction and another binding molecule binds another particular subsequence in another stoichiometrically free interaction, then any interaction *between* the two binding molecules will be a consequence of stoichiometrically free second-order interactions. (Even steric hindrance at this level will be a stoichiometrically free interaction, on account of relations resulting from the interaction of the lower-level polymer and sequence-specific binding molecules.) Now we have a situation in which interactions between two binding molecules are mediated by relations among subsequences in the polymer and second-order interaction specificities of particular binding molecules. The tertiary relation between the binding molecules carries "surplus information" about the relations of subsequences in the polymer and conversely.

The stoichiometric freedom of the sequence property of the linear "genetic" polymer in the elaborated chemoton is the chemical basis for a potentially biological property—a specific interaction among two other molecules "represented" in the polymer. Conversely, we can say that the polymer sequence—specifically the relations between subsequences—"codes for" the tertiary interaction. Insofar as the latter interaction has a "feed-down" effect on the template reproduction of the polymer, and template reproduction has a stoichiometric relation to metabolism and membrane growth and division, this biologically emergent property can potentially serve a regulation and control function in the chemoton.

This is by no means to suggest a coherent model for the emergence of a genotype/phenotype separation in the elaborated chemoton, but it does point to a more fundamental feature of emergent biological properties: the way in which biological emergence depend on freedom from purely chemical constraints. It should not escape notice, however, that the system of relations manifest in modern living systems with genotype/phenotype separation depends on at least tertiary relations among molecules analogous to those described here. The amino acids tethered to tRNAs, which bind mRNA, facilitated at the ribosome, engage in stoichiometrically free tertiary interactions.

22.4 Conclusion

We have shown that the chemoton model, together with Gánti's life criteria, provides a fruitful platform for theoretical and empirical modeling and simulation of various infrabiological and protochemoton systems. Although the chemoton theory has not been applied to actual cases of artificially synthesized chemical systems, it has great potential for such application because it offers a coherent set of life criteria together with a well-specified model, the chemoton, which satisfies the criteria. The chemoton theory sheds light on the project of artificial synthesis of living systems as well as the evolutionary origins of life in three ways: (1) by identifying significant synthetic challenges (such as the side reaction problem), (2) by identifying the full chemoton grade of organization (stoichiometrically coupled metabolism, genetic molecule, and membrane) as a minimal living system, and (3) by distinguishing infrabiological and protochemoton pathways from chemical to living organization. These pathways may present greater or lesser synthetic challenges as well as more or less likely evolutionary paths to modern living systems. Far from being a mere historical curiosity, chemoton theory is very much relevant to present attempts to fulfill a long-standing dream of mankind: the synthesis of artificial living systems.

Acknowledgment

This work was supported by the Hungarian Scientific Research Fund (OTKA T047245), the National Office for Research and Technology (NAP 2005/KCKHA005), and the COST D27 action (Prebiotic chemistry and early evolution).

References

Békés, F. (1975). Simulation of kinetics of proliferating chemical systems. *BioSystems*, *7*, 189–195.

Berlin, I. (1958). *Two concepts of liberty*. Oxford: Clarendon.

Bernal, J. D. (1967). *The origin of life.* London: Weidenfeld & Nicolson.

Breslow, R. (1959). On the mechanism of the formose reaction. *Tetrahedron Letters*, *1* (21), 22–26.

Butlerov, A. (1861). Bildung einer zukerartigen Substanz durch Synthese. *Liebigs Annalen Der Chemie*, *120*, 295.

Chakrabarti, A. C., & Deamer, D. W. (1992). Permeability of lipid bilayers to amino acids and phosphate. *Biochimica et Biophysica Acta Biomembranes*, *1111*, 171–177.

Christensen, W. D., & Hooker, C. A. (2000). Autonomy and the emergence of intelligence: Organized interactive construction. *Communication and Cognition—Artificial Intelligence*, *17* (3–4), 133–157.

Crick, F. (1968). The origin of the genetic code. *Journal of Molecular Biology*, *38*, 367–379.

Csendes, T. (1984). A simulation study on the chemoton. *Kybernetes*, *13*, 79–85.

Dawkins, R. (1976). *The selfish gene*. New York: Oxford University Press.

Deamer, D., & Fleischaker, G. (Eds.) (1994). *Origins of life: The central concepts.* Boston: Jones and Bartlett.

Decker, P., & Speidel, A. (1972). Open systems which can mutate between several steady states ("bioids") and a possible prebiological role of the autocatalytic condensation of formaldehyde. *Zeitschrift für Naturforschung B*, *27*, 257–262.

Eigen, M. (1971). Self-organization of matter and the evolution of biological macromolecules. *Naturwissenschaften*, *58*, 465–523.

Feldman, F. (1992). *Confrontations with the reaper: A philosophical study of the nature and value of death.* New York: Oxford University Press.

Fernando, C., von Kiedrowski, G., & Szathmáry, E. (2007). A stochastic model of nonenzymatic nucleic acid replication: "Elongators" sequester replicators. *Journal of Molecular Evolution*, *64*, 1–15.

Ferris, J. P. (2002). Montmorillonite catalysis of 30–50 mer oligonucleotides: Laboratory demonstration of potential steps in the origin of the RNA world. *Origins of Life and Evolution of the Biosphere*, *32*, 311–332.

Fleischaker, G. R. (1994). A few precautionary words concerning terminology. In G. R. Fleischaker, S. Colonna, & P. L. Luisi (Eds.), *Self-production of supramolecular structures: From synthetic structures to models of minimal living systems* (pp. 33–41). Dordrecht: Kluwer.

Folsome, C., & Brittain, A. (1981). Model protocells photochemically reduce carbonate to organic carbon. *Nature*, *291*, 482–484.

Fraser, C. L., & Folsome, C. E. (1975). Exponential kinetics of formation or organic microstructures. *Origins of Life*, *6*, 429–433.

Gánti, T. (1966). *Forradalom az élet kutatásában* (*Revolution in life research*). Budapest: Gondolat.

Gánti, T. (1971). *Az élet princípuma* (*The principle of life*). Budapest: Gondolat.

Gánti, T. (1975). Organization of chemical reactions into dividing and metabolizing units: The chemotons. *Biosystems*, *7*, 15–21.

Gánti, T. (1978). *Az élet princípuma* (*The principle of life*) (2nd rev. ed.). Budapest: Gondolat.

Gánti, T. (1979). *A theory of biochemical supersystems and its application to problems of natural and artificial biogenesis*. Baltimore: University Park.

Gánti, T. (1983). The origin of the earliest sequences (in Hungarian). *Biológia*, *31*, 47–54.

Gánti, T. (1987). *The principle of life*. Budapest: OMIKK.

Gánti, T. (1997). Biogenesis itself. *Journal of Theoretical Biology*, *187* (4), 583–593.

Gánti, T. (2003a). *The principles of life, with a commentary by James Griesemer and Eörs Szathmáry*. Oxford: Oxford University Press.

Gánti, T. (2003b). *Chemoton theory. Volume 1: Theoretical foundations of fluid machineries. Volume 2: Theory of living systems.* New York: Kluwer Academic/Plenum.

Gilbert, W. (1986). The RNA world. *Nature*, *319*, 618.

Goldacre, R. J. (1958). Surface films: Their collapse on compression, the shapes and sizes of cells, and the origin of life. In J. F. Danielli, K. G. Pankhurst & A. C. Riddiford (Eds.), *Surface phenomena in biology and chemistry* (pp. 12–27). New York: Pergamon.

Grey, D., Hutson, V., & Szathmáry, E. (1995). A re-examination of the stochastic corrector model. *Proceedings of the Royal Society of London B*, *262*, 29–35.

Griesemer, J. (2000a). Development, culture and the units of inheritance. *Philosophy of Science*, *67*, S348–S368.

Griesemer, J. (2000b). Reproduction and the reduction of genetics. In P. Beurton, R. Falk & H.-J. Rheinberger (Eds.), *The concept of the gene in development and evolution: Historical and epistemological perspectives* (pp. 240–285). Cambridge: Cambridge University Press.

Griesemer, J. (2000c). The units of evolutionary transition. *Selection*, *1*, 67–80.

Griesemer, J. (2003). The philosophical significance of Gánti's work. In T. Gánti, *The principles of life, with a commentary by James Griesemer and Eörs Szathmáry* (pp. 169–194). Oxford: Oxford University Press.

Haldane, J. B. S. (1929). The origin of life. *The Rationalist Annual (1929)*. Reprinted in D. Deamer & G. Fleischaker (Eds.) (1994). *Origins of life: The central concepts* (pp. 73–81). Boston: Jones and Bartlett.

Hargreaves, W. R., Mulvihill, S. J., & Deamer, D. W. (1977). Synthesis of phospholipids and membranes in prebiotic conditions. *Nature*, *266*, 78–80. Reprinted in D. Deamer & G. Fleischaker (Eds.). (1994). *Origins of life: The central concepts* (pp. 205–207). Boston: Jones and Bartlett.

Huber, C., & Wächtershäuser, G. (1997). Activated acetic acid by carbon fixation on (Fe,Ni)S under primordial conditions. *Science*, *276*, 245–247.

Joyce, G. (1994). Foreword. In D. Deamer & G. Fleischaker (Eds.), *Origins of life: The central concepts* (pp. xi–xii). Boston: Jones and Bartlett.

Kauffman, S. (1986). Autocatalytic sets of proteins. *Journal of Theoretical Biology*, *119*, 1–24.

King, G. (1982). Recycling, reproduction, and life's origin. *BioSystems*, *15*, 89–97.

Kole, R., & Altman, S. (1981). Properties of purified ribonuclease P from Escherichia coli. *Biochemistry*, *20*, 1902–1906.

Korányi, P., & Gánti, T. (1981). Coenzymes as remnants of ancient enzyme ribonucleic acid molecules (in Hungarian). *Biológia*, *19*, 107–124.

Kruger, K., Grabowski, P. J., Zaug, A. J., Sands, J., Gottschling, D. E., & Cech, T. R. (1982). Self-splicing RNA: Autoexcision and autocyclization of the ribosomal RNA intervening sequence of Tetrahymena. *Cell*, *31*, 147–157.

Lehman, N. (2003). A case for the extreme antiquity of recombination. *Journal of Molecular Evolution*, *56*, 770–777.

Leigh, E. G. (1983). When does the good of the group override the advantage of the individual? *Proceedings of the National Academy of Sciences of the United States of America*, *80*, 2985–2989.

Luisi, P. L. (1998). About various definitions of life. *Origins of Life and Evolution of the Biosphere*, *28*, 613–622.

Maturana, H., & Varela, F. (1980). *Autopoiesis and cognition: The realization of the living*. Boston: D. Reidel.

Maynard Smith, J. (1986). *The problems of biology*. New York: Oxford University Press.

Maynard Smith, J. (1987). How to model evolution. In J. Dupre (Ed.), *The latest on the best: Essays on evolution and optimality* (pp. 119–131). Cambridge, MA: MIT Press.

Maynard Smith, J., & Szathmáry, E. (1995). *The major transitions in evolution*. Oxford: W. H. Freeman Spektrum.

Maynard Smith, J., & Szathmáry, E. (1999). *The origins of life: From the birth of life to the origins of language*. Oxford: Oxford University Press.

Michod, R. (1999). *Darwinian dynamics: Evolutionary transitions in fitness and individuality*. Princeton: Princeton University Press.

Monnard, P. A., & Deamer, D. W. (2001). Nutrient uptake by protocells: A liposome model system. *Origins of Life and Evolution of the Biosphere*, *31*, 147–155.

Morowitz, H. (1992). *Beginnings of cellular life: Metabolism recapitulates biogenesis*. New Haven: Yale University Press.

Morowitz, H., Heinz, B., & Deamer, D. (1988). The chemical logic of a minimum protocell. *Origins of Life and Evolution of the Biosphere*, *18*, 281–287.

Morowitz, H., Kostelnik, J., Yang, J., & Cody, G. (2000). The origin of intermediary metabolism. *Proceedings of the National Academy of Sciences of the United States of America*, *97* (14), 7704–7708.

Oberholzer, T., Wick, R., Luisi, P. L., & Biebricher, C. K. (1995). Enzymatic RNA replication in self-reproducing vesicles: An approach to a minimal cell. *Biochimica et Biophysica Research Communications*, *207*, 250–257.

O'Connor, T., & Wong, H. Y. (2005). Emergent properties. In E. N. Zalta (Ed.), *The Stanford encyclopedia of philosophy* (Summer 2005 ed.). Accessed March 14, 2007, at: http://plato.stanford.edu/archives/sum2005/entries/properties-emergent/.

Oparin, A. I. (1924). Proiskhozhdenie zhizny (The origin of life). Moscow. Izd. Moskovshii. Reprinted (Ann Synge, trans.) in D. Deamer & G. Fleischaker (Eds.). (1994). *Origins of life: The central concepts* (pp. 31–71). Boston: Jones and Bartlett.

Orgel, L. (1968). Evolution of the genetic apparatus. *Journal of Molecular Biology*, *38*, 381–393.

Orgel, L. (2000). Self-organizing biochemical cycles. *Proceedings of the National Academy of Sciences of the United States of America*, *97* (23), 12503–12507.

Penny. D. (2005). An interpretive review of the origin of life research. *Biology and Philosophy*, *20*, 633–671.

Rebek, J. (1994a). Synthetic self-replicating molecules. *Scientific American*, *271* (1), 48–53, 55.

Rebek, J. (1994b). Extrabiotic replication and self-assembly. In G. R. Fleischaker, S. Colonna, & P. L. Luisi (Eds.), *Self-production of supramolecular structures: From synthetic structures to models of minimal living systems* (pp. 75–87). Dordrecht: Kluwer.

Ruiz-Mirazo, K., Peretó, J., & Moreno, A. (2004). A universal definition of life: Autonomy and open-ended evolution. *Origins of Life and Evolution of the Biosphere*, *34*, 323–346.

Santos, M., Zintzaras, E., & Szathmáry, E. (2004). Recombination in primeval genomes: A step forward but still a long leap from maintaining a sizable genome. *Journal of Molecular Evolution*, *59*, 507–519.

Shapiro, R. (1986). *Origins: A skeptic's guide to the creation of life on Earth*. New York: Bantam.

Shapiro, R. (1988). Prebiotic ribose synthesis: A critical analysis. *Origins of Life and Evolution of the Biosphere*, *18*, 71–85.

Singer, S., & Nicolson, G. (1972). The fluid mosaic model of the structure of cell membranes. *Science, 175* (4023), 720–731.

Smith, E., & Morowitz, H. (2004). Universality in intermediary metabolism. *Proceedings of the National Academy of Sciences of the United States of America, 101* (36), 13168–13173.

Sober, E. (1987). Comments on Maynard Smith's "How to model evolution." In J. Dupre (Ed.), *The latest on the best: Essays on evolution and optimality* (pp. 133–145). Cambridge, MA: MIT Press.

Stillwell, W. (1980). Facilitated diffusion as a method for selective accumulation of materials from the primordial oceans by a lipid-vesicle protocell. *Origins of Life, 10*, 277–292.

Stillwell, W., & Rau, A. (1981). Primordial transport of sugars and amino acids via Schiff bases. *Origins of Life, 11*, 243–254.

Szabó, P., Scheuring, I., Czárán, T., & Szathmáry, E. (2002). In silico simulations reveal that replicators with limited dispersal evolve towards higher efficiency and fidelity. *Nature, 420*, 360–363.

Szathmáry, E. (1989). The emergence, maintenance, and transitions of the earliest evolutionary units. *Oxford Surveys in Evolutionary Biology, 6*, 169–205.

Szathmáry, E. (1990). RNA worlds in test tubes and protocells. In J. Maynard Smith & G. Vida (Eds.), *Organizational constraints on the dynamics of evolution* (pp. 3–14). Manchester: Manchester.

Szathmáry, E. (1995). A classification of replicators and lambda-calculus models of biological organization. *Proceedings of the Royal Society of London Series B, 260*, 279–286.

Szathmáry, E. (1999). Chemes, genes, memes: A revised classification of replicators. *Lectures on Mathematics in the Life Sciences, 26*, 1–10.

Szathmáry, E. (2002). Units of evolution and units of life. In G. Pályi, L. Zucchi, & L. Caglioti (Eds.), *Fundamentals of life* (pp. 181–195). Paris: Elsevier.

Szathmáry, E. (2003). The biological significance of Gánti's work in 1971 and today. In T. Gánti, *The principles of life, with a commentary by James Griesemer and Eörs Szathmáry* (pp. 157–168). Oxford: Oxford University Press.

Szathmáry, E., & Demeter, L. (1987). Group selection of early replicators and the origin of life. *Journal of Theoretical Biology, 128*, 463–486.

Szathmáry, E., Santos, M., & Fernando, C. (2005). Evolutionary potential and requirements for minimal protocells. *Topics in Current Chemistry, 259*, 167–211.

Szostak, J., Bartel, D., & Luisi, P. (2001). Synthesizing life. *Nature, 409*, 387–390.

Varela, F. (1979). *Principles of biological autonomy*. New York: North-Holland/Elsevier.

Varela, F. (1994). On defining life. In G. R. Fleischaker, S. Colonna & P. L. Luisi (Eds.), *Self-production of supramolecular structures: From synthetic structures to models of minimal living systems* (pp. 23–31). Dordrecht: Kluwer.

Varela, F., Maturana, H., & Uribe, R. (1974). Autopoiesis: The organization of living systems, its characterization and a model. *BioSystems, 5*, 187–195.

von Kiedrowski, G. (1986). A self-replicating hexadeoxynucleotide. *Angewandte Chemie International English Edition, 25*, 932–934.

von Kiedrowski, G. (1999). Molekulare Prinzipien der artifiziellen Selbstreplikation. In D. Ganten (Ed.), *Gene, Nerone, Qubits & Co. Unsere Welten der Information* (pp. 123–145). Stuttgart: S. Hirzel.

Wächtershäuser, G. (1992). Groundworks for an evolutionary biochemistry: The iron-sulphur world. *Progress in Biophysics and Molecular Biology, 58* (2), 85–201.

White, H. B. (1976). Coenzymes as fossils of an earlier metabolic state. *Journal of Molecular Evolution, 7*, 101–104.

Wimsatt, W. C. (1987). False models as means to truer theories. In M. Nitecki & A. Hoffman (Eds.), *Neutral models in biology* (pp. 23–55). London: Oxford University Press.

Woese, C. (1967). *The genetic code*. New York: Harper & Row.

Zintzaras, E., Mauro, S., & Szathmáry, E. (2002). "Living" under the challenge of information decay: The stochastic corrector model versus hypercycles. *Journal of Theoretical Biology, 217*, 167–181.

23 Viral Individuality and Limitations of the Life Concept

David C. Krakauer and Paolo Zanotto

23.1 Minimality

One of the outstanding problems in research into the origins of life has been deriving a definition of life that satisfies two criteria—one inclusive and one exclusive. The exclusion criterion is that the definition should dichotomize the physical universe into living and nonliving systems, whereas the inclusion criterion is that the definition should accommodate all compelling empirical examples of life—from bacteria to protists, algae, flies, and mammals. Based on common observed properties of biological life, there is agreement that all living entities should be capable of exploiting chemical energy stores in order to replicate with high fidelity and should be capable of doing so in a range of conditions. And, given the tendency of molecules to diffuse in R^3 space, they should also have some means of concentrating essential reactive chemical species. This typically takes the form of a semipermeable compartment. Hence, replication, adaptability, metabolism, and cellularity have emerged as *minimal* necessary conditions for life (De Duve, 1991; Orgel, 1998; Rasmussen et al., 2003).

One of the problems with the life concept is that it is categorical, or quasi-categorical. Systems are said to be alive or dead—or at least not alive. Cases such as the mobile catalytic parasites known as viruses are often located in a sort of life-limbo, neither living nor dead, but alternating between these two states over the course of a single life cycle (Haldane, 1929). Applying the minimality conditions filter to viruses, we see they possess simple cellularity through the capsid or envelope and are capable of adaptive proliferation in appropriate environments, but never anabolic metabolism. As they are in partial fulfillment of life criteria, they have been hard to place, and some would classify them as unique hyperparasites, providing little insight into the rest of biology.

We suggest that viruses expose a fundamental scientific limitation of our current life concept, which strives to exclude physical systems from biology based on rather artificial requirements derived ex post facto from a subset of the full range of biology we observe on Earth.

It remains unstated, but frequently assumed, that replication, adaptability, metabolism, and cellularity need to be encoded in the sequence and structure of a single entity. Functionally, however, these four properties need only be highly correlated in space and time. Ensuring that all four derive from a single source—*evolutionary autonomy*—is one way of increasing coordination among members of a network, but it is not the only way. Biology presents us with a continuum of autonomy, with highly autonomous systems typically chosen as illustrative examples of life at one end of the spectrum and viruses at the other. Even large mammals have a need for essential amino acids they are incapable of synthesizing; these include tryptophan, lysine, methionine, phenylalanine, threonine, valine, leucine, and isoleucine.

In this chapter, rather than discuss living and nonliving, we attempt to identify a continuum of physical phenomena, some range of which we have identified as biological. This program is useful for several reasons. It represents an approach that seeks to reveal continuities among physics, chemistry, and biology. It requires that we provide an extended conceptual framework for thinking about adaptive matter; the central concept we favor is autonomy. And it should provide a more inclusive theory for the full range of biology observed on Earth and not exclude lineages based on values we assign to correlate biology.

One virtue of making viruses *exemplary* biosystems is that they make explicit the spatially and temporally distributed network structure of adaptive systems, and suggest an alternative to the view that organisms (tightly correlated autocatalytic networks) are alive while biochemistry is dead, in favor of a more inclusive definition of individuality. For example, there is little sense in the question *is a worm alive?* without an appreciation of the network of regulatory and metabolic dependencies that enable sustained, adaptive diversification. A single worm instantly transported into space is not alive, even though its essential physiological processes remain transiently ordered. However, there is something that allows us to single out a worm as an *individual*.

The question arises: What do we mean by individuals, and why have individuals provided the focus for questions relating to the origin of life? From the network point of view, individuals represent tightly correlated clusters of autocatalytic activity embedded sparsely within larger networks of dependencies. Individuals are often physically distinguishable as tight correlations enabled by physical boundaries and physical linkages serving to concentrate activity and minimize interference. To use the terminology of network complexity measures, individuals are partially segregated subnetworks weakly integrated into a larger network of dependencies. In other words, individuals are *modules* (Brandon, 1999; Wagner, 1996). Viruses are a perfect model for this definition (figure 23.1) as the "very essence of the virus is its

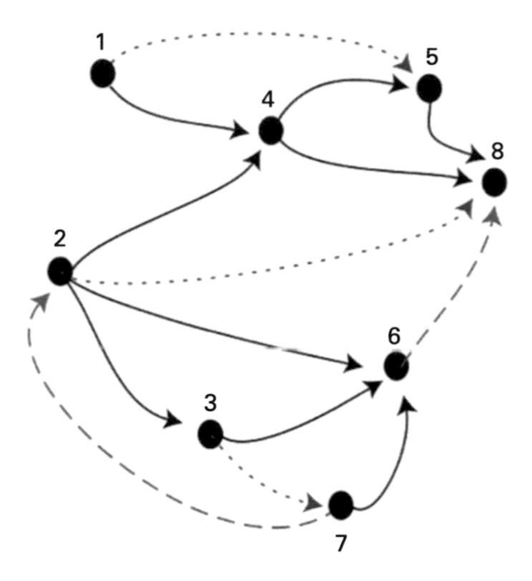

Figure 23.1
Host-facilitated virus regulatory network. Virus-encoded proteins are numbered 1 through 8. Edges connecting nodes are a function of the host environment and arrows represent biochemical production. Solid edges represent interactions catalyzed by viral enzymes once within the cell. Dashed and dotted edges are interactions catalyzed by host factors. The reactions that transpire (dashed or dotted) frequently depend on the cell type infected by the virus. For example, viral protein 1 contributes directly to the production of viral protein 4, whereas viral protein 1 requires species-specific host factors to produce viral protein 5.

fundamental entanglement with the genetic and metabolic machinery of the host" (Lederberg, cited in Morse, 1994).

23.2 Sufficiency

Few life forms are wholly autonomous. This implies that most living systems are not capable of surviving independently from all externally derived macromolecules. This observation leads us to a view of a living system as a concentrated network of synthetic pathways embedded within larger *distributed* networks of statistically predictable and essential components. Some fraction of molecules within these networks provides specialized mechanisms that decrease the uncertainty of access to components. These we typically associate with structural elements such as membranes, compartments, and chemical backbones. The question naturally arises about the identity of individuals within the larger network of dependencies. Individuals are identified within these larger networks as those minimal subnetworks that, when present, provide sufficient information to predict their own representation and proliferation into the future. In other words, we are suggesting that biological individuals are those networks that constitute *sufficient statistics* of their futures (Kullback, 1997).

What is a sufficient statistic? Assume that we have samples from a distribution. We do not know what form the distribution takes, but we do know that the distribution in question derives from a set of distributions characterized partly or wholly by a parameter vector p. A statistic is said to be sufficient to make inferences about p if and only if the values of any sample from that distribution give no more information about p than does the value of the statistic on that sample. In other words, a sufficient statistic is a minimal representation of the distribution that is as good as any sized sample at describing the distribution. More formally, consider a series of n observations or random variables, $X_1, X_2, \ldots, X_n$. Suppose that the distribution of these variables depends on the parameter vector p. The statistic $U = h(X)$ is sufficient for p if U contains all the information about p contained in any permissible value of X.

To clarify these remarks, consider a simple phage genome. The genome physically consists of a string of nucleotides surrounded by a protein capsid. We suggest that for a virus, this sequence constitutes the parameter vector defining the sufficient statistic. Most of the proteins encoded in the phage genome are unable to react among themselves. The phage is, internally, reactively inert. Once within the bacterial host cell, reaction networks are established and synthetic pathways are opened up. Within the host cell, the virus-encoded proteins constitute a small fraction of the larger synthetic networks required to generate replicas of the phage genome and capsid protein (virion). Hence, the individuality of the phage as a *physical* entity is diluted within the cell reaction pathways. However, the phage genome is *informationally* the best predictor of future phage genomes, and not the host factors or host genome sequence contributing to the construction of new virions. This is not equivalent to stating that host factors are unimportant. It is just that they are not the informative parameters predictive of the coarse-grained degrees of freedom defining the virus genome. Different phages can infect the same host, making use of essentially similar host factors and pathways yet generating very different daughter virions. The information that makes a difference is carried by the virus.

Individuality is the property that allows the phage to become subject to ongoing selection pressures. Genetic variation in host genes is not typically the best predictor of variation in phage genes. This allows for extensive host ranges and diverse cell tropisms. Put in the strongest terms, the information required to predict the phage resides in the phage when all else is held constant. In many cases we are aware of, this strong definition of individuality is violated. Many host genes covary uniquely with phage genes, and without these the virus would be unable to propogate. This requires that the individuality of the phage be extended to include a subset of host genes, even though these genes are not encoded in the genome transmitted by the virion. Hence, our sufficient statistic, U, formerly defined only in terms of the virus genome, is expanded to include a proper subset of host genes: $U_c = G_v + g_h$ defining

$g_h \subset G_h$, where G_h is the host genome. The subset g_h is predicitive of specific sequence variation in the virus genome and does not include all the host genes necessary for producing virions capable of supporting a wide range of genetic variants.

23.2.1 Individuality, Macrostates and the Levels of Selection

Individuality, thought about in terms of statistical sufficiency, makes a natural connection to the physical concept of a macrostate. Thermodynamic macrostates are typically systemwide extensive variables or bulk averages, such as temperature and pressure (Amit and Verbin, 1995). These are arrived at from more microscopic kinetic description through a process of contraction or coarse graining. Shalizi and Moore (2003) have suggested that macrostates are partitions of phase space that possess a Markovian property, such that they represent the minimal number of variables successfully accounting for all regularities in a temporal sequence at a preferred level of analysis. This argument makes use of the concept of the causal state (Crutchfield and Shalizi, 1999). Say we observe a temporal stochastic process, which we arbitrarily divide into past observations and future observations about one time point. Two histories belong to the same causal state if they are equivalent for predicting the future time series. Shalizi and Moore offer the following formal account: Consider a stationary time series. Write $\overleftarrow{S}$ for the set of all histories and $\overrightarrow{S}$ for the set of all futures. The objective is to predict one future $\overrightarrow{s}$ from one past $\overleftarrow{s}$. Prediction methods treat histories as equivalent by including only some finite number of partitions of past events k and designating these histories as equivalent when the k observations back in time are identical. Two histories $\overleftarrow{s}$ and $\overleftarrow{s}'$ are causally equivalent if and only if they give the same conditional distribution for futures:

$$\overleftarrow{s} \sim_\epsilon \overleftarrow{s}' \quad \text{iff} \quad Pr(\overrightarrow{s}|\overleftarrow{s}) = Pr(\overrightarrow{s}|\overleftarrow{s}')$$

The relation $\sim_\epsilon$ divides the set $\overleftarrow{S}$ of all pasts into equivalence classes. The causal state of a history is a compressed or reduced representation of equivalent histories:

$$\epsilon(\overleftarrow{s}) \equiv \{\overleftarrow{s}' || \overleftarrow{s}' \sim_\epsilon \overleftarrow{s}\}$$

The causal state captures the dependence of the future on the past,

$$Pr(s_{t+1}|\overleftarrow{s_t}) = Pr(s_{t+1}|\epsilon(\overleftarrow{s_t})).$$

The causal state lumps together, in compressed form, all the information from the past that is predictive of the future.

Shalizi and Moore demonstrate that causal states form optimal partitions of histories by showing that an optimal partition maximizes the mutual information between the past and future, and thereby constitutes a sufficient statistic of the future. We suggest that biological individuals behave analogously to macrostates, as they represent partitions of phase space of a minimal statistical complexity predictive of their success into the future. For highly recombinant organisms, the appropriate level of selection has been hypothesized by some to be the gene (Dawkins, 1989; Hamilton, 1964; Williams, 1996), since only the gene persists for long enough to come under persistent selection pressure. For tightly bound genes in linkage disequilibrium, whole genomes, organisms, or populations can constitute relevant macrostates (Gould, 2002).

An additional value of the macrostate approach to the levels of selection question is that it makes clear its anthropic character, which arises through the choice of a particular coarse-graining. It is not that the genic view is intrinsically better then a genomic or even population partition. It simply depends on the question we wish to answer. Having chosen a partition, we can then determine its legitimacy according to its adherence to a Markovian property. Partitions that fail to be good predictors (i.e., that are insufficient) are, on formal grounds, poor candidates. As we alluded to earlier, the minimality conditions provide some of those physical conditions promoting sufficiency. This means that there will be problems for which the entire virus genome does not need to be specified, merely some subset or, alternatively, some union over essential genes distributed over a population of virus particles infecting the same cell. This introduces a subjective element into establishing levels of individuality, but this is something we view as unavoidable and an implicit property of the concept responsible for some of the confusion raging over the levels of selection.

23.2.2 Complexity

Because causal states are optimal predictors of the future, the complexity of a process or sequence is equal to the statistical complexity of the causal states. For our purposes, statistical complexity is an extensive measure of individual autonomy. The greater the number of causal states, the greater the statistical complexity, and thereby autonomomy, of the individual. In other words, the complexity or autonomy is given by

$$A = -\langle log_2 Pr(\epsilon(\overleftarrow{s_t}))\rangle.$$

The trend toward increasing individual complexity or autonomy—for example, an increased coding capacity of a genome—can be viewed as one means of increasing correlations among codependent molecules and higher-order structures in a network. A good example of this trend is the internalization and enslavement of symbiotic organisms into reduced organelles (Margulis, 1993). Organelles—sources of essential substrates—become tightly coupled to the set of chemical species constituting the nuclear genome. Division of cells is coupled to the division of organelles, thereby minimizing resource uncertainties. From the organelle perspective, what was once a free-living unicell dependent on an unpredictable environment has become a reduced unicell dependent on the more constant environment of the cell. An analogous security is obtained from the perspective of the cell.

Viruses do not possess a true cell wall, nor do they encode metabolic pathways, and they are thereby often excluded from life. Adopting the distributed, codependency network framework, we see that this is a rather arbitrary exclusion that does not provide a better understanding of those lineages we do deem living. We suggest it is better to view viruses as occupying one end of the spectrum of delocalized control over metabolic and regulatory networks. Viruses can function by virtue of the extreme regularity of their host environment. More specifically, viruses depend on hosts to provide the complement to their own genomes for effective replication, and the stability of this evolutionary strategy is a result of the effectively deterministic nature of the virus-host environment. Extrapolating from the viruses, we can think of biology more broadly in terms of its varying autonomy as a function of the degree of predictability of the components of extended support networks. The greater the environmental stochasticity, the greater the need to endogenize essential components of the dependency network. This uncertainty is likely to scale with system size, hence, organisms requiring a greater energetic throughput will have to rely to a reduced degree on the network of interactions specified outside of their own genomes.

23.2.3 Degeneracy

One reason that we are able to make a statement about sufficiency is that a variety of strains or types of virus are able to infect the same host cells. If only a single sequence was ever observed, we would not be able to separate, informationally, the virus-encoded genes from those of the host. We can physically do this as a result of the mobile virion. The diversity of viruses infecting a single host allows us to factor out the host contribution when defining the individual. Obligately symbiotic associations do not have this property. Mitochondria and chloroplasts divide with, and are under the control of, the cell (Reid and Leech, 1980). It is the morphological correspondence and phylogenetic similarity with algae that allude to the ancestral individuality of organelles.

Degeneracy is the property of having a diversity of viruses all capable of performing the same function—proliferating on the same host genetic background. It arises out of the combinatorial sequencing of nucleotides into genomes that are all energetically equally stable and biosynthetically equally accessible. Degeneracy is a feature of singular importance for biology, because its makes the prediction of a genome sequence a function of its contingent history in addition to physicochemical constraints.

Conditioned on a given host background and choice of nucleic acid genome, extant viruses are likely to constitute, from the genome architecture point of view, neutral alternatives with respect to proliferation potential. Variation among architectures will have arisen through the usual mutation, recombination, and gene transfer operations available to viruses.

23.3 Viral Diversity: Phylogenetic Relationships

Biologists traditionally classify living systems into three scales: individuals, species, and ecosystems. Each higher scale is assumed to contain the lower, and the role of natural selection is assumed to weaken toward the higher scales. In terms of autonomy, this reflects a reduction in the predictability of a spatial sequence at more inclusive levels of organization. Thus, Darwinian dynamics features individuals and genes as its arguments, whereas ecological dynamics features species. This is a matter of convention rather than formal necessity.

Species are classified according to two criteria: a grouping criterion and a ranking criterion (Mishler and Brandon, 1987). The grouping criterion identifies characteristics according to which individuals should be placed together. This can be interbreeding, phenetic or genetic similarity, or functional analogies. The ranking criterion establishes a threshold beyond which individuals are no longer deemed members of a single species, and is typically derived from genetic data informative of descent. Hence, species grouped according to some trait vector V must also constitute a monophyletic group (they share a recent common ancestor).

Viruses are most frequently grouped by function or genome. Hepatitis viruses all infect the cells of the liver. However, hepatitis viruses violate the phylogenetic ranking criterion (they are paraphyletic). A similar violation of ranking occurs when we classify viruses according to genome organization.

We shall see that viruses differ from most evolving lineages in terms of the quantity of genetic material they appropriate from other individuals as opposed to material they evolve themselves through genetic duplication and expansion events. The question relates to the origin of evolutionary novelties. For viruses, the majority of novelties were genetically plagiarized from host genomes and, in many cases, remain resident in the host genome and exploited through spatial propinquity following infection.

The extensive genome plagiarism has led to two alternative sets of hypotheses for the origins of viruses (see Morse, 1994). One set consists of the escaped transcript hypotheses, which state that viruses are of cellular origin and derive from some form of retrograde devolution. The best known of these is the Green-Laidlaw hypothesis (Green, 1935; Laidlaw, 1938) that views viruses as degenerate organisms. A variant is the Galatea hypothesis (Andrewes, 1967) that envisages a bacterial nucleoprotein as the viral precursor. The second set views viruses as relics of a primitive form of life, or prebiotic nucleic acid from the RNA world that became co-adapted to primitive, and subsequently, more recent, forms of cellular life. The two sets of theories differ with respect to the antiquity of viruses and their mechanism of origination.

23.3.1 The Origin of RNA Viruses

Early evolution is likely to have employed self-replicating RNAs to extend and stabilize metabolic functions (Doolittle and Brown, 1995). RNA possesses both autocatalytic properties (ribozymes) and primitive translation capabilities. RNA viruses could have emerged independently and opportunistically once autocatalytic self-replicating RNA molecules were able to harness the energy surplus from metabolic replicators in the RNA world. However, although there is evidence that life progressed from a RNA-based progenote to a cenancestor with a DNA genome, there is no definitive evidence that present-day RNA viruses share key metabolic features (ribozyme activity and self-replication) with RNA-based early life.

More likely, RNA viruses originated and then organized their replication strategies based on cellular mRNA. In this case, RNA viruses have a more recent and recurrent cellular origin. All replication strategies of RNA viruses make use of positive-stranded RNA, either as genome or as template for negative-strand genomes.

The RNA genome retroviruses are also likely to have descendended from cells, since reverse transcriptase (RT) is homologous with cellular telomerase, and retrovirus replication strategies resemble those of bona fide cellular transposable elements.

A positive-strand RNA molecule (like a mRNA) encoding RT function suffices as a retrotransposon, which, if able to move from cell to cell (with the assistance of encoded structural proteins), would constitute a virus.

23.3.2 The Origin of DNA Viruses

One hypothesis is that DNA viruses evolved by host cell reduction. This would entail cell-like structures reducing autonomy to the extent that the resulting genome would require additional cell factors to replicate. In support of this hypothesis are large viruses that possess many cellular genes. The poxviruses (smallpox) are one example, with genomes one-third the size of a small bacterial genome and a large virion (around 450 nm) with dozens of active enzymes. Recently, a huge DNA virus was isolated from *Acantamoeba polyphaga*—the mimivirus—with an 800-nm virion (the size of a small bacterium) with a genome of 1.2 Mbp (the size of small bacterial genome) encoding 1,260 genes, with the metabolic capacity to synthesize 150 proteins, including chaperones and DNA repair enzymes.

An alternative hypothesis derives from the widespread incidence of episomal DNA (e.g., bacterial plasmids). Horizontally transmitted episomal DNA may have provided the mobility mechanism that spawned complex viruses. DNA viruses cluster

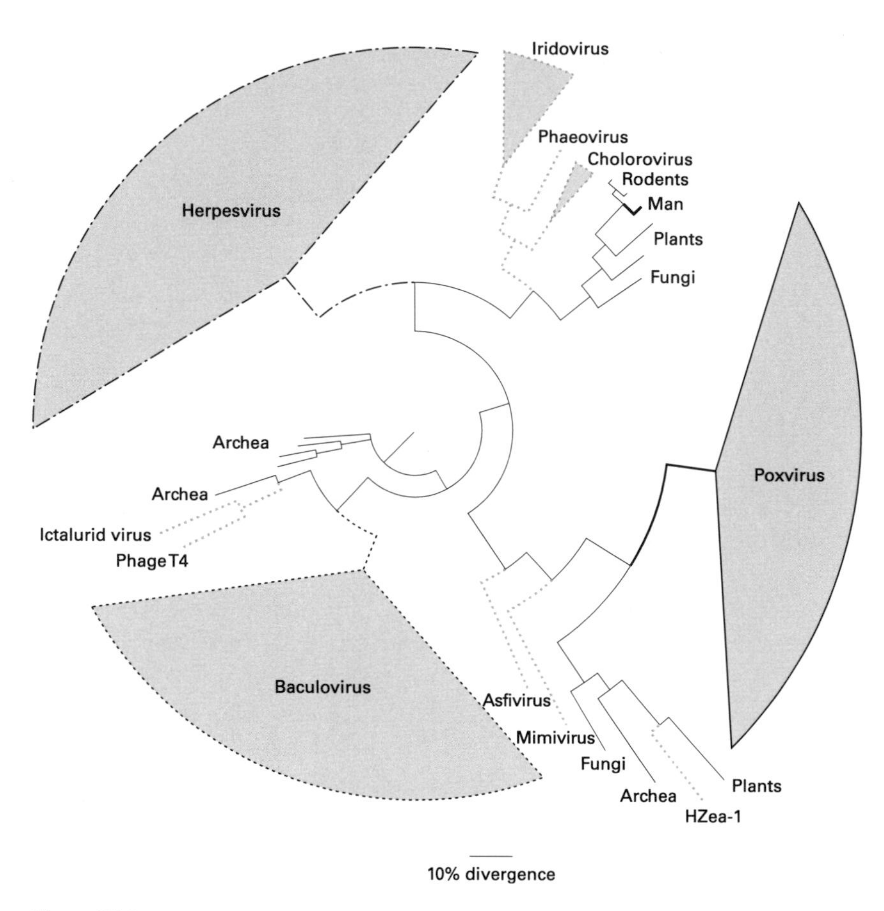

Figure 23.2
Phylogeny based on the DNA-dependent DNA polymerase (DdDp) illustrating clusters of the major DNA viruses and cellular DdDp. Representatives of distinct viral lineages, such as the mimivirus, asfivirius, HZea-1, ictalurid virus, phaeovirus, chlorovirus, and the four large DNA virus families of herpesvirus, poxvirus, iridovirus, and baculovirus do not form a monophyletic group, but radiate from Archea or eukaryotic cell lineages. This supports a view of viral history as arising through recurrent escape events.

into families that have either type B (of eukaryotic origin) or type A (of prokaryotic origin) DNA-dependent DNA polymerases (DdDp). For four large DNA viral families (Zanotto, Kessing, and Maruniak, 1993), a strong agreement (i.e., congruence) is observed among phenetic clusters derived from complete genomes and maximum likelihood phylogenies obtained from the DdDp (see figure 23.2). The adenoviruses have a type A DdDp, related to those found in bacteriophages, suggestive of a possible prokaryotic origin. The large DNA viruses from the remaining three families studied (*Poxviridae*, *Herpesviridae*, and *Baculoviridae*) possess type B DdDp, with orthologues in eukaryotic cells (Braithwaite and Ito, 1993; Garcia-Maruniak et al.,

2004). This strong agreement supports the idea that a core set of genes (including the replicase) in each of the virus lineages has had congruent phylogenetic histories.

The hypothesis of a conserved, coevolving core set of genes is further supported by the gene content of DNA viruses. Baculovirus gene content studies show that all baculoviruses share a core set of 30 genes, some of which, including helicase ac96, 38K(ac96), maintain similar relative positions (Herniou et al., 2004). Half of them (15) have known orthologues outside the *Baculoviridae*. One-third (10) of these conserved genes are involved either in replication (4) or transcription (6), whereas the remaining are structural proteins (9), auxiliary (1), or have unknown functions (10). This suggests that most conserved functions shared among baculovirus are related to the replicative/multiplicative core, which in turn can be traced back to host genes.

Whole genome trees (reconstructed by including the full genome sequence) agree with phylogenies of the baculovirus based on single genes (Zanotto, Kessing, and Maruniak, 1993) or with clusters of orthologues (Garcia-Maruniak et al., 2004; Herniou et al., 2004). Interestingly, a large DNA virus (Hz-1) persistently infecting *Heliothis zea* (*lepidoptera*) cell lines shares a few genes with baculovirus that also infect *lepidoptera*. However, this virus has a DdDp not related to that of the baculovirus.

The maximum parsimony reconstruction (MPR) of gene content in 27 genomes of representatives of the family *Baculoviridae* shows that, for the 663 genes shared by all baculovirus, there were seven to nine times more independent gene gain events (from host cells and other viral families) than gene losses along the radiation of the virus family.

Baculovirus have exchanged large numbers of genes with their host cells and appear to be accumulating gene content in time, all the while preserving the same core set of replication-related genes. Similarly, in the poxviruses, a core set of 34 genes involved in replication and viral assembly is found to be present in all 20 genomes investigated. The orthopox genomes have an additional set of 52 shared genes (McLysaght, Baldi, and Gaut, 2003). It has been proposed that poxvirus genomic evolution takes place under heterogeneous rates of gene gain and loss along distinct lineages, whereas a core set of 34 genes defines the functional or metabolic identity of the virus family.

In the case of the herpesvirus, the core set of genes has been studied in great detail. Like baculoviruses and poxviruses, the largest core set of genes—57% of the 14 COGs including all herpes genomes (except that of the ictalurid herpesvirus)—are related to DNA metabolism (Montague and Hutchinson, 2000). Furthermore, an additional study of the frequency distribution of distinct functional classes of 19 herpes genomes suggests that most of the core set of shared genes are involved in nucleotide metabolism and DNA repair, structural functions (capsid and tegument) and replication (Alba et al., 2001). In each of these case studies, it can be observed that a core set of genes and functions typically related to replication is conserved whereas auxiliary functions, possibly of contextual adaptive value, are gained and lost in time as viruses coevolve with their specific hosts.

23.3.3 Cores and Satellites

The picture that these data are painting is of a core set of highly conserved genes ultimately derived from a host cell, combined with satellite genes prone to rapid loss and gain, providing the local genetic variation required by the diversity of host environments.

Why should a set of genes remain closely linked through evolutionary time? A principle of organization according to which a stable replicative metabolic core (plus those elements essential for horizontal transmission), which exhibits continuity and coherence in evolutionary time, is Eigen and Schuster's hypercycle organization (Eigen and Shuster, 1979). The hypercycle refers to that set of proteins constituting a closed catalytic network. By invoking the hypercycle structure of networked genetic dependencies, we can go some way toward explaining the finding that the complete genomes of viruses have histories that are well represented by their replicative core, the central hub of which is polymerase. By treating the whole genome as a quantitative variable, thereby deriving a genomic phenotype, we can observe epistatic effects among genes that lead to linkage disequilibrium among members of the genomic core. This core cannot be perturbed without rendering the virus defective, and hence, a large group of genes exhibit concerted evolution.

Virus cores are informative in much the same way that eukaryotic phenotypes are informative: They indicate tightly coupled genetic dependencies, which we think of as individuality. These dependencies can be used, as we have done here, to reveal the nonoverlapping evolutionary histories of complete virus genomes. In particular, these histories include the independent origin of the adenoviruses from ancestral prokaryotic host populations, and the independent origin of the baculoviruses, herpesviruses, and poxviruses from eukaryotic ancestral populations. These data support the escaped transcript hypothesis, which posits distinct virus lineages as having independent origins. We have gone further to suggest that multiple genes escape simultaneously, centered on a polymerase catalytic core.

23.4 Viral Disparity: Genomic Architectures

Viruses to do not represent a homogeneous group of organisms, but rather a diverse set of replicative strategies varying in the degree of autonomy and control exerted over extended environmental networks. The standard classification of viruses is based on genomic architecture and was conceived by Baltimore as a means of specifying the causal flow of information over the course of a virus life cycle. The classifi-

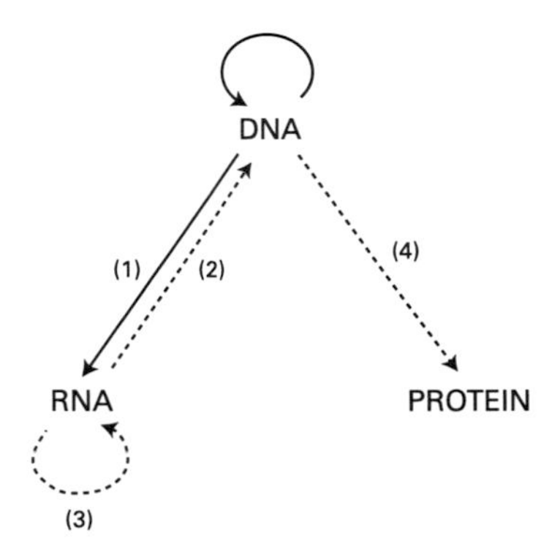

Figure 23.3
The central dogma of molecular biology. Arrows represent canonical transfers of information between macromolecules.

cation derives from Francis Crick's hypothesis of the central dogma of molecular biology. In a famous statement made in 1958, and expanded into a paper in 1970, Crick expounded the central dogma, which in concise form states that once (sequential) information has passed into protein it cannot get out again. The reasons for this irreversibility were suggested to be stereochemistry (moving from a folded, three-dimensional object to a one-dimensional sequence) as well as the lack of any machinery comparable to that used to go from RNA to protein. Crick expressed the central dogma in diagramatic form reproduced in figure 23.3.

Crick spoke of DNA → DNA, DNA → RNA, and RNA → Protein as general transfers and RNA → RNA, RNA → DNA, and DNA → Protein as special transfers. The group of unknown transfers that would shake the whole intellectual basis of molecular biology includes all decodings from proteins: Protein → Protein, Protein → DNA, and Protein → RNA.

Baltimore recognized that viruses represent strategies that exploit the predictable properties of the specific and general forms of the central dogma in order to maximize their representation within the gene pool. Referring back to figure 23.1, the numbered edges represent mechanisms of decoding from one biochemical alphabet to another. Alphabets vary in cardinality, combinatoriality, stability, and activity. The first two refer to logical properties of the coarse-grained degrees of freedom we describe chemically as nucleotides. The second two are functional attributes of sequences conferred by attractive forces among atoms.

At different stages of the virus life cycle, various combinations of these properties are preferred. During persistent infection, stability is required and many viruses make use of DNA genomes or DNA intermediates to provide long-term genetic memory. During ingress and egress, when immune surveillance is high, mechanisms for generating genetic and functional diversity are required, and many of the more

Table 23.1
Properties of biochemical coding alphabets

	Cardinality	*Combinatoriality*	*Stability*	*Activity*
RNA	4	4^L	Low	Intermediate
DNA	4	4^L	High	Low
Amino acid	20	$20^{L/3}$	High	High

successful disease-causing viruses have RNA genomes. The functional differences between these are captured in table 23.1, which summarizes the basic properties of biochemical control alphabets.

23.5 Control Architectures

Each of the different genome coding alphabets, when filtered through protein synthesis pathways of the host cell, generates a different sequence of events culminating in virion production. The virus life cycle is thereby closely linked to the underlying structure of the virus genome. We illustrate the relation of form to function for the positive and negative sense single-stranded RNA viruses. In order to do so, we present a minimal description of the virus in terms of genome, host factors, and polypetides. In the following, we use the logic symbols $\oplus$ and $\ominus$ to refer to the polarity of the genome.

23.5.1 Minimal RNA Virus Elements

1. $\oplus m_i$ for positive sense RNA fragment i
2. $\oplus G$ for $[f(\sum_i \oplus m_i)]$ linked positive sense RNA genome
3. $\ominus m_i$ for negative sense RNA fragment
4. $\ominus G$ for $[g(\sum_i \ominus m_i)]$ linked negative sense RNA genome
5. R for host encoded ribosomes
6. p_i for virus protein translated from viral fragment $\oplus m_i$

23.5.2 Life Cycle of $\oplus$ RNA Virus

1. $\oplus G \stackrel{r\oplus}{\rightarrow} \oplus G + \oplus G$: host encoded replication into minus strand copy
2. $\oplus m_i \stackrel{tR}{\rightarrow} p_i$: genome fragements translation into protein or $\oplus G \stackrel{tR}{\rightarrow} \sum_i p_i$: whole
3. genome translation
4. $\ominus G \stackrel{r_{\ominus}p_j}{\longrightarrow} \oplus G + \oplus G$: virus encoded replication into positive strand copy
5. $\oplus G + \sum_i p_i \stackrel{k}{\rightarrow} \oplus y$: virion production

23.5.3 Life Cycle of ⊖ RNA Virus

1. $\ominus G \xrightarrow{r_{\ominus} p_j} \ominus G + \oplus G$: virus encoded replication into positive strand copy
2. $\oplus G \xrightarrow{r_{\oplus}} \oplus m_i$: host encoded transcription into messenger fragments
3. $\oplus m_i \xrightarrow{tR} p_i$: translation into protein
4. $\sum_i \oplus m_i \xrightarrow{r_{\ominus} p_j} \ominus G + \sum_i \oplus m_i$: virus encoded replication into linked negative strand copy
5. $\ominus G + \sum_i p_i \xrightarrow{k} \ominus y$: virion production

These kinetics statements are easily translated into dynamical equations from which we can derive a number of results (Krakauer and Komarova, 2003). For the positive sense virus, the RNA genome serves three functions: a genome for replication, a messenger RNA for translation, and a genome for packaging into the virion for egress from the cell. Furthermore, all of these processes are self-starting, since the virus is effectively indistinguishable from host cell mRNA. Once the genome is present, translation is initiated. Translation yields proteins required for genome replication and genome egress as virions.

The sequence of events leading to the production of daughter virions follows directly from the polarity of the genome. The positive genome is rapidly translated according to host-dependent factors. The positive genome quickly synthesizes viral proteins with minimal delay. However, because the translated sequence also serves as a genome for packaging, viral proteins can prematurely assort with the genome for egress from the cell, thereby curtailing infection. This leads naturally to a tension between two levels of selection, or alternatively, two partitions of sequence space. Within the cell, those sequences free of the encapsidating genes are the most effective predictors of their own future. Packing proteins cause sequences to be exported from the cell and undergo selection at the superordinate cellular level. Since there is no reason to expect the selection process to be identical within and among cells—they constitute different environments—the underlying genome experiences frustration simultaneously in favor of purging genes and maintaining these genes for cellular proliferation (Krakauer and Komarova, 2003).

For the negative sense virus, the genome is of the incorrect polarity for translation. Consequently, a positive sense genome must be synthesized. Because the virus cannot yet have made the necessary protein (RNA-dependent RNA polymerase), this needs to be carried within the virion. Translation proceeds with use of the positive sense fragments. Generation of the linked negative genome then proceeds by concatenating the positive subgenomic fragments rather than generating a full-length transcript.

With negative genomes, translation is initially rate-limited by the virus-encoded RNA-dependent RNA polymerase carried by the virus in the virion. Hence, the host assumes control of the translational portion of the life cycle only once the positive subgenomic fragments are synthesized. The virus then assumes control over replication as the negative genome is assembled from the positive subgenomic fragments with the assistance of a virus-encoded protein—the nucleocapsid protein.

Hence, the negative virus exerts greater control over all stages of the life cycle than the positive virus. Control is exerted over both the start of the life cycle and the assembly of a genome for egress from the cell. The negative virus has slightly increased its autonomy and gained greater control over its life cycle. However, the negative virus has had to carry additional genes in order to exercise this control and suffers a replicative handicap in the race to synthesize viral proteins. This shift from positive to negative, with an attendant increase in life cycle complexity, reduces the uncertainty that a virus faces when infecting a host cell.

23.6 Self-Assembly and Selection

Perhaps one of the most important debates in recent evolutionary theory is that between the view of life as the outcome of a series of contingent symmetry-breaking events, and the view that a complete understanding of the physicochemical boundary conditions will be sufficient to generate the essential features of living systems. The first position emphasizes various forms of environmental feedback and filtering constituting natural selection. The second seeks to minimize the role of historically unpredictable events in favor of a physical uniformitarianism, which attempts to derive elements of biological regularity from physical law.

The virus contributes to this debate in a number of ways. First, for a virus, the host genome evolves relatively slowly, thereby establishing the ecological regularities and boundary conditions that come to be reflected in the complexity of virus genome and behavior. To a virus, the host stands in the same relation as does the environment to a multicellular species. Hence, the effective *laws of nature* influencing virus evolution are slow evolutionary lineages (host lineages) in addition to the chemistry and physics. Second, as a result of the spatial and temporal scales at which viruses operate, developmental processes for these organisms are largely physical and driven by the minimization of free energy rather than elaborate multistep genetic programs. The first, *host invariance*, allows us to explore evolution in the presence of complicated boundary constraints; the second, *developmental physics*, reduces the amount of work selection needs to perform to create order.

23.6.1 Host Invariance and Maxwell's Demon

The most general statement about the origins of ordered states of matter was captured in the 1867 thought experiment of James Clerk Maxwell that has come to be known as Maxwell's demon. Consider two containers, A and B, filled with the same volume of gas at equal temperature (see figure 23.4I). A small trapdoor connects the two containers and is attended by a small individual who monitors the position and

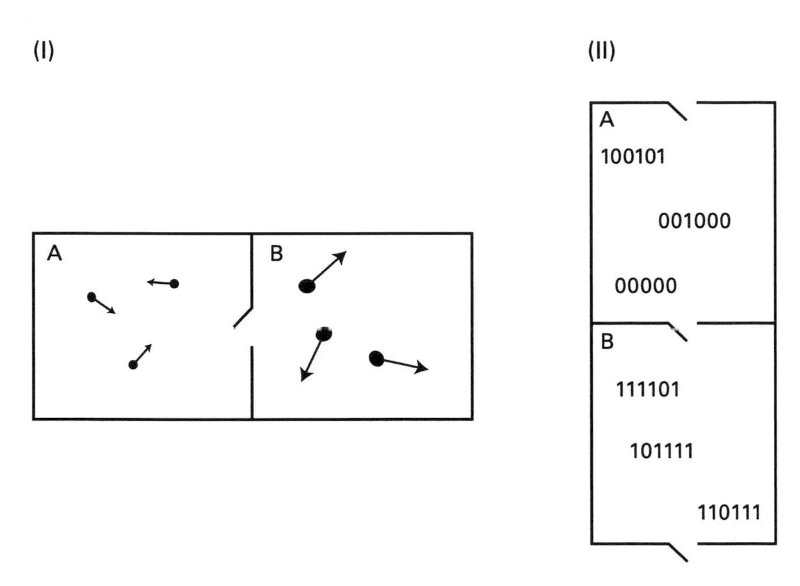

Figure 23.4
The correspondence between Maxwell's demon (I) and natural selection (II). Maxwell's demon is able to detect molecule position and velocity, allowing faster molecules to pass from compartment A into B. Natural selection is able to detect preferred genome sequences and allow them to pass from generation A to generation B. In both cases, the reduction in entropy of B is paid for by an increase in the entropy of the demon. Natural selection confers order through an increase in environmental thermodynamic dissipation.

velocity of molecules in each container and preferentially allows faster molecules to pass from A into B. In this way, container B increases in temperature. For Maxwell, this outcome runs counter to the second law of thermodynamics whereby an ordered state in a closed system has arisen without expenditure of work from outside of the system.

In 1929, Szilard suggested that the act of assaying molecular velocities required a dedicated mechanism of detection and that this information acquisition would involve the expenditure of free energy. Treating the demon and the gas in the two containers as a closed system, the entropy increase in the demon during information processing comes to exceed the ordered state associated with lowering the entropy of the gas. In this way, the second law is not violated.

In 1961, Landauer suggested that if the demon could store information about particle velocities in memory, without erasing information about past observations, then there need be no consequent increase in the entropy of the demon. This derives from *Landauer's dissipation principle* which states that the erasure of n bits of information must always incur a cost of $k \ln n$ in thermodynamic entropy.

In 1982, Bennett exorcised the demon arguing that, at some point, the memory of past events would have to be erased in order to store new particle velocities. In this way, neither the Landauer principle nor the second law are violated.

In an evolutionary context, Maxwell's demon (MD) is nothing other than Darwin's principle of natural selection (NS) (figure 23.4II). Like the demon, natural selection acts as a filter or sieve, letting through (from one generation to the next) adaptive variants and eliminating nonadaptive variants. The increase in the ordered state of an evolving lineage must be paid for by an increase in the thermodynamic entropy of the environment.

For a virus, the environment is the host cell, and the demon, all those interfaces within and between the cells that the virus depends on for proliferation. Evolution of a virus necessitates some reduction in the ordered state of the cell. This is typically measured biologically as virulence or host morbidity.

The natural selection–Maxwell's demon correspondence makes clear the physical roots of adaptive matter and the need for a more continuous treatment for the origin of adaptive matter and individuality. The key differences between NS and MD reside in: (1) the complexity of the boundary conditions and (2) the degeneracy of the states of order.

Why is life any more complex than a Ranque-Hilsch vortex tube? The vortex tube is a heat pump without moving parts, that produces cold air at one end and hot air at the other when supplied with compressed air; in other words, a Maxwell's demon apparatus. In biology, the boundary conditions we generically designate as selection pressures are able to operate independently or semi-independently over a much larger number of degrees of freedom. Typically, these are the nucleotides forming the linear chain of an individual genome. The critical chemical requirement (met by RNA and DNA) is that the target macromolecule should have a vast number of thermodynamically equivalent sequences. Each nucleotide is then potentially an information-bearing degree of freedom, and the ability of selection to target a single nucleotide can be measured, for example, by its degree of conservation. This is the approach of Adami, Ofria, and Collier (2000). A single site i in a linear genome can adopt four configurations with probabilities p_j^i, where j indexes the four nucleotides. The entropy of a site once we have aligned multiple homologous genomes is given by

$$H_i = -\sum_j p_j^i \, log \, p_j^i,$$

where we assume log base four, and the information per site is

$$I_i = H_{max} - H_i,$$

where H_{max} is given by the equiprobable distribution over nucleotides. For a genome with l base pairs, the strength of selection, and thereby order or statistical complexity of the genome, assuming independence among sites, can be given by

$$C = l - \sum_i H_i.$$

Complexity measures agreement among aligned sites, and thereby, the strength of selection acting at a site. Assuming a host cell synthesizes proteins that are also encoded in the virus genome, there is no selection on virus nucleotides to maintain the protein. The redundancy between the virus host-cell selection and the host-environment selection for the shared polypeptide allows the virus genome to random walk in sequence space (see hierarchical demon in following section) at this gene locus. The complexity measure will reflect this lack of concensus by declining. This reduces the information that the virus environment needs to preserve (and subsequently erase) and it reduces the entropy increase in the host cell. The more statistically complex or autonomous the virus genome, the more work host-cell selection needs to perform to preserve order, and the greater the dissipative consequences.

23.6.2 Developmental Physics

Development is the process that generates adaptive phenotypes from sequence information and maternal factors (Carrol, Grenier, and Weatherbee, 2004). In metazoans, development is a highly regulated process involving a shared set of transcription factors constituting the homeobox. For a virus, development makes use of few genes other than those constituting the structural genes and the temporal order of their expression. This would be like a mammal synthesizing structural genes and then allowing these to self-organize into the phenotype without developmental programs. Clearly, for a virus, development is a relatively simple process. In fact, it is very close to the physics and chemistry of protein folding, and consequently, highly dependent on the underlying sequence of the genome. In this respect, it closely resembles developmental folding of RNA (Fontana, 2002). Viruses implement a developmental physics whereby low-level processes within the cell construct virions with minimal regulation. This physics is a second source of order more fundamental to that provided by natural selection. The best example of this is the construction of the capsid. In the hepatitis B virus, weak protein-protein interactions drive self-assembly of the capsid. The capsid is constructed from 120 copies of the capsid protein arranged into $T = 4$ icosahedral symmetry. The construction is entropy-driven by the hydrophobic contacts among capsid dimers (Ceres and Zlotnick, 2002).

23.7 Robustness, Information Flow, and Genetic Dissipation

In this short chapter, we have used the example of the virus to emphasize some general properties of adaptive lineages and stressed important continuities with physical theory, in particular, with statistical physics. From our perspective, the life concept offers little intellectual leverage. By stressing discontinuity (living versus nonliving), the search for the origin of life requires discovering a profound emergent transition from physics. This chapter offers the alternative concepts of individuality (and its formal exposition in statistical sufficiency) and autonomy (measured, for example, in terms of the statistical complexity of causal states). Autonomy will be increased through realization of the minimality conditions (metabolism, cellularity, etc.) typically associated with life. Hence, the extensive investigation of these minimality conditions contributes to the study of autonomy.

The pressure toward increased autonomy is increased certainty. In variable environments, greater autonomy ensures greater statistical predictability. This is nothing other than a statement of the robustness of lineages. Mechanisms of robustness are those that increase the persistence of some adaptive feature in the face of perturbations (Krakauer, 2005). We have recently demonstrated that robustness places a lower bound on a measure of network complexity (Ay and Krakauer, 2007). Hence, increasing robustness can act as a driving force toward networks capable of processing larger amounts of information. The argument, put simply, is that robustness is ultimately derived through some form of redundancy, which, in networks, derives from correlated activity among nodes. In the computational context, these correlations are required for integration; hence the link between robustness and complexity.

We have, through the correspondence of Maxwell's demon and natural selection, stressed a further information processing perspective on evolution. Natural selection, acting as a demon, must select among alternative genome configurations through thermodynamic dissipation in order to increase order in viral lineages. For a virus, the demon is the physical environment of the cell, and the virus can store only as much information as the host cell can acquire about the virus. When the host cell synthesizes factors contributing to viral proliferation, which the virus could encode (e.g., tRNAs), redundancy disables selection in one individual. Typically, the virus loses autonomy, since the selection process is no longer able to exert any influence over the redundant viral sequence. This takes us a little beyond the simple two-compartment Maxwell's demon *gedanken* experiment. We should rather think of a recursive or hierarchical architecture in which compartments are nested (see figure 23.5). At one level, the host cell acts as selection (or the demon) to the virus; at another, the host environment act as demon to the host. When two states of order are preferred simultaneously by both selection mechanisms, one ceases to be required. Imagine a sequence 111 preferred over 110 by the virus demon but also by the host demon for host viability. The host demon, constituting an effectively constant background for a virus, has already selected in favor of 111. When the virus infects, the virus's contribution of this sequence will go largely unnoticed (redundant) by the host demon and will be free to drift. The host and environmental demons do not have to duplicate their efforts.

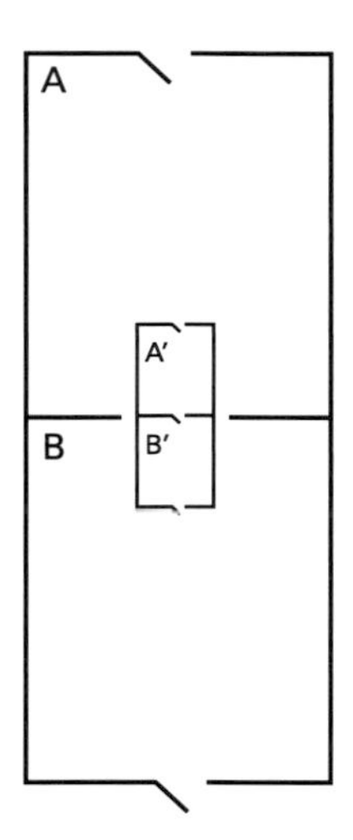

Figure 23.5
The hierarchically structured selective demon. Selective processes are nested with two demons selecting redundandantly. Here, the larger compartments represent two generations of virus, and the smaller compartments, two generations of host. The same information passes through two demons. As the messages have already been sorted by the host environment demon (A′ → B′), the requirement to identify the message going from A → B is lost. Dissipation is thereby experienced by the host environment rather than the cell.

This demonic-redundancy shifts thermodynamic dissipation from one selective environment to another. The genetic dissipation of the host increases the greater the autonomy of the virus, or stated differently, the less mutually redundant the host and virus genomes. This can be conceived of in terms of mutual information. When the mutual information between virus and host is high, the virus diminishes, because virus and host share a common selective environment. When the mutual information is low, the virus is required to expand its unique information content and does so by expending the energy reserves of the host cell.

This raises a natural question about the status of niche construction, construed in the literature as the ability of organisms to modify their selective environments (Odling-Smee, Laland, and Feldman, 2003). Through niche construction, organisms become responsible for multiple demons, constituting an effective hierarchy. The essential novelty arising from this strategy is that the more inclusive constructs are able to act on multiple individuals concurrently. This points toward mutualistic interactions among individuals and species, promoting a coordinated shaping of shared environments and a collective dissipation of free energy.

Having militated against the life concept, there remains the intuitive sense of life and death as expressed at the end of a life. There is, after all, something categorical about a biological individual dying. But this is little more than a statement that those chemical reactions predictive of future proliferation have been terminated. This is dramatic in highly autonomous lineages (metazoans for example) but far less so for prokaryotes. We suspect that this is an example of the general tendency of human brains to transform continuous variation into categorical differences.

23.8 Conclusions Bearing on Protolife Research

Having played down life and played up individuality and autonomy, grounded as they are in statistical theory and quantitative formalisms, where does this leave us in origin of life research? As previously stated, biology possesses several enhancements over mechanisms traditionally found in purely physical systems:

1. A variety of isoenergetic combinatorial macromolecules (degeneracy),
2. Physical and regulatory means of generating macromolecular forms (developmental physics),
3. Complex selective boundary conditions with a hierarchical character (recursive demons), and
4. Chemical means of proliferating macromolecules (replication).

In both virtual ("soft" artificial life) and laboratory settings, manipulating selective complexity and developmental physics have been the chief obstacles to progress in synthesizing adaptive lineages. Creating energy fluxes capable of supporting simple macromolecular order has met with greater success. Since complex individual order comes about through a combination of development and selection, we need new theories of these processes. Von Neumann (1949, described in Burks, 1966) tackled aspects of development in his theory of universal construction but assumed rather idealized conditions. The physics of computation community has provided novel insights into Maxwell's demon. Viruses offer a rich empirical system for exploring some combinations of these ideas, and offer some suggestions for how they might be extended.

References

Adami, C., Ofria, C., & Collier, T. (2000). Evolution of biological complexity. *Proceedings of the National Academy of Sciences of the United States of America*, *97*, 4463–4468.

Alba, M. M., Das, R., Orengo, C., & Kellam, P. (2001). Genome wide function conservation and phylogeny in the herpesviridae. *Genome Research*, *11*, 43–54.

Amit, D. A., & Verbin, Y. (1995). *Statistical physics: An introductory course*. London: World Scientific.

Andrewes, C. (1967). *Viruses and evolution*. Birmingham: Birmingham University.

Ay, N., & Krakauer, D. C. (2007). Geometric robustness theory and biological networks. *Theory in Biosciences*, *2*, 93–121.

Bennett, C. H. (1982). The thermodynamics of computation—A review. *International Journal of Theoretical Physics*, *21*, 905–940.

Braithwaite, D. K., & Ito, J. (1993). Compilation, alignment, and phylogenetic relationships of DNA polymerases. *Nucleic Acids Research*, *21*, 787–802.

Brandon, R. (1999). The units of selection revisited: The modules of selection. *Biology and Philosophy*, *4*, 167–180.

Brown, J. R., & Doolittle, W. F. (1995). Root of the universal tree of life based on ancient aminoacyl-tRNA synthetase gene duplications. *Proceedings of the National Academy of Sciences of the United States of America*, *92*, 2441–2445.

Burks, A. (Ed.). (1966). *Theory of self-reproducing automata by Von Neumann, J.* Urbana: University of Illinois Press.

Carroll, S. B., Grenier, J. K., & Weatherbee, S. D. (2004). *From DNA to diversity: Molecular genetics and the evolution of animal design.* Malden, MA: Blackwell Science.

Ceres, P., & Zlotnick, A. (2002). Weak protein-protein interactions are sufficient to drive assembly of hepatitis B virus capsids. *Biochemistry*, *41*, 525–531.

Crick, F. (1970). The central dogma of molecular biology. *Nature*, *227*, 561–563.

Crutchfield, J. P., & Shalizi, C. R. (1999). Thermodynamic depth of causal states: Objective complexity via minimal representations. *Physical Review E*, *59*, 275–283.

Dawkins, R. (1989). *The selfish gene.* Oxford: Oxford University Press.

De Duve, C. (1991). *Blueprint for a cell: The nature and origin of life.* Burlington, NC: Patterson.

Eigen, M., & Schuster, P. (1979). *The hypercycle: A principle of natural self organization.* Berlin: Springer Verlag.

Fontana, W. (2002). Modelling evo devo with RNA. *BioEssays*, *24*, 1164–1177.

Garcia-Maruniak, A., Maruniak, J. E., Zanotto, P. M. A., Doumbouya, A. E., Liu, J. C., Merritt, T. M., & Lanoie, J. S. (2004). Sequence analysis of the genome of the Neodiprion sertifer nucleopolyhedrovirus. *Journal of Virology*, *78*, 7036–7051.

Gould, S. J. (2002). *The structure of evolutionary theory.* Boston: Harvard University Press.

Green, R. (1935). On the nature of filterable viruses. *Science*, *82*, 443–445.

Haldane, J. B. S. (1929). The origin of life. *Rationalist Annual.*

Hamilton, W. D. (1964). The genetical evolution of social behaviour I. *Journal of Theoretical Biology*, *7*, 1–16.

Herniou, E. A., Olszewski, J. A., Cory, D. J. S., & O'Reilly, R. (2003). The genome sequence and evolution of baculovirus. *Annual Review of Entomology*, *48*, 21134.

Krakauer, D. C. (2005). Robustness in biological systems: A provisonal taxonomy. In T. Deisboeck (Ed.), *Complex systems science in biomedicine* (pp. 183–210). New York: Kluwer Academic.

Krakauer, D. C., & Komarova, N. L. (2003). Levels of selection in positive-strand virus dynamics. *Journal of Evolutionary Biology*, *16*, 64–73.

Kullback, S. (1997). *Information theory and sufficient statistics.* New York: Dover.

Laidlaw, P. P. (1938). *Virus diseases and viruses.* Cambridge: Cambridge University Press.

Landauer, R. (1961). Irreversibility and heat generation in the computing process. *IBM Journal of Research and Development*, *3*, 183–191.

Margulis, L. (1993). *Symbiosis in cell evolution: Microbial communities in the Archean and Proterozoic eons.* New York: Freeman.

McLysaght, A., Baldi, P. F., & Gaut, B. S. (2003). Extensive gene gain associated with adaptive evolution of poxviruses. *Proceedings of the National Academy of Sciences of the United States of America*, *100*, 15655–15660.

Mishler, B. D., & Brandon, R. N. (1987). Individuality, pluralism and the phylogenetic species concept. *Biology and Philosophy*, *2*, 397–414.

Montague, M. G., & Hutchison III, C. A. (2000). Gene content phylogeny of herpesviruses. *Proceedings of the National Academy of Sciences of the United States of America, 100*, 15655–15660.

Morse, S. (Ed.). (1994). *The evolutionary biology of viruses.* New York: Raven Press.

Odling-Smee, J., Laland, K. N., & Feldman, M. W. (2003). *Niche construction: The neglected process in evolution.* Princeton: Princeton University Press.

Orgel, L. E. (1998). The origin of life—A review of facts and speculations. *TIBS, 23*, 491–495.

Rasmussen, S., Chen, L., Deamer, D., Krakauer, D., Packard, N., Stadler, P., & Bedau, M. A. (2004). Transitions from nonliving to living matter. *Science, 303*, 963–965.

Reid, R. A., & Leech, R. M. (1980). *Biochemistry and structure of cell organelles.* New York: John Wiley & Sons.

Shalizi, C. R., & Moore, C. (2003). What is a macrostate? From subjective measurements to objective dynamics. Last accessed April 20, 2007 http://arxiv.org/abs/cond-mat/0303625.

Szilard, L. (1929). The interpretation of thermodynamic entropy as an information metric. *Zeitschrift fur Physik, 53*, 840–856.

Wagner, G. P. (1996). Homologues, natural kinds and the evolution of modularity. *American Zoologist, 36* (1), 36–43.

Williams, G. C. (1996). *Adaptation and natural selection: A critique of some current evolutionary thought.* Princeton: Princeton University Press.

Zanotto, P. M. A., Kessing, B. D., & Maruniak, J. E. (1993). Phylogenetic interrelationships among baculoviruses: Evolutionary rates and host associations. *Journal of Invertebrate Pathology USA, 62* (2), 147–164.

24 Nonlinear Chemical Dynamics and the Origin of Life: The Inorganic-Physical Chemist Point of View

Jerzy Maselko and Maciej Maselko

24.1 Introduction

Chemical systems that are sustained far from thermodynamic equilibrium have exhibited a broad spectrum of temporal and spatial behaviors hitherto seen only in biological systems. The researchers in the field of nonlinear chemical dynamics have developed many theoretical and experimental tools to describe and study such phenomena. We will use those tools to discuss a possible mechanism for the origin of life, and to address possible origins of artificial, nonbiological protocells.

The transition from nonliving to living matter has many theoretical and experimental problems (Deamer, 1997; de Duve, 2003; Eigen and Schuster, 1979; Fox, 1973; Luisi and Varela, 1989; Morowitz, 1987; Orgel, 1998; Rasmussen et al., 2003; Rasmussen et al., 2004). The gap between biological and chemical structures is huge, and very little is known about the transition between chemical/abiotic and the first biological/biotic systems when life began. A major obstacle is that, today, even the simplest biological cell consists of thousands of chemicals precisely organized in time and space, fulfilling many functions. The lack of a universally accepted definition of life makes the design of experimental systems an even more complex and poorly defined task.

The idea of an early cell ancestor was proposed to account for the fact that the entity bridging the nonliving and living matter transition must have been simpler than today's cells. (These are sometimes called *protocells* in the literature, but here they are called *early cell ancestors* for terminological consistency with this book.) Over time, this cell ancestor would have acquired, in a stepwise fashion, the functions that make up biological cells today. The time period for each improvement and the sequence of steps remain unknown.

The experimental synthesis of a protocell is a very challenging task simply because we have no idea what the composition of the original cell ancestors was. Close mimicking of structures and functions of existing biological cells is not the best experimental approach because of the great evolutionary distance from these original ancestors. In this chapter, we explicitly assume that early cell ancestors were formed before the genetic code for protein synthesis and before RNA and DNA emerged. Therefore, a different mechanism of evolution and control of complex metabolism must have existed.

24.2 Precision of Temporal and Spatiotemporal Organization

At the heart of the challenge facing the research community is that biological cells exhibit complex spatiotemporal organization. We cannot reproduce the complexity of cellular organization by simply mixing all the cellular components at the right concentrations. In life, the concentrations of a multitude of chemical compounds are organized in both space and time. Furthermore, the concentrations of different chemicals are continuously changing, and this spatiotemporal organization is incredibly precise. The possibility of successfully assembling all of the chemicals in their proper positions and correct concentrations is almost nonexistent. The difference in the precision of spatiotemporal organization of chemical versus biological systems is an important distinction between the two.

Let us consider the following chemical experiment: In two beakers we can put a solution of a simple inorganic compound. After some time, the water will vaporize and crystals will appear on the bottom in patterns that are distinguishable between the beakers. We have only two processes in this example, vaporization and crystallization, and only one chemical compound. Now compare this with the development of identical twins. Their development involves many chemical compounds and countless chemical reactions spanning many years. However, in looking at the "products" of those reactions we will not see much difference. This incredibly precise spatiotemporal organization is simply mind-boggling for any experimental chemist.

The determinism in chemical kinetic systems originates from the large number of interacting molecules. The stochastic nature of Brownian motion of simple molecules is averaged by the number of molecules, which is on the order of 10^{20} (mM). This is not the case in biological systems. The number of molecules is sometimes very small, and might even be just a few. In this case, the meeting of two molecules might take a few seconds or a few days, depending on the size of a cell. Many biochemical reactions should not behave deterministically. However, because of the careful transfer of many compounds and precisely juxtaposed enzyme complexes, biological systems do not leave chemical meetings to chance. Therefore, one of the issues associated with the synthesis of a prebiological entity is understanding and achieving the precise temporal and spatial chemical organization necessary for biological systems to function.

Our knowledge of complex temporal organization in chemical systems comes from the study of complex chemical oscillations and chaos in the CSTR (the continuously stirred tank reactor). The reactants are pumped into the reactor and thoroughly

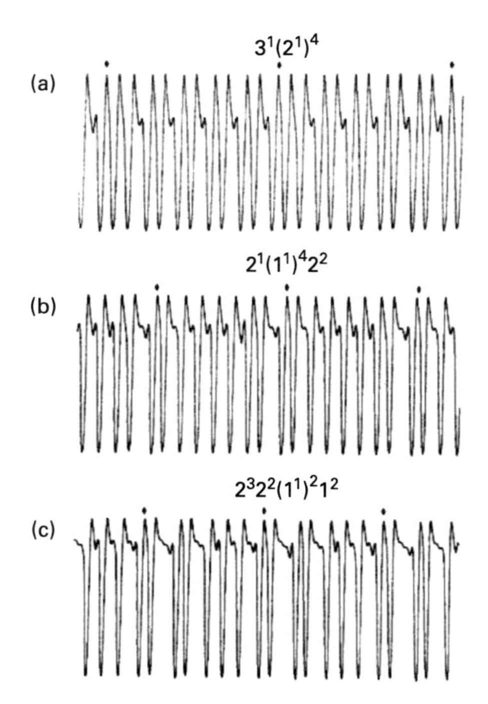

Figure 24.1
Complex temporal organization in chemical systems (Maselko and Swinney, 1986). Complex periodic oscillations are assembled from simpler blocks. The assembly is governed by Farey's arithmetic. The y axis is the potential on Pt electrode that corresponds to the concentration of catalyst. The x axis is time. The average period is about one minute. Temporal and spatiotemporal complexity are common in nature, but with regard to the origin of life, only functional complexity is important.

mixed by stirring rod. They react, and the products are pumped out so the constant flow of resources into the reactor and constant flow of products from the reactor is ensured. The most complex oscillations detected experimentally are presented in figure 24.1. The temporal pattern is periodic and built from simpler oscillations (Maselko and Swinney, 1986). The origin of the pattern can be understood by considering a phase locking on a four-dimensional torus. Unfortunately, these oscillations exist for only a very narrow range of parameters and can be easily destroyed. The chaos is even more complex but, because of the sensitivity to initial conditions, it cannot be the source of precise temporal organization.

To mimic complex temporal organization in biological systems, it has been suggested that one consider the analogies between complex networks of chemical reactions and the structure of computer algorithms (Gánti, 2003; Maselko, 1979; Ross, 1995). Different reactions can be connected differently, and in this way various time organizations can be achieved. Such ideas have been discussed only theoretically, and their experimental realization has not been achieved. Algorithmic chemistry—the chemistry in which many chemical and physical processes are precisely organized in space and time forming a chemical entity with the required properties, similar to biological systems—is just at its beginning (Maselko, 2003).

Even more difficult than temporal organization is the problem of spatial organization. Chemical systems, far from thermodynamic equilibrium, can form Turing structures (Quyang and Swinney, 1991; Turing, 1952), where chemical compounds are organized in space. Unfortunately most of the spatial organization is rather simple, for example, formation of dots of different concentrations in symmetrical arrangements (Lee et al., 1994). Little is known about complex spatiotemporal organization that is not chaotic (Pearson, 1993).

Interestingly, numerical simulations show that Turing structures can be formed in a system with only 100 molecules (Hasslander, Kapral, and Lawniczak, 1993). The precision of spatial organization, however, decreases substantially with decreasing numbers of molecules. In a biological cell, we know that the stochastic nature of Brownian motion is often substantially constrained and the molecules' movement in space is precisely organized.

Although we realize that the CSTR is imperfect as a model for a protocell, we will nonetheless use it to look at the issue of temporal organization. It has already been shown that many important properties of cells can be modeled by considering kinetics alone (Novak and Tyson, 1993). By using the well-developed and understood nonlinear chemical dynamics, important progress can be made in understanding the origin, development, and properties of protocells.

24.3 The Formation of a Chemical Cell

One suggested approach for learning about the formation of early cell ancestors was to remove particular parts of the cell and look at the functioning of the rest. This is similar to the process of taking parts from a 747 in hope that one will produce the Wright brothers' plane. When looking at the 747, it is impossible to imagine that it has "descended" from the Wright's plane.

However, it is obvious that some parts or subsystems in both the 747 and the Wrights' planes must fulfill the same functions, even if carried out by completely different structures and mechanisms. Following this analogy, we should look at what could have provided lift, rather than how to fashion a wing from aluminum without modern technology, or for forms of propulsion other than a jet engine. The next step would be to study the "evolution" from the Wrights brothers' plane to the 747, which in this case was achieved by human design and implementation, including a lot of trial and error and refinement. Let us use this approach to explore what functions must have been performed by the biological equivalent of the Wright brothers' plane.

1. The protocell must be capable of continuously existing far from thermodynamic equilibrium. Systems that are far from thermodynamic equilibrium are usually produced by stressing the system by putting it between two reservoirs with different free energetic potentials, such as reservoirs at different temperatures. The flow of energy or matter through the system is a necessary condition for forming structures. Examples might include the Bernard structures or chemical oscillations and chaos in a CSTR. In chemical cellular systems such as CSTR, the chemicals diffuse into the cell, react, and the products diffuse out of the cell. It is an analog of simple one-celled organisms in which waste products diffuse out of the cell and useful products remain inside. Of course, the definition of waste and useful products depends on the function of particular chemicals.

The spontaneous formation of cellular chemical systems that exist far from equilibrium, and into which reactants diffuse, react at enhanced rates due to catalysts with products diffusing out, was recently achieved in a simple inorganic system (Maselko and Strizhak, 2004).

2. Perhaps the two most important properties of living systems are the multiplication of cells and the possibility of evolution. Studies suggest that a necessary condition for evolution may be that the diverging daughter cells have different concentrations of enzymes (Morowitz, 1987; Szostak, Bartel, and Luisi, 2001). Szathmáry suggests that information can be carried forward as analog (as opposite to digital) information, in the form of the concentrations of enzymes, which is maintained or is able to reestablish itself following the division process (Szathmáry and Maynard Smith, 1997). This is in direct contrast to the digital information carried by genes.

3. The complexity of protocells should increase in time, allowing them to adapt to changes in environment.

The spontaneous formation and subsequent multiplication of protocells has been studied for many years (Deamer, 1997; Hanczyc and Szostak, 2004; Luisi and Varela, 1989; Mavelli and Luisi, 1996; Rasi, Mavelli, and Luisi, 2003; see also chapters 2 and 3). The simplest case is the system in which the membrane building material is in a solution. An initial single body is created and then additional building material attaches itself to the membrane that comprises the body, thereby increasing the surface of the membrane, and the body multiplies. In the next, more complex system, the reactant diffuses into the body and reacts with a catalyst, and the products become part of the membrane. These systems are very simple, and are unable to either evolve or increase in complexity.

Therefore, in addition to the possibility of multiplying, the cell must contain some mechanism that will allow it to increase in complexity. This continuous increase in complexity from selective exposure to disruptive environments leads to the incredibly complex biological systems that we observe today.

24.4 The Viable System Model and Protocells

Stafford Beer (1985) analyzed viable systems in biology and in the economic sector. The question was: Why do some species and companies survive, grow, evolve, and propagate in spite of the perturbations from the environment that caused others to disappear? He developed the viable system model, which aims for an integrated model of both systems and their environment. Following is a short summary of the viable system model approach to modeling a protocell, clarifying how it compares with the current approach.

The model divides the system into operation (metabolism) and control (enzymes and information systems that govern operation). A key concept is variety, or complexity, introduced by Ashby (1956). The variety of the system is defined by the number of its states or components. In chemical and biological systems, the variety can be measured by the number of chemical compounds or steady states. The Ashby Law of Requisite Variety states that control can be obtained if the variety of controller is at least as great as the variety of the situation to be controlled. In practice, it is effectively impossible for biological systems to obtain a variety equal to that of the environment. However, an increase in variety is a necessary property of viable systems. This is done by mutation and evolution. Another requirement is that the variety of controller should be as great as the variety of metabolism. One issue that Beer does not discuss in his work is that not just any variety will do. The new states in a system must be compatible with the changes in environment.

For a protocell to be viable and evolve in a continuously changing environment, it has to continuously increase the number of chemical compounds that can increase the size of a protocell's basin of attraction or the number of steady-state attractors (by forming a new basin of attraction). In the following section is a discussion of the formation of chemical systems that fulfill this requirement, based on our knowledge of the dynamics of nonlinear chemical systems.

24.5 Chemical Dynamics of a Protocell

Thermodynamically open chemical systems can be generally described, in the case of a system consisting of two compounds X and Y, by the following set of chemical reactions (Tyson and Light, 1973):

(a) $A \rightarrow X$

(b) $A \rightarrow Y$

where A represents different reactants diffusing in from the environment and whose concentrations are constant in time.

The kinetics term is described as $dX/dt = d_1 A$ or $dY/dt = d_2 A$, where d_i are determined by diffusion of different A compounds.

(c) $X \rightarrow$

and $Y \rightarrow$

represent diffusion of X and Y from a cell to the environment. These have the form $dX/dt = d_x X$ and $dY/dt = d_y Y$

(d) $X \rightarrow Y$

(e) $Y \rightarrow X$

(f) $A + Y \rightarrow X$

(g) $A + X \rightarrow Y$

Kinetic terms for reactions (f) and (g) have the form $dX/dt = k_5 AY$ and $dY/dt = k_{15} AX$, respectively. The four additional reactions are listed below.

(h) $X + A \rightarrow 2X$ autocatalytic reaction

(i) $X + Y \rightarrow 2Y$ autocatalytic reaction

(j) $Y + A \rightarrow 2Y$ autocatalytic reaction

(k) $X + Y \rightarrow 2X$ autocatalytic reaction

The dynamical behavior of this system is given by the following set of differential equations:

$$\frac{dX}{dt} = d_1 A + k_1 AX - k_2 XY - k_3 AX + k_4 XY + k_5 AY$$

$$- k_6 X + k_7 Y + k_8 Y^2 - 2k_9 X^2 - k_{10} X - d_x X \quad (1a)$$

$$\frac{dY}{dt} = d_2 A + k_{11} AY - k_{12} XY - k_{13} AY + k_{14} XY + k_{15} AY$$

$$- k_{16} Y + k_{17} X + k_{18} X^2 - 2k_{19} Y^2 - k_{20} Y - d_y Y \quad (1b)$$

The open chemical system with two variables, described by the preceding equations, will have a maximum of four steady states. We consider separately the two cases that the system has either an even number or an odd number of steady states. The number of steady states can change by two in the process of bifurcation. Any chemical system with an even number of steady states will not have a global attractor. This

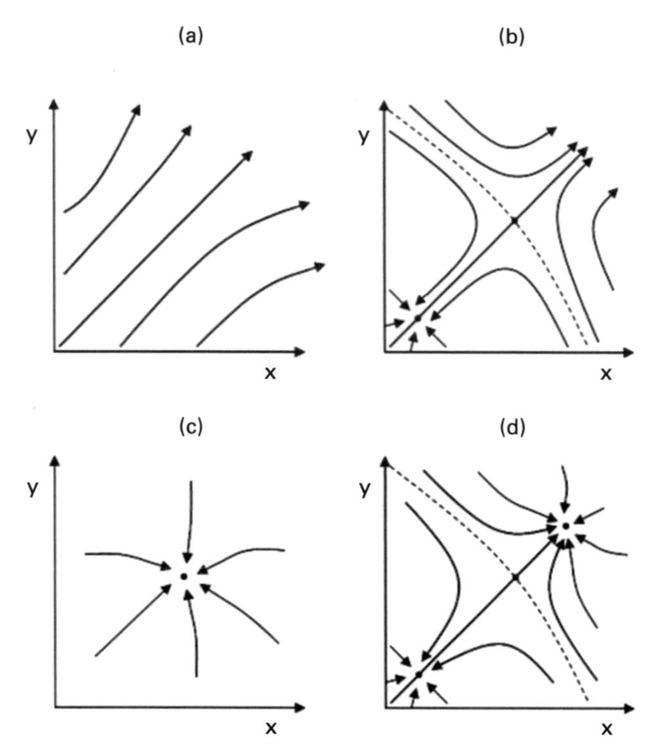

Figure 24.2
Schematic representation of the phase portraits for simple two-component chemical systems. (a) and (b) show systems with an even number of steady states, 0 and 2, respectively. In (a), all trajectories go to infinity, whereas in (b) there is one stable steady state and one saddle point. The broken line indicates a separatrix. All trajectories that start left of the separatrix will end in a stable steady state. All trajectories that start right of the separatrix will go to infinity; (c) and (d) show systems with an odd number of steady states, 1 and 3, respectively. In (c) there is just a stable steady state, whereas in (d) the system has two stable steady states and one saddle point located on the separatrix that separates the two basins of attraction.

means that there are initial conditions in which the trajectory will go on to infinity (see figure 24.2a and b). In contrast, a system with an odd number of steady states will always have a global attractor (see figure 24.2c and d). Adding more variables will increase the number of steady states by a factor of 2 for each new variable. The system with three variables may have a maximum of eight steady states, and the maximum number of possible steady states is equal to 2^d, where d is the number of chemical components. The maximum number of *stable* steady states or attractors related to them is equal to 2^{d-1}. By changing parameter A, we can switch the system from one attractor to another. The relation between parameters and steady states is schematically presented in figure 24.3. The system may also be switched from one attractor to another by a perturbation. By changing system parameters, we can change the properties and concentrations in a cell, but the system is in the same

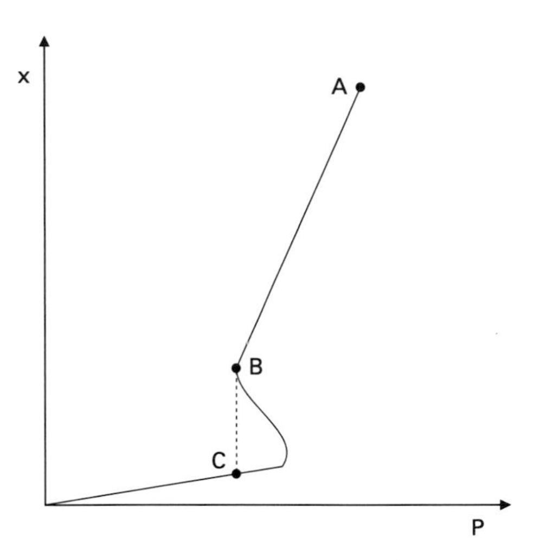

Figure 24.3
The dependence of the concentration of compound X in the steady state on the value of system parameter P. Between points A and B a system is in the stable steady state and concentration changes smoothly with changes in the system parameter. At point B, the system dramatically switches to a different attractor. To the left of point C, the system also has one stable steady state, but a different one. To the right of point C, two attractors coexist (see figure 24.2d). The system may be drawn to either of the two attractors, depending on the initial conditions.

steady state. The changes may be smooth or dramatic. In dramatic changes, systems can switch abruptly to another attractor (see figure 24.3).

Therefore, by increasing the number of reacting chemical compounds, we increase the number of possible steady states for a protocell, and therefore the complexity, with possible different emerging properties. Some attractors make cells more suitable for changing environment than others.

24.6 Transition to Self-Control

The dividing protocell may create two cells that will later compete for resources, and Darwinian evolution begins (Morowitz, 1987; Szostak, Bartel, and Luisi, 2001). In our model of cells, the cells are in a stable steady state surrounded by the basins of attraction. If there are no concentration perturbations, cells will remain in the same steady state with the same concentrations.

The transition from one attractor to another can be caused by a strong perturbation if the system exhibits multistability, or by changing the system parameters. Assume the system parameters of equation 1 are represented by A, corresponding to concentrations of compounds in the surroundings. The possibility is very slim that changes in A will cause a switch to a new steady state that will substantially increase viability of the system. Therefore, some other mechanism must exist that allows other steady states to be created, and the system to switch to them. Networks of reactions that will produce new compounds must exist. By producing new compounds, we can increase the number of possible stable steady states, and a few of them may increase the viability of the chemical protocell. A few of the new compounds produced may be enzymes that perform a functional catalytic role. If a reaction is catalyzed, the initial rate constant k_i is replaced by $k_i'E$, where E is the concentration of catalyst. In this way, the concentration of catalysts becomes an internal controlling parameter. In the language of Beer's analogy between biological and economic systems, the enzymes belong to management.

The enzyme E, for example, may catalyze the following reaction:

(k) $X \rightarrow Y$ with a rate described by $dY/dt = k_{17}X$

in which case this reaction becomes

(l) $X + E \rightarrow Y + E$ with a rate described by $dY/dt = k_{17}'EX$.

The value of E becomes the controlling system parameter similar to A. By changing the concentration of E, one may switch the system from one steady state to another.

The existence of self-control is probably the most crucial step in the formation of protocells. The protocell is no longer completely controlled by the environment; it responds to environmental changes and increases its viability.

The mechanism that will continuously produce new compounds is necessary for chemical systems to increase their complexity and to remain viable. We call this kind of system a *constructor*, and will discuss it in the following section. In biological systems, mutations and the creation of new species occur internally through the mutation of DNA, mostly in the process of duplication. The new DNA may produce new catalysts.

24.7 Constructors

The constructor, and particularly the universal constructor, is an analog of the universal computer introduced by Turing (1936). The universal constructor was introduced by von Neumann (Burks, 1970). We define a constructor as a system that can spontaneously produce the entire spectrum of complex structures. The constructor consists of a set of simple building blocks and rules. Depending on the building blocks, rules, and initial conditions, a wide spectrum of structures can be formed.

A fundamental constructor consists of protons, neutrons, electrons, and the rules of quantum mechanics. The three kinds of particles constitute all forms of matter including living systems. Another basic constructor consists of the letters of the alphabet that can create every word. The words, by rules of grammar, can form text in every book. Similarly, one can construct all music from the eight musical notes. Drexler (1986) defined an assembler as a molecular machine that can be programmed to build virtually any molecular device or structure from simpler building blocks. This is probably the best definition of a molecular chemical constructor. The building blocks of the protein constructor are 20 standard amino acids that may produce billions of proteins, and the building blocks of the DNA constructor are four nucleic acids that produce different DNA sequences. Therefore, in a protocell we may need either a chemical constructor able to construct different chemicals, or several different constructors.

In nature, a constructor produces a constructor; for example, protons, neutrons, and electrons form one constructor that produces all of the chemical elements that constitute a second constructor. This constructor produces chemical compounds that can produce cells that, in turn, produce biological organisms. In this way, the hierarchy of constructors is established. The biological cellular constructor produces billion of different organisms.

24.7.1 Mathematical Models of Constructors

There are several mathematical models dealing with constructors. Here we briefly mention a few of them for future reading. Mathematical models of constructors include the L-systems (Lindermayer, 1975), free grammars, context dependent grammars (Kauffman, 2000), and cellular automata. The simplest cellular automata (CA) have two different cells (with the internal states 0 or 1) and different rules of development. CA have been classified into four types (Wolfram, 2002) based on their qualitative dynamics, and for our purposes the most interesting is the type 4 CA. Their time evolution is a process of development in which structures in the CA state space increase in complexity. The development is impossible to predict without complete simulation of the dynamics. The type 4 CA apparently have structure that may develop continuously as time goes to infinity (so-called open-ended systems).

24.7.2 Examples of Molecular Chemical Constructor Model

We know very little about the chemical constructors. An important paper was written by Farmer (Farmer and Kauffman, 1986). This and subsequent papers concern the dynamics of the polymerization system that begins with a few monomers. In this system, a large spectrum of polymers is created:

$$A + B = AB$$

$$AB + A = ABA$$

$$AB + B = ABB$$

$$ABB + ABA = ABBABA\ldots,$$

where A and B represent monomers, and AB, ABA, and so on represent polymers. And so the process continues, creating successively longer and longer polymers. New compounds can catalyze the formation of other compounds and soon a fully connected network of reactions can be formed, which is reminiscent of the intricate and complex maps of biochemical reactions.

The theoretical considerations by Farmer and coworkers led to the hypothesis that the formation of an autocatalytic network is highly probable. It is an important model, delivered by a group of theoretical scientists without the participation of an experimental chemist. Unfortunately, the experimental realization of spontaneous generation of such networks has never been achieved.

Another example of an open-ended chemical system was observed during the calculation of pattern formation in a multicellular chemical system (Maselko, 2000). Chemical cells are arranged in two-dimensional grids, and every cell communicates by diffusion with four neighbors. The chemical mechanism is very simple, with only two chemical components. The kinetic constants and diffusion coefficients are chosen so that the system may exhibit Turing patterns.

Growth begins from a perturbation of a single cell. Billions of structures may develop in this simple two-component chemical system. The growth of structure may terminate by forming a stationary structure limited in space, or it may be periodic in space and develop to infinity, or be an analog of type 4 CA, developing continuously with increasing complexity. The final state, if it exists, can be obtained only by iterating the initial conditions through time.

The study of constructors proposed here differs from traditional research. In traditional research, after discovering some new phenomena, the main focus is on finding the physical laws that govern the latter and then mathematically modeling similar phenomena based on physical laws or their hypothesized form. Here the approach is different, and reminiscent of the exploration of Conway's Game of Life. The rules (laws) are known from the beginning and the research is devoted to the search for new phenomena and new properties starting from different initial conditions. Even though the Game of Life has been known for more than 30 years, the theory for it has not been developed. In complex systems, the search for the emergence of new and surprising phenomena is the goal of the research.

24.7.3 Initial Chemical Constructor

In biological systems there are many compounds, including lipids and the many intermediates of the metabolic pathways. These compounds were probably formed before the synthesis of RNA and DNA or even proteins. Therefore, we may assume the existence of a primary initial constructor (IC) that produced all these chemicals. The main question is, What is this initial constructor and how was it created?

In early experiments, the formation of simple organic compounds with increasing complexity from simple inorganic compounds was considered (Chyba and McDonald, 1995; Cody, 2004; Miller, 1998; Morowitz, 1987; Wächtershäuser, 1990). Unfortunately, the experimental realization of this scenario has not been completely successful. Many compounds—even complex organic ones—were created, but only at very low concentrations.

The initial constructor has to produce an essentially never-ending collection of new compounds with increasing complexity and also reproduce itself. In biological systems, the creation of new catalysts and enzymes from mutated DNA is mostly synchronized with the multiplication of a cell (Novak and Tyson, 1993). However, this requires very complicated mechanisms that were not available in early cell ancestors. In the beginning, the creation of new catalysts and division of cells were probably not coordinated.

The multiplication of cells can take a few hours, whereas the formation of organic compounds without a catalyst might take months. Before the formation of DNA, the synthesis of new compounds and evolution of complexity was a deterministic process defined by the initial constructor, which itself was formed by a deterministic process. Considering the slow synthesis of organic compounds without enzymes, this process might have lasted a few months, years, or much longer.

The production of new compounds by the initial constructor could have many implications in terms of the resulting chemistry. However, from the persistence point of view, it can have only three implications: detrimental, essentially indifferent, or beneficial. In most cases the new species has no function and will use only matter and energy, that is, it is less fit for persistence. Alternatively, the new chemical compound may fulfill some function that will increase the chance of persistence.

When a new compound is produced by the IC, the number of possible steady states may increase. The system may move from one stable steady state to another, thereby exploring the network of steady states. Some transitions will lead to the thermodynamic steady state that corresponds to the death of a cell. Some transitions will lead to a system with an attractor that has a larger area basin, thereby increasing its ability to survive. The evolution of a protocell occurs through its existence on a network that expands with time.

Detailed studies of chemical constructors are only in their beginning, and continuous transition to protocells is very far from experimental realization.

24.8 Functional Approach

Fundamental to the transition between nonliving and living matter is the emergence of functions. There are no functions in chemistry or physics, but functions are of fundamental importance in biology. The idea that the existence of functions marks the difference between chemistry and biology is instilled into every biology student with the phrase *form follows function.*

The following are major structures and functions in the biological cell, as described in Miller's (1978) excellent book *Living Matter*:

1. Reproducer: the ability to produce a similar cell
2. Boundary: something that separates the cell from its environment
3. Subsystems that process mass and energy: ingestor, which brings in matter and energy; distributor, which carries substances to different places in cell; converter, which changes substances into products more useful to the cell; extruder, which transmits substances out of the system; motor, which moves the systems or its parts; and supporter, which maintains proper spatial relations between components
4. Subsystems that process information: input transducer, internal transducer, channel and nets, decoder, associator, memory, decider, encoder, and output transducer

The protocells we consider are more primitive and, as such, will have a simpler list of functions and structures. In summary, protocells should have the structures and mechanisms that will

- produce the compartment;
- guarantee that the system will stay far from thermodynamic equilibrium (this function is secured by metabolism);
- store/use/pass on information to reproduce the protocell and its functions; and
- increase complexity and provide pathways for emergent phenomena.

For our chemical protocell to survive in a changing environment, the complexity of the system must increase and be compatible with changes in the environment. In other words, it must increase in functional complexity. Therefore, the number of steady states of the chemical compounds must increase continuously. The formation of a single entity that will constantly increase its variety or complexity in a way that will always be compatible with changes of the environment is very difficult in the chemical, technological, or biological world. The complexity of this entity must reach that of the environment, and if the environment is composed of other such entities, the complexity of the entities as well as that of the environment goes to infinity. It is very probable that this entity will be switched to a nonviable state and be annihilated. There must be some feedback from the environment that, in protocells and biological systems, can be realized by the formation of many new different cells and the annihilation of those that are incompatible with the environment. However, the first necessary step is an increase in the number of steady states or chemical compounds. Thus, the protocell must have a constructor that has the capability to produce new compounds. The functional requirements for a protocell to successfully multiply and survive are

$$A_1 + K \Leftrightarrow M + K, \tag{1}$$

namely, compound A_1 diffuses into a cell from outside, reacts with catalyst K, and produces compound M, which will be incorporated into a membrane, thereby causing the cell to increase its volume and multiply.

$$A_2 + K \Leftrightarrow 2K, \tag{2}$$

namely, another species A_2 reacts with the catalyst in an autocatalytic reaction, thereby increasing the number of catalyst molecules.

$$A_3 + Me \Leftrightarrow 2Me, \tag{3}$$

namely, compound A_3 reacts with the compounds running the metabolism, thereby producing copies of themselves.

$$A_4 + C \Leftrightarrow Z, \tag{4}$$

namely, compound A_4, fed either from outside the cell or from the metabolism, is used to produce a new compound, Z. The C represents the set of the compounds and is part of a constructor. The final requirement is that

$$A_5 + C \Leftrightarrow 2C, \tag{5}$$

namely, compound A_5 is used to create a copy of the chemical constructor. (Note that for each of these requirements, the listed compounds could each represent either a single compound or a combination of multiple compounds.)

The original artificial protocell will consist of all the compounds produced in steps 1 through 3 and step 5. Hence, in running through those steps, it will produce a copy of itself. In addition, it will produce a new compound Z, which on cell division, would end up in one of the two daughter cells. Depending on the compounds that made up original protocells and their environments, the daughters may or may not be viable. Producing a new and more suitable steady state corresponds to the daughter having increased viability. Thus, the protocell can multiply and those that are compatible with the environment will survive. With the constructor functionality of the protocell, the daughter with the new compound may have an additional means of interacting with the environment or its own compounds, in theory having an added capability and thus the ability to be more compatible with the environment, greater viability, and higher fitness. We therefore argue that the existence of a constructor is necessary for the collection of cells to be vital in the face of competition or changes in the environment.

24.9 Cellular Chemistry: The Experimental Search for "Other Biology"

The quest to understand life has been highly focused on how it originated on Earth some 3.5 to 4 billion years ago. However, the formation of complex organizations and a continuing increase of complexity are not unique to biological systems but have also been observed in physical, chemical, and geological systems. There must be general laws governing these processes, and it is likely that different geochemical systems may create "life," forming different biologies apart from what is found on Earth. These highly organized structures found outside of life should not be viewed as idiosyncrasies, but rather as evidence for underlying tendencies toward increasing localized complexity and order. These tendencies must be studied in order to understand how life began and where else it may be found.

One relevant field of study is cellular chemistry, that is, study of the formation and properties of chemical cells or the community of cells. Chemical cells are defined as compartments in which chemicals are separated from their surroundings. The external chemicals diffuse or are transported into the cell, react (metabolism), and the products may then diffuse or be transported out of the cell. The entrance and exit of the chemicals and products may be selective and depend on the properties of the membrane or the transport system if it exists. The membrane may be formed as a part of the metabolism. The study of CSTRs can contribute to the understanding of cellular chemistry and biology.

The spontaneous increase of complexity and the formation of cellular and even multicellular structures may even be observed in very simple inorganic chemical systems, such as in Ca^{2+}-silicate, Cu^{2+}-phosphate, and similar systems. We observe spontaneous formation of cells separated from surroundings by a semipermeable membrane. The size of the cells/cell communities is from micrometer to centimeter, just as in biology. In our experiments, an effort was made to study the shapes and properties of inorganic cells.

In a simple experiment, a volume becomes surrounded by a semipermeable membrane that is spontaneously created by submerging a pellet of calcium chloride in a solution of sodium carbonate. The membrane is semipermeable, and small chemicals can diffuse through it. In a somewhat more complex experiment, the pellet is a mixture of calcium chloride and copper chloride and the solution is a mixture of sodium carbonate, sodium iodide, hydrogen peroxide, and starch. The iodide and hydrogen peroxide diffuse into the spontaneously formed cellular structure where the following reactions occurred:

$$Cu^{2+} + I^- \Leftrightarrow Cu^+ + \tfrac{1}{2}I_2$$

$$Cu^+ + H_2O_2 \Leftrightarrow Cu^{2+} + OH\cdot + OH^-$$

Figure 24.4 (color plate 21)
A cell made from a semipermeable calcium carbonate membrane surrounded by a purple solution resulting from the formation of a complex of iodine and starch. The green-blue color inside the cell is caused by the copper catalyst.

Figure 24.5
A monocellular chemical structure in a Fe^{2+}-silicate system.

$$OH\cdot + I^- \Leftrightarrow OH^- + \tfrac{1}{2}I_2$$

The I^- reactant diffuses into the cell volume and reacts with a catalyst Cu^{2+}. The product I_2 diffuses out. The catalyst is later regenerated by the reaction with H_2O_2. The product I_2 reacts with the starch in the external solution, forming the purple color that can be detected (see figure 24.4, color plate 21). In this way, the cellular system may permanently sustain itself far from thermodynamic equilibrium, which is a necessary thermodynamic requirement for living systems. Similar to the biological world, the cells in chemical systems are not always uniform spheres but may sometimes obtain much more elaborate structures, as is shown in figure 24.5.

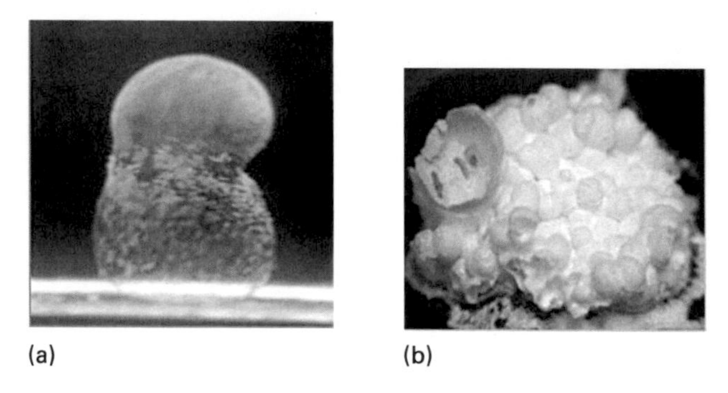

Figure 24.6
(a) The formation and development of a cell that is created by pumping 5.0-M $CaCl_2$ into a 25% saturated silicate and 0.1-M carbonate solution. (b) The growth of a colony of chemical cells formed by dripping a 5-M $CaCl_2$ solution into a 0.33-M silicate and 1.0-M carbonate solution. The image is actually that of the "fossil" of a multicellular chemical system after it has been dried out.

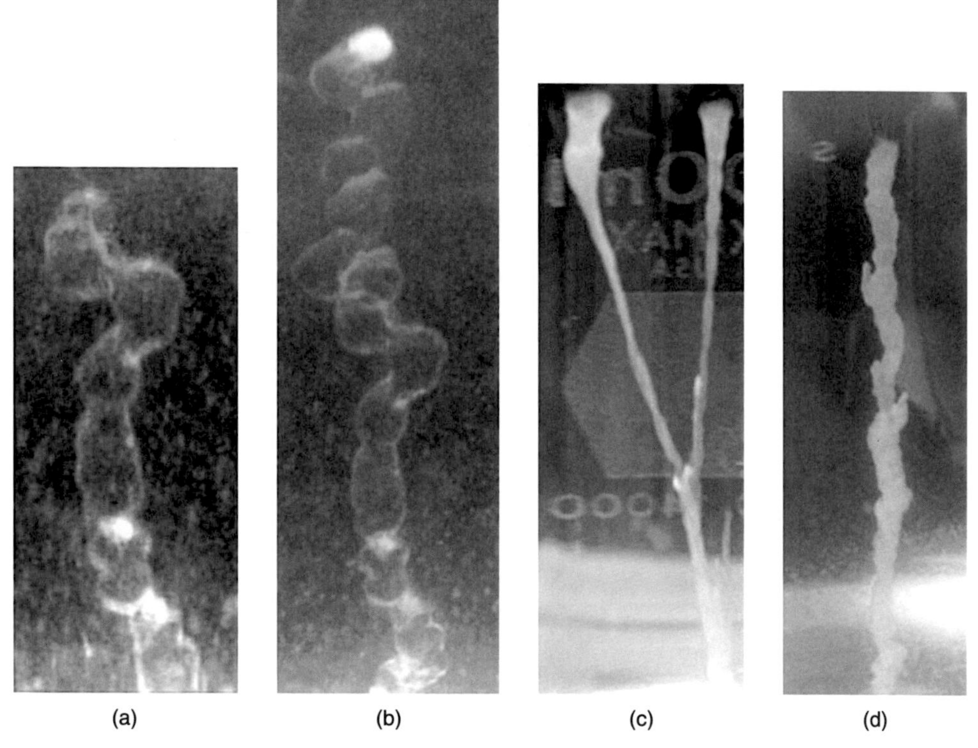

Figure 24.7
Panels (a) and (b) show the growth of a multicellular stem formed by injecting a 1.7-M $CaCl_2$ solution up from the bottom into a silicate solution that is 50% saturated. (c) Branching of a stem formed by injecting a 1.0-M $AlCl_3$ solution up from the bottom into a silicate solution that is 50% saturated. (d) A zigzag stem. After a new branch is formed, the old branch stops growing. The structure was formed by injecting a 1.0-M $AlCl_3$ solution up from the bottom into a silicate solution that is 75% saturated.

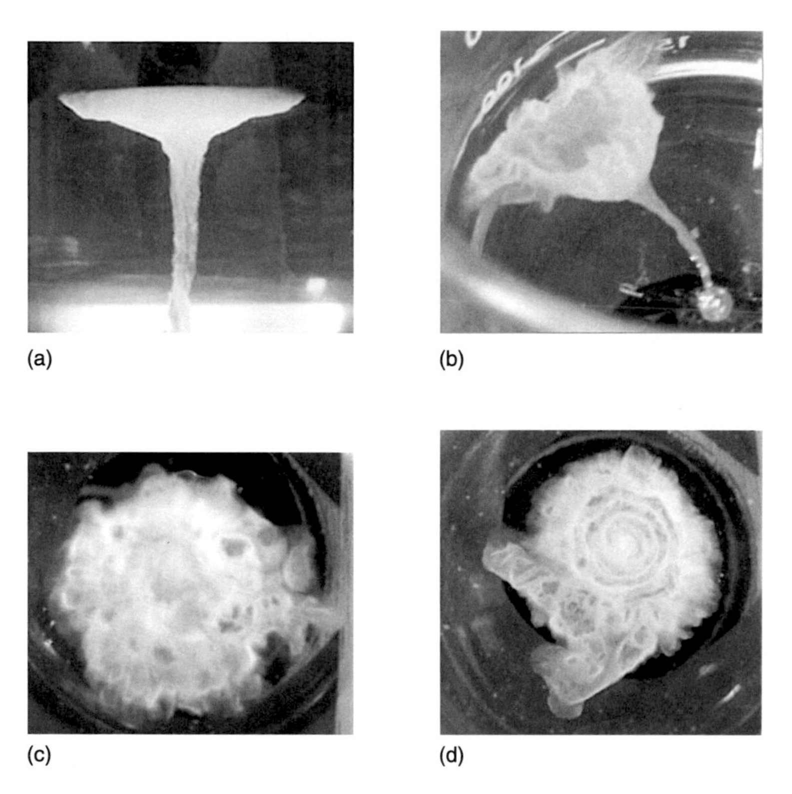

Figure 24.8
Inorganic structures that resemble flowers. (a) The side view; Height 5.2 cm, $AlCl_3$ 1.0 M, silicate 10%. (b) The inside of the flower is empty and resembles a chalice. (c) View from the top. The inside of the flower is more complex and shows internal structure. (d) View from the top. "Pellets" form a spiral similar to that seen in roses.

In another very simple chemical system, we have observed multicellular chemical structures in which cells form different and sometimes very complex morphologies (see figures 24.6 through 24.8 and 24.9, color plate 22). All these experiments were done with the liquid pumped into the tank from the bottom. The cells were often found to form multicellular structures with different morphologies. The multicellular stems can grow from the top end of the cell or from the bottom by continuously pushing up the higher parts of the structure. A stem can form branches directly from the initial stem, or new branches can successfully bifurcate. The most elaborate structures are probably the chemical flowers seen in figure 24.8. The structures presented in figures 24.5 and 24.6 through 24.8 were obtained in two component chemical systems. Increasing the number of reactants will increase the number of possible structures. An example is presented in figure 24.9 (color plate 22), where the addition of glycine changes the morphology.

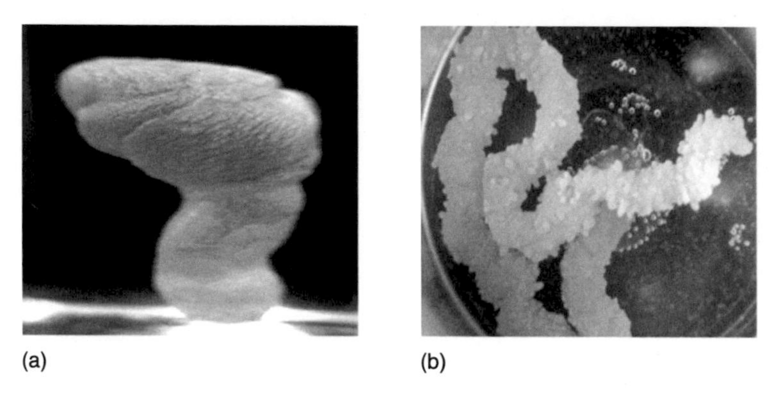

Figure 24.9 (color plate 22)
Influence of an additional compound on cell formation. In (a) the cell developed in a Cu^{2+}-silicate system, whereas in (b) the structure developed in a Cu^{2+}-silicate-glycine system. The additional structures seen on this cell's surface results because glycine is both a complexing agent and a surfactant able to form its own cells.

The multicellular chemical structures in which cells are connected by diffusion were also studied experimentally and theoretically. The properties of multicellular chemical systems were explored by numerical simulations of hundreds or thousands of cells in a square geometry, where every cell was connected by diffusion to four neighbors. The diffusion between cells was chosen to enhance the formation of Turing patterns. The chemistry of each cell was described by the Gray-Scott model (Gray and Scott, 1985). It is probably the simplest two-component model of chemical kinetics that can realize a broad spectrum of temporal and spatial structures in chemical systems. Initially, all cells were in the same stable steady state. The development of structures was then triggered by perturbation of a central cell. The developing structures were found to exhibit a surprising richness of behavior, where *behavior* is defined as a response of the systems to its surroundings. Most of these behaviors have until now been seen only in biological systems. The stem structures that develop from the initially disturbed cell may be solid or fenestrated. Stems behave differently after reaching a nonpermeable wall, depending on the system's parameters. Stems may stop growing or change direction and continue to grow. These entities also have self-repairing properties, and after the stem is cut, the same stem may be regenerated, the new thinner stem may grow, or there may be no regeneration. The growing entity may also influence surrounding territory and annihilate the growth of other remaining entities (Maselko, 1996a, 1996b; Maselko and Anderson, 1999), as seen in biological systems. Figures 24.10 and 24.11 present a simulation model of multicellular chemical structures.

It is important to notice that very simple, two-component chemical systems may create such complex structures even without a complicated information mechanism

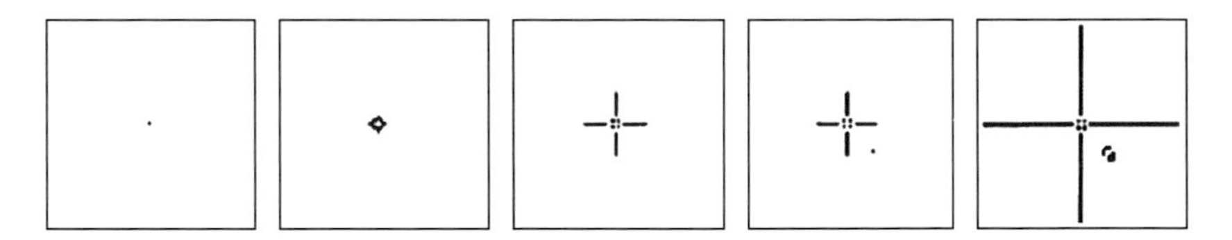

Figure 24.10
Simulation of pattern formation in a 2-dimensional array (100 × 100) of chemical cells. The chemistry is described by the Gray-Scott 1988 model. The parameters are chosen to assure Turing pattern formation. The first frame on the left shows initial perturbation; the next two frames show pattern development. On the next frame the second perturbation has been implemented. The last frame shows the final stationary structure. The second perturbation does not develop to a full structure. This is an example of environment control by the first developed structure.

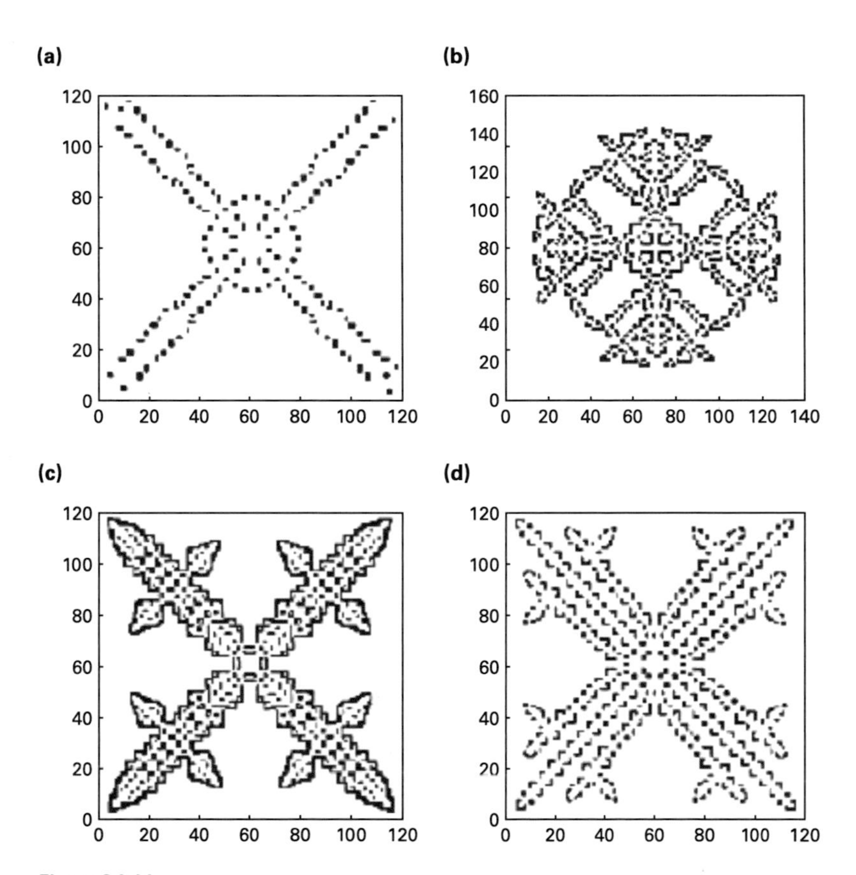

Figure 24.11
Numerical simulation of pattern formation in multicellular chemical systems. Parts (a), (c), and (d) represent different open-ended chemical systems; (b) represents a pattern that is stationary and does not develop further.

such as DNA in biological systems. As can be seen from figure 24.6b, the multicellular chemical structures may also be obtained in simple inorganic systems.

Though these studies are only preliminary, they suggest that spontaneous formation of complex forms and increased complexity are common and that different biologies may be possible. In our experimental systems, by changing systems parameters, we may create different structures with different emerging behaviors. Some properties of inorganic cells are different from properties of protocells and biological cells. The variety of the cells produced depends on the pumped chemicals, which in turn are controlled by the experimentalist, in contrast to biological systems, where the constructor forms the chemicals controlling the cell's properties. Thus far, experimentally obtained cellular and multicellular chemical structures do not have the capacity to evolve, and substantial experimental effort will probably be required to implement this capability. However, in many applications of biological systems, like genetic engineering, the search for new species with required properties is not subject to natural evolution. In these cases, mutations, changes in the system's controlling parameters, and selections are made by humans. A similar human control/selection can also be done in our "different biology." In both cases, the job at hand is to develop the needed information to get the desired property and store it within the system in a usable form. The information will, of course, have to be specific to the environment the system lives in. If this information is truly unknown beforehand, the only way to get there is by trial and error.

24.10 New Technology of Self-Constructing Chemical Systems

The recreation of life is probably the most intellectually challenging problem in science. Related studies, and particularly the study of multicellular chemical structures, may lead to a technological revolution that should be compared to the nanotechnology revolution.

In the contemporary technology, all complex structures are built by producing parts and then assembling them. Biological technology is very different. A seed is immersed into a special solution, and incredibly complex and very different forms can develop spontaneously from it. If we learn how to control these processes and incorporate them into engineering practice, we could (in theory) have cars, buildings, and almost any technological object that would grow spontaneously. Of course, controlling these systems will be a very difficult task, and many experiments and the development of a predictive theory of developing structures are necessary.

Initial calculations (Maselko, 2000) indicate that we do not need an information-control apparatus that is nearly as complicated as DNA. Even in a very simple two-variable multicellular chemical system we may observe billions of possible structures depending on only a few parameters. These parameters are concentrations in solution, concentrations in pellets, or the pumping of particular enzymes into the internal solution. To control the synthesis of the products generated, it is necessary to experimentally study the phase diagram. That diagram may be very complicated, but it is the primary source of our information about the system and its dynamics. By choosing different concentrations of different chemicals, we can choose different developing structures. As we observed in modeling and experiments, these structures may have holes or produce tubes. But the most interesting of these simple chemical structures would be the ones that show self-repair capabilities or different controlled responses to changes or obstacles in their environment.

Even very simple two-chemical structures may learn and mark territory (Maselko and Anderson, 1999). Just as in the Game of Life, a change in initial composition can lead to a broad spectrum of new emerging behavior. As mentioned before, the chemical cellular system is probably the universal constructor, which means that it can, in principle, produce all cellular structures that can be built, including existing biological structures.

24.11 Conclusion

During biological evolution, a continuous increase in complexity and the continuing emergence of new phenomena are observed. Many living forms have been created and many of them are incredibly complex. Therefore, we may seek to create protocells that can spontaneously undergo evolution and that will lead to such complexity.

In discussing the origin of the protocell, we should concentrate on an entity that will have the ability to create many different cells and explore the network of possible protocells. To form protocells we must now learn more about chemical constructors. We should study chemical constructors both theoretically and especially experimentally.

Biological systems are dynamic systems existing far from thermodynamic equilibrium. Most of the effort in attempting to create a protocell or synthesize life has been devoted to the reconstruction of chemical compounds and their paths of synthesis. The dynamics of protocells has been recognized as crucial but has been investigated only marginally (Eigen and Schuster, 1979; Kaneko, 2003; Stadler and Stadler, 2003). Here, we are proposing an approach to the protocell based on nonlinear chemical dynamics. Studying the complex kinetics of chemical systems far from thermodynamic equilibrium has developed this approach. The nonlinear chemical dynamic approach to the origin of life is just at its beginnings, from both the theoretical and the experimental points of view. It raises the following important questions:

1. What conditions are needed to form open-ended chemical systems, particularly in experimental studies? What are the properties of open-ended chemical systems? For

life to emerge, it is probably necessary for the universal constructor constantly producing new compounds to cooperate with the self-organizing phenomena seen far from thermodynamic equilibrium.

2. When and how does the transition from a deterministic to a probabilistic chemical system occur, or more specifically, when does the transition to Darwinian evolution occur? In the beginning, we consider formation of new compounds by the chemical constructor as a deterministic process.

The Darwinian evolution with its exponential growth occurs once one has the capacity for self-catalysis, or perhaps more accurately, self-driven reproduction, with evolution occurring as a result of occasional errors in the self-reproduction process, with the defective product still having the ability to self-reproduce. Competition for resources coupled with the environment itself will, of course, provide the subsequent selection.

3. What types of functions are related to the protocell?

The study of chemical cells will probably be a huge and rapidly developing area of chemistry. The initial experimental results, in simple inorganic systems with only two components, show great potential. The approach based on cellular chemistry described here will allow the creation of many complex systems that, in contrast to biological systems, will have initial conditions and survival defined by humans. These systems might not be alive from the biological point of view, but might well perform many tasks and be the beginning of a technological revolution comparable with the computer or nanotechnology revolutions.

Acknowledgment

This chapter could not have been completed without the assistance of P.-A. Monnard, S. Rasmussen, and H. Ziock (to whom we are especially grateful).

References

Ashby, W. R. (1956). *An introduction to cybernetics.* London: Chapman & Hall.

Beer, S. (1985). *Diagnosing the system for organizations.* New York: John Wiley & Sons.

Burks, A. (1970). *Essays on cellular automata.* Urbana: University of Illinois Press.

Chyba, C. F., & McDonald, G. D. (1995). The origin of life in the solar system: Current issues. *Annual Review of Earth and Planetary Science*, *23*, 215–249.

Cody, G. D. (2004). Transition metal sulfides and the origin of metabolism. *Annual Review of Earth and Planetary Science*, *32*, 569–599.

Deamer, D. (1997). The first living systems: A bioenergetic perspective. *Microbiology and Molecular Biology Review*, *June*, 239–261.

De Duve, C. (2003). A research proposal on the origin of life. *Origins of Life and the Evolution of the Biosphere*, *33*, 559–574.

Drexler, K. E. (1986). *Engines of creation.* New York: Anchor Press.

Eigen, M., & Schuster, P. (1979). *The hypercycle.* New York: Springer-Verlag.

Farmer, D. J., & Kauffman, S. A. (1986). Autocatalytic replication polymers. *Physica D*, *22*, 50–67.

Fox, S. W. (1973). A theory of molecular and cellular origin. *Nature*, *205*, 328–340.

Gánti, T. (2003). *Chemoton theory.* New York: Kluwer Academic.

Gray, P., & Scott, S. K. (1985). Sustained oscillations and other exotic pattern behavior in isothermal reactions. *Journal of Physical Chemistry*, *89*, 22.

Hanczyc, M., & Szostak, J. (2004). Replicating vesicles as models of primitive cell growth and division. *Current Opinion in Chemical Biology*, *8*, 660–664.

Hasslacher, B., Kapral, R., & Lawniczak, A. (1993). Molecular Turing structures in the biochemistry of the cell. *Chaos*, *3*, 7.

Kaneko, K. (2003). Recursiveness and evolvability in a mutually catalytic reaction system. *Advances in Complex Systems*, *6*, 79–92.

Kauffman, S. (2000). *Investigations.* New York: Oxford University Press.

Lee, K. L., McCormick, W. D., Swinney, H. L., & Pearson, J. E. (1994). Experimental observation of self-replicating spots in a reaction-diffusion system. *Nature*, *369*, 215–218.

Lindermayer, A. (1975). Developmental algorithms for multicellular organisms: A survey of L-systems. *Journal of Theoretical Biology*, *54*, 3–22.

Luisi, P., & Varela, F. (1989). Self-replicating micelles: A chemical version of a minimal autopoietic system. *Origin of Life and Evolution of Biosphere*, *19*, 633–643.

Maselko, J. (1979). On applications of oscillation reactions in chemical technology processes. *Material Science*, *5*, 155.

Maselko, J. (1996a). Growth of patterns in a multicellular chemical systems and their responses to boundaries. *Journal of the Chemical Society, Faraday Transactions*, *92*, 2879–2881.

Maselko, J. (1996b). Self-organization as a new method for synthesizing smart and structured materials. *Materials Science and Engineering C*, *4*, 199–204.

Maselko, J. (2000). Large scale structures in multicellular chemical systems. *Polish Journal of Chemistry*, *74*, 311–319.

Maselko, J. (2003). Patterns formation in chemical systems. *Advances in Complex Systems*, *6*, 3.

Maselko, J., & Anderson, M. (1999). Some properties of multicellular chemical systems: Response to the environment and emergence. *InterJournal Complex Systems*, article no. 227. Accessed May 15, 2007 at: http://www.interjournal.org/.

Maselko, J., & Swinney, H. (1986). Complex periodic oscillation in the Belousov-Zhabotinskii Reaction. *Journal of Physical Chemistry*, *85*, 6430.

Maselko, J., & Strizhak, P. (2004). Spontaneous formation of cellular chemical system that sustains itself far from thermodynamic equilibrium. *Journal of Physical Chemistry*, *108*, 4937.

Mavelli, F., & Luisi, P. L. (1996). Autopoietic self-reproducing vesicles: A simplified kinetic model. *Journal of Physical Chemistry*, *100*, 16600–16607.

Miller, J. G. (1978). *Living systems.* New York: McGraw-Hill Book Company.

Miller, S. L. (1998). The endogenous synthesis of organic compounds. In A. Brack (Ed.), *The molecular origin of life: Assembling pieces of the puzzle* (pp. 59–85). Cambridge: Cambridge University Press.

Morowitz, H. J. (1987). *Beginnings of cellular life.* New Haven, CT: Yale University Press.

Novak, B., & Tyson, J. (1993). Modeling the cell division cycle: M-phase trigger, oscillations and size control. *Journal of Theoretical Biology*, *165*, 101–134.

Orgel, L. E. (1998). Polymerization on the rock: Theoretical introduction. *Origins of Life and the Evolution of the Biosphere*, *28*, 227–234.

Pearson, J. (1993). Complex patterns in simple systems. *Science*, *261*, 189–192.

Quyang, Q., & Swinney, H. L. (1991). Transition from an uniform state to hexagonal and striped Turing patterns. *Nature, 352*, 610–612.

Rasi, S., Mavelli, F., Luisi, P. L. (2003). Cooperative micelle binding and matrix effect in oleate vesicle formation. *Journal of Physical Chemistry B, 107*, 14068–14076.

Rasmussen, S., Chen, L., Deamer, D., Krakauer, D., Packard, N., Stadler, F., & Bedau, M. (2004). Transition from nonliving to living matter. *Science, 303*, 963–965.

Rasmussen, S., Chen, L., Nilsson, M., & Abe, S. (2003). Bridging nonliving and living matter. *Artificial Life, 9*, 269–316.

Ross, J. (1995). Implementation of logic functions and computations by chemical kinetics. *Physica D, 84*, 180–193.

Stadler, B. M. R., & Stadler, P. F. (2003). Molecular replication dynamics advancement. *Complex Systems, 6*, 47–78.

Szathmáry, E., & Maynard Smith, J. (1997). From replicators to reproducers: The first major transition leading to life. *Journal of Theoretical Biology, 187*, 555–571.

Szostak, J., Bartel, D., & Luisi, P. L. (2001). Synthesizing life. *Nature, 409*, 387–390.

Turing, A. M. (1936). On computable numbers, with the application to the Entscheidungsproblem. *Proceedings of the London Mathematical Society, 2* (42), 230–265.

Turing, A. M. (1952). On the chemical basis of morphogenesis. *Philosophical Transactions of the Royal Society of London B, 237*, 37–72.

Tyson, J., & Light, J. (1973). Properties of two-component bimolecular and trimolecular chemical reactor systems. *Journal of Chemical Physics, 59*, 4164–4173.

Wächtershäuser, G. (1990). Evolution of the first metabolic cycles. *Proceedings of the National Academy of Sciences of the United States of America, 87*, 200–204.

Wolfram, S. (2002). *A new kind of science.* Champaign, IL: Wolfram Media Inc.

25 Early Ancestors of Existing Cells

Andrew Pohorille

25.1 Introduction

Recent rapid advancements in structural, molecular, and cellular biology have for the first time facilitated efforts to integrate the functions of containment, metabolism, and heredity in simple, laboratory-built, cell-like structures, which by many standards will be considered to be alive. Most of these efforts proceed along two lines. One is to use existing simple cells as templates and simplify them in various ways. Another is to explore as building blocks a wide variety of materials and structures that are not necessarily related to any known life, to explore a broad range of alternative life forms.

In this book, the structures that might emerge from both lines of research are collectively called *protocells*, and in the literature some of them are referred to as *minimal cells*, *artificial cells* or *synthetic cells*. Creating any of these structures in the laboratory is a formidable task, and one that can be assisted by summarizing our current knowledge about the only experiment in creating life that has so far been successful—the one carried out by nature when life first originated. This chapter provides such a summary. In this context, it is only natural to focus on universal principles that must be shared by both naturally evolved and manmade protocells. Specific chemical or environmental realizations serve only as examples to illustrate these general principles.

The central position on the evolutionary pathway connecting inanimate and animate matter is occupied by the earliest ancestors of cells. (These are sometimes called protocells in the literature, but here they will be called *early cell ancestors* for terminological consistency with this book.) Even though there is no fossil evidence or living example of an early cell ancestor, it is widely accepted that they must have existed on the early Earth. It is also generally believed that early cell ancestors were evolving, heterogeneous structures. This means that initially cell ancestors were endowed with very few functions and separated from the environment by boundary

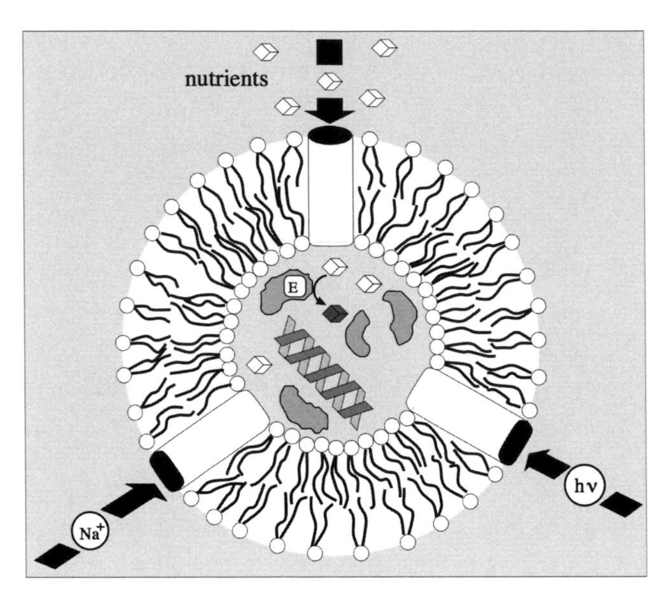

Figure 25.1
Schematic of an early cell ancestor capable of carrying out essential cell functions: metabolism (chemical reactions catalyzed by protoenzymes), information transfer, energy transduction, and transport of ions, nutrients, and waste products across cell walls. Similarity of symbols in the figure to conventional drawings of biomolecular structures in contemporary cells (e.g., a double helix) is used only to help the reader in identifying different functions. Molecules and mechanisms used to carry out these functions in cell ancestors might have been different from those in contemporary terrestrial cells.

structures (see figure 25.1). Bilayers built of amphiphilic molecules that form vesicles are protobiologically relevant examples (Deamer and Pashley, 1989; Dworkin et al., 2001). Over a period of thousands or millions of years, they acquired complex structural and functional characteristics of modern cells. Furthermore, at any given time, not all early cell ancestors were identical. The primary goal of research on cell ancestors is to explain how they self-organized, performed their essential functions given only the simple material available in the environment, and evolved in a continuous manner toward structures that are often called the last common ancestor. Any viable explanation of these processes must be firmly rooted in our understanding of universal laws of physics and chemistry, as they operated in protobiological milieu.

Although the conceptual distinction between protocells (or artificial cells) and minimal cells is somewhat blurry, each concept is useful because each is related to different scientific objectives. Most broadly, minimal cells are defined as membrane-bound constructs that can self-maintain, self-replicate, and evolve (Luisi, 2003). Commonly, they are considered cells with the minimal number of genes that still allow for retention of these properties (Gil et al., 2004; Hutchison et al., 1999; Koonin, 2003). The agenda for studying minimal cells is to start with the simplest contemporary organisms and successively delete nonessential genes or, possibly, substitute a group of genes with a smaller set of natural or synthetic genes (see chapters 2, 3, and 16). Although ideas for different routes to the reduction of the genome might come from studies on the origin of life, there is no requirement or guarantee that the minimal cell represented a point on the path from nonliving to living matter.

Protocells or artificial cells are simple, laboratory-made, cell-like constructs that exhibit some but not necessarily all traits of existing living cells (Noireaux et al., 2005; Pohorille and Deamer, 2002). As Pohorille and Deamer (2002) have argued, many functional components needed for their construction might already have been built or extracted from cells, and the main challenge is to put them together so that they work in concert. Clearly, efforts along these lines might be helpful for understanding how early cell ancestors functioned. However, since the development of artificial cells is most often motivated by needs in biotechnological, biomedical, and pharmaceutical fields (Pohorille and Deamer, 2002), no effort is made for compatibility with protobiological conditions. One more reason why they should not be considered as examples of early cell ancestors is that, in general, the components of artificial cells are not well balanced in their evolutionary complexity.

The study of early cell ancestors has been mostly confined to the field of the origin of life. Unfortunately, this field is not nearly as vigorous, rapidly developing, or well funded as molecular biology or biotechnology. Not surprisingly, progress on building laboratory and theoretical models of cell ancestors has been relatively slow. Recently, however, this line of study acquired a new context by becoming a part of the growing field of astrobiology NASA initiated nearly a decade ago. Astrobiology is defined as a discipline concerned with the origins, evolution, distribution, and future of life in the universe. It is still in a natural and healthy phase of identifying itself through research of the participating scientists, but at least so far its agenda has been largely guided by NASA (2003). One of its explicitly stated goals is to move beyond the specifics of our own origins to create a broader field of *universal biology*. Origins of cellularity and protobiological systems are identified as an inherent part of this field, with particular focus on the origins and early coordination of key cellular processes such as metabolism, energy transduction, and information transfer between consecutive generations. Again, these investigations are to be carried out without special regard to how life actually emerged on Earth. A clear implication of this point of view is that early cell ancestors are more than precursors of cells on the early Earth—they were (and perhaps still are) universal intermediates that bridged nonliving and living matter in the universe. This, in turn, brings to center stage several questions that are not new but have so far been considered peripheral to the problem:

What is universal about early cell ancestors and what are the reasons for this universality if it exists? That is, if an early cell ancestor were found on another planet, would it be recognized as such on the basis of present knowledge of terrestrial chemistry and biochemistry?

These questions cannot be fruitfully addressed without connecting them to a long-standing debate about the fundamental nature of processes that led to the emergence of life. One view holds that the origin of life is an event governed by chance, and the result of so many random events is unpredictable. Monod (1971) eloquently expressed this view in his book *Chance and Necessity*, in which he argued that life is a product of "nature's roulette." Although his arguments mostly dealt with the genomic domain, he also reasoned that the pregenomic domain was equally unpredictable. In an alternative view, the origin of life is considered a deterministic event. Its details need not be deterministic in every respect, but the overall behavior is predictable. An elegant exposition of this view was made by Morowitz (1992). A corollary to the deterministic view is that the emergence of life must have been determined primarily by universal chemistry and biochemistry, rather than by subtle details of environmental conditions, such as narrow temperature range, precise composition of the atmosphere, or a very specific suite of chemicals available for biogenesis. Since these details are unknown and, most likely, unknowable, the origins of life could not be reconstructed by scientific means and therefore we would be forced to accept Monod's view.

This preliminary discussion brings us to the main objectives of this chapter. We argue that, indeed, early cell ancestors are likely candidates for universal ancestors of living systems. We further argue that their self-organizing and functional capabilities were strongly constrained by fundamental principles of chemistry. As a consequence, cell ancestors can form in only a subset of possible environmental conditions and will all exhibit certain common features. This, however, does not mean that we fully embrace the deterministic point of view. In fact, we argue that the processes underlying the emergence of life are stochastic and, therefore, can be described only in probabilistic terms. Their outcome is predictable, however, although not in full detail.

In the next section, we briefly review the main hypotheses about the origin of the biosynthesis of early cell ancestors. The focus is on mechanisms of information transfer between generations and universal constraints implied by different hypotheses. The picture is painted with a broad brush, but the reader can find many details in other chapters of this book. We then move to examining the emergence of structure and function at a molecular level, again emphasizing universal constraints. Membrane-bound peptides are used as an example. Since molecular structure and function cannot be treated separately from environmental conditions, the kinds of solvents suitable for supporting life must be considered.

25.2 Early Cell Ancestor Biosynthesis

25.2.1 Preliminaries

An essential condition for the emergence of early cell ancestors is the formation of barriers that separate their interior from the environment. In contemporary cells, these barriers are typically built of phospholipid bilayers. Phospholipids, however, are not likely to play the same role in cell ancestors because no primitive biosynthetic pathway for their synthesis appears to exist. Instead, ancestral cell walls might have been formed by other amphiphilic molecules, such as long-chain alcohols, carboxylic acids, and fatty acids. Indeed, it has been shown that under appropriate conditions amphiphilic material extracted from the Murchison meteorite or obtained in laboratory simulations of interstellar or cometary material can yield vesicles (Deamer and Pashley, 1989; Dworkin et al., 2001; Mautner, Leonard, and Deamer, 1995). This indicates that vesicle-forming amphiphilic molecules might be fairly common in space and might have been delivered to planets in the solar system or beyond to seed life. Thus, it is plausible that vesicles or related structures built of amphiphilic material were universal precursors of cell ancestors.

By themselves, vesicles were nothing more than envelopes that contained molecules needed for chemical reactions supporting the self-maintenance, growth, and evolution of cell ancestors. This leads to the question: What were those chemical reactions? Taking clues from modern biosynthesis, two significant classes of reactions may be distinguished: intermediary metabolism, responsible for the synthesis of small molecules, and the synthesis of biopolymers from monomers. The assumptions made about the order in which these metabolisms emerged and how they were executed lead to very different scenarios for the early evolution of cell ancestors.

Valid assumptions must be selected on the basis of several fundamental criteria. One is that chemical reactions constituting a metabolism must be permitted under protobiological conditions. In the language of physical chemistry, these are thermodynamic and kinetic constraints. Another important criterion is that an early cell ancestor system cannot undergo combinatorial explosion in the number of its components. Combinatorial explosion occurs when a very large number of possible combinations can be created as the number of entities that can be combined increases. A hypothetical "random chemistry," in which every compound in the original mixture can react with every other compound, can serve as an example. Combinatorial explosion is a threat whenever increasingly complex systems are constructed from a variety of building blocks. This threat applies equally to polymer synthesis and networks of chemical reactions. Once it occurs, almost all resources are expended on synthesizing molecules that play little or no constructive role in the overall system. Thus, scenarios for early cell ancestor evolution must make provisions for taming combinatorial explosion.

25.2.2 Early Cell Ancestors with Genetic Polymers

Historically, most approaches have concentrated on biopolymer synthesis. The paradigm for this approach is the RNA world hypothesis (see chapters 17 and 26). The discovery that RNA is capable of catalysis led to the suggestion that the present world of nucleic acids and proteins was preceded by another "world," wherein RNA molecules alone acted as both catalysts of biochemical reactions and information storage systems (Gestland, Cech, and Atkins, 1999; Gilbert, 1986; Joyce, 1996). The hypothesis is supported by the fact that a variety of RNA enzymes have been created in the laboratory using in vitro selection (Wilson and Szostak, 1999), and by the recent discovery that the decoding and peptidyl transferase centers of ribosomes are composed entirely of RNA (Nissen et al., 2000).

According to the RNA world hypothesis, selection of improved combinations of nucleic acids obtained through random mutations drove evolution of biological systems from their inception. This evolution can be described by adapting concepts from Eigen's theory of early genomic systems (Eigen, 1971). It suggests that template-directed replication of nucleic acids provides a natural mechanism for preventing combinatorial explosion through autocatalytic processes (Eigen and Schuster, 1977). In this theory, the pace of evolution is determined by noise (mutation rates) in the system. This pace has its limits. Above a certain error threshold, all information is lost and the system experiences an "error catastrophe." Estimates of error rates for nonenzymatic synthesis of RNA imply that polymers longer than 100 nucleotides will undergo such a catastrophe. To increase the fidelity of replication, enzymes are needed. This leads to an apparent paradox: No replicase can exist without a large genome, and no large genome can exist without a replicase. Eigen and Schuster (1977, 1979) proposed an elegant solution to this paradox—error rates can be markedly reduced in the presence of short polymers only if they are organized into hypercycles. The hypercycle is a self-reproducing macromolecular system, in which RNA molecules and enzymes cooperate in a cyclic manner, which ensures its stability. In this view, hypercycles provide coupling between inheritance and metabolism and, for this reason, can be considered as the essence of early cell ancestors.

Bringing the concept of the RNA world to its logical conclusion, Szostak, Bartel, and Luisi (2001) proposed a hypothetical construct that can be viewed as "the minimal RNA cell." This construct consists of only two ribozymes encapsulated in a vesicle. One of them is a replicase capable of copying both ribozymes. The second ribozyme catalyzes the synthesis of the membrane-forming molecules from their precursors. These molecules can become incorporated into the membrane. As a result, the cell would grow in size and eventually reach thermodynamic instability leading to its spontaneous division. Thus, at least in principle, the system could self-maintain, self-replicate, and undergo evolution through mutations in the nucleic acid composition of the ribozymes. However, the apparent simplicity of this construct is somewhat deceiving—the number of components is truly minimal, but it is highly unlikely that either of them would be simple. In particular, it is not clear whether small replicases exist. It would be difficult therefore to consider the minimal RNA cell as an RNA cell ancestor, because a large replicase would be a subject of error catastrophe, mentioned earlier.

The RNA world hypothesis has another, well-known weakness—it requires the presence of activated monomers for RNA polymerization. However, synthesis of nucleotides, particularly the ribose part, is quite difficult to achieve under conditions compatible with the protobiological environment. To circumvent this difficulty, it has been proposed that transitional genetic polymers, whose monomers were simpler to synthesize under primitive conditions, might have preceded RNA. These polymers might have had, for example, backbones containing the pyranose form or ribose (Eschenmoser, 1999) or a peptide unit (Wittung et al., 1994). In this case, however, it is also necessary to assume that the emergence of a genetic polymer in cell ancestors was preceded by a "pregenomic world," in which the required monomers were synthesized through networks of chemical reactions. This view seems to gain acceptance even among supporters of the RNA-first hypothesis (Orgel, 2000).

25.2.3 Nongenomic Cell Ancestors

The considerations presented at the end of the previous subsection lead to the *Chekhovian argument*. Anton Chekhov, the famous nineteenth-century dramatist, once remarked that in a well-staged play if a rifle is hanging on the wall in the first act somebody should be shooting it in the third act. In the context of the evolution of cell ancestors, this argument can be paraphrased as follows: Synthetic routes leading directly from simple organic molecules available in the prebiotic milieu to genetic polymers do not appear to be simple. Routes leading, for example, to amino acids and possibly peptides do not seem to be more complicated, and in fact may be simpler (Leman, Orgel, and Chadiri, 2004; Rode, 1999; Weber, 2001). Thus, if RNA or its analogs existed in the cell ancestor milieu, then peptides should also have existed. Since they have a considerable catalytic potential, they might have played a role in early cell ancestor evolution as protoenzymes, possibly organized into catalytic networks.

These kinds of arguments raise a possibility that both intermediary metabolism and proteins were early inventions in the evolution of cell ancestors. This hypothesis has a long intellectual tradition, but its developments are considerably more diverse than the developments of the RNA hypothesis. This is hardly surprising. In the absence of template-directed replication mechanisms, providing plausible evolutionary scenarios that prevent combinatorial explosion is not a simple or unique task. This task has been approached in several ways, leading to different pictures of early cell ancestor metabolism.

One paradigm for the emergence of metabolism is based on the existence of random reaction networks in the early cell ancestor environment. The pioneering theoretical work in this area was done by Kauffman (Kauffman 1986, 1993) and then extended by others (see, e.g., Bagley and Farmer, 1991). Kauffman addressed this question: What is the probability that a set of nonreplicating, catalytic polymers contains a subset that is reflexively autocatalytic (i.e., capable of self-repeating)? He showed that as the polymer set becomes large enough to reach a certain threshold, this probability rapidly increases to nearly 1. Dyson (1982, 1999) proposed another model for the self-organization of metabolism. In this model, monomers exist in either catalytically active or catalytically inactive states. The activation reaction is reversible, which under certain conditions prevents combinatorial explosion and, instead, allows the system to reach a steady state. Lancet and his collaborators pursued related ideas in a more quantitative fashion (Sergé and Lancet, 1999, 2000; Sergé et al., 2001; see also chapter 11 for an extended discussion). They showed that compositions of cell ancestors persist over generations, thus forming "compositional genomes." This demonstrates that information transfer in simple chemical systems may proceed through mechanisms other than template-directed replication. Sergé and Lancet (1999) provided an excellent review of random reactions concepts.

Morowitz, Kostelnik, Yang, and Cody (2000) presented a different approach to limiting chemical diversity. They applied a small set of physical and chemical constraints relevant to prebiotic conditions to 3.5 million chemical compounds listed in the Beilstein database and emerged with 153 molecules. Among them were all 11 members of the reductive citric acid cycle, which they argued provided support for the strongly deterministic hypothesis that this autocatalytic cycle was at the origin of cell ancestor biosynthesis (Morowitz et al., 2000; Smith and Morowitz, 2004). This approach was criticized on two grounds. First, since the enzymes that catalyze the reactions in the cycle are not produced in the process, the cycle is truly autocatalytic only if it proceeds without the need for enzymes. This may be unlikely (Orgel, 2000). Second, it was pointed out that some of the selection criteria were somewhat arbitrary, and the selection process might be biased (Orgel, 2000).

The second argument is correct in principle, but it seems to miss the main conceptual idea behind the Morowitz approach—prebiotic chemistry is not random because it was very strongly constrained both thermodynamically and kinetically. Once these constrains are taken into account, combinatorial explosion in the number of chemical compounds may not be a serious issue. Weber (2002, 2004) has developed this idea in a systematic fashion. He points out that chemical reactions needed for primi-

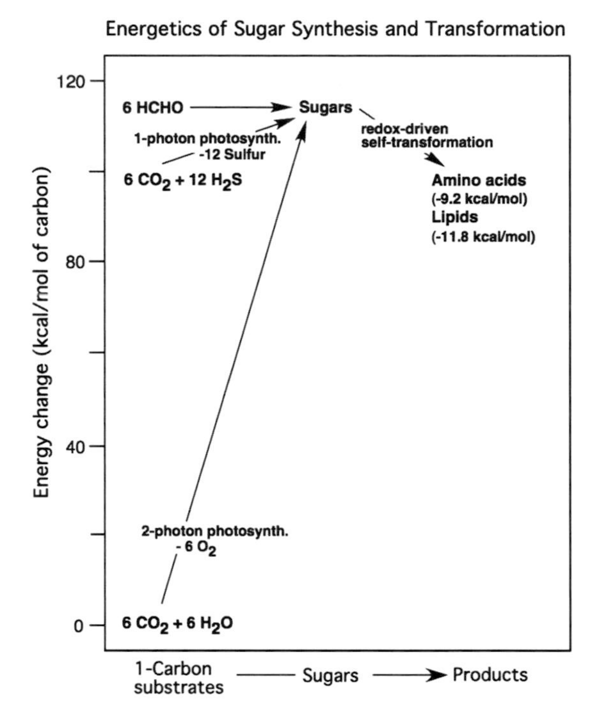

Figure 25.2
Schematic representation of simple metabolism leading from simple carbon substrates to amino acids and its energetics (courtesy of A. Weber).

tive metabolism under mild aqueous conditions must proceed stepwise through a chain of reactions, which produces intermediates separated by a single electron-pair transfer. He further argues that, in the absence of powerful contemporary enzymes, primitive cell ancestor metabolism must have consisted of a continuous series of chemical transformations that had favorable free energies and rate constants. A thorough analysis of free energies and kinetics of carbon group transformations (Weber 2002, 2004) leads to the conclusion that the initial set of prebiotically possible chemical reactions is quite constrained, with sugars playing a central role in early biosynthesis (Weber, 1999). Within this framework, shown schematically in figure 25.2, amino acids appear much more likely to be biosynthetic intermediates than nucleotides (Weber, 2001).

The synthetic limitations of early metabolism would be relaxed with the emergence of high-energy compounds that could capture some of the free energy released in downhill reactions and subsequently be used to drive uphill reactions. Even with these innovations, the diversity of chemical transformations remained restricted by the requirements that only ~10 kcal/mol of carbon was available for biosynthesis, and energy-rich molecules could be synthesized only by irreversible reactions with large, favorable free energies.

Building on earlier theoretical ideas and some recent experiments, Pohorille and New have proposed a model for nongenomic evolution of cell ancestors that combines the emergence of metabolic networks with the emergence of their protein catalysts (New and Pohorille, 2000). Although proteins do not replicate, stochastic mechanisms still exist that can lead to reproduction and improvement of their catalytic functions *in a population*. In fact, self-replication of macromolecules may not have been required during the early stages of evolution; the reproduction of cellular functions alone might have been sufficient for the self-maintenance of cell ancestors. The precise transfer of information between successive generations of the earliest cell ancestors was unnecessary and could have impeded the "discovery" of cellular metabolism. Many proteins with unrelated amino acid sequences could have performed each biochemical function at an evolutionarily acceptable level (Keefe and Szostak, 2001). Such proteins can be found in libraries of randomly synthesized amino acid chains with probabilities comparable to those of finding ribozymes in random RNA libraries (Keefe and Szostak, 2001). As evolution progressed, however, proteins must have performed their functions with increasing efficiency and specificity. These proteins constituted a progressively decreasing fraction of all proteins and, at some point, the likelihood of generating them through noncoded synthesis was so small that further evolution was not possible without storing information about their sequences. Beyond this point, further evolution required the coupling of proteins and informational polymers that is characteristic of modern life.

The starting point for the proposed nongenomic mechanism of cell ancestor reproduction and evolution is the emergence of protoenzymes that catalyze the formation of peptide bonds (ligases) and, therefore, generate new peptides. It is known that even some dipeptides exhibit ligating activity, and a simple ligase has been developed experimentally by Ghadiri and coworkers (Severin et al., 1997). Although most of the peptides synthesized with the aid of these poor catalysts would be nonfunctional or only weakly functional, a few of them might be better ligases than the peptides that generated them. They, in turn, might ligate even more peptide bonds and, by doing so, increase the repertoire of peptides in the cell ancestor system. As a consequence, the likelihood of finding an even better ligase might increase as might the likelihood of finding proteins that catalyze other chemical reactions, which would be subject to the constrains discussed by Weber. Some of the peptides generated by ligases act as proteases and hydrolyze the already formed peptide bonds. Proteases require water for their function. Peptide bonds are disordered and, therefore, nonfunctional or poorly functional molecules are more likely to be exposed to water than bonds in structured peptides. This means that proteases preferentially cleave nonfunctional peptides, thus reducing their inventory.

Preliminary computer simulations based on this model (Pohorille, unpublished results) indicate that, although most early cell ancestors exhibit little catalytic activity, some encapsulate metabolisms composed of a series of consecutive chemical reactions, which occasionally organize into autocatalytic cycles. Even though the underlying processes are highly stochastic and the mathematical formulations of the model are fully probabilistic, several concepts inherent to Darwinian evolution, such as the "species" (defined as similar metabolic networks), fitness to the environment, and inheritance appear to hold for the population but not for individual cell ancestors.

A novel feature of the model is the introduction of balance between constructive (peptide synthesis) and destructive (preferential hydrolysis of unstructured, nonfunctional peptides) processes as a mechanism for limiting combinatorial explosion. Considering that this balance is a universal phenomenon acting at different levels on all living systems, it is likely that it also played an important role in the beginnings of life.

Some biologists might feel uncomfortable with the idea that "metabolism first" models do not take self-replication of a genome, which they consider the essence of life, as the first step in its origins. We simply observe that, with the exception of replication and transcription, information transfer in cellular systems, from protein folding to transmission of neural signals, proceeds through structural recognition. More generally, self-organization and self-maintenance in a natural and social world rarely require linear information storage and transfer, which are highly advanced ideas (e.g., an alphabet or digital computers) invariably preceded by simpler concepts.

25.3 The Origin of Cell Ancestor Self-Organization and Function at a Molecular Level

25.3.1 The Role of Solvent in Promoting Biological Structures and Functions

The discussion so far has not referred directly to molecular mechanisms underlying the evolution of early cell ancestors. This appears to be in sharp contrast with current trends in biology, which emphasize the molecular basis of cellular processes. Much research along these lines is aimed at elucidating intricate interactions between specific, complex molecules, but there is also a large body of work devoted to understanding fundamental principles underlying biomolecular structures and interactions. It has long been recognized that, to gain such an understanding, it is necessary to consider explicitly the role of solvent in promoting biological organization (Kauzmann, 1959; Tanford, 1982). For terrestrial biology, this solvent is water. This immediately raises a question: Is water unique in this respect, and if not, would protobiology be very different in alternative solvents?

The question is hardly new, and several interesting contributions have been made to this subject recently (Bains, 2004; Benner, Ricardo, and Carrigan, 2004;

Schulze-Makuch and Irvin, 2004). An interesting speculation from recent studies is that alternative biochemistries are possible, in principle, in solvents other than water (Bains, 2004; Benner, Ricardo, and Carrigan, 2004). It is not our intention to repeat or contest these considerations. Instead, we offer a complementary point of view that goes beyond organic chemistry.

A cell, or even an early cell ancestor, is a complex system whose structure and function are largely modulated by noncovalent interactions. Indeed, noncovalent interactions determine, for example, self-assembly of boundary structures, protein folding, ligand-enzyme interactions, ion transport across membranes, and regulation of gene expression. To be effective, these interactions must be in the right energy range. They have to be sufficiently strong that inherent thermal noise does not raise havoc in the system. However, they cannot be too strong. Otherwise, regulation of local thermodynamic equilibria could not be accomplished without considerable expenditure of energy. In other words, biomolecular interactions would become irreversible for practical purposes. Furthermore, the system should exhibit sufficient stability that it could function properly over the temperature range existing in the environment. These desiderata are imposed by biology rather than organic chemistry.

One implication of these desiderata is that electrostatic interactions between biomolecules cannot be too strong. This means that the solvent for life should be characterized by a high dielectric constant. Another expectation is that interactions sufficiently strong to organize nonpolar molecules or groups should be present. In these respects, water is an excellent solvent for life—it has a high dielectric constant, it promotes hydrophobic interactions between nonpolar species (for recent views on the hydrophobic effect see Chandler, 2005; Paulaitis and Pratt, 2002; Pratt and Pohorille, 2002) and its macroscopic parameters most relevant to supporting life remain virtually unchanged over a wide range of temperatures. In fact, the balance between hydrophilic (electrostatic) and hydrophobic interactions, which are quite often of the same order of magnitude on the energy scale, is commonly exploited in biomolecular structures, interactions, and regulation.

The fact that water is a very good solvent for biology does not mean that other pure or mixed solvents that might be present elsewhere in the universe are not endowed with the same capability. A host of solvents—for example, formamide and dialcohols—have high dielectric constants. Some can promote the formation of vesicles resulting from solvophobic effect (Huang et al., 1997; McDaniel, McIntosh, and Simon, 1983). However, very few solvents have all the desired properties simultaneously. One candidate liquid that has not been sufficiently explored yet is formamide. At any rate, any solvent with the desired characteristics for supporting life would be similar to water in the sense that it would promote similar types of biomolecular interactions.

25.3.2 The Origin of Membrane-Mediated Self-Organization and Functions

We will illustrate several universal molecular principles underlying self-organization of biomolecules that lead to the emergence of functions in an example of membrane proteins. This may seem an odd choice, considering that most metabolic and information transfer processes take place inside cellular compartments. This choice, however, is not accidental. First, membrane proteins mediate functions that are essential not only to all cells but also to their early ancestors. These functions include transport of ions, nutrients, and waste products across cell walls, capture of energy and its transduction into the form usable in chemical reactions, transmission of environmental signals to the interior of the cell, cellular growth, and cell volume regulation. Second, no other macromolecules implicated in the origin of cellular life appear to be capable of performing similar functions. In particular, attempts to evolve RNA molecules that mediate such functions, carried out mostly through in vitro evolution, have so far failed, yielding only molecules that attach to and disrupt membranes, causing unspecific ionic leakage (Vlassov, Khvorova, and Yarus, 2001). In fact, these efforts may never succeed because of poor hydrophobic matching between nucleic acids and lipid tails. This, in turn, suggests that proteins must have been present in early cell ancestors and played a functional role as soon as there was a need for communication (controlled and selective transport of chemical species or signals) between the ancestor's interior and the environment (Pohorille, Schweighofer, and Wilson, 2005).

Considering that contemporary membrane channels are large and complex, both structurally and functionally, a question arises as to how their presumably much simpler ancestors could have emerged, performed functions, and diversified in early cell ancestor evolution. Remarkably, despite their overall complexity, structural motifs in membrane proteins are quite simple, with α-helices being the most common. This suggests that these proteins might have evolved from simple building blocks already present in early cell ancestors. This suggestion is supported by analyses that indicate the ability of simple membrane proteins to self-organize and function is largely a consequence of just a few general principles (Pohorille, Schweighofer, and Wilson, 2005; Popot and Engelman, 2000, White and Whimley, 1999), which operate at interfaces between water (or, possibly, a similar solvent) and the membrane-forming material (Pratt and Pohorille, 2002). This is schematically shown in figure 25.3.

It has been demonstrated both experimentally and theoretically that peptides with the amino acid sequences able to form α-helices in which hydrophobic and hydrophilic residues are located at opposite faces readily fold at water-membrane interface (for a recent review, see Pohorille, Schweighofer, and Wilson, 2005). Such peptides are called amphiphatic and are quite common among small membrane proteins in contemporary cells. The match between the polarities of the peptide and its environment

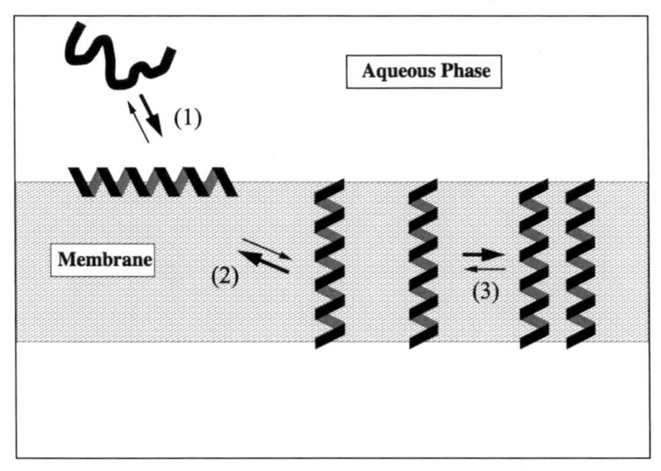

Figure 25.3
Self-organization of peptides proceeds in three stages, shown from left to right. Peptide folding (1), their insertion into the membrane (2), and self-assembly (3) are largely driven by the nature of the interface between water (or possibly another polar solvent) and membrane-forming material.

renders the amphiphatic helices particularly stable. The specific identity of the amino acid residues appears to be less important. This is a desirable protobiological property because neither a precise protein synthesis mechanism nor the full suite of amino acids would be required for the formation of amphiphatic helices. Considering that these helices had to fulfill only very modest sequence constraints, their presence in the early cell ancestor environment should not have been rare.

The peptides folded at the interface could have become inserted into the membrane so that they spanned the lipid bilayer. This process is thermodynamically unfavorable, as can be understood by referring, again, to the hydrophobic effect. Even if all amino acids are hydrophobic, a peptide still contains hydrophilic groups in its backbone. Their transfer from water to a nonpolar phase requires expending free energy.

Stability of transmembrane peptides can be regained through their specific recognition and association into larger assemblies. Examples of such assemblies are associations of four to seven α-helices into ion-transporting channels. In contrast to folding and insertion, this process is strongly sequence-dependent. In fact, the ability to associate was probably one of the selection mechanisms operating on cell ancestor transmembrane peptides. It appears that the specific nature of the lipid environment plays a lesser role in stabilizing helix associations. Many membrane proteins or complexes of helical peptides remain stable in other nonpolar solvents, such as detergents, or can be reconstituted in nonnative lipid environments (Popot and

Engelman, 2000). Again, this is a desirable property from the early evolution perspective because the composition of cell ancestor membranes might have been considerably more heterogeneous than the composition of membranes in contemporary cells. This property, however, does not necessarily extend to complex channels of contemporary cells, which are often sensitive to membrane composition. This may point to coordinated evolution of membranes and integral membrane proteins.

Despite their simple structure, membrane peptides could spontaneously form functional assemblies endowed with properties that, at the first sight, appear to require markedly more complexity. One contemporary example is a 25-amino acid fragment of the M2 protein of the influenza virus, which forms a tetrameric, α-helical channel capable of efficient and selective transport of protons (Pinto et al., 1997; Pinto, Holsinger, and Lamb, 1992). Another example is a 16-residue peptaibol, antiamoebin, which is a simple structural analog of the voltage-gated potassium channel, KvAP, again built from α-helices (O'Reilly and Wallace, 2003). Properties of these simple proteins can be subtly modulated by local modifications to the sequence rather than global changes in molecular architecture. This is a convenient evolutionary solution because it does not require imposing conditions on the whole amino acid sequence.

The example of membrane proteins illustrates a few universal principles that operate at a molecular level. First, the interface between water (or possible other water-like solvents) and a nonpolar phase has a powerful organizing effect on biomolecules. Second, the specific nature of the side chains is critical to function for only a limited number of residues. Otherwise, preserving polarity and perhaps size suffices. In other words, the mapping between sequence and function exhibits considerable degeneracy. Since many sequences can perform similar functions, this provides yet another mechanism of taming combinatorial explosion, but this time at the functional level. Perhaps most remarkably, a small number of very simple structural domains have sufficiently universal architectures that they can readily adapt to the diverse functional demands arising during cell ancestor evolution. This, in turn, suggests that, with careful attention to universal principles governing interactions in chemical and biological systems, self-organization and evolution of early functional macromolecules can be understood at the molecular level.

25.4 Conclusion

Early cell ancestors are good candidates for universal intermediates that bridge inanimate and animate matter. At first sight, it would appear that the number of ways in which they can self-organize and evolve is enormous, perhaps prohibitively large for fruitful analysis. This would render their universality meaningless. It seems, however, that this diversity is greatly reduced once universal constraints that must have acted on cell ancestors in protobiological environments are enforced. The constraints act at different levels, including reducing the number of possible chemical reactions and imposing rules on their self-organization into networks, and forming a molecular basis for the self-assembly of macromolecular structures into functional units. These constraints apply equally to cell ancestors on Earth and elsewhere in the universe. In fact, if we found alien cell ancestors, they might turn out to be surprisingly (and perhaps disappointingly) similar to hypothetical terrestrial cell ancestors. This does not mean that they would be endowed with the same genetic material, genetic code, or the same suite of catalytic residues (e.g., the same amino acids), but the underlying organizational principles of metabolism and other functions would be similar.

Several theoretical models have been put forward to explain the self-organization and early evolution of cell ancestors. The best-known models assume that template-directed replication is required for cell ancestor evolution. However, the conclusion that genetic polymers should be placed at the origins of cell ancestors is far from certain, and models based on alternative ways of information transfer exist. Unfortunately, all theoretical models have only circumstantial experimental support. Since only a few efforts so far have been devoted to building models of cell ancestors (see, e.g., Hanczyc and Szostak, 2004; Luisi, Rasi, and Mavelli, 2004; see also chapters 2, 3 and 5), it is unlikely that the nature of information transfer throughout cell ancestor evolution will be resolved in the near future.

Manmade protocells that bridge living and nonliving matter can leap over evolutionary barriers that cannot be surmounted by naturally evolving systems and therefore can differ markedly from early cell ancestors, but the universal principles highlighted in this chapter apply to both types of structures. For example, the ability to undergo Darwinian evolution is considered by many to be an essential attribute of life. This in turn implies that any cell-like structure, either natural or made in the laboratory, must be endowed with mechanisms to control combinatorial explosion and excessive information loss during reproduction. Further, studies of cell ancestors point to the fundamental roles of thermodynamic and kinetic constraints associated with energy flow in the system and self-organization driven by solvent-mediated spatial segregation of chemical groups, molecules or cellular components. Implementing these principles in specific constructs will unquestionably be a major challenge. This challenge can be met only through collaborative research of experimentalists and theorists.

Acknowledgments

This work was supported by the grants from NASA Astrobiology Institute and NASA Exobiology Program. The author thanks Michael Wilson for critically reading the manuscript, and Lawrence Pratt and Arthur Weber for illuminating discussions.

References

Bagley, R. J., & Farmer, J. D. (1991). Spontaneous emergence of a metabolism. In C. G. Langton, C. Taylor, J. D. Farmer, & S. Rasmussen (Eds.), *Artificial life II, SFI studies in the science of complexity*. Reading, MA: Addison-Wesley.

Bains, W. (2004). Many chemistries could be used to build living systems. *Astrobiology*, *4* (2), 137–167.

Benner, S. A., Ricardo, A., & Carrigan, M. A. (2004). Is there a common chemical model for life in the universe? *Current Opinions in Chemical Biology*, *8* (6), 672–689.

Chandler, D. (2005). Interfaces and the driving force of hydrophobic assembly. *Nature*, *437* (7059), 640–647.

Deamer, D. W., & Pashley, R. M. (1989). Amphiphilic components of the Murchison carbonaceous chondrite: Surface properties and membrane formation. *Origins of Life and Evolution of the Biosphere*, *19*, 21–38.

Dworkin, J., Deamer, D., Sandford, S., & Allamandola, L. (2001). Self-assembling amphiphilic molecules: Synthesis in simulated interstellar/precometary ices. *Proceedings of the National Academy of Sciences of the United States of America*, *98*, 815–819.

Dyson, F. J. (1982). A model for the origin of life. *Journal of Molecular Evolution*, *18*, 344–350.

Dyson, F. J. (1999). *Origins of life*. Cambridge: Cambridge University Press.

Eigen, M. (1971). Self-organization of matter and the evolution of biological macromolecules. *Naturwissenschaften*, *58* (10), 465–523.

Eigen, M., & Schuster, P. (1977). The hypercycle: A principle of natural self-organization. Part A: Emergence of the hypercycle. *Naturwissenschaften*, *64* (11), 541–565.

Eigen, M., & Schuster, P. (1979). *The hypercycle: A principle of natural self-organization*. Berlin: Springer.

Eschenmoser, A. (1999). Chemical etiology of nucleic acid structure. *Science*, *284* (5423), 2118–2124.

Gestland, R. F., Cech, T. R., & Atkins, J. F. (Eds.) (1999). *The RNA world*. Cold Spring Harbor, NY: Cold Spring Harbor Laboratory Press.

Gil, R., Silva, F. J., Pereto, J., & Moya, A. (2004). Determination of the core of a minimal bacterial gene set. *Microbiology and Molecular Biology Reviews*, *68* (3), 518–537.

Gilbert, W. (1986). The RNA world. *Nature*, *319*, 618.

Hanczyc, M. M., & Szostak, J. W. (2004). Replicating vesicles as models of primitive cell growth and division. *Current Opinions in Chemical Biology*, *8* (6), 660–664.

Huang, J. B., Zhu, B. Y., Zhao, G. X., & Zhang, Z. Y. (1997). Vesicle formation of a 1:1 catanionic surfactant mixture in ethanol solution. *Langmuir*, *13* (21), 5759–5761.

Hutchison, C. A., Peterson, S. N., Gill, S. R., Cline, R. T., White, O., Fraser, C. M., et al. (1999). Global transposon mutagenesis and a minimal Mycoplasma genome. *Science*, *286* (5447), 2165–2169.

Joyce, G. (1996). Ribozymes—Building the RNA world. *Current Biology*, *6*, 965–967.

Kauffman. S. A. (1986). Autocatalytic sets of proteins. *Journal of Theoretical Biology*, *119*, 1–24.

Kauffman, S. A. (1993). *The origins of order—Self-organization and selection in evolution*. Oxford: Oxford University Press.

Kauzmann, W. (1959). Some forces in the interpretation of protein denaturation. *Advances in Protein Chemistry*, *14*, 1–63.

Keefe, A. D., & Szostak, J. W. (2001). Functional proteins from a random-sequence library. *Nature*, *410*, 715–718.

Koonin, E. V. (2003). Comparative genomics, minimal gene sets and the last common ancestor. *National Review of Microbiology*, *1*, 127–136.

Leman, L., Orgel, L., & Ghadiri, M. R. (2004). Carbonyl sulfide-mediated prebiotic formation of peptides. *Science*, *306* (5694), 283–286.

Luisi, P. L. (2003). Autopoiesis: A review and a reappraisal. *Naturwissenschaften*, *90*, 49–59.

Luisi, P. L., Rasi, P. S., & Mavelli, F. (2004). A possible route to prebiotic vesicle reproduction. *Artificial Life*, *10* (3), 297–308.

Mautner, M., Leonard, D., & Deamer, D. (1995). Meteorite organics in planetary environments: Hydrothermal release, surface activity, and microbial utilization. *Planetary and Space Science*, *43*, 139–147.

McDaniel, R. V., McIntosh, T. J., & Simon, S. A. (1983). Nonelectrolyte substitution for water in phosphatidylcholine bilayers. *Biochimica Biophysica Acta*, *731*, 97–108.

Monod, J. (1971). *Chance and necessity*. New York: Alfred A. Knopf.

Morowitz, H. J. (1992). *Beginnings of cellular life*. New Haven, CT: Yale University Press.

Morowitz, H. J., Kostelnik, J. D., Yang, J., & Cody, G. D. (2000). The origin of intermediary metabolism. *Proceedings of the National Academy of Sciences of the United States of America*, *97* (14), 7704–7708.

NASA. (2003). *Astrobiology roadmap*. Available at: http://astrobiology.arc.nasa.gov/roadmap (accessed October 2006).

New, M. H., & Pohorille, A. (2000). An inherited efficiencies model of non-genomic evolution. *Simulation Practice and Theory*, *8*, 99–108.

Nissen, P., Hansen, J., Ban, N., Moore, P. B., & Steitz, T. A. (2000). The structural basis of ribosome activity in peptide bond synthesis. *Science*, *289* (5481), 920–930.

Noireaux, V., Bar-Ziv, R., Godefroy, J., Salman, H., & Libchaber, A. (2005). Toward an artificial cell based on gene expression in vesicles. *Physical Biology*, *2*, P1–P8.

O'Reilly, A., & Wallace, B. A. (2003). The peptaibol antiamoebin as a model ion channel: Similarities to bacterial potassium channels. *Journal of Peptide Science*, *9*, 769–775.

Orgel, L. (2000). Self-organizing biochemical cycles. *Proceedings of the National Academy of Sciences of the United States of America*, *97* (23), 12503–12507.

Paulaitis, M. E., & Pratt, L. R. (2002). Hydration theory for molecular biophysics. *Advances in Protein Chemistry*, *62*, 283–310.

Pinto, L. H., Dieckmann, G. R., Gandhi, C. S., Papworth, C. G., Braman, J., Shaughnessy, M. A., et al. (1997). A functionally defined model for the M_2 proton channel of influenza A virus suggests a mechanism for its ion selectivity. *Proceedings of the National Academy of Sciences of the United States of America*, *94*, 11301–11306.

Pinto, L. H., Holsinger, L. J., & Lamb, R. A. (1992). Influenza virus M_2 protein had ion channel activity. *Cell*, *69*, 517–528.

Pohorille, A., & Deamer, D. (2002). Artificial cells: Prospects for biotechnology. *Trends in Biotechnology*, *20*, 123–128.

Pohorille, A., Schweighofer, K., & Wilson, M. A. (2005). The origin and early evolution of membrane channels. *Astrobiology*, *5*, 1–17.

Popot, J. L., & Engelman, D. M. (2000). Helical membrane protein folding, stability, and evolution. *Annual Review of Biochemistry*, *69*, 881–922.

Pratt, L. R., & Pohorille, A. (2002). Hydrophobic effects and modeling of biophysical aqueous solution interfaces. *Chemical Reviews*, *102*, 2671–2692.

Rode, B. M. (1999). Peptides and the origin of life. *Peptides*, *20*, 773–786.

Schulze-Makuch, D., & Irwin, L. N. (2004). *Life in the universe: Expectations and constraints*. Berlin: Springer.

Segré, D., & Lancet, D. (1999). A statistical chemistry approach to the origin of life. *Chemtracts—Biochemistry and Molecular Biology*, *12*, 382–397.

Segré, D., & Lancet, D. (2000). Composing life. *EMBO Reports*, *1*, 217–222.

Segré, D., Shenhav, B., Kafri, R., & Lancet, D. (2001). The molecular roots of compositional inheritance. *Journal of Theoretical Biology*, *213*, 481–491.

Severin, K., Lee, D. H., Kennan, A. J., & Ghadiri, M. R. (1997). A synthetic peptide ligase. *Nature*, *389*, 706–709.

Smith, E., & Morowitz, H. J. (2004). Universality in intermediary metabolism. *Proceedings of the National Academy of Sciences of the United States of America, 101* (36), 13168–13173.

Szostak, J. W., Bartel, D. P., and Luisi, L. P. (2001). Synthesizing life. *Nature, 409*, 287–390.

Tanford, C. (1982). *The hydrophobic effect: Formation of micelles and biological membranes.* New York: John Wiley & Sons.

Vlassov, A., Khvorova, A., & Yarus, M. (2001). Binding and disruption of phospholipid bilayers by supramolecular RNA complexes. *Proceedings of the National Academy of Sciences of the United States of America, 98*, 7706–7711.

Weber, A. L. (1999). Sugars as the optimal biosynthetic carbon substrate of aqueous life throughout the universe. *Origins of Life and Evolution of the Biosphere, 30*, 33–43.

Weber, A. L. (2001). The sugar model: Catalysis of amines and amino acid products. *Origins of Life and Evolution of the Biosphere, 31*, 71–86.

Weber, A. L. (2002). Chemical constraints governing the origin of metabolism: The thermodynamic landscape of carbon group transformations under mild aqueous conditions. *Origins of Life and Evolution of the Biosphere, 32* (4), 333–357.

Weber, A. L. (2004). Kinetics of organic transformations under mild aqueous conditions: Implications for the origin of life and its metabolism. *Origins of Life and Evolution of the Biosphere, 34* (5), 473–495.

White, S. H., & Whimley, W. C. (1999). Membrane protein folding and stability: Physical principles. *Annual Reviews in Biophysics and Biomolecular Structure, 28*, 319–365.

Wilson, D. S., & Szostak, J. W. (1999). In vitro selection of functional nucleic acids. *Annual Reviews of Biochemistry, 68*, 611–648.

Wittung, P., Nielsen, P. E., Buchardt, O., Egholm, M., & Norden, B. (1994). DNA-like double helix formed by peptide nucleic acid. *Nature, 368* (6471), 561–563.

26 Prebiotic Chemistry, the Primordial Replicator, and Modern Protocells

Henderson James Cleaves II

26.1 Introduction

Modern organisms are essentially lipid-encased systems of biochemicals that propagate themselves through a complicated system of gene transcription and translation, using precisely folded protein catalysts (enzymes). The folding of linear sequences of polymers into discrete catalytic three-dimensional structures conducive to the propagation of the catalyst polymers is the essence of biochemistry. Water and environmentally available carbon, nitrogen, sulfur, phosphorus, and a few other elements, together with energy sources, such as light, organic molecules, metals, sulfur, and nitrogen species, are used to produce new biological materials for replication.

Although an all-encompassing definition of life is difficult, all terrestrial biological organisms share three basic properties: a genetic storage mechanism, universally based on DNA, an encapsulating membrane, which also functions in energy transduction (generally composed of either isoprenoid ethers or fatty acid esters), and a metabolism used to convert environmental resources into monomer subunits for propagation and multiplication (mediated by both RNA and proteins).

Though all three components are found in modern organisms, it seems unlikely that all three were simultaneously involved in the origin of life. Although models have been proposed for *gene-first*, *encapsulation-first*, and *metabolism-first* (Wächtershäuser, 1988a, 1988b) origins of life, among these, only the gene and encapsulation models have been shown experimentally to be able to replicate themselves (Orgel, 2000), and the gene-first model seems unique in its ability to evolve.

There are two complementary approaches to the study of the origin of life (figure 26.1): the *prospective* and the *retrospective*. The prospective, or prebiotic chemistry, approach attempts to assemble plausible primitive building blocks into more complex structures capable of evolvable replication (Miller, 1998), whereas the retrospective, or comparative biochemistry, approach attempts to deconstruct living organisms into their common fundamental components and processes through comparative biochemistry (Benner, Ellington, and Tauer, 1989). Both are likely to be useful in solving the daunting problem at hand.

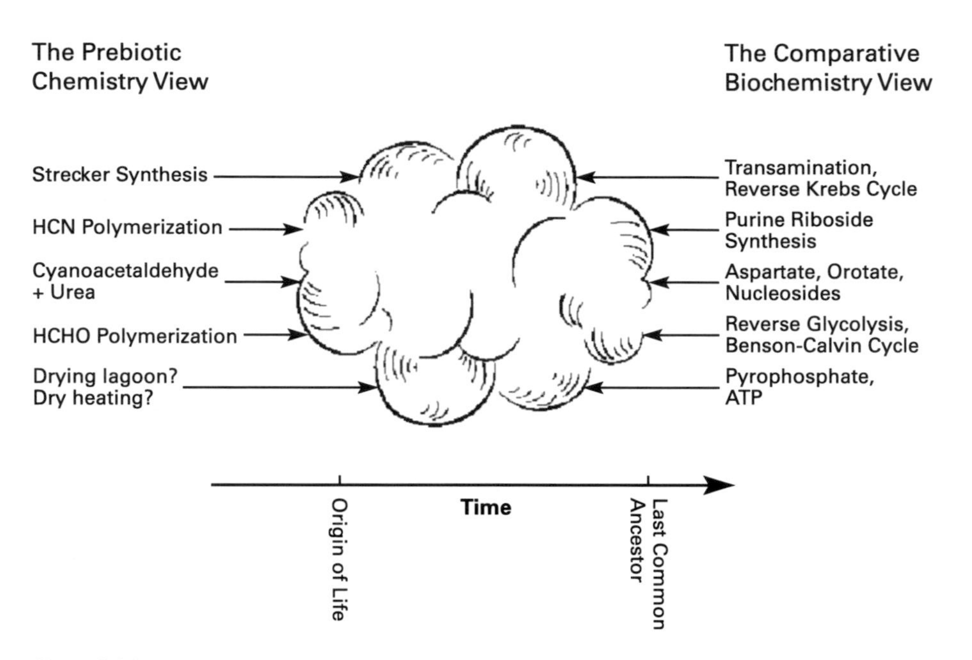

Figure 26.1
The prospective and the retrospective approaches to the study of the origin of life.

Oparin was the first to propose that the first organisms were assembled from organic compounds provided by abiotic environmental syntheses. These organisms were thus heterotrophs, and later developed the various sophisticated mechanisms for assembling their components from simpler compounds by backwards assembly of the metabolic pathways. This is known as the *heterotrophic hypothesis* (Horowitz, 1945) or *Oparin-Haldane* hypothesis (Haldane, 1929; Oparin, 1924, 1938).

Because of the numerous shared qualities of modern organisms, all life on Earth is assumed to be descended from a Last Universal Common Ancestor (LUCA), which was itself derived from a simpler ancestor (Doolittle, 2000; Forterre, 2002; Woese, 2000). Unfortunately, the properties of the LCA can be inferred only from those properties shared by all modern organisms, and the LCA must have already been a highly evolved organism, as it was certainly cellular and already possessed the mechanisms of DNA replication, modern protein coding mechanisms, as well as a fairly sophisticated metabolism.

Organic chemistry appears to follow similar pathways throughout the universe (Ehrenfreund and Charnley, 2000; Ehrenfreund and Menten, 2002). Despite the underlying simplicity of this chemistry, it allows a great multitude of possible structures and an even greater multiplicity of possible interactions. The combinations of these simple atomic units pales in comparison with the possible higher-order combinations, and it is in this larger set that the complex interactions generating the phenomenon of life arise. It is the goal of origins of life research to provide a plausible and compelling reconstruction of this process.

The solar system is thought to have formed from the condensation of atomic and molecular species generated by previous cosmic and galactic processes to generate further combinations that were segregated based on their physical properties (Wetherhill, 1990). The rocky planets are thus mainly composed of high boiling point compounds such as iron, nickel, and silicates, although these are relatively rare cosmically, whereas the outer planets are largely composed of low boiling point helium, H_2, and the hydrides of higher elements (H_2O, NH_3, phosphine, etc.).

Prebiotic chemistry has shown a remarkable overlap between identifiable chemical species produced by abiological processes and the components of modern biochemistry. It has been known for some time that small organic molecules could be generated by the interaction of energy sources and simple gases (Löb, 1913); however, not until Miller's pioneering 1953 experiment was it demonstrated that such a process could have provided the raw material for the origin of life (Miller, 1953). According to the model at that time, the early Earth's atmosphere was reducing, which allowed for the synthesis of simple reactive intermediates such as HCN and HCHO, which subsequently reacted in the oceans to form bio-organic compounds such as amino acids (Urey, 1952). These then polymerized to form catalysts that somehow attained the ability to self-replicate. In 1953, it was also shown that the molecular structure of DNA could explain physical bases for the long sought principle of genetic inheritance (Watson and Crick, 1953), which allowed for further refinement of origin of life models.

The modern synthesis of the gene-first theory holds that self-replicating genetic polymers arose spontaneously. Their self-assembly was driven by the inherent chemical properties of their monomers, that is, the properties embodied by monomer-template interactions, with polymerization driven by either externally provided free energy sources or the tendency of the monomers to polymerize (Joyce, 1987). Combinatorial pools of these polymers, which were constantly being nonenzymatically synthesized and degraded, accumulated in the environment.

Once a pool of polymers had been established, natural selection allowed those capable of catalyzing their own replication to predominate, and, as they exhausted the raw materials available in their environment, these mutated to produce new variants capable of catalyzing the needed reactions. From here the system self-organized through numerous adaptive and selection events to give rise to the LCA (Lazcano and Miller, 1994; see figure 26.2).

In all living organisms, biological information flows from DNA to RNA to protein. RNA acts as a central player in the form of mRNA, tRNA, and the ribosome. Retrospective analyses thus hypothesized the existence of an RNA world (Crick,

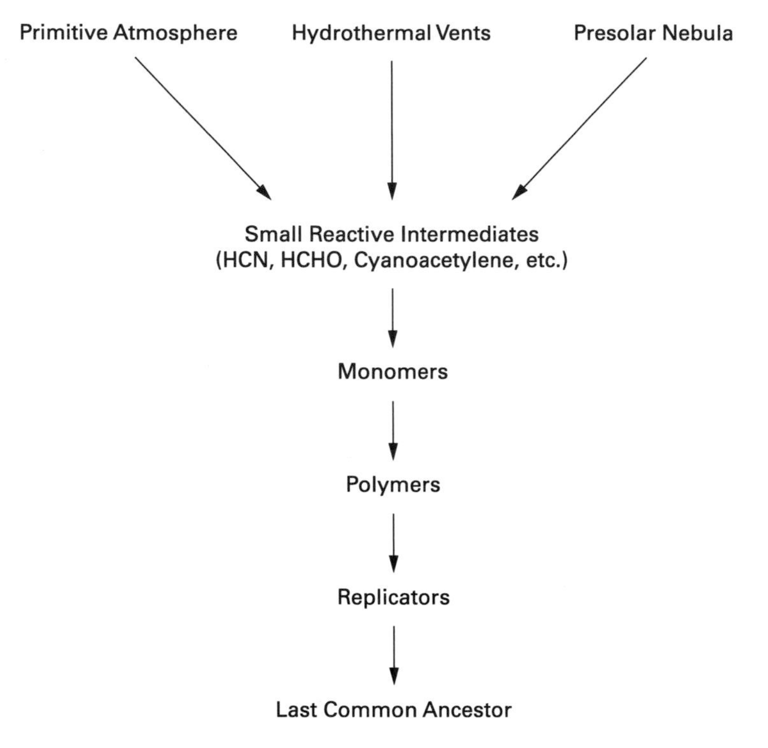

Figure 26.2
Schematic of the modern *genes first* theory of the origin of life.

1968; Gesteland, Cech, and Atkins, 1999; Gilbert, 1986; Orgel, 1968), where RNA could have carried out both the catalytic and informational roles now played by protein and DNA, respectively. RNA seemed to be the answer to the long sought "Holy Grail" of the origin of life problem, and the prebiotic synthesis of its components became the focus of research. Subsequently, syntheses of the nucleosides and nucleotides were demonstrated, highlighting both the potential facility and difficulty of the prebiotic synthesis of RNA.

This model has been validated by the discovery of ribozymes in biological systems (Cech, Zaug, and Grabowski, 1981; Guerrier-Takada et al., 1983) as well as by laboratory models that have shown the prophesied ability of RNA molecules to serve as evolvable catalysts (Forst, 1998; Johnston et al., 2001; Rimmele, 2003).

The RNA world hypothesis suggests that DNA inherited some of RNA's duties, while proteins inherited others. If such "genetic takeovers" were possible (Cairns-Smith, 1977), then it is also possible that other replicating molecules had passed their information to RNA, and these earlier replicators were perhaps more easily synthesized prebiotically (Joyce et al., 1987). These molecules were subsequently discarded by evolving chemical systems.

If this were possible, in much the same way that DNA and RNA are structurally similar and share the properties of molecular recognition and self-assembly provided by Watson-Crick base-pairing mediated by weak noncovalent interactions, then these same properties might have been embodied in the hypothesized earlier molecules. The idea of a "pre-RNA world" was thus introduced (Joyce et al., 1987).

Such precursor molecules must be fairly simple. This model has led to numerous advances in both theoretical understanding of molecular evolution and practical applications of this technology (Bowser, 2005). Such structures should obey the same universal weak-intermolecular chemical bonding self-assembly rules found in biology and found to facilitate nonbiological macromolecular self-assembly (Park, Feng, and Rebek, 1992; Rebek, 1991). It is possible that the emergence of evolvable chemical systems may depend on something besides molecular recognition, but little experimental evidence has been given for other theories, and many people believe that the greatest gains in the future are likely to come from investigation of the RNA world principle.

26.2 Prebiotic Synthesis

Although much of the following discussion focuses on chemistry likely to have taken place on the primitive Earth, extraterrestrial input may also have supplied the early environment with a significant amount of organic material. Most of the discussion here would also apply, although, of course, the types of chemistry that may have occurred in space environments may be distinct in many respects (Muñoz-Caro et al., 2002).

Until 1969, no examples of authentic prebiotic organic chemistry were known. The Murchison meteorite, which fell in Australia that year, serves as the Rosetta stone of plausible prebiotic chemistry. Many of the compounds found in the Murchison meteorite match those produced in prebiotic simulations quite closely (Peltzer et al., 1984; Wolman, Haverland, and Miller, 1972; see table 26.1), although there are exceptions.

"Prebiotic" reaction mixtures and their natural analogs, such as the Murchison meteorite, contain complex mixtures of organic molecules. The characterization of their components has often been carried out to show that simple pathways between important modern bio-organic compounds and primitive solar system syntheses exist.

There may have been numerous steps between modern biochemistry and the compounds originally supplied by the primitive environment, as the complexity of the hypothesized LCA suggests. Some of these may have been more chemically conducive to the synthesis of self-replicating biochemical systems than modern biochemicals. These would be easily overlooked without a compelling theoretical reason for searching for them, since many compounds are present in fairly trace amounts (Muñoz-Caro et al., 2002).

Table 26.1
Relative abundances of amino acids detected in the Murchison meteorite and a spark discharge experiment

Amino Acid	Murchison	Electric Discharge
Glycine	****	****
Alanine	****	****
α-Amino-*n*-butyric acid	***	****
α-Aminoisobutyric acid	****	**
Valine	***	**
Norvaline	***	***
Isovaline	**	**
Proline	***	*
Pipecolic acid	*	<*
Aspartic acid	***	***
Glutamic acid	***	**
β-Alanine	**	**
β-Amino-*n*-butyric acid	*	*
β-Aminoisobutyric acid	*	*
γ-Aminobutyric acid	*	**
Sarcosine	**	***
N-Ethylglycine	**	***
N-Methylalanine	**	**

Adapted from Wolman, Haverland, and Miller, 1972.

The primitive Earth may have harbored a variety of atmospheres as it evolved, ranging from highly reducing (CH_4/NH_3; see Urey, 1952), to neutral (CO/N_2—although high CO atmospheres could have existed only transiently; see Miyakawa et al., 2002c), to oxidizing (CO_2/N_2; see Kasting, 1993). The nature of the primordial atmosphere would have depended on the manner and rate at which the Earth formed (Wetherhill, 1990).

High-temperature or high-energy reactions in the gas phase are thought to proceed through radical intermediates, which recombine and "freeze-out" when exposed to lower temperature or energy regimes. The resulting nonequilibrium species, such as HCN, HCHO, acetylene, cyanoacetylene, and acrylonitrile, may be considered to store energy that can be used for subsequent reactions. The degree to which these species are generated is a function of the nature of the starting gases, their partial pressures, and relative partial pressures, and the energy source acting on them, as well as the sources and sinks for the species involved (Miller, 1998).

Though reducing gas mixtures are relatively efficient at producing small reactive intermediates such as HCN and HCHO, oxidized gas mixtures are not (Stribling and Miller, 1987). It is now widely held that the primordial atmosphere was not particularly reducing (Kasting, 1993). However, the yields of reactive intermediates are still significant under oxidizing conditions (Stribling and Miller, 1987) and, although not as well studied, these should not be neglected. Paleoatmospheric models are still somewhat poorly constrained, and the plausibility of a reducing atmosphere has recently been resurrected (Tian et al., 2005). If the atmosphere was not reducing, it is possible that extraterrestrial input (Chyba and Sagan, 1992) or hydrothermal organic material was more significant, and this might change the subsequent chemistry to some extent. For example, comets contain significant amounts of HCN, NH_3, and HCHO (Chyba et al., 1990; Oró, 1961; Oró and Lazcano, 1997), whereas vents are possible sources of NH_3, but not HCN. HCHO is still produced efficiently from CO_2 atmospheres by UV irradiation (Pinto, Gladstone, and Yung, 1980).

Small reactive intermediates hydrolyze and photolyze over geological timescales. For example, ammonia is readily photolyzed in the gas phase (Kuhn and Atreya, 1979). HCN hydrolyzes to formamide, formic acid, and ammonia. Steady states of HCN, which depend on atmospheric composition and oceanic composition, have been estimated at 4×10^{-12} M and 2×10^{-5} M at 100° C and 0° C, respectively, at pH 7 (Miyakawa, Cleaves, and Miller, 2002a). These steady-state concentrations would affect the degree to which subsequent reactions could occur.

Various energy sources were available on the primitive Earth. These include ultraviolet (UV) radiation from the early sun, electric discharges, γ-rays, energy from radioactive decay, and energy from shock waves generated by thunder as well as meteor and comet impacts (Miller, 1998). The likely contributions of each around the time of the origin of life 3.5 to 4 billion years ago can be estimated based on current fluxes (table 26.2). Not all energy sources are equally effective at producing small reactive intermediates. For example, electric discharges are considerably more efficient than UV light at producing HCN (Miller, 1998).

Another possibility is that molecules outgassed from the crust-mantle boundary were reacted over mineral catalysts on the upward journey through what has been termed the *Fischer-Tropsch type* synthesis (Anders, Hayatsu, and Studier, 1973) to produce organic compounds (Gold, 1992). It is difficult to detect such processes operating today because of the abundance of contaminating biologically derived organic compounds. The extent to which this synthetic mechanism might have occurred on the primitive Earth certainly merits further experimental investigation.

26.3 Monomer Synthesis

The bulk of prebiotic research has focused on the synthesis of compounds important in modern biochemistry (Miller and Orgel, 1974), and the maximization of the yields of such syntheses. There are, of course, numerous potentially interesting, as yet

Table 26.2
Fluxes of energy from various sources on the present Earth

Source	Energy (cal cm^{-2} yr^{-1})	Energy (J cm^{-2} yr^{-1})
Total radiation from sun	260,000	1,090,000
Ultraviolet light < 300 nm	3400	14,000
Ultraviolet light < 250 nm	563	2360
Ultraviolet light < 200 nm	41	170
Ultraviolet light < 150 nm	1.7	7
Electric discharges	4.0[a]	17
Cosmic rays	0.0015	0.006
Radioactivity (to 1.0 km)	0.8	3.0
Volcanoes	0.13	0.5
Shock waves	1.1[b]	4.6

a. 3 cal cm^{-2} yr^{-1} of corona discharge and 1 cal cm^{-2} yr^{-1} of lightning.
b. 1 cal cm^{-2} yr^{-1} of this is the shock wave of lightning bolts and is also included under electric discharges.
Note: Fluxes on the primitive Earth may have differed, in particular the flux of shorter wavelength UV light, especially if there was increased UV output by the early sun, and in the absence of ozone and molecular oxygen in the primitive atmosphere. The majority of the solar flux incident on the upper atmosphere is presently in the form of longer wavelength radiation (l > 300 nm) of lower energy per photon, which is unable to engage in useful atmospheric prebiotic chemistry. Adapted from Miller and Orgel (1974).

unidentified side products of any of the numerous possible combinations of small reactive intermediates.

Important prebiotic syntheses of simple biological molecules are briefly reviewed here, to lend insight into the synthesis of alternative replicators. It is important to bear in mind that most of these syntheses occur extremely rapidly on geological timescales. Miller's original experiment was conducted in one week, and adenine can be detected in concentrated HCN solutions within hours of their preparation. At the other extreme, biomolecules, and indeed most organic compounds, degrade rapidly at high temperature (Larralde, Robertson, and Miller, 1995; Levy and Miller, 1998; White, 1984), thus low temperatures are generally more conducive to the synthesis and accumulation of organic compounds.

26.3.1 Amino Acids

Adolf Strecker demonstrated the first abiological synthesis of an amino acid in 1850, when he produced alanine by reacting aqueous acetaldehyde, ammonia, and cyanide. Miller showed that the amino acids produced in his 1953 experiment were probably formed by Strecker's mechanism by simple compounds generated in the gas phase reacting in the aqueous layer below (Miller, 1957). This was suggested by the racemic nature of the products detected, as well as by the presence of both α-amino acids and α-hydroxy acids in the product mixture (figure 26.3).

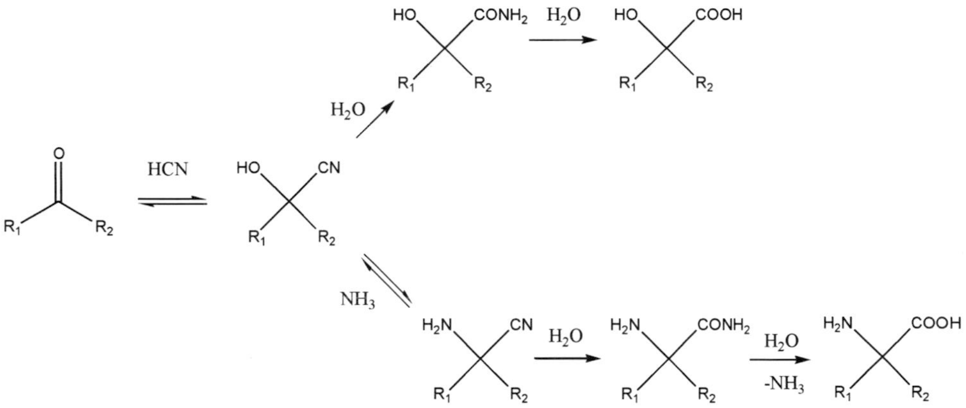

Figure 26.3
The Strecker and cyanohydrin mechanisms for the formation of amino- and hydroxy-acids from ammonia, aldehydes and ketones, and cyanide.

The equilibria and kinetics of these reactions were investigated in some detail (Schlesinger and Miller, 1973), and it was calculated that this synthesis could have occurred at extremely low dilution, even in the primitive open oceans. This is not true for most of the steps believed to be important for nucleic acid monomer and polymer synthesis, which might need to occur in more specialized environments.

A more limited set of amino acids is generated from aqueous NH_4CN solutions by different mechanisms, and various β- and γ-amino acids can be generated using other reactive precursors (Peltzer et al., 1984). This illustrates the fact that any given monomer type may have multiple possible synthetic mechanisms.

26.3.2 Purines

Oró first demonstrated the remarkable reaction of aqueous solutions of HCN to produce adenine (Oró, 1960; Oró and Kimball, 1961), which is formally a pentamer of HCN. The mechanism of synthesis and the kinetics of reaction were later investigated in considerable detail (Sanchez, Ferris, and Orgel, 1967), and syntheses of all of the biological purines, as well as some not known in biology, were accomplished (Sanchez, Ferris, and Orgel, 1968). Suggested mechanisms for these syntheses are shown in figure 26.4.

The polymerization of HCN or NH_4CN also produces a large quantity of insoluble and intractable polymer. This is true of most prebiotic reactions, and the bulk of the organic carbon in Murchison is also in the form of a heterogeneous polymer.

HCN polymerization in solution occurs only when the concentration of HCN is 10^{-2} to 10^{-3} M or greater (Miyakawa et al., 2002a; Sanchez, Ferris, and Orgel,

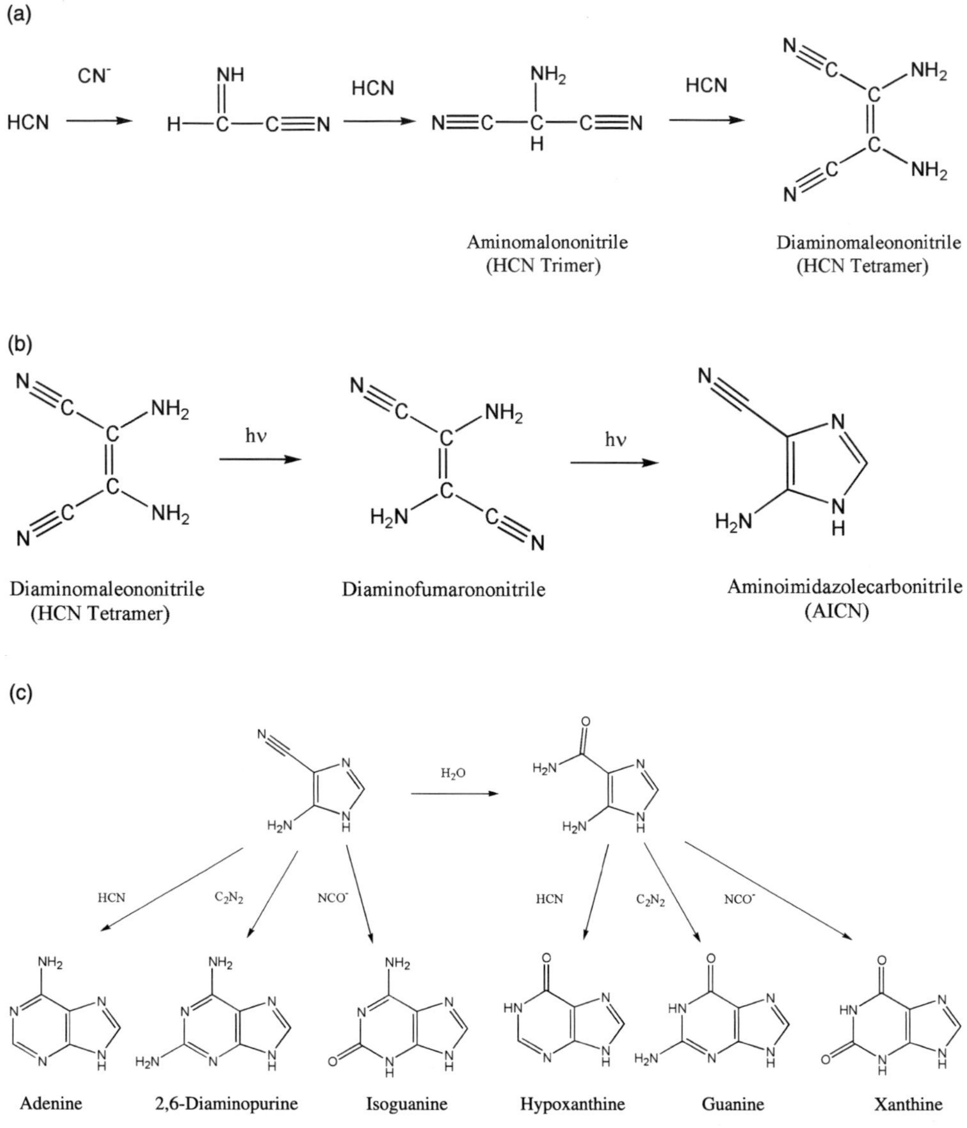

(a)

$HCN \xrightarrow{CN^-} H{-}C({=}NH){-}C{\equiv}N \xrightarrow{HCN} N{\equiv}C{-}CH(NH_2){-}C{\equiv}N \xrightarrow{HCN}$ Diaminomaleononitrile

Aminomalononitrile (HCN Trimer)

Diaminomaleononitrile (HCN Tetramer)

(b)

Diaminomaleononitrile (HCN Tetramer) $\xrightarrow{h\nu}$ Diaminofumaronitrile $\xrightarrow{h\nu}$ Aminoimidazolecarbonitrile (AICN)

(c)

AICN $\xrightarrow{H_2O}$ aminoimidazole carboxamide

HCN, C_2N_2, NCO^-

Adenine, 2,6-Diaminopurine, Isoguanine, Hypoxanthine, Guanine, Xanthine

Figure 26.4
Proposed mechanisms of synthesis of purines from HCN.

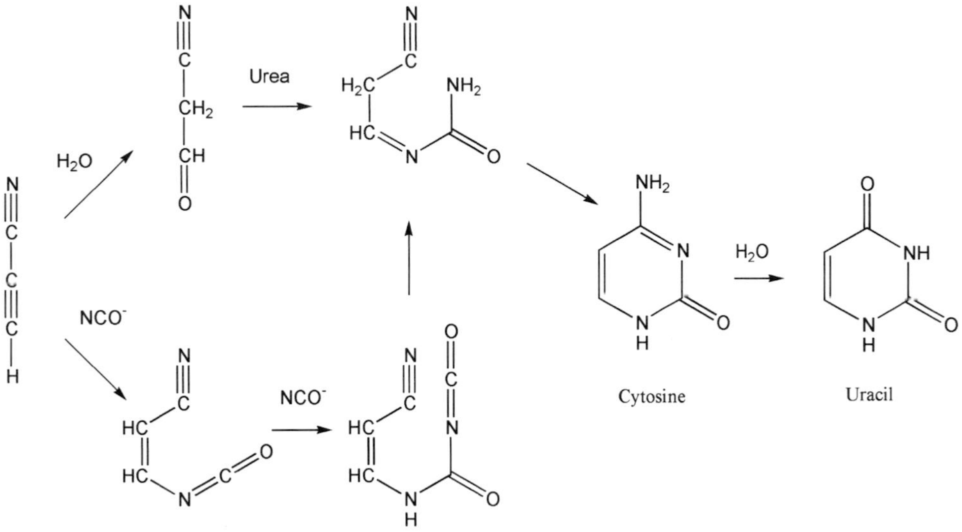

Figure 26.5
Possible mechanisms for the prebiotic synthesis of the biological pyrimidines.

1966a), and as mentioned earlier these appear to be unlikely prebiotic steady-state oceanic concentrations. However, when dilute solutions of HCN are frozen, concentrated pockets of HCN brines are produced as the ice matrix forms, and purine synthesis could have occurred in specialized environments where freezing occurred at least occasionally (Levy et al., 2000; Miyakawa, Cleaves, and Miller, 2002b; Schwartz, Joosten, and Voet, 1982). Purines have also been identified in the Murchison meteorite (Hayatsu et al., 1975; van der Velden and Schwartz, 1977).

26.3.3 Pyrimidines

Cyanoacetylene (CA), generated from spark discharges acting on reduced gases (Sanchez, Ferris, and Orgel, 1966b), and cyanate react in solution to form the biological pyrimidines (Ferris, Sanchez, and Orgel, 1968). It was later shown that guanidine or urea and cyanoacetaldehyde, produced from the hydration of CA, react under drying conditions to give high yields of the biological pyrimidines (Robertson and Miller, 1995). The various schemes that have been proposed to account for this are shown in figure 26.5.

Uracil has also been produced from the UV photolysis of 5,6-dihydrouracil (Schwartz and Chittenden, 1977) and the hydrolysis of HCN polymers (Voet and Schwartz, 1982). On hydrolysis, HCN polymers also yield substantial amounts of orotic acid, the biochemical precursor to the pyrimidines (Ferris et al., 1978). Thymine is produced from the reaction of HCHO, HCOOH, and uracil (Choughuley et al., 1977).

26.3.4 Sugars and Polyols

In 1861, Butlerov showed that aqueous basic solutions of HCHO produced a sweet-smelling substance he called "formose" (Butlerow, 1861), which was later shown to include a wide variety of the biological sugars, including ribose in low yield, and their isomers (Decker, Schweer, and Pohlmann, 1982; Reid and Orgel, 1967). The mechanism of synthesis has been elucidated in some detail since then (Matsumoto, Komiyama, and Inoue, 1980; Shigemasa et al., 1983). Figure 26.6 presents a possible mechanism accounting for the observed products.

Polyols such as glycerol are also produced in these reactions by cross Cannizzaro reactions. These compounds have been detected recently in the Murchison meteorite (Cooper et al., 2001).

Sugars are extremely unstable on geological timescales (Larralde, Robertson, and Miller, 1995; Reid and Orgel, 1967; Shapiro, 1988), leading to questions of whether they would have been available for prebiotic synthesis. It was recently shown, however, that borate can stabilize ribose considerably (Ricardo et al., 2004). Another interesting recent discovery is that simple chiral compounds such as L-proline are able to enantioselectively catalyze cross aldol reactions (Córdova, Notz, and Barbas, 2002).

26.3.5 Other Small Bio-Organic Molecules

Prebiotic syntheses have been described for a number of other important biochemical building blocks including fatty acids, isoprenoid alcohols, quinones, porphyrins, and cofactors. These may be important in other theories of the origin of life that do not depend on genetic mechanisms; for this reason, however, they are not detailed herein. A number of other small bio-organic molecules, such as lysine, pyridoxal, and thiamine, have so far resisted attempts at synthesis. These compounds may have been products of an already very sophisticated functioning biological system, and perhaps are not needed for the origin of life.

26.4 Dehydration Reactions

Many important biochemicals are produced by condensation reactions, which result in the elimination of a water molecule. These include membrane lipids, nucleosides, nucleotides, polypeptides, and nucleic acids. Dehydration reactions are difficult in aqueous solution, and most successful prebiotic syntheses have used either extremely concentrated solutions of reactants, reaction in the dry state as might have been

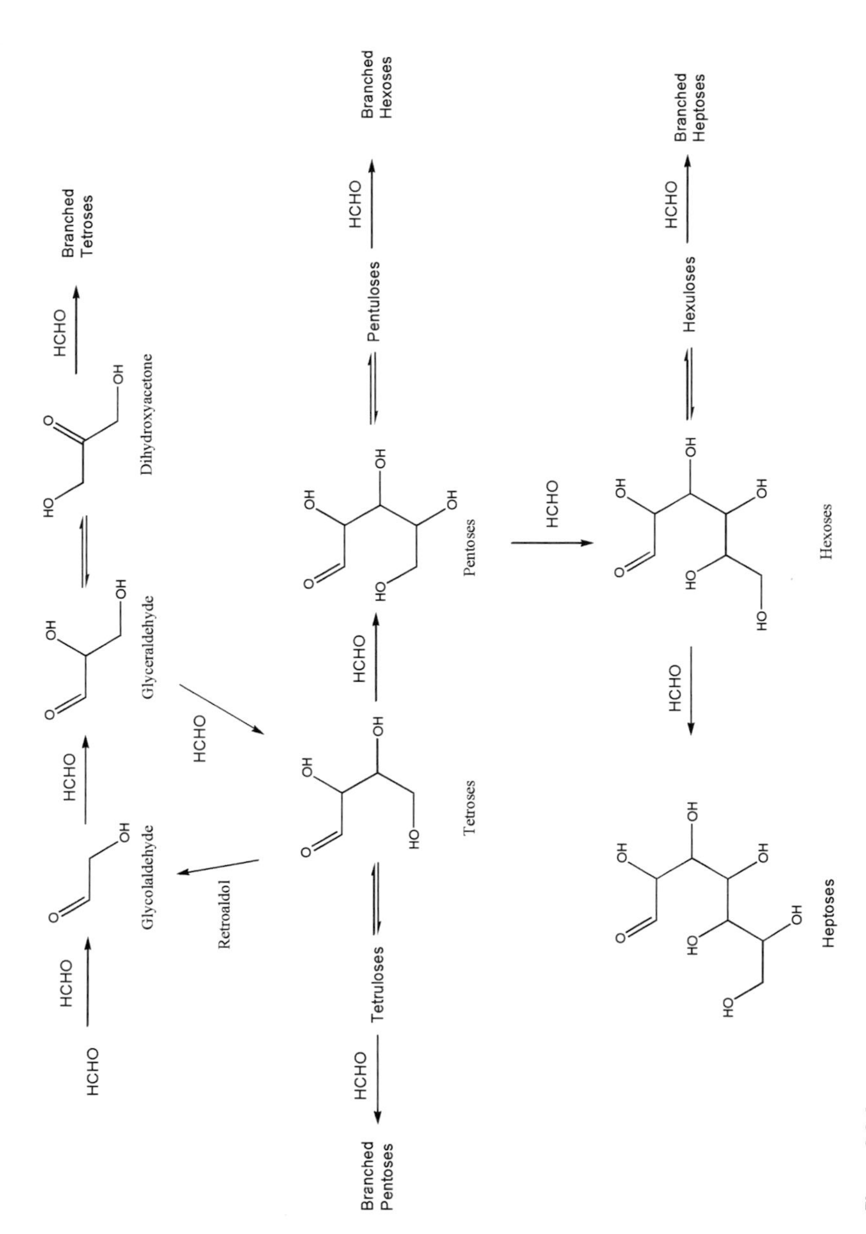

Figure 26.6
A simplified scheme of the formose reaction.

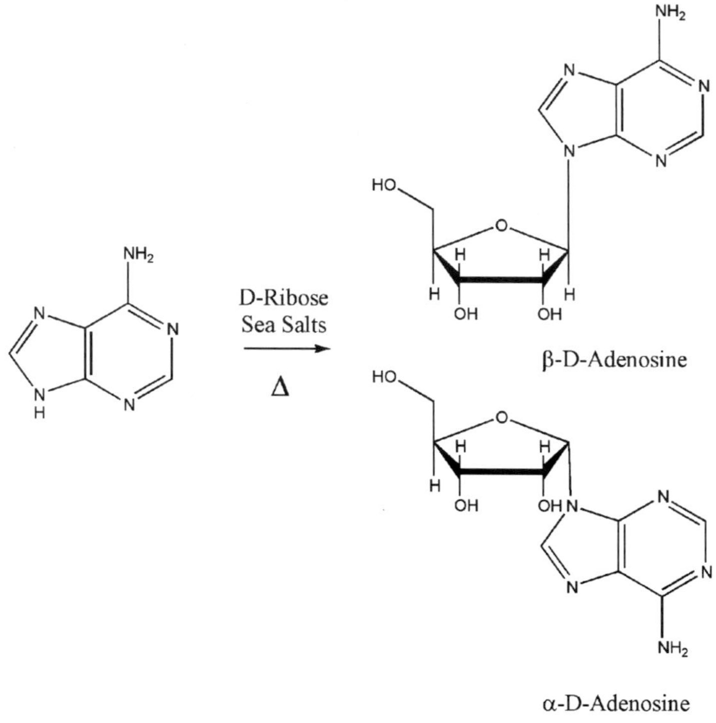

Figure 26.7
Ribonucleosides can be generated by heating dry mixtures of ribose, common sea salts, and purines.

achieved in drying beach or lagoon environments (Nelson, Robertson, Levy, and Miller, 2001), or activating agents that, for the most part, are prebiotically implausible because of their lability.

26.4.1 Nucleosides

When solutions of ribose or deoxyribose and purines are dried and heated with various inorganic components of modern seawater, small but significant amounts of purine ribosides or deoxyribosides are formed, although both the α- and β-isomers are produced (Fuller, Sanchez, and Orgel, 1972; see figure 26.7).

It seems likely that this reaction would work with other sugars; thus, various nucleosides would likely be produced if sugars were produced by formose chemistry. The analogous reaction has not been demonstrated for the pyrimidines, although various more complicated schemes have been devised (Ingar et al., 2003; Sanchez and Orgel, 1970). It is possible that the first replicating polymer used only purine nucleosides (Wächtershäuser, 1988a; Zubay, 1996).

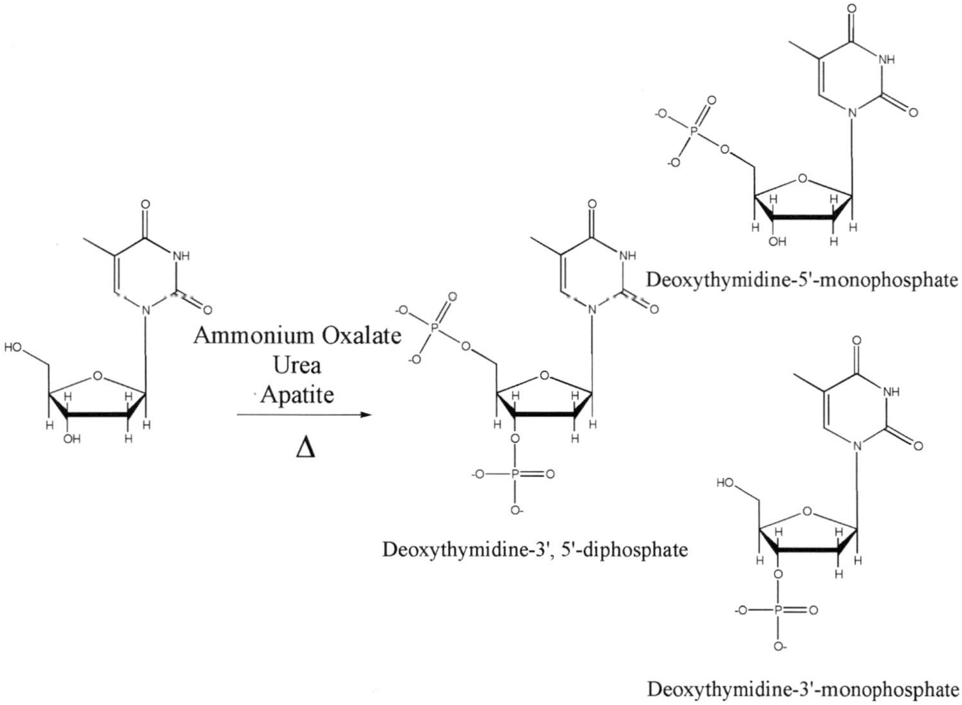

Figure 26.8
Nucleosides and deoxynucleosides can be phosphorylated by heating dry mixtures of ammonium oxalate and urea over common naturally occurring phosphate minerals.

26.4.2 Nucleotides

Heating dry mixtures of nucleosides with phosphate or phosphate minerals and urea or formamide results in synthesis of nucleotides (Lohrmann and Orgel, 1968, 1971; Schoffstall, Barto, and Ramos, 1982; Schwartz et al., 1975), phosphorylated variously at all of the possible hydroxyl groups (figure 26.8).

The cyclic phosphate compound trimetaphosphate, which has been shown to be a component of modern volcanic emissions (Yamagata et al., 1991), is also an efficient phosphorylating agent for α,β-diols including ribonucleosides (Schwartz, 1969; Yamagata, Inoue, and Inomata, 1995). Riboside $2',3'$-cyclic phosphates are obtained in high yield from many of these schemes (figure 26.9); their importance will be elucidated later.

Some authors have pointed out, however, that phosphate is unlikely to have been prebiotically abundant and thus, despite reasons given for its use in biological systems (Westheimer, 1987), it may not have been a component of the first biopolymers (Keefe and Miller, 1995).

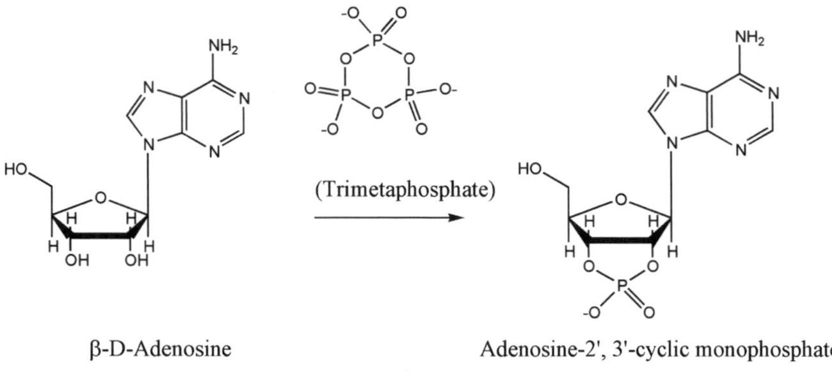

(Trimetaphosphate)

β-D-Adenosine Adenosine-2', 3'-cyclic monophosphate

Figure 26.9
The cyclic compound trimetaphosphate has been shown to phosphorylate nucleosides in aqueous solution. Among the products are cyclic nucleoside monophosphates.

26.4.3 Polypeptides

Ghadiri and coworkers have shown that peptides are capable of self-replication (Lee et al., 1996), however, it is unclear whether such chemistry could have been important for the origin of life. Because peptide nucleic acids have been offered as possible solutions to the problems facing RNA, a brief discussion of peptide formation is included here. Fox and coworkers showed that polypeptides could be formed from dry heated mixtures of amino acids (Harada and Fox, 1958). Meggy (1953) had shown earlier that polyglycine could be produced under similar conditions, and it was later shown that diketopiperazines are possible intermediates in this reaction (Nagayama et al., 1990). Polypeptides degrade rapidly through the loss of N-terminal dipeptide units in the reverse of this reaction (Steinberg and Bada, 1983), thus the peptides produced by this mechanism are essentially in equilibrium with the cyclic dimer (figure 26.10).

26.4.4 Oligonucleotides

Orgel and coworkers showed that polynucleotides were produced from the 2′,3′-cyclic phosphates of the nucleosides analogously through the reverse reaction of RNA hydrolysis (Komiyama and Yoshinari, 1997; Verlander, Lohrmann, and Orgel, 1973) (figure 26.11).

Equilibrium monomer-polymer mixtures were not deemed to be of sufficient length to make this mechanism significant; nevertheless, it serves as a model for the primordial generation of combinatorial libraries of polymers. The numerous likely congeners make this synthesis somewhat unlikely, and as stated earlier, phosphorylation reactions are relatively nonspecific and there would most likely have been a mixture of purines and sugars. Thus, a resulting homogeneous RNA polymer would have been unlikely.

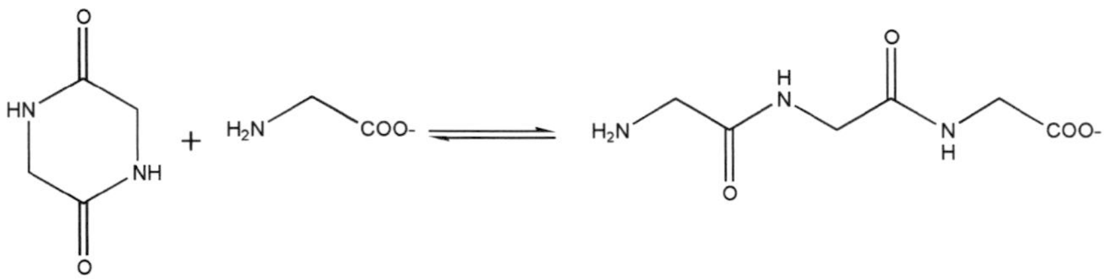

Figure 26.10
α-Amino acids can be elongated in concentrated solution through reaction of the cyclic dimer (a diketopiperazines). The reverse reaction also occurs in solution.

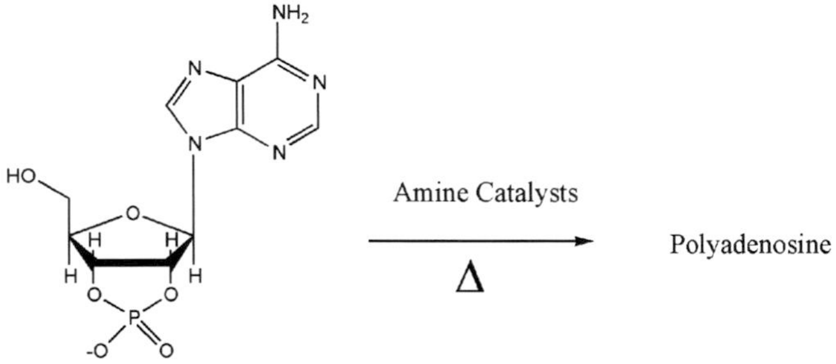

Adenosine-2', 3'-cyclic monophosphate

Figure 26.11
Nucleoside monophosphates can be oligomerized in the dry state in the presence of suitable amine catalysts.

26.5 Combinatorial Libraries

The human genome is $\sim 3 \times 10^9$ nucleotides long. Each human being is genetically unique, and even though some 6×10^9 humans are alive today, all of the possible sequences for the human genome have not been sampled in the history of humanity, nor are they ever likely to be. Indeed, only a very small fraction of these sequences would code for something even remotely human. Even given rapid generation times, the total possible sequence space of relatively small genes has not yet been sampled during the geological history of life on Earth. Interestingly, however, it has been shown that enzymes may catalyze reactions at up to 10^{18} times the uncatalyzed background rate, and that some enzymes appear to be optimal catalysts with respect to substrate diffusion rates (Miller and Wolfenden, 2002).

How large do catalytic polymers need to be? A length of 30 to 60 nucleotides has been suggested by Schöning and coworkers (2000). To take a more tractable and cogent example, a 45-nucleotide RNA molecule includes some 10^{27} possible sequences, or $\sim 2 \times 10^7$ g. For comparison, it has been suggested that some 1.4×10^9 g yr^{-1} of

HCN was delivered to the Earth by comets around the time of the origin of life (Chyba and Sagan, 1992). Only a fraction of these sequences would be expected to fold meaningfully, and many less into functional molecules with any given catalytic activity, although the total number that might fold into functional motifs is difficult to estimate.

Sequence space is a function of the number of monomer types and the length of the polymer. Although the degree of activity required to identify an aptamer as "catalytic" is difficult to define, sequence space grows as n^x, where n is the number of monomers and x is the polymer length. Longer polymers are undoubtedly able to adopt more complex three-dimensional shapes, and the total number and type of catalysts in sequence space is likely to grow as the space itself grows. The frequency of catalysts in sequence space may, however, increase, decrease, or remain relatively constant.

26.6 Alternative Polymers

If RNA is not a particularly attractive prebiotic molecule, alternative molecules should be considered. Biological nucleic acids are of obvious interest as models of self-recognition polymerization chemistry. Nucleic acid monomers possess three essential functionalities: two points of attachment and a recognition surface for a complementary polymer (figure 26.12). By idealizing a monomer in this way, the possible chemistry that could be used for attachment and recognition can be analyzed, which may help identify and construct such analogs.

Various types of covalent linkages could be used to construct such an analog. Assuming that a polymer must have a *regularly* repeating linkage type, the possible

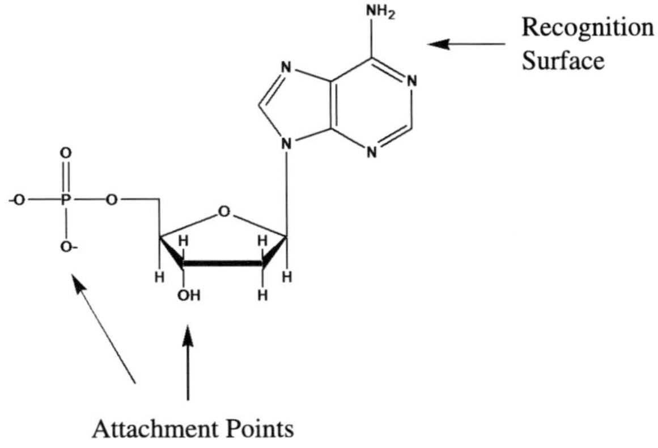

Recognition Surface

Attachment Points

Figure 26.12
An alternative nucleic acid monomer should include two attachment points for incorporation into a polymer, and a recognition surface for complementary strand binding.

linkages, such as ethers, sulfates, ester, sulfonates, phosphonates, phosphates, and anhydrides, can be enumerated and investigated. Only those that could be generated under plausible geochemical conditions need be investigated with respect to the origin of life.

A variety of stable or transiently stable weak interactions might mediate molecular recognition, polymerization, and replication. Molecular recognition for polymerization can be understood as essentially a method to increase the "effective concentration" of reactants to allow reactions to occur at dilutions at which reactive groups would otherwise be unlikely to interact. Modern nucleic acids use a combination of weak forces such as solvophobic effects, H-bonding, and p-stacking to attain their shapes and mediate recognition (Kool, 1997; Kool, Morales, and Guickian, 2000; Saenger, 1984). Forces such as ionic bonding could also be used. With the preceding discussion of the prebiotic synthesis of RNA in mind, a brief review of alternative nucleic acids of potential interest in the origin of life is presented.

Numerous alternative nucleic acid structures are possible, given the constraints of a monomer with two attachment points and a recognition surface. If it is assumed that the RNA bases have been retained from prebiotic chemistry, then only modifications of the backbone need to be considered. These monomers must be prebiotically synthesizable, sufficiently stable to accumulate enough to allow polymerization, and capable of forming structures compatible with molecular recognition.

Because of modern interest in antisense and gene therapies, a large number of possible nucleic acid analogs have been synthesized in the laboratory (Rimmele, 2003;

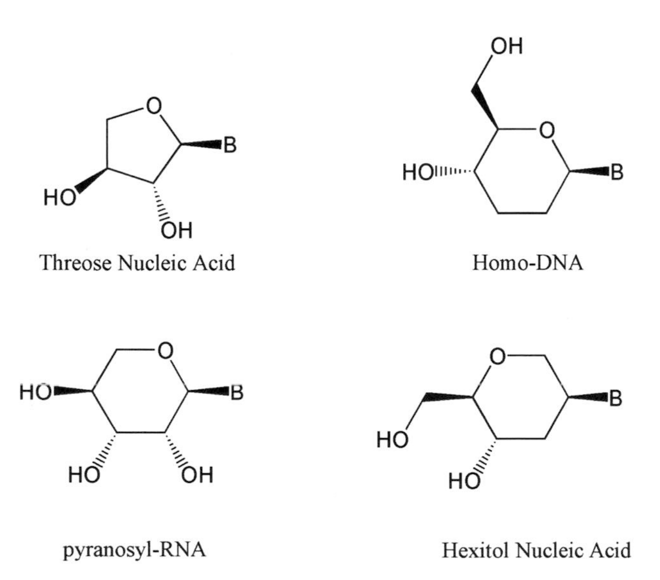

Figure 26.13
Some carbohydrate-based alternative backbones described in the chemical literature.

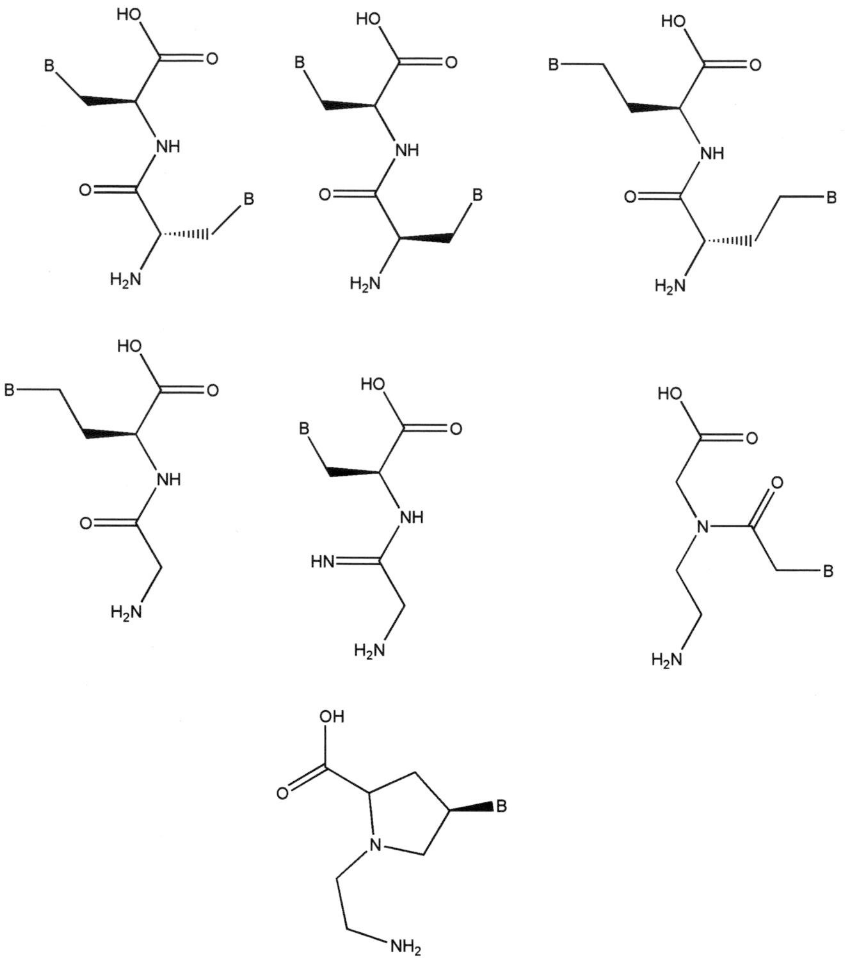

Figure 26.14
Some peptide nucleic acid analogs described in the chemical literature. In some cases, the stereochemistry of these has been shown to be crucial, and this may be an important constraint on the plausibility of these structures as candidates for the origin of life.

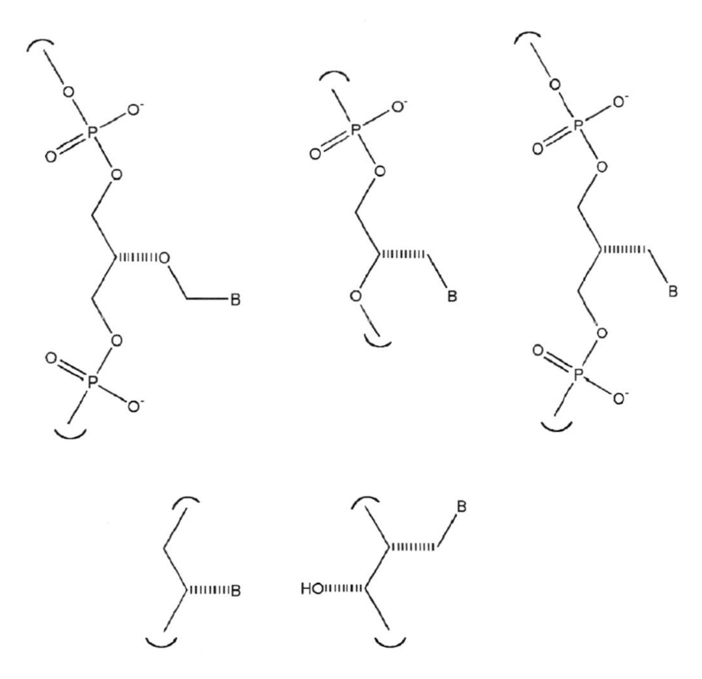

Figure 26.15
Some acyclic nucleic acid analogs described in the chemical literature.

Summerton, 2004). These include the large variety of carbohydrate-based backbones described by Eschenmoser and colleagues and others (Pitsch et al., 1995; Pitsch et al., 1993; Schöning et al., 2000; see figure 26.13), numerous peptide nucleic acids (Lenzi et al., 1995; Lohse et al., 1996; Muller et al., 1990; Nielsen et al., 1991; Weller et al., 1991; see figure 26.14), acyclic nucleoside analogs (Nelsestuen, 1980; Pitha and Pitha, 1970; Pitha, Pitha, and Ts'o, 1970; Schneider and Benner, 1990; Schwartz and Orgel, 1985; Tohidi and Orgel, 1989; Visscher et al., 1989; Zhang, Peritz, and Meggers, 2005; see figure 26.15) and variations using modified linkers (Harada and Orgel, 1990; Stirchak, Summerton, and Weller, 1987; Wu and Orgel, 1991; see figure 26.16).

Carbohydrate-based analogs are likely to suffer from the same problems as RNA, namely, the nonselectivity of sugar synthesis through the formose reaction, the instability of sugars, the difficulty of nucleoside formation, the great number of possible stereoisomers, and the limited availability of phosphate.

Although several peptide nucleic acids (PNA) had been described previously (Buttrey, Jones, and Walker, 1975; Muller et al., 1990), Nielsen and coworkers were the first to prepare the molecules based on N-acetic acid substituted nucleobases attached to an aminoethylglycine (aeg) backbone (aeg-PNA) (Nielsen et al., 1991).

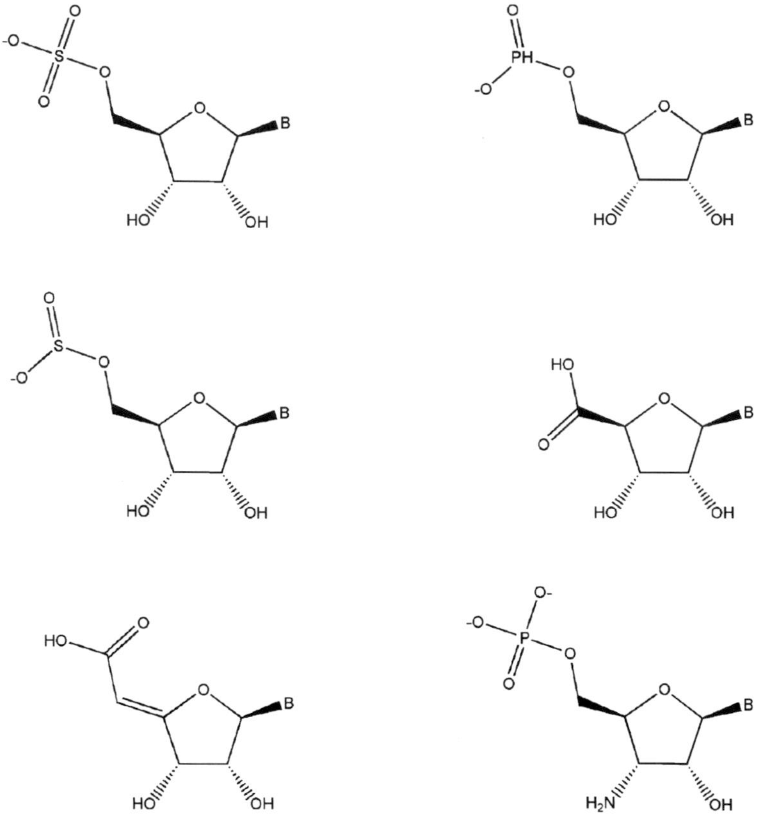

Figure 26.16
Some linker modifications of RNA that have been described in the literature. It should be noted that many of these are unlikely to have plausible prebiotic syntheses, and may be chemically unstable.

These bind complementary aeg-PNA, RNA, and DNA strands extremely well, and are achiral. The prebiotic synthesis of the monomers of aeg-PNA has been demonstrated (Nelson, Levy, and Miller, 2000; see figures 26.17 and 26.18); however, it appears that the construction of polymers may be difficult because of facile rearrangement reactions (Schmidt, Nielsen, and Orgel, 1997). See chapter 15 for a more detailed discussion of the possibilities of utilizing PNA as the information carrier in protocells.

The prebiotic synthesis of the monomers is significant in that it is the first demonstration of the synthesis of the monomers of an alternative structure. Such investigations allow evaluation of the relative difficulty of prebiotic RNA synthesis. Many alternative PNA polymers are possible (figure 26.14), and it seems likely that many of the alternatives could also be synthesized prebiotically.

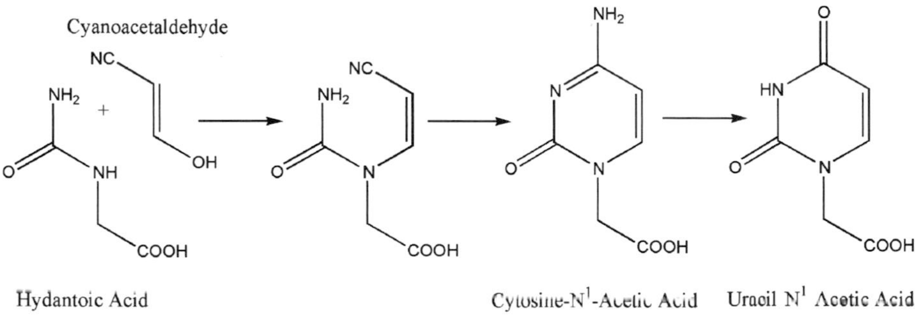

Figure 26.17
The suggested mechanism of synthesis of the aeg-PNA pyrimidine monomers from plausible prebiotic reactants. The backbone was synthesized in remarkably high yield through the Strecker mechanism from ethylene diamine, HCN, and HCHO.

Acyclic monomers (figure 26.15) have the possible advantage that in some cases they may avoid the possible problems associated with adoption of chirality, which has been shown to interfere with oligomerization of nucleotides (Joyce, 1987), and are structurally simple. Despite remarkable success with their higher-order chemistry (Schwartz and Orgel, 1985; Zhang, Peritz, and Meggers, 2005), little attention has been paid to their prebiotic synthesis.

Both Orgel and Benner (Harada and Orgel, 1990; Hutter, Blaettler, and Benner, 2002) have pointed out the apparent importance of charge in a polynucleotide, for both strand repulsion and polymer solubilization. Function, stability, and reactivity considerations could help pare down the most plausible of these possibilities.

Perhaps the simplest template-directed information polymers proposed in the origins of life discussion are based on π stacking of simple aromatic copolymers (Ehrenfreund et al., 2006). A prebiotic synthesis of the involved monomers is evident since simple aromatic compounds are the most abundant organic molecules in the universe. However, experimental evidence of their polymerization and subsequent replication is still lagging.

26.7 Conclusion

The idea that life began with a self-replicating genetic polymer is extremely attractive based on the success of template-directed oligomerizations and SELEX technology (Forst, 1998; Joyce, 1987; Rimmele, 2003). Although there is an immense number of possible modifications of the backbone or bases, only a small minority of them may be prebiotically synthesizable, some are undoubtedly more stable than others, some are probably more easily oligomerized nonenzymatically, and some may be more adept at molecular recognition, catalysis, and ultimately replication. Thus,

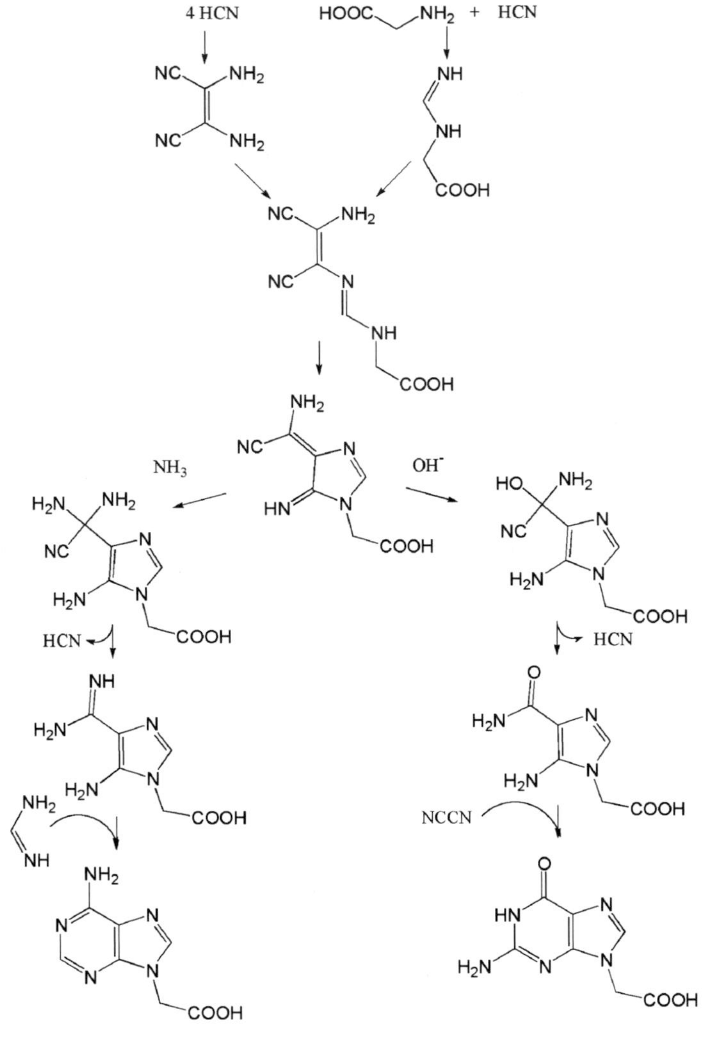

Adenine-N^9-Acetic Acid

Guanine-N^9-Acetic Acid

Figure 26.18
The suggested mechanism of synthesis of the aeg-PNA purine monomers from plausible prebiotic reactants.

although there is potentially a huge structural space to examine experimentally, the number of structures worth investigating can be carved down considerably based on "first principles" examination of the prebiotic chemistry of the monomers.

Biological molecules such as RNA are relatively easy to work with experimentally because of the enzymatic tools available. Nonnatural systems suffer from the difficulty of their manipulation, thus progress in this field may be slow in the foreseeable future. Given sufficiently sensitive screening methods, it seems plausible that other polymers might also produce catalysts with the same facility as RNA. Nothing in principle indicates that an alternative set of monomers could not also produce versatile catalytic structures capable of replication. Thus, the earliest polymers could have been significantly different, and synthetic biochemistries should bé possible, but they may be difficult to demonstrate experimentally.

It is possible that the emergence of complex replicating systems requires the interaction of an encapsulating chemical system with a replicating polymer chemistry to form a primitive protocell and that the conjunction of these two results is a more sophisticated system than could be achieved by either alone. These, unfortunately, would be even more complicated systems to work with experimentally, since there are many more variables and it is difficult to identify catalytic activities in either system alone. This chemical complexity issue is the largest challenge the protocell community currently faces.

References

Anders, E., Hayatsu, R., & Studier, M. H. (1973). Organic compounds in meteorites. *Science*, *182*, 781–789.

Benner, S. A., Ellington, A. D., & Tauer, A. (1989). Modern metabolism as a palimpsest of the RNA World. *Proceedings of the National Academy of Sciences of the United States of America*, *86*, 7054–7058.

Bowser, M. T. (2005). SELEX: Just another separation? *Analyst*, *130* (2), 128–130.

Butlerow, A. (1861). Formation synthetique d'une substance sucree. *Comptes Rendus de l'Academie des Sciences*, *53*, 145–247.

Buttrey, J. D., Jones, A. S., & Walker, R. T. (1975). Synthetic analogues of polynucleotides XIII: The resolution of DL-a-(thymin-1-yl) alanine and polymerization of the DL-a-(thymin-1-yl) alanines. *Tetrahedron*, *31*, 73–75.

Cairns-Smith, A. (1977). Takeover mechanisms and early biochemical evolution. *BioSystems*, *9* (2–3), 105–109.

Cech, T. R., Zaug, A. J., & Grabowski, P. J. (1981). In vitro splicing of the ribosomal RNA precursor of tetrahymena: Involvement of a guanosine nucleotide in the excision of the intervening sequence. *Cell*, *27*, 487–496.

Choughuley, A., Subbaraman, A., Kazi, Z., & Chadha, M. (1977). A possible prebiotic synthesis of thymine: Uracil-formaldehyde-formic acid reaction. *BioSystems*, *9* (2–3), 73–80.

Chyba, C., & Sagan, C. (1992). Endogenous production, exogenous delivery and impact-shock synthesis of organic molecules: An inventory for the origins of life. *Nature*, *355* (6356), 125–132.

Chyba, C., Thomas, P., Brookshaw, L., & Sagan, C. (1990). Cometary delivery of organic molecules to the early Earth. *Science*, *249*, 366–373.

Cooper, G., Kimmich, N., Belisle, W., Sarinana, J., Brabham, K., & Garrel, L. (2001). Carbonaceous meteorites as a source of sugar-related organic compounds for the early Earth. *Nature*, *414*, 879–883.

Córdova, A., Notz, W., & Barbas, C. F. (2002). Direct organocatalytic aldol reactions in buffered aqueous media. *Chemical Communications*, *24*, 3024–3025.

Crick, F. H. C. (1968). The origin of the genetic code. *Journal of Molecular Biology*, *38*, 367–379.

Decker, P., Schweer, H., & Pohlmann, R. (1982). Identification of formose sugars, presumable prebiotic metabolites, using capillary gas chromatography/gas chromatography-mass spectrometry of n-butoxime trifluoroacetates on OV-225. *Journal of Chromatography*, *225*, 281–291.

Doolittle, W. F. (2000). The nature of the universal ancestor and the evolution of the proteome. *Current Opinion in Structural Biology*, *10* (3), 355–358.

Ehrenfreund, P., & Charnley, S. B. (2000). Organic molecules in the interstellar medium, comets, and meteorites: A voyage from dark clouds to the early earth. *Annual Review of Astronomy and Astrophysics*, *38*, 427–483.

Ehrenfreund, P., & Menten, K. M. (2002). From molecular clouds to the origin of life. In G. Horneck & C. Baumstark-Khan (Eds.), *Astrobiology* (pp. 7–23). New York: Springer-Verlag.

Ehrenfreund, P., Rasmussen, S., Cleaves, J., & Chen, L. (2006). Tracking the experimental steps of the origins of life: The aromatic world. *Astrobiology*, *6*, 482–512.

Ferris, J., Joshi, P., Edelson, E., & Lawless, J. (1978). HCN: A plausible source of purines, pyrimidines, and amino acids on the primitive Earth. *Journal of Molecular Evolution*, *11*, 293–311.

Ferris, J., Sanchez, R., & Orgel, L. (1968). Studies in prebiotic synthesis III: Synthesis of pyrimidines from cyanoacetylene and cyanate. *Journal of Molecular Biology*, *33*, 693–704.

Forst, C. V. (1998). Molecular evolution: A theory approaches experiments. *Journal of Biotechnology*, *64* (1), 101–118.

Forterre, P. (2002). The origin of DNA genomes and DNA replication proteins. *Current Opinion in Microbiology*, *5* (5), 525–532.

Fuller, W., Sanchez, R., & Orgel, L. (1972). Prebiotic synthesis VII: Solid-state synthesis of purine nucleosides. *Journal of Molecular Evolution*, *1* (3), 249–257.

Gesteland, R., Cech, T., & Atkins, J. (Eds.). (1999). *The RNA World* (2nd ed.). Cold Spring Harbor, NY: Cold Spring Harbor Laboratory Press.

Gilbert, W. (1986). The RNA world. *Nature*, *319*, 618.

Gold, T. (1992). The deep, hot biosphere. *Proceedings of the National Academy of Science*, *89*, 6045–6049.

Guerrier-Takada, C., Gardiner, K., Marsh, T., Pace, N., & Altman, S. (1983). The RNA moiety of ribonuclease P is the catalytic subunit of the enzyme. *Cell*, *35*, 849–857.

Haldane, J. (1929). The origin of life. *The Rationalist Annual*, 3–10.

Harada, K., & Fox, S. W. (1958). Thermal condensation of glutamic acid and glycine to linear peptides. *Journal of the American Chemical Society*, *80*, 2694–2697.

Harada, K., & Orgel, L. E. (1990). Template-directed oligomerization of 5′-deoxy-5′-nucleosideacetic acid derivatives. *Origins of Life and Evolution of the Biosphere*, *20* (2), 151–160.

Hayatsu, R., Studier, M., Moore, L., & Anders, E. (1975). Purines and triazines in the Murchison meteorite. *Geochimica et Cosmochimica Acta*, *39*, 471–488.

Horowitz, N. H. (1945). The evolution of biochemical syntheses. *Proceedings of the National Academy of Sciences of the United States of America*, *31*, 153–157.

Hutter, D., Blaettler, M. O., & Benner, S. A. (2002). From phosphate to bis(methylene) sulfone: Non-ionic backbone linkers in DNA. *Helvetica Chimica Acta*, *85* (9), 2777–2806.

Ingar, A., Luke, R., Hayter, B., & Sutherland, J. (2003). Synthesis of cytidine ribonucleotides by stepwise assembly of the heterocycle on a sugar phosphate. *ChemBioChem*, *4* (6), 504–507.

Johnston, W., Unrau, P., Lawrence, M., Glasner, M., & Bartel, D. (2001). RNA-catalyzed RNA polymerization: Accurate and general RNA-templated primer extension. *Science*, *292*, 1319–1325.

Joyce, G. F. (1987). Non-enzymatic template-directed synthesis of informational macromolecules. *Cold Spring Harbor Symposium on Quantitative Biology*, *52*, 41–51.

Joyce, G., Schwartz, A., Miller, S., & Orgel, L. (1987). The case for an ancestral genetic system involving simple analogs of the nucleotides. *Proceedings of the National Academy of Sciences of the United States of America*, *84* (13), 4398–4402.

Kasting, J. (1993). Earth's early atmosphere. *Science*, *259*, 920–926.

Keefe, A., & Miller, S. (1995). Are polyphosphates or phosphate esters prebiotic reagents? *Journal of Molecular Evolution*, *41* (6), 693–702.

Komiyama, M., & Yoshinari, K. (1997). Kinetic analysis of diamine-catalyzed RNA hydrolysis. *Journal of Organic Chemistry*, *62* (7), 2155–2160.

Kool, E. T. (1997). Preorganization of DNA: Design principles for improving nucleic acid recognition by synthetic oligonucleotides. *Chemical Reviews*, *97*, 1473–1487.

Kool, E. T., Morales, J. C., & Guckian, K. M. (2000). Mimicking the structure and function of DNA: Insights into DNA stability and replication. *Angewandte Chemie International Edition in English*, *39*, 990–1009.

Kuhn, W., & Atreya, V. (1979). Ammonia photolysis and the greenhouse effect in the primordial atmosphere of the earth. *Icarus*, *37* (1), 207–213.

Larralde, R., Robertson, M., & Miller, S. (1995). Rates of decomposition of ribose and other sugars: Implications for chemical evolution. *Proceedings of the National Academy of Sciences of the United States of America*, *92*, 8158–8160.

Lazcano, A., & Miller, S. (1994). How long did it take for life to begin and evolve to cyanobacteria? *Journal of Molecular Evolution*, *39* (6), 546–554.

Lee, D. H., Granja, J. R., Martinez, J. A., Severin, K., & Ghadiri, M. R. (1996). A self-replicating peptide. *Nature*, *382* (6591), 525–528.

Lenzi, A., Reginato, G., Taddei, M., & Trifilieff, E. (1995). Solid phase synthesis of a self-complementary (antiparallel) chiral peptidic nucleic acid strand. *Tetrahedron Letters*, *36*, 1717–1718.

Levy, M., & Miller, S. L. (1998). The stability of the RNA bases: Implications for the origin of life. *Proceedings of the National Academy of Sciences of the United States of America*, *95*, 7933–7938.

Levy, M., Miller, S., Brinton, K., & Bada, J. (2000). Prebiotic synthesis of adenine and amino acids under Europa-like conditions. *Icarus*, *145*, 609–613.

Löb, W. (1913). Behavior of formamide under the influence of the silent discharge nitrogen assimilation. *Berichte*, *46*, 684–697.

Lohrmann, R., & Orgel, L. (1968). Prebiotic synthesis: Phosphorylation in aqueous solution. *Science*, *161* (3836), 64–66.

Lohrmann, R., & Orgel, L. (1971). Urea-inorganic phosphate mixtures as prebiotic phosphorylating agents. *Science*, *171* (3970), 490–494.

Lohse, P., Oberhauser, B., Oberhauser-Hofbauer, B., Baschang, G., & Eschenmoser, A. (1996). Chemie von a-Aminonitrilen XVII: Oligo(nukleodipeptamidinium)-Salze. *Croatia Chemica Acta*, *69*, 535–562.

Matsumoto, T., Komiyama, M., & Inoue, S. (1980). Selective formose reaction catalyzed by diethylaminoethanol. *Chemistry Letters*, *7*, 839–842.

Meggy, A. B. (1953). Glycine peptides I: The polymerization of 2,5-piperazinedione at 180° C. *Journal of the Chemical Society, Abstracts*, 851–855.

Miller, B. G., & Wolfenden, R. (2002). Catalytic proficiency: The unusual case of OMP decarboxylase. *Annual Review of Biochemistry*, *71*, 847–885.

Miller, S. (1953). A production of amino acids under possible primitive Earth conditions. *Science*, *117*, 528.

Miller, S. (1957). The mechanism of synthesis of amino acids by electric discharges. *Biochimica et Biophysica Acta*, *23*, 480–489.

Miller, S. (1998). The endogenous synthesis of organic compounds. In A. Brack (Ed.), *The molecular origins of life: Assembling pieces of the puzzle* (pp. 59–85). Cambridge: Cambridge University Press.

Miller, S., & Orgel, L. (1974). *The origins of life on the Earth.* Englewood Cliffs, NJ: Prentice Hall.

Miyakawa, S., Cleaves, H., & Miller, S. (2002a). The cold origin of life. A: Implications based on the hydrolytic stabilities of hydrogen cyanide and formamide. *Origins of Life and Evolution of the Biosphere, 32* (3), 195–208.

Miyakawa, S., Cleaves, H., & Miller, S. (2002b). The cold origin of life B: Implications based on pyrimidines and purines produced from frozen ammonium cyanide solutions. *Origins of Life and Evolution of the Biosphere, 32* (3), 209–218.

Miyakawa, S., Yamanashi, H., Kobayashi, K., Cleaves, H., & Miller, S. (2002c). Prebiotic synthesis from CO atmospheres: Implications for the origin of life. *Proceedings of the National Academy of Sciences of the United States of America, 99* (23), 14628–14631.

Muller, D., Pitsch, S., Kittaka, A., Wagner, E., Wintner, C. E., & Eschenmoser, A. (1990). Chemie von alpha-Aminonitrilen (135). *Helvetica Chimica Acta, 73*, 1410–1468.

Muñoz Caro, G. M., Meierhenrich, U. J., Schutte, W. A., Barbler, B., Arcones Segovia, A., Rosenbauer, H., et al. (2002). Amino acids from ultraviolet irradiation of interstellar ice analogues. *Nature, 416* (6879), 403–406.

Nagayama, M., Takaoka, O., Inomata, K., & Yamagata, Y. (1990). Diketopiperazine-mediated peptide formation in aqueous solution. *Origins of Life and Evolution of the Biosphere, 20* (3–4), 249–257.

Nelsestuen, G. (1980). Origin of life: Consideration of alternatives to proteins and nucleic acids. *Journal of Molecular Evolution, 15* (1), 59–72.

Nelson, K., Levy, M., & Miller, S. (2000). Peptide nucleic acids rather than RNA may have been the first genetic molecule. *Proceedings of the National Academy of Sciences of the United States of America, 97* (8), 3868–3871.

Nelson, K., Robertson, M., Levy, M., & Miller, S. (2001). Concentration by evaporation and the prebiotic synthesis of cytosine. *Origins of Life and Evolution of the Biosphere, 31* (3), 221–229.

Nielsen, P., Egholm, M., Berg, R., & Buchardt, O. (1991). Sequence-selective recognition of DNA by strand displacement with a thymine-substituted polyamide. *Science, 254* (5037), 1497–1500.

Oparin, A. (1924). *Proiskhozhedenie zhizni* (Mosckovskii Rabochii, Moscow). Reprinted and translated in J. D. Bernal. (1967), *The Origin of Life*. London: Weidenfeld and Nicolson.

Oparin, A. (1938). *The origin of life.* New York: Macmillan.

Orgel, L. E. (1968). Evolution of the genetic apparatus. *Journal of Molecular Biology, 38*, 381–393.

Orgel, L. E. (2000). Self-organizing biochemical cycles. *Proceedings of the National Academy of Sciences of the United States of America, 97*, 12503–12507.

Oró, J. (1960). Synthesis of adenine from ammonium cyanide. *Biochemica Biophysica Research Communications, 2*, 407–412.

Oró, J. (1961). Comets and the formation of biochemical compounds on the primitive Earth. *Nature, 190*, 442–443.

Oró, J., & Kimball, A. (1961). Synthesis of purines under primitive earth conditions. I: Adenine from hydrogen cyanide. *Archives of Biochemistry and Biophysics, 94*, 221–227.

Oró, J., & Lazcano, A. (1997). Comets and the origin and evolution of life. In P. J. Thomas, C. F. Chyba, & C. P. McKay (Eds.), *Comets and the origin and evolution of life* (pp. 3–27). New York: Springer.

Park, T. K., Feng, Q., & Rebek, Jr., J. (1992). Synthetic replicators and extrabiotic chemistry. *Journal of the American Chemical Society, 114*, 4529–4532.

Peltzer, E., Bada, J., Schlesinger, G., & Miller, S. (1984). The chemical conditions on the parent body of the Murchison meteorite: Some conclusions based on amino-, hydroxy-, and dicarboxylic acids. *Advances in Space Research, 4*, 69–74.

Pinto, J. P., Gladstone, G. R., & Yung, Y. L. (1980). Photochemical production of formaldehyde in Earth's primitive atmosphere. *Science, 210* (4466), 183–185.

Pitha, J., & Pitha, P. M. (1970). Preparation and properties of poly-9-vinyladenine. *Biopolymers, 9*, 965–977.

Pitha, J., Pitha, P. M., & Ts'o, P. O. P. (1970). Poly (1-vinyluracil): The preparation and interaction with adenosine derivatives. *Biochemica et Biophysica Acta, 204*, 39–48.

Pitsch, S., Krishnamurthy, R., Bolli, M., Wendeborn, S., Holzner, A., Minton, M., et al. (1995). Pyranosyl-RNA (pRNA): Base-pairing selectivity and potential to replicate. *Helvetica Chimica Acta, 78*, 1621–1635.

Pitsch, S., Wendeborn, S., Juan, B., & Eschenmoser, A. (1993). Why pentose and not hexose-nucleic acids? Part VII: Pyranosyl-RNA ("p-RNA"). *Helvetica Chimica Acta, 76*, 2161–2183.

Rebek, J. (1991). Molecular recognition and the development of self-replicating systems. *Experientia, 47*, 1096–1104.

Reid, C., & Orgel, L. E. (1967). Synthesis of sugars in potentially prebiotic conditions. *Nature, 216*, 455–456.

Ricardo, A., Carrigan, M. A., Olcott, A. N., & Benner, S. A. (2004). Borate minerals stabilize ribose. *Science, 303*, 196.

Rimmele, M. (2003). Nucleic acid aptamers as tools and drugs: Recent developments. *ChemBioChem, 4* (10), 963–971.

Robertson, M. P., & Miller, S. L. (1995). An efficient prebiotic synthesis of cytosine and uracil. *Nature, 375*, 772–774.

Saenger, W. (1984). *Principles of nucleic acid structure.* New York: Springer-Verlag.

Sanchez, R., Ferris, J., & Orgel, L. (1966a). Conditions for purine synthesis: Did prebiotic synthesis occur at low temperatures? *Science, 153*, 72–73.

Sanchez, R., Ferris, J., & Orgel, L. (1966b). Cyanoacetylene in prebiotic synthesis. *Science, 154*, 784–785.

Sanchez, R., Ferris, J., & Orgel, L. (1967). Studies in prebiotic synthesis II: Synthesis of purine precursors and amino acids from aqueous hydrogen cyanide. *Journal of Molecular Biology, 30*, 223–253.

Sanchez, R., Ferris, J., & Orgel, L. (1968). Studies in prebiotic synthesis IV: The conversion of 4-aminoimidazole-5-carbonitrile derivatives to purines. *Journal of Molecular Biology, 38*, 121–128.

Sanchez, R., & Orgel, L. (1970). Studies in prebiotic synthesis V: Synthesis and photoanomerization of pyrimidine nucleosides. *Journal of Molecular Biology, 47* (3), 531–543.

Schlesinger, G., & Miller, S. (1973). Equilibrium and kinetics of glyconitrile formation in aqueous solution. *Journal of the American Chemical Society, 95* (11), 3729–3735.

Schmidt, J., Nielsen, P., & Orgel, L. (1997). Information transfer from peptide nucleic acids to RNA by template-directed syntheses. *Nucleic Acids Research, 25*, 4797–4802.

Schneider, K., & Benner, S. (1990). Oligonucleotides containing flexible nucleoside analogs. *Journal of the American Chemical Society, 112* (1), 453–455.

Schoffstall, A. M., Barto, R. J., & Ramos, D. L. (1982). Nucleoside and deoxynucleoside phosphorylation in formamide solutions. *Origins of Life, 12*, 143–151.

Schöning, K., Scholz, P., Guntha, S., Wu, X., Krishnamurthy, R., & Eschenmoser, A. (2000). Chemical etiology of nucleic acid structure: The alpha-threofuranosyl-(3′–2′) oligonucleotide system. *Science, 290*, 1347–1351.

Schwartz, A. (1969). Specific phosphorylation of the 2′- and 3′-positions in ribonucleosides. *Journal of the Chemical Society D: Chemical Communications, 23*, 1393.

Schwartz, A. W., & Chittenden, G. J. F. (1977). Synthesis of uracil and thymine under simulated prebiotic conditions. *BioSystems, 9*, 87–92.

Schwartz, A. W., Joosten, H., & Voet, A. B. (1982). Prebiotic adenine synthesis via HCN oligomerization on ice. *Biosystems, 15*, 191–193.

Schwartz, A., & Orgel, L. (1985). Template-directed synthesis of novel, nucleic acid-like structures. *Science, 228*, 585–587.

Schwartz, A. W., van der Veen, M., Bisseling, T., & Chittenden, G. J. F. (1975). Prebiotic nucleotide synthesis-demonstration of a geologically plausible pathway. *Origins of Life and the Evolution of the Biosphere, 6*, 163–168.

Shapiro, R. (1988). Prebiotic ribose synthesis: A critical analysis. *Origins of Life and the Evolution of the Biosphere*, *18*, 71–85.

Shigemasa, Y., Matsumoto, H., Sasaki, Y., Ueda, N., Nakashima, R., Harada, K., et al. (1983). Formose reactions part 20: The selective formose reaction in dimethylformamide in the presence of vitamin B1. *Journal of Carbohydrate Chemistry*, *2* (3), 343–348.

Steinberg, S. M., & Bada, J. L. (1983). Peptide decomposition in the neutral pH region via the formation of diketopiperazines. *Journal of Organic Chemistry*, *48* (13), 2295–2298.

Stirchak, E. P., Summerton, J. E., & Weller, D. D. (1987). Uncharged stereoregular nucleic acid analogues I: Synthesis of a cytosine-containing oligomer with carbamate internucleoside linkages. *Journal of Organic Chemistry*, *52*, 4202.

Stribling, R., & Miller, S. (1987). Energy yields for hydrogen cyanide and formaldehyde syntheses: The hydrogen cyanide and amino acid concentrations in the primitive ocean. *Origins of Life and the Evolution of the Biosphere*, *17* (3–4), 261–273.

Summerton, J. E. (2004). Morpholinos and PNAs compared. *Letters in Peptide Science*, *10*, 215–236.

Tian, F., Toon, O. B., Pavlov, A. A., & De Sterck, H. (2005). A hydrogen-rich early Earth atmosphere. *Science*, *308* (5724), 1014–1017.

Tohidi, M., & Orgel, L. (1989). Some acyclic analogues of nucleotides and their template-directed reactions. *Journal of Molecular Evolution*, *28*, 367–373.

Uhlmann, U., & Peyman, A. (1990). Antisense oligonucleotides: A new therapeutic principle. *Chemical Reviews*, *90*, 543–579.

Urey, H. (1952). On the early chemical history of the Earth and the origin of life. *Proceedings of the National Academy of Sciences of the United States of America*, *38*, 351–363.

Van der Velden, W., & Schwartz, A. (1977). Search for purines and pyrimidines in the Murchison meteorite. *Geochimica et Cosmochimica Acta*, *41* (7), 961–968.

Verlander, M. S., Lohrmann, R., & Orgel, L. E. (1973). Catalysts for the self-polymerization of adenosine cyclic 2′,3′-phosphate. *Journal of Molecular Evolution*, *2* (4), 303–316.

Visscher, J., Bakker, C. G., Van der Woerd, R., & Schwartz, A. W. (1989). Template-directed oligomerization catalyzed by a polynucleotide analog. *Science*, *244*, 329–331.

Voet, A., & Schwartz, A. (1982). Uracil synthesis via hydrogen cyanide oligomerization. *Origins of Life*, *12* (1), 45–49.

Wächtershäuser, G. (1988a). Before enzymes and templates: Theory of surface metabolism. *Microbiological Reviews*, *52*, 452–484.

Wächtershäuser, G. (1988b). An all-purine precursor of nucleic acids. *Proceedings of the National Academy of Sciences of the United States of America*, *85*, 1134–1135.

Watson, J. D., & Crick, F. H. C. (1953). A structure for deoxyribose nucleic acid. *Nature*, *171*, 737–738.

Weller, D. D., Daly, D. T., Olson, W. K., & Summerton, J. E. (1991). Molecular modeling of acyclic polyamide oligonucleotide analogues. *Journal of Organic Chemistry*, *56*, 6000–6006.

Westheimer, F. H. (1987). Why nature chose phosphates. *Science*, *235*, 1173–1178.

Wetherill, G. (1990). Formation of the Earth. In A. Albee & F. Stehli (Eds.), *AnnualRreview of Earth and Planetary Sciences*, *18*, 205–256.

White, R. (1984). Hydrolytic stability of biomolecules at high temperatures and its implication for life at 250°C. *Nature*, *310* (5976), 430–432.

Woese, C. R. (2000). Interpreting the universal phylogenetic tree. *Proceedings of the National Academy of Sciences of the United States of America*, *97* (15), 8392–8396.

Wolman, Y., Haverland, W., & Miller, S. (1972). Nonprotein amino acids from spark discharges and their comparison with the Murchison meteorite amino acids. *Proceedings of the National Academy of Sciences of the United States of America*, *69* (4), 809–811.

Wu, T., & Orgel, L. E. (1991). Disulfide-linked oligonucleotide phosphorothioates: Novel analogues of nucleic acids. *Journal of Molecular Evolution*, *32*, 274–277.

Yamagata, Y., Inoue, H., & Inomata, K. (1995). Specific effect of magnesium ion on 2′,3′-cyclic AMP synthesis from adenosine and trimetaphosphate in aqueous solution. *Origins of Life and Evolution of the Biosphere*, *25* (1–3), 47–52.

Yamagata, Y., Watanabe, H., Saitoh, M., & Namba, T. (1991). Volcanic production of polyphosphates and its relevance to prebiotic evolution. *Nature*, *352* (6335), 516–519.

Zhang, L., Peritz, A., & Meggers, E. (2005). A simple glycol nucleic acid. *Journal of the American Chemical Society*, *127*, 4174–4175.

Zubay, G. (1996). Arguments in favor of an all-purine RNA first. *Chemtracts: Biochemistry and Molecular Biology*, *6*, 251–260.

27 Cell-like Entities: Scientific Challenges and Future Applications

John M. Frazier, Nancy Kelley-Loughnane, Sandra Trott, Oleg Paliy, Mauricio Rodriguez Rodriguez, Leamon Viveros, and Melanie Tomczak

27.1 Introduction

In the last few decades, scientists have learned how to manipulate the basic components of life, how to design biomolecular networks, how to evolve biomolecules with unique characteristics, and how to direct and control cellular processes at the molecular level. Using this knowledge as a foundation, it is theoretically possible to conceive of designing biological constructs, which we refer to as cell-like entities (CLEs), that use custom engineered biological machinery to accomplish specified tasks. The practical challenge is: Can we fabricate biological constructs for specific purposes using the same principles and components found in natural biological systems? For example, can CLEs be designed to detect very low levels of specific chemicals and fluoresce to indicate their presence, can they be designed to synthesize functional chemicals on demand, or can they use engineered metabolic pathways to detoxify environmental pollutants? The building of CLEs that could accomplish these goals would result in fundamental breakthroughs in exploiting our understanding of cellular control systems and biochemical information processing.

The concept of the CLE is based on understanding how the biochemical reactions network that we call a living cell functions, how is it controlled, and what the key processes and components are that will provide unique functions when integrated. The goal is not to create a living organism, but to create a biological construct that uses the essence of living systems to provide a wide array of solutions to current technological and medical challenges.

27.1.1 Definition of Nonliving Versus Living

The emergence of cellular life is one of the major transitions in history. The critical factor was the establishment of a closed boundary encapsulating catalytic reactions and genetic information in a well-defined compartment. This co-containment allowed for the parallel evolution of biochemical processes and the information that defines those processes. Although there are significant differences of opinion as to the definition of what constitutes "living," there are several characteristics of living organisms that all agree are components of life. First, living organisms are open dynamical systems—energy and matter flow across the physical boundary of the system in both directions. The internal components of the organism are not in equilibrium and can be characterized as a quasi-steady state, but are subject to major state transitions. Compared to that of macrochemical systems, the behavior of cellular reactions at the molecular level is unique in several ways. Processes such as diffusion, binding, and catalysis occur in parallel within a confined three-dimensional microvolume. Many of these processes are stochastic in nature, and this stochastic behavior often contributes to functionality (McAdams and Arkin, 1997). Second, living organisms can reproduce. Their physical structure and information content are passed on from generation to generation, and in theory the lineage of any given cell can be traced back to its primordial ancestor. The physical structure of a living organism persists in time, although it may undergo significant transformations. On the other hand, the informational content is relatively stable on a macro scale. Somatic mutations and stochastic phenomena result in individual variability in a population of cells whereas genomic mutations result in population drift over time. This variability permits the exploration of novel solutions at the edges of the available reaction space in response to environmental perturbations. Finally, an important characteristic of living systems is the ability to perform useful (meaningful) work. The definition of "useful" must ultimately relate to ensuring the survival of the species.

As a consequence of the physical chemical nature of the biomolecular networks that constitute the living cell, living cells exhibit several unique properties. Although the basic complement of cellular components is transmitted from parent to daughter cells at the time of cell division, the full structures of the daughter cells require additional assembly and growth. This is accomplished as a consequence of the self-assembling properties of the basic molecular components of the cell combined with a supply of external energy and global control exerted by regulated gene expression. Furthermore, if the cellular structure is physically or chemically damaged, the nature of the damage is recognized and repair mechanisms are activated. The capacity to self-repair allows the cell to exist and maintain normal functions for periods of time that exceed the lifetime of individual biomolecular components, such as proteins and lipids. Another unique feature of cellular systems results from the modular nature of expression of gene sets in response to both internal and external stimuli. The genome contains specific subroutines that are run when biochemical switches are activated. This allows for adaptive responses and reprogramming of cellular functions to ensure survival in a chaotic environment. The responsiveness of cellular systems is mediated by a wide spectrum of receptor molecules that detect environmental signals, transduce the message in the context of the current state of the cellular system, and activate appropriate responses. This capability is facilitated by the exquisite sensitivity and selectivity of biomolecular reactions. All of these features combine to allow cells to behave in an autonomous yet collective manner and exhibit emergent behaviors that are not manifested by the individual components. The complex regulation of cellular performance results in a robust and stable system that exhibits the properties of a "living" organism.

27.1.2 Useful Components and Functions of Cells

When exploring the properties of biological systems to identify unique capabilities that can be exploited for human purposes, several additional features of biological systems are attractive. Cells are the fundamental unit of living systems and can be thought of as low energy micro-scale building blocks. Their small size is important in constructing intricate microstructures that are highly adapted for efficient utilization of particular niches in the environment. The low energy requirements of cells needed to support internal biochemical and biophysical reactions are provided by a small number of high-energy biochemicals (e.g., ATP, GTP, NADH, NADPH) that are derived from a diverse range of environmental resources. Another important aspect of biological systems is that all information processing is accomplished through biochemical reactions. Unlike silicon-based computer hardware that requires the input of external software programs to perform useful functions, biological systems are run by wetware—the functional program is a fundamental consequence of the properties of the biomolecular components and their interactive biochemical network. Finally, biological systems are capable of interacting with the nonliving world through communication channels that use common message exchange media—photons, electrons, and chemicals. These features can be manipulated to perform useful functions and form the basis of new technologies.

Through evolution, biological systems have solved a wide variety of engineering problems including nano-/micro-scale detectors, mechanical effectors, logic networks, and efficient energy harnessing systems. What are some of the useful functions biological systems perform? They can detect electromagnetic radiation from low-energy infrared to high-energy ionizing radiation. They can detect and respond to physical factors, such as osmotic pressure, mechanical tension, pH, and heat, as well as biochemical factors such as nutrients and signal molecules. Cells have developed selective membranes with associated proteins to control the flux of chemicals ranging from water and ions to macromolecules and, in fact, continuously transport specific molecules in and out of the cell to control the internal environment. At higher levels of organization, cellular membranes consisting of layers of cells with defined gaps filter large volumes of fluids at the molecular level. Semipermeable membranes and biomolecular transport systems can create ionic potentials and serve as batteries. Biological organisms can collect sunlight and produce useful chemical energy with high efficiency. They use flagellar motors to move themselves or to move fluids across their surface. If they are not free-swimming organisms but are confined to the solid phase, they can still move by systematic rearrangements of their internal scaffolding. Cells can generate photons as luminescence. They can communicate with each other through chemical signals both in the aqueous environment and in the air. They can hibernate or sporulate to protect themselves and survive life-threatening conditions, propagate when conditions are good, and initiate programmed cell death if appropriate. Living organisms have many more amazing capabilities yet to be discovered, all of which may be exploited to solve difficult technical challenges.

27.1.3 Key Features of Cellular Control Systems

As described, living cells are robust autonomous micro-scale agents that possess amazing capabilities to self-organize, self-repair, and evolve new functionalities. From a systems engineering point of view, cells consist of a complex set of nested, nonlinear control systems that, taken together, can ensure survival in the face of large perturbations from "normal" conditions in a dynamic environment. To this end, cells are constantly monitoring their environment with a multitude of sensors and sensor strategies. The inputs from these sensors are integrated into an overall control strategy to survive adverse perturbations and maintain essential functions. Cells are extremely complex, high-performance systems.

The more complex the system is, the more critical the issues concerning control are. The real world is highly chaotic from the perspective of the cell. External perturbations can have catastrophic consequences without adequate control mechanisms. Hence, through evolution, biological systems have explored and exploited novel solution spaces to provide robust biochemical control systems. These biological control networks are dominated by the segregation of information and feedback loops over multiple levels of molecular organization. The stochastic nature of reactions in the intracellular space requires a robust control system that operates in a highly noisy environment. In this situation, the concept of robustness does not mean maintaining the system in a fixed steady state, but rather maintaining a dynamic nonequilibrium system in a condition that is recognizable as a particular phenotype.

Cells must make appropriate decisions to successfully respond to challenges in the environment. There are many examples of specific control systems in biological systems. The metabolic response of *E. coli* to changes in nutrients through the lac operon (Ozbudak et al., 2004) involves changes in the regulation of gene expression that ultimately control the energy metabolism in the cell. The decision to divide and subsequent regulation of the cell cycle to successfully generate daughter cells (Pomerening, Sontag, and Ferrell, 2003) require the synchronized expression of a spectrum of genes in response to internal signals. At a more fundamental level is the interaction between bacteria and viruses (bacteriophages). The interplay between the genome of the bacteria and the bacteriophage determines not only the survival of the cell but the behavior of the virus (see, e.g., the interaction of *E. coli* and the λ-

phage; Arkin, Ross, and McAdams, 1998). In all of these cases, decisions within cells are achieved by biochemical switches.

Research into complex cellular control systems is providing new insights into the molecular mechanisms of biocontrol processes. The study of "minimal cells" provides clues as to which genes and processes are the most fundamental to a viable cell (Hutchison et al., 1999; see also chapters 3 and 16). Another approach to understanding the complexity of biological systems employs perturbation analysis that is accomplished by combining genetic manipulations to perturb the system and techniques to measure global gene expression in response to the perturbation (Ideker et al., 2001; Rao and Arkin, 2001). Using this research approach, researchers discovered that specific changes in the expression of genes in a single pathway have repercussions on gene and protein expression of almost every major biochemical pathway in the cell, emphasizing the interconnectedness of cellular reactions and the importance of distributed control systems to maintenance of the system in its entirety.

27.2 The Cell-Like Entity

27.2.1 Concept of CLEs

The basic concept of the CLE is that of a biological construct engineered on the principles of cellular systems, but not an engineered cell. CLEs will have an artificial genome that contains the blueprints of the construction and functioning of the CLE. It will have synthetic systems to transcribe genes and translate mRNA to produce functional proteins. But it will not in fact be a living organism; that is, once constructed, the CLE cannot reproduce or evolve. By this definition, the CLE falls well into the nonliving world (figure 27.1), no different from many products based on biomolecules. CLEs can be engineered using molecular techniques to integrate a spectrum of biological components and processes that provide signal transduction, signal fusion, and decision-making capabilities based on the nested control strategy

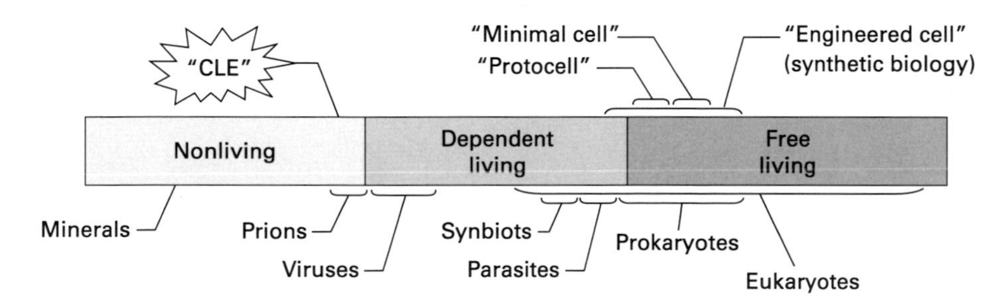

Figure 27.1
The location of bioengineered products on the spectrum of nonliving to living. Synthetic biology and the minimal cell are top-down approaches (reengineering of living organisms). The cell-like entity (CLE) and protocells are bottom-up approaches.

of cells. Custom engineering of extracellular adhesion molecules on the surfaces of CLEs will enable programming of complex network architectures for sensing and control functions. CLEs have the potential to serve as the interface between silicon- and carbon-based technologies. The vision of the fully operational CLE will be a multifunctional, robust, self-organizing bioengineered entity that can be integrated into a wide range of systems and devices and at the same time be affordable, self-repairing, and sustainable.

The strategy to build CLEs from the bottom up was selected to overcome some of the disadvantages associated with using engineered living cells. First, the behavior of cells is highly dependent on the environment in which they live. For a given organism, growth, survival, and functional properties are maintained in a "normal" range only when environmental physical/chemical conditions are controlled within a limited range. Extremeophiles can thrive in conditions that are adverse to most other cells; however, they perform poorly under more normal conditions. Few cells have large dynamic ranges and therefore have limited robustness with respect to device engineering. Second, it is difficult to express foreign genes that are toxic to biochemical machinery within cells. These products of gene expression often trigger unforeseen reactions that are detrimental to the cell and limit the design space for engineering functionality. Third, many engineered cells exhibit genetic instability with respect to the engineered components. Even when the construct is relatively stable, there is a large range of functionality in supposedly identical constructs. This diversity is beneficial to free-living organisms, allowing subsets of the population to use environmental noise, but the efficiency of the system is less than the theoretical maximum from an engineering point of view. Fourth, the design of synthetic circuits is complicated by the requirement that control elements should not cross-talk with normal cellular control processes. This limits the spectrum of possible control elements that can be used in any particular cell type. And finally, there are ethical issues concerning the use of engineered organisms in the general environment. The concern is that such organisms can impact the stability of natural systems and have unintended consequences that are socially unacceptable. The CLE concept does not necessarily solve all of these problems, but if it can be demonstrated that CLEs have an advantage over living organisms in any of these problem areas, then they will have an operational niche.

A basic tenant of this project is that the necessary components needed to construct a CLE for a given function can be found in nature. The concept that one can artificially integrate the molecular scale interactions of cellular components to achieve a defined function is relatively recent (Pohorille and Deamer, 2002). However, the feasibility of such a concept has been demonstrated already by the work of Libchaber and Noireaux (Noireaux and Libchaber, 2004; Noireaux et al., 2005; Yu et al., 2001) and Ishikawa and cowokers (2004). Noireaux and Libchaber constructed an encapsulated bioreactor, an initial step toward developing a CLE. The system consisted of a plasmid containing the enhanced green fluorescence protein (eGFP) gene encapsulated in a unilamellar lecithin vesicle that contains a commercial in vitro transcription translation system. The researchers measured the formation of eGFP in the vesicles as an indicator of functionality. They addressed the requirement of nutrient and energy availability by incorporating the α-hemolysin gene into the plasmid. The protein product from this gene spontaneously inserts in the encapsulating membrane, forming pores to allow influx of substrates. In another study, Ishikawa and colleagues (2004) constructed a genetic network by expressing a transcriptional activation cascade (SP6 RNA polymerase makes T7 RNA polymerase that transcribes GFPmut1-His6) in a cell-free transcription-translation system. This system was encapsulated in liposomes and its functionality demonstrated by observing the increase in fluorescence emission at 545 nm (excitation at 488 nm) from the reaction mixture. Both of these studies demonstrate the feasibility of the CLE concept—to use biological machinery at the molecular level to perform specific tasks—and provide insights into the creation of self-assembling synthetic pathways.

27.2.2 Components of CLEs

The basic components of a fully functional CLE are illustrated in figure 27.2, and the key features of these components are described in table 27.1. The first five components (vesicle, artificial genome, synthetic system, energy converting system, and input system) are essential to the basic CLE platform. The vesicle is the functional

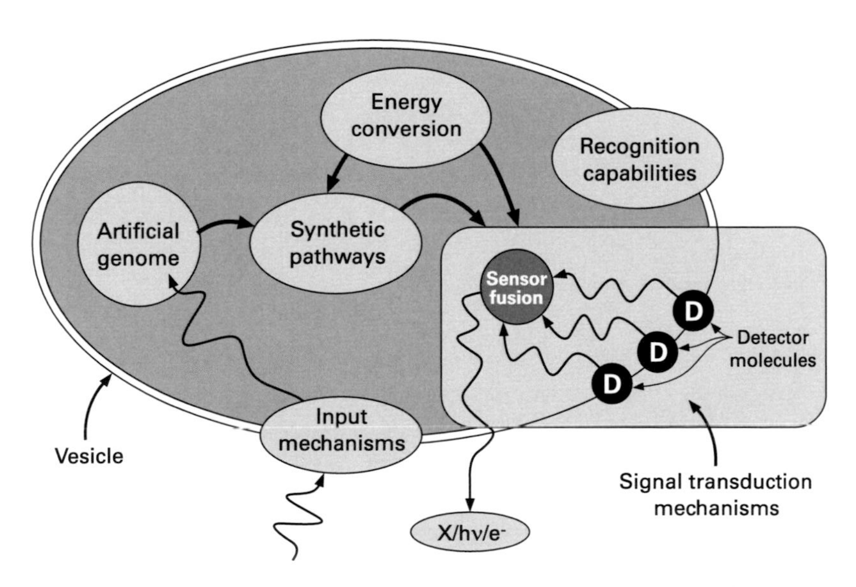

Figure 27.2
Schematic diagram of a CLE illustrating basic components and an integrated functionality (signal detection).

Table 27.1
Basic components of a cell-like entity (CLE)

Component	Description
Artificial genome	Permanently stores information required to construct CLEs and enable them to accomplish specific functions.
Vesicle	The envelope that encapsulates the components of the CLE and regulates transport of molecules between the external and internal environments.
Energy conversion system	Provides required energy in a suitable chemical currency to self-assemble and maintain structure of CLE.
Synthetic systems	Synthetic capability to take external substrates and synthesize molecular components of CLE.
Input system	Capability to activate specific functional cassettes of the artificial genome to program CLE functionality.
Output systems	Spectrum of processes to output information captured by CLE to operator.
Recognition systems	Provide capability to organize CLEs into higher-order structures, either in 2 or 3 dimensions.
Signal transduction system	Augments basic CLE with the capability to detect environmental signals and communicate that information through the output system.

container for the individual CLE. The artificial genome contains the blueprints for the structure and function of the CLE. The information content of the artificial genome is converted by the synthetic system into operational components of the CLE that self-assemble into the functional CLE using energy provided by the energy generation system. The input system allows for reprogramming the CLE from one functionality to another in a way analogous to how cytokines reprogram cells for specific functions. Given the basic CLE platform consisting of these five components, additional capabilities can be introduced to detect environmental signals, organize CLEs into hierarchical structures, and output information collected by the CLE to the device operator. These systems must be integrated to generate a micro-scale entity that will conduct useful work.

27.3 Status of CLE Development

27.3.1 Research Plan

The CLE concept is to use the principles and processes inherent in cellular control systems to engineer a unique entity that possesses the functionalities required for autonomous control of physical devices. It is the understanding of the nature of these molecular processes that will enable the development of novel capabilities. The objectives of the program are to demonstrate feasibility, prototype functionality, and control capability of CLEs. Following are the major efforts of the research program:

1. Component engineering: Identify and exploit existing biochemical/physiological processes that will provide essential components of a CLE (lipid vesicle, artificial genome, energy conversion systems, and synthetic pathway to assemble and repair vesicles).
2. Systems integration: Develop a proto-CLE, a simple organic (liposome-like encapsulated) system based on known biochemistry and physiology of single-celled organisms and demonstrate proto-CLE functionality, for example, signal transduction or synthetic function.
3. Advanced development: Optimize basic components and integrate new functionalities to produce fully functional CLE platform and demonstrate CLE control of electromechanical component or subsystem.
4. Manufacturing and fabrication: Develop microfabrication techniques to manufacture large quantities of functional CLEs at low cost for systems applications.

The necessary research and development of CLEs require a staged effort to accomplish. The initial goal is to develop and demonstrate the proto-CLE, consisting of an encapsulated in vitro transcription-translation system with a defined synthetic function. In essence, the proto-CLE will be a vesicle construct that is on intensive care. All substrates and energy (e.g., ATP and GTP) will be provided externally. This construct will serve as the benchmark for future CLE development. The next step is to evolve the CLE from the proto-CLE to a minimal CLE, which consists of a system that functions with minimal life support, that is, it will generate its own chemical energy from an external carbon source and synthesize and self-assemble the majority of its synthetic and functional components. The third stage will produce the fully operational CLE platform. This bioconstruct will self-assemble all of its biomolecular components, harvest energy from external resources, and self-organize into higher-order architectural structures. Along the way, it is anticipated that various spinoff products will be developed from CLEs at different levels of evolution.

To execute this research and development plan, a series of research efforts must be undertaken. Each effort—encapsulation, artificial genome, energy conversion, synthetic systems, and functional systems—addresses the development of one of the major components of the CLE. The current status of these efforts is summarized in the next section. It should be noted that most of these issues are also critical for the assembly of minimal cells (see, e.g., chapter 16).

27.3.2 Encapsulation—Confinement of Components

Development of an encapsulation system for a CLE presents a technological hurdle, since this encapsulation system must allow substrates, chemical energy, and monomer subunits (i.e., amino acids and ribonucleotide triphosphates) to cross the membrane while maintaining the proper barrier to retain CLE components and exclude undesired foreign components. As a critical step in the development of a CLE, the main focus is on capturing the in vitro cell-free transcription/translation (CFTT) system within liposomes (see, e.g., discussion in chapter 3). Alterations to the encapsulation process or the lipid components of the liposomes provide strategies to increase the efficiency and functionality of encapsulation. After the CFTT is determined to have been captured within the liposome, expression tests will determine whether the synthetic system is still actively transcribing DNA and translating proteins inside liposomes.

A large technical gap exists between incorporating the CFTT system into an artificial membrane system and having a functional CLE. The major challenge is to move necessary substrates or precursor components across the membrane faster than they are consumed in the interior of the liposome. Lipid membranes are permeable to small molecules, including water and ions, to varying degrees (Chakrabarti and Deamer, 1992; Paula et al., 1996). To increase the permeability, multiple lipid species can be used and, because of the slight mismatch in packing between lipid species (i.e., saturated versus mono- or diunsaturated), the permeability of the liposome membrane can be manipulated (Mouritsen and Zuckermann, 1987). As the lipid complexity increases, macroscopic membrane domains will form, the boundaries of which are considerably mismatched (Leidy et al., 2001). Such a system will probably be necessary when large components must cross the membrane or when integral membrane proteins are incorporated into the system. At the outset, two or three lipid component systems may be sufficient to achieve the appropriate flux across the membrane. Another technical hurdle to be addressed is whether the liposomes fuse with each other during the course of CLE operations. Using a low percentage of lipids that have a charged head group could induce the appropriate repulsion between adjacent liposomes. As an alternative, giant unilamellar vesicles (GUVs) can be used. GUVs are on the order of tens of micrometers in diameter, are effectively "flat" because their radius of curvature is relatively large and, thus, have little affinity to fuse (Fisher, Oberholzer, and Luisi, 2000). By using a combination of these techniques, it is possible to optimize the membrane for incorporation and efficient operation of the proto-CLE.

In addition to the uptake of essential "nutrients," the CLE vesicle will have to include physical pathways to dispose of biochemical waste and reaction byproducts, the accumulation of which may limit the efficiency of synthetic reactions and render the CLE nonfunctional. Porins and other active and nonactive transport proteins incorporated as part of the encapsulation could potentially assist in eliminating unneeded chemicals. In addition, similar mechanisms can be used to construct CLEs capable of responding to stimuli by exporting chemical signals into the surrounding milieu and stimulate synergistic responses in other CLEs by quorum sensing circuitry. Pohorille, Schweighofer, and Wilson (2005) have recently discussed the adaptability of membrane channels to diverse functional demands, which can be applicable to the design of CLEs.

CLE encapsulation systems must be tested under a variety of environmental conditions to provide experimental data as to which systems perform best. Under real-world conditions, temperature and humidity are not regulated. Lipid membrane fluidity is affected by temperature, causing it to undergo transitions from liquid crystalline to gel at temperatures around the phase transition temperature (T_m). In the gel state, the membrane permeability is considerably decreased. However, during the phase transition, the membranes are considerably more permeable than either above or below the phase transition temperature (Cruzeiro-Hansson and Mouritsen, 1988; Hays et al., 2001). Furthermore, the temperature range in which the membrane will be stable is dictated by the lipid composition and characteristics of any membrane proteins present (Oberholzer, Albrizio, and Luisi, 1995; Oberholzer et al., 1995). There are extremophiles that live in hot, cold, and dry environments, and understanding how their membrane components are adapted for such environments and mimicking those components in a CLE membrane system would increase the range in which CLEs could be used (Kiran et al., 2004; Macalady et al., 2004). Another environmental stress that CLEs may encounter is dehydration. Under functional conditions, lipid membranes must be hydrated and the associated water molecules help to maintain the fluidity of the membrane. There are techniques that can "trick" the membrane into thinking that it is still hydrated, and the most common of these is to add the disaccharide trehalose to the membrane suspension before it is dehydrated. The hydrogen bonding between the trehalose and the lipid headgroups mimics that between the lipid headgroups and water, so that the trehalose effectively traps the lipids in the fluid phase when dehydrated and maintains its integrity (Crowe, Carpenter, and Crowe, 1998; Crowe, Reid, and Crowe, 1996).

One drawback of lipid-based membranes for encapsulating CLEs is that they are biologically based, and therefore are labile to environmental stresses. Alternative membrane systems, termed polymerosomes, are based on diblock copolymers. Diblock copolymers have hydrophilic moieties on both ends and a hydrophobic region in the center of the molecule. These diblock copolymers spontaneously form vesicles, similar to liposomes, consisting of a monolayer of copolymers with a hydrophilic compartment in the center sandwiched between the polar ends. Diblock copolymer vesicles can be made and tested for their permeability to various ions and CLE components (Discher and Eisenberg, 2002; Taubert, Napoli, and Meier, 2004). Depending on their permeability properties, polymerosomes can be employed as the encapsulation system for CLEs and tested for efficient performance compared to the liposome-based encapsulation system. Polymerosomes have the advantages that they are mechanically stronger than liposomes, are not as biologically labile, and may not be affected as much as lipid-based membranes under extreme temperature conditions.

Optimal performance of the CLEs will depend directly on the establishment of adaptive metabolite transport and transmembrane signal transduction coupled to the desired responses. Although useful applications with semipermeable membranes based on mixtures of lipid components will be achieved with the CLEs, in many cases, these will prove insufficient and more complex asymmetric membranes will be necessary. As the CLE system develops and becomes increasingly more complex, technical advances in membrane systems will continue to be evaluated with the goal of optimizing membrane components. A combinatorial approach can be taken to determine which lipids, polymers, or hybrid combinations are the most efficient for optimal CLE functionality. In addition, it will be necessary to incorporate a spectrum of proteins into the encapsulation membrane to transport a variety of cofactors and substrates across the membrane. It is yet to be determined which membrane transport proteins are appropriate for the efficient loading of the CLEs, and they will have to be tested to determine whether they can be incorporated into various vesicle systems. Recent studies have been concerned with the activity of integral membrane proteins in polymer-based vesicles, and these will be used as a basis to test the use of polymerosomes in our CLE system (Haefele, Kita-Tokarczyk, and Meier, 2006; Nallani et al., 2006; Ranquin et al., 2005).

27.3.3 Artificial Genomes—Genes and Control of Gene Expression

The artificial chromosome serves as the blueprint for self-assembly of the operational CLE. It consists of the set of genes required to provide the basic functions of the CLE as well as the control mechanisms at multiple levels of molecular organization—genome, transcriptome, proteome, and metabolome—to efficiently regulate expression of genes in biomolecular constructs. Specific technical challenges to designing and engineering the artificial chromosome are

- identifying the minimal gene set to provide the CLE with sufficient basal activity to self-assembling given a starter set of substrates and components,
- synthesizing an artificial chromosome that contains the required gene set, and
- developing biomolecular switches to regulate gene expression at multiple levels of biological organization.

Successfully meeting these challenges requires a coordinated research effort.

A circular bacterialike chromosome is the genome of choice for the CLE because of its higher stability. Linear DNA molecules are sensitive to exonuclease degradation and therefore would require additional features for the protection of the DNA ends (Bendich and Drlica, 2000; Ishikawa and Naito, 1999). The CLE genome must include all necessary genes to maintain the CLE and also integrate genes for specific functions depending on the application of the CLE. The "application" genes will encode proteins involved in performing the CLE function. Several attempts have been made to determine the minimal gene set for bacteria by experimental approaches (Akerly et al., 2002; Forsyth et al., 2002; Kolisnychenko et al., 2002; Sassetti, Boyd, and Rubin, 2001) as well as by comparative analysis of prokaryotic genomes (Gil et al., 2004; Islas et al., 2004; Koonin, 2000). By comparing the genomes of two simple prokaryotic organisms, *Haemophilus influenza* (about 1,700 genes) and *Mycoplasma genitalium* (about 470 genes), Mushegian and Koonin (1996) identified a set of 255 genes (Koonin, 2000) that were orthologous between the two species. Because *M. genitalium* and *H. influenzae* belong to two different bacterial lineages, the genes that are conserved in these two bacteria were thought to be essential for the functioning of a modern-type cell even under the most favorable environmental conditions (i.e., abundance of nutrients and lack of competition or stress). This set was hypothesized to constitute a minimal gene set of a living cell. In a more recent study, Gil and coworkers (2004) suggested a minimal gene set consisting of about 206 genes that encode proteins involved in transcription, translation, DNA replication and repair, cell division, protein folding and secretion, protein and RNA processing and degradation, metabolite transport, energy metabolism, and maintenance of the cell membrane. Since CLEs are not intended to divide, the genes for DNA replication and cell division would not need to be included in the CLE genome. The number of genes can be further reduced by using only one codon per amino acid, so that fewer genes encoding aminoacyl-tRNA-synthetases and tRNA would be necessary (Luisi, 2002). Depending on the nutrients provided to the CLE, the basic maintenance genes might vary and individual genes will have to be added or deleted (see discussion of minimal genome size in chapter 3).

Once it has been decided which genes should be part of the CLE genome, the artificial chromosome must be synthesized. In recent years, multiple attempts have been made to not only improve but also reduce time and cost of de novo DNA synthesis. Different error-correcting techniques such as mismatch-binding protein MutS and enzymatic cleavage of mismatches have been developed to diminish the number of errors in synthesized DNA fragments (Carr et al., 2004; Fuhrmann et al., 2005). Smith, Hutchison, Pfannkoch, and Venter (2003) demonstrated the assembly of a whole genome (5,836 bp) from ∮X174 bacteriophage starting from synthetic oligonucleotides. Further synthesis methods for large DNA fragments have been described in the literature (Kodumal et al., 2004; Xiong et al., 2004).

The third challenge is to establish an in vitro gene expression system that can be controlled at the transcriptional, translational, and protein degradation level. A key operational parameter is how rapidly functional gene expression circuits can be shut down after challenging the system with a control signal, and how fast they can restart again when the signal is removed. Our hypothesis is that each individual control mechanism will inhibit the gene expression to a certain degree. All three inhibitory mechanisms combined should be able to exert stringent control of the expression of the proteins of interest and shorten the time delay of the response of the system after exposure to the external signal. Our long-term interest in controllable genetic elements is to incorporate them into the artificial genome to regulate the reading of the blueprints for CLEs.

27.3.4 Energy Conversion Systems—Providing Energy

A critical issue to be resolved is that of how to design these constructs to extract energy from their environment and how to program them for energy conservation. Energy coupling will be required for the functional catalysis of biochemical reactions inside the CLE and to establish communication between its interior and the environment. The processes to be carried out by these microdevices must obey the thermodynamic principles of energy conversion. Two types of mechanisms have been identified to drive the energetics of biological constructs: an internal substrate-level phosphorylation (i.e., ATP synthesis through other high-energy intermediates) and generation of an electrochemical proton potential coupled with ATP synthesis (e.g., respiration and photosynthesis). Extensive literature provides insight into how to tackle this issue (Anthony, 1980; Harold, 1986; Nichols and Ferguson, 1992). In bacteria, for example, the oxidation of an aldehyde by NAD drives the formation of a high-energy acyl phosphate. Acyl phosphates serve as phosphoryl donors for ATP formation. Electron-transport–coupled phosphorylation, on the other hand, functions as a free energy converter. Under aerobic conditions in bacteria, the electron-transport chains consist of a dehydrogenase and a reductase, and this redox process is coupled to ATP synthesis by proton translocation across the membrane. In photophosphorylation, electron-transport and proton translocation are driven by light (Lengeler, Drews, and Schlegel, 1999). Both approaches will need to be explored and adapted to the CLE platform for specific applications.

27.3.5 Synthetic Systems—Generating Components

To convert the information stored in the artificial chromosome into functional elements, proteins and metabolites, a set of synthetic pathways is required for the CLE. The first step is the development of these synthetic systems. This involves the establishment of an in vitro transcription and translation system in which genes contained in the artificial genome can be expressed. Protein biosynthesis outside an intact cell has been studied for several decades (Betton, 2003; Spirin et al., 1988; Swartz, 2001; Voloshin and Swartz, 2005). Development of procedures to separate endogenous mRNA from the ribosome was a key discovery that allowed for diverse applications of in vitro transcription and translation systems (Spirin, 2002). Extracts from *E. coli*, rabbit reticulocyte, and wheat germ are available from commercial vendors that allow protein synthesis of virtually any exogenous mRNA or DNA template (Spirin, 2002). Although these synthetic systems are versatile, many are not energy efficient and, because of the crude nature of the separation from other cellular material, contain many side reactions unrelated to transcription and translation.

Incorporation of well-defined transcription and translation components into the CLE will improve energy efficiency and eliminate extraneous reactions that may interfere with CLE function. One such defined system has been developed by Shimizu and coworkers (2001). The "protein synthesis using recombinant elements," termed the PURE system, includes 32 individually purified components and has been shown to produce functional proteins. The PURE system uses as a template either mRNA or DNA, allowing the genetic flexibility needed for the CLE. The components have been designed with hexa-histidine tags and GST fusion peptides in order to achieve one-step purification of a desired protein, which would allow for simplified quantification and quality control testing of CLE protein products (Shimizu, Kanamori, and Ueda, 2005). Ueda and coworkers have constructed a system without inhibitory factors such as nuclease, proteases, and other enzymes that hydrolyze nucleotides as well as compensatory reactions for energy regeneration and sulfur bond formation. Well-defined control of the essential processes (transcription, translation, and energy regeneration) of the CLE are crucial for its functionality.

The CLE will include other synthetic processes, and a spectrum of several genes that can be expressed by T7 RNA polymerase under the control of the T7 promoter is being investigated. One case study is focused on demonstrating the self-assembly of the glutathione (GSH) synthetic pathway in an in vitro CFTT system. The key goals are to

- construct a functional self-assembling in vitro metabolic pathway;
- obtain quantitative measurements of system behavior to define operational parameters; and
- investigate and optimize the behavior of the in vitro synthetic system.

The glutathione synthetic pathway was selected for this demonstration because GSH plays an important role in modulating the oxidation-reduction status of the cell. GSH is a ubiquitous tripeptide that participates in diverse biological functions, including gene expression, apoptosis, signal transduction, metabolism, and membrane transport (Sies, 1999). GSH is synthesized in bacteria through a two-step enzymatic process involving γ-glutamylcysteine ligase (GCL) and glutathione synthetase (GS) (Kelly, Antholine, and Griffith, 2002). In the nomenclature of *E. coli* genes, the two enzymes correspond to the two genes *gshA* and *gshB*, respectively. Both GCL and

GS depend on ATP and Mg^{2+} for catalytic activity. In bacteria, GS is functionally active as a homotetramer (Yamaguchi et al., 1993). The rate-limiting step of GSH synthesis is controlled by GCL, which is controlled through feedback inhibition by GSH. Plasmids containing *gshA* and *gshB* have been constructed and the functional gene products have been expressed in both *E. coli* and several commercial in vitro transcription/translation systems. The GCL and GS protein products have been constructed with a hexa-histidine tag and S-tag, respectively, in order to simplify detection, quantitation, and purification for subsequent kinetic parameterization. Formation of reaction products, γ-glutamylcysteine and glutathione, and consumption of amino acid substrates over time are measured by a rapid and sensitive high-performance liquid chromatography (HPLC) method for amino acid detection (a modification of the method of Henderson et al., 2001; and Nardi, Cipollaro, and Loguercio, 1990). ATP utilization and ADP formation are also measured as indicators of GCL and GS enzymatic activity. Such quantitative data will support the mathematical modeling of this biomolecular network and help build the biomolecular engineering tools to enhance CLE design and engineering.

Measuring the rates of transcription, translation, and glutathione synthesis in this system allows for the investigation of this synthetic system under defined conditions. Protein levels, transcripts, all amino acids, γ-glutamylcysteine, glutathione, and all nucleotides are quantitatively measured. These time course data provide the needed reaction parameters to quantitatively predict the behavior of this self-assembling synthetic network model. The next phase of the program will include encapsulating the self-assembling GSH synthetic network into liposomes (see section 3.2). Real-time monitoring methods are under investigation to determine protein production (using a GCL-fluorescent protein fusion product) in the enclosed system. Additional synthetic pathways will be developed for the CLE to provide desired functionality.

27.3.6 Functionality

Building a proto-CLE by successfully integrating the major components (first four components in table 27.1) to provide the basic operational capabilities is an enormous challenge. However, such a basic CLE is not very useful on its own since it lacks any input-output systems and cannot perform useful work apart from its own repair. The next objective is to use the proto-CLE as a basis for design and manufacture of a more complex functional CLE capable of unique actions and functions. If such capabilities can be engineered as separate interchangeable genetic modules, they can be easily added to a basic proto-CLE to produce a variety of functions.

To build a CLE module with a specific function, three separate components are usually required. The first component is a sensor/receptor. The ability to sense environmental stimuli is an integral part of almost any biological organism, and living cells use a variety of sensors. Receptors for a variety of stimuli, including metal ions, environmental stress (temperature, osmolarity, UV damage, gamma irradiation, oxidative damage), and organic and inorganic compounds are known (Nivens et al., 2004). These sensors can be engineered into a CLE functional module so that the module gains the ability to detect a corresponding stimulus and respond accordingly. Alternatively, several classes of molecules such as RNA ribozymes and aptamers can be designed to bind to a variety of chemicals and thus serve as sensors (Bayer and Smolke 2005, Breaker 2002). Yet another approach is to use computational tools to redesign an active site of an existing receptor to bind a different molecule (Looger et al., 2003).

The second component of a functional module involves signal transduction. This component serves as a conduit of information flow from the sensor to the output mechanism. In many cases, signal transduction is an integral part of receptor regulation (Bayer and Smolke, 2005), as is the case for the two-component signal transduction systems (Stock, Robinson, and Goudreau, 2000) and bacterial transcriptional regulation. In many cases, a link between sensor and output can be engineered to create different receptors with the same output function. Alternatively, different output functions can be coupled to the same receptor (Bayer and Smolke, 2005).

The third component of the functional module is the output function. As in previous cases, many choices are available for the output function. Examples include reporters that produce a measurable quantitative signal (fluorescent proteins, luciferases, enzymes participating in a colorimetric reaction) and proteins that can participate in biodegradation of a specific compound (bioremediation of organic pollutants, heavy metals, etc.). Not all three components of a functional module must to be present at the same time; one can, for example, design an "always-on" module that will not require any sensing or signal transduction but will have only an output function. For example, a genetic oscillator can periodically produce a chemical and thus serve as a molecular time clock (Elowitz and Leibler, 2000).

Current research is focused on an effort to combine several recent advances in signal transduction and gene regulatory circuits to synthesize a sophisticated biosensor module in a bacterial system with the ultimate goal of incorporating the module into the CLE. The proposed module is designed to sense the presence of bivalent mercury ions (a toxic water contaminant) and translate the signal into an output in the form of a fluorescent protein. Bivalent mercury binds to a Mer repressor protein, which then becomes an activator of the Mer promoter. Transcription from the Mer promoter produces several proteins: yellow fluorescent protein, which serves as a measurable real-time output; mercury reductase, which converts bivalent mercury into the neutral species that can diffuse from the environment; and a T7 RNA polymerase, which activates the expression of the second operon. Once the signal (bivalent mercury ions) is dispersed and no new contamination is introduced, the operon returns to its original repressed state and thus ceases to produce yellow fluorescent protein output. However, T7 RNA polymerase transcribed from the Mer promoter will jump-start a second autoinducible operon consisting of a cyan fluorescent protein and another gene for T7 RNA-polymerase. This operon will continuously produce a persistent, "mnemonic" output to document the past presence of the contaminant. A version of such a module can potentially be used in CLEs as a mercury biosensor and bioremediator.

At the current stage of the program, we have successfully built the first operon that was shown to respond to the addition of mercury chloride, with the magnitude and length of the response (as measured by the total fluorescence of bacterial cultures) dependent on the initial concentration of mercury. Experiments are under way to implement the second operon and to combine both operons into one functional module.

27.3.7 Outstanding Issues

At this stage of development of the CLE technology, several technical challenges remain. Many potential applications of the CLEs will require that they be functional situations other than water-based applications. In these cases, the application will require that gases and fluids be able to interact with CLEs in order to generate the desired responses. These responses will be more efficient and controllable if CLEs are constructed using microfluidic channels on solid or semisolid supports. In addition, microfabrication would allow arraying CLEs of multiple functionalities capable of transducing signals across the microchannels and increase their versatility. Several groups have established protocols for microfabrication of gels and other polymers within microfluidic channels; progress in this field has recently been reviewed (Peterson, 2005). Other examples of sophisticated methods to control microfluidic applications both temporarily and spatially have been developed, as well as a renewed understanding of the ability of microfluidics to use the particular properties of fluids at the micro-scale, where viscosity and surface tension are most important (Atencia and Beebe, 2005). These technologies can be merged with that of the CLEs to harness the full potential of these hybrid systems. It should be noted that the same trend using programmable microfluidics is seen in the development and possible programming of protocells (see chapter 12).

One final issue is the mass production of CLEs. Fabrication techniques will undoubtedly be affected by microfluidic technologies. Recent developments have been made using microfabricated flow cytometers that can detect, analyze, and sort cells or particles, through an integration of microfluidics, optics, and electronics (Huh et al., 2005). These high-speed devices will allow for the analysis and separation of CLEs following encapsulation of the molecular components.

27.4 Applications

The CLE program seeks to create novel technologies for civilian and military applications, including smart biosensors that are able to respond to changes in the environment, tracking devices to locate personnel, and monitoring devices to detect physiological change with built-in response systems to modulate any adverse conditions that may be detected. CLEs, being the biological equivalent of an integrated circuit, can provide multifunctional capabilities that support combined detection and response functionalities or "smart sensors." Proposed applications for this technology include sensing of physical, chemical, and biological events in the environment; monitoring of human health and fatigue status; decision making on the basis of incongruent, incomplete, or corrupted information (fuzzy logic); and control of micro-/nanoelectromechanical devices (MEMS, NEMS).

Even without exploiting the potential of a gene system for sensing or responding, the CLE may serve as a platform for nanoscale motors or similar devices. The potential for producing customized hybrid cell-like platforms with specific molecule adhesion and recognition properties offers new immobilization options that may survive extreme environments better than engineered cells.

27.4.1 Remote Operations Sensing

CLE-based multidimensional sensing (CBMS) exploits cellular signal transduction components, processes, and systems to perform as real-time surveillance and reconnaissance, to decrease information processing time, and to improve integration and decision-making capabilities based on multidimensional information (heat, light, chemical, biological, ionizing radiation, and vibration signatures). Remediation can be integrated with the detection functions, resulting in numerous applications. The utilization of cells in detection systems is not new (DeBusschere and Kovacs, 2001; Gilchrist et al., 2001; Gray et al., 2001; Pancrazio et al., 1998; Park and Shuler, 2003; Rainina et al., 1996; Stenger et al., 2001). However the options that will be available for CLE components provide a much greater range of applications, storage and operational capacity. CBMS sensors could

- monitor activities in an area of interest such as detection and analysis of movement of animals or persons, vehicle emission detection, or environmental impact assessment;
- go where current macro-scaled technology cannot and significantly reduce risk of personnel exposure to toxic or hazardous environments;
- offer stealthy surveillance and reconnaissance potential, since these systems are chemically rather than electronically based, and thus do not radiate telltale signals unless interrogated or activated;

• be embedded into materials with the ability to be formed into a variety of shapes and features that provide natural camouflage for sensor networks;

• assist with rapid decision making based on real-time data gathered from multidimensional sensors;

• integrate bio-based constructs with mechanical devices to act as the central controller for autonomous movement and activity; and

• replace heavy, bulky equipment with ultra-lightweight sensors and systems for space applications.

27.4.2 Medical Applications

CLEs have a wide range of potential applications in medical diagnosis and treatment. Implants containing multifunctional CLEs could be designed to provide continuous monitoring of physiological biomarkers that indicate degradation in health or performance and rapidly modulate the effects with appropriate drugs or antibiotics before clinical symptoms develop. As a survival tool, CLEs can be designed to augment digestive capability, allowing enzymatic digestion of complex biomolecules such as cellulose (converting plant fibers to absorbable sugars).

In principle, CLEs could replace many treatments based on stem cells or gene therapy currently being evaluated. Instead of introducing a pluripotent stem cell and hoping the specific progeny develops in a timely fashion in the right place, the mature function could be designed into CLEs and implanted. Genes for producing insulin and clotting factors have been known for years and have been prime targets for both stem cell and gene therapy. CLEs could combine the production of these substances with the capacity to monitor the physiological status of the patient, making them an obvious target for this application (Sapir et al., 2005; Shi et al., 2005).

CLE implants could provide cytokine-based cell communication in a more cost effective manner than current treatments. Specific cytokine and chemokine cocktails could be engineered for immune system modulation for cancer treatments, tissue regeneration, and more. Incorporated into synthetic skin for burn patients, CLEs could activate on application and accelerate the integration of the material as well as producing adhesive, bacteriostatic, and immune-inducing compounds.

27.4.3 Other Applications

Much discussion has taken place over micro-scale and nano-scale motors. Such devices require integration into larger-scale platforms and present a challenge for control and functional support. The CLE platform has the potential to solve this problem by providing the capability for autonomous control of microelectromechanical devices. The size, durability, low energy requirements, and multifunctionality of CLEs provide advantages over other possible solutions.

CLEs also have some advantages in the area of biocomputing. The rate of advances in biomolecular computing is rapidly increasing. Functions such as inverters and AND gate toggle switches have been developed in *E. coli* from naturally occurring proteins (Feng et al., 2004; Weiss, 2001; Weiss et al., 2003; Yokobayashi et al., 2003). CLEs can be engineered to carry out any logic function that is possible in bacteria with two advantages: The simplicity of the CLE relative to natural bacteria reduces the chance of accidental interference between the genetic circuit and critical cellular functions, and CLEs will be able to survive better in environmental extremes than natural bacteria. In addition, the CLE membranes and input/output functions can be specifically designed to optimize bio-silico communication.

27.5 Conclusion

The concept of the cell-like entity is enticing because it uses the basic principles of living organisms (biomolecular reaction networks that couple genome to function) without actually qualifying as living. Thus, the CLE is on the boundary between nonliving and living. The technical challenges that remain are formidable, but recent research in the published literature has demonstrated the feasibility of the concept. Over the next several years, the big issues—synthesizing large artificial genomes, preparing defined transcription/translation systems, designing and fabricating encapsulating systems, and engineering functional modules—will be addressed and resolved. CLEs will not solve every technological problem, but they will provide unique options for solving a wide range of problems and very likely open up yet to be discovered application avenues.

References

Akerley, B. J., Rubin, E. J., Novick, V. L., Amaya, K., Judson, N., & Mekalanos, J. J. (2002). A genome-scale analysis for identification of genes required for growth or survival of Haemophilus influenzae. *Proceeding of the National Academy of Sciences of the United States of America*, *99*, 966–971.

Anthony, C. (1980). *Bacterial energy transduction*. London: Academic Press.

Arkin, A., Ross, J., & McAdams, H. H. (1998). Stochastic kinetic analysis of developmental pathway bifurcation in phage lambda-infected Escherichia coli cells. *Genetics*, *149*, 1633–1648.

Atencia, J., & Beebe, D. J. (2005). Controlled microfluidic interfaces. *Nature*, *437*, 648–655.

Bayer, T. S., & Smolke, C. D. (2005). Programmable ligand-controlled riboregulators of eukaryotic gene expression. *Nature Biotechnology*, *23*, 337 343.

Bendich, A. J., & Drlica, K. (2000). Prokaryotic and eukaryotic chromosomes: What's the difference? *BioEssays*, *22*, 481–486.

Betton, J. M. (2003). Rapid translation system (RTS): Promising alternative for recombinant protein production. *Current Protein and Peptide Science*, *4*, 73–80.

Breaker, R. R. (2002). Engineered allosteric ribozymes as biosensor components. *Current Opinions in Biotechnology*, *13*, 31–39.

Carr, P. A., Park, J. S., Lee, Y. J., Yu, T., Zhang, S., & Jacobson, J. M. (2004). Protein-mediated error correction for de novo DNA synthesis. *Nucleic Acids Research, 32*, 162.

Chakrabarti, A. C., & Deamer, D. W. (1992). Permeability of lipid bilayers to amino acids and phosphate. *Biochimica et Biophysica Acta, 1111*, 171–177.

Crowe, J. H., Carpenter, J. F., & Crowe, L. M. (1998). The role of vitrification in anhydrobiosis. *Annual Review of Physiology, 60*, 73–103.

Crowe, L. M., Reid, D. S., & Crowe, J. H. (1996). Is trehalose special for preserving dry biomaterials? *Biophysical Journal, 71*, 2087–2093.

Cruzeiro-Hansson, L., & Mouritsen, O. G. (1988). Passive ion permeability of lipid membranes modeled via lipid-domain interfacial area. *Biochimica et Biophysica Acta, 944*, 63–72.

DeBusschere, B. D., & Kovacs, G. T. (2001). Portable cell-based biosensor system using integrated CMOS cell cartridges. *Biosensors and Bioelectronics, 16*, 543–556.

Discher, D. E., & Eisenberg, A. (2002). Polymer vesicles. *Science, 297*, 967–973.

Elowitz, M. B., & Leibler, S. (2000). A synthetic oscillatory network of transcriptional regulators. *Nature, 403*, 335–338.

Feng, X., Hooshangi, S., Chen, D., Li, G., Weiss, R., & Rabitz, H. (2004). Optimizing genetic circuits by global sensitivity analysis. *Biophysical Journal, 87*, 2195–2202.

Fisher, A., Oberholzer, T., & Luisi, P. L. (2000). Giant vesicles as models to study the interactions between membranes and proteins. *Biochimica et Biophysica Acta, 1467*, 177–188.

Forsyth, R. A., Haselbeck, R. J., Ohlsen, K. L., Yamamoto, R. T., Xu, H., Trawick, J. D., et al. (2002). A genome-wide strategy for the identification of essential genes in Staphylococcus aureus. *Molecular Microbiology, 43*, 1387–1400.

Fuhrmann, M., Oertel, W., Berthold, P., & Hegemann, P. (2005). Removal of mismatched bases from synthetic genes by enzymatic mismatch cleavage. *Nucleic Acids Research, 33*, 58.

Gil, R., Silva, F. J., Pereto, J., & Moya, A. (2004). Determination of the core of a minimal bacterial gene set. *Microbiology and Molecular Biology Review, 68*, 518–537.

Gilchrist, K. H., Barker, V. N., Fletcher, L. E., DeBusschere, B. D., Ghanouni, P., Giovangrandi, L., & Kovacs, G. T. A. (2001). General purpose, field-portable cell-based biosensor platform. *Biosensors and Bioelectonics, 16*, 557–564.

Gray, S. A., Kusel, J. K., Shaffer, K. M., Shubin, Y. S., Stenger, D. A., & Pancrazio, J. J. (2001). Design and demonstration of an automated cell-based biosensor. *Biosensors and Bioelectronics, 16*, 535–542.

Haefele, T., Kita-Tokarczyk, K., & Meier, W. (2006). Phase behavior of mixed langmuir monolayers from amphiphilic block copolymers and an antimicrobial peptide. *Langmuir, 22*, 1164–1172.

Harold, F. M. (1986). *The vital force: A study of bioenergetics.* New York: Freeman.

Hays, L. M., Crowe, J. H., Wolkers, W., & Rudenko, S. (2001). Factors affecting leakage of trapped solutes from phospholipid vesicles during thermotropic phase transitions. *Cryobiology, 42*, 88–102.

Henderson, J. W., Ricker, R. D., Bidlingmeyer, B. A., & Woodward, C. (2001). Rapid, accurate, sensitive and reproducible analysis of amino acids. (Publication # 5980-1193EN, pp. 1–10). Palo Alto, CA: Agilent Technologies.

Hutchison, C. A., Peterson, S. N., Gill, S. R., Cline, R. T., White, O., Fraser, C. M., et al. (1999). Global transposon mutagenesis and a minimal Mycoplasma genome. *Science, 286*, 2165–2169.

Huh, D., Gu, W., Kamotani, Y., Grotberg, J. B., & Takayama, S. (2005). Microfluidics for cell cytometric analysis of cells and particles. *Physiological Measurement, 26*, R73–R98.

Ideker, T., Thorsson, V., Ranish, J. A., Christmas, R., Buhler, J., Eng, J. K., et al. (2001). Integrated genomic and proteomic analyses of a systemically perturbed metabolic network. *Science, 292*, 929–934.

Ishikawa, F., & Naito, T. (1999). Why do we have linear chromosomes? A matter of Adam and Eve. *Mutation Research, 434*, 99–107.

Ishikawa, K., Sato, K., Shima, Y., Urabe, I., & Yomo, T. (2004). Expression of a cascading genetic network within liposomes. *FEBS Letters, 576*, 387–390.

Islas, S., Becerra, A., Luisi, P. L., & Lazcano, A. (2004). Comparative genomics and the gene complement of a minimal cell. *Origins of Life and Evolution of the Biosphere, 34*, 243–256.

Kelly, B. S., Antholine, W. E., & Griffith, O. W. (2002). *Escherichia coli* γ-glutamylcysteine synthetase. *Journal of Biological Chemistry, 277*, 50–58.

Kiran, M. D., Prakash, J. S., Annapoorni, S., Dube, S., Kusano, T., Okuyama, H., et al. (2004). Psychrophilic pseudomonas syringae requires trans-monounsaturated fatty acid for growth at higher temperature. *Extremophiles, 8*, 401–410.

Kodumal, S. J., Patel, K. G., Reid, R., Menzella, H. G., Welch, M., & Santi, D. V. (2004). Total synthesis of long DNA sequences: Synthesis of a contiguous 32-kb polyketide synthase gene cluster. *Proceedings of the National Academy of Sciences of the United States of America, 101*, 15573–15578.

Kolisnychenko, V., Plunkett III, G. C., Herring, D., Feher, T., Posfai, J., Blattner, F. R., & Posfai, G. (2002). Engineering a reduced *Escherichia coli* genome. *Genome Research, 12*, 640–647.

Koonin, E. V. (2000). How many genes can make a cell: The minimal-gene-set concept. *Annual Review of Genomics and Human Genetics, 1*, 99–116.

Leidy, C., Wolkers, W. F., Jorgensen, K., Mouritsen, O. G., & Crowe, J. H. (2001). Lateral organization and domain formation in a two-component lipid membrane system. *Biophysical Journal, 80*, 1819–1828.

Lengeler, J. W., Drews, G., & Schlegel, H. G. (Eds.) (1999). *Biology of the prokaryotes.* Stuttgart, New York: Blankwell Science.

Looger, L. L., Dwyer, M. A., Smith, J. J., & Hellinga, H. W. (2003). Computational design of receptor and sensor proteins with novel functions. *Nature, 423*, 185–190.

Luisi, P. L. (2002). Toward the engineering of minimal living cells. *Anatomical Record, 268*, 208–214.

Macalady, J. L., Vestling, M. M., Baumler, D., Boekelheide, N., Kaspar, C. W., & Banfield, J. F. (2004). Tetraether-linked membrane monolayers in Ferroplasma spp: A key to survival in acid. *Extremophiles, 8*, 411–419.

McAdams, H. H., & Arkin, A. (1997). Stochastic mechanisms in gene expression. *Proceedings of the National Academy of Sciences of the United States of America, 94*, 814–819.

Mouritsen, O. G., & Zuckermann, M. J. (1987). Model of interfacial melting. *Physics Review Letters, 58*, 389–392.

Mushegian, A. R., & Koonin, E. V. (1996). A minimal gene set for cellular life derived by comparison of complete bacterial genomes. *Proceeding of the National Academy of Sciences of the United States of America, 93*, 10268–10273.

Nallani, M., Benito, S., Onaca, O., Graff, A., Lindemann, M., Winterhalter, M., et al. (2006). A nanocompartment system (synthosome) designed for biotechnological applications. *Journal of Biotechnology, 123*, 50–59.

Nardi, G., Cipollaro, M., & Loguercio, C. (1990). Assay of gamma-glutamylcysteine synthetase and glutathione synthetase in erythrocytes by high-performance liquid chromatography with fluorimetric detection. *Journal of Chromatography, 530*, 122–128.

Nichols, D. G., & Ferguson, S. J. (1992). *Bioenergetics.* London: Academic Press.

Nivens, D. E., McKnight, T. E., Moser, S. A., Osbourn, S. J., Simpson, M. L., & Sayler, G. S. (2004). Bioluminescent bioreporter integrated circuits: Potentially small, rugged and inexpensive whole-cell biosensors for remote environmental monitoring. *Journal of Applied Microbiology, 96*, 33–46.

Noireaux, V., Bar-Ziv, R., Godefroy, J., Salman, H., & Libchaber, A. (2005). Toward an artificial cell based on gene expression in vesicles. *Physical Biology, 2*, P1 P8.

Noireaux, V., & Libchaber, A. (2004). A vesicle bioreactor as a step toward an artificial cell assembly. *Proceeding of the National Academy of Sciences of the United States of America, 101*, 17669–17674.

Oberholzer, T., Albrizio, M., & Luisi, P. L. (1995). Polymerase chain reaction in liposomes. *Chemical Biology, 2*, 677–682.

Oberholzer, T., Wick, R., Luisi, P. L., & Biebricher, C. (1995). Enzymatic RNA replication in self-reproducing vesicles: An approach to a minimal cell. *Biochemical and Biophysical Research Communications, 207*, 250–257.

Ozbudak, E. M., Thattai, M., Lim, H. N., Shraiman, B. I., & Van Oudenaarden, A. (2004). Multistability in the lactose utilization network of *Escherichia coli. Nature*, *427* (6976), 737–740.

Pancrazio, J. J., Bey, P. P. Jr., Cuttino, D. S., Kusel, J. K., Borkholder, D. A., Shaffer, K. M., et al. (1998). Portable cell-based biosensor system for toxin detection. *Sensors and Actuators B Chemical*, *53*, 179–185.

Park, T. H., & Shuler, M. L. (2003). Integration of cell culture and microfabrication technology. *Biotechnology Progress*, *19*, 243–253.

Paula, S., Volkov, A. G., Van Hoek, A. N., Haines, T. H., & Deamer, D. W. (1996). Permeation of protons, potassium ions, and small polar molecules through phospholipid bilayers as a function of membrane thickness. *Biophysical Journal*, *70*, 339–348.

Peterson, D. S. (2005). Solid supports for micro analytical systems. *Lab on a Chip*, *5*, 132–139.

Pohorille, A., & Deamer, D. (2002). Artificial cells: Prospects for biotechnology. *Trends in Biotechnology*, *20*, 123–128.

Pohorille, A., Schweighofer, K., & Wilson, M. A. (2005). The origin and early evolution of membrane channels. *Astrobiology*, *5*, 1–17.

Pomerening, J. R., Sontag, E. D., & Ferrell, J. E., Jr. (2003). Building a cell cycle oscillator: Hysteresis and bistability in the activation of Cdc2. *Nature Cell Biology*, *5*, 346–351.

Rainina, E., Efremenco, E., Varfolomeyev, S., Simonian, A. L., & Wild, J. R. (1996). The development of a new biosensor based on recombinant *E. coli* for the direct detection of organophosphorus neurotoxins. *Biosensors and Bioelectronics*, *11*, 991–1000.

Ranquin, A., Versees, W., Meier, W., Steyaert, J., & Van Gelder, P. (2005). Therapeutic nanoreactors: Combining chemistry and biology in a novel triblock copolymer drug delivery system. *Nano Letters*, *5*, 2220–2224.

Rao, C. V., & Arkin, A. P. (2001). Control motifs for intracellular regulatory networks. *Annual Review of Biomedical Engineering*, *3*, 391–419.

Sapir, T., Shternhall, K., Mwivar-Levy, I., Blumenfeld, T., Cohen, H., Skutelsky, E., et al. (2005). Cell-replacement therapy for diabetes: Generating functional insulin-producing tissue from adult human liver cells. *Proceedings of the National Academy of Sciences of the United States of America*, *102*, 7964–7969.

Sassetti, C. M., Boyd, D. H., & Rubin, E. J. (2001). Comprehensive identification of conditionally essential genes in mycobacteria. *Proceedings of the National Academy of Sciences of the United States of America*, *98*, 12712–12717.

Shi, Y., Hou, L., Tang, F., Jiang, W., Wang, P., Ding, M., & Deng, H. (2005). Inducing embryonic stem cells to differentiate into pancreatic beta cells by a novel three-step approach with activin A and all-trans retinoic acid. *Stem Cells*, *23*, 656–662.

Shimizu, Y., Inoue, A., Tomari, Y., Suzuki, T., Yokogawa, T., Nishikawa, K., & Ueda, T. (2001). Cell-free translation reconstituted with purified components. *Nature Biotechnology*, *19*, 751–755.

Shimizu, Y., Kanamori, T., & Ueda, T. (2005). Protein synthesis by pure translation systems. *Methods*, *36*, 299–304.

Sies, H. (1999). Glutathione and its role in cellular functions. *Free Radical Biology and Medicine*, *27*, 916–921.

Smith, H. O., Hutchison III, C. A., Pfannkoch, C., & Venter, J. C. (2003). Generating a synthetic genome by whole genome assembly: phiX174 bacteriophage from synthetic oligonucleotides. *Proceeding of the National Academy of Sciences of the United States of America*, *100*, 15440–15445.

Spirin, A. S. (Ed.) (2002). *Cell-free translation systems*. London: Springer-Verlag.

Spirin, A. S., Baranov, V. I., Ryabova, L. A., Ovodov, S. Y., & Alakhov, Y. B. (1988). A continuous cell-free translation system capable of producing polypeptides in high yield. *Science*, *242*, 1162–1164.

Stenger, D. A., Gross, G. W., Keefer, E. W., Shaffer, K. M., Andreadis, J. D., Ma, W., & Pancrazio, J. J. (2001). Detection of physiologically active compounds using cell-based sensors. *Trends in Biotechnology*, *19*, 304–309.

Stock, A. M., Robinson, V. L., & Goudreau, P. N. (2000). Two-component signal transduction. *Annual Review of Biochemistry*, *69*, 183–215.

Swartz, J. R. (2001). Advances in *Escherichia coli* production of therapeutic proteins. *Current Opinions in Biotechnology*, *12*, 195–201.

Taubert, A., Napoli, A., & Meier, W. (2004). Self-assembly of reactive amphiphilic block copolymers as mimetics for biological membranes. *Current Opinions in Chemical Biology*, *8*, 598–603.

Voloshin, A. M., & Swartz, J. R. (2005). Efficient and scalable method for scaling up cell free protein synthesis in batch mode. *Biotechnology and Bioengineering*, *91*, 516–521.

Weiss, R. (2001). *Cellular computation and communications using engineered genetic regulatory networks*. Ph.D. thesis. Massachusetts Institute of Technology.

Weiss, R., Basu, S., Hooshangi, S., Kalmbach, A., Karig, D., Mehreja, R., & Netravali, I. (2003). Genetic circuit building blocks for cellular computation, communication and signal processing. *Natural Computing*, *2*, 47–84.

Xiong, A. S., Yao, Q. H., Peng, R. H., Li, X., Fan, H. Q., Cheng, Z. M., & Li, Y. (2004). A simple, rapid, high-fidelity and cost-effective PCR-based two-step DNA synthesis method for long gene sequences. *Nucleic Acids Research*, *32*, e98.

Yamaguchi, H., Kato, H., Hata, Y., Nishioka, T., Kimura, A., Oda, J., & Katsube, Y. (1993). Three-dimensional structure of the glutathione synthetase from *Escherichia coli* at 2.0 Å resolution. *Journal of Molecular Biology*, *229*, 1083–1100.

Yokobayashi, Y., Collins, C. H., Leadbetter, J. R., Weiss, R., & Arnold, F. H. (2003). Evolutionary design of genetic circuits and cell-cell communications. *Advances in Complex Systems*, *6*, 1–9.

Yu, W., Sato, K., Wakabayashi, M., Nakaishi, T., Ko-Mitamura, E. P., Shima, Y., et al. (2001). Synthesis of functional protein in liposome. *Journal of Bioscience and Bioengineering*, *92*, 590–593.

28 Social and Ethical Issues Concerning Protocells

Mark A. Bedau and Emily C. Parke

28.1 Introduction

Protocells are microscopic compartmented entities that spontaneously assemble from simple organic and inorganic compounds. They are not natural but artificial, coming into existence only through the experimental efforts of human scientists and engineers. Protocells are sometimes referred to as *artificial cells*, but this phrase is also used to designate such things as artificial red blood cells, and these lack the essential properties of the living state. By contrast, protocells embody the essential minimal conditions for genuine life. They grow by harvesting nutrients and energy from their environment and converting them into their component molecules. This metabolic process is modulated by genetic information, typically through the production of catalysts that control metabolism. At a specific point during the growth process the microscopic compartments can undergo division into two or more similar but not identical daughter cells. Because of this variation, populations of protocells are subject to selective pressures and can thereby evolve. These are the fundamental properties of the simplest living organisms, and the term *protocell* emphasizes that they can be viewed as prototypes for the simplest single-celled forms of life. But protocells are simpler than existing bacteria, and furthermore exist only because of intentional human activity. At least this is what protocells *will* be like when they exist, for they do not exist today.

However, as this book illustrates in detail, a number of scientific teams are working to create fully functional protocells, and incremental progress is yielding a stream of increasingly lifelike systems of molecular assemblages (see chapters 2 through 4). For example, experimental achievements include spontaneously reproducing vesicles (Takakura, Toyota, and Sugawara, 2003; see also chapter 4) and vesicles containing polymerase enzymes that replicate encapsulated DNA (Oberholzer, Albrizio, and Luisi, 1995; see also chapters 3 and 4) or RNA (Walde et al., 1994; see also chapters 3 and 4). When the rich supply of complex biological structures extracted from living bacteria is encapsulated in vesicles, it has been possible to transcribe and translate cascading genetic networks (Ishikawa et al., 2004; see also chapters 3, 4, and 7), and expressing ever more complex genetic networks inside vesicles is only a matter of incremental progress. It is not unreasonable to expect that fully functional protocells integrating these kinds of capacities will be created in the next decade or two.

The race to create protocells is already starting to capture public attention. New companies for creating artificial life forms are being founded in the United States and Europe, and an increasing number of multimillion-dollar grants are funding protocell research in the United States, Europe, and Japan. Thus, it is no surprise that this activity is attracting increasing notice in the popular media. The accelerating pace of breakthroughs in protocell science will heighten public interest in their broader implications, and this interest is significantly related to the fact that the creation of protocells raises a number of pressing social and ethical issues. There are ample reasons to believe that protocells could benefit human health and the environment and lead to new economic opportunities, but they could simultaneously create new risks to human health and the environment and transgress cultural and moral norms. As living systems created from nonliving matter, protocells will be unlike any previous technology and their development will lead society into uncharted waters.

This chapter surveys some of the main social and ethical issues that surround the prospect of creating protocells. The survey is not exhaustive and does not aim to resolve all the controversies discussed. More extensive discussion of these issues from a wide variety of perspectives can be found elsewhere (Bedau and Parke, 2009). The aim of this chapter is to convey some of the complexity of the issues and facilitate constructive and responsible reflection and discussion about them, so that stakeholders can become informed and involved in the discussion. Because the creation of protocells promises to profoundly change the world in which we live, we all are likely to be stakeholders in the long run.

28.2 Potential Risks and Benefits

Two main motivations drive protocell research. One is pure science, characterized by curiosity-driven research. The ability to synthesize life in the laboratory is a scientific milestone of immense proportions. Achieving it will mark a profound understanding of the biochemical systems that embody life, and produce fundamental scientific insights. Our ability to create such systems represents a critical test of the extent to which we really understand how they work (this motivation is further discussed in chapter 5).

But protocells also have practical uses. Natural cells are much more complex than anything yet produced by technology, and none of the current productions of artificial intelligence or artificial life would be taken for more than a moment as something genuinely alive (e.g., Brooks, 2001). The next watershed in devising intelligent machines could well depend on first bridging the gap between nonliving and living matter. Perhaps the dreams of self-reproducing robots populating factories on distant planets (e.g., Farmer and Belin, 1992) will be possible only once we have gained the deep understanding of living systems that will come from creating protocells. In these and related ways, making protocells that organize and sustain themselves and adapt to their environment will open the door to creating technologies with the impressive capacities of living systems. The promise of harnessing those capacities for social and economic gain is similar to that of recombinant DNA in the 1970s, which has now produced multibillion-dollar industries. The reason is that protocells are a threshold technology that opens the door to qualitatively new kinds of applications (see chapter 27 and the introduction to part IV).

Protocell research will produce scientific breakthroughs in our fundamental understanding of the chemistry of life, and these breakthroughs can be counted on to revolutionize medical science, as have all similar breakthroughs in the past. Pohorille and Deamer (2002) note numerous pharmacological and medical diagnostic functions that protocells could perform. One application is drug-delivery vehicles that activate a drug in response to an external signal produced by target tissues. Another is microencapsulation of proteins, such as protocells that carry enhanced hemoglobin or special enzymes in the bloodstream. A third application is multifunction biosensors with activity that can be sustained over a long period of time. After reviewing these examples, Pohorille and Deamer (2002, p. 128) conclude that

> Protocells designed for specific applications offer unprecedented opportunities for biotechnology because they allow us to combine the properties of biological systems such as nanoscale efficiency, self-organization and adaptability for therapeutic and diagnostic applications.... [I]t will become possible to construct communities of protocells that can self-organize to perform different tasks and even evolve in response to changes in the environment.

It is easy to expand the list of hypothetical protocell applications. Two possibilities are populations of protocell "factories" that synthesize pharmaceuticals or degrade environmental toxins. We can also imagine protocells designed to defend against bioterrorism, and even a cure for heart disease involving circulating protocells in the bloodstream that remove atherosclerotic plaque. Protocells could also lead to applications as self-repairing materials, such as protocell films that protect against corrosion, or extremely sensitive microscopic chemical sensors, or even nano-scale self-assembling logical units in molecular computers. When conventional engineering produces devices that are highly complex, they tend to be nonadaptive, brittle, and costly to redesign. In contrast, living entities exhibit self-repair, open-ended learning, and flexible adaptability to changing environments. The creation of living technology should dramatically help the future engineering of autonomous, robust, adaptive systems, and protocells will be a striking concrete step down this road.

Notwithstanding all these potential benefits, protocells also raise significant social and ethical concerns. Ethical issues related to creation of artificial forms of life have a long history, dating back at least to the artificial production of urea in 1828, the first manmade organic compound synthesized from inorganic materials. Concerns about nanostructures proliferating in natural environments were expressed in the nanotechnology community more than a decade ago (Merkle, 1992), and a piece by Bill Joy in *Wired* about the combination of nanotechnology with genetic engineering (Joy, 2000) brought similar issues to the attention of a large audience, sparking extensive commentary on the Web. Similar public concerns have surfaced over the research of J. Craig Venter, Hamilton O. Smith, and colleagues aimed at putting a entirely manmade genome inside a free-living microorganism (Hutchinson et al., 1999), which some in the popular press have dubbed "Frankencells." This public outcry prompted Venter to halt research and commission an independent panel of ethicists and religious leaders to review the ethics of synthesizing protocells. When this panel subsequently gave a qualified green light to this line of research (Cho et al., 1999), Venter and Smith announced the resumption of their protocell project (Gillis, 2002). This move attracted quick commentary on editorial pages (e.g., Mooney, 2002). Events like these are increasingly bringing the social and ethical implications of self-reproducing nano-entities to the attention of the general public.

One of the most widespread worries about protocells is their potential threat to human health and the environment. Joy's *Wired* article (2000) worried about molecular machines that could reproduce and evolve uncontrollably. Referring to the dangers of genetic engineering and Eric Drexler's (1986) warnings about the dangers of self-reproducing nanotechnology, Joy concluded that our species, "by its own voluntary actions, has become a danger to itself—as well as to vast numbers of others," and he described one key problem thus:

> "Plants" with "leaves" no more efficient than today's solar cells could out-compete real plants, crowding the biosphere with an inedible foliage. Tough omnivorous "bacteria" could out-compete real bacteria: They could spread like blowing pollen, replicate swiftly, and reduce the biosphere to dust in a matter of days. Dangerous replicators could easily be too tough, small, and rapidly spreading to stop—at least if we make no preparation. We have trouble enough controlling viruses and fruit flies.

To the health and environmental risks of protocells, Joy added the threat of new and vastly more lethal forms of bioterrorism.

Most scientists would probably conclude that Joy's speculations are naive and uninformed, at least with respect to what can be achieved today and in the forseeable future. The widespread sentiment in the protocell research community is that protocells will be extremely fragile and for many years will require massive human experimental support to remain alive. However, in the long run protocells still pose potential dangers, stemming from two key properties. First, since protocells self-replicate, any danger that they pose has the potential to be magnified on a vast scale as they proliferate and spread in the environment. Second, because they evolve, their properties could change in ways that their creators never anticipated. For example, they could evolve new ways of competing with existing life forms, and evading any eradication methods. This potential for open-ended evolution makes the long-term consequences of creating protocells unpredictable.

One can envision strategies for coping with these risks, of course. One is simply to contain protocells in confined areas and not let them escape into the environment. This is a familiar way of addressing dangerous natural pathogens such as the Ebola virus and smallpox. Another method is to take advantage of the fact that they are artificially created and engineer their chemistry in such a way that they do not interact with existing forms of life. Similarly, protocell engineers could build in mechanisms that cripple or control them. For instance, they could be made dependent on a specific form of energy or raw material that is normally unavailable in the environment, so that they would survive only under controlled conditions. Protocells could also be designed to have a strictly limited lifespan, so that they die after performing their intended function. There is evidence that magnetic fields can be used to turn genes on and off (Stikeman, 2002), so it might be possible to engineer protocells to remain alive only on receiving regular external signals or to die when an externally triggered switch is flipped. A further form of crippling would be to block their ability to evolve, perhaps by preventing reproduction (e.g., the *cell-like entities* discussed in chapter 27). Merkle (1992) has also proposed encrypting artificial genomes in such a way that any mutation would render all of the genetic information irretrievable. A further suggestion is to put a unique genetic "bar code" inside each protocell, so that any protocell that creates damage can be traced back to the responsible parties.

No containment method is perfect, of course, and effective containment becomes increasingly expensive as risks are reduced. Another cost is that containment significantly hampers research, thus impeding research on the protocells' potential beneficial uses. Many of these benefits involve protocells inhabiting our environment or even our bodies, and such applications would be impossible if protocells were to be isolated inside strict containment devices. Furthermore, methods for crippling or controlling protocells could well be ineffective. When humans have introduced species into foreign environments, it has often proved difficult to control their subsequent spread. More to the point, viruses and other pathogens are notorious for evolving ways to circumvent our methods of controlling or eradicating them, and protocells would experience significant selection pressure to evade our efforts to cripple or control them. Another social cost of crippling protocells is that doing so would defeat many potential benefits of living technology. The appeal of living technology includes taking advantage of life's robustness and its flexible capacity to adapt to environmental contingencies; crippling life would sacrifice this capacity.

Creating protocells will dramatically alter our world, and like any powerful new technology, the potential risks and benefits are both substantial. In time, protocells could enable many impressive benefits for human health, the environment, and defense, and could dramatically accelerate basic science. But they also could create new risks to human health and the environment and enable new forms of bioterrorism. In addition, creating protocells will fundamentally reshape public perception about life and its mechanistic foundations, thereby undermining some entrenched cultural institutions and belief systems. Given the potential for such significant consequences, it is appropriate to start asking now how society should navigate these issues.

28.3 Social and Ethical Issues

In this section we discuss six clusters of social and ethical issues concerning protocells: (1) questions about what lessons we can learn from our previous experience with related social controversies, including genetic engineering, cloning and stem cells, nanotechnology, and nuclear power; (2) worries about violating the natural order or the sanctity of life; (3) worries about playing God, even for those who do not believe in God; (4) questions about what kind of say in determining our future with protocells should be given to different interested bodies (the general public, protocell scientists, private industry, professional ethicists, nongovernmental organizations, or NGOs, and governmental officials); (5) issues about the proper applicability of the standard methods for making social policy decisions on related issues (risk analysis, cost-benefit analysis, and the precautionary principle); and (6) the light that creating protocells will shed on our self-understanding and sense of our place in the universe. The aim here is not to settle controversies but to raise issues, present multiple perspectives, and see where threads in the discussion lead.

28.3.1 Learning from History

Protocell technology is too new to have yet generated a substantial literature on social and ethical issues, so today's investigations of these issues are, of necessity, pioneering. One strategy for getting traction is to make comparisons with related technologies that *have* received significant public scrutiny. Social values and preconceptions about life have already felt significant pressure from a number of new biotechnologies. So, one can ask how to draw lessons about protocell technology from the track record of those analogous technologies. This kind of argument from analogy can be criticized (Boniolo, 2008), but it can also be defended if it is done with appropriate care.

Genetic engineering, genomics, stem cell research, cloning, biomedical scicnce, and agricultural biotechnology, for example, are increasingly changing the nature of the life forms that exist, and this has prompted people to reconsider our attitudes toward simple life forms and their pragmatic implications. Each of these new technologies has brought in its train a heightened public awareness of and sensitivity to new and potentially risky technologies. These related biotechnologies provide a mirror in which we can glimpse how a future with protocells might appear. For example, synthetic biology today is designing and constructing new biological parts and systems and reconfiguring existing simple life forms to perform specific functions, and the social and ethical implications of synthetic biology are generating significant attention (e.g., ETC Group, 2007; Maurer, Lucas, and Terrell, 2006). This debate could provide useful models for those concerned about the social and ethical implications of protocells.

28.3.2 Violating Nature

The prospect of protocells will unsettle some people for the simple reason that they are unnatural forms of life, and thus seem to violate the sanctity or wisdom of life or upset the balance of nature. It must be admitted that some kinds of protocells will indeed be quite unlike any existing form of life. For example, the minimal protocell described in chapter 6 (also known as the *Los Alamos bug*) is often described as being based on micelles and having no DNA or RNA or proteins, unlike any existing form of life.

However, it is far from clear that unnatural life forms are unequivocally bad. The growth of modern civilization has relied directly on domesticating plants and animals, and these in turn have involved extensive artificial selection. Human hands have altered the genes of virtually all of the vegetables, grains, or animal products available in modern supermarkets today. So those who accept with equanimity the production and incorporation of these genetic "monsters" into daily life but simultaneously worry that protocells would be violations of nature should explain the relevant difference between these two cases (Bedau and Triant, 2009; Johnson, 2009). Of course, some people might object to protocells for the same reason that they object to modern supermarket foods, but that objection is not a worry specifically about protocells.

The example of the protein-free micellar Los Alamos bug illustrates that protocells might well significantly differ from domesticated plants and animals. The key question is to determine whether those differences are ethically or morally relevant. For example, one might worry that protocells would upset the balance of nature in a way that existing forms of life do not, but it is still an open question whether being based on micelles and containing no proteins would have such an effect.

28.3.3 Playing God

Concerns about violating nature are sometimes connected to the worry that creating protocells would be playing God. The idea behind this worry is sometimes that only

God *can* create wholly new forms of life, but that view seems to be patently falsified by the forward march of genetic engineering and synthetic biology. As the persecuted but devout Christian scientist Galileo observed four hundred years ago in his *Letter to the Grand Duchess Christiana*, it would be an unwise theology that let itself be hostage to the contingencies of empirical observations. More serious is the worry that only God *should* create new forms of life. At this juncture, one must deal with the diversity of religious perspectives (Bedau and Triant, 2009) and diverse interpretations of each. Opinions based merely on individual articles of faith would be unpersuasive to those who do not share that faith, which is why religious dogma is generally considered to be an unstable foundation for social policy in modern liberal democracies with heterogeneous subcultures.

Humans undeniably possess the capacity to create new forms of life, and this capacity is developing and increasing with advances in synthetic biology and protocell science. Thus, the issue of exercising this capacity wisely becomes increasingly urgent, because life's autonomous regeneration, repair, and adaptation and its capacity for open-ended evolution make it both especially powerful and especially difficult to predict and control. At this juncture, a pragmatic and secular form of the playing God worry arises. In a nutshell, the concern is whether humans have the understanding and wisdom necessary to properly and judiciously exercise the capacity to create new forms of life. Exercising this capacity might well be appropriate for an entity with godlike discernment of the consequences of such actions, godlike wisdom about the proper ends for which to exercise this capacity, and godlike power to correct any unanticipated problems that might arise. The worry is that it is hubris to think that mere mortals possess such godlike capacities. This is an extreme form of the challenge that those in a position of power and responsibility always face. It is a matter of ongoing debate whether and, if so, how best to apportion and structure the responsibility of deciding what role, if any, protocells should have in society's future.

28.3.4 Who Should Have a Voice

A fourth issue arises at this juncture. If there is to be a debate about the proper place of protocells in scientific research and, eventually, commercial development, who should have a voice in this discussion and what role should these different parties have in the decision-making process? Candidates for participating in this process include the scientists engaged in protocell research, entrepreneurs and commercial interests attracted to the commercial potentials of protocells, professional ethicists, governmental regulatory agencies, international NGOs such as Greenpeace, and the general public themselves. Various issues arise about each of these candidates.

Scientists have often viewed their work as ethically neutral. They often presume a division of labor in which the scientists' task is to unlock the secrets of nature, leaving it to policymakers to decide how the new knowledge should be deployed. However, some now argue that this traditional separation of scientific and ethical realms is no longer sensible, and that scientists themselves must grapple with the ethical complexities and implications of their research (Khushf, 2009).

Some protocell scientists anticipate with dread the establishment of uninformed regulatory burdens on their research. Social policy should be deeply informed by this scientific expertise because the scientists themselves possess the most up to date and accurate understanding of the actual and likely future capabilities and risks of protocell research. But letting protocell scientists simply police themselves would saddle them with a conflict of interest that might prove untenable to the scientists and unacceptable to other concerned parties.

Protocell entrepreneurs and commercial interests have a somewhat similar informed but conflicted position, and it is further compromised by the profit motive. Contemporary society refuses to accept commercial development of a very few extremely dangerous technologies, even though there would be a market for them; nuclear and biological weapons are obvious examples. When a technology is sufficiently powerful, society restricts what private enterprise can do. So when protocells exist and start to acquire significant powers, the appropriate governmental regulatory agencies might need to exercise oversight on protocell commerce and perhaps even academic protocell research.

It is uncontroversial that the general public deserves a say in how protocells are to become part of our world, but their voices are usually expressed indirectly, either through elected officials and their agents (e.g., regulatory agencies), or through voluntary participation in NGOs. A number of NGOs have been active in previous controversies concerning new biotechnologies, including genetic engineering, cloning, stem cells, and synthetic biology. There is every reason to expect this for protocells, too, once they become a reality. However, it is important to ensure that the public's actual voice is heard, rather than just the voices of interest groups who purport to represent the public but might actually be responding more to their own private agendas (Durodié, 2009).

Society could consider supplanting narrow interest groups like NGOs with alternative social institutions. For example, in response to the historically unprecedented capacity for humans to destroy global ecosystems, Arne Naess (1989), the inventor of the phrase *deep ecology*, proposed creating a new social institution in which scientists would regularly explain their research projects, plans, and possibilities to diverse collections of independent amateurs who are intelligent and engaged with social issues. After discussing issues and expressing concerns, the scientists and amateurs together would make recommendations about social policies concerning the scientific activities.

Another possible response to the difficulty of navigating an appropriate future with protocells would be to engage the assistance of professional ethicists. A precedent for this is the small fraction of the human genome project budget that was earmarked for studies of the ethical, legal, and social implications of sequencing the human genome, and a number of professionally staffed institutes and centers for social and ethical implications of nano- and biotechnologies now exist. But ethical advice offers no simple resolution of ethical dilemmas, nor can it absolve scientists from the need to make ethically difficult decisions. It is important to realize that enlisting the help of professional ethicists will not eliminate the ethical problems raised by protocells, and might even magnify them (Gjerris, 2009; Häyry et al., 2006).

28.3.5 Assessing Risks and Exercising Caution

Contemporary advanced societies typically weigh complex and difficult questions of public policy with the aid of risk analysis (Ropeik and Gray, 2002; Wilson and Crouch, 2001). So an understanding of the risks and benefits presented by creating new forms of life is one key to addressing and evaluating concerns about protocells. With the potential for great benefit comes the potential for great misuse and harm, so balancing these considerations is not easy. In addition, because fully functional protocells have not yet been created, we can only speculate about their actual risks. There is also the related but different and complex matter of how society perceives risk (Slovic, 2000). We might be able to learn something about the perceived acceptability of protocell technology and the proper way to manage the accompanying risks from our history with the chemical and genomics industries (Cranor, 2009).

Any analysis and evaluation of the risks and benefits of protocells is especially complex, and those who must grapple with these issues are in an especially difficult position. Because life is so unpredictable, the complex of good and bad consequences of creating new forms of life is virtually impossible to gauge. Thus, the traditional methods of risk analysis offer scant help for deciding how to act when we are in the dark about the ultimate consequences of our decisions (Bedau and Triant, 2009). Analyzing risks and benefits while lacking firm ground on which to base these analyses is a genuine challenge. Making wise and responsible decisions in the dark will require the development of a new paradigm of risk assessment.

There is a growing sentiment today that society's decision-making tools need to be augmented with the precautionary principle (Raffensberger and Tickner, 1999; Tickner, 2003), which states that precaution should be an overriding concern when weighing the adoption of new technologies and practices. Partly because of its growing popularity, the precautionary principle has become a lightning rod for criticism (Morris, 2000; Parke and Bedau, 2009). Before using the precautionary principle as a framework for evaluating critical policy issues involving new technologies, one should review the various arguments for and against the principle, consider its different formulations, and determine their plausibility and applicability to issues specifically about protocells. Some have argued that following the precautionary principle would be too risky, because society cannot afford to forego all the potential benefits of new technologies like protocells (Bedau and Triant, 2009). It would be instructive to examine recent history and tabulate in which cases precautionary and nonprecautionary decision making have actually led to good and bad decisions.

28.3.6 Our Place in the Universe

One final issue is worth mentioning briefly. Success at creating new and potentially very unusual forms of life from nonliving materials will resolve one of the remaining fundamental mysteries about creation and our place in the universe. The prospect of discovering extraterrestrial life has both fascinated and repelled those people who take the possibility seriously, in part because of the unpredictable implications it could have for our understanding of ourselves and the world we live in. The practice of creating alien forms of life from scratch could well provoke a similar sense of wonder. It could also lead to feelings of dismay and hostility, if its implications violate entrenched cultural preconceptions. Not everyone would appreciate or embrace a brave new world revealed by the scientific achievement of protocells. Although people might disagree about how to negotiate such cultural sensitivities, presumably nobody would suggest that scientists should hide or obscure what they learn about nature. However, some people might council that society should be consulted before it is burdened with the need to refashion its sense of itself. Others will embrace the opportunity to learn something fundamental about what we are and why we exist.

28.4 Conclusion

Scientists are rapidly advancing in the quest to make increasingly lifelike systems, and an especially striking example will be fully functioning protocells. The idea of protocells soon becoming a reality generates unprecedented social and ethical issues. The issues are complex, and they raise many open questions about how best to proceed. This chapter has discussed some of these issues in a brief and preliminary fashion. The aim has been to identify rather than answer questions, with the ultimate goal of raising awareness of these issues and initiating responsible and informed discussion of them.

References

Bedau, M. A., & Parke, E. C. (Eds.) (2009). *The prospect of protocells: Moral and social implications of creating life in the laboratory*. Cambridge, MA: MIT Press.

Bedau, M. A., & Triant, M. (2009). Social and ethical implications of creating artificial cells. In M. A. Bedau & E. C. Parke (Eds.), *The prospect of protocells: Social and ethical implications of creating life from scratch*. Cambridge, MA: MIT Press.

Boniolo, G. (2009). Methodological considerations about the ethical and social implications of protocells. In M. A. Bedau & E. C. Parke (Eds.), *The prospect of protocells: Social and ethical implications of creating life from scratch*. Cambridge, MA: MIT Press.

Brooks, R. (2001). The relationship between matter and life. *Nature*, *409*, 409–411.

Cho, M. K., Magnus, D., Caplan, A. L., McGee, D., & The Ethics of Genomics Group (1999). Ethical considerations in synthesizing a minimal genome. *Science*, *286*, 2087.

Cranor, C. (2009). The acceptability of the risks of protocells. In M. A. Bedau & E. C. Parke (Eds.), *The prospect of protocells: Social and ethical implications of creating life from scratch*. Cambridge, MA: MIT Press.

Drexler, K. E. (1986). *Engines of creation: The coming era of nanotechnology*. New York: Doubleday.

Durodié, B. (2009). Ethical dialogue about science in the context of a culture of precaution. In M. A. Bedau & E. C. Parke (Eds.), *The prospect of protocells: Social and ethical implications of creating life from scratch*. Cambridge, MA: MIT Press.

ETC Group. (2007). Extreme genetic engineering: An introduction to synthetic biology. Available at: https://idl-bnc.idrc.ca/dspace/bitstream/123456789/129/1/82813.pdf (accessed May 2007).

Farmer, D., & Belin, A. (1992). Artificial life: The coming evolution. In C. G. Langton, C. Taylor, J. D. Farmer, & S. Rasmussen (Eds.), *Artificial life II* (pp. 815–840). Redwood City, CA: Addison-Wesley.

Gillis, J. (2002). Scientists planning to make new form of life. *Washington Post*, November 21, A01.

Gjerris, M. (2009). This is not a hammer: On ethics and technology. In M. A. Bedau & E. C. Parke (Eds.), *The prospect of protocells: Social and ethical implications of creating life from scratch*. Cambridge, MA: MIT Press.

Häyry, M., Takala, J., Jallinoja, P., Lötjönen, S., & Takala, T. (2006). Ethicalization in bioscience: A pilot study in Finland. *Cambridge Quarterly of Healthcare Ethics*, *15*, 282–284.

Hutchinson III, C. A., Peterson, S. N., Gill, S. R., Cline, R. T., White, O., Fraser, C. M., et al. (1999). Global transposon mutagenesis and a minimal Mycoplasma genome. *Science*, *286*, 2165–2169.

Ishikawa, K., Sato, K., Shima, Y., Urabe, I., & Yomo, T. (2004). Expression of a cascading genetic network within liposomes. *FEBS Letters*, *576*, 387–390.

Johnson, B. (2009). New technologies, public perceptions and ethics. In M. A. Bedau & E. C. Parke (Eds.), *The prospect of protocells: Social and ethical implications of creating life from scratch*. Cambridge, MA: MIT Press.

Joy, B. (2000). Why the future does not need us. *Wired* 8 (April). Available at: http://www.wired.com/wired/archive/8.04/joy.html (accessed May 2007).

Khushf, G. (2009). Open evolution and human agency: The pragmatics of upstream ethics in the design of artificial life. In M. A. Bedau & E. C. Parke (Eds.), *The prospect of protocells: Social and ethical implications of creating life from scratch*. Cambridge, MA: MIT Press.

Maurer, S. M., Lucas, K. V., & Terrell, S. (2006). From understanding to action: Community-based options for improving safety and security in synthetic biology, draft 1.1. Richard & Rhonda Goldman School of Public Policy, University of California, Berkeley. Available at: http://gspp.berkeley.edu/iths/Maurer%20et%20al._April%203.pdf (accessed May 2007).

Merkle, R. (1992). The risks of nanotechnology. In B. Crandall & J. Lewis (Eds.), *Nanotechnology research and perspectives* (pp. 287–294). Cambridge, MA: MIT Press.

Mooney, C. (2002). Nothing wrong with a little Frankenstein. *Washington Post*, December 1, B01.

Morris, J. (Ed.) (2000). *Rethinking risk and the precautionary principle*. Oxford: Butterworth-Heinemann.

Naess, A. (1989). *Ecology, community and lifestyle: Outline of an ecosophy*. Translated by D. Rothenberg. Cambridge: Cambridge University Press.

Oberholzer, T. M., Albrizio, M., & Luisi, P. L. (1995). Polymerase chain reaction in liposomes. *Current Biology*, *2*, 677–682.

Parke, E. C., & Bedau, M. A. (2009). The precautionary principle and its critics. In M. A. Bedau & E. C. Parke (Eds.), *The prospect of protocells: Social and ethical implications of creating life from scratch*. Cambridge, MA: MIT Press.

Pohorille, A., & Deamer, D. (2002). Protocells: Prospects for biotechnology. *Trends in Biotechnology*, *20*, 123–128.

Raffensperger, C., & Tickner, J. (Eds.) (1999). *Protecting public health and the environment: Implementing the Precautionary Principle*. Washington, DC: Island Press.

Ropeik, D., & Gray, G. (2002). *Risk: A practical guide to deciding what's really safe and what's really dangerous in the world around you*. Boston: Houghton Mifflin.

Slovic, P. (2000). *The perception of risk*. London: Earthscan Publications.

Stikeman, A. (2002). Nanobiotech makes the diagnosis. *Technology Review*, May, 60–66.

Takakura, K., Toyota, T., & Sugawara, T. (2003). A novel system of self-reproducing giant vesicles. *Journal of the American Chemical Society*, *125*, 8134–8140.

Tickner, J. (Ed.) (2003). *Precaution, environmental science, and preventive public policy*. Washington, DC: Island Press.

Walde, P., Goto, A., Monnard, P.-A., Wessicken, M., & Luisi, P. L. (1994). Oparin's reactions revisited: Enzymatic synthesis of poly(adenylic acid) in micelles and self-reproducing vesicles. *Journal of the American Chemical Society*, *116*, 7541–7547.

Wilson, R., & Crouch, E. A. C. (2001). *Risk-benefit analysis*. Cambridge, MA: Harvard University Press.

Glossary

Activated nucleotide A standard nucleotide monomer with modifications of the phosphate group that allow the nucleotide to polymerize during nonenzymatic RNA replication.

Acyl A chemical compound of the form RCOO- where R is an organic group.

Alanine A nonpolar amino acid, one of the 20 biological amino acids that compose proteins.

Amino acid The basic molecular component of proteins, which are polymers of amino acids linked through peptide bonds.

Amphipathic peptide A peptide that folds into secondary structures in which polar (hydrophilic) and nonpolar (hydrophobic) side chains are spatially segregated. Such peptides are amphiphiles that can interact with lipid bilayer membranes.

Amphiphile An organic compound with both hydrophilic and hydrophobic groups. The groups are separated within the molecule to produce hydrophilic and hydrophobic poles, commonly referred to as heads and tails, respectively. All lipids are amphiphiles, and some lipids, such as phospholipids and cholesterol, can self-assemble into lipid bilayers. Amphiphiles compose the permeability barrier of biological and artificial membranes.

Anion A negatively charged ion such as chloride (Cl-), phosphate, or carboxyl groups (-COO-). Macromolecules such as DNA are negatively charged because of their content of anionic monomers.

Anisotropy Something is anisotropic if its properties or characteristics vary depending on the direction of measurement or observation.

Antisense RNA A single-stranded RNA that is complementary to a messenger RNA strand transcribed in a cell.

Apoptosis Programmed death of cells. Apoptosis is a natural process that is essential to the formation of morphological structures during embryogenesis and in the development of the nervous system.

Aptamer A DNA, RNA, or peptide molecule that has been selected from a large pool of molecules for its ability to bind to a specific molecular target.

Astrobiology The study of the origin, distribution, and evolution of life in the universe.

Attractor A subset of a dynamical system's state space, the set of points that orbits approach asymptotically (in the limit of long times). The set of points whose orbits approach the attractor is called the attractor's *basin of attraction*. Not all dynamical systems possess attractors; only those that have dynamical rules that shrink state space volumes, at least in some parts of the state space, can have attractors.

Autocatalysis A form of catalysis in which one of the products of a reaction serves as a catalyst for the reaction.

Autocatalytic reaction network A collection of reactants and products, each of which is able to catalyze the next reaction. An autocatalytic cycle is an autocatalytic reaction network that goes in a loop, thereby catalyzing its own reproduction.

Autonomous living system A cohesive, self-producing, self-maintaining system that regulates its own behavior so as to sustain its defining organization. In the case of chemotons, the defining relations are the chemical reactions that must be regulated so that the stoichiometric coupling of subsystems is maintained. They do not depend directly on other living systems in the informational sense. Their chemical environment can be rich, but does not need to contain informational macromolecules (nucleic acids or proteins).

Autopoiesis Literally, self-creation; a term invented by Varela and Maturana. The classic example of an autopoietic system is a living cell, which is able to continually regenerate and produce its own components using energy and nutrients derived from its environment.

Autotroph An organism that synthesizes its own organic substances from inorganic compounds, that is, carbon dioxide. Plants and some bacteria are autotrophs.

Bacteriophage A virus that infects a bacterial cell.

Bilayer lamellar membrane The canonical biological membrane consists of two layers (leaflets) of amphiphiles, with proteins either inserted into and through the bilayer or attached to the surface. Artificial bilayer membranes, called liposomes, lack protein components and are used as experimental models of cellular compartments.

Binary fission A form of asexual reproduction that results in a living cell dividing into two daughter cells that live independently.

Biogenesis The process by which living organisms reproduce themselves, in contrast to the origin of life arising from nonliving components.

Biosynthesis The process by which complex biochemical compounds are produced from simpler reactants (nutrients and metabolites) within a living cell. Biosynthesis is an end result of metabolism.

Bottom-up approach An approach to creating protocells that involves fabrication of cellular compartments by self-assembly, using basic materials to build simple models of living system.

Brownian motion The random motion of minute particles in fluid. Mathematical models that describe this random motion precisely were first developed by Albert Einstein.

Calvin cycle A series of biochemical reactions that takes places in the stroma of the chloroplasts of photosynthetic organisms. The cycle is a metabolic pathway in which carbon enters as CO_2 and exits as reduced organic compounds that can be used by the organism either as a source of chemical energy or as metabolites and building blocks for more complex structures. One such compound is glucose, which is the basic monomer of polymeric cellulose and starch.

Capsid The shell of a virus particle, surrounding the nucleic acid or nucleoprotein core.

Catalysis The process by which the rate of a chemical reaction is increased. Catalysts such as enzymes are substances that accelerate the rate of a given reaction by decreasing its activation energy but are not changed or consumed by the reaction.

Cation A positively charged ion such as sodium (Na^+) and the amine groups of amino acids ($-NH_3^+$). Proteins such as histones are positively charged because of their content of the amino acids lysine and arginine, which bear positively charged amine groups.

Cellular automaton (CA) A model studied in mathematics and theoretical biology, consisting of a collection of cells, or squares, on a grid. Each cell is in one of a finite number of states, and evolves through a number of discrete time steps according to a set of rules based on the states of its neighboring cells.

Cenancestor The last common ancestor of all life (also called the last universal common ancestor [LUCA]).

Chemical supersystem A system in which qualitatively different chemical systems are stoichiometrically or kinetically coupled (or both). Such systems can have emergent properties.

Chemoton A chemical supersystem invented by Tibor Gánti that is also a biological minimal system. It consists of a metabolic network, template replication and a boundary system, each of which are autocatalytic. A chemoton (short for "chemical automaton") in which one or more of the subsystems is catalytic but not autocatalytic is called a protochemoton.

***Cis* reaction** A molecular reaction in which the catalytic molecule modifies itself.

Citric acid cycle One of the three stages of cellular respiration, also known as the Krebs cycle or tricarboxylic acid cycle. The citric acid cycle is the metabolic pathway in which carbohydrates and lipids (fatty acids) serve as a source of reduced compounds that, in turn, provide electrons for mitochondrial or bacterial electron transport. The energy generated as the electrons ultimately react with oxygen is used to produce ATP, the primary energy currency of all cellular life.

CLE Cell-like entities (CLEs) are biological constructs engineered to accomplish specific tasks. CLEs are engineered on the principles of cellular systems, but are not engineered artificial cells.

Combinatorial chemistry A technology for rapidly testing large numbers of different molecules for desired properties. Applied primarily in the pharmaceutical and materials discovery industries, combinatorial chemistry allows researchers to create a virtual library of possible compounds and select subsets of the library for large-scale laboratory synthesis.

Combinatorial explosion A phenomenon that occurs when a very large number of possible combinations is created as the number of building blocks increases exponentially. An example is the hypothetical "random chemistry," in which every compound in the original mixture can react with every other compound.

Compartment world A hypothesis proposing that compartments (i.e., cell membranes) were the first biologically relevant structure to appear on the early Earth, which later incorporated catalysts and genetic material to become the first living cells.

Compositional information Information encoded in a system of associated molecules, in which information is specified by the composition of the system rather than the sequence of monomers in polymeric components.

Composome A dynamic state of a molecular assembly, spanning several cycles of growth and division, in which the composition of the assembly is constant.

Connection degree The number of connections (or edges) for each node in a network.

Conway's Game of Life A two-dimensional cellular automaton rule, a dynamical system whose state space is the set of configurations of 0s and 1s over a two-dimensional square lattice. One state goes to another with application of a local rule applied synchronously to every site, based on the values of the site and its eight nearest neighbors: A 1 changes to 0 if fewer than two neighbors have the value 1 or more than three neighbors have the value 1, otherwise it remains 1; a 0 changes to 1 if it has exactly three neighbors with value 1, otherwise it remains 0.

Covalent bond A chemical bond formed by sharing pairs of electrons between two components of a given molecule.

Critical micelle concentration The concentration at which a solution of amphiphilic molecules undergoes a transition from individual molecules in solution to aggregates of amphiphiles, referred to as micelles.

Cycle stoichiometry A special branch of chemical stoichiometry designed to handle cyclic (catalytic) processes. At its center is the cyclic process sign and the associated operational rules.

Cytokine A signaling compound secreted by the immune system used for intercellular communication (similar to a neurotransmitter).

Dinucleotide A single piece of genetic material (DNA or RNA) that is two nucleotides long.

Directed assembly This term is used here to describe processes like transcription and translation, in which the sequence of nucleotides in a nucleic acid directs the sequence of amino acids in a protein that is synthesized using that information.

Directed evolution Also known as *in vitro evolution*, a method used in protein engineering to evolve proteins or RNA with desirable properties not found in nature.

Dissipative particle dynamics (DPD) A computational method for modeling chemical reaction systems that uses second-order equations of motion, with explicit conservation of momentum (in contrast to Langevin or Brownian dynamics).

Dynamic light scattering (DLS) A technique that correlates the patterns of scattered light to estimate the average size of particles in a sample.

Dynamical system A mathematical construction consisting of a *state space*, each point of which typically represents the state of the system under study, and a *dynamical rule* that specifies how the system progresses from one state to another with the passage of time. Beginning with a starting state, or *initial condition*, the set of points generated by repeated application of the dynamical rule form an *orbit*. There is a wide variety of types of dynamical systems—for example, iterated maps, ordinary differential equations, certain types of partial differential equations—and they are used to model a wide range of phenomena, from physical to chemical to ecological systems to economic systems.

Encapsulation The process by which lipid vesicles capture solutes within their internal volume.

Endosymbiont An organism that lives within the cells or body of another organism.

Ergodicity The idea that, as a system explores its microscopic configurations, it repeatedly visits configurations and transitions between configurations that are representative, in the sense that measurements correspond to averages over the representative configurations and transitions.

Ester An organic compound in which an acid and an alcohol are covalently linked when a water molecule is removed, a chemical reaction called condensation. Important examples of ester bonds include those that combine fatty acids and glycerol into fats, and the phosphodiester bonds that link nucleotide monomers into long chains of nucleic acids.

Eukaryote An organism with cells containing membrane-bound nuclei. Animals, plants, and fungi are eukaryotes, in contrast to prokaryotes such as bacteria, whose cells lack nuclei.

Exonuclease An enzyme that catalyzes the hydrolysis of single nucleotides from the end of a DNA or RNA molecule.

Extremophile An organism that thrives in extreme environments characterized by high temperatures, high salinity, and very acidic or basic pH ranges. Typically unicellular bacteria, extremophiles are of particular interest in the field of astrobiology because such organisms could potentially survive in environments like those of other planets.

Fatty acid An amphiphile consisting of a carboxylic acid (hydrophilic head) with a long hydrocarbon tail (hydrophobic group).

Flow field-flow fractionation (FFF) A technique for separating and characterizing particles, which relies on laminar flow to fractionate a population of particles based on size.

Fluid mosaic All biological membranes exist in a fluid rather than gel state, with embedded proteins forming a kind of mosaic within the lipid bilayer. The fluid mosaic structure is the conventional model of biological membranes.

Fluorescence resonance energy transfer (FRET) A technique used to quantify the distance between two fluorescent dyes. This technique can be used to estimate the surface density of dyes embedded in a vesicle membrane.

Fokker-Plank equation An equation that describes the time evolution of the probability density function of position and velocity of a particle.

Fusogenic Capable of producing fusion between membranous compartments. For instance, a fusogenic liposome has a lipid composition that promotes fusion of the liposome with other liposomes or with a living cell.

GARD (graded autocatalysis replication domain) model A conjectural model for prebiotic evolution based on molecular assemblies capable of capturing energy and "nutrients" from the environment to undergo a kind of growth and division. Such systems can pass compositional information between generations.

Genetic takeover A concept first promoted by Graham Cairns-Smith, in which the composition of an inorganic structure such as a clay surface is used to guide the sequential order of monomers in an organic polymer. This order may happen to be useful as genetic information with biological relevance. Thus, the heritable information of the clay surface is "taken over" by the genetic order required for life.

Genome reduction Process of gene content diminution resulting from a relaxed selection on unnecessary genes under certain conditions, such as the transit from a free-living lifestyle in bacteria to parasitism or permanent endosymbiosis within an eukaryotic host that provides a chemically rich environment.

GNA An artificial simplified nucleic acid containing glycerol as a linking monomer, rather than ribose or deoxyribose of RNA and DNA.

Gram-negative bacteria Bacteria that do not retain a violet dye during the Gram stain process.

Gram-positive bacteria Bacteria that retain a violet dye during the Gram stain process. They have a thicker cell wall than gram-negative bacteria.

Hereditary system (limited/unlimited) Such systems transfer some characteristics from a parent to offspring. Systems with unlimited heredity can, in principle, store and transmit many more states than can be realized in real, finite populations. All possible states of limited hereditary systems can be realized in finite populations.

Heterotroph An organism that obtains its energy from organic substances produced by other organisms. Animals and fungi are heterotrophs.

Homeostasis The process by which organisms regulate their internal environment to maintain a stable condition.

Hoogsteen pair A minor variation of base-pairing in nucleic acids, in which two nucleobases on each strand are held together by hydrogen bonds in the major groove.

Hydrolysis A chemical reaction in which the addition of water to chemical linkages such as ester bonds break down larger molecules into small components. An example of hydrolysis is the enzyme-catalyzed reaction in which proteins are hydrolyzed to smaller peptides and amino acids during digestion.

Hydrophilic Literally "water-loving," this refers to the end of an amphiphilic molecule that associates with water in an aqueous environment.

Hydrophobic Literally "water-fearing," this refers to the end of an amphiphilic molecule that tends to be excluded from water. The result is that molecular aggregates of the amphiphiles produce stable structures such as micelles and bilayers in aqueous environments.

Hydrophobic effect The tendency of nonpolar molecules or groups to aggregate in such a way that they minimize their contacts with aqueous solvent. This effect is responsible for many self-organizing phenomena in biological systems, such as the formation of cellular boundary structures, protein folding, and higher-order protein assembly. A signature of the hydrophobic effect is its unusual temperature dependence—hydrophobic interactions often become stronger as temperature increases (typically up to 50–70°C).

Hypercycle A self-replicating macromolecular system, in which RNA molecules and enzymes cooperate in a cyclic manner.

Infrabiological system Infrabiological systems are chemical supersystems that include some, but not all, component subsystems of living systems. Any combination of two of the three subsystems of a chemoton—template, metabolism, and membrane—would constitute an infrabiological system.

Intermediary metabolism Enzyme-catalyzed processes within cells that extract energy from nutrient molecules and use that energy to construct cellular components.

In vitro In biology, in vitro refers to a scenario in which molecules are taken out of their normal biological contexts and manipulated in a controlled or artificial setting, usually in a defined chemical environment.

In vivo This refers to conditions in a living cell or organisms (in contrast to the artificial conditions referred to as in vitro).

Ion channels Proteins or protein assemblies that mediate transport of ions across membranes.

Ion gradient Because lipid bilayers are barriers to the free diffusion of ions, concentration gradients of ions such as sodium, potassium, and protons can be maintained across a membrane. Such gradients are essential energy sources for biological processes. For instance, proton gradients drive ATP synthesis in mitochondria and chloroplast membranes, and sodium-potassium gradients are required for excitability and action potentials in neurons.

Kinetics In chemistry, kinetics is the study of reaction rates in a chemical reaction.

Langevin equation A stochastic differential equation describing Brownian motion in a potential.

LANL bug A particular model system of an artificial life form proposed by researchers at Los Alamos National Laboratory (LANL). The model guides an ongoing effort to construct an artificial cell (see chapter 6 for further explanation).

Lattice artificial chemistry (LAC) An approach to computational autopoiesis.

Lattice molecular automaton A computational tool for simulating self-organization of molecular systems. The model uses a cellular automaton environment to formulate molecular dynamics and self-assembly processes.

Leaflet One of two monolayers of amphiphiles that form a typical bilayer membrane.

Lennard-Jones potential A mathematical model that represents two interactive forces that neutral atoms and molecules are subject to: an attractive force and a repulsion force.

Life criteria (absolute/potential) According to Gánti, all living systems must obey the absolute criteria at every moment they are alive. The potential criteria must be satisfied by units that can build up a biosphere (a living world). The absolute criteria are inherent unity, metabolism, inherent stability, surplus information, regulation, and control. The potential criteria are growth and multiplication, hereditary change and evolution, and mortality.

Ligase An enzyme that catalyzes the formation of a chemical bond linking two relatively large biological polymers. For example, a DNA ligase catalyzes the synthesis of a phosphodiester bond to link two separate strands of DNA.

Ligation A type of chemical reaction by means of which two polymers are joined to form a single longer one (see ligase).

Lipid Naturally occurring amphiphilic molecules such as fat (triglyceride), cholesterol, and phospholipids. All lipids tend to be soluble in organic solvents and spread as monolayers at air-water interfaces. Some lipids—phospholipids and cholesterol—can self-assemble into bilayer membranes.

Lipid world A possible scenario for the emergence of cellular life, which suggests that life may have started from a collection of amphiphiles forming compartments that incorporate a network of reactions characterized by mutual catalysis.

Lipidomics The experimental and theoretical study of lipids.

Lipophilic Literally "lipid-loving," this refers to materials that attract nonpolar organic compounds, such as oils, fats, and greases. Lipophilic materials are used in chemical separation processes to remove nonpolar from polar compounds.

Liposome A spherical supramolecular structure with an aqueous interior formed by a bilayer lamellar membrane consisting of amphiphiles. Liposomes are also referred to as vesicles.

Marshall replicators Faster-replicating RNA parasites of an in vitro evolution reaction that had the rare characteristic of longer genomes than their parental RNA templates.

Membrane asymmetry The composition of one leaflet of a lipid bilayer membrane differing in composition from the other leaflet.

Mesobiotic A term to describe the transition state from "prebiotic" or "nonliving" to "biotic" or "living" molecular systems.

Metabolic network A conceptual model of metabolism that describes the set of catalysts, reactions, and products involved in cellular biochemistry.

Metabolism The sum of chemical reactions that occur within an organism to maintain life, which include the biochemical changes that occur as nutrients are taken into a cell from the environment, and the biosynthesis and breakdown of complex macromolecules such as proteins and nucleic acids.

Metabolite A substance (usually a small molecule) formed during, or essential for, metabolism.

Metabolome The set of small-molecule metabolites found within an organism. The metabolome is a large network of metabolic reactions, in which outputs from one enzymatic chemical reaction are inputs to other chemical reactions.

Metazoan A member of the largest division of the animal kingdom, that is, a multicellular animal with differentiated tissues.

Micelle A spherical supramolecular structure with no aqueous interior formed by amphiphiles, in which their hydrophobic tails reside in the center and hydrophilic heads are in contact with the surrounding solvent.

Michael additions A conjugate addition involving the nucleophilic addition of a carbanion to an unsaturated carbonyl compound.

Minimal cell A membrane-bounded construct that can self-maintain, self-replicate, and evolve. This term is often used to describe a cell with the minimal number of genes that is still able to retain these properties.

Minimal genome A repertoire of genes that is necessary and sufficient to support cellular life, that is, to allow the cell to maintain a minimal metabolic network within a boundary, reproduce, and evolve under the most favorable conditions.

Mitosis A kind of cell division whereby a cell separates its duplicated genome, resulting in two daughter cells, each with the same number and type of chromosomes as the parent cell.

Molecular chaperone A protein that aids in the correct folding of a second protein following synthesis, and repairing potential damage caused by misfolding.

Molecular dynamics (MD) A computational method for modeling interactions between components in small chemical reaction systems.

Molecular ecology A complex network of chemical reactions in which individual types of molecules behave similarly to species in an ecosystem by competing with each other for resources (i.e., input material).

Molecular template A molecule that must be specifically recognized to direct a particular chemical reaction. For example, DNA sequences act as templates for the replication reaction, which produces a copy of the DNA template.

Monomer A molecule that can chemically bond to other molecules to form a polymer. For example, amino acids are the monomers of proteins, and nucleotides are monomers of nucleic acids.

Monte Carlo method A class of computational algorithms for simulating the behavior of physical or mathematical systems.

Multiangle laser light scattering (MALLS) A technique that correlates the patterns of scattered light at multiple angles to estimate the average size/shape of particles in a sample.

Myristoleic acid A natural fatty acid amphiphile consisting of 14 carbons, one double bond, and a carboxylic head group.

Network diameter The length of the longest pathway among all path lengths in a network.

Nongenomic evolution A hypothetical stage in the evolution of early cell ancestors, which proceeded in the absence of a genome.

Nucleophilicity The nucleophilicity of an electron-rich functional group is a measure of its reactivity with electron-deficient functional groups, leading to the formation of a covalent bond.

Nucleoside A compound consisting of a purine or pyrimidine base linked to a sugar. When a nucleoside has a phosphate group attached to the sugar, it is called a nucleotide. For example, adenine and cytosine are purine and pyrimidine bases, and when they are attached to deoxyribose they form nucleosides called adenosine and cytidine. With a phosphate group attached, they are nucleotides called adenosine monophosphate and cytidine monophosphate, two of the four monomers of DNA.

Nucleotide The chemical compounds that are basic structural units of DNA and RNA, consisting of a nucleoside linked to one or more phosphate groups.

Nucleotide triphosphates (NTPs) A nucleotide with three phosphates linked together, used by a cell to synthesize DNA and RNA polymers.

Oleic acid A natural fatty acid amphiphile consisting of 18 carbons, one double bond, and a carboxylic head group.

Oligomer A polymer containing a relatively small number of repeating monomer units. Examples include short, single-stranded DNA fragments and peptides with two or more amino acids. Peptides containing 50 or more amino acids, such as the insulin molecule, are classified as proteins.

Operon A unit made of linked genes that regulate other genes responsible for protein synthesis.

Orthologous genes Homologous genes (i.e., derived from a common ancestor) present within different species, usually exhibiting very similar or identical functions.

Osmotic pressure The pressure produced across a semipermeable membrane that separates different concentrations of a solute such as salt or sucrose. The pressure is produced when water molecules diffuse through the membrane from the lower to the higher concentration of solute.

Outer Helmholtz plane An electrochemical term used to describe the effective plane of charge. The excess ions in solution line up in one plane (Helmholtz plane) very close to a charged surface. Ions in the outer Helmholtz plane are about two solvent-molecule diameters away from the charged surface.

Path length The number of connections (or edges) in the shortest pathway between two nodes in a network.

Peptide A short chain of amino acids connected though peptide bonds. Ordered peptides typically contain only one element of secondary structure (e.g., a -helix). Naturally occurring peptides are often synthesized nongenomically, but can also be produced by hydrolytic cleavage of a larger protein molecule.

Permeability Lipid bilayer membranes represent the primary barrier to free diffusion of solutes into and out of a cell. Solutes cross the membrane barrier at different rates, and the property of the bilayer related to the rate is referred to as its permeability to a given solute.

Permeability coefficient A mathematical equation relates the rate of flux of a given solute across a membrane to the concentration gradient of the solute and a constant that is specific for that solute. The constant is referred to as the permeability coefficient. For example, the coefficient for water is in the range of 10^{-4} cm/s, and that for potassium ions is 10^{-11} cm/s. From this we can infer that water crosses a lipid bilayer 10 million times faster than potassium ions.

Phase transition The transition from one phase to another (i.e., change in physical properties) in a system, often referring to transitions between gas, liquid, and solid phases.

Phospholipid A natural amphiphile consisting of a head group containing phosphate as a hydrophilic component, which is linked to a glycerol molecule with two long-chained fatty acids attached by ester bonds.

Phosphorylation The introduction of a phosphate group into a molecule or compound.

Plasmid A DNA molecule in a cell that can replicate independent of the chromosomes. Plasmids occur in bacteria and some eukaryotic organisms, and are often used in laboratory manipulation of genes.

PNA Peptide nucleic acid is a chemical that structurally mimics DNA or RNA, but differs in the composition of its backbone. PNA is synthesized and does not occur naturally in any known living organisms.

Polymer A molecule whose structure consists of a number of subunits (monomers) connected by covalent chemical bonds.

Polymerase An enzyme that catalyzes the synthesis of polymers, for example, nucleic acids.

Polymerization The chemical reaction during which monomers are bound together to form a polymer.

Polyols Chemical compounds containing multiple hydroxyl groups.

Prebiotic An era of the Earth's history before the emergence of life.

Primary endosymbiont A bacterium with a permanent eukaryotic host-associated lifestyle, such as some Proteobacteria permanently associated to insects like aphids.

Primer A strand of nucleic acid (or similar material) that serves as a starting point for DNA replication.

Prokaryote An organism whose composite cell or cells lack a nucleus and membrane-bound organelles. Bacteria are prokaryotes (in contrast to eukaryotes, which include plants and animals).

Protease An enzyme that catalyzes hydrolysis of peptide bonds in proteins.

Proteome The entire complement of proteins expressed by a cell, tissue, or organism.

Proteomics The study of the structures and functions of protein.

Protocell A cell-like construct fabricated experimentally that exhibits some, but not necessarily all, traits of living cells. (Some sources have defined protocells as precursors to living cells, i.e., things that have some but not all features of minimal living cells.)

Regiospecificity A regiospecific reaction is one in which bonds form or break preferentially at one of two or more functional groups on a molecule.

Replicase An enzyme that catalyzes the synthesis of a complementary nucleic acid molecule from a nucleic acid template.

Replication (holistic/modular) The act or ability to make a copy. Chemically, it always rests on autocatalysis. In modular replication, the copy is assembled module by module, usually on a template surface: It is apparent when replication is half complete, for example. Holistic replication can consist of any number of qualitatively different steps, and replication is apparent only at the end as a net result.

Replicator Generic term for any entity that is capable of replication or that is replicated in a suitable environment. In particular, living organisms are (usually) replicators.

Reverse transcriptase DNA polymerase that transcribes single-stranded RNA into double-stranded DNA.

Ribosome A minute particle in the cytoplasm of cells that synthesizes proteins from activated amino acids. Ribosomes consist of RNA and proteins, with a ribozyme-like active site composed of RNA. They bind messenger RNA and transfer RNA to produce peptide bonds linking amino acids into proteins.

Ribozyme Shortened from *RNA enzyme*, ribozymes are functional catalytic sites on RNA molecules.

RNA world The phrase coined in 1986 by Walter Gilbert to describe a hypothetical stage in the origin of life in which RNA molecules carried genetic information and catalyzed chemical reactions.

rTCA cycle The reverse tricarboxylic acid (rTCA) cycle (also known as the reverse citric acid cycle or reverse Krebs cycle) is a sequence of chemical reactions some bacteria use to produce carbon compounds from carbon dioxide and water. The reaction is essentially the TCA cycle in reverse.

Scale-free network A network where the connection degree of nodes follows a power-law distribution, that is, $P(x) = Kx^{-a}$, where $P(x)$ is the frequency of the connection degree of node x, and K and a are constants. In other words, there is no typical connection degree nor normal distribution of connection degrees, but few nodes have many connections (hubs) whereas many other have few connections.

Self-assembly Spontaneous organization of molecules into more complex structures. An important example is the ability of phospholipids to self-assemble into lipid bilayer membranes.

Selfish parasite A molecule that is capable of being replicated by a member of a hypercycle replication system, but does not replicate any other member in turn.

Self-replication In biology, this term is most commonly used to describe the process in which a biopolymer (e.g., nucleic acid) or a set of organic molecules makes copies of itself. This definition can be extended to other, not necessarily biological, objects.

Self-reproduction In many instances, this term is synonymous with self-replication. Sometimes it is used to describe processes in which larger structures make copies of themselves without templates.

Sequence space The abstract space that contains all and only the possible genomes of some set of organisms.

Shortcut parasite A replicator that bypasses its original target template and replicates a downstream member of a hypercycle, thereby short-circuiting the hypercycle and resulting in information loss.

Simplex A mathematical space composed of vectors whose entries are non-negative and sum to unity.

State space A mathematical space that is used to describe the state of system. For example, a population can be described by a vector of frequencies that represent the abundance of different types of organisms.

Stereochemistry The study of the relative spatial arrangement of atoms within molecules, and the effect of this arrangement on chemical reactions.

Stoichiometric freedom A property of a chemical system is stoichiometrically free if and only if it is determined by chemical processes but does not depend on mass balance relations of the reactions that realize the property. An example is the sequence of nucleotides in a polynucleotide. Although the chemical composition of the molecule depends on mass-balance relations, the order (sequence) of nucleotides in a polymer of given composition does not; every ordering of a given composition is stoichiometrically correct.

Stoichiometry The calculation of quantitative relationships of the reactants and the products of a chemical reaction. Stoichiometry also refers to the relative number (usually an integer) of a particular molecule in proportion to another molecule in some environment in which the molecules interact.

Substrate The specific reactant on which enzymes act to enhance the rate of a reaction. For example, glucose is the substrate of the enzyme glucose oxidase.

Surfactant A wetting agent that lowers the surface tension of a liquid.

Syntrophy When one species lives off the products of another species.

TCA cycle The tricarboxylic acid (TCA) cycle (also known as the citric acid cycle cycle, or Krebs cycle) is a series of chemical reactions of central importance in all living cells that use oxygen as part of cellular respiration.

Thermophile A bacterium or microorganism that thrives at higher than normal temperatures. A thermophile is a kind of extremophile.

TNA Like PNA, threose nucleic acid is a synthetic nucleic acid structurally similar to DNA or RNA, but with a backbone containing the sugar threose in place of ribose.

Top-down approach An approach to fabricating protocells that starts with existing life forms (or components of existing life forms) and modifies them to create new and typically less complex life forms.

Transcription The process by which genetic information (i.e., a DNA sequence) is copied into newly synthesized molecules of RNA, with the DNA acting as a template.

Transcriptome The set of all messenger RNA molecules, or "transcripts," produced in a cell or group of cells.

Transesterification The process of exchanging the alkoxy group of an ester compound with another alcohol. This type of reaction is typical in making and breaking bonds in nucleic acid polymers.

Translation The process, following transcription, by which a messenger RNA molecule is "decoded," giving rise to a specific sequence of amino acids to synthesize a polypeptide or protein.

Transmembrane proteins Proteins located in membranes such that they span the width of the lipid bilayer. Their functions include facilitating transport of ions, nutrients, and waste products across membranous barriers.

Transposon A sequence of DNA that can move to different positions in the genome of a cell, causing mutations in the genome.

***Trans* reaction** A molecular reaction in which the catalytic molecule modifies a different molecule.

Van der Waals interactions An attractive force between atoms and molecules is produced by transient dipoles that occur when the molecules are in close proximity. Such forces stabilize molecular aggregates when stronger forces such as hydrogen bonds and electrostatic interactions are not involved. For instance, the molecules of a hydrocarbon wax such as paraffin are held together in solid form through van der Waal's interactions.

Vesicle A larger spherical supramolecular structure with an aqueous interior formed by a bilayer lamellar membrane consisting of amphiphiles. Vesicles are also referred to as liposomes.

Virion The complete form of a virus outside a host cell, with a capsid and core of genetic material.

Watson-Crick pair The major form of base-pairing in nucleic acids such as RNA and DNA, stabilized by hydrogen bonds formed between adjacent bases. In DNA, adenine (A) forms a base pair with thymine (T), as does guanine (G) with cytosine (C). In RNA, thymine is replaced by uracil (U).

Zwitterion From the German word *zwitter* (between). A molecule that has both positive and negative ionic groups. All amino acids and some amphiphiles are zwitterions.

About the Authors

James Bailey is a technical staff member at Los Alamos National Laboratory. He received his PhD from the University of Victoria, Victoria, Canada, in 1991. Recent research activities include studies of the structure-function relationship in cytochrome oxidases, particularly with regard to ligation events and coupled electron/proton transfer, and investigation of the structure and dynamics of biologically inspired hybrid materials.

Mark A. Bedau (PhD Philosophy, University of California, Berkeley, 1985; Professor of Philosophy at Reed College; Visiting Professor at the European School of Molecular Medicine) studies complex adaptive systems. He has coedited many books, including *Emergence: Contemporary readings in philosophy and science* (MIT Press) and *The prospect of protocells: Social and ethical implications of recreating life* (MIT Press). He is editor-in-chief of the journal *Artificial Life*, cofounder of ProtoLife Srl., and cofounder of the European Center for Living Technology.

James Boncella is currently a staff member at Los Alamos National Laboratory. Prior to coming to Los Alamos in 2003, he was a professor of chemistry at the University of Florida. He received his PhD in Chemistry from the University of California, Berkeley, in 1984 and spent two years as a postdoctoral fellow at Oxford University, Oxford, UK, before starting his faculty position at the University of Florida in 1986.

Irene A. Chen grew up in San Diego and received a bachelor's degree in chemistry from Harvard. She received a doctorate in biophysics working with Dr. Jack Szostak on physical aspects and emergent behaviors of protocells. She is currently completing a medical degree at Harvard.

Liaohai Chen is a molecular biologist and group leader in the Biosciences Division at Argonne National Laboratory, and an associate professor at Rush University Medical Center. He got his PhD in chemistry from the University of Rochester, and his current research focus is on systems biology and synthetic biology with emphasis in building a minimal cell or protocell experimentally.

Henderson James Cleaves II obtained a BA in molecular, cellular and developmental biology at the University of California, Santa Cruz, and a PhD in chemistry and biochemistry from the University of California, San Diego (UCSD), working with Professor Stanley Miller. Later he was a postdoctoral fellow at the Scripps Institution of

Oceanography, working with Professor Jeffrey Bada, and a lecturer in chemistry at UCSD. He is presently a Senior Staff Associate at the Carnegie Institution of Washington Geophysical Laboratory working with Dr. Robert Hazen. His research concerns the synthesis, transformations, and self-organization of organic compounds in planetary environments and the synthesis of novel biopolymers, as well as the development of novel, sensitive life detection techniques.

Stirling Colgate is associate senior physicist and senior fellow at Los Alamos National Laboratory. He obtained his BS and PhD at Cornell University (1952) and has worked in physics, plasma physics, and astrophysics. In this project he has been concerned with the minimum information for the origin of life and the association between planet formation and the necessary diversity of chemical precursors. He is a member of the National Academy of Sciences.

Gavin Collis received a PhD in organic chemistry from the University of Western Australia. He has held positions as a scientist and assistant director at the Nanomaterials Research Centre in New Zealand, and collaborated in multidiscipline programs while at Los Alamos National Laboratory. He recently joined CSIRO in Australia as a Research Scientist.

Shelley D. Copley is Professor of Molecular, Cellular and Developmental Biology at the University of Colorado, Boulder. Her research interests center on the molecular evolution of catalysts in two primary areas: (1) the evolution of protein enzymes to serve new functions, and (2) the role of simple catalysts on the early Earth before the emergence of macromolecules.

David Deamer is Research Professor of Chemistry and Biochemistry at the University of California, Santa Cruz, and Acting Chair of the Department of Biomolecular Engineering. His research interests are nanopore analysis of DNA and self-assembly of amphiphilic compounds into membrane structures.

Michael DeClue received his PhD in organic chemistry from the University of California in San Diego in 2003. He is currently a postdoctoral fellow in the Materials, Physics and Applications Division at Los Alamos National Laboratory. His research activities include synthetic life, biocatalysis, mechanistic enzymology, multistep total synthesis, photorefractive polymers, and reagent development.

Andrew D. Ellington is the Fraser Professor in the Department of Chemistry and Biochemistry at the University of Texas at Austin. He got his BS in biochemistry from Michigan State University in 1981, received a PhD in biochemistry and molecular biology with Steve Benner at Harvard, and was a postdoctoral fellow with Jack Szostak at Massachusetts General Hospital. He was an assistant professor of chemistry at Indiana University prior to moving to Texas.

Harold Fellermann studied applied system science and molecular biology at the University of Osnabrück, Germany, where he graduated in 2005. He is currently working at the ICREA–Complex Systems Lab (Barcelona, Spain) and the Los Alamos National Laboratory (New Mexico, US). His research interests include complex systems, systems and molecular biology, and origin of life.

John M. Frazier received his PhD in physics from Johns Hopkins University in 1971. Dr. Frazier served on the faculty at The Johns Hopkins University from 1970 to

1994. He moved to the Air Force Research Laboratory in 1994 as a Senior Scientist and retired from the Air Force in 2006. Currently Dr. Frazier is an Air Force Research Laboratory Emeritus.

Toni Gabaldón received his education in biology and biochemistry at the University of Valencia (Spain), where he graduated with an honors degree in 1996. From 1997 to 2000 he researched stress response in yeast at the department of biochemistry and molecular biology of the University of Valencia. In 2001 he moved to work as a junior researcher in the computational genomics group at the NCMLS center of the Radboud University of Nijmegen (The Netherlands), where he applied comparative genomics to study the evolution and function of mitochondria. In 2005 he obtained his PhD degree from that university under the guidance of Martijn A. Huynen. He is an EMBO postdoctoral fellow in the bioinformatics department of the Príncipe Felipe Research Center in Valencia.

Rosario Gil got her PhD in pharmacy at the University of València (Spain) in 1991. After several years of work in yeast genetics, Dr. Gil moved to evolutionary genomics in 2001. Since then, she has been involved in the genome sequencing of several endosymbiotic bacteria of insects; her main goal is to understand the role of the different bacteria in their respective symbiosis. As a consequence of the acquired knowledge about reduced genomes, she has also been working on an attempt to define the core genes that must be present on a minimal genome of a modern-type bacterial cell. She is a recipient of a contract in the *Ramón y Cajal* Program from the *Ministerio de Ciencia y Tecnología*, Spain.

Goran Goranović is currently assistant professor at the University of Southern Denmark and guest researcher at the University of Copenhagen. He is working on the EU integrated project PACE (EU-IST-FP6-FET-002035) on development of microfluidic implementation of both self-replication and coupled protocellular reactions. He graduated in theoretical physics from the University of Zagreb, Croatia, in 1999, and obtained his PhD in theoretical microfluidics from the Technica University of Denmark in 2003.

James Griesemer is professor and chair of the Department of Philosophy, a member of the Center for Population Biology, and a member of the Science and Technology Studies Program at the University of California, Davis. He has been a fellow of the Wissenschaftskolleg in Berlin, a guest of the rector at Collegium Budapest, and a guest of the Max Plank Institut für Wissenschaftsgeschichte in Berlin. His current research is to develop a general theory of reproduction as the fundamental basis of evolutionary processes and its implications for theory in genetics, development, origins of life, and the evolution of culture. With Eörs Szathmáry, he is a coeditor of Tibor Gánti's seminal work on the chemoton model of living systems, *The Principles of Life*, Oxford University Press, 2003.

Martin M. Hanczyc is chief chemist at ProtoLife Srl. in Venice, Italy. He received a bachelor's degree in biology from Pennsylvania State University and a doctorate in genetics from Yale University His interest in the synthesis of a protocell stems from his training in population genetics and experimental evolution.

Takashi Ikegami is an associate professor of physics at the University of Tokyo since 1994 and teaches complex systems and theoretical biology. He has been working in the field of artificial life for more than 15 years and his current interest is to understand the continuity between life and mind.

Keitaro Ishikawa is a PhD student at the Graduate School of Frontier Biosciences, Osaka University. He has researched in the field of artificial life, and is currently researching the fitness landscape of protein.

Ask F. Jakobsen has an MS in theoretical physics from the Niels Bohr Institute at the University of Copenhagen (2003), and a PhD in theoretical biophysics from the University of Southern Denmark (2006) based on a thesis on dissipative particle dynamics methods applied to membranes. Since July 2006 he has worked as a software developer in seismic inversion at Western Geco/Schlumberger in Copenhagen.

Yi Jiang received her PhD in physics from the University of Notre Dame in 1998. She is a technical staff member at Los Alamos National Laboratory and adjunct professor at Notre Dame. Her primary research interests include biophysics, biomedical modeling, soft condensed matter, pattern formation, and multiscale modeling methods.

Kunihiko Kaneko has been a professor at the University of Tokyo since 1994, as well as a visiting professor at Osaka University, external faculty of Santa Fe Institute, and research director of ERATO Complex Systems Biology Project. Previous positions include Stanislaw Ulam Fellow (1988–89), associate professor (1990–94), assistant professor (1985–90) at University of Tokyo, and visiting assistant professor, University of Illinois at Urbana–Champaign (1987–88). He holds a PhD from the University of Tokyo, Department of Physics (1984).

Nancy Kelley-Loughnane received her PhD in biochemistry from Boston College in 2000. Following her postdoctoral fellowship at Children's Hospital Medical Center in Cincinnati, Ohio, in 2001, she became a technical area coordinator for genomics and proteomics for Air Force and Navy contracts. She is currently the Project Director of the Cellular Dynamics and Engineering Program in the Applied Biotechnology Branch at the Air Force Research Laboratory.

Günter von Kiedrowski studied chemistry at the Westfälische Wilhelms–University of Muenster. He received his PhD degree from the Georg August–University Goettingen under the supervision of Lutz F. Tietze and was a postdoctoral fellow at the Salk Institute in San Diego. He gained his postdoctoral lecture qualification in Goettingen on a self-replicating minimal system before he became professor in organic chemistry at the University Freiburg. In 1996 he moved to Ruhr University–Bochum where he is professor of bioorganic chemistry. His research interests include self-replication in artificial chemical systems, synthetic bioorganic chemistry, supramolecular chemistry, reaction kinetics, theory of evolution, and programmable nanotechnology.

Chad Knutson received graduate training in environmental engineering. His primary research interest is computational modeling of multiphase fluids. He is currently a postdoctorate at Sandia National Lab in Albuquerque, New Mexico.

David C. Krakauer is professor at the Santa Fe Institute in New Mexico. He was educated at the University of London and the University of Oxford. He was a long-term member of the Institute for Advanced Study in Princeton and visiting professor of evolutionary biology at Princeton University. David's work is concerned largely with the evolutionary history of information processing mechanisms in biology, with an emphasis on information transmission, signaling dynamics, and their role in promoting novel, higher-level structures.

Doron Lancet studied chemistry and immunology in Israel, and had postdoctoral training at Harvard and Yale (US). He headed the Department of Membrane Research at the Weizmann Institute of Science, and is now professor at the Department of Molecular Genetics, and Director of Israel's National Center for Genomics. He discovered the molecular basis of smell transduction, and currently studies the genomics and population genetics of human olfaction. He also developed GeneCards, a widely used Web-based gene compendium. In the area of the origin of life, he developed the computer-simulated GARD/Lipid World model, whereby early selection and evolution may occur via compositional information, independent of nucleic acid sequence information.

Amparo Latorre is associate professor of genetics and a member of the Evolutionary Genetics Group of the Cavanilles Institute on Biodiversity and Evolutionary Biology of the University of Valencia. In her major research areas of molecular evolution, and evolutionary and comparative genomics of bacterial endosymbionts of insects she is in charge of supervising several new genomes from aphids, weevils, and cockroaches. Her main goal is to characterize the initial and final stages of the bacterial adaptation to a strict intracellular way of life, and the possible replacement for a new symbiont in the case of an extreme genome reduction. Her teaching activities are related to molecular genetics and genetic engineering for undergraduates.

Matthew Levy received his MS in chemistry from the University of California, San Diego, in 1997, where he studied origins of life chemistry with Dr. Stanley L. Miller. In 2003 he received a PhD in molecular biology from the University of Texas, Austin, where he studied the design and selection of nucleic acid catalysts. He is currently working as a research associate at University of Texas, Austin.

Pier Luigi Luisi is professor of biochemistry at the Third University of Rome. He received a degree in chemistry at the Scuola Normale Superiore of the University of Pisa. Professor Luisi was first *oberassistent* at the Swiss Federal Institute of Technology (ETH) of Zurich, then assistant professor (1973), and full professor of macromolecular chemistry (since 1984). His scientific interests include macromolecular chemistry, enzymology, self-assembly structures as micelles, reverse micelles, liposomes, and giant vesicles. He has recently focused on the chemical implementation of autopoiesis, the origin of cellular complexity, and self-reproduction of supramolecular structures.

Javier Macía is a physicist by training, and received his PhD in physics from the Universitat de Barcelona. He is currently a postdoctorate at the Complex Systems Lab. His fields of research include complex systems, systems and synthetic biology, and biocomputational designs.

Duraid Madina is currently a PhD student at Tokyo University, where his research is in systems and software for large-scale simulation of complex systems.

Jerzy Maselko graduated from the Department of Chemistry, Technical University, Wroclaw, Poland. He is the author of 47 publications in the area of nonlinear chemical dynamics, pattern formation, cellular chemical systems, and complex chemical systems.

Maciej Maselko is a graduate student working toward a doctoral degree in molecular and cellular biology at Oregon State University.

John S. McCaskill graduated from Sydney University, Australia, in 1978 and finished a DPhil in theoretical chemistry at the University of Oxford (1982). He became professor at the University of Jena in 1992, and has been developing biomolecular information processing at the German National Research Center for Information Technology (GMD) near Bonn since 1999. He is currently a visiting professor at the Ruhr University, Bochum, where he heads the research group BioMIP. His multidisciplinary research bridging theory and experiment has brought together the fields of molecular information processing and evolutionary self-organization, and resulted in several spin-off companies, the latest being Protostream GmbH.

Pierre-Alain Monnard received his PhD in natural sciences (colloid and polymer chemistry) from ETH, Zurich, Switzerland, in 1998. He is a technical staff member at Los Alamos National Laboratory in the Earth and Environment Division. His research interests include self-assembly of chemicals, catalysis in heterogeneous media and micro- and nanostructures, and the origins of life. He is an associate editor of *BioSystems*.

Harold J. Morowitz is Robinson Professor of Biology and Natural Philosophy at George Mason University, where he works at the Krasnow Institute for Advanced Study. His research is on biochemical networks and the origin of life. His most recent book is *The Emergence of Everything*.

Fouzi Mouffouk received his PhD in biomolecular chemistry from University of Montpellier II, France, in 2000. Currently he holds a postdoctoral position at Rush University Medical Center, Department of Obstetrics and Gynecology, and at Argonne National Laboratory in Chicago. His research interests are materials science and bioengineering.

Ole G. Mouritsen (PhD, DSc, FRSD) is professor of theoretical biophysics at the University of Southern Denmark and director of the MEMPHYS–Center for Biomembrane Physics. His fields of research cover statistical physics, soft matter, and computer simulation. His current areas of focus are biological membranes, lipids, proteins, and liposome-based drug delivery.

Andrés Moya is professor of genetics and head of the Cavanilles Institute on Biodiversity and Evolutionary Biology (University of Valencia). His research interests include theoretical and experimental studies on population genetics, evolution of RNA viruses, and the evolution of interacting genomes. As part of his research program in biodiversity, he has applied population genetics theory and molecular tools to conservation biology and biological control of several species. His teaching activities are related to introduction of evolutionary theory for undergraduates.

Andreea Munteanu received her BS in physics from University of Bucharest, Romania, and her PhD in astrophysics from the Polytechnic University of Catalunya, Spain, in 2003. Her current research interests are centered on the influence of noise and stochasticity in biological systems, with emphasis on signal transduction and gene regulation.

Giovanni Murtas is currently senior scientist at the University of Roma Tre, Rome (Italy). After completed his degree in biology at University *La Sapienza* in Rome, he moved to John Innes Centre, Virus Department (UK), where he was appointed scientific officer. In 1994 he joined the Molecular Genetics Department, Cambridge Laboratory, JIC, (UK). He was later appointed research fellow at the University of Warwick (UK). In 2005 he was awarded a senior grant from Centro Fermi, Rome. Dr Murtas's scientific interests include genetics, molecular genetics, molecular biology, genomics, and origin of life. He was recently involved in a synthetic biology project to construct a semisynthetic minimal cell as a model for early living cells.

Peter E. Nielsen has been affiliated with the University of Copenhagen since 1980, and he became full professor in 1995. He is one of the inventors of peptide nucleic acid (1991), and has further studied and developed this DNA mimic in the last 15 years. Peter Nielsen is the coauthor of more than 300 scientific papers and 20 patents and patent applications, and he serves on the advisory board of 8 scientific journals. He received his PhD in 1980 from University of Copenhagen and an honorary doctor's degree in 1990. He is the cofounder of two Biotech companies in Denmark. He has received several scientific prizes.

Naoaki Ono received his PhD in theoretical biology from the University of Tokyo. He is currently a researcher in the Department of Bioinformatic Engineering, Osaka University. His research interests include the evolution of biochemical reaction networks, cellular structures, and gene regulatory systems.

Norman H. Packard is cofounder and CEO of ProtoLife Srl (Venice, Italy). He has over two decades of experience in chaos theory, learning algorithms, predictive modeling of complex systems, statistical analysis of evolution, artificial life, and complex adaptive systems. He cofounded Prediction Company in 1991, and served as CEO from 1997 to 2003. He holds a PhD in physics from the University of California at Santa Cruz (1983). He also has a long-standing involvement with the Santa Fe Institute, currently serving on its external faculty.

Oleg Paliy received his PhD in biology from the University of Manchester, UK, in 2001. He was a postdoctoral fellow at University of California, Berkley, until 2004 when he accepted his current position at Wright State University as an assistant professor in the Department of Biochemistry and Molecular Biology.

Emily C. Parke is business manager at ProtoLife Srl. (Venice, Italy). She is coeditor of *The prospect of protocells: Social and ethical implications of creating life from scratch* (MIT Press), and formerly was editorial assistant of the *Artificial Life* journal. She has a BA in philosophy from Reed College in Portland, Oregon, and wrote a thesis in bioethics on the precautionary principle.

Volker Patzke studied chemistry at the Ruhr University–Bochum and received his MS degree in organic chemistry in 1999 on the synthesis of monofunctionalized polyamines. He received his PhD degree from the Ruhr University–Bochum under the supervision of Günter von Kiedrowski on a self-replicating minimal system. In his postdoctoral fellowship he is working in the BioMIP research group at the BMZ in Dortmund on the development of chemical systems toward an artificial cell in microfluidic systems.

Juli Peretó is associate professor of biochemistry and molecular biology at the University of València. He is a member of the Evolutionary Genetics Group at the Cavanilles Institute for Biodiversity and Evolutionary Biology. His research interests include the origin and early evolution of life, with a focus on metabolic pathways evolution. He is also interested in the history of ideas on the origin of life, and public understanding of science. His teaching activities are related to general biochemistry and chemical and biochemical evolution for undergraduates.

Andrew Pohorille received his PhD in theoretical physics from University of Warsaw. In 1996 he joined the staff of NASA Ames Research Center, where he directs the NASA Center for Computational Astrobiology. He is also professor of chemistry and pharmaceutical chemistry at the University of California, San Francisco. His main scientific interests have focused on modeling the origins of life, computer simulations of biomolecular systems, modeling genetic and metabolic networks, and statistical mechanics of condensed phases; other research interests range from the structure of comets to the mechanism of anesthetic action, high-performance computing, and risky decision making.

Steen Rasmussen received his PhD in physics at the Technical University in Denmark in 1985. He is currently a professor and center director at University of Southern Denmark and an external professor at the Santa Fe Institute. For 19 years he was at Los Alamos National Laboratory, the last seven as a team leader, and in 2004–05 he was a visiting professor at the University of Copenhagen. He has a broad interest in self-organizing processes and he has developed new concepts and methods for a wide variety of systems ranging from physicochemical to sociotechnical systems.

Mauricio Rodriguez Rodriguez received his PhD in biochemistry from Texas A&M University in 2005. Among other awards, he has received the UNESCO Fellowship in Biotechnology, a Fulbright Fellowship, and a NRC Research Associateship from the National Research Council, National Academy of Science. He is currently a research scientist at the Air Force Research Laboratory in the Applied Biotechnology Branch.

Josep Sardanyés is a theoretical biologist working at the Complex Systems Lab (PRBB–Pompeu Fabra University). His fields of research include nonlinear dynamical systems theory applied to replicator dynamics of hypercycles, molecular quasispecies and antagonistic ecological dynamics, complex networks, and synthetic biology.

Kanetomo Sato is currently a researcher at the R&D institute, Sekisui Chemical Co., Ltd. He has a PhD in engineering from the Graduate School of Engineering, Osaka University. He is now researching various topics concerning applied liposome technology.

Peter Sazani received a bachelor's degree in biochemistry from SUNY–Binghamton and a doctorate in pharmacology from University of North Carolina–Chapel Hill. His postdoctoral work in the laboratory of Dr. Szostak led to his interest in self-replicating genetic material for use in the protocell. Dr. Sazani is director of discovery research at Ercole Biotech, Inc., in Research Triangle Park, North Carolina.

Anjana Sen received her PhD in biophysical chemistry from University of Calcutta in 1997. She is currently working as a guest researcher at the Panum Institute, Department of Cellular and Molecular Medicine, University of Copenhagen, Denmark. Her primary research focus is on the biophysical aspects of nucleic acids.

Barak Shenhav is a multidisciplinary scholar residing in Israel. He holds a BS in physics and computer science from The Hebrew University of Jerusalem. He has two MS degrees, in computer science from the Hebrew University and in the life sciences from the Weizmann Institute of Science. He was until recently a doctoral student under the supervision of Professor Doron Lancet at the Weizmann Institute. His thesis addresses a computational model (GARD) for the Lipid World scenario for the origin of life. At present, Shenhav is establishing a group as an independent researcher at The College of Judea and Samaria in Ariel, while pursuing an MBA at the Inter-Disciplinary Center, Herzliya.

Andy Shreve is a technical staff member at Los Alamos National Laboratory. He received his PhD in physical chemistry from Cornell University, and his research interests include study of charge and energy transfer in chemical and biological systems, development and application of spectroscopic and optical imaging methods, and study of self-assembled materials.

Zachary Booth Simpson is an engineer, artist, and molecular biology researcher from Austin, Texas.

Eric Smith received a BS degree in physics and mathematics from the California Institute of Technology, and a PhD in physics from the University of Texas at Austin. He is currently a professor at the Santa Fe Institute, interested in self-organization in thermodynamically reversible as well as irreversible physical systems, and applications of related statistical ideas to the origin of life and the emergence of core biochemistry, and to the stability of the Darwinian selection process.

Ricard V. Solé is a biologist and physicist by training. He received his PhD in physics from the Universitat de Barcelona. He is ICREA research professor at the Catalan Institute for Research and Advanced Studies at Universitat Pompeu Fabra, where he is head of the Complex Systems Lab, and external professor of the Santa Fe Institute (New Mexico) and senior member of the NASA–Associate Center of Astrobiology. His fields of research include complex systems, systems and synthetic biology, and network theory.

Bärbel M. R. Stadler studied mathematics in Vienna and is currently affiliated with the Max Planck Institute for Mathematics in Science in Leipzig, Germany. Main areas of research include topological structures in evolutionary theory and dynamics of replicator networks.

Peter F. Stadler studied chemistry in Vienna. He is professor of bioinformatics at the University of Leipzig, Germany, and a member of the external faculty of the Santa Fe Institute in New Mexico. His main areas of research are comparative genomics, RNA bioinformatics, dynamics of replicator networks, artificial chemistry, graph theory, theoretical biology, and molecular evolution.

Pasquale Stano is a research assistant at the University of Roma Tre (Rome, Italy) since 2004. After completing a physical chemistry degree on supramolecular complexes at the University of Pisa, he joined Luisi's group (2001–03) at the Swiss Federal Institute of Technology (ETH). His current research interests involve the properties of supramolecular assemblies and compartments, dynamic light scattering, biospectroscopies, and structural characterization of proteins. He is also developing new approaches for the experimental study of vesicle populations, and the understanding of reactivity within and between vesicles.

Takeshi Sunami is currently a doctoral student at Osaka University in Japan. He has a master's degree in engineering from Osaka University, and is researching several reactions within liposomes.

Eörs Szathmáry (1959) is professor of biology at the Department of Plant Taxonomy and Ecology of Eötvös Loránd University, Budapest, and permanent fellow of Collegium Budapest (Institute for Advanced Study). His main interest is theoretical evolutionary biology and focuses on the common principles of the major steps in evolution, such as the origin of life, the emergence of cells, the origin of animal societies, and the appearance of human language. He has coauthored two books with John Maynard Smith—*The Major Transitions in Evolution* (Freeman, 1995) and *The Origins of Life* (Oxford University Press, 1999)—and has also published numerous papers in various scientific journals.

Jack W. Szostak is an investigator at the Howard Hughes Medical Institute, professor of genetics at Harvard Medical School, and the Alex Rich Distinguished Investigator in the Department of Molecular Biology and the Center for Computational and Integrative Biology at Massachusetts General Hospital. His current research interests are in the laboratory synthesis of self-replicating systems, and the origin of life.

Jeffrey J. Tabor received his BA in biology and biochemistry from the University of Texas in 2001, during which time he studied evolutionary biology with Dr. James J. Bull. He is currently a PhD candidate in molecular biology at the University of Texas studying synthetic biology with Dr. Andrew D. Ellington.

Arvydas Tamulis has held the senior researcher position at the Institute of Theoretical Physics and Astronomy of Vilnius University, Lithuania, since 1996. He received his PhD in 1985 in the field of theoretical and mathematical physics and has published 193 science papers. He is currently working on several quantum mechanical nanoscience and nanotechnology projects that have the common goal of quantum modeling and construction of nanobiorobots using molecular electronics and spintronics-based logical controls.

Melanie Tomczak received her PhD in biochemistry from University of California at Davis in 2000. She was a postdoctoral fellow in the Department of Biochemistry, Queen's University, Kingston, Ontario, Canada, from 2000 to 2003. She is currently a program manager and research scientist at UES, Inc., in Dayton, Ohio, and has collaborated extensively with the Materials Laboratory and Human Effectiveness directorates at the Air Force Research Laboratory.

Bryan Travis has a PhD in applied mathematics, and completed a graduate program in medical science. He has worked in several areas of computational physics during his 32 years at Los Alamos National Laboratory, including flow and transport in porous media, thermodynamics, electromagnetics, microbiology, and neuroscience.

Sandra Trott obtained her doctoral degree from the University of Stuttgart in 2000. Her main areas of expertise are biotechnology and microbiology. Dr. Trott received her postdoctoral training at the Air Force Research Laboratory, Tyndall Air Force Base, Forida, and at Cargill, Inc., Dayton, Ohio. Dr. Trott is a research assistant professor in the Department of Biochemistry and molecular biology at Wright State University and works in close collaboration with the Air Force Research Laboratory at Wright-Patterson Air Force Base.

Sergi Valverde is a computer scientist working on complex networks and the origin of innovation in natural and artificial evolution. He is with the Complex Systems Lab, University Pompeu Fabra, Barcelona, Spain.

Leamon Viveros received his MS degree in applied biology from Bowling Green State University in 1998. He has served in the U.S. Air Force for 21 years as a medical laboratory officer and is a specialist in immunohematology. He is currently the program manager for Cellular Dynamics & Engineering in the Applied Biotechnology Branch at the Air Force Research Laboratory.

Pawel Weronski has a PhD in physical and theoretical chemistry. Since 1992 he has worked at the Institute of Catalysis and Surface Chemistry, Polish Academy of Sciences in Kraków, Poland; currently he is on leave at Los Alamos National Laboratory. His research interests include application of numerical and statistical methods in chemical physics and biophysics, stiff differential equations, colloid and protein adsorption, self-assembling nanostructures, Brownian dynamics, Monte-Carlo methods, molecular dynamics, and multiscale modeling and simulations.

William H. Woodruff received his BA in chemistry from Vanderbilt University in 1962. Following five years as a U.S. Naval Aviator he received MS (1967) and PhD (1972) degrees in inorganic chemistry from Purdue University. He was a National Institutes of Health Postdoctoral Fellow at Princeton University and held faculty positions in chemistry at Syracuse University and the University of Texas at Austin before moving to Los Alamos National Laboratory where he is a laboratory fellow (retired associate). He is also an external professor at the Santa Fe Institute and an adjunct professor of chemistry and biology at the University of New Mexico.

Tetsuya Yomo is currently a professor at the Graduate School of Information Science and Technology, and Frontier Biosciences Department at Osaka University in Japan. He is experimentally synthesizing artificial life, genetic network for cell differentiation, and symbiosis to understand fundamental rules behind biological complex systems.

Paolo Zanotto is a professor at the Microbiology Department of the Institute of Biomedical Sciences at the University of São Paulo in São Paulo Brazil. He was educated at the University of São Paulo, University of Florida in Gainesville, and University of Oxford. He currently does research and teaches genomics and viral genetics and coordinates a network of laboratories researching the genetic diversity of human viruses.

Jinsuo Zhang received a PhD in mechanics in 2001 at Zhejiang University, and is interested in theoretical modeling and numerical simulation in the fields of fluid mechanics and materials science. Dr. Zhang joined the Center for Non-Linear Studies as a postdoctorate in 2001, and has been a staff member of the International Nuclear Engineering System group since 2004.

Xin Zhou received his PhD in physics from Institute of Theoretical Physics, Chinese Academy of Sciences in 2001. He is a postdoctorate at Los Alamos National

Laboratory. His primary research interests include modeling and simulations of biophysics, soft condensed matter, and multiscale modeling methods.

Hans Ziock received his PhD from the University of Virginia in 1980 in the field of intermediate energy physics. He has been a technical staff member at Los Alamos National Laboratory since 1985 and has pursued projects ranging from particle physics, to carbon dioxide mitigation, to information systems. He is currently working on an attempt at Los Alamos to build a protocell using the bottom-up approach.

Index

Printed in the United States
by Baker & Taylor Publisher Services